A Companion to the 'Indian Mutiny' of 1857

A Companion to the 'Indian Mutiny' of 1857

GENERAL EDITOR

P. J. O. TAYLOR

DELHI
OXFORD UNIVERSITY PRESS
BOMBAY CALCUTTA MADRAS
1996

Oxford University Press, Walton Street, Oxford OX2 6DP
Oxford   New York
Athens   Auckland   Bangkok   Bombay
Calcutta   Cape Town   Dar es Salaam   Delhi
Florence   Hong Kong   Istanbul   Karachi
Kuala Lumpur   Madras   Madrid   Melbourne
Mexico City   Nairobi   Paris   Singapore
Taipei   Tokyo   Toronto
and associates in
Berlin   Ibadan

© Oxford University Press 1996

ISBN 0 19 563863 8

Printed in India at Pauls Press, New Delhi 110020
and published by Neil O'Brien, Oxford University Press
YMCA Library Building, Jai Singh Road, New Delhi 110001

PREFACE

'Indian Mutiny', 'The War of Independence', 'The Great Indian Rebellion', 'The Start of the Freedom Struggle' — what we call it is irrelevant: we talk of the momentous events of 1857 in India: they speak for themselves.

EVIDENCE

The task of the historian is bedevilled by two cynical observations, each of which unfortunately contains a measure of credibility: 'The first casualty of War is Truth', and 'The Victor writes the History'. Add to that the misfortune of the serious student of the events of 1857 that over 95 per cent of the documentary evidence emanates from the British side, and it will be seen that the task of looking objectively at the struggle is fraught with difficulty. But it can be done. One aid to impartiality is the determination to be aware of the subtle influence of 'colonial terminology' in that great bulk of documentary evidence. Documents cannot be rewritten or misquoted, but reinterpretation is quite legitimate, as is a change of emphasis: the rebel military victories at Chinhat (30 June 1857), Sassiah (5 July 1857), Kanpur (28 November 1857), Jagdishpur (23 April 1858), etc. deserve, for example, greater attention than the British thought fit to give them. Similarly the overwhelming support given to the rebels in some rural areas cannot be dismissed cavalierly as the action of 'badmashes' or 'dacoits'. Simplistic stereotyping was convenient, but it is misleading *and* inaccurate. 'Great Empires are not made', said an old British Colonel in 1856, 'by seeing the other fellah's point of view.' We can hardly expect military men's reminiscences to be objective, and few are, but to be forewarned is to be forearmed. In practice the evidence of a J.W. Sherer may be preferred to that of a Captain Wolseley, and it will not take the open-minded reader long to see why.

SPELLING

An editor has no right to tamper with the spelling of words in documents from which he *quotes*; but the forms in use one hundred and forty years ago have now, in the process of evolution, become obsolete, and it would not be desirable to revert to them, save *in quotation*. In this book, therefore, Hindu and Hindoo may be found side by side, as will Kanpur and Cawnpore: pedantic attempts like those of Colonel G.B. Malleson to adhere to 'scientifically correct' spelling leads to even greater confusion, obscurity, and worse, irritation. At the same time a genuine effort has been made to use modern spelling for place-names and individuals—*except in quotation*—whenever this is at all possible. But there are still difficulties; there are some words which from ignorance or sheer cussedness the British insisted on pronouncing, and spelling, quite differently from modern ideas: take the Kingdom of 'Awadh' as an example—it appears in contemporary, and subsequent documents and reports as 'Avadh', or, more usually, 'Oude' or 'Oudh'. Now to the non-Hindi speaker 'Awadh' and 'Oude' do not appear to be in any way related, let alone representative of the same place. If therefore the text is interspersed with occasional bracketed alternatives—e.g. 'Awadh [Oude]', the General Editor craves pardon and indulgence: his intention is to avoid anachronism but also to be as helpful as possible. Englishmen in India in the eighteenth century wrought sad havoc with the vowel sounds of 'Hindustani', or 'Moors' as they called it, and that is reflected in the spelling they adopted: it has been suggested that this was partly the influence of Bengali as spoken in Calcutta. But I will leave that to better linguists than myself to argue. In particular the English came to pronounce the long 'a' as 'o': thus *lal sharab* came out as *lol shrob*. Many of these renderings of Indian words persisted into the nineteenth century, *mali* and *dali* were still *molly* and *dolly*. *Cawnpore* is another relic of the distortion; it will not be a surprise to learn that *pani* was *pawnee*, *dak bahangi* was *dawk buhngee*, and *adalat* was *adawlut*. If these distortions are seen in the pages that follow it will be because they are in passages *directly quoted* or in book titles.

MAIN ENTRIES

Entries are in alphabetical order, obviating therefore the need for an index. Names etc. which appear in the body of an entry in **bold type** will be found to have been themselves the subject of a specific entry. Where the reader is being *directed* to such a supplementary source of information the entry to look for appears in CAPITALS, thus: 'He was buried in the Rajpur Cemetery. *See* CEMETERIES'. Each entry is intended to be an aid of first resort, and will normally contain (a) useful information (b) leads and references by which to explore the theme in greater depth; thus: SAUGUR [Sagar] is identified and its major significance in 1857–8 outlined, and, at the end of the entry, reference is made to *The Revolt in Central India, 1857–58*, Simla, 1908, and also **FSUP** (for abbreviations see below) Vol. III, which in turn either quotes from or directs the reader to the *Mutiny Narrative for the Saugur and Nerbudda Territories*. It will be seen that these leads and references are almost invariably monograph items in the Bibliography: this is done deliberately, as providing the most helpful extension to the text. ***The function of this COMPANION is not so much to tell the story of the Indian Mutiny/Rebellion of 1857, as to direct the scholar or the general reader to appropriate sources of information***. The Bibliography is the most comprehensive ever produced on the subject, and, if all else fails, a thorough scrutiny of it should provide a potential source of information on the chosen theme.

ANECDOTES

These appear in the same alphabetical sequence as Main Entries, but are identified by the headwords being in ***BOLD ITALICS***. Thus there might be two entries under **AGRA**; the one containing anecdotal material, or in loose terms, a story rather than a recital of facts, will be headed ***AGRA***. Anecdotal entries will usually be attributed, that is their source will be noted if it is known, but it will not usually have the additional leads and references appended as is the case with Main Entries.

ABBREVIATIONS

To avoid the constant repetition of oft repeated phrases, sources, titles, etc., the following abbreviations are used throughout:

ASW	Andrew Ward
Ball	Chas Ball, *The History of the Indian Mutiny*, London n.d. (1858?)
BL	British Library, London
BM	British Museum
BNI	Bengal Native Infantry
Bod	Bodleian Library, Oxford
Cav	Cavalry
CM	P.J.O. Taylor, *Chronicles of the Mutiny and other Historical Sketches*, Delhi, 1992

CSAS	Centre of South Asian Studies, Cambridge
DNB	*Dictionary of National Biography*
DIB	*Dictionary of Indian Biography*
EIC	East India Company
Forrest	Sir G.W. Forrest, ed., *Selections from Letters, Dispatches and other State Papers of the Military Dept., of the Government of India 1857–58*, 4 vols, Calcutta, 1893–1912
FPD	Foreign Political Department
FPP	Foreign Political Proceedings
FQP	P.J.O. Taylor, *A Feeling of Quiet Power*, HarperCollins, Delhi, 1994
FSC	Foreign Secret Consultations
FSUP	Bhargava, M.L. (ed. with Rizvi, S.A.A.), *Freedom Struggle in Uttar Pradesh*, 6 vols, Source Material, Lucknow, 1957–61
G.O.C.	General Officer Commanding
HEIC	Honourable East India Company, 'John Company'
HM	Her Majesty, in particular 'HM's regiments', meaning Queen's troops not Company's
IA	*Indian Antiquary*
IESHR	*Indian Economic and Social History Review*
ILN	*Illustrated London News*
Ind.Inst	Indian Institute Library, Oxford
Inf	Infantry
INTACH	Indian National Trust for Art and Cultural Heritage
IOL	India Office Library & Records (now incorporated in the British Library)
JASB	*Journal of the Asiatic Society of Bengal*
Kaye	Sir John Kaye, *History of the Sepoy War*, 3 vols, London, 1864–67
Lt.Col	Lieutenant Colonel
Martin	R. Montgomery Martin, *The Indian Empire*, 3 vols, London, 1858–61. Vol. II is concerned with the 'Mutiny'
NIR	*Narrative of the Indian Revolt*, London, 1858. Published weekly, price one penny
NA	National Archives, New Delhi
NWP	North Western Provinces, with Headquarters at Agra
OIOL	ORIENTAL & INDIA OFFICE Collections of the British LIBRARY
PJOT	P.J.O. Taylor (General Editor)
RHA	Royal Horse Artillery
SEN	Surendra Nath Sen, *Eighteen Fifty-Seven*, Delhi, 1957
SHERER	John Walter Sherer, *Daily Life during the Indian Mutiny*, London, 1898
SR	P.J.O. Taylor, *A Sahib Remembers*, HarperCollins, Delhi, 1994
SSF	P.J.O. Taylor, *A Star Shall Fall*, HarperCollins, Delhi, 1993
Thomson	Captain Mowbray Thomson, *The Story of Cawnpore*, London, 1859
Times	The London *'Times'* newspaper
Williams	G.W. Williams, *Depositions taken at Cawnpore etc.*, Allahabad, 1859

ACKNOWLEDGEMENTS

In a work of this nature it is right and proper that there will be a large number of individuals and institutions to whom gratitude and recognition are due. Inevitably also there will be oversight. Where this occurs may I unreservedly apologise: there is no intention on my part to either denigrate or ignore.

My first duty, or rather pleasure, is to thank the contributors to this volume. Though I have not met them all, I feel they are old friends, and that we are united by a bond, a common interest: it is very straightforward—we want scholars, and the general reader, to be helped to a wider understanding of the great events of 1857 in India. We have all written books or learned articles upon the subject; we have all been aware of the human as well as the dispassionately academic story involved; we are all dedicated to revealing truth. In alphabetical order the contributors are: Dr Seema Alavi, Assistant Professor, Centre for Historical Studies, Jawaharlal Nehru University, New Delhi, and author of *Sepoys and the Company: Tradition and Transition in Northern India, 1770–1830*, OUP, Delhi, 1995; Dr Gautam Bhadra, Reader in History, University of Calcutta; Professor B.M. Bhatia, former Vice-Chancellor, Jodhpur University; Ms Neelmani Bhatia M.A., Writer and College Lecturer; Sri Satish Bhatnagar, I.A.S. (ret'd), writer and authority on Lucknow; Dr Katherine Frank, Reader in English, Nottingham Trent University, and author of *Lucie Duff Gordon*, London, 1994; Dr Rosie Llewellyn-Jones, writer, researcher, historian, specialist on Lucknow and author of *A Fatal Friendship: The Nawabs, the British, and the City of Lucknow*, OUP, Delhi, 1985; Lieutenant-General S.L. Menezes PVSM SC, Indian Army (ret'd); Dr Rudrangshu Mukherjee, former academic, now Senior Assistant Editor, the *Telegraph*, Calcutta, author of *Awadh in Revolt, 1857–58: A Study of Popular Resistance*, OUP, Delhi, 1984; Dr Kirti Narain, Senior Lecturer, Department of History, D.G. College, Kanpur; Dr David Omissi, Lecturer in the Department of History, Hull University, and author of *The Sepoy and the Raj: The Indian Army, 1860–1940*, London, 1994; Jane Robinson M.A. (Oxon), writer and lecturer, and author of *Angels of Albion*, Viking Penguin, London, 1996; Dr Tapti Roy, Senior Lecturer, University of Calcutta, and author of *The Politics of a Popular Uprising: Bundelkhand in 1857*, OUP, Delhi, 1994; Andrew Ward, novelist and historian, specialist in Kanpur history, author of *Our Bones are Scattered*, New York, 1996, and *The Blood Seed*, New York, 1985.

To Andrew Ward I owe an especial debt, as he allowed me access to his copious and meticulously organised notes on Kanpur. To Jane Robinson and Katherine Frank I owe the stimulus to start—'if you've got all this in your head, why not produce an encyclopaedia?' To my Indian friends in Agra (Dr Neville Smith and the late Thomas Smith), Delhi (the Bhatia family, Mr R.K. Mehra and the staff of HarperCollins, the editor and staff of the New Delhi *'Statesman'*, Ms Prabha Chandran and others—too numerous to mention), Kanpur (Mr S.P. Mehra, the Mahendrajit Singhs and family, the Subhedar brothers, Mr & Dr (Mrs) Narain, etc.), Lucknow (Ram Advani, *the* bookseller of North India, Satish Bhatnagar, Mr and Mrs Sudhanshu Bushan and Ashok Priyardarshi IAS), and Calcutta (Mr Irani, Ravindra Kumar and Sujata Sen of the Calcutta *Statesman*, and Dr Tapti Roy), I have to say thank you, to you all, for your support and encouragement, and reassurance that the work of a *firanghi* such as myself would not be resented.

No serious student of this subject can ignore the British Library. The facilities and the material available at both the Reading Room at the British Museum, and at the old India Office Library and Records (transferred to the British Library in April 1982, and now known as Oriental and India Office Collections of the British Library) are unique and indispensable; I have to thank the staff for their assistance. Similarly the Centre of South Asian Studies, at Cambridge University, houses significant unpublished material, and I am grateful to the Secretary-Librarian, Dr Lionel Carter for his invariable amiability and assistance. The Bodleian Library, at Oxford, and in particular the staff of the Upper Reading Room, have been most supportive; without the facilities and comprehensive book stock of the Indian Institute Library there might have been no 'Companion': I am most grateful to the staff there whom I regard as personal friends and collaborators in the enterprise—Dr Simon Lawson, Mrs Elizabeth Krishna and Dr Gillian Evison.

I owe particular gratitude to Brigadier K.A. Timbers, the Historical Secretary of the Royal Artillery Historical Trust, based at the Old Royal Military Academy at Woolwich, who not only made available to me the facilities of the Royal Artillery Library, but introduced me to the 'Maude Album'—a collection of photographs dating from 1857–8: the student of the 'Indian Mutiny' will recognise some prints in this album at once, as the well-known work of Felice Beato etc., but what is unique about the collection is that each print carries the hand-written note of Colonel F.C. Maude VC, RA, as to the contents of the photograph and the personalities etc. involved: it is quite clear that Maude himself was present when many of the photographs were taken, and indeed appears himself in some of them. I have to thank the RA Historical Trust for permission to re-photograph, and reproduce, some of these remarkable prints.

Gratitude also to David Higham Associates, on behalf of author Dennis Kincaid and Routledge, Publishers, for permission to quote from *British Social Life in India*, and to include two drawings therefrom. And to Sheil Land Associates Ltd, for their permission to quote from *A Matter of Honour*, by Philip Mason, Jonathan Cape, London, 1974, in particular Philip Mason's masterly explanation, for the layman, of the caste system. Amita Baig, Director General, Architectural Heritage, INTACH (Indian National Trust for Art and Cultural Heritage), New Delhi, could not have been more helpful and co-operative.

There are many illustrations in this book: their quality, in terms of technical excellence, varies considerably; photography was young in 1857, and its exponents in India were few, but better a bad photograph than none at all; and better an inaccurate engraving by an artist who never got nearer to India than the East India Docks Road, than no picture at all? Thus many of the illustrations come from contemporary sources, *Punch, Narrative of the Indian Revolt*, Chas Ball's *History of the Indian Mutiny*, R. Montgomery Martin's *Indian Empire, Chambers' History of the Revolt*, Captain G.F. Atkinson's *Curry and Rice, etc.*, or from books published in Britain and in India during the great swell of interest in the 'Mutiny' that lasted for more than twenty years after 1857; many come from photographs taken by me or given to me by friends. The source of the Colour Plates is invariably stated.

Finally, I have to thank my family, particularly my wife Hazel, for patient acceptance over so many years of my Indophile obsession. The active support and constructive criticism of 'she who must be obeyed' has been invaluable.

Corpus periit, opera manent

P.J.O. Taylor
Lewes, East Sussex

October 1995

CONTENTS

Main Entries	1 – 354
Summary of Major Events	355
Glossary	366
Bibliography	372
Published Works, Indian Languages	406
Novels and Plays, etc.	408
Archival Sources	410
Newspapers, Journals and Periodicals	412

ILLUSTRATIONS

Between pp. 116 and 117
and between pp. 340 and 341

A

ABBAS ALI He was a Risaldar in the **Jodhpur Legion** who, with another native officer named Abdul Ali, protected the Europeans at Erinpura when the Legion mutinied in August 1857: in particular he saved the life and engineered the escape of Lieutenant Conolly of his unit. He was obliged, by *force majeure*, to ride off with the mutineers but afterwards, on the advice of Conolly, communicated with Captain Monck Mason, Commissioner of **Ajmer**, and offered to desert with a large party of his men and the guns and to bring the party into Jodhpur, provided, naturally, that he and his comrades were pardoned and reinstated. The Commissioner had to refuse as he was bound by the bureaucratic Calcutta directive that forbade him to make any agreement with rebels with arms in their hands. Abbas Ali was able to slip away from the mutineers at **Awah**, and to remain in concealment in **Bikaner** until he was duly pardoned - although thanked and congratulated might have been more appropriate! For details of this affair and the story of the Jodhpur Legion *see* I.T.Pritchard, *The Mutinies in Rajputana*, London, 1864, and, *The Revolt in Central India*, Simla, 1908.

ABBOTT, Captain H.D. Officer commanding the 1st Regiment of Irregular Cavalry of the **Nizam's Contingent,** part of which mutinied in June 1857, *see* AURANGABAD.

ABDUL ALI Risaldar of the Jodhpur Legion *see* ABBAS ALI.

ABID KHAN, General *See* PERSONS OF NOTE, APRIL 1859.

ABU, MOUNT The 'Hill station' of **Rajputana**, some sixty miles west of Udaipur. There were a number of non-combatant Europeans there when it was attacked on the morning of 21 August 1857 by men of the **Jodhpur Legion** from Erinpura, who had hitherto shown no signs of mutiny although it contained eight companies of Oude sepoys. For an account of this attack *see* I.T.Pritchard, The *Mutinies in Rajputana*, London, 1864.

ACCIDENT OR TRAP? The following story comes from Lucknow, November 1857: *'Just before sunset a large quantity of loose powder was found in one of the outbuildings, and Colonel Hamilton, who by this time had arrived, directed a party of the 64th Regiment he had brought with him to remove it for safety's sake, and in doing so it unfortunately got ignited (how, it was never known), some thought purposely by a hidden sepoy, but this was not proved. At any rate 13 or 14 men were blown up or scorched, two bodies were never found. One was found dead in the ruins, three died the first night, two have since died, and two more are likely to follow. One poor fellow I met coming out of the wreck was mistaken by a comrade for a Sepoy and shot by him through the chest. He was one of those who died the first night.'* **Dr Francis Collins** in *St George's Gazette,* 31 August 1894.

ADDISCOMBE The fine old red-bricked mansion of that name, set in 57 acres of well-timbered and beautiful grounds, was acquired from Lord Liverpool in 1809 by the Honourable **East India Company.** The purchase was made specifically to provide a military college to train the officers of the Company's army in India - with a particular emphasis on Engineers and Artillery. The first headmaster was a Dr Thomas Andrew, who taught Mathematics and Classics - a common enough combination for academics in those days. His Classics - Latin and Greek, and Ancient History, - would also be regarded then as a necessity for a general education. There is a story of one officer saying the only practical use to which he had been able to put his knowledge of Greek was to send messages in it during the 'Mutiny', secure in the knowledge that none but an English public school man would be able to understand them. There were 60 cadets at first, for whose board and education the Headmaster received £80 each *p.a.*: a sum with which he expressed his total satisfaction. A Mr Glennis taught 'Fortification', assisted by Mr Bordwine: Mr Shakespear taught 'Hindustani': he was assisted by a Mr Hasan Ali. There was also a military plan drawing master, and a Colonel Mudge acted as public examiner. Each cadet was paid £30 *per* half year, to cover the cost of books, materials, pocket money etc. At first they were admitted at 13½ - 16, but the minimum age was soon raised to 15, and the maximum to 18. It seems at first that the students gave little trouble: nothing is recorded against them 'save inattention to the study of Hindustani'. The prizes for successful candidates were appointments to India. So discipline was strict. Cadets were not, for example allowed to go into Croydon, nor even beyond the grounds without the Headmaster's permission; they had always to wear uniform, even when on leave; although not subject to corporal punishment, they could be fined, get extra drill, or be 'put in the BLACK HOLE', there to be fed on bread and water, provided only that the confinement should end each day at 9 o-'clock' . One of the earliest cadets, we read, was **Archdale Wilson**, of Delhi fame, or infamy. The cadets were not pampered. Many names well-known in British Indian days were associated with Addiscombe, mainly as cadets: eg Robert **Napier** (Lord Napier of Magdala), and Robert **Montgomery**, successor to Sir **John Lawrence** in the Punjab; Henry Durand, **Henry Lawrence, James Outram**, Richard Strachey, **Baird-Smith, Sir John Kaye, Sir Frederick**

Sleigh Roberts ('Bobs', the Commander in Chief in India at the end of the century) etc.

Frederick Roberts, aged 20.

The 'Mutiny' in India in 1857, and the demise of the Company as a result, sounded the death knell for Addiscombe, and it was closed: its last batch of officers received their commissions in 1860. With all its shortcomings - and they were many - Addiscombe was yet an unrivalled nursery of military Captains. **PJOT**

ADIL MUHAMMAD KHAN A **jagirdar** of **Bhopal** who took the opportunity presented by the mutiny of the **Bhopal Contingent** to rebel against the Begum. For an excellent and succinct account of these events and the uprisings at Sehore and Berasia, see K.D.Bhargava, *Descriptive List of Mutiny Papers in the National Archives of India, Bhopal*, Vol II, New Delhi, 1963. Adil Muhammad's title was Nawab of Ambapani; he had escaped from **Rahatgarh** when that place was captured by **Sir Hugh Rose** in January 1858. He remained active in rebellion for many months, although he did not recover well from a defeat by Brigadier Wheeler, commanding the **Sagar** District On 27 March 1859. He was again attacked near Basoda on 16 May and in a major fight at Gunapura on 23 June 1859 when he is said to have had a force of 2,000 men including 80 sowars and 700 regular sepoys. He again escaped. *See The Revolt in Central India*, Simla, 1908 p.189.

ADJUTANT GENERAL, The There is only one source for this story: Dr Collins of the 5th Fusiliers, but as he was a non-combatant and a medical man who had care of the wounded whether British or rebel, he may have been in a good position to report. He was in the **Alambagh** in the period between the relief of the **Lucknow Residency** in November 1857 and the final attack on **Lucknow** by the British in March 1858. He witnessed an attack made by the forces of the Begum **Hazrat Mahal**, on the Alambagh on 16 January 1858. '*...the rebels made an impromptu attack upon us, coming suddenly down on our right without any warning... Brasyer at the head of the Ferozepore Regiment (Sikhs) was already doubling off at the head of his men in spite of the enormous odds against him....on this occasion the Sikhs were so quickly upon them that they got amongst them before they had the opportunity of carrying off their dead and wounded. These therefore fell into our hands. Among them was their leader, said to be the Adjutant-General of the Forces, who was dressed up as the Monkey God. It was he who inaugurated the attack, and came riding at the head of his troops, indeed in the line of their skirmishers. He fell from his horse at about the first discharge, a rifle ball having broken his thigh-bone. The Seikhs were on him before he could be removed, and handled him somewhat more than roughly, for they gave him three fearful gashes with their tulwars across his forehead, two of which completely destroyed his eyes, and the third penetrated his skull down to the brain which could be seen pulsating at the bottom of the wound. Of course the fall of their leader hastened the discomfiture of his followers who looked upon him as holding a charmed life and as being invulnerable....After the repulse of the enemy the medical officers of the Ferozepore Regiment brought in the almost lifeless body of the leader of the attack, who was so maltreated, and notwithstanding his desperate wounds, and his resistance to treatment, at any rate in the first instance, he was still living and doing better than could be expected. He tore off the splints and bandages from his legs the first night they were put on, but later on allowed them to be replaced. The terrible gash across his chin had already almost united, and those across the forehead were looking healthy, and in a better state than could be supposed...two months later this man was not only still living, but there was every prospect that he would make a good recovery. His sight however was completely destroyed, and his further history I never heard.*' Reported in the *St George's Gazette*, 30 March 1895.

AGA BAKAR IMAMBARA, Lucknow Said to be the first Imambara ever built in the city, it was razed to the ground during the revolt of 1857, but rebuilt as an act of piety by Mirza Haidar Shikho in 1859. Courtesy, INTACH.

AGAR (or Augor) A town some eighty miles north of **Indore**, where the 5th Infantry **Gwalior Contingent** mutinied on 4 July 1857, killing the Adjutant, Lieutenant O'Dowda, but sparing the other officers (although the doctor and his wife were shot by persons unknown), who, with their families and those who had fled from **Sipri**[Shivpuri], wandered in the jungle for some twelve days before being taken to Hoshangabad by one Lalji, a Jemadar of the Gwalior Police. Lalji was still alive, aged 100, in 1908. *See* **FSUP** Vol III, *see also NEILL, MAJOR JAMES.*

AGRA, BATTLE of This encounter (not to be confused with the Battle of **Sussiah[Sasia]**), took place on 10 October 1857 virtually on the parade ground in front of the Fort. The British column under Colonel **Greathed** from Delhi had marched some forty-four miles in twenty-eight hours and appeared in the early morning under the walls of the fort whence virtually the entire European community observed them. They encamped at about eight a.m. on the parade ground and were eating breakfast at ten-thirty when they were attacked, in a complete surprise, by the rebels of the **Mhow and Indore Brigades**. Both British commanding officers - Greathed and Cotton were not present when the attack was started, and somewhat fortuitously, it was repulsed, and some 2,000 of the rebels reported killed. Then followed a grossly unfair punishment, by burning, of a number of villages around Agra, for 'failing to warn Agra'. Ironically the peasants had also failed to tell the Mhow and Indore rebels of the arrival of Greathed's Column, so, apparently, the surprise on the parade ground was on both sides. There are numerous accounts of the battle, the most succinct being in R.Montgomery Martin, *The Indian Empire*, Vol II., London, 1858.

Agra Fort

AGRA FORT The European and Christian community of Agra spent some months in the Fort at Agra, and many accounts later spoke of the 'siege' of Agra: this was misleading, as food and supplies - even mail - from outside the fort were never cut off, and the fort itself was never attacked by an enemy. The worst privations stemmed from the lack of privacy etc. For statistics of persons taking refuge within the Fort, *see* Letter No 1 in the anecdote entry *AGRA*.

AGRA A large and stately city, capital of the NW Provinces at this time. Situated on the banks of the Jumna[Yamuna] river, 139 miles south-east of Delhi. The troops in the station consisted of one company of artillery (Europeans), H.M.'s 3rd Foot and the 44th and 67th regiments Bengal N.I. The Lieutenant Governor, **John Russell Colvin** took charge of affairs as soon as the news of the *emeute* at **Meerut** became known. Charles Raikes, the Judge at Agra, described Colvin as 'exposed to that rush of alarm, advice, suggestion, expostulation, and threat, which went on increasing for nearly two months, until he was driven nearly broken-hearted into the fort'. Colvin himself wrote to the Governor General on 22 May 1857: 'It is a vitally useful lesson to be learnt from the experience of present events, that not one step should be yielded in retreat, on an outbreak in India, which can be avoided with any safety. Plunder and general license (sic) immediately commence, and all useful tenure of the country is annihilated. **It is not by shutting ourselves in forts in India that our power can be upheld; and I will decidedly oppose myself to any proposal for throwing the European force into the Fort except in the last extremity.'** Ironically he led the entire European and Christian population of Agra into the fort where they remained bottled up until **Greathed's** Column from Delhi had defeated the rebels in the battle of **Agra** on 10 October. *See also* COLVIN, Sir AUCKLAND

AGRA Recently there were discovered two letters, dating from August 1857. They were in the possession of Miss Vere Annesley, a direct descendant of Sir Elijah Impey, the first Chief Justice, under the British, of Bengal, and friend of Warren Hastings; her great grandfather was ***Captain Archibald Impey*** a Bengal Engineer officer who was serving at this time in **Cawnpore[Kanpur], Allahabad and Benares**[Varanasi]. In the same bundle of private family letters addressed by Archibald to his wife, are these two, without explanation or provenance. Neither is signed or addressed, yet they are quite genuine: they were clearly written from the Fort at Agra where the British had shut themselves up, in a panic, at the time of the so-called 'Battle' of Agra (more properly of **Sussiah[Sasia]** on 5 July 1857. These letters had not seen the light of day for nearly 150 years, and are therefore most exciting to an historian. *LETTER NO.1 'Agra 7th August 1857. As a messenger is going back to Benares, I send you a few lines tho' it is some time since you have heard from us direct, yet I doubt not you are* familiar with our position, besides the official and demi-official despatches from Mr Colvin. I have taken pains to send almost daily letters to Havelock containing all the news of any importance relating to us, Delhie etc, the General sends on such portions of intelligence as he thinks expedient. 10 or 15 days after the fight of the 5th July the Neemuch mutineers left Muttra for Delhi on the 18th. Since that we have not been directly threatened from <u>any</u> quarter. Gwalior we have all along felt to be our weak point, the mutineer troops threaten us with a siege on their way to Delhi. Their numbers were greatly exaggerated. Persons who ought to have known represented them 10, 15 and even 100,000. By a later statement rec'd by Major Macpherson from Gwalior, it is shewn that including Indore mutineers the cavalry is under 1,000, the infantry only 4,000 and 1,000 Ghazees, but they are strong in artillery. They have four batteries of 6 guns each, & 7 guns from Indore, but the latter without gunners. They have also a well stocked 3rd Class Magazine, and siege Guns. Scindia however has talked them over to stay till after the rains and after the rains please God we shall be able to turn the tables against them. Meanwhile a regular siege of Agra, which is the most important arsenal in these parts would have constituted a serious diversion in favour of the Delhie mutineers. The possibility of this danger is I trust passing away. The Indore mutineer troops have arrived at Gwalior. We have not had, I said, particulars of what has transpired since the Gwalior and Mhow Mutineers were not expected to pull well together. Scindia might possibly be unable to restrain them. He was every day finding his position more difficult; but we must trust that he will succeed in holding this force in hand. He has paid them large advances but that is no reason to question Scindia's present line of conduct. There are frequent communications but of an informal nature between him and Major Macpherson. At <u>Delhie</u> our position is improving. There was a lull in the enemy's camp and attacks for the last 8 or ten days of July & this was taken advantage of by our troops - strengthening our outworks. The Kumaon Ghoorkhas joined us on the 1st. Regular trains are established for the conveyance of stores etc from Umballa. The road is free from Meerut to Delhie via Baghpat, but the Country South and East is entirely and utterly disorganised. Our Cossids from Meerut and Delhie encounter much risk & several are reported to have lost their lives. Large premiums are paid for rapid deliverance (sic) yet we receive our letters most regularly. It had been remarked that the attacks of the Pandies (as the sepoys are now universally styled) have become feebler & feebler. They have lost heart. The reproaches of the king etc drive them out into the field with vaunting speeches as to dislodging us from the heights - but, the moment we turn on them they fly (as Greathed says) like the wind. The affairs of the 31st and 1st August are quite in accordance with this. A long concerted plan of attacking us in the front and rear utterly failed and a night attack protracted from 5 P.M. to 6 A.M during which the cannonade and musketry was kept up without intermission was resisted by us with very slight loss. On the 2nd an attack was made on our pickets at Subzee Mundee. It was so feeble as not to require us to call up the supports. Unable thus to succeed in offensive operations the Mutineers will find their position entirely altered by the arrival of reinforcement on the 10th. 1,000 Europeans and 3,000 Punjabees will enable us there is very little doubt to resume action attacks, but even with this addition, the enemy are so numerous and have such a weight of Artillery that it seems more than doubtful whether anything will be attempted in a general assault till your reinforcements join the Delhie camp. If Havelock is enabled to dispose of the Oudh Mutineers without much delay we are looking for an early advance this way, and we can send an amount of heavy guns, ammunition & mortars with the column. Delhie once settled, then the country to the north can next be settled also, though a campaign in Bundelkund and Rohilcund may be anticipated in the cold weather. You know of course that from Meerut to Cawnpore & from Nynee Tal to Etawah we have not a single post or officers exercising authority for us. This encourages the enemy to fling out offshoots from Delhie. We can do little in Rohilcund where the Nawab of Rampoor is holding Moradabad (it is believed bonâ fide for us). Between this and Meerut a petty Nawab Wallee Dad Khan professing to hold a patent from the King of Delhi, set up at a fort called Malaghar and from such beginnings raised a numerous band of followers & assumed the Soobah from Delhie to Allahabad. He has had the arrogance to send a naib soubah to Alligurh who within a week has with the countenance of part of the Mahomedan population established administration, raised levies, and commenced collecting the Revenue. I hope it may be possible for us to raise an expedition to chase the fellow away and put in 2 faithful Talookdars (who have already received a warrant for the purpose) in charge of this city. The Malaghar rebel has rec'd a serious check in attempting to extend his rule towards Meerut by the Militia and a party of the Carbineers & Rifles who carried his entrenchment, took his guns killing 92 cavalry. The guns were manufactured of telegraph post piles with bits of the wire for grape. It is possible we may make some rigorous attempt of this kind now, under the rigorous (sic) auspices of Col.Cotton, but we have no cavalry except militia and only some 700 European soldiers, we have plenty of guns but not well off for gunners either officers or men, and our artillery shed was perfectly crippled at the battle. The Nana extended his rule up to our very doors, sending purwannahs of soobecarries to Etawah, but as he is gone Etawah is now ready to receive us, but no European Officer can go without troops. At Futteghur (=Fatehgarh) our pensioner the Nawab Raees has entertained the Seetapoor mutineers & threatens to march on Cawnpoor. Havelock and Neil will no doubt give a good account of him. The armed boats from Calcutta will be of great use on the Ganges & Jumna. If we had only one or two as a guard we might send down a fleet of boats with the women and children. The prospect of large reinforcements in the cold weather is most gratifying.

Census of Persons in the Fort 26th July 1857:

Europeans --- ---- 1689
Eurasians---------- 1541
Native Xtians----- 858
Hindoos --------- 1157
Mahomedans ----- 299

5,544

Provisions well laid in for 6 months & to spare.'

LETTER NO.2 'We have again rapid intelligence from Delhie up to the 7th and 8th (August '57). Mr Colvin is sending extracts of these to Genl Neill which you will see, but I may recapitulate here. In a letter of the 7th Mr Greathed says that the rebels had sent a force from Delhie to Jujher (Jhajjar?) to enforce a demand of 5 lacs. Another body of 800 foot and 150 horse and two guns had been sent into the Doab to Malagarh to raise money. Murwallies called in for forced loans, the Ghazees (fanatics who have sworn to kill Kafirs or Europeans) are going home, they get nothing to eat. The attack of the 6th had died out at 2 PM when an attack on the Metcalfe picket was abandoned on the first show of resistance by us. Nicolson had arrived and the reinforcements were expected from 13th to 15th. They must have been about at Kurnaul. They had heard through Lahore of the messages of the 27th June regarding reinforcements from home. The 1st rumour of your victory between Cawnpore and Lucknow had reached the city with the addition that your troops had sacked Lucknow for three days. The letter of the 8th relates that great confusion had been caused the day before in the city, whereby 500 artificers and all the stock in hand of sulphur & saltpetre were blown into the air. The sepoys suspected Hukkeem Hussin Ullah Khan of being at the bottom of it and plundered his house. There had been no attack to speak of since the 2nd. Besides the Doab force, one had been sent towards **Muzzaffanagar** to levy contributions. The same perhaps as beforementioned as having gone to Jujjher. They have taken now to attacking our position with heavy field pieces behind the cover of their houses but their fire is ineffectual, they tried a heavy piece on the 7th but it was soon silenced. On the 8th all was quiet, health good and cholera ceased. <u>All are looking forward to the day when the approaches will be made</u>. You will see by Mr Colvin's letter to Genl Neill that he regards as an unfavorable symptom the ability of the enemy to be detaching bodies of troops to beat up for money. No doubt it shows that they have plenty of men, besides that their heart is broken for fighting at Delhie. I think that the words I have underlined are so full of promise and hope as quite to outweigh that consideration. In another letter of the 5th when speaking of the expectation held out by Col.Tytler of your marching towards Delhie from Lucknow on 29th he adds 'I don't think that the capture of Delhie depends now upon his coming'. From all which I gather that when the reinforcements arrive active operations and offensive measures will be recommenced, regular approaches made & please God a successful assault. When matters begin to mend at Delhie we shall afford to laugh at the detachments sent to the East & West for money - otherwise, of course their outward movements are not to be despised. F.Williams - Commissioner of Meerut writes on the 4th that spies had been discovered taking plan of the intrenchments - two were hung, and he thinks an attack there in force not at all improbable, especially as it is his opinion that the rebels are likely on evacuating Delhie to pass over into Rohilcund. As our siege train closes upon Delhie I should think that all ideas of attacking Meerut would be abandoned. The rebels appear to be strengthening themselves on the left bank, and to be throwing up intrenchments on the Eastern side of the bridge of boats, and Williams adds, that he cannot send out a party strong eno' to defy even 2 regts with guns - in addition to Wally Dad Khan and his Goujurs, the latter since his defeat at Malagam has got an addition of 500 men. Fancy the arrogance of the king and his advisers - if they had succeeded in their night attack, the king was to have held his Eed in Camp, and the termination of our reign proclaimed. His Majesty of Delhie has composed a couplet to the effect, that the mighty English who boast of having vanquished Rooss and Iran, have been overthrown in Hindustan by a single <u>cartridge</u>. The consciousness of power had grown up in the army - the cry of the cartridges brought the latent spirit of revolt into action. Greathed writes I fully believe this to have been the cause with most of the sepoys. Nothing new from Gwalior. 13. Locally our position remains the same as when I last wrote '(the letter I sent)'. We could take a portion of Delhie any day, but to do this we should have to leave our camp empty, and to risk the failing of our ability to keep the town after an assault. When the moveable column arrives (Nicholson's) you may expect to hear of our walking into Delhie - we shall not wait for Havelock. all well cholera left and barring the 75th the Europeans have not suffered from sickness. '   This letter ends without signature, and is neither dated nor addressed. Until one reads the penultimate sentence, internal evidence seems to point to it having been addressed to General Havelock himself, but this cannot be since he is referred to in the third person; ditto his chief of staff Colonel Tytler (the man who 'promised' to march on to Delhie after subduing Lucknow!). *See* 'Impey Letters' in the collection of **PJOT.**

AHMAD ULLAH SHAH Otherwise known as the 'Moulvie'. *See* MAULVI AHMADULLAH SHAH.

AHMADULLA SHAH *See* MAULVI AHMADULLAH SHAH.

AHMEDABAD A graphic description of the execution of eighteen mutineers at Ahmedabad is given in the *London Times* of December 3rd 1857, p.7. Ten were hanged, five shot from guns and three shot by firing squad. Until the arrival in January 1858 of their distinguished correspondent **W.H.Russell**, the material published by the *Times* tended to be sensationalist, little attempt being made to verify facts, and emphasis being laid upon gruesome atrocity stories.

AHMED YAR KHAN 'We were at Kaemgunge some fifteen or twenty days [until 25 September 1857?]. A deal of revenue was collected. Ahmed Yar Khan had four guns with a party of sowars (I heard they came from Sagar) and some matchlock-men. Subadar Bhoondoo Singh had ammunition and elephants, which Ahmed Yar Khan demanded from him, stating that as they belonged to the Sirkar, the subadar had no right to keep them, but he refused to give them up. On this, there was a disagreement. Bhoondoo Singh had about 500 men under his command; Ahmed Yar Khan 200 sowars and 200 matchlockmen. The next day was the Nishan Pooja. The 17th Native Infantry had their colours. Bhoondoo Singh arranged with his men that when volleys were fired in honour of the colours, they should load with ball, fire at Ahmed Yar Khan and his party, and make a rush at the four guns to capture them....... The next morning before the sepoys had assembled, Ahmed Yar Khan, with his sowars, seized all the ammunition, elephants and horses belonging to Bhoondoo Singh: the sepoys then forsook him and joined in plundering the treasure. The subadar and his eldest son were made prisoners, and the sowars took away his two daughters. They were girls about 14 or 15 years of age. No, they were not disgraced; the Nawab Ahmed Yar Khan interfered, and they were given up, and with the subadar and his son, turned out of camp. Some five or six sepoys of his own corps remained faithful to him. I heard that they all returned to Futtehgurh, and lived in the Lal seraie in the city. After this, two nawabs came from Shumshabad Mhow and took command. I do not know what became of Ahmed Yar Khan.' *See Deposition of John Fitchett.* **Williams**, Vol. III. pp l-lxxii.

AHSAN ALLY, General *See* PERSONS OF NOTE, APRIL 1859.

AITWURYA He was a man 'who a month subsequently was proved to be one of the principal butchers, or executioners, employed by the **Nana** in dragging away the bodies of the poor European victims, and who was foremost in stripping the dead of their garments and taking as booty all he could get upon their persons ...was hanged on the surree tree in the court-yard of the very building where he had so short a time back been the principal actor in dragging the dead bodies of the unfortunate women and children and throwing them into that horrible well while life was yet warm in some of them.' This refers to the massacre of the women and children in the **Bibigurh**. *See* W.J.Shepherd. *Personal Narrative. etc.*, Lucknow 1886, *see also* SARWAR KHAN.

AJIT GARH *See* MUTINY MEMORIAL.

AJMER A British controlled district, surrounded by independent States, in the centre of **Rajputana.** There was a magazine there, garrisoned by a company of the 15th Bengal N.I., and a company of the Mhairwarra Battalion was also posted for duty in the city. There were no British troops in the whole of Rajputana at the beginning of the outbreak; **Colonel George Lawrence**, when he heard on 19 May of the events at **Meerut** telegraphed to **Deesa** in the territory now known as Gujerat, for a Light Field Force to be sent to his assistance. This was sent and consisted of 3 Horse Artillery guns (European), a squadron of Bombay cavalry, 400 European soldiers and 200 Bombay N.I.. Their arrival, on 12 June at **Nasirabad** near Ajmer, did not completely prevent an outbreak in the area but it did much to calm the situation. One hundred of the British infantry, from HM's 83rd Foot came on to Ajmer. The city and the arsenal there were commanded by Taragarh Hill, on the summit of which was a Muslim shrine and a fort. The chief priests there offered their services for the defence of this important post, their offer was accepted and they performed this duty regularly and efficiently until all danger was past. The only 'outbreak' at Ajmer seems to have occurred on 9 August when about fifty of the prisoners escaped from the gaol: Lawrence himself started in pursuit accompanied by some of the leading Muslims of the city, and met the civil mounted police on their way back - they had killed fifteen of the prisoners and recaptured twenty-five. *See* I.T.Pritchard, *The Mutinies in Rajputana*, London, 1864

AJODHYA PRASAD, or Jotee Pershad There was a well-known 'portrait' of **Nana Sahib** circulating throughout India - rather like a Wild West 'Wanted' picture - for some years after 1857. The British wanted to catch him and bring him to trial, but the picture they distributed was ***not*** that of the Nana Sahib. Incredibly the same, mistaken, portrait was used to 'identify' **Rajah Kunwar Singh, or Koer Singh.** The picture of 'Nana Sahib' that they used is in fact a portrait of a banker called Ajodhya Prasad (or Jotee Pershad) originally of Meerut who also acted as contractor to the British (eg victualling Agra Fort during the 'siege'). Sometime before the mutiny/rebellion he gave his portrait to John **Lang**, a London barrister in gratitude for his successfully prosecuting a case for him. Lang in London in summer 1857, may have lent the portrait to the Illustrated London News to be engraved, as a joke.

Certainly this 'portrait of the Nana' was circulated throughout India at a later date, and Lang was in 'a real stew' in case his banker friend ever found out that it was he who had 'leaked' it to the Press.

AJOODEEA PERSHAD, Before the outbreak in Kanpur rumours abounded among the sepoys, and they were undoubtedly spread by those seeking to rouse them to mutiny. The case of the *CASHIERED SUBALTERN* illustrates this: no one was in fact killed in the incident, but : 'Soon after, I forget the date, one night, an officer shot a sowar; I know not for certain whether the latter died or survived, but I heard that he died, and that no justice had been done him. This circumstance led the sepoys to imagine that the English were displeased with them. I was also informed that the officers had ordered the sepoys to come into the entrenchment unarmed to receive their pay, and that this increased their suspicion.' See Deposition of Ajoodea Pershad. **Williams.** See also The *CASHIERED SUBALTERN*.

ALAMBAGH, THE Between the relief of the **Lucknow Residency** in November 1857, and the final assault on Lucknow in March 1858 by **Sir Colin Campbell**, the British Commander-in-Chief, the only British troops in **Oude[Awadh]** were in the Alambagh, some three or four miles from the city on the Kanpur road. They were under the command of **Sir James Outram**. Their presence within the Kingdom of Awadh was an offence to the **Begum Hazrat Mahal** who frequently upbraided her forces for their lack of bravery in not driving the foreigners from Awadh, even threatening that if they did not attack the British she would make terms herself with them. Stung by her reproaches the army did mount a major attack on the well-fortified Alambagh on 13 January 1858. The only detailed accounts of this attack are unfortunately all on the British side, but there is one objective account, by a non-combatant (Dr Francis Collins): *'Yesterday, by order of the Begum, the Pandies made a general attack upon us. It is by no means the first time that such an order has been issued, and though they have marched into the trenches, they have never had the pluck to show themselves in front of them. The General (Outram) had information, by means of spies that this time the attack would come off, so he had his troops prepared to meet it. Orders were issued that all Regiments were to be under arms by day break, and on the assembly sounding each brigade was to fall in on its own parade, and wait there for further orders. We never expected that the enemy would do more than just show themselves in the distance. At 8a.m. we heard their assembly sound, and shortly afterwards ours followed. The men fell in at once, eagerly and impatiently waiting for orders to do something more than stand to arms. There being a large wide and open plain between the camp and the enemy's entrenchment, we watched the Pandies marching down in thousands, through their lines, on our front and left front, and very shortly they opened fire on our pickets. This was quickly responded to by our men, for we could easily distinguish the sharp crack of the Enfields from the duller sound of the old Brown Bess, with which our opponents are armed. All were shortly hidden from us by the smoke from the guns of both parties, but we were afterwards told that the rebels threw out their skirmishers, which were supported by large bodies of infantry, and these again by their cavalry, while they threw into our batteries shot and shell from their own artillery without intermission. On they came, yelling like demons, and as they approached a picket, consisting of a subaltern and ten men of the Madras Fusiliers the latter withdrew his men from a tope they occupied, quietly and slowly, firing on the rebels who kept advancing, until he reached a second tope, immediately in front of the left village picket. The rebels now rushed forward to the first tope, so soon as they saw it was evacuated, expecting no doubt to carry everything before them, and anticipating an easy victory, particularly as he had retired on his supports in the left village. The rebels at once followed and took possession of the second tope, in which is a small ruined mud village, and established themselves there. This tope is only 180 yards from our left front village picket, where we had two guns masked and in position. As soon as they were fairly settled here, these hidden guns were opened upon them, and at the same time those of our heavy gun battery were turned on them; for the moment* *they stood in amazement, being quite taken aback, and the 78th and 90th regiments, which were drawn up in the village in their front, with a hearty cheer sent in volley after volley amongst them, completely discomforting them, so that they took to their heels as quickly as they could run, and sought the protection of their entrenchments. Before they got away they had into their midst six rounds of grape from the two masked guns independent of the fire from our heavy guns and from the rifles of the 78th and 90th regiments.'* The attack turned out to be a failure but an analysis of the cause again reveals the lack of leadership and military know-how on the part of the rebels: they lacked planning rather than courage. Probably 30,000 men were brought into the field by the Begum's generals but of those only 10,000 were actually used: the British used fewer than 500 to repulse them, plus the guns mentioned. From spies the British learned that the rebels were much dispirited by their repulse, and that they suffered a loss of about 400 killed, against only three (Sikhs) wounded on the British side. The attack had lasted from 8.30 am to 12.30pm about four hours in all. In terms of equipment the rebels were handicapped by not having the Enfield rifle (which they had effectively rejected *see* THE CARTRIDGE QUESTION), and in relying too much on round shot and having fewer shells (which did far more damage) to fire from their artillery. Dr Collins account is given in the *St George's Gazette* of 28 February 1895.

The Gateway to the Alambagh

ALAMBAGH, The Three miles south of Lucknow on the Kanpur road, the Alambagh or 'World's Garden' was the pre-1856 summer residence of the late Queen-mother of Awadh: it consisted of a fine large mansion, very strongly built, a mosque close by, an Imambara for the celebration of the Islamic festival of the Mohurrum (the Royal family of the Kingdom were Shiahs), and various other buildings, situated in the midst of pleasure-grounds, walled in with stone bastions at the angles. It was therefore, in the military sense, an extremely strong place, capable of fortification and with a sufficient force of guns and infantry capable of withstanding severe attack. When therefore, after the withdrawal of the British garrison from the **Lucknow Residency** in November 1857, **Sir Colin Campbell**, the British Commander-in-Chief, was seeking a strong point in which to leave a British force until he was powerful enough to attack and subdue Oude[Awadh], he chose the Alambagh, and left there a considerable force, under the command of Sir James **Outram** the Chief Commissioner-designate of Oude [Awadh]. The Alambagh was kept supplied by regular convoys from Kanpur, it being found impossible to obtain supplies from any local people. The presence of a British force in the Alambagh was a source of very considerable annoyance to the **Begum Hazrat Mahal** and her Durbar, for in a sense it negatived the victory they had achieved in driving Sir Colin Campbell from Lucknow in November 1857, and it represented a permanent threat to the kingdom, which was realised in March 1858 when the British army, reinforced, returned. So, in the three months November 1857 to February 1858, Outram was attacked no fewer than six times, on 22 December, 12 and 16 January, and 15, 21 and 25 February. The first attack was made by 4,400 men, and 4 guns, and they tried to get at the south of the position by circling round the right flank; but Outram drove them to flight by turning their rear and threatening their line of retreat. Although employed on civil duties Outram was a soldier by origin - indeed held the rank of Major General - and was unlikely to be out-manoeuvred by the forces of the Begum. The second attack was in great force, perhaps thirty thousand men, in two bodies, attacking simultaneously the right and the left of the Alambagh position: they were allowed to come somewhat nearer, and were then received with such a powerful fire of artillery, and musketry, as to drive them back at some speed. Twice they attacked, twice they were repulsed; they then attacked the centre of the Alambagh position but were driven off again by a very heavy and concentrated artillery fire. The third attack was similar to the second, but fewer numbers were involved, and consequently they were more heavily punished. The fourth attack, led apparently by the **Maulvi Ahmadullah**, though comparatively weak, and desultory, lasted longer than the earlier attacks and was pressed home with greater conviction, with the result that posts much closer to the British position were occupied: thence they could attack with greater suddenness. Again however they were defeated. The fifth attack was on the front of the position *and* on both faces, but before it came Outram had, providentially for him, been reinforced by additional troops, especially cavalry of which he had been deficient, and while containing the frontal attack, mainly by artillery fire, he launched his cavalry on the two flank attacks, who were thus assailed from their own flanks, and they suffered very severely. The sixth and last attack was the strongest of all, but Outram was well prepared for it. Checking the main body in the usual way by heavy artillery and musketry fire, he sent out large bodies of cavalry and light infantry both to the right and left of the enemy on his right flank, and caused them to fly precipitately, inflicting heavy loss: the remainder of the Begum's forces, seeing their comrades so badly defeated, also withdrew, in some confusion. This proved to be the final action, no other attempt was made to overrun the Alambagh position, and thenceforth it was used simply as a scene for the concentration of the various columns and troops from the south and west who were massing for the final attack upon Lucknow. There is little left today of the Alambagh, save the gateway, which is preserved by the Government of India Archaeological Survey Department, and just inside, the grave of Major General **Henry Havelock**, who died of dysentery soon after the second relief of Lucknow in November 1857.

ALI MUHAMMAD KHAN This was the former name, or alias, of **Mammu Khan.**

ALI NUCKY KHAN The Chief Minister of the ex-King of **Oude**, **Wajid Ali Shah**. He was an extremely able man, though it would appear he used his abilities rather to feather his own nest than to serve his master. He accompanied Wajid Ali Shah into exile in Calcutta in 1856, and was arrested with him and placed as a prisoner in Fort William on 15 June 1857, on suspicion of being in communication with the mutineers, although this now seems to be unlikely. Buried in his palace on the **Gomti** at **Lucknow** was a great fortune in treasure, some of which was used by the **Begum Hazrat Mahal** to pay her soldiers. Colonel **Sleeman** knew him well, and described him as 'one of the cleverest, most intriguing, and most unscrupulous villains in India; who had obtained influence over his master by entire subservience to his vices and follies, and by praising all he did, however degrading to him as a man and a sovereign.' See W.H.Sleeman, *A Journey through the Kingdom of Oude in 1849-50*, 2 vols, London, 1858.

ALIGHAR An important city some fifty miles north of **Agra**. It was significant especially for the fact that it spanned the line of communication up country. In 1857 it was garrisoned by four companies of the 9th N.I., who on the outbreak being first notified from Meerut, behaved steadily. On 19 May a *faquir* appeared and incited the sepoys to mutiny, but was seized by two of them and carried to the officer in command; he ordered an instant court-martial, and the native officers who constituted the court sentenced the prisoner to death. The next day the troops were paraded and the *faquir* hanged, no opposition or displeasure being shown. But a *Brahmin* sepoy later stepped from the ranks and abused his comrades for having betrayed 'a holy man who came to save them from disgrace in this world and eternal perdition in the next'. The troops listened, wavered and then enthusiastically declared for mutiny and a march to Delhi to join their comrades; before leaving they broke open the treasury, opened the gaol, burnt the bungalows, but neither hurt nor abused any Europeans, civil or military. The latter retreated to Hathras or Agra; Lady Outram (wife of **Sir James Outram**) arrived in Agra on 23 May footsore, as she had fled from Alighar without shoes.

ALIPUR (Or Allipore) The town 'one march from Delhi' where the army of the Commander in Chief **General Barnard**, from **Ambala** via Karnal, joined forces on 7 June 1857 with the Meerut Brigade commanded by Brigadier **Archdale Wilson**, which had just fought the battle of the **Hindun River**. The combined force then provided the nucleus of what was to be called the **Delhi Field Force** which took up its position on the **Ridge** to the north-west of Delhi and eventually, on 14 September, assaulted the city.

ALIPUR A writer signing his name T.E.H. who claimed to have traversed the Delhi district during the Mutiny wearing nothing but a pair of white trousers, answered **Judex** with testimony of his own. 'Everyone who was in the camp at Alipore, 35 miles from **Delhi**, on the 5th of last June,(1857), may remember the trial of a zemindar for the rape and subsequent mutilation of an English lady escaping from Delhi.' He had found the lady 'wandering in the fields almost exhausted, she, poor girl, having walked nearly 40 miles under a burning sun — thermometer 110 deg. in the sun (shade?). He then stripped her and made her sit under a tree in front of the village for all the villagers to stare at, and after two hours of this torture, led her away to a clump of trees, where, after enduring every indignity which could be heaped upon a woman, she was murdered. The brute, when brought before the Court, not only did not deny the charge, but boasted of the manner in which he had slain the innocent.' *See* London *Times*. April 1, 1858. p. 7. This is another example of what might be called 'pre-Russell' news from India in the *Times,* being uncorroborated and sensationalist.

ALL SOULS CHURCH All Souls at Kanpur, otherwise known as the Memorial Church, was designed by Walter Granville of the Eastern Bengal Railway. There is apparently no foundation for the delightful and highly plausible story that the church (which certainly looks Italianate) was designed for a village near Florence, but that the architect's plans were sent to India 'by mistake', leaving us with the pleasant thought that perhaps some Italian peasants are still wondering how a Victorian Gothic church was erected in their village! Cost 200,000 Rs. to build, its marble floor a gift of the Maharaja of Jodhpur. Built in 1875.

The design 'was especially intended to resist the intense heat of the summer's sun, but when the summer came, it was discovered that the absence of ventilation and the character of the material employed enhanced rather than reduced the temperature.' The building was later ventilated and from then on services could be held in all seasons. The chancel floor is of Minton's glazed tiles, and all the interior design owed its good taste to the District Engineer, Mr. Cleburne. A handsome brass lectern was donated by Mrs. W.C.Plowden, and a font of pure white Carrara marble was erected by the Officers of the 2nd Light Cavalry. There appears to be no truth in the report that **Queen Victoria** personally donated the funds for the font. *See* H.G.Keene, *Handbook for Visitors etc.*, Calcutta, 1896. p. 30.

ALLAHABAD MOVABLE COLUMN   This was the official name for the force under **Brigadier General Henry Havelock** which attempted to relieve **Kanpur** in July 1857, but which arrived too late to save the garrison or the women and children killed in the **Bibigurh**.

ALLAHABAD   'The City of God', built by the Emperor Akbar. At the confluence of the rivers Ganges and Yamuna (Jumna). It is on the Grand Trunk Road (Calcutta to Peshawar). The lofty and very strong fort dominates both rivers. The troops in Allahabad at the beginning of the struggle were (with the exception of seventy-four invalid artillerymen) all Indian, in particular the 6th regiment N.I. At first the latter enthusiastically averred their intention of remaining loyal to government and begged to be led against the mutineers. But on 5 June news reached them of the events in Benares[Varanasi], and they came to believe that the 37th N.I. after being disarmed had been faithlessly massacred by the Europeans. Their agitation was fanned by the Maulvi of Allahabad (**Liaquat Ali**). The mutiny of the 6th N.I. was accompanied by unusually brutal scenes, many of the officers being killed, including eight unposted ensigns, aged 16, who were bayonetted in the mess-room: one (Ensign Cheek) survived for a few days and his story, graphically related in the Narrative of one 'Gopinath Nundy' became well-known in England. *See* The *Times* 7 September 1857. The Sikhs of Lieutenant **Brasyer** remained staunch to the British and they and the remaining Europeans retired to the Fort and held out until the arrival of Col **James Neill** and the Madras Fusiliers a few days later altered the position. The story of Allahabad at this time makes remarkable reading: Disease, drunkenness and insubordination among the Europeans and Sikhs and their indiscriminate looting of the property of both friend and foe, added to the hatred already felt here for the English, and Neill's policy of destruction of suburbs and 'making a severe example by laying the city under the heaviest possible contribution, to save it from destruction also' must have aggravated this hatred. For a detailed account of the events in Allahabad at this time *See* R.Montgomery Martin, *The Indian Empire,* Vol II, London (n.d.) p292-303, *and* **FSUP**. *See also* WATKINS, SERGEANT MAJOR.

ALLAHABAD ATROCITIES   The British were responsible for many acts of appalling cruelty and injustice, 'justified', some claimed, by the stories from **Meerut** and **Delhi**. Nowhere was this seen with greater harshness than in the march of **Neill's** Madras Fusiliers up country through **Benares[Varanasi]** and **Allahabad**. 'Scouring through the town and suburbs, they caught all on whom they could lay their hands — porter or pedlar — shopkeeper or artisan, and hurrying them on through a mock-trial, made them dangle on the nearest tree. Near six thousand beings had thus been summarily disposed of and launched into eternity. Their corpses hanging in twos and threes from branch and sign-post all over the town, speedily contributed to frighten down the country into submission and tranquillity. For three months did dead-carts daily go their rounds from sunrise to sunset, to take down the corpses which hung at the cross-roads and market places, poisoning the air of the city, and to throw their loathsome burdens into the Ganges.' ..... 'It mattered little whom the red-coats killed, the innocent and the guilty, the loyal and the disloyal, the well-wisher and the traitor, were confounded in one promiscuous vengeance. To 'bag the nigger' had become a favourite phrase of the military sportsmen of the day. 'Pea-fowls, partridges, and Pandies rose together, but the latter gave the best sport. Lancers ran a tilt at a wretch who had taken to the open from his covert.' In those bloody assizes, the bench, the bar, and jury were none of them in bland humour, but were bent on paying off scores by rudely adminstering justice with the rifle, sword, and halter. — making up for one life with twenty. 'The first spring of the British Lion was terrible, its claws were undiscriminating.' Bolanauth Chunder, *Travels of a Hindoo*. London, 1869, 2 vols.

Allahabad Fort

ALMORA   The chief town of the Kumaon District in Rohilkhand, memorable as the site of the decisive contest with the Gurkhas in 1815. Garrisoned originally by the 66th BNI (Gurkhas). The British refugees from **Naini Tal** retreated to this place in July 1857. **Khan Bahadur Khan** from **Bareilly** sent a force to take Almora, but as they camped in the freezing hills Major Ramsay sneaked in with 'thirty gentlemen and the twenty-five faithful sowars' and attacked the sepoys as they lay bundled in their blankets against the cold. Ramsay and his men killed 114 rebels and dispersed the rest, who refused ever to attack Almora again. *See* William Butler, *Land of the Veda*, New York, 1872 pp. 290-1.

ALNWICK CASTLE   Houses the regimental museum of the Royal Northumberland Fusiliers, formerly the 5th Fusiliers who took part in the campaigns in India in 1857-58, notably at **Arrah** and in the final attack upon **Lucknow** in March 1858. The museum has a digest of the services of the regiment, but, of greater interest, the diary of **Dr Francis Collins**, an Assistant Surgeon with the regiment: although technically a noncombatant Collins was naturally very well-informed as to the day-to-day events of the campaign and his account includes some revealing anecdotes.

AMAR SINGH   Brother of **Kunwar Singh**, and co-leader with him of the Bihari rebels. He continued to fight and to defy the British long after his brother's death. For the best account of his activities *See* K.K.Datta, *Biography of Kunwar Singh and Amar Singh,* Patna, 1957

AMAZONIAN SURVEILLANCE   'One night, (June 1857, in **Wheeler's Entrenchment** at Kanpur) eleven prisoners were placed together in the main-guard, and as all available strength was required in action, the wife of a private in her Majesty's 32d Regiment, Mrs. Widdowson, volunteered to keep guard over them with a drawn sword. They were only secured by a rope, which fastened them wrist to wrist, but they sat motionless upon the ground for more than an hour, under the Amazonian surveillance to which they were subjected. Presently, when the picket returned, and they were placed under masculine protection, *they all contrived to escape*. Whatever was the influence that restrained them while under their female warder, it must be confessed that Mrs.Widdowson was as much blessed with great courage, as she was distinguished by rare physical strength.' **Mowbray Thomson.**

AMBALA   (or, as it was usually referred to at the time, 'Umballah'). Apparently it was here that some of the earliest discontent and distrust excited by the new **cartridges** had been most marked.. This was a military station 55 miles north of Karnal, 120 miles NNW of Delhi and about 90 miles SW of Simla. As at **Meerut** there were so many European troops (2,290 to 2,890 Indians) that no anxiety was felt among the government authorities as to the possible spread of the mutiny/rebellion. But in fact the 5th and 60th N.I. regiments were poised to mutiny, and only the presence of so many British troops prevented the explosion. The Commander-in-Chief, General **Anson**, came down from Simla , arriving at Ambala on 15 May, and held a Council of War. Incredibly none of the five senior officers there present lived long enough even to see the capture of Delhi, Generals Anson, Barnard, Brigadier Halifax and Colonel Mowatt died of cholera, Colonel Chester the Adjutant General, was killed in action. Hence there is considerable uncertainty as to what actually did happen and what was decided at Ambala. No official account was ever published of the Ambala *emeute*. *See* GENERAL ANSON. *See* also article entitled *The Bengal Mutiny* in Blackwood's Magazine, Edinburgh, 1858, p.387 letter in the *Times,* September 18th 1857, **Ball, Martin,** *and* F.Cooper, *Crisis in the Punjab*, Lahore, 1858, Henry Mead, *The Sepoy Revolt,* London, 1857

AMERICAN MISSIONARIES   'At the time of the Mutiny there were several Americans scattered around the Doab, most if not all of them engaged in missionary work. By 1850 five American boards of missions were operating in India, most in the Punjab or the south, but many Indians had no idea where America was and regarded Americans as a variety of Englismen. The Board of Foreign Missions of the Presbyterian Church in the United States (distinct from the American Board of Foreign Missions) was the largest in North India, and its headquarters at **Allahabad** included a printing press and orphan asylum that were entirely destroyed in the uprising. Reverend J.Owen was one of the few to protest the retributive excesses at Allahabad, and to deplore Colonel Simpson's ineffectual command. Another major mission was founded at **Farrukhabad** to care for orphans of the famine of 1837-8. By 1844 it included a tent factory that was famed throughout North India, and by the time of the Mutiny the married orphans of Farruckabad had established schools, workshops and a large and bustling 'Christian village' that were the envy of their British counterparts. In 1857 the orphanage and tent factory were in the hands of John E. and Elizabeth Freeman, the high school was run by Albert and Amanda Osborne, the 'Christian Village' was overseen by James and Sarah McMullin, all under the guidance of David and Maria Campbell. All of them were to die with the Campbell's two children, in the first flotilla of civilians that fled neighbouring **Fatehgarh** on 3 June 1857, only to be massacred at **Cawnpore[Kanpur]** nine days later. Their bravery and devotion was exemplary. The handsome, square-jawed Freeman, 'his eyes full of tears', had to be commanded to leave his mission, and when the flotilla was stranded on an island and fired upon by the rebels, he and Campbell led the others in a prayer service. 'My dear friends,' said Freeman as cannon balls ploughed into the sand around them, 'it is my belief that this is our last day on earth. Let us, before doing anything else, prepare to die'. He was only slightly mistaken; they were not killed until the next day in the massacre at the **Savada Kothi** that made **Nana Sahib** infamous weeks before the massacres at **Satichaura** and the **Bibigurh**. Another American who figured in the mutiny was Reverend William Butler, a Methodist missionary who had been refused permission by **Coverley Jackson** to establish a station at **Lucknow**. He eventually opened a mission at **Bareilly**, where Magistrate Robertson was more welcoming. In early June of 1857, Butler wisely declined his Presbyterian brethren's invitation to join them on their doomed exodus to Cawnpore[Kanpur], choosing instead to flee with his pregnant wife to **Naini Tal**. He left in the nick of time. Shortly afterwards the Bareilly rebels hanged Robertson and destroyed Butler's mission, burning his library of 1,000 volumes. After a hair-raising journey through rebel territory and up precarious mountain paths, Butler and his wife found refuge at Naini Tal (where another American he did not name volunteered his services as a Kentuckian expert in the uses of the long rifle). Butler later visited Lucknow to preach to the remnants of **Havelock's** troops - 'one of the highest privileges of my life' - and returned to Bareilly in time to visit the imprisoned **Khan Bahadur Khan**, the rebel 'Nawab of **Rohilkhand'**, on the eve of his execution. Butler's attempt to bring Khan Bahadur Khan to Jesus was less than successful. 'The Koran was enough', wrote Butler disappointedly, 'he wanted nothing more, and wished to hear nothing else', (indeed he was satisfied to have killed 'a thousand Christian dogs,' Butler heard him declare from the gallows, 'and I would kill a thousand *more* now if I had the power'). Another American missionary named Woodside made a similar visit to the prison cell of the Emperor of Delhi, **Bahadur Shah**, who was somewhat more polite. ('The old man assented to the general excellence of the Gospel,' Woodside told Butler, 'but stoutly declared that it was abrogated by the **Koran**.') American missionaries were generally admired for their tact, but Woodside was an exception. In September the Magistrate at **Saharanpur** threatened him with imprisonment for haranguing the Muslims of the station. In 1858 Butler toured the ruins of the Farrukhabad mission. 'All was destroyed and desecrated now', he wrote, 'and these dear missionaries and their wives were numbered 'among the noble army of martyrs' '. The American Missions lost about Rs 3,50,000 worth of property in the Mutiny, but outraged American Christians raised a great deal more, and the missions at Farrukhabad and Allahabad were eventually reestablished. It was not until the American Civil War increased the demand for Indian cotton that Indians began to understand where Americans came from.. Butler wrote of his experiences in *The Land of the Veda,* 1872. Other sources include Reverend Fullerton's and Ishuri Dass's accounts of the Farrukhabad mission in Sherring, *The Indian Church in the Great Rebellion,1859;* Alter, In the Doab, 1986; and the archives of the Presbyterian Church (U.S.A.) Department of History in Philadelphia, Pennsylvania.' **ANDREW WARD.** *See* Acknowledgements, and Bibliography.

AMETHIE Lal Madho Singh of Amethie had played an ambiguous role in the struggle. His fort had undoubtedly sheltered English men, women and children when in danger *viz* Mrs Goldney, Mrs Block and Mrs Stroyan *et al.*, and forwarded them in safety to Allahabad, and he seems to have been very badly treated by the British 'If I am rightly informed the authorities, without any proof, took it for granted that he was a rebel, and seized upon several lacs of rupees which he had at Benares; and, to his application for redress, he received, in reply, a summons to come in and surrender himself' (**W.H.Russell** in the *Times* 21 December 1858). This did not prevent his fort being invested on 11 November 1858 by **Lord Clyde** in person, but hostile operations were stayed by the submission of the raja, who protested against the decree for the disarmament of his followers and the surrender of his arms. He was eventually 'rehabilitated', but can have had little faith in British justice or gratitude. It seems likely that he was confused with **Beni Madhoo Sing**, who remained an honourable but implacable enemy of the British to the end.

AMNESTY When Government, in the **Queen's Proclamation** of 1 November 1858, made it clear that large numbers of the rebels could expect favourable treatment should they decide to surrender, some of the local British officials, even of high rank, 'chose to evince their repugnance to the amnesty in a most inexcusable manner'. **Mr Russell** gives a case in point: 'it will be credited with difficulty that a very distinguished officer of the government, whose rank in the councils of the Indian Empire is of the very highest, actually suggested to one of the officers charged with the pacification of Oude, *that he should not send the proclamation till he had first battered down the forts of the chiefs;* and yet he did so. Had a military officer so far contravened the orders of his superior, nothing could save him from disgrace and the loss of his commission. A more disgraceful suggestion could scarcely have been made to a man of honour; one more ruinous to our reputation, more hurtful to our faith, certainly could not be imagined.' If this attitude were widespread it would go some way to explain the reluctance of many leading rebels to surrender - they just did not trust the British officials to whom they would have to submit. *See The Times,* December 21 1858.

AMRITSAR The holy city of the Sikhs in the Punjab. There was the possibility that the Sikhs, shaken by the recent annexation of the Punjab by the British might make common cause with the mutineers. In the event the reverse was the case, and not only the warrior class of Sikh but also the agricultural class of **Jats** in the neighbourhood espoused the cause of the British with enthusiasm. The Deputy Commissioner of the Punjab, **Frederick Cooper**, had his base at Amritsar and appears to have had a good relationship with the predominant Sikhs. The evening of 14 May 1857 saw an alarm in the city as a rumour spread that the disarmed sepoys from **Lahore** might decide to move on the fort of **Govindghar** adjacent to the town. *See Ball. See also* SIKHS.

AMY HAINES The name by which **Amy Horne** was originally known in the initial accounts of her return to Anglo-Indian Society.

AMY HORNE *See* BENNETT, AMELIA.

ANDAMAN ISLANDS The Penal Colony established on the islands received many mutineers/rebels sentenced to transportation for life. Captain C.D. Campbell of the Indian Navy wrote to **Beadon** that on 18 May, 1858, 81 mutineers had been hanged for trying to escape from Port Blair in the Andaman Islands. Several had returned voluntarily, and many more escapees were killed by the hostile aborigines. Twenty were dead from disease and 'from the savages, who had attacked several navy crews with bows and arrows'. The Hon. J.P. Grant, President in Council, ordered the hangings to stop and prescribed corporal 'or more severe punishment.' Almost a third of all the convicts, 228 in March and April alone, had escaped within 51 days of arriving at Port Blair. Grant made 'no charge of inhumanity' against the superintendent, Dr. Walker, but 'deeply deplored the result.' Escape was 'a natural impulse, to which it is impossible to attach moral guilt.' A ban on native boats and a 'good flogging would have answered all purposes.' **NA**. HDP. 19 June, 1858, p 262. Called 'Kala Pani' (Black water) by its convicts, and dreaded for many reasons. 'Up to 3,000 mutineers' were sent there. *See* L.P.Mathur, *Kala Pani, History of Andaman and Nicobar Islands, with a study of India's Freedom Struggle*, Delhi, 1985.

ANANT RAO *See* DAMODAR RAO.

ANDERSON, George and SUBADAR, Manilal Bhagandas Authors of *The Development of an Indian Policy, last days of the Company: a source book of Indian History 1818-1858*, London, 1918. Particularly useful for analysis of permanent policy results of the mutiny/rebellion of 1857.

ANJOOR TEWARRI *See* SPIES, BRITISH.

A Spy - perhaps Anjoor Tewarri - Reporting

ANSON, Maj-General George (1797-1857) British Commander-in-Chief in India on the outbreak of the rebellion. He would appear to have been quite unfit for the post. He arrived first in India in 1856 and his military experience before that had been very limited indeed. The *Times* of 19 July 1857 carried a letter from one Major-General

The *Anticlimax* : an Artist's Impression of the Relief of Lucknow

Tucker which said 'I venture to say it will be found on enquiry that he was quite unequal to the occasion; and painful as it is to point to the weakness of one who was talented, amiable and gentlemanly, it is yet due to the country, and to those whose sons and daughters, and kith and kin, are being sacrificed in India, to expose the favouritism which in high places has led to many such appointments.' *The Daily News* went even further on 5 August: 'General Anson's death saved him from assassination. He was hated by the troops and they burnt his tents. He was quite unfitted for the post. Horses and gaming seem to have been his pursuits, and as a gentleman said, no court pet flunky ought to come to India. Everyone gave a sigh of relief when they heard he was gone. **Pat Grant** is come over from Madras, to head the army till orders come from England.' Raikes the judge in Agra, reported that a 'native journal' (unspecified) had in ***February*** 1857 declared: 'Now is the time for India to rise, with a governor-general who has had no experience of public affairs in this country, and a commander-in-chief who has had no experience of war in any country.' Anson died of cholera at **Karnal** on his way down from **Simla** to take command of the troops marching down to **Delhi**. He is said to have been the author of a popular book on the playing of Whist. *See* **DNB**.

ANTICLIMAX In November 1857, **Sir Colin Campbell** relieved the British garrison in the **Lucknow Residency**, and the occasion was not as wonderful in reality as in prospect. Nor does the following bear out the elaborate graphic displays which the imagination of artists (not present) subsequently produced: 'What an entry compared with the one we had promised ourselves!' an officer wrote to his parents. 'We expected to march in with colours flying and bands playing, and to be met by a starving garrison, crying with joy; ladies waving handkerchiefs on all sides and every expression of happiness; but instead of all that we entered as a disorganized army, like so many sheep, finding the whole of the garrison at their posts, as they always remained, and a few stray officers and men only at the gate to meet us.' *See* London *Times,* 1 February, 1858, p 8.

AONG, BATTLE OF **Havelock** and the **Allahabad Column** were advancing on **Kanpur** and had already beaten the forces of **Nana Sahib** at **Fatehpur.** They then had to fight two actions in one day to continue the advance, 14 June 1857. ' Early on the 15th, we marched on towards Aoung, where the General had been led to suppose he should meet with considerable resistance. His information was perfectly correct,— the village was occupied in strength. The enemy had intrenched themselves across the road, not indeed in a very formidable manner, but the village offered great cover in the walled gardens, thickly grown with trees, which flanked it on either side. From this shelter, a steady fire of musketry was kept up for a considerable time. It was in this engagement that the enemy's cavalry made more than one attempt to get round our force and cut off the baggage. Once or twice, they regularly charged, but as soon as the bullets of the baggage guard began to fly amongst them, they pulled up and galloped away in quite a ludicrous fashion. After a struggle of some little endurance, the village of Aoung was taken, and as it was supposed the enemy would try and injure the bridge over the Pandoo Nuddee, the General pushed on. The rebels had placed two heavy guns on the bank on the opposite side of the Pandoo, one a 12-pounder, and the other an old carronade, I think, of large calibre. These were fired straight down the high road, but Enfield riflemen were sent on through the fields to the river bank, and from that position, very soon dislodged the gunners. and the whole body then made off, leaving the guns. Some miserable attempts had been made to blow up the bridge, but quite ineffectual in their nature, and our troops marched across with perfect ease, and occupied the opposite bank. ....The battle of Aoung was fought early in the forenoon of the 15th, and the Pandoo Nuddee was forced to the best of my recollection, by about eleven o'clock the same day. There was therefore ample time for news of the repulse, and the steady advance of the British troops to have reached Cawnpore early in the afternoon. There is every reason therefore to suppose, that the fate of the unhappy captives was immediately made the subject of discussion. The decision arrived at is now known and execrated throughout the civilized world. It was decided that the captives should be put to death. The order was carried into execution about sun-down. There were four gentlemen, three of them of the Futtehgarh party, who by some mischance, or for some especial reason, had been reserved from the fate which had already fallen upon their male companions. These were first taken out of the Beebeegurh, and murdered on the high road. Then the general massacre commenced. It seems probable, that vollies (sic) were first fired into the doors and windows, and then that executioners were sent in to do the rest with swords. If the work was any thing like completed, it must have taken a considerable time. At length, the doors were closed, and night fell upon what had happened. The hotel, where the Nana had his quarters, was within fifty yards of this house, and I am credibly informed that he ordered a *nautch* and passed the evening with singing and dancing.' **Sherer.** *See also* BIBIGURH.

APOPLEXY 'A Highlander had covered his face with a grimy handkerchief and lay down from fatigue in the plain - his face a deep blue, contrasting with a fierce red beard gave him the most dreadful appearance. When they tried to rouse him he was found to be dead, although there was not a mark upon him.' From a letter by J.W.Sherer, Magistrate of Cawnpore, 17 July 1857.

APPA SHASTRI The **Nana Sahib** had a particular grudge against all who had taken the part of his adoptive father (**Baji Rao II**) 's widows against him; when British authority was swept away in June 1857 in **Kanpur**, he took a swift revenge: 'After this, the Nana blew away from a gun **Goordeen,** the Baee Sahib's agent, and his family; also confined me, Lalla Ram, and Appa Jee Punth, at the thanah, and put fetters on us; and, until the re-establishment of the British rule, we remained in that state, together with a number of other persons connected with the Baee'. *Deposition of Appa Shastri.* **Williams**.

ARCHAEOLOGICAL SURVEY OF INDIA This department is responsible for the maintenance of a large number of historical monuments, including many connected with the Revolt of 1857. As an example here is the list in respect of Lucknow: *2785 Cemeteries on La Martiniere road; 2786 Cemetery at Alambagh; Cemetery at Dilkusha; 2791 Cemetery at Fort Machhi Bhawan; 2800 Kaiser Bagh gates; 2805 Monument of 93rd Highlanders; 2808 Neil's Gate; 2809 Old Palace at Dilkusha; 2811 Residency Buildings; 2814 Sikandar Bagh buildings; 2829 Old Mariaon cantonment;* Courtesy INTACH.

ARCOT 'At the siege of Arcot in 1751, where, outnumbered and exhausted, Clive's garrison of 80 British and 120 native troops prevailed over the French-backed claimant to the Carnatic, the sepoys first proved themselves. Trained, drilled and led by British Officers they were to prove invincible in the wars conducted by the HEIC'. Philip Mason. *A Matter of Honour*, Jonathan Cape, London, 1974, p 36.

ARMY, *See* FORCES WITH REBEL GOVERNMENT AT LUCKNOW; *See also* STATISTICS.

Rustic Recruitment to the British Army 1857

ARMY REORGANIZATION, 'As might be expected, the events of 1857-58 had a lasting impact on the Indian Army. The changes which accompanied and followed the rebellion meant that the army of the 1860s differed significantly from that which had mutinied in 1857. Four major changes stand out -- in recruiting policy, in the army's ethnic balance, in its weaponry and in its discipline. In each of these areas, measures were taken whose chief purpose was to make a future mutiny less likely to occur and easier to contain. The changes in recruiting policy chiefly affected the Bengal Army, the largest of the three main military forces in colonial India, and the one from which most of the mutinous regiments were drawn. Before the mutiny, the Bengal army recruited mainly from higher castes (Rajputs and Brahmins) of Bihar and the United Provinces -- especially Awadh. As the events of 1857 showed, this policy left the discipline of the Bengal Army vulnerable to any discontent affecting this very narrow range of communities, who might all share similar grievances at the same time. Changes in recruiting policy after 1857 made this collective dissidence less likely. To suppress the mutiny, the British raised many new units, particularly from the recently-conquered province of Punjab, where the disbandment of Ranajit Singh's armies after 1849 had created a large pool of military unemployment. Their loyalty proven in 1857-58, many of the Punjabi units were retained after the mutiny, whereas most of the mutinous high-caste regiments of the old Bengal Army were disbanded and not re-raised. A large Punjabi element was therefore permanently introduced to the British military establishment in Northern India. (The composition of the Bombay and Madras Armies remained largely unaffected.) The shape of the Indian Army as it emerged from the mutiny was largely accidental, but it was consistent with a policy of 'divide and rule' and was later justified on those grounds. According to this strategy, the Army was divided into four 'watertight compartments' of roughly the same size, but of differing regional composition -- the Armies of Madras and Bombay, the new levies from the Punjab, and the remnant of the old Army of Bengal. These four very different elements were unlikely to share the same grievances at the same time, or to combine in mutiny. Any dissent in one army could therefore be crushed by loyal troops from the others. The broad outlines of recruiting policy remained the same until the mid-1880s, when more and more Punjabis and Pathans began to be enlisted in the search for the so-called 'martial races'. The second major new safeguard involved adjusting the numerical balance between British and Indian troops in favour of the former. Before the mutiny, the ratio of Indian troops to British in India had been around five or six to one. When the revolt began, small British garrisons in many cantonments were overwhelmed or were hard pressed to hold their own against more numerous sepoys. Following the recommendations of the Peel Commission of 1858-9, the number of British troops was increased and the number of Indian troops reduced. From the 1860s until 1939, the Army in India maintained an approximate peacetime ratio of no more than two Indian soldiers to every one British. The chief aim of this cautious policy was to ensure there would always be plenty of British troops at hand to deter or to crush another mutiny among Indian units. The third area of reorganization was in military hardware. The sepoys of the East India Company had traditionally carried the same weapons as their British counterparts. (Indeed it was the issue to the Indian troops of the new, more efficient, Enfield rifle - whose cartridges, bitten open during the loading process, were thought to cause ritual pollution - that had occasioned the outbreak of the mutiny.) The Indian Army also contained its own artillery units, including horse and foot batteries with light and field pieces. Many of these weapons were turned, often with deadly effect, against the British in 1857. Never again. After the mutiny, the Indian Army was kept permanently inferior in fire power to the British. The sepoys were stripped of nearly all their artillery, keeping only the light mountain guns. Almost all the remaining heavy weapons were kept in British hands until the 1930s. Furthermore, until 1914 the Indian infantry were normally armed with the model of rifle which the British Army had just discarded in favour of a newer and more deadly type. This deliberate imbalance in firepower was intended to secure a slight British advantage in combat should the Indian troops mutiny once more. One last change worked in the sepoys' favour. Although the mutiny was suppressed, the sepoys arguably gained bargaining power by making British military authorities nervous of another outbreak. Relations between officers and men suggest this is true. In ways difficult to measure, the tone of Indian Army discipline subtly changed after 1857, becoming more sensitive to the cultural concerns of the other ranks. For example, sepoys were less likely to be squeezed into uniforms of European cut; instead more and more regiments adopted turbans and loose pantaloons of indigenous style. Above all, much more care was taken not to upset or impede the troops' customary religious observances. After the mutiny, British officers tended to treat the religious feelings of the troops with increasingly scrupulous - and fearful - respect. The mutiny was a useful, if chilling, reminder to the British that the loyalty of the troops, so vital to the Raj, had to be earned, not taken for granted.' **Dr David Omissi, Lecturer in Imperial & Military History, University of Hull.**

ARRAH HOUSE 'Up till the end of July 1857 the government in Calcutta was pre-occupied with sending military aid - pitifully small and late - up country towards Allahabad, Kanpur and Lucknow via the new railway that ran just a few hundred miles to the northwest, and by the great arteries of the river Ganges and the ancient Grand Trunk Road that ran (and still runs) from Calcutta to Peshawar on the N.W.Frontier. It is just as well for the British the Europeans in Arrah did what they did. Without their stand, Bihar would have fallen to the rebels: the road from Calcutta, whence came all British reinforcement to the area most affected by mutiny/rebellion ie the North-west Provinces from the Punjab to Allahabad, would have been cut. It is for this reason that Giberne Sieveking speaks of the Siege of Arrah House as 'A turning point in the Indian Mutiny'. Arthur Littledale was the senior civil official present, though he clearly was not the leader. He was the Judge, appointed to the Saran District of Bengal. He was aged about 38. Herewald Wake, Magistrate, Bengal Civil Service, commonly known as 'Wideawake', was the leading spirit. About 35. Richard Vicars Boyle was the resident railway engineer in Arrah. He was then 35, but lived to be 86 and had quite a distinguished career building railways, especially in Japan. It was his idea to fortify a building, which he did alone, using his own materials on one of his own buildings. Next comes Mr Coombe, Bengal Civil Service, the officiating Collector. We have to thank John James Hall B.A.(Cantab) FRCS, Assistant Surgeon at the civil station of Arrah for the best contemporary account of the affair : he called it *Two months in Arrah in 1857'*, and it was published in 1860. Mr Colvin, Bengal Civil Service, Assistant Magistrate, bore a famous name : he was the son of the Lieutenant Governor of the North West Provinces of India, who, within a month or so, was to die in the Red Fort of Agra, not by the bullet of the mutineers but sick and worn out with work. His tomb is still there in the Fort. There was also a Mr Field, whose title was 'Sub Deputy Opium Agent'. Not all this band was British. There was a most interesting character called M.Delpeiroux, a Frenchman, another railway worker. He 'worked, fought, and talked with the buoyant vivacity peculiar to his French extraction.' Then there were two somewhat shadowy figures, Eurasians of whom we know nothing but their names and that they were clerks : Mr Godfrey and Mr Da Costa. And then, in some ways head and shoulders above the rest, was Jemadar Hukum Singh commanding the fifty Sikh policemen, 'Rattray's Sikhs', that the Commissioner at Patna, Mr **Tayler**, had sent a few days before the actual siege began. He was 6ft 2 ins tall. His favourite expression was 'Kooch purwa nahin!' (no harm done, no matter!). Syed Azimoodin Hossein, the elderly Deputy Collector of Arrah threw in his lot with the British although at great personal risk, for his family and possessions were at the mercy of the mutineers. On 8 June 1857 a letter arrived from Mr Tayler, the Commissioner at Patna, saying that mutiny was expected at Danapur at any moment, and that it was likely that the sepoys would depart thence, in a hurry, and head for Arrah. The Europeans of Arrah all gathered therefore at the Judge's house and spent the night there. The following morning all the men moved on to the Magistrate's house for a council of war: ironically there was not a soldier among them. The next day, 10 June, all the non-officials departed, some on horseback, others in carriages: they were a formidable party, and carried with them a veritable arsenal of double barrelled guns and revolvers. They headed for Danapur, and go out of our story, but not before perhaps we relate a somewhat discreditable anecdote, attributed to William Tayler the Commissioner of Patna: 'Some of the gentlemen of the railway who fled to Dinapore from Arrah are said to have done so in woman's clothes. An ekka was stopped and challenged by a sentry with the question 'quon hie' ('who is there?'), a voice replied 'ham aurat hie' ('we are women'), when the sentry not having heard so gruff a voice before issuing from fair female's lips, lifted up the curtain and found a burly red-faced Englishman.' We are not told what happened to him. On 16-17 June Mr Boyle, the railway engineer seems to have taken a decision of his own. Earlier it had been suggested that one house be fortified that all might enter if the situation proved critical. It had been negatived by the majority, on what had seemed sound enough grounds at the time. Fortunately for them all, Boyle had the will and the means to think differently. He had an immense store of new bricks destined for railway construction: he used cartloads of them

Arrah House Today

to build up the verandah arches of a two-storeyed building originally destined as a billiards room, 60 yards from his own house. No mortar was used but the walls were 'artistically constructed' and were sufficient defence against musket balls. The low arches beneath were entirely bricked up with the exception of a large loophole, and on the upper floor between the pillars a kind of breastwork was constructed, with numerous sandbags. Boyle put in a large supply of rice, grain, biscuits and water, and a small quantity of beer and brandy. He then proposed, not unnaturally, that the party should move in, or at least move in to his residence nearby: Objections were made to the idea; it was thought that the concentration of the Europeans in this one fortified place would cause panic in the town. On 20 June a letter came from Mr Tayler saying that the treasure at Danapur was about to be moved: this clearly implied criticism of the loyalty of the Danapur sepoys was almost certain to provoke the long-expected mutiny. He said also that there was a strong rumour to the effect that Kunwar Singh was 'tampering' with the sepoys of the 40th Native Infantry Regt., many of whom came from villages on his own estates. On the same day also a letter came from Buxar saying that sepoys from mutinous regiments to the west and north were flocking into Shahabad from that side. On the evening of 25 June, the following laconic epistle directed 'by express urgent' was received by the judge:- 'To the Judge or Senior Civil Officer, Arrah. Dinapore 25th July 1857. Sir, A revolt among the native troops at Dinapore is expected to occur this day. Stand prepared accordingly. Your Obed't Servant, W.Lydiard. Major. A.A.General.' Despite all their planning and preparation the news when it finally came to the Europeans in Arrah was a shock. There was good news too: Tayler thoughtfully sent fifty Sikh policemen under Jemadar Hukum Singh to reinforce the little garrison: they had arrived on 23 July. Then the last stragglers came in - Mr Cock, Messrs Godfrey and Da Costa left their homes in the town, so did Syed Azimoodin Hoosein; two servants came in also. On 26 July,

Defending Arrah House

a Sunday, information was received that the rebels had indeed headed for Arrah and that they were crossing the Son river: they actually fired on the messenger who brought the news to Wake. About 10 a.m. the last two Europeans, Messrs Delpeiroux and Boyle, came spurring into the judge's compound. They reported that the sepoys were now over the river in force and were burning railway works and bungalows. The judge and the others now decided that with dignity and without haste they should all go to Boyle's 'fortification'. It must have given Boyle much satisfaction to be vindicated thus. Wake in particular had been scathing about the 'billiard room fort': we must wonder if he now avoided Boyle's eye. They loaded a dog-cart with arms, and escorted by the Sikhs and a few of the Europeans on horseback they went to Boyle's compound. The local population seemed 'indifferent and careless spectators'. They went at first to Boyle's dwelling house for a few hours, and wrote letters and made last minute preparations for what they thought was to come. Many still thought they were in need only of a very temporary refuge while the sepoy horde swept through Arrah. The doctor had just had a delivery of six dozen cases of port and sherry: possibly because he did not want them to be looted by the rebels he had them conveyed, still not unpacked, into the fort at the eleventh hour. They were to contribute not a little to the health and good spirits of the garrison. The Sikhs also took in a supply of water. Halls does not count the two servants, but they were there also, making a garrison of 67 men. They were to face upwards of 5,000 rebels. That night they went into the fortified billiard room and bricked themselves up. The Siege of Arrah House was about to begin. On Monday 27 July the sepoys arrived. They first went to the gaol and released the prisoners; they were joined by the gaol guard and many of the townspeople. Next they raided the Treasury and looted it of the remaining government cash, approximately Rs 70,000. It seems that the three regiments of Native Infantry, the 7th, 8th and 40th, had just calmly walked out of **Danapur**, through the apathy of General **Lloyd**, and in the face of 4 large guns and 600 British troops who, metaphorically, watched them go. Major General Lloyd was subsequently dismissed from his command for this 'gross inefficiency'. The sepoys next assembled on rising ground 600 yds from Arrah House and a formidable sight they must have been to the defenders. They came on steadily, in disciplined order, by sound of trumpet, to within 200 yds, and then the charge was sounded. They came on shouting and firing. Arrah House returned the fire, the Europeans with their double-barrelled guns and the Sikhs with their carbines. Many sepoys were knocked over, and the charge was halted. Most of the attackers went behind a large house only 60 yds from the 'fort', and others took skirmishing order on flanks and rear sheltered by trees and outhouses: they kept up a continual fire all day and occasionally in the night. That first assault was the most fearful that the garrison had to endure: they had decided to sell their lives as dearly as possible. Perhaps this became known in some way to the attacking sepoys; certainly the latter did not press home their enormous numerical advantage. They could have kicked down or scaled the defences with ease, but they would have had many casualties in so doing and this they knew, and it held them back from all-out assault. They never again exposed themselves in a charge like the one on that first day: they had lost perhaps 30 men, and in the whole of the rest of the siege they probably lost another thirty, all picked off by rifle fire and sharp-shooting by the garrison. The Europeans contemplated a sally to spik a gun that was brought up: the problem was that they could not find a suitable nail or other spike to do the job. The nearest thing was an instrument in the doctor's bag: it turned out to be a surgical tool for withdrawing body fluid in cases of dropsy. Dr Halls offered it, but it was thought that its use would be so entirely unprofessional, it was refused. On Wednesday 29 July also, towards midnight, they all heard, with great excitement, the sound of regular volleys of musketry. It seemed to be not more than a couple of miles away, and they concluded that the relief they were expecting had now arrived. Some were not so sanguine as the night was pitch black and the sound of the firing did not seem to get any closer: indeed it got fainter as time went on. This was Captain **Dunbar's** ill-fated relief column that was decimated by the sepoys in a well-laid ambush among the mango groves just outside Arrah. One of the Sikh policemen who had accompanied the column slipped through the sepoy sentries by a stratagem , and calling to the garrison from outside was drawn up by ropes into the stronghold. As the little garrison was pressed for water by this time they decided to dig a well within Arrah House. In twelve hours it was complete, 18ft deep. Everyone had a bath. We can imagine that the sight of water running down the outside of Arrah House, from an improvised waste outlet, must have caused some astonishment to the sepoys. The latter now tried to smoke out the British whom they guessed were already in great discomfort from the heat, dust, and lack of fresh air, by burning a mound of capsicums to windward. Sure enough a terrible cloud of acrid smoke came pouring towards the house, but the wind changed and the sepoys themselves caught the full brunt of the smoke. This evening also a sally was made to bring in four goats - for food, and two cage-birds - from pity: they had not been fed or watered since the siege began and were in a pitiable state. On 30 July, a Thursday, after dark there was a great hullaballoo raised and deafening shouts of Maro! Maro! (Kill! kill!), but no all out attack came. Instead there were attempts made to bribe the Sikhs to give up the Europeans: Rs 500 each was offered. In the last two or three days there were only a few rifle shots fired although the cannonade was constant. On the night of 2 August they were almost despairing. There was no news of a relief column, and they guessed rightly that Patna and Danapur had given them up for lost. The mine they thought must by now be underneath the foundations and might be exploded at any time. The miserable conditions in which they existed were beginning to tell on their spirits. But the end, when it came, was altogether unexpected: a voice was heard calling them from outside. It said not to shoot, there was good news. Two men were standing close to the house behind a tree, they were allowed to approach the house, with great caution for a trap was thought likely. They certainly had news, great news. The two strangers came under the walls and said that the sepoys had been defeated 6 miles off, towards Buxar, at Bibiganj, by British troops under Major **Vincent Eyre** (the 'Cabool man'). The garrison sallied out but no sepoys were found: they had all gone, fled. The two guns, and powder, were brought in. The mine was examined: it was found to have been all ready: the powder was in thirteen large earthenware jars beneath Arrah House, the train was laid. Why was it not blown? It needed only a match to the train from the outside entrance of the mine for the garrison to have been blown sky-high. The Europeans were still not sure that all was well, so they remained on watch until 7 a.m. when two of the volunteer cavalry, men they knew from Patna, rode in waving their hats. The garrison gave three belated but hearty cheers. Soon the compound was alive with townspeople and the Europeans' servants: they came protesting their joy that the British had won; they brought in arms of all sorts, some wounded sepoys and others described as 'traitors'. Two more cannon which had just been mounted on the doctor's buggy wheels were also brought in. They heard almost with incredulity how the 'miracle' had happened. Major Eyre with three guns and a few artillerymen had been on his way up the Ganges by steamer, with orders to report to the General at Allahabad. He had reached Buxar where he had stopped for the night, and found 150 men of H.M. Fifth Fusiliers and one officer. The latter had expressed himself as quite happy to accompany Eyre on an attempt to rescue the Arrah garrison, provided he had it in writing from Eyre that he was ordering him to do so. With considerable moral courage Eyre agreed. If the expedition had failed, he would have been a ruined man, for he had no business turning aside from his main duty, or disregarding his orders which were to get to Allahabad as soon as possible. It seems that friends of the Arrah garrison at Buxar had persuaded him to make the attempt. The little column had had a hard fight of it but had put to flight the sepoys: these were now streaming away from the vicinity of Arrah and heading for **Jagdishpur**, the home of Kunwar Singh. The sepoy prisoners and others were promptly tried by drumhead court-martial composed of Littledale and Wake, with Major Eyre as President. Littledale and Wake often asked for mercy for the non-sepoy prisoners on the grounds that they had probably had no choice when the sepoys were in charge, but Eyre overruled them, and they were all hanged, in the compound of Arrah House. All the Europeans' houses were found to have been wrecked, but in some cases the servants had saved some of the valuables by burying them. Major Eyre was now reinforced by 200 of H.M. 10th Regiment, from Danapur, and proceeded in pursuit of the sepoys, and to the attack of Jagdishpur. 'Few events were more productive of immediate and tangible benefit to the British rule in India than the successful defence at Arrah'. So says the *'Friend of India'* on 14 January 1858. ' *See also* **SSF**. **PJOT**

ARRAH The town of Arrah is in the district of Shahabad, near the junction of the River Ganges (10 miles) and the River Son (8 miles). It is 24 miles from **Danapur**, which in turn is close to the provincial capital, **Patna**. In 1857 Arrah seems to have been best known as an administrative centre for the district, and for the possession of an unusually large gaol, in which there were 3-4,000 convicted prisoners.

ARTILLERY On balance the advantage where artillery was concerned lay heavily with the British, both in terms of trained gunners, good quality guns of heavy calibre, and ammunition. Time and again the rebels were defeated by the influence of the artillery, siege guns, field guns and, perhaps most significantly of all, horse artillery which was expertly handled and could keep up with cavalry in the pursuit of rebel soldiers. The rebel artillery was reported as a 'miscellaneous collection of old brass and iron pieces, chiefly of small calibre, but some have been taken carrying 12-pound shot, and others of larger size are reported to be in their possession'. The rebels' supply of ammunition was never good. At the siege of **Lucknow** they had large mortars but no shells for them so instead fired great logs of wood eighteen inches long by 12 inches diameter, or even large stones. Their capture and use however of the eight-inch howitzer at the Battle of **Chinhat** had dramatic consequences since it caused the death of the leader of the British within the Residency, **Sir Henry Lawrence**. *See Further Papers No.8 relative to the Mutinies in the East Indies,* Inclosure 31 in No.2 pp41-2, *and* Gubbins M.R. *An Account of the Mutinies in Oudh,* 1858, p 369. *See also* **FQP**. *See also* Julian R.J.Jocelyn, *History of the Royal and Indian Artillery in the Mutiny of 1857,* London, 1915. A truly detailed *military* history of the revolt with particular emphasis on the role of the artillery.

ARTILLERYMEN Casualties among artillerymen, particularly in static or siege warfare, were always exceptionally high, since the prime target for the enemy's guns would be your guns. Nowhere was this more apparent than in the siege of **Wheeler's Entrenchment** at Kanpur : 'The whole of the artillerymen at the station, 59 in number, were killed at their guns, except four or five, who survived till the embarkation. Our force of European troops at the time of the outbreak consisted of 59 artillerymen, 75 of the 32nd Regiment (invalids), 15 of the 1st Madras Fusileers, and about 50 of the 84th Foot, about 200 in all. I am sure that 100 of these fell during the siege.' Mowbray Thomson's answer, dated 8 September, 1858, to Mrs. Murray's account. London *Times*, 11 September, 1858.

ARTILLERY, post 1857 'Whatever the artillery force may be,' wrote Charles Raikes, 'one precaution seems indispenable. All and every sort of great gun should be kept in the hands of Europeans, and Europeans alone. ...***The possessor of the strongest artillery is the master of India.'*** see Charles Raikes, *Notes on the Revolt in the North-western Provinces of India,* London, 1858, pp 130-1.

ARTILLERY HOSPITAL This was a building some 500 yards from **Wheeler's Entrenchment** at **Kanpur**, and was the place assigned by Wheeler for the use of the loyal sepoys for whom he considered there was no room in the Entrenchment. It proved to be of little use: 'When I and Seeta Ram Pandey reached the Entrenchment, Major [Hillersdon] ordered us to convey all the arms and accoutrements of the sick, and of men on furlough, loaded on the artillery carts, to the entrenchment. When we reached the lines of our regiment, we, with those who had not run away, put the arms, &c., on the carts, and drove them to the Entrenchment. The whole party accompanied the carts without arms. Arrived there, the Major ordered us to unload the carts, but occupy the hospital barracks (there being no room in the Entrenchment) with a promise that arrangements would be made for our subsistence. Soon after Rs. 400 were distributed amongst us by the Major's orders'. *See Deposition of Ram Buksh.* **Williams.** Those who wished to remain were not allowed to come into the entrenchment but were told if they still desired to serve the Govt they could hold the Artillery Hospital which was about 500 yards from the entrenchment. Mowbray Thomson, Cawnpore: Report to the Commander-in-Chief: 9 March, 1858. 'We said in such a building we could not manage to save our lives, as the round shot would reach us from all sides. The General telling us there was no fear, and recommending us to look after the rear of the building, returned with the Major to the entrenchment. (He came, I omitted to say, and inspected the barrack.) A short time after this, we were sent for, and ordered to bring our men down from the Magazine, a distance of about half a mile from the barrack. We went there, and breaking the lock, loaded seven carts with ammunition, and brought them back to the entrenchment; after this we returned to the barrack. About this time, Kullunder Singh, 6th company, 53rd Native Infantry, who was on paymaster's treasure guard, came to the General, and reported that the Nana had plundered the treasury.' *Deposition of Bhola Khan.* **Williams**.

ARUNGABAD A small town between **Mohammadi** and **Sitapur** in **Oude[Awadh]**, and the scene of a massacre of Europeans fleeing from **Shahjahanpur** etc. *See* **SSF**. *see also* ORR, CAPTAIN PATRICK

ASHE, Lieutenant 'A few days afterwards, Lieut. Ashe, of the Bengal Artillery, arrived from Lucknow with a half battery, consisting of two nine-pounders and one twenty-four-pounder howitzer. These ten guns were all the artillery that could be brought to the position, and they constituted our sole means of defence by artillery; and the poor little mud wall our only bulwark.' Speaking of **Wheeler's Entrenchment**. Mowbray Thomson, *The Story of Cawnpore,* London, 1859, pp 30-1.

ASHE'S MISCALCULATION One of the regiments at **Kanpur** was believed to be more likely to remain loyal to government than the other three; this was the 53rd BNI. In a bizarre incident on 5 June 1857 when the 2nd Cavalry had mutinied as had the 1st BNI and the 56th was about to go also, the 53rd were fired upon by the British artillery (commanded by Ashe), using 9pdr guns from within the Entrenchment. 'The Oudh gunners though supposed to have too thoroughly implicated themselves in our favour nevertheless at this time [6 June] went off to join their mutinous brethren,' **Thomson** wrote in a passage he later deleted from his acount for the Commander-in-Chief, probably because of its implicit criticism of his brother officer Ashe. Ashe's calculation obviously appalled Thomson, who still believed his men (the 53rd) would have been staunch. To Thomson's way of thinking, by firing on the loyal 53rd Ashe had traded the loyalty of an entire infantry regiment for the shaky and, as it turned out, very temporary allegiance of a handful of **golundazes.** 'The gunners belonging to Lieut. Dempster's Battery were told by General **Wheeler** that if they did not wish to serve the Government any longer they were at liberty to depart and away they all went and commenced firing on the entrenchment immediately afterwards.' See Mowbray Thomson, *Cawnpore: Report to the Commander-in-Chief: 9 March, 1858.* **ASW**.

ASPIRANT FOR A CB 'Gong-hunting', that is the deliberate pursuit of some action that will lead, with luck, to the award of a medal for valour or meritorious service, is not a new phenomenon. The **Victoria Cross** had only recently been instituted in the British Army, and such was the prestige of its possessor that men would go to great lengths to obtain it, even foolishly exposing themselves to excessive danger: it is said that that is the way that Colonel **Neill** met his death; *see also* Frederick Sleigh Roberts's account of his efforts (successful) to get the coveted award, in F.S.Roberts, *Letters Written during the Indian Mutiny*, London, 1924. An officer who took part in the pursuit of **Tatya Tope** wrote - 'Each fresh commandant who took the field fancied he could catch Tantia; prodigious marches were made, officers and men threw aside all baggage, even their tents, and accomplished upwards of forty miles a day - the rebels did fifty. The end was, all our horses were sore-backed, and the halt of a week or ten days rendered absolutely necessary. *Then came a new aspirant for a CB* (Commander of the Order of the Bath) and Tantia's head, who brought fresh troops and camels into the field. He perhaps had not only to chase Tantia, but to keep clear of other forces commanded by a senior in rank to himself. It was wonderful the amount of energy that was thrown into the pursuit, and the hundreds of dead camels strewn over every jungle track: roads were no object, or rivers either, to pursued or pursuers. On they went till they were dead beaten...we had the very worst of information, even in the territories of professedly friendly rajas. The sympathy of the people was on their side.' See *The Revolt in Central India,* Simla, 1908.

ASSEMBLY ROOMS In Kanpur. Highly significant for functions, and an essential element in the social life of the Europeans in 1857 pre-outbreak. The building, much of which remains, is now a Bank. 'Built by private subscription the Assembly Rooms fulfilled the function of a club. They were used for dances and suppers, and there were billiard and card smoking rooms. The ballroom had the distinction of a boarded dance floor. It was a huge room with two long rows of pillars, brilliantly lit by chandeliers and wall shades, comfortably furnished with sofas and decorated with potted palms.' Zoë Yalland. *A Guide to the Kacheri Cemetery*. London, 1985, p 17. Oddly enough, while the Europeans were prepared to pay, by public subscription for Assembly Rooms, Theatres, Race Courses and schools for their children etc (in other words they considered enter-tainment and education entirely their responsibility), they were not prepared to pay for a Church - which they considered Government should provide: by some strange logic the care of souls was regarded as an official function, to be attended to by the proper authorities.

Assembly Rooms Today - the State Bank Building

At the end of November 1857 the **Gwalior Contingent** successfully attacked Kanpur, and 'The Assembly Rooms,' wrote an officer, 'with all they contained, consisting of 11,000 rounds of Enfield cartridges, the mess plate of some four Queen's corps, paymasters' chests and baggage untold of officers and men, fell into the hands of the Philistines, who dislodged our men, but whom we could not in return dislodge, though not more, certainly, than 450 yards from our batteries. Between the Assembly Rooms and church (which they also occupied), the enemy placed several guns in position, that worried us for several successive days.' See London *Times,* 27 January, 1858, p 9. 'We are blowing up the Assembly Rooms of Cawnpore in order to clear the ground in front of the guns of our entrenchment, and billiard-rooms and ball-rooms are flying up in fragments to the skies. Is not that a strange end for all Cawnpore society to come to?' **W.H.Russell** in the London *Times,* 13 April, 1858, p 9.

From Kincaid's 'British Social Life in India'

ATHARAH SAU SATTAVAN Hindi, literally, eighteen fifty-seven, a way of referring to the events of that year

ATROCITIES There were atrocities perpetrated on both sides, frequently the result of panic and ignorance, and as a result of the spreading of wild and unfounded rumours of ill-conduct by the other side. 'Meanwhile at Tirhoot: martial law has been proclaimed throughout the district; and Holmes at Segowley is hanging right and left, mostly sepoys returned from the scene of action laden with booty. Those that belong to the insurgent regiments will be hanged unless they are away on leave. This has had a very salutary effect.....Holmes is stringing up the fellows like a brick. Venables at Azimgarh is at the head of 150 soldiers (natives) and doing good service in the disturbed villages. He has hung a lot of rebels and sent an indent for twenty new ropes a few days ago. Give full stretch to your imagination- think of everything that is cruel inhuman infernal and you cannot then conceive anything so diabolical as what these demons in human form have perpetrated....We burnt seven villages on the road and hung seven lumberdars. One of these wretches had part of a lady's dress as a cummerbund.- he had seized a lady from Delhi, stripped her, violated and the murdered her in the most cruel manner, first cutting off her breasts. He said he was sorry he had not had the opportunity of doing more than he had done. We found a pair of boots, evidently those of a girl of six or seven years of age with the feet in them. They had been cut off just above the ankle. We hung many other villains and burnt the villages as we came along. A man who witnessed the last massacre in Delhi...'little children were thrown up in the air and caught on the points of bayonets or cut at as they were falling with tulwars.' Some journals, notably, eventually, the *'Times'*, were prepared to be restrained and temperate in their language and indeed to report what they considered British excesses, even at the height of the ill-feeling towards the 'mutineers'. The *'Times'* special correspondent, Wm Russell *q.v.*, wrote thus of the activities of Major Renaud: 'In two days 42 men were hanged on the road side, and a batch of 12 men were executed because 'their faces were turned the wrong way' when they were met on the march. All the villages in his front were burned when he halted. These 'severities' could not have been justified by the Cawnpore massacre *because they took place before that diabolical act.* The officer in question remonstrated with Renaud on the ground that if he persisted in this course *he would empty the villages and render it impossible to supply the army with provisions.'* One might hope that some humanitarian thoughts also entered into the argument! Brigadier General Neill's action was also condemned, though the criticism is more muted: - 'A shameful thing, although a good deal of excuse for him. *6,000 natives lost their lives and most of them were innocent.* Villages were fired over the heads of old women and little children, hanging parties went out, devising new ways of strangling with a rope, an elephant or a mango tree; amateur sportsmen 'peppered the niggers' with shotguns. Another Bloody Assize, an orgy of hatred and injustice.' Once again it is the English periodicals that produced the most exaggerated stories, and we are left to wonder at their motivation. Was it just sensationalist journalism, aimed at increasing the readership, or something more sinister *eg* preparing the public for a more intense effort to convert Indians to Christianity? ' Atrocities - Burning family alive - Killing by inches - cutting off noses, ears, fingers, toes; they violated mothers in the presence of their children - Bareilly bungalow fired and 40 killed as they ran out - Shahjehanpore church - 'all' murdered at Sunday service and heads and feet of the women and children strewn about the road. Jhansi women publicly violated then murdered and hacked to pieces. Delhi six European women in one room; one hid under the sofa. Others violated and beheaded, blood trickled under sofa - she screamed with terror, dragged out and sent to King of Delhi's harem. Little children 1 year old thrown into the air and caught on bayonets. The Beresfords father, mother and 6 babes murdered: the throats of the children cut with pieces of glass to increase their suffering. At Raee a wretch seized a lady from Delhi, stripped her, violated and then murdered her brutally, first cutting off her breasts. Another hiding under a bridge ditto. Party of fugitives from Delhi found a pair of boots evidently of a girl 6 or 7 years old - with the feet still in them. We select these facts at random from the Indian newspapers and the private correspondence published in London, and could add other details as incredible but unfortunately as true.... And shall there be mercy to such fiends?' - *Illustrated London News,* 8 August 1857: no attempt seems to have been made to check the 'truth' of these accounts. ' It is right to say that not only out of regard to decency but to avoid publishing that which would miserably

Colonel Neill fires Villages between Allahabad and Kanpur

and uselessly harrow the nerves of readers, we and doubtless many of our contemporaries abstain from publishing the most horrible of the crimes of these miscreant revolters.' - ILN as above, 15 August 1857. 'Cawnpore: the wives and children of the officers and soldiers consisting of 240 persons were taken into Cawnpore and sold by public auction when after being treated with the highest indignities they were barbarously slaughtered by the inhabitants. ' For God's sake, let us at them !' is in every European soldier's mouth.' - ILN as above, 5 September 1857. But *See also* Edward J. Thompson, *The Other side of the Medal*, London, 1925. He deals with English atrocities against Indians during the rebellion, and claims that native memories of the Mutiny are a heavy obstacle to friendship and understanding. Goes too far in his denunciation to be considered truly objective, but his book is valuable, and, almost, unique in its approach....One non-combatant (**Dr Francis Collins**) observed an incident near to the **Secunderbagh** in November 1857 at the time of the relief of the **Lucknow Residency**: *One wretched old man came out of the village opposite, not long after it was set on fire, he was unarmed, but how far he was or was not rebel I know not. When some of the soldiers standing near saw him, one said 'that's a sepoy'. The wretched old man sat down on the ground about thirty yards in front of them, with his back turned to them and a poor helpless creature he looked. One man then said 'I'll have a shot at him', and up went his rifle which, standing close by as I was, I knocked up, as also that of a second and third, and I believe also a fourth, when a man behind, fired at the poor creature, missing him and then another, and then I think another before the poor creature was hit, but this came at last in spite of any opposition I could make. Indeed the blood of the soldiers, was so up that I almost expected that they would have turned on me, being a stranger to them, none belonging to my regiment.* From *St Georges' Gazette*, 31 August 1894. Regretfully it was 'men of the cloth' who were most given to spreading sensationalist accounts without taking much trouble to verify for accuracy. Here is one of the worst examples: 'Neither age, sex, nor condition has been spared. Children have been compelled to eat the quivering flesh of their murdered parents, after which they were literally *torn asunder* by the laughing fiends who surrounded them. Men in many instances have been mutilated, and before being absolutely killed, have had to gaze upon the last dishonour of their wives and daughters previous to being put to death. But really we cannot describe the brutalities that have been committed.' He has already done so and no doubt this was deliberate. *See* Reverend Hollis Read, *India and Its People*. Columbus, 1858. p. 63. *See also* JUDEX

AUGIER PAPERS These are held in the Centre of South Asian Studies at Cambridge University, and contain a number of interesting documents, including two that relate to the story of the murder of *Major Neill* at Augur in 1887 (son of Colonel **James Neill**); and a typescript account of the mutiny at **Nasirabad**.

AURANGABAD   Once the capital of the dominions of the Nizam of Hyderabad, it was still, in 1857, a city of great importance, in the western region of the Deccan, and well away therefore from the main centres of the rebellion. A portion of the Hyderabad Contingent force (constituted by treaty with the British), consisting of the 1st Regiment of the Nizam's irregular cavalry, was, in May and the beginning of June 1857 stationed at Aurangabad, under the command of **Captain H.D.Abbott.** The latter had no reason to doubt the loyalty of the corps until the news of the revolt in the NW Provinces began to arrive. This caused much agitation among the Muslim sowars in the regiment, and they apparently persuaded many of the Hindus to join them in declaring that they would not march against the rebels in the north (rumours had been spreading that they were to be led against the mutinous regiments of the Bengal Army), nor would they fight against their King, ie the King of Delhi. That was on 13 June; apparently by 15 June matters had considerably changed, for the Hindus had separated themselves from the die-hard Muslims who were bent on mutiny, and indeed many of the Muslims, under Rissaldar Abdul Rhyman Khan, also expressed regret for their action and said they were now willing to obey orders. Abbott was instructed to arrest the ringleaders, try them by court-martial and carry the sentence into execution: a moveable column under General Woodburn was, he was told, on its way to help in the coercion. On the regiment being paraded under menace of guns, some men refused to stay, and when they mounted they were fired upon 'and a few were killed, and about a dozen or so were cut down by the dragoons'. One sowar stepped forward and fired his pistol at Abbott but it misfired and he was arrested, tried by drumhead court-martial and hanged the next morning. Thirty or forty of the cavalrymen were put on trial together with a subadar of the 2nd Infantry Hyderabad Contingent and some *golandauzes*. The affair ended with two men blown from guns, seven shot by dragoons, 'several' hung, 30-40 transported, 100 disbanded and turned out of the station, 50-60 flogged or otherwise punished. There was no further mutiny among the Hyderabad Contingent. It is worth remembering that the **Nizam,** and his chief minister **Salar Jung,** remained firmly on the British side throughout the struggle. For a detailed account of this affair *See* Charles Ball, *History of the Indian Mutiny,* London, (n.d.) p 425 *et seq.*

AUSAN SINGH   Reported as 'Captain Ausan Singh, formerly a havildar in the artillery attached to Captain Bunbury's regiment before the annexation, subsequently a subadar in the 2nd Oude Military Police under Captain John Hearsey, took a prominent part in the insurrection and raised a Cavalry Corps at Lucknow for the Begum'; *See* PERSONS OF NOTE, APRIL 1859.

AVA   'Wreck of the Ava. All contents lost, all passengers saved.' *See* ILN No. 907. 20 March, 1858, p 294.   ' The Ava has sunk, and with her my last letter from India. The second was merely leaves of my note-book, and some loose pages written to save a mail, on my arrival at Cawnpore,' *see* W.H.Russell, *My Diary in India,* London, 1860, Vol I, p 191. The Ava was also a disaster for future historians. It took **Gubbins**'s journals, **Thomson**'s notes, and the letters of **Edmund Vibart** to J. **Powers,** the Magistrate of **Mainpuri.** *See Mss.Eur.F18/11* in the IOL. The Ava set off from Calcutta on 10 March 1858, paused at Madras, and then struck a rock just off the coast of Ceylon on the afternoon of 16 March. As water poured into the engine room 'the crew were ordered to clear away boats to land the passengers, which was done without any accident. An officer was put in charge of each boat, and all the ladies and children, and then the gentlemen were embarked, with orders to lie by the ship until daylight.' The surf was too high on the nearby beach, and so the next morning they made for Trincomalee, where they all arrived safely. Thus **Thomson** and **Delafosse,** who were passengers, added another high adventure to their prodigious experience. *See* London *Times,* 7 April, 1858, p 10.

AVADH   *See* OUDE.

AWAH   A town in **Rajputana** where an action took place on 18 September 1858. Rebels had retired into the town and strengthened its fortifications. Brigadier General **Lawrence** had assembled a force at Beawar to co-operate with the Marwar troops but the latter were defeated by the rebels at Pali; Lawrence determined to attack the latter therefore at Awah but he could make no impression against the fortifications which were stoutly defended, so had to withdraw, thus giving the rebels the semblance of a victory. *See The Revolt in Central India,* Simla, 1908. *See also* Iltudus T.Pritchard, *The Mutinies in Rajpootana, being a personal narrative of the mutiny at Nusseerabad,* London, 1864.

AWADH, The Last Days of the Kingdom of: the Annexation of 1856   'At the close of 1855 Company troops assembled at the cantonment of **Cawnpore,** prior, it was rumoured, to the annexation of Awadh. When worried officials of King **Wajid Ali Shah** questioned the British Resident, Lieutenant General Sir **James Outram,** he said he could not account for their presence unless it was to quell a local disturbance on the Nepalese frontier. By January 1856 Outram was told the real reason (if he did not already suspect it), by the Governor-General, **Lord Dalhousie,** who had summoned him to Calcutta. The Court of Directors in London had given **Dalhousie** carte-blanche for annexation and the Resident was now briefed on how to effect this. 'The Instructions to Outram, the letter to the King, and the proclamations have been finished', wrote Dalhousie on 20 January

Lucknow, in Nawabi Days

1856, 'the Council agreed to them all on Friday. Either the King must give us the administration or we shall take it.' Outram left Calcutta three days later, with the papers, and arrived back in **Lucknow** on 30 January, 'anxious and desponding as to the success of the negotiations'. The Residency surgeon, Dr **Joseph Fayrer**, had already been asked to reassume his other job as Post Master, Lucknow 'at the approaching emergency', and most of the new administrative functions for the assumption of British rule were in place. Outram carried two proclamations with him, the first to be issued if the King agreed without demur to the Treaty of annexation; the second, if he refused, was couched in diplomatic terms, but its purport was clear. The East India Company would assume the government of the territories of Awadh 'for ever', whether the King liked it or not. He was given less than three days to decide. Outram had been told by Dalhousie to offer the King an annual pension of Rs 15 lakhs initially, but had the authority to increase the amount to Rs 18 lakhs, if he thought extra money would sway the King.

Both the Queen Mother and the King's brother pleaded against annexation, to no avail, with the Resident. The King himself, recognising the *fait accompli*, then sorrowfully took off his turban and placing it in the Resident's hands, said, 'Treaties are necessary between equals only'. Now that his titles, rank and position were all gone, it was not for him to sign a Treaty. Thus the Kingdom of Awadh passed into British hands at 9 a.m. on 7 February 1856. A new Governor-General, **Canning**, was appointed five weeks later, and Dalhousie sailed for England, having finished this last task. The annexation of Awadh, threatened for more than fifty years, had now been successfully completed, more in sorrow than in anger, of course, but at least it was done. The irritant of Awadh, the only remaining independent state in the great north eastern swathe of ceded and conquered provinces that stretched from Bengal to the Punjab, had been resolved. The Company officials were now in a frenzy of organisation. There was so much to do, so many long-conceived plans to be put, at last, into operation. Awadh was divided into four Divisions, each with a Commissioner, and the Divisions further split into twelve Districts, with Deputy Commissioners. James Outram, the last Resident, became the first Chief Commissioner. The Director General of the Post Office in India decreed that Lucknow would be the chief sorting office for letters in Awadh, and Dr Fayrer was busy setting up lines of communication to the stations selected for the new Commissioners. His staff of letter peons sped along them, bearing instructions, newspapers and law books. The **electric telegraph**, which the King had wanted to set up (and which had been refused by Dalhousie the previous year), was quickly extended from Calcutta to Lucknow, so quickly in fact that the engineers must have been standing by with their galvanised iron rods, copper wires and wooden supports on 7th February. Only a month later, Calcutta was complaining about the number of messages tapped out by eager Company staff in Lucknow 'upon matters of the smallest importance and of no urgency....and there is reason to doubt whether the messages thus sent have always been sent with the knowledge of the Chief Commissioner'. Surveying and road building was a priority, for 'without a Survey it will be impossible to organise the Police of the City, to lay down roads connecting the Military posts with the Civil Station, or to devise those measures for developing the [re]sources of the country which are expected at our hands and which are as politically expedient as they are necessary for the improvement of the province.' The Cawnpore road, 'our only communication with the province is rapidly falling into disrepair and requires a fresh layer of **kunker** along the entire line'. An Engineers Department was needed for 'the formation of Cutcherries, Treasuries, Jails and other public establishments of the Capital', and a Public Works Department was mooted by the Civil Engineer, Captain J.Anderson. Newly appointed Company officials inspected the 'public' buildings that they had inherited. The Judicial Commissioner, Mr **Ommanney**, found some prisoners 'in cages' in one of the Nawabi jails and the Kotwali jail, with 250 men in confinement ' in a condition of squalid filth'. He proposed that all four existing jails be closed and a new District Jail established that would hold 1,000 prisoners. More schools were needed, too, both for the many children of 'the poorer classes of Europeans in the City', and for the Indian children in the villages. But at the same time, a College founded by the King's father for teaching the Shi'a faith, with 205 pupils, was abolished. It had been established in the Asafi Imambara, but after annexation the King no longer paid the Rs 2,600 monthly costs, so the teachers had to be pensioned off. Indeed the financial burdens that the Company had to bear were much greater than they could have anticipated.

Provision had been made for the necessary costs of annexation, the salaries of the numerous officers and their staff, the new buildings and the roads, but what became increasingly apparent as 1856 unfolded, were the enormous debts that the King had left unpaid, sometimes for years. Every day seemed to bring forth new claims. The King had left Lucknow on 13 March to travel to Calcutta, where he hoped that his pleas to reverse the annexation would meet with a more sympathetic response. He had made no provision for the many staff who worked in his palaces, the gardens and the stables of Lucknow. By May his servants were 'greatly in arrears and clamorous for payment. These poor wretches were absolutely starving....' There were also hereditary pensioners, relatives, descendants of deceased kings, and a host of people about the Court, including muezzins, artists, umbrella carriers, dancing girls and attendants of the royal menagerie, not to mention the animals themselves, which cost Rs1,100 to feed *per* day. The animals were in fact the easiest to deal with - they were auctioned. But the total liabilities of the King, including unpaid loans, were estimated at Rs 40 lakhs, which the Company agreed, unenthusiastically, to meet. The departure of the King had led not only to financial hardship for his dependants, but the removal of judicial protection for them too. This was something else that Company officials had not foreseen. A summons for non-payment of a debt to a local trader, Ali Mohamed, against the widow of a former King had been served by the new Assistant Commissioner. Tactlessly he had used the familiar 'tum' and 'tumhari' forms of address and had omitted her full titles. The summons was returned and a debate then ensued on how far the King's authority still extended in judicial matters. Were his relatives and dependants now 'to be made amenable to our Courts', pondered Outram. The Governor-General ruled that 'the process of our Courts will not run' in the palaces of **Kaiserbagh,** the **Dilkusha** and the Park, but the King was asked to submit a list of residents there. No one who was not on this list could claim legal immunity, and the arrangements were only to last during the King's lifetime. (People could still, in certain circumstances, be 'extradited' from the two palaces) On the King's unexpected departure for Calcutta, however, the order was rescinded, so that no-one retained any special privileges.   The King's troops were another source of anxiety. After correct muster rolls were drawn up, they were found to number 60,349 men, most in arrears for the whole of the previous year and many for former years as well. Rs 12 lakhs went to them in back pay, but only half these troops could be accommodated in new Company Regiments, and in the Police Battalions and City Police. This left 31,000 discharged men, with pensions and gratuities roaming about Awadh, bearing arms which were their own property. Outram reported that 'They moreover openly display their sympathy for the late King and are ready to a man to join any standard which would hold out a prospect of restoring them to their former life of rapine. The spirit of dissatisfaction evinced by this large body of armed men, now let loose upon the country....is a very serious consideration.' These disaffected men were the most visible warning that the annexation of Awadh was not, after the first shock, to be accepted without protest. Other grievances were mulled over as people saw physical changes in the towns and countryside and established status overthrown by their new masters who believed financial recompense could turn the Indian heart. One of the most poignant pre-Mutiny events must surely be the three weeks that the deposed King spent in Cawnpore on his journey to Calcutta. He had arrived in Mr Brandon's mailcoach, followed by 300 or so followers, some of whom were then taken back to Lucknow by the Magistrate on Outram's orders. They included the Chief Eunuch, the physician, the 'Provider of the Table', the 'Provider of the Lights', the 'Master of the Wardrobe' and the 'Chamberlain'. With hindsight we can imagine how the inhabitants of Cawnpore, including the **Nana Sahib** , viewed their unfortunate neighbour's predicamant.'   **Dr Rosie Llewellyn-Jones.**

AYODHYA   A sacred city in Oude[Awadh], the ancient capital of the Hindu kingdom, associated with the God-King Rama whose story is told in the Ramayana. It played no significant part in the mutiny/rebellion save in its propinquity to **Faizabad** and **Shahganj**.

AZEEZUN,[Azizan]   Azeezun is one of the legendary heroines of the rebellion, and if legend is thought to be half way between myth and history then her reputation is secure. Unfortunately there is very little fact, though much conjecture, known about her. If **Ulrica Wheeler** and **Amy Horne** are the best known of the British heroines of Kanpur, then their counterparts on the mutineers' side are Azeezun [Azizan] and **Husainee[Hosaini]**. Attempts have been made, more than once, to give the two latter women a glamorous and significant role in the drama , either as scene stealing patriot or power behind the throne, but always the shadow outweighs the substance. The in-filling of evidence by guesswork has to be stretched to the limit - and only the very loving or the blinkered can find the detail plausible. No matter, we know they lived and that they played a part which they believed in. There is no evidence that these two women knew each other, or even met. In 1857 they both lived in Kanpur, and belonged to the world's oldest profession and they must at least have heard of each other. For they had more in common than the strangely addictive resonance of their names. Their link is that they were 'instigators of mutiny and rebellion, and perpetrators of the most fiendish acts of cruelty', or, if you prefer, 'Muslim women freedom-fighters': it depends on your point of view. Before the sepoys broke out into open revolt in Kanpur on 4 June 1857, there was much secret coming and going of conspirators. Exactly who these were is difficult to prove: successful conspirators are keen to advertise their involvement; it is not so when the plot fails. Some say the Nana Sahib was one of them, and that Azimullah Khan, his vakil, was the arch planner. A

Subhedar or senior native officer of the 2nd Cavalry was certainly a leader. A trooper of the same regiment, by name Shamshoodeen was in the plot: his role was limited to providing a house in Kanpur where the plotters could meet; but he certainly 'built up his part' in the drama to impress his girl-friend: this was the courtesan Azeezun. Shamshoodeen told her, not long before the Mutiny broke out, that already the Nana Sahib held the Magazine and the Treasury, and in a day or two 'would be paramount in all things, and then...then she would see... he would fill her house with gold mohurs'. He was in liquor at the time and doing his best to show off, but his words are well attested and probably reflect the truth, *ie* that the Mutiny in Cawnpore was planned for 4 June, and that the Nana Sahib was involved. The train was laid, all it needed was the match. Azeezun far from concealing things actually boasted of the wealth that would soon be hers. At this time both General Wheeler and Mr Hillersdon the Magistrate trusted Nana Sahib implicitly: they had appealed to him for help and he had responded by sending 300 troops, plus guns, to guard the Treasury at Nawabganj and the great Magazine containing a great deal of warlike stores, on the banks of the Ganges. There was one man who did not trust the Nana, and was sure that a conspiracy existed and that the Nana was involved in it. This was **Nanak Chand**, a lawyer of Kanpur, loyal to Government and, providentially (or conveniently) keeping a diary right through this period. His record of detail is meticulous, as is his naming of names. It is this latter habit that causes some to cast doubt on what he says for there are many names which, as a lawyer with past scores to settle, he might wish to see brought to the attention of an avenging Government. It is his pre-occupation with the name of Azeezun that is intriguing. It is as though he cannot put her name aside for long: he is fascinated, if not by the personality and the events, then perhaps by the beauty of the courtesan herself. Whether she was really so important as he would have us believe is debatable. The 'dastardly rebel' theorists would say not, the 'beautiful freedom-fighter' lobby would have us believe all, and more, that Nanak Chand tells us. Who then was Azeezun? She lived in the street of the brothels in Kanpur, and by most accounts was a favourite of the cavalry, although she later denied this, claiming that she belonged to one man only and was not permitted to see others. Colonel Williams in 1858 painstakingly took down the evidence of many of Kanpur's citizens when Government was trying to see what really happened in May - August 1857: this evidence is still available for us to read, but tantalises rather than informs. So many witnesses had axes to grind, scores to settle or deeds to conceal: there is truth there, but there is also fear, and with it come lies, or what is almost worse from the historian's point of view, a string of 'facts' that the witness thinks the Government wants to hear and so obliges. We have little else to go on: without Colonel Williams's summary of evidence almost nothing would be known of Azeezun. Kunhya Pershad, mahajun of Kanpur, made this statement in his evidence : 'Azeezun was born in 1832, left motherless when very young, and brought up in the house of Umrao Jan Ada, the famous courtesan, at Satrangi Mahal, Lucknow. She moved to Kanpur and when the flag was raised against the British, she appeared on horseback in male attire, dressed in the uniform of her favourite regiment, armed with pistols and decorated with medals. I saw her as thousands of others did.' So far so good: we have other witnesses that testify to the basic facts, but hardly to the same effect. For example, one modern interpretation reads as follows 'Kanpur was sacked by the freedom fighters on 4 June 1857. Leading them was Azeezun Begum, a charming danseuse who set an example of organizational skill and patriotism. This fair damsel was born in Lucknow in 1832 and her parents were Husain Khan and Hamida Begum. She lived in the house of the famous courtesan Umrao Jan Ada in Satrangi Mahal. Her mother died soon after her birth. In response to the call of Nana Sahib, Azeezun collected a party of young women and set about doing her duty.' Sounds interesting, but we have no details. In the 1860s Sir George Trevelyan in his excellent account of Kanpur's troubles, seems to be quoting Kunhya Pershad when he speaks of Azeezun :

Indian Cavalry, 1857

'She appeared on horseback amidst a group of her admirers, dressed in the uniform of her favoured regiment, armed with pistols, and decorated with medals.' But he adds his own summary: 'Azeezun was the Demoiselle Theroigne of the revolt'. Let us see what Nanak Chand says of her, remembering that his is very much a contemporary account: his 'Narrative of Events at Cawnpore' is based closely upon the entries in his diary. He tells us that Azeezun chose to make one of the gun batteries her headquarters. This was the battery to the north of Wheeler's entrenchment, between the racket

Court and the Chapel of Ease: it fired shot and shell into the entrenchment almost from the first day of the siege, and must have killed many . 'It shows great daring in Azeezun,' he writes, 'that she is always armed and present in the batteries, owing to her attachment to the cavalry, and she takes her favourites among them aside, and entertains them with milk etc on the public road'. Nanak Chand's pre-occupation with the girl is intriguing: the fascination she exercises over him is obvious. He may have been physically attracted to her, or she may have refused him her favours, or perhaps it is a combination of the two: not many pages of his narrative pass without some mention of her. He too reports the story of Shumshoodeen the trooper boasting of the wealth he would shower on her when the Nana 'was paramount'. He heard the tale on 26 May, and immediately drew up a petition to Government declaring that a conspiracy existed and that the Nana was a traitor, but, he declares, no one would listen, no one would hear a word against the Nana, he claimed afterwards. Hillersdon the magistrate told him bluntly to desist: Nanak Chand's animosity to the Nana was well known, he was told; was there not ample evidence of this from the numerous court cases he had brought at the instigation of Bajee Rao's widows who claimed the Nana held them prisoner and had stolen their wealth? Nanak writes 'I told Mr Hillersdon that the Nana had long harboured enmity to the government, and a great number of rascals belonged to his party, that he (the Magistrate) would remember my caution, and that I had obtained certain intelligence from the men of the Nana's household. But the Magistrate would not listen.' As the latter did not survive the Kanpur massacre, we do not know if he regretted his brusque dismissal of Nanak's petition, or indeed whether he ever received it. After the Mutiny was put down in Kanpur, Government was very interested in the story of the 'raising of the green flag'. When the troops had mutinied and the Europeans had all retreated into Wheeler's entrenchment, someone sought to rouse the Muslim citizens of the city by appealing to their religion and declaring the jihad or Holy War by the raising of the green, Mahumdie, flag. It was thought that this 'someone' might still be in Kanpur and could be brought to book by Government. It was difficult to pin the deed on any one person, but Colonel Williams did his best, and as his interrogations continued so more light was shed, by coincidence on Azeezun. Let us look deeper into Kunhye Pershad's evidence: Question: 'Do you know whether Aseesun, a prostitute, lived in the city?' Answer: 'Yes, one Aseesun, a prostitute, who was in the service of Kulloo Mull, lived in the Lurkee Mahil, in Oomrao Begum's house. She was very intimate with the men of the 2nd Cavalry, and was in the habit of riding armed with the sowars. the day the flag was raised she was on horseback in male attire decorated with medals, armed with a brace of pistols, and joined the crusade ....I was on bad terms with the 2nd Cavalry as four troopers had been in prison on complaint from me and I feared vengeance. I knew that many of the troopers frequented the house of the prostitute named Aseesun, and bought over her servant Emambux. She informed me that the sowars of the 2nd Cavalry were plotting with the Nana.' (He then goes on to corroborate the story that Shumshoodeen had promised Azeezun a houseful of gold coins). Then comes the naming of names, that Colonel Williams was no doubt hoping for. 'The Kazi Waseeoodeen and the Moulvie Salamut Oollah one day concerted together, got ready a green Mohumdie flag and set it up on the open ground near Moulvie Salamut Oollah's, near the old Assembly Rooms. The Mussulmans congregated round it from the early morning and some measures were concerted. But in the afternoon more than 4,000 men were collected and went by way of the Filkhana Gate to attack the entrenchments, were defeated and fled back' (Another source says that a single cannon-ball dispersed them). What was Azeezun doing ? Thomas Smith, the late doyen of Agra Indian journalists has no doubts. 'The company of fair warriors under Azizan fearlessly went around cheering the men in arms, attended to their wounds and distributed arms and ammunition. They also acted as informers and messengers. Azizan was an active member of the revolutionary high command. She enjoyed the confidence of Tatya Tope, Nana Sahib, Azimullah Khan and other leaders...She in fact became a legendary figure and her beauty defended her. After the fall of Kanpur she was brought before Sir Henry Havelock who was so struck by her comeliness that he could hardly believe what she was accused of. To all appearances Havelock could have given in to his heart and spared her life, had she only compromised but Azizan knew how her weakness at this juncture would affect the morale of her compatriots. She boldly retorted, 'I stand committed to destroy the British, lock stock and barrel', and as the bullets of the firing squad hit Azizan, she once more cried out 'Nana Sahib ki jai !' (Long live Nana Sahib!), sending shivers down the spines of the British.' That is one account, stirring and romantic and perhaps we might wish it to be true. But, alas, facts point otherwise: there is no evidence that the strict Baptist General Havelock ever met Azeezun, let alone was smitten by her charms. More unfortunate still for this version is that long after Havelock himself was dead, Azeezun was still alive. We must return to the indefatigable Colonel Williams, for proof of this, and highly dramatic proof it is: 'No.28. Deposition of Azeezun, prostitute and resident of Cawnpoor. Question. Before the mutiny, what sowars of the 2nd Cavalry were in the habit of paying you visits? Answer. I was in the keeping of Kulloo Mull, mahajun; therefore no sowars of the 2nd Cavalry were allowed to visit me. Question. About two days before the mutiny Shumshuddeen Khan, sowar, came to your house and said that in one or two days more the Peshwa's rule would be established and the Nana would be supreme? Answer. Before the mutiny Shumshuddeen Khan sowar never came to my house, I don't know the man. He may have come without my knowing it. Question. What do you know about the religious flag raised at Cawnpoor? Answer. I heard that it was raised by Azeemoollah Khan, who took Moulvi Salamut

Oollah with him. The Moulvie in vain attempted to resist. He also took all the residents of the city, and said if you don't come I will blow you all from the mouth of cannon. He must also have taken Kazi Waseeoodeen. The sowars collected all the people and took them to a house near the canal and they took me also. There were about 1,000 persons, men and women, collected there. The Nana and Azeemoollah ordered the people to attack the entrenchments. Moulvie Salamut Oollah and the people said 'You first attack them, then we will'. They then sent the people away and I also returned home. I remember seeing the Nunne Nawab, Azim Ali Khan, Daroga, Agha Meer Shah Alie, Ruar Ali, Moulvie Salamut Oollah, Bakir Ali, Kazi Waseeoodeen, and Ahmed Ali Khan, Vakil, Moulvie Ubdul Rulman, Hoolas Sing, Kotwal, and Rehim Khan native doctor, and all the Government officials were present, and a good many from the city.' This sounds like the statement of a very frightened girl - but who can blame her?' *See* Chapter 6 of **SSF**. **PJOT**.

'I asked about the prostitute Azizun, and was told that she and a quarter of a jagir in Bundelkhand had been a present to Baji Rao from one Govind Punt, when the Peshwa was at his peak, and that she had accompanied him to Bithur. Baji Rao peaked between his coronation in 1796 and 1801, when Holkar began menacing his territories, so assuming that Azizun was a child of, say, six, in 1801, she could hardly have been the voluptuous young revolutionary some have portrayed her as during the Mutiny, but may have been a woman of perhaps sixty. (another informant declared that he had always imagined she was advanced in years, for only a woman of advanced years would have been likely to exercize such influence over the rebels.) As of 1818 Baji Rao was not permitted to own any property beyond his Bithur jagir, so I imagine the latest he could have been presented with Azizun would have been 1817. If she was, say, thirteen when she was given to Baji Rao, she would have been born in 1804, and would have been no younger than 53 at the time of the mutiny and may by this time have been more of a madame than a prostitute.' **ASW**.

AZIMGARH An account of the mutiny/rebellion in Azimgarh District, between 12 May and 10 October 1857, is to be found in the papers of Charles Horne, the Magistrate and Collector which are preserved in the Oriental and India Office collections of the British Library, *see MSS.Eur. D.533*. The city was also the scene of one of the triumphs of **Kunwar Singh** who took possession of it on 25 March 1858, repulsing an attempt by Colonel Dames to re-take it. He also hampered the force under **Sir Edward Lugard** (despatched by **Sir Colin Campbell** from **Lucknow**) by destroying the bridge over the river **Gomti**, but was eventually, after another sharp engagement in which he acquitted himself well, obliged to quit Azimgarh on 13 April and head for his hereditary dominions at **Jagdishpur**.

AZIMGARH PROCLAMATION *See* NATURE OF THE REVOLT, **K**.

AZIMOODIN HOSSEIN, SYED Deputy Collector of **Arrah**. He threw in his lot with the British, especially in the defence of *ARRAH HOUSE*, although at great personal risk, for his family and possessions were at the mercy of the rebels.

AZIMULLAH 'A deputation of babus was selected from those imprisoned in the kotwali and brought before the **Nana**, who commanded them, on pain of death, to cease all communication with the British and never again to write in English. Before the babus were released, **Azimullah** delivered a long speech to a grand assembly of **Cawnpore** citizens declaring that he had been to Britain and knew it to be very scantily populated. 'If it were not so,' he wanted to know, 'why would not more soldiers have been sent long ago? ...What fools, then, we natives have made ourselves, so quietly to surrender our country to a handful of tyrannical foreigners, who are trying in many ways to deprive us of our religion and privileges!' He commanded them to annihilate their enemies 'root and branch, from the face of all India. Let not a soul escape, let not the name of a Christian be ever named in Hindoostan.' He assured them that the rebel forces were strong and numerous, but that if they needed assistance he would call upon his friends, 'the Plenipotentiaries of France, Russia,' and other states, who were 'willing and ready to do anything' for him. The speech convinced most people since Azimullah had actually crossed the oceans and visited Britain, and a good many natives heretofore loyal to the Company 'joined the rebel cause.' The babus, however, 'were carefully watched.' W.J.Shepherd, *A Personal Narrative etc.*, Lucknow, 1886.

AZIMULLAH KHAN The secretary or adviser of the **Nana Sahib** of **Bithur**. It was he that was sent to London before the outbreak to attempt to change the decision of the East India Company not to recognise his master as **Peishwa**, or to continue the pension that had been paid to his (adoptive) father **Baji Rao II**. It is believed also that he was a leading figure in the decisions that were taken in the Nana's council in June 1857 to throw in their lot with the mutineers, and, subsequently to set up the ambush that became known as the massacre at **Satichaura Ghat**. He died at Bhutwal in October 1859, probably from the fever of the **Terai** which had already claimed **Bala Sahib** and, some think, his brother Nana Sahib also. There is another theory: According to A. Busteed, Azimullah was supposed to have died of small pox in the company of 'But-du-Hind, surviving brother of the Nana' (Baba Bhut?) while trying to make his way into Calcutta to ask a native banker about property belonging to the Nana Sahib. *See* note by A. Busteed written at the back of the India Office Library's copy of Mowbray Thomson's *The Story of Cawnpore*. Fiend incarnate? Or hero of the Indian struggle for independence? Or perhaps a colourful adventurer, endowed as are most of that ilk , with charisma but not much scruple? It will depend on your point of view.For a full account of his career *See* chapter on *'Darling Azimullah'* in **SSF**. For interesting details of his sojourn in London *see* Katherine Frank, *Lucie Duff Gordon,* London, 1994.

AZIMULLAH KHAN IN ENGLAND

'Azimullah Khan came to England in January 1854 as the agent of Nana Sahib. His mandate was to petition the Directors of the East India Country to restore the pension of Nana Sahib's late adoptive father, Baji Rao II. Although he ultimately failed in his mission, Azimullah enjoyed a great personal and social success in London. This was largely due to his generous host, Sir Alexander Duff Gordon, and even more to Sir Alexander's beautiful, witty and accomplished wife Lucie. The Duff Gordons were neither rich nor powerful (Alexander had a clerkship at the Treasury and Lucie translated books from German and French), but they were one of the most glamorous and popular couples in London in the 1850s and numbered among their friends Dickens, Thackeray, Tennyson, the Carlyles and the Brownings. Azimullah probably came to lodge with the Duff Gordons through their close friend John Stuart Mill who was a clerk at the East India Office. (Mill's father, James Mill, was the author of A History of British India, published in 1818.) When Azimullah arrived on the Duff Gordons' doorstep in Esher, twelve miles from central London, in Surrey, he was accompanied by two retainers (one of whom, Mohammed Ali Khan would be executed by the British four years later at Lucknow) and half a dozen trunks full of treasures, including a bolt of exquisite kincob cloth woven of rose silk and real gold thread and a necklace of rare emeralds, rubies and pearls - both gifts for Lucie from Nana Sahib. Soon Lucie was writing to her friends about Azimullah whom she described as 'a very grand-looking and very amiable, charming young Mussulman Mahratta [sic].' For the next fourteen months Azimullah lived with the Duff Gordons as one of the family and was a great favourite of their two children, especially twelve-year-old Janet Duff Gordon. Azimullah also accompanied them when they travelled to Brighton, Wales and France. His great social success - even notoriety - however took place in the upper echelons of London society. With the Duff Gordons and later on his own, Azimullah moved through British drawing rooms, exclusive clubs, and country houses, charming the aristocracy with his handsome looks, fluent and cultivated English, exotic Indian dress and deferential and delicate manners. It helped that Azimullah came to England well-endowed in other ways as well: Nana Sahib had sent with him the extravagant sum of £50,000. Lucie Duff Gordon presided over an eclectic salon at their Esher home that included nobility, philosophers, academics and politicians as well as writers and artists. Lucie was especially close to the great Whig peer, Lord Lansdowne and Lansdowne entertained Azimullah at formal balls at Lansdowne House in London and at house parties at his magnificent country seat, Bowood in Wiltshire. Lucie also entreated Lansdowne - unsuccessfully as it turned out - to arrange for Azimullah to be presented to the Queen. Victoria declined, presumably because of Azimullah's claim against the East India Company. But Azimullah still got a glimpse of her from afar because Lansdowne secured him a ticket for the opening of Parliament. The artist Henry Phillips painted Azimullah's portrait, the poet Samuel Rogers invited him to his famous breakfasts in St. James's Square; the writer Caroline Norton had him to dinner. Lucie frequently took Azimullah to dine at her mother-in-law's, Lady Duff Gordon's home in Hertford Street in Mayfair and then they would go on together to the opera. In the summer old Lady Duff Gordon came to stay with her son's family in Esher and on the long, warm afternoons, sat in the garden talking with Azimullah who had his trunks brought down from his rooms and showed Lucie and Lady Duff Gordon his vast collection of beautiful Indian clothes. In the evenings, after everyone else had gone to bed, Lucie and Azimullah would sit up late before the fire talking. To Lansdowne, Lucie described Azimullah as her protégé: 'You would be amused at the incessant questioning that goes on [with Azimullah] - I have gone through such a course of political economy and social sciences . . . and get so many books for my pupil to devour that I feel growing quite solemn and pedantic.' But Lucie, clearly learned as much from Azimullah as he from her: about his employer Nana Sahib, about the East India Company, about Islam and India itself. Lucie and Azimullah, in fact, became intimate friends. Gossip that they were also romantically involved was unfounded. Lucie was a happily married matron nearly ten years his senior. Azimullah, however, most likely did have bona fide romantic entanglements with other English ladies, married and unmarried. Certainly there were numerous reports to this effect and in particular one concerning a fiancée in Brighton. Undoubtedly Azimullah broke at least one English heart when he returned to India in the spring of 1855. His journey home was by way of Malta, Constantinople and the Crimea where he acquired a low opinion of the English army. By the time he arrived at Cawnpore, there may well have been a letter from Lucie waiting for him. Though no letters between the two have survived, it is highly likely that they corresponded until the outbreak of the Mutiny. Lucie was an inveterate and fluent letter writer and she was too fond of Azimullah to lose touch with him after he left England. Frederick Roberts found a stash of Azimullah's correspondence when he occupied Nana Sahib's palace in December 1857. Writing to his sister, Roberts reported with barely suppressed rage, 'while searching over the Nana's palaces at Bithur . . . we found heaps of letters directed to that fiend Azimula [sic] Khan by ladies in England, some from Lady -----, ending 'your affect. Mother." (Letters Written During the Indian Mutiny) Lucie Duff Gordon, who often referred to herself as

Azimullah's 'English Mother', is almost certainly the 'Lady----' whose letters Roberts discovered at Bithur. And Roberts' weren't the only eyes to see these letters. In his diary Arthur Moffatt Lang describes how the letters were passed around and read by Roberts' soldiers. With the outbreak of the Mutiny, all communication between Lucie and Azimullah, of course, must have ceased. Lucie's view of the uprising was, as she put it, 'singular' and very different from that held by the majority of her countrymen. While all about her public feeling was being whipped up to a frenzy and alleged 'atrocities' perpetrated against the English dominated the press, Lucie's sympathies, as she wrote to a friend, were with 'the natives who are . . . harried by the English. . . . Who will pity the poor, helpless mass of [Indian] people? What a vista of disaster and hatred is before us and them! I execrate the tone of everybody in England on the whole affair.' She steadfastly refused to believe that Azimullah could have been implicated in the massacre at Cawnpore. In 1862, suffering from tuberculosis, Lucie went to Egypt where she lived for the next seven years until she died in 1869. She is remembered today as the author of an enchanting book about her life there, Letters From Egypt. Azimullah died in 1859, ten years before Lucie. He became in the English imagination one of the most reprehensible villains of the Mutiny. It is only in Lucie Duff Gordon's unpublished letters that a very different vision of the man in all his complexity and charm comes down to us'. **Dr Katherine Frank, Reader in English, Nottingham Trent University.**

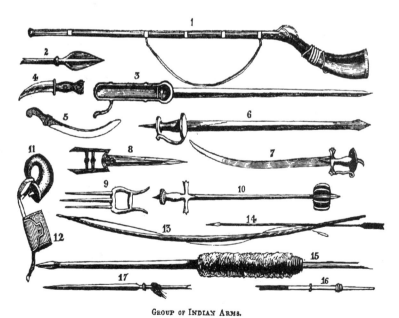

GROUP OF INDIAN ARMS.

1. Matchlock. 2. Head of a Hunting-spear. 3. Potta. 4. Creece. 5. Knife. 6. Hunting Tulwar. 7. Common Tulwar. 8. Kundeer. 9. Kundeer. 10. Ballagondeeka. 11. Powder-horn. 12. Pouch for balls. 13. Bow. 14. Arrow. 15. Borsee Spear—carried before chiefs, &c. 16. Bottom end of a Spear. 17. Head of common Spear.

B

BABA BHUTT He preferred to be called Baba Rao. He was the elder brother of the Nana and of Bala Rao, but unlike them had *not* been adopted by **Baji Rao II**. After his brief spell of being a (very severe) judge under the **Nana** in **Kanpur**, he fades somewhat from the headlines. He was 55 years old, tall and fair and well made, with a high forehead and long thick nose and large eyes behind heavy spectacles. He was believed to have been suffering the early stages of leprosy which left some marks on his face, but they may have merely been birthmarks. He was of course much older than most of the protagonists, and therefore his role, after the movement from Kanpur, seems to have been confined to being Nana's 'Chief personal attendant', although he was reported to be with **Tatya Tope** at **Kalpi** in January 1858. (**FPP** 30 Dec.1859 Supplement Cons. No.620, pp 40-41, **NA**). We know that he accompanied the Nana to **Shahjahanpur** after the fall of **Lucknow** in March 1858 and they then went to **Bareilly** to join **Khan Bahadur Khan**. It seems he followed the Nana in the last unhappy phase of the revolt when they crossed over into Nepal. The '*Bengal Hurkaru*' of 26 August 1858 reports that **Bala Rao** and Baba Bhut were both with the Nana in the jungle of bamboos 18 miles from Dhorghuree. It would seem certain that he died in the **Terai** of fever with his two brothers. He was never captured and never surrendered (all three brothers would have known there was no chance whatsoever of their escaping the noose if they had given themselves up). There is no official record of his having been presumed dead other than a magistrate's undetailed statement in 1864. He had a son Ramrow, Rammoo or Ram Rao, who was taken prisoner by the British, also in 1864. *See Kanpur Collectorate Mutiny basta.*

BABA BHUTT 'I cannot help mentioning here that an order had gone out some time ago, that a severe punishment would be inflicted upon any body who dared to utter the name of 'Baba Bhutt,' that in future his name was to be simply 'Baba Sahib.' Now Bhutt is a Maharatta word signifying mendicant, or one who derives his support from charity, and this fellow was such by birth. I am told that he is the eldest of three brothers. When Bajee Rao, the late Peishwa of Poonah and Sittarah, adopted his two younger brothers (Nana and Bala as related elsewhere), this Baba Bhutt (alias Neroo Punth) was rather overgrown and ugly-looking; he was therefore allowed to remain with his own father, and to follow the same pursuit, till the death of Bajee Rao, when the Nana asked him to come and live with him, though he was still called by his proper name 'Baba Bhutt;' but, of course, it would not do now that he was a 'Chief Commissioner' to be still called a Bhutt !!' *see* W.J.Shepherd, *Personal Narrative etc.*, Lucknow, 1886.

BABER (or Babur, the Tiger). Zahir-ud-din Muhammad, (1483-1530). Descended from Timur on his father's side, and Chingiz Khan on his mother's, he revived Timur's claims on Hindustan, fought and won the battle of (yet again) Panipat against Sultan Ibrahim Lodi, and the battle of Khanua against the Rajput confederacy led by Rana Sangrama Singh of Mewar. His empire stretched from the Oxus to the Bengal frontier and from the Himalayas to Gwalior, but he died young and did not live to conquer the west, south and east of the sub-continent. He was the father of the Mughal Emperor Humayun, and may be called perhaps the first of the Mughals really to rule India. *See* S.M.Edwardes, *Babur: Diarist and Despot,* London, n.d. (1935?).

BACSA The British Association for Cemeteries in South Asia. A prestigious organisation founded after Indian Independence to work for the preservation of Christian cemeteries and graves in the sub-continent and beyond, for which purpose it raises money and disburses to local organisations prepared to effect the work. It has however sponsored a wide variety of published material - some related to the struggle of 1857-8 which is of interest to students of the period. The association's magazine, produced twice a year, is called *Chowkidar*. The Hon Sec is Theon Wilkinson M.B.E., 76½ Chartfield Avenue, London, SW15 6HQ.

BADLI-KE-SERAI The site of a battle on the Grand Trunk Road just north of Delhi. At 1 a.m.on the morning of Monday 8 June 1857 the combined British forces from **Meerut** and **Kurnaul** under the command of Major-General **Sir Henry Barnard**, the new Commander-in-Chief advanced from **Alipur** towards Delhi and after marching for about four miles, came upon a body of the rebels, numbering about 3,000, strongly posted in an entrenched position, defended by twelve guns, at Badli-ki-Serai. The rebels' artillery was extremely well served and caused many British casualties (including Colonel Chester the Adjutant General of the Army), but a bayonet charge on the guns by H.M.'s 75th regiment gave victory to the British who drove their enemy within the walls of Delhi, and occupied the old cantonment area on the Ridge, from the old Signal Tower on the left to **Hindu Rao's House** on the right. From this time the so-called 'siege' of Delhi may be said to have begun. *See* **Kaye, Sen, Ball, Martin** etc. *See also* A.Llewellyn, *The Siege of Delhi,* London, 1977; Harvey Greathed, *Letters written during the Siege of Delhi,* London 1858; Charles Metcalfe, *Two Native Narratives of the Mutiny of Delhi,* London, 1890.

BADLI-KI-SERAI, BATTLE OF 'Headquarters, Delhi Cantonments, June 12th (1857), We are now encamped on the Flagstaff Hill, having dislodged the insurgents on the 8th. They fought most obstinately, and disputed the ground inch by inch; but British courage, and I may add, ferocity, forced all before them . *In the enemy's camp was found a European actually laying the guns! He was literally cut to pieces by the enraged soldiery.* It is suspected there are others in the city. Three hundred of the mutineers tried to escape by getting into a Seraie, but they were seen by some of our fellows, and a party went to dislodge them; they fired from the windows, when our fellows burst open the doors, and, rushing in, killed every one. Not one was left alive'

BADSHAH BAGH The King's Garden, now Lucknow University. An important stronghold of the revolutionaries during the Uprising of 1857. The British gave it to Maharaja Harminder Singh of Kapurthala. In 1905 the State government purchased the estate and handed it over to the authorities of Canning College. Courtesy INTACH.

BAGPIPES A new use for the bagpipe (On **Havelock's** entry to Lucknow in September 57): on this occasion the Highlanders' piper, who had lost his way, suddenly found one of the enemy's cavalry, sabre in hand, about to cut him down. His rifle had been fired off, and he had no time to use his bayonet. 'A bright idea' said he afterwards, when relating the story, 'struck me. All at once I seized my pipe, put it to my mouth and gave forth a shrill tone, which so startled the fellow, that he bolted like a shot, evidently imagining it was some infernal machine. My pipe saved my life.'

BAHADUR SHAH ZAFAR The Sepoys are said to have divulged their plan to the King of Delhi, Bahadur Shah, himself 'You are the King of both worlds - terrestrial and spiritual. The whole of India is under your sway and every announcement is preceded by 'God is the Master of Creation, Order belongs to the King and the Command is that of the Company' . The English have been ruling on your behalf. We have come to you with our grievances. We expect justice from you. We are the servants of the English. It is we who have conquered the whole territory of 1400 Kos extending from Calcutta to Kabul for the English, because they did not bring any English army with them from England. Now when they have subjugated the whole of India, they have changed their mind and want to convert every one of us into Christians and thus meddle with our religion. They applied this policy first in our army. After careful deliberation and due consideration they thought of a new plan. A new type of gun was invented for which a cartridge greased with the fats of certain animals was used. God only knows as to which were those animals whose fats were used. Both the Hindoos and Mussulmans among us refused to obey the order. The Hindoos pleaded that they belonged to high castes - Brahmin and Kshattriyas etc and that they did not take meat at all. The Muslims also objected saying they did not take the meat of those animals which were not killed in accordance with the Muslim religious rites.....We fought them. In the morning we started towards Delhi and with forced marches reached this place after covering a distance of 30 Kos. Your Majesty should extend your kind and sympathetic hand, for we have done all this in the cause of religion.' *See* Dastan-i-Ghadar, Zahir Dehlvi, pp 46-53. Perhaps it was this address that prompted Bahadur Shah to compose his couplet: 'Na Iran ne kiaya, na Shah Russ ne,- Angrez ko tabah kiya Kartoosh ne.' - 'the mighty English who boast of having vanquished Russia and Persia, have been overthrown in Hindoostan by a single cartridge.'

Bahadur Shah Zafar

The veteran King of Delhi, Bahadur Shah Zafar, is dismissed at the time as a 'pantaloon emperor'. Indeed he may unwittingly have been the cause of the failure of the rebels, for it is said that the ambitions of the Delhi faction to reinstate the Mughal dynasty alienated from the outset most of the other potentates and kept them neutral or pro-British. On 4 January 1862 the *Illustrated London News* reported: 'The King of Delhi died at Rangoon on 11 Nov 1861 and was buried the same day. Little interest was exhibited by the Mohammedan population of Rangoon' *See* Husain, Mahdi, *Bahadur Shah II and the War of 1857 in Delhi,* Delhi, 1958. *See also* BOWRING, Lewis Bentham. In June of 1859, William Howard **Russell** visited the deposed old King in his prison cell, and found him a most unlikely leader of men. 'That dim wandering-eyed, dreamy old man, with feeble hanging nether lip and toothless gums,' Russell wondered, '...was he, indeed, one who had conceived that vast plan of restoring a great empire, who had fomented the most gigantic mutiny in the history of the world, and who from the walls of his ancient palace, had hurled defiance and shot ridicule upon the race that held every throne in India in the hollow of their palms?' When the King spoke, 'Alas! it was to inform us that he had been very sick, and that he had retched so violently that he had filled twelve basins.' Russell noted that as a poet Mohammed Shah was 'rather erotic and warm in his choice of subject and treatment, but nevertheless, or may be therefore, an esteemed author of no less than four stout volumes of meritorious verses, and he is not yet satiated with the muse, for a day or two ago he composed some neat lines on the wall of his prison by the aid of a burnt stick.' See W.H.Russell, *My Indian Mutiny Diary*, London, 1860. Bahadur Shah II was said to be 82 years old in 1857, but this may have been lunar years in which case he was 78 by the western calendar. As a young man he had been renowned for his athleticism, especially in archery. He is said to have never drunk wine but there are conflicting stories about his addiction, or otherwise, to opium. Certainly the story that he had a fixed belief that he could transform himself into a fly or gnat and that he could in this guise convey himself to other countries and learn what was going on there, implies addiction to some drug, as do the well-known portraits of him lying out at full stretch inhaling from his hookah. *See* Sayyid Ahmed Khan, *Rissalah Asbab-e-Bhagawat-i-Hind*, transl. Graham and Colvin, Benares, 1873.

BAISWARA (or Byswara) A District of Oude[Awadh] which strongly resisted the British. Among others **Rana Beni Madhoo** came from here (**Shankerpur**). On 20 November 1858 the Chief Commissioner of Oude reported to the Governor-General as follows: 'While it is a subject of regret that this rebel Chief (Beni Madhoo) of the Baiswara clan has thus obstinately refused the royal grace so mercifully extended by Her Majesty the Queen, still the Chief Commissioner considers it as a matter of great congratulation that the country of Baiswara whose people are the most warlike, and the most worthy to cope with our arms has been thus cleared of the rebel armies without bloodshed, and whatever may be the final determination of the Government regarding the Estates and persons of the fugitive chief and his followers, there is satisfaction in the knowledge that the utmost forbearance, and most lenient offers have been proffered by the British Government, and almost thrust on their acceptance. The rebels having counted the cost of refusing such terms, have, with wilful obstinacy, sealed their own ruin.....Civil Officer reports that the *Zemindars* of Baiswara are rendering ready submission and the breaking up of the Baiswara Talooqdars, some of whom have fled with their Chief, will relieve from much oppression the subordinate landholders', see Foreign Political Consultations, 30 December 1858, No.200.

BAJI RAO II The last **Peishwa**. Having no natural heirs of the body, he adopted, by due and formal Hindu rite, in 1827, one Dhondu Rao Punt (**Nana Sahib**) who, when Baji Rao died in January 1851, claimed from the government the continuance of the annual pension of 8 lakhs of rupees (=£80,000) granted to his adoptive father when the latter abdicated his authority, by treaty with the British, in 1818. The question is not one of adoption (as at **Jhansi**), which was not disputed; even had Baji Rao left issue of his own body it is doubtful indeed whether the pension would have been continued after his death, since the treaty did not stipulate that it should. This is, ironically, all the more likely in that Baji Rao had enjoyed his very considerable pension for far longer than anyone could have anticipated. In !818 he was considered to be a man of feeble constitution and dissolute habits, far advanced in years - *yet he lived to enjoy his pension for 33 years!* He had certainly saved a great deal by his death and is said to have left the sum of no less than Rs16,00,000 to his heir, the Nana Sahib. The failure to inherit the pension, the title, the prestige of his adopted father was what drove Nana Sahib to rebellion.

BAJI RAO VERDICT 'In public life, even if his many faults are born in mind, one cannot but feel that he was particularly unfortunate. In the early years of his career, hardly was a Peshwa more ill-served. His enmity towards Nana Fardnavis and his associates deserves the strongest condemnation, but at the same time it should be remembered that it had been Baji Rao's fate to move in an atmosphere that was politically vicious, and there was hardly any reason for his feeling grateful to his father's enemies. ' .... 'the dissolution of the Maratha Empire had set in before Baji Rao's time, and a man of far superior qualities would have found it equally impossible to arrest the decay.' P.C.Gupta, *Baji Rao and the East India Company,* Oxford, 1939, p 206.

BAJI RAO II's WILL His will was drawn up in 1839. 'Strictly it should be said that he adopted three sons and a grandson. His will says, 'That Doondoo Punt, Nana, my eldest son, and Gungadhur Rao, my youngest and third son, and Sada-Sheo Punt Dada, son of my second son, Pundoo Rung Rao, my grandson; these three are my sons and grandson. After me Doondoo Punt, Nana, my eldest son, Mookh Perdan, shall inherit and be the sole master of the Guddee of the Peishwa, &c'. *See* Sir John Kaye, *A History of the Great Revolt.* Vol I, p 101 *et seq.*

BAKT or BUKT KHAN, Mohd. A one-time Commander-in-Chief of the rebel forces in Delhi. His former battery commander in the Company's artillery described him as a 'fat and lazy subordinate'. In September 1857 when the British assaulted Delhi, the Commander-in-Chief of the forces of the King of Delhi Bahadur Shah Zafar, was Bakt Khan. He had been appointed to the command of the King's troops almost literally because there was no one else eligible. For a time one of the King's sons had been in command, but he had no military knowledge whatsoever, and

Bakt Khan - A Sketch by Capt.Maisey

was soon discredited - and humiliated - in the eyes of the sepoys. They wanted, needed, a leader in the military as well as the political sense, so turned to Bakt Khan, who had been a Subhedar in the Company's army, and commanded the respect of a substantial and influential part of the King's forces - the Bareilly Brigade as it was called. The British did not share this respect: they lampooned Bakt Khan, said he was too fat to mount his horse unaided, and lacked courage. Whether this was true, or just the usual morale-boosting denigration of the enemy's leader, is unclear. Bakt Khan was not exactly *persona grata* with a large number of influential Indians - even those nominally on his side. In 1857 in the North of India the whole country when in open revolt was apparently acting with one national will under separate leaders, but in reality this may have been far from the case: Just before the fall of Delhi, in September 1857, Bakt Khan left the city at the head of those troops whom he managed to keep together in some reasonable state of order. He went to **Farrukhabad**, on the Ganges north of **Kanpur**, where he expected to be welcomed. On the contrary, he and his men were by no means well-received, and were called on to pay their way: they seem to have found considerable difficulty in obtaining even the minimum of supplies for survival. On the advance of the British forces Mahomed Bakt Khan joined **Moosun Alli Khan**, one of the nazims of the **Nawab of Farrukhabad**, in opposing the British, and, on being defeated at **Khudaganj**, he was making the best of his way across the Ganges, but still at the head of a small but disciplined force. It was at this moment of all others that the Nawab of Farrukhabad, who must have foreseen with the most unpleasant clarity the capture of his capital city by the British - he did indeed flee from it himself within three days - chose to issue, on 25 December 1857, the following order to Likha Singh, talukdar of Allahganj: 'Mahomed Bukt Khan, with his companions in arms, being in possession of five or six lacs of rupees, is fleeing before the Europeans, therefore, to Us the all-powerful, this appears right that you plunder all the cash and personal effects of Bukt Khan and his companions. You will not be called to account for this pillaging, but the guns, magazine, and elephants attached to his force you are to send in to Us, that We being pleased, it may be to your advantage. This is written as an injunction. You are to act strictly as you are ordered.' Likha Singh, rightly guessing who was going to win the struggle, sought to make his peace with the British by sending in this letter to them, hastily adding that he had not acted upon it. It is not clear what Bakt Khan had done to make him the enemy of the Nawab of Farrukhabad. But good soldier or not he remained true to the cause of rebellion, for we next hear of him, as 'General Bukht Khan (the Bareilly Subadar)' joining the Begum Hazrat Mahal in July 1858, long after the fall of Lucknow, and when she was in retreat towards Nepal. He is said then to have had with him 200 sepoys, 15 sowars, and one gun.' For comments and brief assessments of the rebel leaders, see *Chambers Journal* No.20 1930, p 549 *et seq*.

BALA RAO The brother of Dhondu Pant, the **Nana Sahib**, adopted, with his brother, by **Baji Rao II**, the last **Peishwa** of the Marathas. He was a leading member of the Nana's council, and at first had some pretensions to being a military leader, but was badly wounded in the right shoulder in the engagement with **Havelock's** forces at Fatehpur on 12 July 1857. It may be that it was he who most strongly pressed upon the Nana the desirability of slaughtering the European woman hostages. Certainly in the early days of the Nana's success in Kanpur Bala Rao was a man of considerable influence, although this is not later so noticeable. In 1857 he was about 28 years old, dark, tall and lean. With a long pock-marked face with round eyes and a crooked nose. His beard made him look Muslim. He was missing his front teeth and had a scar on his chest from his wound at **Fatehpur**. The Commissioner of **Gorakhpur** reported that Bala Rao and his family, acompanied by 500 Bhojepur sepoys, entered the Nepal terai on 6 January 1859, intending to claim the protection of the Nepalese authorities and dispersing to their homes should protection be refused by **Jung Bahadur.** *See* Edmonstone *Memorandum* dated 12 January, 1859. **NA. FDP.** *Despatch to Secretary of State*, 4 March, 1859. Bala Rao died in the **Terai,** probably of fever, possibly about the same time as his brother, although nothing is certain. *See* the *Times* 21 January 1860.

BALA RAO'S DEATH 'The Nana', **Ramsay** reported, 'who had come up from his own camp to see his brother when he heard of his being so seriously ill was with him when he died.' But the Commissioner of Gorakhpur placed little confidence in reports of sickness in the rebel camp, and believed the Gurkhas were making so much money selling the rebels provisions 'that they were now protecting them from capture.' A letter from Sirdar Siddhuman

Singh, Raja of Bhundair, stated that he died on Thursday, 30 June, 1859, and 'none of his Ranees have become Suttees.' **NA. FDP.** FC. 30 December, 1859, Nos. 742-55.

BALDEO SINGH, THAKUR

In June 1857 he assisted some European fugitives from **Gwalior**, including Brigadier Ramsay, taking them by by-paths to avoid the *ghazis* who lay in wait for them and escorted them safely to the banks of the Chambal, on the opposite side of which some troops of the Rana of Dholpur were waiting with elephants: they were then escorted safely to Agra. Baldeo Singh's motivation appeared to be some personal regard he had for the Resident at **Scindia's** court, Major **MacPherson.** His son, Rissaldar-Major Gopal Singh, became ADC to the Viceroy.

BALL, Charles

The author of a standard history of the struggle, produced before the final curtain and therefore having the merit at least of freshness and contemporaneity. A very readable, profusely illustrated, but highly jingoistic account. The author was not an historian but he could write, and he knew his public. An interesting feature is that he quotes from many letters written by protagonists in the great struggle. One of the standard histories, but somewhat dated as it makes little effort to be objective. *See* Chas Ball, *History of the Indian Mutiny* (2 vols) London, n.d. (1858?).

A Typical Fort in Oude

BAMBOO

The **Talukdars** of Oude[Awadh] surrounded their fortifications with vast stands of bamboo so resilient and impenetrable that a six-pounder ball would merely snag or rebound with a clatter. The Resident, Colonel Sleeman, had them razed.

BANDA

In the year of the Great Rebellion Banda was a relatively unimportant place in central India, but it was to play a considerable part in events. The district area amounted to about 3,000 square miles, with a population, mainly Rajputs, of some half a million. The land was studded with rocks set amongst dry basaltic black soil, unfertile unless irrigated. There were few large estates or rich landholders, but at Banda itself there lived a mediatised prince, the Nawab Ali Bahadur, of Maratha origin, whose family had seized power during the anarchy of the eighteenth century, and had been converted to Islam. The town is on the right bank of the river Ken, a tributary of the Jumna[Yamuna], and is 95 miles southwest of Allahabad. The chief civil officer in May 1857 was a man of probity and energy named F.O.Mayne, nicknamed 'Foggy' Mayne. He was not of great intellectual ability, nor indeed was his judgement always sound, but he was a strong character, with considerable influence over those he met. His *'Narrative'* and J.W.Sherer's *'Daily Life etc'* are the chief sources, on the British side, for what happened in Banda. As the news from Delhi and Meerut arrived so Mayne took steps to strengthen the position of the government agencies: he recruited more police, and stationed them particularly at district outstations. He put an embargo on the Jumna ferries in an attempt to keep 'troublemakers' out; he set up horse patrols in and around the town; English officers personally visited the police posts by night; some of the gentry and richer traders were encouraged to recruit armed men for their protection; there were regular troops also - three companies of the 1st BNI - who for the time being remained loyal. Tranquillity was preserved but it was fragile: Mayne himself believed it was only a matter of time before law and order collapsed. He thought the local population had been 'ruined' by over-assessment for land tax, were 'half-starving', and would take the first opportunity to rebel. We are talking, be it noted, of a general rising not just a military mutiny. Indeed the first outbreaks when they came were unconnected with the sepoys: some Tahsils were attacked and records destroyed by villagers 'in order that no record of their liabilities might remain to the new government'. Mayne's deputy, Cockerell, was out at Kirwi and was even more isolated. The government officials, the amlah, remained staunch, and some even died in the defence of government property and records. But their task was hopeless because, unlike the situation in many other areas, there was no civilian feeling of goodwill towards government. The sepoys of the 1st NI were trusted, and even used to guard and escort treasure, although 'mutinous talk' was reported from their lines. On 8 June there came the first real scare. A body of horsemen was reported approaching the bridge of boats over the Jumna at Chilatara. Almost as though they had been waiting for a signal the some of the townspeople broke out into an orgy of looting, but Mayne was equal to that and used his police effectively to stop the plundering; he also sent the small number of English ladies of the station for safety into the Nawab's palace. Ironically the horsemen turned out to be Europeans, led by the magistrate J.W.Sherer, fleeing from Fatehpur. This party arrived in Banda the same evening. The situation was still perilous so Mayne actually began to fortify the Nawab's palace. On the night of 12 June two bungalows in the station were burned; then, on 14 June the sepoys heard of the mutiny of their regiment at Kanpur, and broke out in revolt: an attempt to disarm them by using the Nawab's armed retainers failed dismally - in fact the guards cheerfully joined the sepoys in rebellion. All the Europeans now became refugees and sought safety in the Nawab's palace. They did not stay; that same evening Mayne led them, hampered by the presence of seventeen women and eight children out of Banda for friendly territory to the east and eventually reached **Mirzapur**. But what of the Nawab of Banda? He now attempted to take charge of the district - indeed he claimed that Mayne had left the district to his charge. Sometimes the sepoys obeyed him, sometimes they opposed. One of the first things to happen was that all remaining Christians in the district were hunted down and murdered. The Nawab was probably glad indeed when the sepoys departed on 19 June, for Kanpur, taking with them all the ammunition and treasure they could lay their hands on. The Nawab's attempts to rule the district, at the behest of the British and presumably in anticipation of their eventual return, were clouded by the fact that although he tried to enter into correspondence with Mayne, the latter refused to reply. It seems that Mayne, unlike Sherer, had decided the Nawab had thrown in his lot with the rebels. That he did so at last, despairing of British help or recognition for his efforts, now seems the inevitable conclusion. For this he paid, not with his life, but with his wealth and estates, being given a pension of only Rs36,000 p.a. on which to live. There seems to be no evidence from any source that the Nawab was anything but a weak and vacillating man: he was certainly not an inspired leader of the Great Revolt. - **PJOT.** In September 1857 there was a great concourse of rebels at Banda, especially when **Kunwar Singh** arrived with 2,000 followers on 29 September. For details *see Revolt in Central India, 1857-58*, Simla 1908.

BANDA, Battle of This took place on 18 April 1858 between the British troops of the Madras army commanded by Major-General **Whitlock**, and some 6,000 rebels, with an estimated 3,000 in reserve, the whole commanded in person by Ali Bahadur the **Nawab of Banda**. The result was an overwhelming victory for the British, the rebels retreating in disorder towards **Hamirpur,** and the Nawab himself reported as weeping bitterly as he withdrew. Banda was then occupied and immense quantities of booty taken, of which the prize agents eventually took possession. The Nawab and his troops reformed at Jalalpur and then marched on **Kalpi** where they joined the **Rao Sahib** and the **Rani of Jhansi**. For full details of the action *See The Revolt in Central India,* Simla, 1908, pp172-176 and numerous entries under **Banda** in **FSUP** Vol III.

BANDA, NAWAB of Rules for government of Banda; the **Nawab Ali Bahadur** appears not to have been an inspirational individual. In his writing desk were found rules for the government of Banda, after the breakaway from British rule. Emphasis is placed on remaining friendly with the Nana Sahib and agreeing total obedience to the Emperor in Delhi. There is one interesting and revealing codicil: 'The English form of government should be continued as it has two advantages - *less expenditure and large receipts*'. *See Banda Collectorate Pre-Mutiny Records,* Central Records Office, Allahabad, quoted in **FSUP** Vol I p 437.

BANGARMAU A town thirty miles west of **Oonao[Unnao]** and 48 miles west of **Lucknow** in **Oude**[Awadh]. For some time after the defeat of the rebels in Lucknow, numerous bands controlled rural Oude: one such was commanded by **Prince Firoz Shah**, of the Royal House of Delhi. For details *see* **FSUP** Vol II p 399 et seq., and Rudrangshu Mukherjee, *Awadh in Revolt,* Delhi, 1984.

BANPUR, Raja of Mardan Singh, Raja of Banpur, was in a position similar in some respects to that of the **Rani of Jhansi** for he considered that he had legitimate grievances against the British government in respect of the tenure of his land: no doubt also, in rebelling, he had hopes of regaining, on the expected overthrow of the British, the entire kingdom of **Chanderi** the ancient possession of his ancestors. However he played at first a double game, professing loyalty to the British while in reality heading the rising. He established his authority in the Chanderi District and remained in possession of the northern part of the **Saugur [Sagar]** District until **Sir Hugh Rose** advanced into the territory in January 1858. He was wounded at the Battle of **Barodia** on 31 January 1858. He gave himself up to Thornton, the Assistant Superintendent at Maraura on 5 July 1858, but was shortly afterwards allowed to return to the **Chanderi** district where he continued intriguing and collecting revenue until early in August 1858 when he finally surrendered and with the **Shahgarh** raja was sent under escort to **Gwalior**. For details of his career *see* **FSUP** Vol III and *The Revolt in Central India, 1857-58,* Simla, 1908.

BANPUR A town some thirty miles south of **Jhansi** on the **Saugur[Sagar]** road in Central India. **Sir Hugh Rose** and his force arrived here on 10 March 1858 and found the strong fortified palace of the Raja had been abandoned; in it were found a complete manufactory for casting guns and mortars as well as clothing and other supplies. The palace was then destroyed. The town was quite empty. For details *see The Revolt in Central India,* Simla, 1908 p105, *and* Thomas Lowe, *Central India during the Rebellion of 1857-58,* London, 1860.

BARAITCH A district of Oude[Awadh], and a large town of the same name. The two British civil officers, Cunliffe and Jordan were both murdered, in mysterious circumstances, together with Lieutenant Clarke the officer i/c two companies of the Oude 3rd Irregular infantry, who mutinied. Apparently the three men tried to reach the hills via Nanpara but were denied entrance to that place because of the *be-duk-ilee* or dispossession grievance that caused much disaffection throughout **Oude[Awadh]**. They then returned to Baraitch and disguising themselves in native clothes set out for Lucknow; at Byram Ghat which was guarded by the **Secrora** mutineers, they were discovered and one account says they 'were tried by the rebels for the murder of **Fuzil Ali** and shot'. *See Mutiny of Bengal Army, an Historical Narrative, by One who served under Sir Charles Napier*, London 1857; *see also* L.E.R.Rees, *A Personal narrative of the Siege of Lucknow*, London, 1858.

BAREILLY BRIGADE This was a considerable body of men, made up of the 18th N.I., 68th N.I., the 8th Irregular Cavalry Regt, 6th Coy Bengal Native Artillery, a total of some 2,974 Indian trained soldiers, with no more than 35 European officers of all ranks. The mutiny at Bareilly took place on 31 May 1857 and the vast majority of the Brigade were swept enthusiastically by what has been called 'an irresistible impulse'. They marched to Delhi and provided a notable reinforcement of Bahadur Shah's forces, under their leader **Bakt Khan,** celebrating their arrival, it is said, by their bands playing 'God Save the Queen.'

BAREILLY A large city in the province of Rohilkhand which lay between **Oude[Awadh]** and the river Ganges which separates it from the Doab. It is 152 miles from Delhi. The military commander of the (almost exclusively Indian) forces of the government was Brigadier Sibbald, who, somewhat surprisingly in view of later events, reported most favourably on the reaction of his soldiers to the news of mutiny at **Meerut** and **Delhi**. With hindsight he might be to some extent justified for throughout Rohilkhand disorganisation in the civil government appears to have preceded military mutiny in the cantonments. A lineal descendant of the last ruler of the country was also a government pensioner as a retired Principal *Sudder Amin*: his name was **Khan Bahadur Khan**. The mutiny when it did occur, on 31 May, followed the usual pattern: Sibbald and many officers were killed, others got away to **Naini Tal**, but there was one unusual event: the gaol held an exceptionally large number of convicts - 4,000 - and when they were released it appears that the scramble for loot between them, the sepoys and some of the townspeople of the city, caused a considerable loss of life. The other unusual occurrence is that five of the civil officers of the city *viz* Robertson and Raikes (judges), Dr Hay, Wyatt (Deputy Collector) and Dr Buch (Principal of Bareilly College), were formally 'tried' before Khan Bahadur Khan as judge, and with witnesses and jury, were found guilty and publicly hanged in front of the gaol. On 5 May 1858 was fought the Battle of Bareilly at which, despite the desperate courage of some *Ghazis*, the British Commander-in-Chief, Sir Colin Campbell, defeated the rebel forces and re-occupied Bareilly. Khan Bahadur Khan held out in the *Terai* until the close of 1859 when he was captured by **Jung Bahadur** and handed over to the British: his life was considered forfeit by the murders committed in Bareilly. *See* Wm Edwards, *Personal Adventures etc*, London, 1858; *FSUP* Vol II; *Mutiny of the Bengal Army, by One who served under Sir Charles Napier*, London, 1857; *Further Parliamentary Papers*, 1857; *and* Chas Raikes, *Notes on the Revolt in the NW Provinces*, London, 1858.

BARGI One of the few places to the south of the river Narbada to have been the scene of active rebel activity. The names of Dalganjan Singh and **Debi Singh Guntia** are connected with this activity. The area was finally settled and pacified in November 1857.

BARJUR SINGH The name of the Raja of **Bilayan**, a town on the road between Orai and **Jhansi** in Central India. He was a man of considerable courage and resource and implacable in his hatred of the British. *See* BILAYAN *and also The Revolt in Central India*, Simla, 1908, *and* Thomas Lowe, *Central India during the Rebellion of 1857-58*, London, 1860, *and* **FSUP** Vol III.

BARNAGAR In some parts of Central India the rebellion appeared to be welcome to many as providing an opportunity for lawlessness and looting, in addition to more patriotic motives in some quarters. But whereas the rebels usually had the support of the local people, this was not always the case, particularly in the towns where people had more to lose from a breakdown in law and order. Thus in November 1857 on entering Barnagar, which had just been evacuated by the rebels, Captain H.O.Mayne of the Intelligence Department reported to his superiors: '*On our coming into the main street every shop was shut, and only a few loiterers with gloomy faces were visible. But on our reaching the Kotwali and announcing our errand, the populace swarmed up as bees from a hive, and in one minute the whole place was densely crowded, as if by magic, and it was difficult to pass through the concourse*'.

BARNARD, GENERAL Sir Henry William Barnard (1799-1857) became the Commander-in-Chief of the British army in India on the death of **General Anson** on 27 May 1857. He himself also died, of cholera, on 5 July in the same year (dying within the space of one day). His want of experience in Indian warfare had told against him as a commander; and his brief tenure of power hardly gave opportunity for a fair judgement to be formed of his military capacity, but by all accounts he was 'a high-minded and true-hearted gentleman'. He did, before he died, continue the

General Barnard

build-up of British forces for the assault upon Delhi. He was buried in the Rajpur Cemetery. *See* CEMETERIES. *See also* **DNB.**

BARODIA A small town on the river Bina, close to **Rahatgarh** in Central India, between **Saugur[Sagar]** and **Bhopal.** The scene of an important battle on 31 January 1858, between the British **Central India Field Force**, commanded by **Sir Hugh Rose**, and the rebel force which contained the troops of the **Raja of Banpur**. The rebels were defeated and the result was that communications between Saugur and the west were now re-opened for the British, and a number of strong fortified places abandoned; guns also were abandoned, and the **Relief of Saugur** made possible. *See The Revolt in Central India*, Simla, 1908, p 92-5, *and* Thomas Lowe, *Central India during the Rebellion of 1857-58*, London, 1860.

BARONNE see MUSABAGH.

BAROWEN see MUSABAGH.

BARRACKPORE 'Barrack-town' is on the Hooghly river, some sixteen miles (25 km) from Calcutta. The Governor General had a residence there, standing in a very large park of some 250 acres, well laid out. Job Charnock is said to have had a bungalow built here before the site of Calcutta was finally decided upon. It was an important military cantonment also. The officer in charge of this Bengal Division was **Major-General Hearsey**, a man of some talent and good judgement. In a letter on the **Cartridge question,** as early as 23 January 1857 he represented to government the extreme difficulty of eradicating the notion which had taken hold on the mind of the Indian soldiers, and urged, as the only remedy, that despite the trouble and expense it would occasion, the sepoys should be allowed to obtain from the bazar the ingredients necessary to grease the cartridges. He also pointed out the influence that was 'probably' exercised by a Brahminical association called the **Dhurma Sobha.** Barrackpore saw the first bloodshed of the rebellion/mutiny when the affair of **Mangal Pande** took place. For an account of this matter, *see* R.Montgomery Martin, *The Indian Empire*, Vol II p131 *et seq.*

BARTRUM, Dr William *'General Outram despatched a party of 250 men to effect a junction with Colonel Campbell of the 90th, and assist him in bringing in the wounded. He ultimately succeeded in this object, but some idea of the dangers they had to encounter may be formed by the following narrative: the doolies containing the wounded remained during the night of the 25th September (1857) in the passage in front of the Motee Munzil Palace, without attracting the attention of the enemy. On the morning of the 26th the mutineers opened fire upon them, and numbers were killed by shot and shells. The surgeons in charge of the wounded behaved with the greatest coolness and courage. One of them, an assistant surgeon in the Artillery, requested one of his brother officers to assist him in an operation. On their way to the spot they were exposed to constant fire. 'Will Bartrum', said one of them, 'I wish I could see my way out of this', 'Oh', said the other, 'there's no danger whatever'. The words had scarcely passed the lips of the assistant surgeon when he was shot down. Two minutes before he was speaking of his wife and child, and the pleasure he would have in meeting them in the Residency. They were never destined to meet in this world.'* Presumably this 'Will Bartrum' is Assistant Surgeon Robert Henry Bartrum of the Artillery whose wife **Mrs Katherine Bartrum** survived the siege of the Residency with her child. This account comes from the *St George's Gazette*, 30 November 1893.

BARTRUM, Mrs Katherine A survivor of the siege of the **Lucknow Residency**, although her husband, Surgeon Robert Henry Bartrum was killed. She had been living at Gonda at the beginning of June 1857 but was called into Lucknow on the orders of **Sir Henry Lawrence** on 6 June. She was as incompetent in the performance of the menial daily tasks of living as were all her class, and found herself in great 'hardship' without servants to wait upon her, particularly to cook. Her story is a particularly sad one however as she was both widowed and lost her only child. Some contemporary letters of hers and the manuscript copy of her diary, giving an account of life in Lucknow during the siege and her journey to Calcutta after the death of her husband are preserved in the Oriental and India Office collections of the

British Library, *see MSS.Eur.A.67; see also* Mrs Katherine Bartrum, *A Widow's reminiscences of the Siege of Lucknow*, London, 1858. NB the original manuscript diary, more complete than the published version is owned by Mrs Dulcie Boyle of Weymouth in Dorset.

BASHIRATGANJ (or Busserut Gunj) A large walled village some nine miles north of **Oonao**[Unnao] on the **Kanpur** to **Lucknow** road. The scene of three battles (29 July, 6 August, and 12 August) between General **Havelock** and the rebels (including, it is said, the Nana Sahib) who were attempting to bar his way to relieve the **Lucknow Residency.** He won pyrhhic victories on each occasion, for his force was very small (perhaps 1,000 effectives) and facing up to 30,000 men with fifty guns - not mutinied sepoys but men determined to defend their homeland, the Kingdom of Awadh. Havelock was compelled to retreat to Kanpur, to await reinforcements. The succinct report of his chief of staff (**Colonel Tytler**) to the Commander-in-Chief tells the story well: 'The whole transaction was most unsatisfactory, only two small iron guns (formerly captured by us, and destroyed in our ideas) being taken. It became painfully evident to all that we could never reach Lucknow; we had three strong positions to force, defended by fifty guns and 30,000 men. One night and a day had cost us, in sick and wounded, 104 Europeans, and a fourth of our gun ammunition; this does not include our killed and dead - some ten men. We had 1,010 effective Europeans, and could, consequently, parade 900 or so; the men are cowed by the numbers opposed to them, and the endless fighting. Every village is held against us, the zemindars having risen to oppose us; all the men killed yesterday were zemindars.' For the best account of these engagements *see* J.C.Marshman, *Memoirs of Major General Sir Henry Havelock,* 1860. For details of the actions at Bashiratganj *see* **FSUP** Vol II, numerous entries.

BAUNDI A large village in Bahraich in Oude[Awadh], on the left bank of the Gogra river. It was here, in a fort belonging to the Raja of Baundi, that the **Begum of Oude** stayed in July 1858, reviewing her position. She was attended by some 15,000 men, of whom 1,500 were cavalry, and 500 mutinied sepoys, the rest being *nujeebs* or followers. It was said that she refused to accept into her force any fugitive sepoys, saying in effect 'You went to fight and ran away, don't come here'. *See Foreign Secret Consultations,*30 July 1858, No 89.

BEADON, Sir Cecil (1816-1881) A career civil servant of considerable ability and reputation. At this time he was Secretary to Government in Calcutta, and a close adviser to **Lord Canning** the Governor-General. On more than one occasion however the measures he advocated were flawed, and this was in part due to his lack of experience 'up-country'. Beadon was 42 years old at the time of the Mutiny/Rebellion. For six years Sir Cecil Beadon's rocket had smouldered and sputtered as he attended to the usual district duties, but with his appointment as Undersecretary to the Governor of Bengal in 1843 his ascendancy was rapid. 'Gracious, approachable, phlegmatic,' Beadon was famously courteous; Emily Eden pronounced him the most perfectly mannered man she had ever met. But as Home Secretary Beadon shared in Canning's unpopularity, and later fell victim to a recurrent fever that exiled him to Darjeeling at crucial periods. His absence from Calcutta during most of the famine of 1860, and his failure to see it coming, made him the subject of a scathing report by a commission of inquiry which was ratified with unusual harshness by the Governor General in Council. Beadon left India in 1868, retired to England and died 13 years later. *See* **DNB**.

BEARDS Well groomed men of regular habits grew accustomed to a certain slovenliness. 'At such times one did not think even of meals,' wrote William Muir of the crisis at Agra, 'and the only thing that flourished was the beard. With me, as with many others, it was the beginning of that luxury.' Muir in his memoirs of the Mutiny at Agra. William Muir, ed., *Records of the Intelligence Department of the Government of the North-West Provinces of India during the Mutiny of 1857.* 2 vols, Edinburgh, 1902.

The Baillie Guard,1858.Photo Felice Beato

BEATO, Felice A photographer whose early work includes a number of photographs of 'Mutiny' sites in Delhi, Lucknow and Cawnpore[Kanpur], dating from 1858 and possibly some from 1857. These are frequently reproduced, and are characterised as 'sombre views', but since the subjects are invariably war and mayhem they could hardly be anything else. There are portfolios etc of his photographs in a number of places, notably the Oriental and India Office collections of the British Library; but the album which was once the property of **Colonel F.C.Maude V.C.,R.A.** and which was presented in 1934 by his daughters to the **Royal Artillery Institution** is of particular interest because each photograph is annotated by Maude, and the implication is that he was with the photographer, especially in Lucknow after the final assault by the British in March 1858. The album may be viewed by appointment in the Library of the Royal Artillery Institution, Woolwich Arsenal, S.E. London.

BEEBEE GURH *See Bibigurh.*

BEES A major 'defeat' for the British Commander-in-Chief was reported by **W.H.Russell** on 17 May 1858: 'On arriving at our camping ground at Tilhour tope, a large swarm of bees, irritated by the smoke of camp fires which ascended the trees to their nests made a furious charge at us, routed Sir Colin, General Mansfield and the officers ingloriously, and forced me to draw the curtains of my dooly, and remain in a state of semi-suffocation till their rage abated.'

BEGUM OF OUDE On 25 January 1858 another determined attack was made on the British position at the **Alambagh** just outside Lucknow, but unfortunately for the attackers the British had the night before been reinforced by a considerable body of cavalry, and it was the latter that was crucial on this day. The rebels' objective was probably to take the fort of **Jellalabad** which was being used by the British as a storehouse and magazine for the coming assault on Lucknow: if these supplies and powder had been lost a considerable blow would have been delivered to the invading force. But the presence of the cavalry, 1,200 strong and consisting of the 7th Hussars, **Military Train**, Volunteer Cavalry and **Hodson**'s and **Wales's Horse**, was enough to turn events in favour of the British who drove back their enemy away from the Alambagh and Jellalabad. A non-combatant who watched the proceedings from the roof of the Alambagh reported: *'The rebels were now totally dispersed, and our cavalry nearly succeeded in securing the Begum, who on this occasion came out to encourage her troops, fully expecting to see them at the least in possession of Jellalabad, if not of the camp also. She managed to escape on an elephant back from a village, just before our cavalry entered it'* **Dr Francis Collins** in *St George's Gazette*, 30 April 1895.

Said to be Hazrat Mahal, Begum of Oude

BEGUM OF OUDE[Awadh] 'Oude was an independent Kingdom once, until the British destroyed and annexed it, in 1856; ironically it was the British who had created it, fifty years before. It was **Hazrat Mahal's** ambition to restore the Kingdom and to place her son upon the throne. She is the Begum of Oude. 'Begum' is a title, rather confusing to foreigners: it means 'lady of high rank', perhaps a Queen, perhaps a King's mother-in-law: they are all begums. So when we are told that 'the Begum of Oude was well regarded even by her enemies, the British, for the manner in which she campaigned against them', we may well ask which Begum? It is confusing even today, but in 1857 many mistakes were made in identifying her. The great Sir William Russell, the world's first War Correspondent, got it wrong. Whoever it was that he saw in Lucknow, it was not *the* Begum of Oude. Anna **Madeleine Jackson**, one of the **Captives of Oude**, was equally deceived perhaps. In an account she wrote many years later for her two sons, she says: 'Our friend brought his wives to see us. The Begum or Queen-Mother also came to see us. He was her daroga. She had a purdah put round her when he came to speak to her - she was a strong masculine looking woman.' At the time Miss Jackson was imprisoned in Lucknow the capital of Oude, the British had not yet recaptured the city, and the Begum - our Begum - was all-powerful: as the mother of the young King, **Birjis Qadr**, she would be thought of as the Queen-Mother. Yet the description of the Begum as 'masculine looking' rules out Hazrat Mahal: she was most beautiful, and entirely feminine. The confusion becomes more understandable when we read that the last King of Oude, **Wajid Ali Shah**, deposed by the British in 1856, and taken to Calcutta where he lived as a virtual prisoner, had had sixty wives and concubines by whom he is said to have had seventy-two children. Hazrat Mahal was one of the wives left in Lucknow; she does not seem to have resented it, indeed she devoted herself to the royal cause with perhaps renewed energy, released as she was from the demands of the royal harem. Her motive would not seem to be to help Wajid Ali her husband so much as to promote the claim of their son Birjis Qadr to the throne. The royal family made an excellent rallying point for the rebellion, but there was more than one candidate for the 'vacant' throne. The sepoys had numbers, arms and ammunition, the support of the countryside and the looted contents of government treasuries; what they did not have - and were in desperate need of - was effective leaders commanding the loyalty and obedience of all. Hazrat Mahal was helped by a raja, **Jai Lal Singh** who belonged to the old, much aggrieved Court faction, that had lost more, in some ways, than the King himself by annexation. The only other strong candidate, supported by the cavalry who were Muslim for the most part, was the brother of the ex-King, one Sulaiman Qadr, and he, perhaps foreseeing the return of the British, refused the honour. With the Hindu infantry sepoys Rajah Jai Lal Singh had most influence; they took little persuading that he was the best military leader available, and that they should support the crowning of Hazrat Mahal's son, Birjis Qadr, as King of Awadh. The boy was only twelve years old: it was assumed, though not explicitly stated, that Hazrat Mahal should rule in his name. She did, and from that moment she had great power. Things went badly for the British at first, at Lucknow as elsewhere. The small garrison, after a disastrous sally a few miles up the road to **Chinhat**, was driven by overwhelming numbers of rebels into the Residency compound which the Chief Commissioner, Sir **Henry Lawrence**, had fortified against just such an event. The 'Siege of the **Lucknow Residency**' then began. Lucknow, and indeed the whole of the **Kingdom of Oude**, suddenly found itself free of the foreigner, and rejoiced mightily in that freedom. In August 1857 Birjis Qadr was accepted by the sepoys and the citizens and was enthroned. There is no record however that the ex-King Wajid Ali in Calcutta was ever asked for his opinion. If he had been it is highly probable that he would have been very put out by the idea: he was still refusing to sign a document of abdication. All the other 'Begums' of Wajid Ali were asked for their opinion, and apparently gave their consent unanimously: considering the jealousies, the rivalries and the deep hatreds that existed in the harem, it is a marvel that they all agreed. Maybe some instinct told them that their best chance of recovering opulence and status lay in handing all power to the ablest of their number, Hazrat Mahal. She seemed all powerful, but the soldiers had not just cheered for their new King, and thrown their hats into the air: they had imposed tough enforceable conditions on the new ruler. For example:- orders from the King of Delhi were to be obeyed by her immediately and without question; the Wazir, or Chief Minister, was to be chosen by the army; the regimental officers were not to be appointed without consultation with the soldiers themselves; sepoys were to get double the pay they had received from the British; the army was not to be interfered with in its pursuit of those whom it believed were friends of the British. This actually worked to the advantage of the Begum and many of the Court party, in the long run: they were able to put the blame for all excesses on the army, and claim their own hands were clean. The old King of Delhi apparently legitimised the whole thing by asking the young Birjis Qadr to rule Oude as his representative. Everyone happily looked back to pre-British times in settling relationships. This adds substance to the claim of many that, in Oude at least, the rebellion was a popular one and a genuinely national rising: Indians looked back with pride and nostalgia to the days before the British had come, and sought to recreate the glory that was the Mughal Empire. **General Havelock** relieved the British garrison in the Lucknow Residency - or so the textbooks would have us believe: in fact all he did was march in and swell the numbers of the garrison. He could no more get out than they had been able to do. Then in November 1857 comes the second 'relief': this time **Sir Colin Campbell**, the Commander-in-Chief, marched in and with a much bigger force was able to evacuate the garrison. But he could not hold Lucknow and had to retreat down the road to Kanpur, leaving a small part of his army at a palace garden in the South of the city, called the **Alumbagh**. In open fight the British, despite being outnumbered many times, were almost always victorious, and this began to depress the rebels. Early in the new year there were many desertions, especially by the old nobility, the landed gentry of Oude, and their armed followers. The Begum was equal to the situation. She called a durbar, or Council, and addressed the chiefs. There is almost an echo of Queen Elizabeth addressing her troops at Tilbury: she displays the same courage, but her message is full of reproach instead of encouragement. Hazrat Mahal said to them 'Great things were promised from the all-powerful Delhi and my heart used to be gladdened by the communications I received from that city, but now the King has been dispossessed and his army scattered; the English have bought over the Sikhs and the rajahs, and have established their government West East and South, and communications are cut off; the Nana has been vanquished; and Lucknow is endangered; what is to be done? The whole army is in Lucknow, but it is without courage. Why does it not attack the Alumbagh? Is it waiting for the English to be reinforced and Lucknow to be surrounded? How much longer am I to pay the sepoys for doing nothing? Answer now, and if fight you won't, I shall negotiate with the English to spare my life.' It is said that the chiefs retorted, 'Fear not, we shall fight, for if we do not we shall be hanged one by one;

we have this fear before our eyes.' A brave speech, a vainglorious reply. **Raja Man Singh**, one of the powerful talukdars or landed gentry of Oude, now showed every sign of going over to the British. Others followed suit. In revenge the Begum confiscated Man Singh's estates, and others, and in the mini - civil war that followed she was much more successful than against the British. Pacification by the latter was a long process. No one stood in the way of a British column as it marched through the land, but as it passed on the rebels fell in again behind: there was a genuine people's revolt in Oude[Awadh]. She was soon to have a rival. The **Maulvi of Faizabad** appeared on the scene: ambitious and astute, he was a Muslim religious leader with pretensions to kingship. The Delhi sepoys saw in him a leader who, because of the divine support he persuaded them he had, could not fail. He certainly displayed more military genius than the Begum's generals. March 1858 saw the British at last successfully attack Lucknow. To the end the Begum was giving fiery exhortations to her troops and maintaining their morale and her own by showing that they were fighting for their King, her son. They were pushed out first to the suburbs of the city, at **Musabagh**, not before the sepoys had proved their gallantry many times over: the courage was not all on the British side. With thousands of her army dead, and many guns lost to the British, she fled to the North-East. Hazrat Mahal sought help from **Jung Bahadur**, the ruler of Nepal, but that wily man was still counting the loot he had gained from co-operating earlier on with the British.

General Sir Hope Grant

The Maulvi joined her however, and they had a combined army of about 18,000 men, a formidable force. In June 1858 they were attacked by the British under General Sir **Hope Grant** at **Nawabganj**: they were compelled to retreat to the North. The problem for the Begum was that she was herself completely without military knowledge or talent: indeed why should it be otherwise? There are not many Joans of Arc or Ranis of Jhansi in history. Fighting was men's work: she had courage and determination, a plan, and a viable strategy, but when it came to commanding in the field, she needed the help of a man, and the only one who might have succeeded was her deadly rival, the Maulvi. So often as you read the story of the Mutiny you marvel at the ineptitude of the rebels' leaders: this was the greatest asset the British had. Her most constant male companion was the quite worthless courtier, **Mammu Khan**, whom she had promoted to the highest posts - far beyond his talent and ability. How often this happens! Women of great personality and intellect get lumbered with great useless oafs on whom they waste their affections and attentions: it is as though a great woman seeks always a child to be her man, whom she can mould and coax and cosset, and eternally dominate. Too late, the talent of another talukdar, Rana Beni Madho, emerged. He too was determined never to submit to the British, but he came too late - the cause was all but lost. Still the rebels continued to plan as though for final victory. The Begum showed sound judgement both of people and politics. For a time it looked almost as though she might prevail. Lucknow was not Paris: the fall of the capital did not mean the end of resistance; this was truly a people's revolt in Awadh. The pursuit by the British troops of the forces of the rebels lasted many months, and we have many first-hand English accounts to tell us how wearisome it was. The noblest, the bravest and the most able of the rebels were thus pursued, including the Begum with a small band of devoted Rajput chiefs, plus Prince **Firoz Shah** of Delhi, and **Khan Bahadur Khan**. With none of these would the British have found serious disagreement: they were honest and honourable adversaries. But now they were joined by others who were thought to be beyond all thoughts of mercy: the Nana Sahib and Azimullah Khan from Kanpur joined the rebel band, and although their soldiers helped to swell the numbers, they were not really welcome. The cruel treachery of the Nana's troops at Kanpur was denounced by the Begum as having brought a curse on the rebel cause. The months dragged on, the rebel army was still formidable and had not been decisively defeated or dispersed. There was hope. Hazrat Mahal's writ still ran in much of the old Kingdom of Awadh, and she proved to be a talented administrator. Even the British admitted that had she been the ruler instead of her husband there would have been scant excuse for the annexation. On 1 November 1858 amid much carefully orchestrated firework displays and military parades, the British played what they planned to be their trump card. A proclamation from Queen Victoria was read out in all the military stations of India declaring that the rule of the East India Company was over, and that the British Government now ruled India directly, under the Queen. Significantly, pardon was offered to all who would lay down their arms and could show they had had no hand in the murder of Europeans. Religious toleration was promised. Ancient treaties and customs would be respected. The British hoped this would be the end of the Mutiny, and congratulated themselves on finding a happy solution to the problem of the anomaly of the Company: they wanted to believe that the solution was popular with Victoria's newly created Indian subjects, and so they told themselves that it was, and reported to the Home government the proclamation had been well received. They gave considerably less prominence to a counter-proclamation issued, with equal pride but much less pomp, by the Begum of Oude. Hazrat Mahal's proclamation is worth looking at because it tells us what, in the eyes of so many Indians, the mutiny/rebellion was all about. She starts by reminding the people of Awadh that the British were not to be trusted, and would not keep their word: they treated the Indians, she said, as inferior beings incapable of doing anything more than manual labour. She had a point. She asked above all what there was in the supersession of the power of the East India Company by that of the Crown which could benefit the people of Hindustan, seeing that the laws of the Company, and their English officials, were all unchanged? She had another good point there. Then she attempted to prove that interference with religion and caste of the people of Hindustan had originated the rebellion, and quoted the ill-treatment of native princes, the violation of treaties and the pernicious nature of Christianity. In December 1858 with Sir Colin Campbell (now Lord Clyde) hard on her heels in the neighbourhood of Bareitch, she was offered asylum, a pension and her life in return for surrender. She was tempted and might have accepted; we shall never know for certain; her chiefs got wind of the possibility, struck camp and fled, taking her and her son as virtual hostages. Soon she had bounced back into her defiant mood, and addressed a letter to **Rajah Man Singh** demanding to know what terms **Queen Victoria** would grant her, a fellow queen, if she thought fit to lay down her arms. She got a pretty abrupt answer: Queen Victoria would not be asked such a question, and the only thing that Hazrat Mahal could now expect was her life would be spared. She, and the remaining chiefs crossed the border into Nepal and asked Jung Bahadur for help. It was again sternly refused: at first he even threatened to hand her over to the British, but later relented and allowed her and a small entourage to stay. Many sepoys lingered for months in the borderlands and in small parties gave themselves up: at the border they were given two rupees and told to make their way home. Even Mammu Khan - rejected at long last by the Begum, for his cowardice, gave himself up. The months passed, the years passed, and the Begum refused to surrender. The '*Times'* in London briefly chronicled her history. At the end of 1858 it was saying 'Like all the women who have turned up in the insurrection she has shown more sense and nerve than all her generals together. ' In January 1859 it reported 'Mummoo Khan has surrendered himself having been previously dismissed the service of the Begum (for want of courage and devotion)'. On 30 Jan 1860 its readers were told that only 'The Prince (Feroze Shah of Delhi, a cousin of the last King of Delhi) and the Begum were still at liberty at the close of 1859. The Begum has less than 1500 adherents, half-armed, half-fed and without artillery'. Hazrat Mahal never did surrender, and she died in Nepal in 1879.What do we know of Hazrat Mahal the person? Surprisingly little. The Dictionary of Indian National Biography calls her ' heroine of the First War of Independence', but tells us little else. It seems she came from a poor family, and that her surname was Iftikharun-nisa, and she was born and brought up in Faizabad. She had 'irresistible physical charms', an 'inborn genius for organisation and command' and was 'fearless and calm in danger'.'Privately trained in music and dancing she received customary education after arrival in harem'. In other words she was a very beautiful dancing girl who caught the eye of the King , Wajid Ali Shah, who not only took her into his harem but, when she gave birth to a son raised her to the rank of one of his wives, under the title Hazrat Mahal. All indications are that the true father of her son Birjis Qadr was that same Mammu Khan of whom we have spoken. He was her lover when she was a dancing girl, and in fact he never left her, despite her marriage, though how he got through the cordon of the 'amazons', if they existed, we do not know. He certainly stayed with her after her husband's banishment to Calcutta, and they raised their son to the throne at the earliest opportunity. The last mention we can find on the family is in a pamphlet issued in 1961 which shows that a grandson of Hazrat Mahal and a son of Birjis Qadr, by name Mehar Quder, lived on in India, and was called by some the Prince of Oude.' **PJOT**.

BELL, Thomas Evan Maharaja Holkar of Indore was suspected, despite his protestations, of being disloyal to the British in 1857. **Sir Henry Marion Durand** in particular doubted Holkar's 'trustworthiness', although he claimed that appearances were against him as his troops had rebelled and he had been unable to bring them back under control. *See* T.E.Bell, *Holkar's Appeal: papers relating to his conduct during the Mutiny*, London, 1881.

BENARES Now known of course as Varanasi. A great city, holy to the Hindus, on the Ganges, and the Grand Trunk Road, 460 miles (736 km) from Calcutta and 83 miles (136 km) from Allahabad. Called Varanashi from the two streams Vara and Nashi so termed in Sanskrit: the Muslims pronounced the name 'Benares', a corruption followed by the English. The Commissioner in 1857 was Henry Tucker, who, in his public life was just, patient and understanding, but who was known also as an actively proselytising Christian. All seems to have been quiet in the city and cantonment where there were three regiments of the Bengal Army plus 150 men of HM's 10th regiment (from Danapur) and about 30 British artillerymen, until the arrival of Colonel **James**

Benares in 1857

Neill of the Madras Fusiliers. His action precipitated an outbreak that might not have taken place: the 'fiery Neill was instrumental in lighting flames which he was compelled to stay and extinguish at the cost of leaving **Sir Hugh Wheeler** and his companions (at **Kanpur**) to perish' - one of many serious charges made against Neill and his conduct. For a detailed account of the mutiny of the 13th Native Cavalry and the 37th N.I. and (part) of the Loodhiana Seikhs, *see* H.Mongomery Martin, *The Indian Empire*, Vol II pp 282-290. *See also. Daily News* 30 October 1857 in which there is quoted a significant letter from **Major-General Lloyd** to his brother, attacking Neill's actions.

BENARES

'So Colonel Neill assumed command and proceeded to the most summary measures, cutting off whole regiments. All the ladies were crowded into one room, with wounded and dying men, and from the window the sight that greeted [an officer's] eye was a row of gallowses, on which the energetic colonel was hanging mutineer after mutineer, as they were brought in.' London *Times*, August 25, 1857. Colonel Neill was putting Barnard to shame, for with 174 men he established order in Benares and punished three regiments of native soldiers 'while General Barnard with some hundreds of European troops at his beck dared not to cut up the mutinous 5th and 60th.' From the *Friend of India* quoted in the London *Times,* 15 August 1857.

BENGAL ARMY, Composition of

While the exact numbers of soldiers enlisted in the Bengal Army in 1857 are relatively easy to ascertain, or at least to compute fairly accurately, the last breakdown of enlisted men, *by caste or creed* dates from 1853: in that year the Bengal native army numbered in all 83,946 men. Of these, 70,079 were infantry. Of the composition of the cavalry the returns are silent (although it is known that a high proportion of them was Muslim), but the infantry was thus classified: 'Brahmins, 26,893; Rajpoots, 27,335; Hindoos of inferior castes, 15,761; Mahommedans, 12,699; Christians 1,118; Sikhs, 50'. *See Daily News,* 8 August, 1857, quoted in B.W.Noel, *England and India,* London, 1859, p 86.

BENI MADHO (BAINEE MADHOO) He was the Rana of Shankerpur in Oude[Awadh], son-in-law of **Kunwar (Koer) Singh** of **Jagdishpur**. One of the most loyal followers of the **Begum Hazrat Mahal** and her son **Birjis Qadr**. Displayed considerable ability both in civil and military matters. Eventually killed in Nepal when **Jung Bahadur** marched against the rebels in the **Terai**. He was a rebel from the beginning, in the sense that he was a man of principle who believed that his loyalty lay to the ruler of Awadh to whom he had sworn allegiance. Even when offered a free pardon and the restoration of his estates, he, after some hesitation, refused, and took up arms again in what was by then a hopeless cause. Admired by the British for his loyalty and competence. See W.H.Russell in the *'Times'*, 17 January 1859. *See also* R.Montgomery Martin's *The Indian Empire*, Vol II, *and* **FSUP** Vol II.

BENI MADHO SINGH - Proclamation by Birjis Qadr. 'As I am fully bent upon populating the land, securing all conveniences for its people and betterment of its inhabitants I have, therefore, decided to exterminate the cruel, ill-behaved Kaffir (the unbelieving) Feranghis from my hereditary dominions, both old and new. Consequently, I have nominated the brave Rajah Beni Madho Singh for the administration of the Ilaqas of Jaunpur and Azimgarh and order that in obedience to the instructions of the said Rajah, you should capture, put to the sword and annihilate the entire group of these perverted unbelievers and make every effort to extirpate them from this country. Considering the said Rajah a permanent Amil (administrator) of this part of the country, you should do your very best for the collection of revenue and betterment of the ryot. You will consequently, be awarded with favours. Dated 26th Zilhijja 1273 A.H. (17th July 1857).' *Mutiny Basta No 6 Trial of Rajah Jey Lal Sing*. Lucknow Collectorate Records.

BENI MADHOO, Rana

The British entertained strong hopes that the **Queen's Proclamation**, issued on 1 November 1858 would bring to an end the mutiny/rebellion by its offer of reasonable terms to those who gave up the struggle. In particular they hoped that **Rana Beni Madhoo** of Shankerpur would submit. The Commander-in-Chief, Lord Clyde was present in person on 15 November some three miles from Shankerpur, and messengers were sent in to the Rana calling upon him to surrender. No direct reply was received, but a most extraordinary letter arrived which no one has been fully able to explain since: a messenger arrived from Shankerpur with the following, professedly from a son of Beni Madhoo: 'I have received Your Excellency's *purwannah*, and with it the Proclamation. I beg to say that I was formerly *qabuliyatdar* of this *ilaqa*, and am still in possession of the same; and if the government will continue the settlement with me, I will turn out my father, Bainie Madhoo. He is on the part of **Birjies Kuddr**, but I am loyal to the British Government, and I do not wish to be ruined for my father's sake.' The comunication, although from the son, was believed to be the composition of Beni Madhoo himself, who also sent in a letter to the Raja of Tiloi, saying that one king was all he could serve and that he had already pledged his fealty to Birjis Qadr.

BENNETT, Mrs Amelia   Also better known as 'Amy Horne'. 'Among those who entered Wheeler's entrenchment in June 1857, and who are known to have survived, there are eight other women, besides **Ulrica Wheeler**, most of whom were of mixed race. Their names are Eliza Morrison, Hannah Spiers, Isabella Spiers, Elizabeth Spiers, Amelia Bradshaw, Ellen Bradshaw, Caroline Letts and Amelia (or Amy or Emma) Horne. We know a great deal about Miss Horne even though by writing at least two, conflicting, accounts of her experiences, she manages to confuse the picture. Her reasons for this are altogether understandable: in the account she gave soon after her rescue she was clearly suffering from shock, and was most unwilling to admit to having been the object of rape: when, in old age, she wrote a full and frank account of what had happened to her she did not spare her own feelings, nor mince her words. Amy's father was Captain Frederick William Horne R.N., of whom she was intensely proud but whom she had not known: he died when she was only one year old. It was **William Howard Russell** the *'Times'* correspondent , who put into words the difference between the two girls : whereas Miss Wheeler was described as 'an unfortunate young lady', Amy Horne was 'the daughter of a clerk, and is, I believe an Eurasian, or has some Eurasian blood in her veins. It would be cruel to give her name, though the shame was not hers'. Amy had been in Kanpur for only a few months in May 1857. In old age she could hardly credit that she had indeed lived through those times and seen so many die. It was on 25 June that the siege of **Wheeler's Intrenchment** came to an end. Amy Horne was the first to see the woman coming towards the ramparts of the entrenchment, carrying a white flag. It was not Mrs Greenway as so many have said, it was Mrs Jacobi who came with the white flag. She was Eurasian also and married to the watchmaker Henry Jacobi. She brought an offer from the Nana to let them all go to **Allahabad** by boats on the river. The story is well known. Two days later they all left the entrenchment. Early in the morning of the 27th the doomed garrison moved off towards the river, over a mile away. The wounded were in palanquins, the able bodied men marched, and the women and children were taken in bullock carts, and some on elephants. The boats were apparently ready for them and were lined up at the **Satichaura Ghaut**.

Massacre in the boats

They were not big boats, normally only six people at most would have travelled on them, now twice that number or more were packed in. It started with a bugle call. As though they had been waiting for the signal, all the boatmen leapt from the boats and scrambled ashore and at once the firing began. The thatch of each boat was set on fire by the boatmen before they left and this added to the panic which had already started. Many fell headlong into the water to be sabred by the troopers who came riding into the river to stop anyone getting back on shore. Sepoys swarmed on to the boat that Amy was in, intent at first more on plunder than with killing. Amy describes what happened. But here her first account begins to differ radically from the one she wrote in her old age. Her first account talks of her being rudely seized and thrown overboard, of crawling ashore and dragging herself on hands and knees for half a mile and of hiding herself in a thicket. She says she then saw the 'well-known face and form' of Ulrica Wheeler, that after an hour a party of the rebels surprised them both, and dragged them away in different directions. She was taken four miles to **Bithur** (it is actually twelve miles from Kanpur), nearly nude, and dragged along unceremoniously by men. In Bithur an AFRICAN interceded for her, gave her clothes, food, a bed, and protected her, although she saw him no more till she went to **Lucknow**: it transpires that he was a eunuch in the employ of the King of Oude[Awadh] and had been sent to Kanpur with despatches from the **Maulvi of Faizabad** to the **Nana Sahib**. She speaks frequently of this eunuch, calling him 'my sable benefactor', but makes no reference whatsoever to her captor or captors. Her second, much more detailed account, written in old age (July 1913), when perhaps she saw no further point in concealing the truth, is very different: ' A trooper rode into the shallows and levelled his musket at me from no more than two feet away and yelled that I was to jump from the boat. I could not, would not move: The next thing I knew I was struck a terrible blow on the side of the head with a musket - wielded by a sepoy already in the boat. I went overboard half unconscious but remember being seized by the hand by the trooper who had first bellowed at me to jump, and found myself half walking and being half dragged against his horse's flank and hauled ashore. I saw Ulrica Wheeler too had been 'captured' in similar fashion: I was not allowed to rest. Two hours later, still held in the firm grip of the trooper, I was still stumbling along beside the horse: we must have covered about three miles, moving all the time away from Cawnpore. I knew from his uniform that he was in the 2nd Cavalry, but that is all I knew: he said absolutely nothing throughout that afternoon. I stole a glance at him from time to time: he was quite young, tall and thin, with a black beard which only partly obliterated the pockmarked sallow face beneath a tall white turban. His piercing black eyes were cruel and without any redeeming spark of humanity. He seemed indifferent to the state I was in - soaking wet, barefooted bareheaded, shivering with terror, and at times only just conscious'. She was dragged in to a large tent, and made to squat on the carpet at the front of the dais. Nothing was said or done for what seemed an eternity, and then she realised that there was a third stranger in the tent, that he stood behind her and that he held in his hand a razor sharp sword with its blade immediately above her head. At last one of the two maulvis began to speak, and to her surprise he spoke in English, excellent English, not the lilting patois of the bazar, and then she knew who he was, although the knowledge was of no comfort to her. It was **Mohammed Liaquat Ali** who a few weeks ago had roused the people of Allahabad and the surrounding countryside and briefly ruled the city before **General Neill** arrived and the rebels fled in terror from his very name. The other maulvi was his brother, a man called Ameerun Ali. What he said to her was plain and to the point. The trooper desired her but she was a Christian, and thus defiled and unclean, and before he could touch her she would have to be converted to Islam. But the choice was hers: if she refused then her life was forfeit and the man standing behind her would despatch the matter immediately. Amy was terror-stricken and knew only that she wanted to live; she had an overwhelming instinctive desire to survive. She sank down before the maulvi and bowed her head in silent assent. Liaquat Ali divided a pomegranate into two halves: it was a blessed fruit, Amy was told, and she must eat it as the first step in her conversion. She did so, and forced down also some sherbet she was handed. Then the maulvi prayed: what he actually said, Amy could not remember, but prayers they were, and the name of God was called upon frequently. They lasted but a short time and then she was taken by a female servant to a kind of tank behind the tent, stripped of her borrowed clothing and allowed to sink into luke-warm clean water: this was not to be for pleasure, it was a part of the process of purification. She was then dressed in a long cotton jacket or kurta, with a white muslin shawl over her head and shoulders and loose pyjama-like trousers gathered in at the ankles. She was to all intents and purposes a Muslim woman in appearance at least. The woman servant told her that the name of the trooper who had brought her thus to the Maulvi: he was **Mohammed Ismail Khan** of the 2nd Cavalry. At long last the trooper returned, the door opened, and she saw him silhouetted against the sun low in the west. She does not describe what happened then: it is enough to say that she was sixteen, a virgin, of what was called then 'of delicate upbringing', in other words she knew little of the facts of life. The horror of the rape remained with her all her life. No wonder that at first she tried to pretend, even to herself that it had not taken place. Mohammed Ismail Khan, her captor, appears never to have treated her with tenderness, and to the end she remained terrified of him. There was to be no 'happy' ending as with Ulrica Wheeler. Amy came to believe that one of the reasons for his cruelty to her was that he feared her almost as much as he desired her. By the middle of July plans were changed, abruptly. The rumour sped around the camp like wildfire: the British had stormed Delhi and recaptured the city. Amy had to pretend to ignorance, indifference, or feign dismay at the stories; but her heart was full of joy. What matter that the news was premature? What was of great moment was that the maulvi and his band believed it, and were shocked beyond measure that their great army had been defeated. She was conveyed to Lucknow that very night in a covered litter, accompanied as always by Mohammed Ismail Khan. The litter or dooly was stopped on the way by a picket of native soldiers, and in reply to the challenge, the trooper simply said that it contained the ladies of his zenana: he had cunningly provided for just such an emergency by placing a native servant in the dooly with Amy, and on the reply to the challenge being made she thrust a brown arm out through the curtains to corroborate his statement. When they reached Lucknow she was concealed in a dyer's hut, quite close to the **Residency** where the British garrison was besieged. It was dirty, dark and uncomfortable, and was to be her 'home' for two whole months. Her misery was the greater because she was so near to the British but had no chance of reaching them. There was hardly room to move about in the hovel: it was large enough only to accommodate a charpoy, or rough stringed bed. There was no ventilation and the heat was so oppressive Amy thought she must surely be suffocated. She was allowed out for only a few minutes each night to take the air, but this saved her life; it certainly saved her sanity. Eventually her captor had to move her, not from compassion or even with regard for her health, but because some women had discovered her existence: they did not believe she was truly a Muslim and they were threatening to report the trooper to the authorities. Their malevolence did Amy a good turn for the place to which she was taken was like a palace compared to the den in which she had been forced to live for the previous two months. It had been the residence of Capt. Simpson, and was also called the 'Observatory'. Amy was in complete despair, and begged her captor to take her to the **Maulvi of Faizabad**. One maulvi had already saved her life, perhaps she could persuade this one to do likewise. He was all powerful in Lucknow, the sepoys believed that nothing could harm him, that he was a saint and was invulnerable from shot and shell. The trooper agreed to take her and she obtained an interview. By now she was highly proficient in the Hindustani language, and was also highly practised in deception and dissimulation. She was still alive and was determined to survive; it is not for us to judge her by the standards of the Victorian drawing-room. The result was that she managed to impose sufficiently on the credulity of the Maulvi for him to agree that she had some claim upon his protection. He made her take an oath to observe all the Muslim religious practices; he also made her share a sweetmeat with him of which he ate half and handed her the other: this made her his 'moreed' or disciple. She became bound to follow him wherever he went. As a devout Christian all her life she claims that it was with repugnance and loathing that she went through this ceremony, 'but I was young and life is sweet to the young'. She did not know whether other, unmentioned, demands would be made of her, as his disciple. In the event there was unexpected advantage in her new position: the attentions of her original captor, the trooper Ismail Khan, became mercifully less frequent as she had to take up residence with the Maulvi. He was then living in **Ali Nucky Khan**'s palace, situated on the banks of the river **Gomti**: she stayed there for a month. The rebels now became greatly concerned by the approach of the British troops. Many deserted and went off to **Farrukhabad**: others stayed loyal to the **Begum** and the Maulvi of Faizabad; a few decided that they had had enough of all fighting and resolved to retire to their homes although they knew the risks that this entailed. Among this last group was Amy's captor. They set off on what was to be a very long march indeed, twenty days, walking all the way, and she was weak with fever. In Rai Bareilly she was detained by the local zemindar who clearly thought she was no true convert : fortunately the guard was slack and after two days, with the help of her captor, she escaped at night. They came at last to Guthni, very close to Allahabad. It appeared that this was Ismail Khan's home and here he intended to stay. Once again Amy felt she was very near to salvation, but yet so far: the English were firmly in control in Allahabad now, and she had no doubt that her uncle was still living there. Her captor told her that she should be free to go if she gave him a solemn undertaking to act as his advocate and obtain a free and full pardon for the part he had played in the uprising. She most willingly agreed to this, but suspicious of her good faith he made her reduce her verbal promises to writing: he enlisted the services of a *Munshi* to translate to him word for word all that she had written. Her ten months' of captivity was coming to an end. She walked to the nearest *Thannah*, or police-station and demanded a dooly to take her to Allahabad, and within three hours she had presented herself to her uncle and aunt. So great a change had taken place in her appearance as a result of her terrible experiences that they at first failed to recognise her, particularly as her proficiency in her own mother tongue had sadly dwindled. Indeed, the *'Bengal Hurkaru'* reported her as being 'a lady in great distress of mind, often in tears, who has forgotten much of the English language and looks prematurely aged'. For some months, with relatives in Calcutta, she suffered severely from melancholia, and must have appeared a mental and physical wreck. After about a year she met and married William Bennett, a railway surveyor, and spent the rest of her life trying to be the wife and mother that she had thought Providence had intended that she should be. William, a good and thoughtful man, died before her, - it was not his fault that he could leave her very little upon which to live in her widowhood; her children all went their own ways, as nature intends shall be, and she was left, an old woman with memories.' **PJOT.** *See SSF*; *see also* Mrs Amelia Bennett, *Ten months' Captivity etc.*, in the *Nineteenth Century* (periodical), June 1913.

BERHAMPOOTER This was the name of a small steamer that in July 1857 was sent upstream from Allahabad on the Ganges, with 100 men of the Madras Fusiliers under the command of Captain Spurgin, in support of General Havelock's land advance along the Grand Trunk Road towards Kanpur. General Neill, who despatched the steamer had the idea that it could actually relieve the investment of Wheeler's Intrenchment at **Kanpur**, but the latter had already fallen to the rebels before the steamer left Allahabad. 'At noon a number of native boys and others announced in great astonishment that a jahauz (a ship) had just arrived in the Ganges. This was the steamer Berhampootur, having a hundred soldiers and several guns on board, and as a vessel of this kind had never yet proceeded so far up the country as Cawnpore, the sight of her both astonished the natives, and gave them confidence in the resources of the British, for the idea of utter destruction of nearly all Europeans had been so instilled by the Nana and his adherents in the minds of the ignorant people, that they were doubtful if the handful of British troops who had so bravely reconquered Cawnpore, could keep their own very long. The object of this vessel, I was informed, was to take the rebels at Cawnpore by surprise and rescue the poor European women and children in bondage: but, alas! it had met with many obstacles on the way, the chief of which was the want of sufficient water in the river, besides having to contend with the rebellious zemindars on the Oude bank of the river who had placed their batteries at convenient spots all along the bank. Hence she was unable to effect the noble object which she had in view, and was obliged to make slow progress, acting in consonance with the land forces under General Havelock all the way from Futtehpore to Cawnpore.' *See* W.J.Shepherd, *Personal Narrative etc*, Lucknow, 1866. *See also* J.C.Marshman, *Memoirs of Major General Sir Henry Havelock*, 1860, p 328; *see also* STEAMERS *and 'The Reluctant Heroes' in* **CM** p 31.

BERHAMPORE [Baharampur] This was the scene of the first mutiny by Bengal army troops in 1857. Berhampore was a military cantonment town some 120 miles (nearly 200 km) from Calcutta. The 19th N.I. on 26 February 1857 refused to accept the new **cartridges** (and the percussion caps that accompanied them) on the grounds that they were contaminated. Their Colonel was a martinet by the name of Colonel Mitchell, who was later adjudged to have been quite unfit to have commanded the regiment. When he heard that the cartridges were being refused, he came on parade and very angrily ordered them to accept them 'or I will take you to Burmah or China, where through hardship you will die. These cartridges were left behind by the 7th N.I. and I will serve them out tomorrow morning by the hands of the officers commanding companies'. He gave the orders so angrily that the sepoys were all convinced that the cartridges were greased, otherwise he would not have spoken so. The 19th N.I. was subsequently disbanded as a punishment: to the rest of the Bengal army they appeared as a body of victims and martyrs.

BERNERS PAPERS Held in the **Cambridge Centre of South Asian Studies** and including source material as follows: Letters from **Mrs Wells**, wife of **Dr Walter Wells** dated 5 April 1857 to 8 April 1858: they are most revealing: she speaks highly of, but disparagingly of **Brigadier Inglis** *'a man universally detested.'* Her husband was the doctor suspected by the sepoys of having spat into a medicine bottle. Tells of the 'Flaming arrow' fired at the Residency in Lucknow on 1 May 1857. Papers include a photograph album containing 22 very fine photographs of Indian scenes.

BETWA RIVER, Battle of Sometimes inaccurately referred to as the Battle of Jhansi, this was one of the most decisive engagements of the entire struggle. In March 1858 **Sir Hugh Rose** at the head of the Central India Force was besieging the city and fort of **Jhansi** in which the **Rani** was putting up a spirited defence. She had sent messengers to **Tatya Tope** and the **Rao Sahib** (nephew of the **Nana Sahib**, but now calling himself **Peishwa** although his uncle was still alive) to come to her aid, and she had reason to know that efforts were being made for her relief. A force, called the 'Army of the Peishwa' estimated at 20,000 men and twenty guns crossed the river Betwa on the 31 March 1858, and took up a position in rear of the British camp and lit an immense bonfire, as a signal to Jhansi of their arrival. This was welcomed by salutes from all the batteries of the fort and city, and shouts of joy from the defenders. Unfortunately for them, their joy was premature. Rose decided to split his force, and while maintaining the siege, to invite a general action, despite his numerical weakness. With silence and regularity he drew up his force, after dark, across the road from the Betwa. That night the two armies slept on their arms, opposite each other and not very far apart. The next morning, before daybreak, Tatya Tope advanced to the attack, but was defeated, pursued for nine miles, and driven, on 31 March, across the Betwa, with the loss of 1,500 men, eighteen guns, and large quantities of stores and ammunition. This destroyed the last hopes of the **Rani**, who fled from the fort of Jhansi with some of the garrison, although some remained to defend unto death the city and fort. There are numerous accounts of this battle, in Chas.Ball and **FSUP** etc and in most of the 'heavyweight' versions of the mutiny/rebellion. In particular *see The Revolt in Central India*, Simla, 1908, p110 *et seq., and* Thomas Lowe, *Central India during the Rebellion of 1857-58*, London, 1860.

BEYOND THE PALE Some French Journals had started the cry of 'English cruelty and barbarity in India,' wrote the *Times*. The *Gazette de France*, citing the siege of Delhi, accused England of openly adopting terrorism. 'England is the only nation of the present day capable of slaughtering the entire population of a city without mercy; the spectacle which she has given to the world places her out of the pale of civilization and humanity.' The *Times* dismissed this as 'too absurd' and doubted if the French even believed it themselves. 'What is the handle upon which this very large and sweeping accusation turns?' the *Times* wanted to know. 'It is a statement in the private letter of an English officer, written after the capture of Delhi, to the effect that some 40 or 50 Hindoos who had concealed themselves were shot, though they were not Sepoys, but residents,' an otherwise unsubstantiated incident 'not at all of a piece with, but in complete opposition to, the general conduct of our army. ...With the exception of this one story, all the evidence respecting the capture of Delhi goes to show that none but Sepoys were killed, special mention being made of one woman who was shot by simple accident.' *See* London *Times,* November 20, 1857, p 6.

BHAGALPUR On the Ganges near to **Danapur** and the scene of the defection of the 5th Irregular Cavalry, possibly as a result of the arrival there of **General Outram** on his way up country on 15 August 1857. They believed, or many of them did, that they would be surprised and disarmed during the night, so they mounted and fled. The headquarters of the regiment had been recently moved from Rohni to Bhagalpur in consequence of something that had happened at the former place on 12 June. *See* ROHNI.

BHAGULPORE [Bhagalpur] 'Remarkable incident in Bhaugulpore(sic). Three officers were sitting on a verandah of a bungalow when a detachment of the 32nd NI having risen in rebellion rushed in on them. One of them was the commanding officer of the detachment, Lieut. Cowper who implicitly trusted his men, who was in constant familiar intercourse with them, and who was believed to be an object of sincere attachment to his corps. Another was Lieut. Rannie of the same corps who had never taken any particular pains to please his men and had never appeared to be a favourite with them. The third was Mr Ronald the Asst. Commissioner in the division who was of course unknown to the sepoys. If all our theories had been worth a straw the men would have shot down Lieut. Rannie and Mr Ronald and spared Lieut. Cowper as their friend and their father. But they singled out Lieut Rannie who was not known to have done them any good, called upon him by name to leave the bungalow and suffered him to depart unmolested whilst they remorselessly butchered Lieut Cowper and the stranger by his side. If they had indiscriminately shot down all three officers, things would have been different. Why the recent convulsion? The real truth is that we know little or nothing about the causes of the Mutiny. My own impression has always been that mutiny is the normal state of an Eastern army, and that the marvel is not that after so many years the Sepoys revolted but that they did not revolt before. Pathans, Sikhs, Mahrattas - all mutiny. Do you know any Indian army that has not mutinied again and again? ....cannot help thinking that our Sepoy army revolted not because it was an army of blacks under a white master but simply because it was an oriental army and all oriental armies revolt. We must not think there is any special hatred of British rule - any especial hatred of a foreign yoke. History is full of instances of the barbarities practised by Indian armies upon their own officers - one of the mildest of which was that of tying them on to guns heated almost to red heat. The regularity of English pay and the certainty of an English pension suppressed the eruption during a long period of years.' Anon., (General **Showers**?), *Blackwoods Magazine*, November 1859.

BHAGWAN BAKHSH   *See* PERSONS OF NOTE, APRIL 1859.

BHANG   From a Sanskrit word meaning Indian hemp, the leaves and seedpods of which are chewed or smoked, or sometimes an infusion of them is drunk. Hashish. A common form of narcotic which acts first as a stimulant and then depresses the individual who indulges in it. Many sepoys were addicted to its use; it is said that **Mangal Pande** was intoxicated with it when he attacked the adjutant of his regiment.

BHARATPUR   *See* BHURTPORE

BHARGAVA, K.D.   A former Director of Archives, Government of India. *See* BIBLIOGRAPHY.

BHATTIS & RANGARS   These two rural castes took an active part in the rebellion in the **Delhi** and Haryana areas. For details, particularly of motivation, *see* Eric Stokes, *The Peasant Armed, the Indian Revolt of 1857*, Oxford, 1986, p 120 *et seq.*

BHOLA SINGH   Deserter Bhola Singh was given transportation for life: his story of being on leave from his regiment and then afraid of being hanged if he turned up was not believed. He may have been unfortunate, but there were many like him that made a good thing out of being on furlough at the time of their regiment's mutiny: it is said that many served - particularly in Oude[Awadh] - against the British, then melted away on the latter's success, only to turn up some months' later, clasping their original furlough certificate, and demanding - and getting! - their arrears of pay. In the case of Bhola Singh, his captor was given 30 rupees - blood money: few Indians betrayed important men, like the Nana Sahib, for whom large rewards were offered, but this delicacy did not extend to lesser rebels, many of whom were betrayed to the avenging British.

BHOPAL CONTINGENT   These contingent troops, established by treaty with the ruler of Bhopal, sympathised in part with the rebels, and a number of them mutinied. On the establishment of the **Central India Field Force** one of the first acts of the British was to deal with these mutineers: at **Sehore** they were tried by court-martial, and no fewer than 195 were executed, 274 dismissed and 228 re-enlisted. The Sikh element of the contingent had remained largely loyal to the British and they were now placed under Captain H.O.Mayne and formed the nucleus of the 1st Regiment Central India Horse.

BHOPAL   This was a princely dependent state of Malwa, bounded on the south-west by the territories of **Holcar and Scindia**. The reigning family were Pathans, although the great majority of the populace were Hindus. The ***Begum of Bhopal***, the Sikander Begum ruled the state as regent at this time, and remained fiercely loyal to the British throughout the conflict. The Bhopal Contingent troops consisted in all of about 800 men, including 48 artillerymen, and although offered to the British, many joined the rebels. 200 of them were later put to death (with the Raja of Amjherra) at Indore 'in a manner repugnant to humanity', *see* The *Times*, 25 August 1858. *Also* Khushhalial Srivastava, *The Revolt of 1857 in Central India-Malwa*, Bombay, 1966.

BHOPAL, BEGUM of   The Sikander Begum of Bhopal was one of the strongest allies of the British, never wavering in her support, based no doubt on her overwhelming conviction that the British would be victorious in the end. **Sir Hugh Rose**, after the Battle of **Kalpi** wrote in his despatch 'Her Highness displayed the very best feeling towards the English interests; she did so courageously in the worst times, when the natives in her part of the world thought that rebellion must triumph. Her Highness gave me two 9-pounder guns and a 24-pounder howitzer, with the gunners belonging to them, very good artillerymen, when I marched through Bhopal, which enabled me to complete No.18 Light Field Battery. Her Highness was indefatigable in obtaining supplies for my force when it was very much in want of them.' *See The Revolt in Central India*, Simla, 1908, *and* Thomas Lowe, *Central India during the Rebellion of 1857-58,* London, 1860, *and* **FSUP** Vol III.

BHOPAL Begum of

As Regent of this princely State she took the decision early on to adhere to the cause of the British. She was among the foremost to be rewarded and praised by the Viceroy at the *Durbars* held when the rebellion had been suppressed. One wonders how she felt however when the Viceroy, after praising her actions and character in fulsome language, ended his speech by turning directly to her and saying 'And you a woman, too!'

Sepoys loyal to Government. Photo 1858

BHOWANI SINGH, Subadar-Major

At **Kanpur** it was believed by the Europeans that the 2nd Cavalry were the most likely to mutiny; a high proportion of the **sowars** were Muslim. But the senior native officer, Subadar-Major Bhowani Singh, of the 2nd Cavalry, was wounded in the head and arms by the mutineers; he refused to allow them to take the colours and treasure belonging to his regiment; 'he was a brave and true man; we heard that he said to the mutineers, he would only obey and serve Government: he was afterwards killed in the entrenchments.' *Depositions of Gobind Singh, Sheik Elahee Buksh & Ghouse Mohomed.* **Williams.** Bhowani Singh had a fine record of loyalty to Government; some years before, the 2nd Light Cavalry was disbanded for cowardice; another regiment, (the Eleventh Light Cavalry) had been raised in the place of the Second; and the officers of the latter had been transferred to it bodily. Only one trooper of the Second had been re-enlisted - the Havildar-Major, Bhowany Singh. The Eleventh was renumbered the Second, for its gallantry at Multan. *See* Sir John Kaye. *History of the Sepoy War.* London, 1864-7, Vol II, p 302.

BHURTPORE [Bharatpur]

This was a princely state lying between **Jaipur** and Agra, the young under-age Raja of which remained loyal to the British in company with **Scindia, Holkar** etc. His contingent troops however who were sent to assist the British in **Muttra** were of the same background as the sepoys of the Bengal army, and had in all respects a fellow-feeling for them, and positively refused to march against them. *See* Mark Thornhill, *Personal Adventures and Experiences of a Magistrate etc.*, London, 1884. *See also* **FSUP** esp. Vol III.

BIBIGURH, EVIDENCE

'The evidence of the Christian drummers declares as follows:— 'After the five Europeans had been removed, the woman named **Hosainee Khanum**, or the Begum, who had the superintendence of the ladies, told them the **Nana** had sent orders for their immediate destruction; an appeal was made by one of them to Yousuf Khan, the jemadar of the guard, and, if the statement made by these drummers be correct, these men (the guard) refused to carry out the Nana's orders. Debased and brutal as many of the sepoys had already become, and steeped though their hands were in Christian blood, they yet hesitated to carry out the fiendish order of ...a still greater fiend than themselves.' *See* **Williams.** 'The Begum, it is said, on their refusal, returned to **Noor Mahomed**'s hotel, and shortly re-appeared with five men, two Mahomedans and three Hindoos (others say seven), but most of the witnesses implicate in particular one man of the Nana's guard named **Sarvur Khan** (a lover of the Begum's). A volley is said to have been fired at random by a few sepoys, but the butchery of the women and children was committed by men sent from the Nana's compound, in executing which they were occupied from about 6 P.M. until dark, when the doors of the building were closed for the night.'

BIBIGURH, EXPLANATION

'On **Bala Rao**'s return to **Kanpur** from the field of battle, wounded on the right shoulder by a musket ball, a council was held at **Noor Mahomed's** Hotel, at which a large number are said to have assembled, and over which the **Nana** presided.... They are said to have been unanimous in one fearful resolve, and that was the death of the unoffending and innocent women and children, and the few gentlemen whose lives had hitherto been spared. Two reasons were advanced in favour of this brutal resolve: the one, that it would probably prevent the further approach of the British; the second, that many rebels even now determined to forsake a losing cause and return to their old allegiance, and knowing full well that many amongst the unfortunate prisoners could recognize the leaders, and give important evidence against them, being intimately acquainted with nearly all those implicated in rebel proceedings, and felt that it was positively necessary to destroy all European evidence, as the only chance of evading the condign punishment their crimes so richly merited. Hence was the fate of these unhappy captives to be sealed in blood, and all were to perish in one common lot.' W.J. Shepherd. *Personal Narrative. etc.*, Lucknow ,1886, p 116.

BIBIGURH List of killed In the Bibigurh at Kanpur was found a list of the persons who had been held there immediately prior to the massacre. It is in *Marathi* and was apparently compiled by a native doctor who had been given (medical) charge of the prisoners some days before by the **Nana Sahib**. There are 210 names, and apart from Colonels Goldie and Smith, and Mr Thornhill, who were taken out and killed earlier, and possibly Mr Greenway, Mr Jacobi and Mr Carroll, there are only women and children on the list. Certainly the latter (even allowing for duplication of 'Moore' and 'Peter') contains the names of over 200 women and children. Their bodies were thrown into a dry well until it was full, but the probability is that, as the well was of no great size, some were dragged down to the Ganges and thrown into that river. The names on the doctor's list are as follows (from Kanpur, total 163): Mr and Mrs Greenway; F.Greenway, Martha and Jane Greenway; Mr Jacobi, Henry, Lucy and Hugh Jacobi; Mrs Tibbett; Miss Peter; Mrs Cocks; Mrs Brothrick; Grace, William and Charlotte Kirk; Mrs White; Miss Macmullen; Mrs Sinclair; John and Mary Greenaway; Lizzie Hornet; Mrs Sheridan,W.Sheridan and baby Sheridan; Mrs Wrexham, Clara and Drummond Wrexham; Eliza Bennett; Mrs Probett, Stephen Probett; Catherine, Jane and Thomas Willup; ? Reid, Susan, James, Julia Reid, C.Reid, Charles and baby Reid; Mrs Gillie; Henry Brett; Mrs Doomey; Henry Duncan; Mrs, James, and L.Levy; Henry Simpson; Miss Colgan; Mrs Keirseile, Mary and Willie Keirseile; Mrs O'Brien; Mrs, and Edward Green; Mrs Crabb; John Fitzgerald; Mrs Jenkins; Mrs Peel and George Peel; Mrs Moore; Marian Conway; F.C.Weston; Mrs Carroll; Mrs Butler; Mrs Johnson; Jane Morpet; Mrs Paterson; Miss Burn; Miss H.Burn; Mrs Dallas; W.O.Connor; Harriet Pistol; Elizabeth Simpson; George Casey and G Casey; Lucy and William Stake; Joseph Conway; James Lewis; Elizabeth West; W.N0ck; Henry Watkins; Jemima Martindall; Weston Darden; William James; Jane Gill; James Conseus(?); Mrs and James Peter; Mrs and Philip Baines; Mrs Harris; Mrs Guthrie and Catherine Guthrie; Mrs White; Mrs Wollen and Fanny and Susan Wollen; Mrs Cooper; Mr and Mrs Carroll and two ayahs; Mrs Sanders and William Sanders; Margaret, Mary, Tom and Ellen Fitzgerald; Mrs Bell and Alfred Bell; Mrs Berrill;; Mrs Murray; Mrs Jones; Mrs Russell and Eliza Russell; Mrs Gilpin, William, Harriet, Sarah, Jane, and F.Gilpin; Mrs Walker; Mrs Coymar; Emma Weston; Mrs Frazer; Mrs Derby; Miss Williams; Mrs Parrott; Mary Peter; Arthur and Charlotte Newman; Mrs Bowling; Mrs Moore; Miss White; Mrs Probert, Joanna, Willie, Emma, and Louisa Probert; Mrs Seppings, John and Edward Seppings; Mrs Dempster, Charles, William and Henry Dempster; Miss Wallet; Mrs Hill; Mrs Basilico; Mrs Lindsay, Frances and Caroline Lindsay; Mrs Scott; Mrs Mackenzie; Mrs Wallis; David Walker; Lucy Lyalls; Mrs Canter. (From Fatehgarh, arrived as refugees, list dated 11 July. Total 49): Mrs Woolyar, Charles and Thomas Woolyar; Mrs Gibbon; Miss Seth; Mrs Tucker, Miss Tucker, Louisa, George, L, and Sutherland Tucker; Mrs Reeve, Mary, Catherine, Ellen, Nelly, Jane, Cornelia and David Reeve; Mrs Thomson; Mr and Mrs Thornhill, Charles and Mary Thornhill; Miss Long; Mrs Maltby, Emma and Eliza Maltby; Mrs West; 'three natives'; Mrs Fatman; Mrs Guthrie; Mrs Heathcote; Godfrey Lloyd, baby Lloyd; Colonel and Mrs Goldie, Mary and Ellen Goldie; Colonel and Mrs Smith; Mrs Rees, Eliza and Jane Rees; Mrs Lewis, Emma and Eliza Lewis.

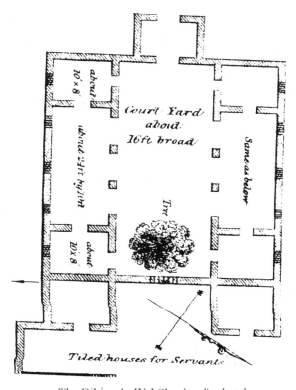

The Bibigurh. W.J.Shepherd's sketch

BIBIGURH, the The name given to the 'House of the Ladies' or the 'House of the Massacre' in **Kanpur** where about two hundred women and children were killed apparently on the orders of **Nana Sahib** on the day before General **Havelock** marched into Kanpur with a British relieving force. Graphically described: 'the Nana had all the ladies whom he had saved from the former massacre - murdered in cold blood. *May God in his mercy, my dear Beadon, preserve me from ever witnessing again such a sight as I have seen this day. The house they were kept in was close to the hotel - opposite the theatre - it was a native house - with a court in the middle, and an open room with pillars - opposite the principal entrance. The whole of the court and this room was literally soaked with blood and strewn with bonnets and those large hats now worn by ladies - and there were long tresses of hair glued with clotted blood to the ground - all the bodies were thrown into a dry well and on looking down - a map of naked arms legs and gashed trunks was visible. My nerves are so deadened with horror that I write this quite calmly. It is better you should know the worst - I am going this very moment to fill the well up and crown its mouth with a mound. Let us mention the subject no more - silence and prayer alone seem fitting. Ever sincerely yours J.W.Sherer'*. Sherer was the newly appointed Magistrate of Kanpur; this extract is from a letter he wrote on 17 July 1857. Major Gordon, who translated the native doctor's note, tried to determine how many captives were alive on the morning of 15 July 1857. He reckoned that 210 prisoners were alive on the 11th, after which 12 died, and therefore 'there must probably have been a hundred and ninety-seven when the massacre took place.' Kaye believes this estimate was correct. 'The well is now bricked over, and there only remains a small circular ridge of brick marking the wall of the well, which was not more than 9 or 10 feet across': Russell writing on February 12, 1858, in the London *Times*. The 'dry well' is still visible in Nana Rao Park in Kanpur. At the time of the Mutiny the Bibigurh was in the compound of Sir George Parker, the Cantonment magistrate. It was occupied as the 'dwelling of a native clerk' but was believed originally to have been the house of an officer's native mistress. *See* Zoë Yalland. *A Guide to the Kacheri Cemetery*. London, 1985, p 26.

BIBIGURH, the KILLING at, The generally accepted view of these events, repeated by most historians, is that orders came to the sepoys of the 6th BNI who were furnishing the guard over the prisoners, to shoot them, and that this the sepoys refused to do; instead they are said to have poked their muskets through the windows of the building and fired upwards into the ceiling, thus hurting nobody (though no doubt petrifying the wretched women and children inside). This version of events has recently been queried, on what seem reasonable grounds: ' My own theory is that they probably refused to enter the Bhibigurh and intended to shoot the women and children from the windows while remaining outside. They probably killed a number of people, but many women and children could have hidden with some security from these lines of fire (the windows were few) behind walls and pillars. Frustrated, probably sickened by now by the cries of the women and children, they may have refused to fire any longer, at which point the butchers were sent in to finish off the survivors. Certainly the butchers would not have had any difficulty massacring them, but it has always seemed to me unlikely that five men armed with swords could have managed to kill 198 people, however young and 'defenceless.' It is more likely that a good many were already dead by the time the butchers entered, and that the long time they took to finish the job (or nearly finish it - there were several victims still breathing in the morning) may have been accounted for by their plundering the bodies of the dead in the gathering gloom. There were many other soldiers besides the sepoys of the 6th to call upon to execute the women with rifles. It seems to me strange that five men with swords would have considered themselves sufficient in number to deal with 200 people. The work they would have been expected to do would have been the work of **jullads**, to finish off the wounded.' **ASW**.....*See also* KULOO.

BIBIS 'It is now very rare and shameful for an officer, civil or military, to live in a state which was normal last generation. The mode of building bungalows has altered. There is now no beebee's house — a sort of European zenana. But yesterday, in again visiting the slaughter-house at Cawnpore, the friend who was with me said he thought it had been a portion of the house of some officer or official, and that the compound had been the residence of some native woman. There are now European rivals to those ladies at some stations. It was the topic of conversation the other day at mess, that the colonel of a regiment had thought it right to prohibit one of his officers from appearing publicly with an unauthorized companion at the band parade; and the general opinion was that he had no right to interfere. But the society of the station does interfere in such cases, and though it does not mind beebees or their friends, it rightly taboos him who entertains their white rivals.' Wm Russell, *My Diary in India*, London, 1860, Vol I, p 188. For all their charms, the bibis were generally hard bitten, and no wonder. 'Their feelings were not involved,' wrote one courtesan, 'because they considered all men worthless. Their affection was mere affectation. If any wretch fell to their wiles, they were the first to pretend they were in love.' Umrao Jan Ada. Nevertheless Burton, who made a study of courtesans, and learned from them the sexual techniques with which he annotated his translations of the *Ananga Ranga* and *Kama Sutra*, admired them for their 'subtlety, ...their wonderful perceptive powers, their knowledge, and their intuitive appreciation of men and things.' Rice. *Captain Sir Richard Burton*. p 53.

BIJNOR   For a History of the Bijnor Rebellion *see* Hafeez Malik and Morris Dembo, transl. *Sir Sayyid Ahmad Khan's History of the Bijnor Rebellion,* East Lansing, 1972.

BIKANER, Maharaja of   Like most of the Princes of **Rajputana** he remained staunchly loyal to the British, and indeed was unusual in that he liked to command his own troops, *eg* at his frontier fortress of Bhadra, in June 1857. He allowed his troops then to be used by the British to assist in the suppression of the rebellion, and some 800 of the Bikaner Horse, under Lieutenant Mildmay served the British well in the Punjab, particularly on 19 August 1857 when they were involved in the repulse of 3,000 sepoys and villagers. But, as with other auxiliary troops from Rajputana they would act loyally in defensive roles but refused to act on the offensive, no doubt because they had more than a passing sympathy for the rebels.

BILAYAN   A town on the Orai-**Jhansi** road. At the end of May 1858 after **Sir Hugh Rose** had defeated the **Rani of Jhansi** and **Tatya Tope** at the battles of Kunch and Kalpi, the Raja of this place, one **Barjur Singh**, attempted with great courage to cut the British communications with Jhansi. On 31 May he was attacked by a British column under the command of a Major Orr, and he and his men, numbering some 250 fought desperately and with discipline: 130 of them were killed and 35 taken prisoner. Barjur Singh lost his horse and standard and got away on foot threading his way through the ravines, having stripped himself of almost all his clothing. For further details *see The Revolt in Central India,* Simla, 1908, p 143 *et seq., and* Thomas Lowe, *Central India during the Rebellion of 1857-58,* London, 1860, *and* **FSUP** Vol III.

BIR BHANJAN MANJHI, Capt.   (or Beer Bhunjun Manjhee). A Nepalese officer in the service of **Jung Bahadur**. He was sent as an emissary to the **Begum** when she was in Nepalese territory, and included in his report is a list of the persons of note who were still with the Begum in March 1859. The list does not include the **Nana** and his brother **Bala Rao**, although they were only a few miles away. The list was briefly annotated in Lucknow and sent on to the Governor General for his information. *See* PERSONS OF NOTE.

BIRCH, Colonel F.M.   Aide de Camp to **Brigadier John Inglis** throughout the siege of the **Lucknow Residency**, *see* INGLIS, Hon.Julia.

BIRD, R.Wilberforce   This gentleman, an army major, gave two lectures to the Southampton Athenæum, 16 February and 30 March 1858, the substance of which was afterwards published as a thirty-seven page tract on sale to the general public. The tract is significant as Major Bird was highly critical of British action in India at a time when it was very unfashionable so to be, and it must have taken considerable courage on his part to put forth his ideas in public. Indeed when he was first asked, in September 1857, he demurred. The lectures themselves are rather tedious and heavy-going (one wonders whether the committee of the Southampton Athenæum lived to regret their insistence that he lecture!), and the charges of misgovernment he makes against the East India Company are neither very original nor convincingly argued eg the doubtful legality of British sovereignty, the failure to establish a policy of neutrality in religious matters, the failure (not very consistent this) to support the employment of native Christians in government service, etc. The most cogent arguments are those dealing with the doubtful legality of the seizure of the **Kingdom of Awadh**. But at least this all shows that even in 1858 there were some English voices prepared to suggest that the British were greater 'offenders' than the Sepoys. *See* R.W.Bird, *The Indian Mutiny: two lectures delivered at the Southampton Athenæum,* London, 1858.

BIRJIS QADR (or BRIJIS KUDR)   He was crowned, aged ten, King of Awadh (Oude) in Lucknow by the rebels, in July 1857, though his (putative) father **Wajid Ali Shah** was still alive (in exile in Calcutta) and there were other candidates for the throne. His mother was **Begum Hazrat Mahal**: her paramour **Mammu Khan** is rumoured to have been his real father. 'The day of the coronation, mutineers, without distinction of rank, crowded into the Palace and spread through the halls, creating the utmost consternation by their disregard of all decorum, and by their noisy comments on the person and appearance of the lad'. After his coronation, government in Lucknow and Awadh was conducted in his name, and numerous proclamations and orders exist bearing his seal: his full name was Mohd Ramzan Ali Bahadur Birjis Qadar. He survived the war and accompanied his mother to exile in Nepal. His son, Mehar Qadar, was alive in 1961, and was called by some the 'Prince of Oude.' *See* **SSF**. *See also* PROCLAMATION OF BIRJIS QADR, and BEGUM OF OUDE, etc.

BITHUR   (or Bithoor) A small town on the right bank of the Ganges some twelve miles (20 kilometres) to the north of **Kanpur**. It was the place of exile selected by or allocated to **Baji Rao II** the last **Peishwa** of the Marathas. It was also the residence of his adopted son Dhondu Pant, **Nana Sahib,** who was unable to persuade the British government, on the death of his father, that he should be allowed to inherit the title of Peishwa and the pension of Rs 8 lakhs *p.a.* The palace of the Nana Sahib was reduced to rubble by the British in December 1857, and the only traces remaining of it are some large well heads. There is a small monument erected on its site. The family of Ramchunder Subadar, the wise and statesmanlike chief minister of Baji Rao II, still live at Bithur, in a house that was built for Baji Rao but which he did not like overmuch so gave to Ramchunder. The follow

Bithur in 1857

ing is taken from a gazetteer:- 'The ancient town of Bithur stands on the banks of the Ganges in 26° 37' N. and 80°16' E., in the extreme northern angle of the tahsil, at a distance of twelve miles above Kanpur. It is connected with the latter by a metalled road, which is constantly threatened by the river and is supplemented by a branch of the Grand Trunk Road, taking off at Sheoli. and also by a branch line of railway which leaves the Kanpur-Achhnera line at Mandhana and has its terminus here: the line is generally known as the Subadar's branch, and was constructed for the needs of the great pilgrim traffic. An unmetalled road goes westwards from Bithur across the Non to Chaubepur and Sheoli. Bithur is not only of the greatest antiquity but is invested with peculiar sanctity in the eyes of the Hindus, for on this spot Brahma celebrated the completion of the creation by a horse sacrifice. The spot is marked by Brahmavarttaghat ghat on the river bank, and a nail of the horse's shoe embedded in one of the steps of the landing-place is still the object of devout worship. At a later date the spot was the residence of Valmiki, the author of the Ramayan, and to him came Sita in her wanderings. The saint gave her shelter, and in his hut were born her twin sons, Lava and Kus, who seized the horse let loose by Rama on the occasion of his aswamedha. By thus accepting the challenge they had to fight the army of Ajodhya, and in the battle they were recognised by their father, with the result that general reconciliation was happily effected. Point is given to the story by the fact that bronze arrowheads have been found in the neighbourhood, while the adjoining village of Ramel to the south is said to be a corruption of Ran-mel, the battle of reconciliation. Similar arrowheads are found at Pariar on the opposite side of the river, which is connected with the same tradition. The residence of Valmiki is still shown on the river bank, and a temple was built in his honour by the Marathas on a mound to the south of the town. There too is the Sita Rasoi and an old temple named Kapaseshwar, probably a. corruption of Kakapaksheshwar, a title of Rama. Bithur afterwards became the capital of a pargana, and from 1811 to 1819 was the headquarters of the district. During that period were built the markets known as Collectorganj and Russellganj, the latter deriving its name from Mr. Claude Russell, the judge, who erected it in 1812. There are four other bazars, known by the names of Raja Bhagmal, Bihariganj, Naubatganj and Kataya Bharamal. In 1819, after the departure of the courts, the place was assigned as a residence to Baji Rao, the deposed Peshwa, who maintained here an almost independent state, attended by a retinue of some 15,000 men, for whose support he was assigned part of Bithur and Ramel in revenue-free tenure, the land being called Arazi Lashkar. The grant was confiscated on the rebellion of the Peshwa's adopted son, the infamous Nana Sahib and was then bestowed for life at a nominal revenue on Narayan Rao, a professed supporter of the British cause. There are still numbers of Maratha Brahmans in Bithur, and the present head of the community is Parsotam Rao Tantia, grandson of Ram Chandra Pant, naib-suba to Baji Rao, and son of Narayan Rao. He is generally known as the Subadar Sahib and is the owner of the Arazi Lasskar, which he purchased in 1895.' H.R.Nevill, Kanpur, a Gazetteer, Allahabad, 1929, pp 258-9. The Subhedar family is still living and well-respected in Bithur.

BITHOOR There is an odd tale told of the first taking of Bithoor[Bithur] by the British on 19 July 1857. On the 16th and the 17th of that month the Nana made a great parade of his newly acquired authority, having gun salutes fired in honour of the King of Delhi, of Baji Rao, of himself, and of his wife and mother - no fewer than *two hundred and eighty guns in all*. But when the British arrived on 19th there were no men at all left in Bithoor, they had all retreated, leaving guns, elephants, bullocks supplies etc. But many of the women of the zenana (including some of the Nana's female relatives with whom he had been quarrelling for some years) fell into the rather embarrassed hands of the British. It is reported that a guard was placed over them for their protection, and *'they were desired to inform their master that they were detained as hostages by the Europeans; and that any indignity offered to English females by his orders, would be retaliated upon their persons.'* It is difficult to understand this story, for the European women and children had already been killed in the **Bibigurh**, and this was only too well known to the British who had arrived in Kanpur only one day after the massacre. A case of bluff or shutting the stable door after the horse had bolted? In any case the Nana's womenfolk were well treated. *See* Chas Ball, *History of the Indian Mutiny*, Vol I, p.385.

BLACK WATCH, The This is a famous regiment of the British army, known at this time as the 42nd Highlanders. They were (nearly all) Scotsmen and were dressed - most unsuitably for the Indian climate - in heavy woollen kilts and padded red jackets and woollen bonnets. They had recently served with **Sir Colin Campbell** in the Crimean War and there was a mutual bond of affection and respect there. Campbell, also a Scot, was often thought to favour the 42nd, though he never hesitated to send them in to the thick of the fighting.

BLOOD MONEY The British frequently offered rewards (very large ones in the case of leaders such as Nana Sahib) for the capture dead or alive of rebels. It is to the great credit of those who were in a position to avail themselves of these rewards that they were very rarely claimed: in fact only two leaders were betrayed - **Tatya Tope** and **Maulvi Ahmadullah Shah**. But in the case of less important rebels 'blood money' was regularly paid out by the British; sepoy stragglers returning to their homes might be given up by villagers not so much for the reward as from enmity that might have existed from time immemorial between different villages or castes. Some rewards were more immediate and tangible than the **'certificates'** that were given to those who helped the British. In particular what we can only describe as 'blood money' was paid promptly to encourage others to inform upon mutineers and assist in their capture: '...regarding the expedition of Lt Metge for the punishment of 5 villages whose inhabitants had become obnoxious to justice, by their barbarous treatment of European officers' ladies during the disturbances; and in reply to sanction the distribution of Rs70 to the 12 Seikhs who distinguished themselves by swimming across the river and seizing 14 villagers, who were endeavouring to escape by boat.' *See* a letter dated 31 July 1858 addressed to the Magistrate of Banda and quoted in **FSUP** Vol IV, p 849.

BOATS On 27 June 1857 the garrison from **Wheeler's Entrenchment** at **Kanpur** marched down to the **Sati Chaura Ghat** on the banks of the Ganges, there to embark on boats provided by the **Nana Sahib**, with the hope, misplaced, that they would be allowed to proceed to **Allahabad**. Originally forty boats had been called for, on Nana's orders, by the Kotwal **Hulas Singh**, but it seems that that number could not be found in the time available. Opinions differ as to exactly how many country boats were eventually supplied, it was not fewer than twenty, and not more than thirty, and the majority opinion would indicate twenty-four. 'I was not acquainted with the names of any of the boatmen until the value of the boats was given, and the money distributed. Munshi Dabeedeen, a boatman, received the money from **Tanteea Topee**, amounting to Rs. 4,467 or 4,465, and distributed it. My brother Lochun was along with me at this time. The following were the owners of the boats belonging to this station: Muheshree Sheo Pershad, son of Dyr Kishen, residing in the city; Baboo Mull, Muheshree; and Jankee Pershad, Ugurwalla; and two boatmen, viz., Banee, residing in the village called Koreean, and Muttra, of the same village. The above five persons were paid in my presence, and the rest of the owners were paid by Munshi Dabeedeen. I do not know who they were.' At least two of the boatmen were subsequently executed by the British, namely Lochin and Hardeo. *See* the *Deposition of Goordial*. **Williams,**

BOB the NAILER This was the nickname given to an African eunuch, formerly in the employ of the King of Oude[Awadh], who, at the siege of **the Lucknow Residency** killed many of the British by his skill as a rifleman. He took up a position outside the perimeter, in Johannes' House which overlooked many of the roads and pathways within the Residency compound. For most of the siege the British casualties averaged something like five a day, killed or mortally wounded, and most of these can be put down to the activities not of the gunners but of the marksmen with musket and rifle who took up positions around the compound, overlooking the British position. There were two of these in particular whom the British had good reason to fear, for they killed many men. One was this Bob the Nailer, alias

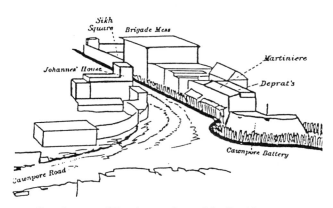

Sketch-map of South-east face of the Residency showing the location of Johannes' House.

Jim the Rifleman, whose headquarters, until it was attacked, was Johannes' House, another, also one of the late King's sharpshooters and reported also as being an African eunuch, made the **Clocktower,** the Lutkun Darwaza, his favourite haunt, and his shots thence were almost always fatal. Bob the Nailer was probably killed in the sortie on 7 July when a little doorway of Johannes House was blasted open, a number of sepoys found sleeping inside were bayonetted, and the rifleman himself, seated at his post in a loopholed turret of the roof, and engaged in returning fire deliberately aimed at him by the garrison to divert his attention, was caught unawares, surrounded and killed.. **PJOT**

BOMBAY TIMES This newspaper got things very wrong when reporting the early stages of the mutiny/rebellion. All its correspondents up-country were, if not *hors de combat,* cut off from any means of communication in the first few months. Thus: The *Bombay Times* reported, at the moment that all the British garrison of **Kanpur** had been slaughtered and the women killed in the **Bibigurh** that 'concerning the Mutiny at Cawnpore we are also in much uncertainty. Part of Her Majesty's 10th Regiment and a battery of European artillery were fortunately on the spot, and a repetition of the tragedies at Bareilly and Shajehanpore prevented. There seems, however, to have been several days' fighting, but the mutineers were eventually driven out of the place. Many on our side were said to have fallen.' *See* London *Times,* 1 August 1857**.**

BONE DUST Among other rumours being spread before the outbreak of the rebellion was one that the British were attempting to defile both Hindu and Muslim (and thus make conversion to Christianity easier) by mixing the ground bones of cattle and pigs with *atta* on sale in the bazar. The story appears to have originated in Meerut near to which, at mills in Bhola, the *atta* was ground before being sent down to Kanpur for sale. Although government had, from first to last, nothing to do with this flour which was in the hands of contractors and merchants, the rumours - and resultant panic - soon spread rapidly and were difficult to contain. See W.H.Carey *The Mahomedan Rebellion*,1857, and Mutiny Narratives, esp. F.Williams, *Narrative of Events attending the Outbreak of Disturbances and the Restoration of Authority in the District of Meerut in 1857-58. See also* **FSUP** Vol I , p 395 *et seq.*

BONE DUST 'On the 8th of March 1857, a Merchant, of the name of Dilowallee Singh, sent down the canal 200 maunds of atta, which had been ground in the mills at Bhola, five miles distant from Meerut. This supply was taken to Cawnpore and there exposed for sale in the market, and sold at once to the retail dealers at the rate of 17 seers per rupee. To the end of the same month other supplies by this and two other merchants came into the bazar from the same mills or others in the neighbourhood, to the extent of 800 maunds more. The first thing that seemed to cause suspicion that this atta was adulterated was its every (sic) cheapness, wheat unground being priced at 18 seers for the rupee. The reason of this cheapness was simply that the first merchant, being aware that other very large supplies were or would be shortly on their way down to the same market, was desirous of clearing himself of the whole of his stock before the arrival of the fresh article, and therefore placed it at a low figure. A sepoy, who had purchased some of the atta from the retail dealer, spread the evil report of its adulteration with bullocks' and pigs' bones ground at the mills. The report soon circulated; it was believed by all who wished to be disaffected. All further sales of atta were suspended, and though every means were used to rebut the senseless and unfounded report that had caused this cessation of sales, for some time all was of no use. A moonshee, of the name of Kurreem Buksh, a Commissariat Gomashta, even went to the mills where the atta was ground, and saw for his own satisfaction the whole process of grinding. He was convinced that there was nothing wrong, and purchased three rupees worth of the article on the spot for his own consumption. Sepoys also, to the number of one hundred, went at various periods to inspect the mills and see the mode of preparing the atta. The moonshee and the sepoys all stated that the report was without foundation. Their statement did thus much, that a small quantity of atta was taken off the dealers' hands at the reduced rate of 28 seers per rupee. No further supplies, however, left the mills, and that which was remaining in the market was spoilt. There were 2,000 maunds ready to be loaded at the mills, but on the 10th of April all operations in this article were closed.'

BONHAM, Lieutenant John He was born in 1834. An officer of the Bengal Artillery 1852-75. He was stationed at **Secrora** when the mutiny broke out and by all accounts was a brave soldier: he was wounded three times and twice mentioned in despatches. He was present at the battle of **Chinhat** and later took part in the siege of the **Residency at Lucknow** where he was responsible for an effective but home-made mortar affectionately known by the garrison as the 'ship'. He was one of the last British surivors of the Mutiny/rebellion, dying in 1928. He wrote a book which is succinct and very readable and contains short accounts of the mutinies at **Lucknow, Secrora, Shahjahanpur, Mohammadi, Chobeypur, Sitapur, Faizabad, Sultanpur, Gonda, and Bareitch**. Thirty years after his retirement Bonham revisited India and had the task of showing the Prince of Wales (later George V) and the Princess of Wales (later Queen Mary) round the Residency at Lucknow. *See* John Bonham, *Oude in 1857,* London, 1928. *See also* Kaye, *History of the Sepoy War,* Vol III, pp 476,507 and 539, *and* **FQP**.

BONNY BYRNE The soubriquet of Amelia Eliza Chandler, neé D'Fountain, of whom **Sir Wm Russell** wrote

'Sherer took me to the native Kotwal, who is most magnificent looking fellow, of great stature, and prodigious abdominous dimensions. The European Kotwal a big English or Anglo/Indian official is almost worthy to compete with the Mahomedan Mayor. He is married to Mrs Byrne, whose husband was killed in the mutinies, and whose life was saved by the Nawab of Furruckabad who *knew her and her mother and received them into his zenana.*' What a very nice way of putting things ! Bonny Byrne, the wife in question, was a notorious woman, whose private life was one of the major scandals of Kanpur. *See* **SSF** Chapter Seven.

BOULTON, LIEUTENANT The 7th Cavalry's Lieutenant Boulton the sole surviving officer after his squadron, encamped northeast of **Kanpur** at Chobeypur, mutinied, galloped into **Wheeler's Entrenchment** on 10 June 1857, leaping his horse over the low mud wall, unconsciously demonstrating how ineffective a barrier it really was. 'At daybreak, spying our position, he rode for it, and cleared our wall at a leap, though, as he had been mistaken for a sowar, he was fired at by our men, and his horse was wounded. He joined the outpicket under Captain Jenkins, and although a great sufferer from the wound in his cheek, he proved a valuable addition to our strength. He lived throughout the siege, and was one of the multitude who perished in the boats.' Mowbray Thomson, *The Story of Cawnpore*, London, 1859, pp 121.

BOURCHIER, Colonel George Even though this army officer was present during the struggle, the account he wrote relies largely upon the work of other writers notably **Henry Norman**'s journal re the siege of Delhi, and **John Walter Sherer**'s story of Kanpur. The book is valuable nevertheless for its succinct account of the mutiny in the Punjab. *See* Colonel G.Bourchier, *Eight Months' Campaign against the Bengal Sepoy Army,* London, 1858.

BOWRING, Lewis Bentham The Private Secretary to the Governor General, **Earl Canning**, 1858-62. While himself playing no active role in affairs he made an interesting collection of autograph letters, seals, portraits and documents of both Indian and British personalities, including an autograph poem by **Bahadur Shah**; the three albums are preserved in the Oriental and India Office collections of the British Library, *see MSS.Eur.G.38.*

BOX WALLAHS An obbrobrious term for one engaged in trade or manufacturing industry. A European so engaged would be regarded as socially inferior to those in military or civil employ in India. Presumably its origin lies in the box or tray that pedlars in the Middle Ages in Europe would use to take round their wares for sale.

BOYLE, Richard Vicars Resident railway engineer in Arrah, superintending the building of the railway to the NW Provinces. In 1857 he was 35 years' old, but he lived to be 86 and had quite a distinguished career building railways, particularly in Japan. Best known for his conduct during the siege of **Arrah House** in Bihar. The house was in fact his own billiard room, specially constructed for that purpose. He was, not without some justification, of the opinion that his part in the affair had been discounted, by the presence of officials like the Judge and the Magistrate, and that he had not received either credit or reward for his efforts. *See* **SSF** Chapter Eight; *see also* a collection of Boyle's letters and press-cuttings preserved in the Oriental and India Office collections of the British Library, Photo.Eur.74.

BRADSHAW, A.F. An Asst Surgeon, 2nd Battalion the Rifle Brigade sent a letter in December 1857 from the deserted and half-ruined town of **Fatehpur**: 'The soldiers are so vindictive against the sepoys that it is almost unsafe to let them even approach prisoners. Most of these luckless gentry are collared (hung) or potted (shot). Our warriors have two dominant ideas *viz* God save the Queen and D... the Darkies; and they are carried out with an energy which startles the Hindoos. Heard a discussion between Russell (*ie* **Sir Wm H.Russell**) and the Chief of Police (*ie* **Major Bruce**) on the subject of the Cawnpore outbreak. Latter has been instituting for some time a very close enquiry at Cawnpore about the murders etc. *I think the result which will be published will prove that the sepoys have been considerably maligned. In a great many of the atrocities attributed to the sepoys the budmashes (scoundrels criminals etc) released from gaols were the sole actors.'* Copies of his letters written between 20 October 1857 and 15 January 1860 are held in the National Army Museum, Royal Hospital Road, Chelsea, London SW3.

BRIND, BRIG-GENERAL Frederick He was commander of the brigade and station of **Sialkot**. He believed passionately in the loyalty of his men and nothing would convince him that they could ever mutiny. On 9 July 1857 he rose early and was drinking a cup of coffee and actually discussing plans with two of his officers about leading his sepoys against some mutineers, who were thought to be coming down from **Jhelum**, when a shot passed through his window. He immediately ordered his horse and said he would go out and quell the disturbance; soon after leaving his bungalow a sowar fired his carbine at him from behind, the ball entering his back close to the spine. The General turned on the man but his pistols had been previously unloaded by his own **khansaman**, and clicked harmlessly. He then seized the barrel of his pistol and riding the trooper down, broke his jaw with the butt end of the weapon. Brind was just able to sit his horse as far as the fort, but had to be carried up the ascent, only to die twelve hours later in great agony. *See* G.Rich, *The Mutiny in Sialkot,* Sialkot, 1924.

BRITISH LIBRARY, The Housed at the moment in the British Museum in Bloomsbury, London, but due to move to new premises at St Pancras in West London, where for the first time most of the Library's collection will be housed under one roof. The Library is probably the finest in the world and as it includes, by statute, a copy of every book published in England (this right is shared with the Bodleian at Oxford) it contains a truly comprehensive collection of published monographs. The separate collection known in the past as the **India Office Library and Records**, and now known as the **'Oriental and India Office Collections of the British Library'**, will eventually also be moved to St Pancras. No student of the mutiny/rebellion can afford to ignore the India Office material: it contains a complete collection of the Official Records of the Government of India at the time. Major series held include *Foreign Department, Proceedings, Political; Foreign Department, Proceedings, Secret; Home Department, Public Proceedings; Military Department Proceedings; Despatches from India and Bengal 1856-58; Political Letters from India, 1856-58; Home Miscellaneous Series 724A-727* ie **Sir John Kaye's** *Mutiny Papers etc.* A complete and detailed list of the material available is included in the invaluable little book by **Rosemary Seton**, *The Indian 'Mutiny' 1857-58,* London, 1986.

BRUCE, J.F. author of an article of scholarship and perception entitled *'The Mutiny at Cawnpore'* . *See Punjab University History Society Journal* April 1934.

BRUCE, Major, C.B. He was apparently a highly regarded and efficient Chief of Police (eventually) in Lucknow under the direct orders of the Chief Commissioner. Was previously responsible for intelligence activities at Kanpur and Lucknow, and for the raising and training of a large body of military police (many of them Sikhs, 'imported' from the Punjab) who were used in the so-called pacification of Oude[Awadh] which followed the military conquest of **Sir Colin Campbell**. While at Kanpur (October 1857) he was particularly highly spoken of by **Sir James Outram**, thus: 'The Intelligence Department by General Outram's wish, is entirely in his hand, as also most of the magisterial and police work of the whole district'. This might have caused severe friction with the newly appointed (August 1857) magistrate of Kanpur, **J.W.Sherer**, had the latter not been of a particularly phlegmatic and philosophic character. *See* F.C.Maude & J.W.Sherer, *Memories of the Mutiny* Vol II London, 1894.

BRYDON, Dr William Dr William Brydon, the army surgeon who was the sole survivor of the retreat from Kabul in the disastrous (for the British) First Afghan War, was one of **Gubbins**'s guests during the siege of the **Lucknow Residency**. He was fortunate to again be a survivor. Seated at dinner one day, with a lady on each side, he was shot in the back by musket ball that somehow penetrated into the dining-room: it went right through his body; the wound, although very painful, turned out not to be dangerous. His wife, Mrs C.M.Brydon, was with him in the Residency and she kept a diary from May to December 1857; the original is owned by Mrs Rachel Blackburn of Criccieth, North Wales.

BUCKLER, F.W. An Historian and Political Theorist who, in 1922, propounded the unpopular (with the British establishment) theory that if there were a mutineer in 1857 it was not a sepoy but rather his master, the Honourable East India Company. *See* his article entitled *Political Theory of the Indian Mutiny* in the *Royal Historical Society's Transactions,* Series 4 Vol V, 1922, and **ASR** p 62 *et seq. See* also a reply to his argument, by Douglas Dewar and H.L.O.Garrett in *Royal Hist.Soc.Transactions,* Vol 7, 1924.

BUDHOO A month previous to the outbreak, he was appointed commissariat contractor of boats, at **Kanpur**, and was the man entrusted with collecting the boats at the **Sati Chaura Ghat** for the coveyance of the Europeans to Allahabad on their surrender at **Wheeler's Intrenchment**. *See* G.W.Williams, *Depositions taken at Kanpur etc,* Allahabad, 1859.

'BUDUKE DASS HUNNOOMAN' Little is known of this man except that he led a brave and determined, but unsuccessful, attack upon the British force holding the **Alambagh** on 16 January 1858, lasting from 10a.m. till 9p.m. The Begum **Hazrat Mahal** had frequently reproached the rebels for their lack of courage in not destroying the small British force at the Alambagh which, from November 1857 to March 1858, constituted the only British presence in the Kingdom of Awadh. This man was described as a 'Hindoo fanatic' by **Major General Outram** commanding at the Alambagh. He was severely wounded in the attack (in which many of his followers were killed) and was finally taken prisoner. Fate unknown, but probably died of his wounds. *See* Lucknow Collectorate *Mutiny Basta.* Possibly to be identified with the *ADJUTANT-GENERAL.?*

BUKT KHAN *See* BAKT KHAN.

BULANDSHAHR A report, came from Bulandshahr, fifty miles from Delhi : ' .....the native population did not rise against the white man, but the moment they thought the white man was powerless they rose against each other, the rival castes and villages plundering and fighting in all directions: the Hindu Gujars raiding the Hindu Jats, and the Mahommedans raiding all Hindus impartially, while the English Magistrate and his young assistant rode about from village to village followed by a few native subordinates and fighting men, vainly trying to keep order and punish the rioters.' - Alfred Lyall in a letter to his mother, 1857. This was by no means the unanimous opinion. ....'The experience of the previous day had not been totally lost sight of. It showed me clearly that the zemindars were one with the lower orders; *that rebellion not plunder alone, actuated the mass of the population'....See* H.Dundas Robertson, *District Duties during the Revolt in the NWP,* London, 1859. *See also* LYALL, SIR ALFRED COMYN.

Bulandshahr in 1857

BULANDSHAHR A large town between Delhi and Agra, which was the headquarters of a man long at enmity with the British whom his family had accused of refusing their lawful rights, one Walidad Khan. After the assault on Delhi the British sent a moveable Column of approximately 2,500 soldiers under Colonel Greathed southwards towards Muttra in pursuit of 5,000 of the rebels. The former left Delhi on 24 September 1857, and on 28 September had to fight a battle at Bulandshahr against the 12th N.I., the 14th Irregular Cavalry and a large body of *burkundazes* and other levies: the rebels were dispersed with a loss calculated at 300 men, and a British loss of 51 of whom the majority were camp followers. Walidad Khan evacuated the nearby fort of Malaghar without firing a shot, and the fort was destroyed. before the Column moved on to Aligarh on 2 October.

BULDEO SING The daroga, or superintendent, of the bridge of boats over the river Jumna[Yamuna], that linked Delhi with Meerut to the north, and the Grand Trunk Road to Calcutta in the south-east, *see* MAINODIN HASSAN.

BULRAMPORE,[Balrampur] Maharajah of He had the confidence of Wingfield the Commissioner of the Bahraich Division that he would remain friendly towards the British, and so it proved. The Maharajah was made KCSI and a member of the Legislative Council of the Governor-General as a reward.. *See* Kaye, *Sepoy War in India,* Vol III, pp 475-6.

BUNDELKHAND IN 1857 'Bundelkhand, a division of the North-Western Provinces, at present in Uttar Pradesh, comprised in 1857 the four British governed districts of Banda, Hamirpur, Jalaun and Jhansi, together with the Lalitpur subdivision. Lying scattered in and around these districts were several independent states of varied size and importance. This region in 1857, witnessed a series of mutinies in the different towns as well as widespread rebellion over a large part of the countryside. Studies on the uprising are numerous. The justification for a fresh appraisal of the actions lies in our disagreement with the way in which the rebellion of 1857 has been problematised, and the available information arranged to produce a certain kind of discourse. A brief review of the existing literature on the subject shows how a pattern and a stereotype evolved to set the standard model for writing the history of 1857. The model served both foreign and Indian scholars even though they set out with quite different perceptions and arrived at quite different conclusions. Stated briefly, casual explanations, by and large, provide the principal tool by which the events of 1857 are reproduced. Such representation, I feel, shifts attention from the 'moments' of violence to the preceding context outside the sphere of the politics of the uprising. The factors enumerated as 'causes' may be relevant equally to areas and people not affected by the rebellion. There could also be instances of rebellion which cannot be accommodated in the casual structure. The objection may be posed at a much more fundamental level. Must the narrative of the rebel actions, the account of the 'moments of violence' always appear between two themes that relate to broader concerns and wider issues? The present description of the uprising in Bundelkhand begins with the actions of the people in rebellion, follows their trajectory through various undulations to their final suppression. Such an exercise throws up unexplored, often overlooked details which in turn help us clarify and rethink some of the received notions about the uprising. The rebellion, for instance, is seen in the existing historiography primarily as a series of negative actions, directed against the colonial state. All positive attempts at organising men and actions are seen to have emerged almost accidentally. The colonial state thereby laid claims to the only positively constituted power which eventually staged a comeback. Can we not suggest a counter proposition that, the rebels of 1857 too, operating at dissimilar levels were involved in a fight for power, in endeavours to capture the apparatus of the state? Months after the restoration of power, rebels set out to build their alternatives to the colonial state and to defend them. Out of these actions emerged patterns of convergence of diverse struggles towards a unity of political action. In order to reconstruct this story, the actions of the people has to be made the starting point of the narrative. We recognise them as they stand at dissimilar social, economic and political levels of society. The extent to which the wider context finds a place in the narrative of rebellion, is to be largely decided by the pohitics of the uprising. One can locate different kinds of actions which British records ascribe to the initiative of different sets of people demarcated by their socio-economic and political standing, These differentiated actions, thus provide the basis on which the narrative can be arranged in terms of the sepoy mutiny, the participation of the rajas, the rebellion of the 'thakurs', and the uprising in the countryside. 'Events' which the official records ascribed to each of them, if put under these heads, will help us sharpen the focus on the complexities in the uprising which are otherwise often overlaid by the standard formula of cause and effect. The mutiny of soldiers in 1857, followed by extensive rebellion among the civilian population, led to the collapse of British rule over large regions of northern and central. India. This is precisely what accounted for the singularity of the conflagration. There had been earlier instances of soldiers' mutinies in the nineteenth century. The difference now in 1857 was a fundamental one. Mutiny in one station graduated into a wide conflagation evoking in the course of its spread not only mutinies in several other stations, contiguous and located far away, but also rebellion among the civilian population. Equally significant was the fact that the mutiny in quite an unprecedented manner, acquired political dimensions. Soldiers actively participated in the process of creating a centralized, supra-local political order, as an alternative to the one they had displaced. This had never happened before and was never to happen again in the nineteenth century. Closely associated with the mutiny of the soldiers were the actions of the rulers and chiefs of small kingdoms and principalities that lay dispersed throughout Bundelkhand. Initially, at least, these potentates rarely ever took the initiative in rebellion. Yet, for the political authority they symbolised and the tradition of power they personified, the rajas became the centre around which the movement in all its phases rallied. In fact it was the British who first called upon them to take command of the situation created by the mutiny and the former's imminent departure. Although they did try to restore the administrative structure, the physical pressure of the mutineers left them with little opportunity for independent action. Ultimately, all those chiefs whom the soldiers supported and rallied around, came to be identified as symbols of resistance to the British rule. In short, till the very end, the rajas and ranis remained victims of a greater political struggle which allowed them only brief moments of independent action. In the countryside, quite a different set of people were seen working out new forms of political articulation and new expressions of dissent. The patterns of alliances, on which they were grounded, were derived from social and ritual equations very dissimilar from those of the soldiers and rajas. The rural actions also operated on a time scale very much their own. There were instances where villages were up in arms even while the British officials ruled in the towns, and quite some time after the mutinies were suppressed resistance in the countryside continued. The more prominent members of the rural community, who revolted in Bundelkhand in 1857, were the thakurs, men enjoying certain superior feudal prerogatives associated with land tenures. They were concentrated in the districts of Jhansi, Lalitpur, Jalaun and Hamirpur and therefore accounts of the civil rebellion here invariably begin with descriptions of the thakurs in arms. While the soldiers and rajas were busy in the towns, widespread thakur insurgency was being reported from the countryside. They also rallied around the soldiers to muster a common resistance to the British counter-insurgency forces. But it was only after the mutiny had been defeated and the major towns reoccupied by British forces that a new phase of rebellion began in Bundelkhand and the thakurs captured centre stage. They mobilized men and resources, attacked the newly-installed revenue and police outposts and drove out the officials and it was not till quite late in 1859 that the British state was able to bring the country around. The thakurs operated essentially within their immediate territorial limits. The springs of their actions, their complaints, grievances, anxieties as well as hopes and aspirations, concerned the world within their reach, and it was in this world that they sought to create alternate domains of power in 1857-58. Thus although the wider political movement against the British absorbed the local and dispersed struggles of the thakurs, they remained unique within it and continued outside it. Finally, there remains the account of those nameless, faceless, nonentities of history, who in the official records are lumped together into categories as unspecific as 'villagers', 'city-badmashes', 'followers', or just 'people', while the space in which they operated became simply 'the village' or 'the country'. Analogous to their status in the established political and economic hierarchy, the resistance of the lesser men was subordinated to that of the dominant groups. Reports of their actions are also, quite predictably, fractured, pieced together only by those moments when these men, through their struggle came to interact directly with the structure of power and thereby to threaten it. Popular response to the insurrectionary situation in 1857 was multifarious. The people also operated within a defined locality, drawing strength from ties of clan or caste. Thus a definite territorial space and collective participation were two crucial factors in their uprising. Fundamental to their reaction to the wider movement were their attempts to negate all visible forms of British power in their immediate vicinity. It was equally common for villages to fall upon neighbouring towns, as yet another means of challenging the contested order. They were able to do so by arming themselves within the means available to them. But it was really in the second phase, that rural resistance proved most difficult to overcome. When the counter-insurgency forces came marching through Bundelkhand, villagers deserted their villages both from fear of retribution and from their refusal to give in to the British order. After the district administrative officials were reinstated, villagers in many instances, drove them out by physically assaulting them, when they went to collect revenue. But the more ubiquitous form of opposition was the constant support, assistance and shelter they extended to the rebel thakurs. One of the reasons why rebellion persisted so long in Bundelkhand was the internal strength of its politics, a product of popular participation and thakur leadership. The rebellion, therefore, has to be reconstructed in terms of the experiences of all those who contributed to make the resistance what it was. Their fears, anxieties, hopes and ambitions have provided the subject matter of this story. It cannot be told in accordance with any preconceived definitions drawn from situations alien to the experiences of the rebels. They had after all defined a political order in their own language only to fulfil their own dreams.' **Dr TAPTI ROY, Historian and Writer**. *See also* Sir George William Forrest, ed., *Selections from the Letters, Dispatches and other State Papers of the Military Department of the Government of India 1857-58*, 4 vols, Calcutta, 1893-1912, Vol IV, *and* Tapti Roy, *The Politics of a Popular Uprising:Bundelkhand in 1857,* Delhi, 1994.

Chillatara Ghat, Bundelkhand

BUREAUCRACY In November 1857, Captain A --- of the 53rd N.I. secured ten horses from the **Prize Agent**, paying him in cash and receiving the Agent's recipts, written on two slips of paper, stationery being at a heavy premium at that time. In 1863, nearly six years' later, being then at **Bareilly**, the bill for the horses was sent to Captain A --- by the Accountant's Department, they having no vouchers to cover the transaction. His assurance that he had paid cash at the time of purchase was deemed insufficient, as well as his statement that all the books and documents of the Regiment were lodged with the station officer at Nagode. The Agent of whom the purchase was made affirmed that as he had no record of payment, the sum could not have been received by him. However, in 1864, seven years after the original transaction, when appointed 2nd in command, 1st Bengal Cavalry, Major A ---, as he had now become, was ordered to Nagode, and after a laborious search among the almost forgotten papers, he found the two slips bearing the Prize Agent's receipt for all the horses purchased. They were worth Rs1,200, besides clearing his character from a nasty stigma. They were forwarded to the Accountant's Department. Regret ? Apology ? Reply even ? - not a word. (*Note.Perhaps Officers in more recent times have similar experiences? In my day there was a very strong rumour obtaining, that said that if the clerks of the Field Controller of Military Accounts at Poona could find any excuse - feeble or not - to query and send back an officer's expenses claim, they received a bonus of one rupee per query. I've often wondered how true was that rumour!* PJOT). When a detachment or complete regiment was disbanded because of suspected disaffection, some of the British officers were so touched at the sight of the shame of those who had lived and fought with them so long, that they laid down their own swords and belts along with the arms of their men.We can be quite sure that the Accountant General accepted no requests for reimbursement for the purchase of replacement swords and belts ! 'On application being made the other day to Government by Staff Officers for compensation for chargers killed in action, they received reply that Govt could not for a moment entertain the idea of allowing compensation for any loss, as the army now was not employed on Field Service, that they were merely quelling a mutiny!'

BURHIYA (or Boorhiya) A village on the Kanpur -Lucknow road some two miles from **Bashiratganj** , which was the scene of an engagement on 11 August 1857 between **Havelock**'s force and the rebels. It was touch and go for some time, the rebels' artillery being particularly well served, but a charge by the Highlanders gave victory to the British. For the best account of this battle *see* J.C.Marshman, *Memoirs of Major General Sir Henry Havelock,* 1860.

BURKUNDAZE A kind of armed auxiliary used in rural areas to support the local *thanah* or *tehsildari*. He was a watchman and was usually armed only with an ancient matchlock and sword. He was of no great use in troubled times but useful to government when the latter was securely established.

BURTON, Major C. He was Political Agent at **Kotah**, [Kota] murdered on 18 October 1857, together with his two young sons, Arthur and Frank, in the Residency at Kota. The troops of the Kotah Maharaja had by now openly joined the rebels, although the Maharaja himself still professed his loyalty to the British government. For details *See The Revolt in Central India*, Simla, 1908 , p 195-6. *See also* Iltudus T.Pritchard, *The Mutinies in Rajpootana, being a personal narrative of the mutiny at Nusseerabad,* London, 1864.

BUXAR A town on the Ganges, on the route from Calcutta and **Patna** to Allahabad. Here **Major Vincent Eyre** heard of the siege of **Arrah** by **Kunwar Singh** and the mutinied Danapur sepoys, and 'commandeering' a detachment of 160 men of the 5th Fusiliers under Captain L'Estrange he marched to the relief of the small British party in Arrah.

An English Artist/Engraver's idea of Indian Sepoys in 1857, from Chamber's *History of the Revolt in India*.

C

CALCUTTA 'Calcutta, the capital of the British Empire in India was in many ways a European settlement. Whatever its original existence may have been, the city in the nineteenth century owed its nature to the founding of the English **East India Company.** Alongside the white colony of the British grew somewhat in its shadow the black town of the natives, comprising both the original inhabitants and migrants from outside. The distance between the two was as much spatial, as racial and psychological. In 1857, when the entire Northern and a good part of Central India were in the grips of a raging mutiny accompanied by civil rebellion, Calcutta remained peaceful though neither unaffected nor undisturbed. The news of the mutiny which filtered in with all the details, stunned and bewildered the white inhabitants of the capital. Shortly, a fear bordering on paranoia spread among the people. There was the dread of the sepoys stationed at **Barrackpore**. A night's march would quite easily bring them to Calcutta, and thereafter overpower the European guards, seize the Fort and massacre the Christian inhabitants. There was also the fear that the 'natives' in the bazaars and suburbs would rise against the white people, release prisoners from jails and feast themselves upon the plunder of the great commercial capital of India. There seemed every possibility that what had happened at **Meerut** and **Delhi** could well be repeated in Calcutta on a larger scale, with worse consequences. Describing the general fear, the *Friend of India*, wrote in its issue of May 28, 'Men went about with revolvers in their carriages, and trained their bearers to load quickly and fire low. The ships and steamers in the rivers have been crowded with families seeking refuge from the attack, which was highly expected, and everywhere a sense of insecurity prevailed, which was natural enough when the character of the danger apprehended is taken into consideration'. As the month wore on, the panic increased. The European community, which included besides the British, French, Americans, Portuguese and the Armenians wanted to be organised, armed and trained to form volunteer corps for the protection of the city. Several public bodies came forward with offers of service. The Traders' Association, the **Masonic** Lodges, the Native Christian Community expressed their sympathies with the cause. They wanted to be mobilised in the manner of soldiers. They also urged the Government to disband the Indian soldiers cantoned in and around Calcutta. For **Canning** and the officials it was a hard decision. To give in to general panic would display the ultimate failure of the state machinery even in Bengal. They had to provide for reinforcements to be sent to **Delhi** and thereby arrest the impending collapse of British rule over large parts of northern India. It was therefore imperative that Bengal remain unperturbed and apparently invincible in the face of all surrounding dangers. Closer home, there were a number of vulnerable pockets which had to be defended - the Fort William with its arsenal, the gun-factory at Kashipur, the powder-manufactury at Ishapur and the artillery school at Dumdum. While providing for the emergency, the officials had to maintain every semblance of normalcy in the city, so that the Indian residents would never be able to doubt the immutability of the Company's government. Canning, therefore, turned down the option of forming a volunteer group or of disbanding the Indian regiments at Barrackpur. The Queen's birthday was celebrated, instead, after the wonted fashion. A grand ball was given at Government House on 25 May and everything went as usual. Canning was particularly keen that nothing should be done to betray any want of confidence in the general loyalty of the people in Calcutta. Lady Canning was thus seen taking her habitual drive in the open suburbs of Calcutta, calm and smiling. Nothing out of the ordinary happened to disturb the month of May, although news from the north did little to assuage the fear in the city. For the paroxysm of fear had spread palpably across the board. Never had the distrust for the whites been so extreme. Canning wrote on 20 May, 'One of the last reports rife in the Bazaar is, that I have ordered beef to be thrown into the tanks, to pollute the caste of all Hindus who bathe there, and that on the Queen's Birthday all the grain shops are to be closed in order to drive the people to eat unclean food.' Even a year later, in April 1858, such rumours were not exceptional. In the wake of an increase in the price of rice, it was widely believed by the Indians in Calcutta, especially the relatively poor, that the government intended to buy up all the rice in the country and afterwards sell it only to those who agreed to become Christians. It was perhaps as a measure to contain the spreading of such panic that a **Press Act** was passed in 1857 whereby news of the rebellion in the north was censored. Thus the *Friend of India* wrote on 5 November 1857, 'Partly from the extreme panic that has prevailed in almost all stations in Bengal, partly from the interruptions of communications, and partly from causes which the Press Act forbids us to describe, the history of the revolt is still but imperfectly understood.' Meanwhile however the town remained peaceful throughout. As precaution, measures were taken to prevent any undesirable excitement. Therefore all liquor shops and punch houses were closed everyday at 5 pm. European sergeants of the Calcutta Police were employed for patrolling round Government House to prevent the sepoy sentries on duty from being assaulted by Europeans, which had become almost a matter of nightly occurrence. The Indians, living on the other side of the town, were fully informed of the events around. The reaction of the educated middle-class Bengali was objective. Regarding the demand for a European private militia, Harish Chandra Mukherjee, the editor of *Hindu Patriot* observed, not without a touch of humour: 'It is stated that the Governor-general is not unwilling to see the inhabitants of Calcutta enrol themselves into a militia. Probably His Lordship wishes them to exhibit an example of loyalty to the rest of Her Majesty's subjects here; for he could scarcely have been so unobservant of the character of Calcutta society as not to have understood that a 'Calcutta Eurppean Militia' is not likely to turn out a more formidable body than the London shopmen formed, when Napoleon's threat of invading England led those estimable citizens to practise the goose-step.' For men like Harish Mukherjee, who was fairly representative of the urban middle-class Bengali, the panic among the 'whites' was as much outside their sphere of experience as the insurgency among the men of the north. Therefore, both were subjects to be written about. 'The burst of mutinous feeling which wrought such lamentable results at **Meerut** and **Delhi** has been succeeded by some weaker explosions at a few other army stations in the Upper Provinces.' Yet this was not the language of the British officials. There was apathy but not hatred for the rebel Indians. On the contrary, the editor seemed concerned at the inevitable reprisals which were bound to follow. 'Heavy as are their crimes, the punishment of the mutineers will be proportionately so. There is one cause of regret, but we fear it is irremediable. The people of Delhi, among whom we doubt not there are many innocent and loyal, will to a great extent have a share in the punishment due to the guilty rebels.' (28 May). About themselves, the editor was equally forthcoming. The mutineers were not one with them, the educated Bengalis who like the Parsis, had outstripped 'the rest of their Indian fellow-subjects in civilisation, knowledge and political progress.' Therefore even in the midst of rebellion the Bengali was seeking changes in the form and policy of the government. That was what agitated him, not the events in Meerut. 'The same system of legislation and government which suffices to rescue a province on the Sutledge or the Indus or the Irrawaddy from anarchy and utter misery is felt as an encumbrance, a dead weight, upon the energies of the Bengali population...To govern all the provinces in British India on the same principles and under the same system would be analogous to enforcing the doctrine of equality by reducing all to one dead level of savage pauperism' (14 May). By the end of the month, the Bengali could no longer remain so complacent. The affairs of the north had become serious business. And for the white men of Chowringhee, Bengalis were primarily Indians like the Jats or Rajputs of the north. This was a rude realisation. The *Hindu Patriot* did not take long to condemn the sepoys as 'brutal and unprincipled, a body of ruffians as ever disgraced a uniform as stained the bright polish of a soldier's sword with the blood of murder.' It added further: 'The people of these provinces both by habit and education are the least likely to swell the ranks of a rebellious soldiery or afford the least countenance or protection to the disturbers of national tranquillity. The Bengallees never aspired to the glory of leading armies to battle or the martyrdom of the forlorn hope. Their pursuits and their triumphs are entirely civil. A strong and versatile intellect enables them to think foresightedly. They are aware that the British rule is best suited to their quiet and intellectual tastes; that under it they might achieve the greatest amount of prosperity compatible with their position as a conquered race.' Harish Mukherjee's position was corroborated by the resolution passed on 22 May 1857 by the British Indian Association, comprising the Bengali landed gentry and the urban middle class of Calcutta. The reolution was passed by a committee of the Association and read: 'The Committee view with disgust and horror the disgraceful and mutinous conduct of the native soldiery at those stations, and the excesses committed by them, and confidently trust to find that they have met with no sympathy, countenance, or support, from the bulk of the civil population of that part of the country, or from any reputable or influential classes among them.....The Committee trust and believe that the loyalty of their fellow subjects in India to the Government under which they live, and their confidence in its power and good intentions, are unimpaired by the lamentable events which have occurred, or the detestable efforts which have been made to alienate the minds of the sepoys and the people of the country from their duty and allegiance to the beneficent rule under which they are placed.' Would all of Calcutta have agreed with such a resolution? What did the mutiny signify to those who in no way could have identified themselves with the members of the British Indian Association, those who could not afford the luxury of intellectual rumination and those who had little to gain from British rule? The unlettered artisan, the semi-literate petty clerk or the Bihari migrant labourer inhabited a world that did not produce an editor or a spokesman who could broadcast their minds. Perhaps the soldier in Delhi or the peasant in Awadh had his unspoken admirer, lost somewhere in the crowd of Calcutta.' **DR TAPTI ROY, Historian and Writer.**

General view of Calcutta from Fort William, 1857

CALPI (or Calpee) *See* KALPI

CAMBRIDGE, Cambridgeshire County Record Office Like many County Record Offices, Cambridge has a few items that relate to the mutiny/rebellion: the most interesting here is the diary of a man who accompanied the **Siege Train** from the Punjab down to **Delhi** in preparation for the final assault upon that city in September 1857. His name was William Tod Brown, and his diary is entitled *'Siege Train Notes.'*

CAMBRIDGE, CENTRE of SOUTH ASIAN STUDIES Situated in Laundress Lane, Cambridge, and ranking as a department of the University, the Centre contains an unrivalled collection of manuscripts relating to the Mutiny/Rebellion. Letters, diaries, sketches, photographs etc provide excellent primary source material. The Centre has produced a South Asian Archive in four volumes (available on microfiche) from which the following are identified as containing material relevant to the revolt of 1857: **Berners Papers; Bishop Papers; Gibbon Papers; Kenyon Papers; Lenox-Conyngham Papers; Monckton Papers; Rickards Papers; Stokes Papers; Wentworth-Reeve Papers; Gore-Lindsay Papers; Gye Papers; Peppé Papers; Showers Papers; Stansfield Papers; Waller Papers; Wilson Papers; Lorimer Papers; Maclagan Papers; Stock Papers; Augier Papers; Campbell/Metcalfe Papers; Coghill Papers; Cra'ster Paper; Fairweather Paper; Gray Papers; Hardy Papers; Hyde-Smith Papers; Macnabb of Macnab Papers; Parsons Papers; Stuart Papers; Perry-Keene Papers.** Details of the contents of these collections are listed under each separate entry.

CAMP FOLLOWERS In Europe this term has tended to become a euphemism for loose women hanging on to an army on the march, but in India it applied to the myriad servants without whom an army, certainly a British army, could not move, and who often had to display great courage and devotion when, unarmed, they met the enemy. In the Bengal Army an ensign had about fifty people working for him alone: a syce, a grass-cutter, a tent-lascar, bullock drivers and coolies etc, and the private soldiers had perhaps four or five servants following each of them, so that a regiment might trail five thousand camp followers and their families, plus hundreds of elephants, camels, bullocks, horses and goats. A column of soldiers on the march might stretch to half a mile in length, but might then be followed by up to five miles of camp followers, particularly vulnerable to flank attacks. Cooks and bhisties in particular were indispensable to the British.

CAMPBELL, SIR COLIN (1792-1863) Son of a Glasgow carpenter. Joined the army as a private soldier, commissioned as Ensign 1808. Served with distinction in the Peninsular Campaign, and was promoted rapidly, for

Sir Colin Campbell

conspicuous gallantry and professional competence. Major General 1854 and commanded the Highland Brigade (with whom he was exceptionally popular) and the First Division in the Crimea, but best known as the Commander-in-Chief in India who succeeded in putting down the Great Indian Rebellion of 1857, or, as the British would term it at the time, 'suppressing the Indian Mutiny'. He had had experience in India (1846-53) and this was to be of the greatest assistance to him. Created Baron Clyde in 1858. Was made Field Marshal - a rare appointment for a commoner - in 1862. Had the reputation of being cautious in risking the lives of his troops, and although this endeared them to him, it led sometimes to unnecessary delays in achieving victory. He has been, erroneously, credited in more than one respectable bibliography of having written an account of the 'mutiny': this has arisen from the somewhat awkward wording of the title of a work that was issued in London, in weekly parts priced one penny, called 'A Narrative of the Indian Revolt from its outbreak to the capture of Lucknow by Sir Colin

Campbell'; he captured, he did not write. *See* **DNB**; *see also* Sir Owen Burne, *Clyde & Strathnairn*, Oxford, 1891.

CAMPBELL/METCALFE PAPERS These are held in the Centre of South Asian Studies at Cambridge University, and consist of Correspondence of and between the Campbell and Metcalfe families 1848-61 in and near Delhi. Box VI contains letters from Sir Edward Campbell describing his involvement in the Mutiny where he fought in the siege of **Delhi**. Box VIII contains letters from **Theophilus Metcalfe**.

CANALS Many reasons have been offered for the outbreak of 'Mutiny' and rebellion in 1857, but one of the oddest adduced has been that the canals built in the Doab had 'poisoned' the land, and that in some way the British must be punished for it. The facts are incontrovertible: a considerable quantity of land watered by the Western Jumna Canal became desert-like and a barren waste, and similar ill-effects were expected of the Eastern Jumna Canal, and even of that mighty engineering project, the Ganges Canal. What was happening was not fully understood by anyone at the time, agricultural science not being sufficiently advanced to explain that the salts held in solution by water from the Himalayas could, in concentrated form, actually harm the land and the crops the canals were intended to assist. Irrigation was all very well, but if the water were full of a superabundance of inorganic material then slime and verdigris, not fertility, resulted. This complaint, and the disaffection that resulted, was loudly voiced in the Saharunpur District, north of Delhi and Meerut, and it is therefore most surprising to hear details of one of the rewards given to 'loyal' villages: 'Lieutenant Hardinge, though he could himself have easily escaped by riding into Saharunpore, preferred linking his fortunes with the sergeants (of the Canal Department) and their helpless families. Though the whole of the surrounding country was in a state of complete anarchy, this party fortunately found themselves in a village, forming one of a rather powerful brotherhood, who had always previously maintained constant feuds with the villagers then in rebellion. This village assembled their brethren, and declared in solemn conclave that the Europeans must be protected; and they kept their word, until Lieutenant Bellew, collecting a sufficient number of elephants to make a rapid and unexpected march, managed, with his usual dashing gallantry, to bring the whole party into Saharunpore, before those in the rebel tract had even heard of his advent. *These villages have since been rewarded through the Canal Department by a free grant in perpetuity of all water required for irrigational purposes from the Eastern Jumna Canal.....this is about as politic a reminder of the results of loyalty as could well be devised.*' Let us hope that by the time they started using the water it had been desalinated or whatever ! NB The saline efflorescence in canals was known as REH.

Lord Canning

CANNING Charles John, Viscount Canning, (1812-1862), Governor General of India, and first Viceroy (From 1 November 1858 when the rule of the East India Company was abolished). Created Earl Canning in 1859. Obtained the soubriquet of 'Clemency Canning' because of his desire to reconcile all parties after the suppression of the mutiny/rebellion. His term of office 1855-62 was comparatively long and extremely eventful, involving war with Persia, the mutiny/rebellion, and the assimilation of Oude[Awadh], annexed in 1856; from 1859-62 he was engaged in the reorganisation of the financial, legal and administrative systems in India. He died at the early age of 50 of a liver complaint. *See* Sir Henry Cunningham, *Earl Canning*, Oxford, 1891; *see also* **DNB**.

CANNING'S CHARACTER Canning hoped for a peaceful administration, but recognized that it would depend 'upon a greater variety of chances and a more precarious tenure than any other quarter of the globe.' Within a year he was at war with the Shah of Persia, and was soon suppressing a revolt in Burma. It was Dalhousie who first horrified Brahmin sepoys by requiring them when commanded to serve overseas, but Canning embraced the change, dismissing his own initial apprehension that the veteran sepoys who believed they would lose their caste by crossing large bodies of water 'might suspect that the change was a first step towards breaking faith with them.' He assigned Sir Henry Lawrence to administer Oude[Awadh]. He was a stickler for details, and fretted over the merest trifles. At first only he could initiate business in any of the departments, and only when the Mutiny threatened to subsume him did he pass small matters off on a cabinet of secretaries.

THE CLEMENCY OF CANNING.
Governor General. "WELL, THEN, THEY SHAN'T BLOW HIM FROM NASTY GUNS; BUT HE MUST PROMISE TO BE A GOOD LITTLE SEPOY."

He was sometimes paralyzed by anxiety, but publicly exuded a sometimes infuriating calm. It took him a while to appreciate the gravity of the crisis. His equanimity infuriated his European constituents. He refused to allow Calcutta businessmen to form a corps of volunteers and delayed ordering disarmament until disarmament was virtually unenforcable and then extended it to the European as well as the native community. He restricted the English press as well as the vernacular, and condemned the retributive excesses of his troops. He seemed cold and reserved, deliberate to a fault, but he was also courageous and generous with his subordinates, and despite numerous and excrutiating provocations never defended himself at the expense of his countrymen. His policy of confiscating the property of rebel talukdars for later disposal as the British government saw fit caused such a hue and cry in parliament that it forced Lord Ellenborough to resign from the presidency of the Board of Control, but Canning survived the storm to become the first Viceroy in 1858 when India redounded to the Crown, and an earl a year later. He oversaw the entire reorganization and reorientation of an ostensibly commercial entity into an imperial government, secured arrangements with the remaining chiefs, including the right of adoption, which he reinstated. He could not, however, extend the right of adoption to himself, and when he died a widower without issue, in 1862, his earldom lapsed. He introduced an income tax, and when Charles Trevelyan published his objections in a newspaper, he dismissed him from the Governorship of Madras. His wife died of jungle fever in late 1861. He died three months later. *Summarised from* DNB *by* **ASW**.

CANNING Canning was 'morbidly (one can almost say) slow and dilatory in his work, and very far indeed from an efficient administrator, quite the contrary. In dealing with the Mutiny his principles were right but his practice very imperfect; and he owes his reputation more to sympathy with the noble stand he made against the Calcutta disposition to excess, and the sobriquet 'Clemency Canning', than to his success in controlling his subordinates. Certainly for a long time Canning did not at all realise the seriousness of the Mutiny. ...'.I feel that the Governor General must to an extraordinary degree have been slow to take in the facts.' Campbell also deplored his authorisation of both military and civil officers to depute Special Commissioners and grant them 'unlimited powers of life and death.' Sir George Campbell, *Memoirs of my Indian Career*, London, 1893, pp 230-1.

CAPTAIN OF THE WELL There was a well, with an unlimited supply of good water, available to the British holed up in **Wheeler's Entrenchment** in Kanpur, but the problem was that it was overlooked by the rebels and drawing water was therefore a most dangerous occupation. 'My friend, John McKillop, of the Civil Service, greatly distinguished himself here; he became self-constituted captain of the well. He jocosely said that he was no fighting man, but would make himself useful where he could, and accordingly he took this post; drawing for the supply of the women and the children as often as he could. It was less than a week after he had undertaken this self-denying service, when his numerous escapes were followed by a grape-shot wound in the groin, and speedy death. Disinterested even in death, his last words were an earnest entreaty that somebody would go and draw water for a lady to whom he had promised it'. Mowbray Thomson, *The Story of Cawnpore*, London, 1859.

CAPTIVES OF OUDE[Awadh] There is no 'official version' of this tale. Some of it comes from letters or statements from survivors and much comes from a typescript manuscript in Box No.1 of the **Haig Papers** kept in the **Centre for South Asian Studies** at Cambridge: the manuscript is the unpublished account, written in 1880 for her sons, of one **Madeline Jackson.** All in all it is a tale that is typical of many of the small stations of the North West Provinces in those turbulent times: mutiny of sepoys and local rebellion, followed by murder, the release of all convicts from the gaol, looting and burning, the pursuit to the death of Christians, the escape and wanderings of a few hapless survivors, helped often by their servants or kindly villagers. But this story is different in that it aroused the imagination and passions of so many who heard only snippets of the whole adventure. Perhaps it was because Madeline and her sister were so beautiful, or well-connected, that they attracted more attention than others. The group of fugitives consisted of the family of **Captain Patrick Orr,** *ie* his wife and daughter, plus Sir **Mounststuart Jackson** and his sister

Sir Mounstuart Jackson

Madeline, Lieutenant Burnes, and Sergeant Morton who was quartermaster of the 10th Regt Oude Irregular Force. The child was little Sophie Christian who, although she was too young to understand it, was already an orphan. Against all odds three of the Captives of Oude survived, together with their benefactor **Wajid Ali Darogah.** *See* LONEE SINGH, RAJA OF MITHOWLI, also *ZUHOOR-OOL-HUSSUN*. For a full account of the adventures of the Captives of Oude *see* **SSF** Chapter 10.

CAREY'S RIDE

Some of the British had lucky escapes when the troops mutinied. One such was a Lieutenant Carey who was caught up by chance with events: he was actually in Kanpur on his way 'up country' to join his regiment, at the end of May 1857. **Captain Fletcher Hayes**, Military Secretary to **Sir Henry Lawrence**, Chief Commissioner of Oude, arrived in Kanpur on an embassy to **Sir Hugh Wheeler**, escorted by a squadron of Oude Irregular Cavalry commanded by Lt Barbor. A Mr Fayrer, brother of **Dr Fayrer**, came along with the cavalry as a volunteer. Leaving Kanpur Hayes and his party got permission to make a reconnaissance up the **Grand Trunk Road** towards **Mynpoorie (Mainpuri)**, and Carey naturally joined the party as it was going in the direction he wanted. A few miles from Mynpoorie they made camp and in the morning one of the native officers of the Irregular Cavalry came to tell the Englishmen that the men had mutinied and that they must fly for their lives. They mounted and set off at a gallop for Mynpoorie, pursued and fired on by some of the troopers. Carey describes it in these words: *'Away we went, with two troops after us like demons, yelling and sending the bullets from their carbines flying all around us. Thank God, neither I nor my horse was hit. Hayes was riding on the side next the troopers, and before we had gone many yards I saw a native officer ride up alongside of him and cut him down from the saddle. It was the work of an instant and took much less time than I have to relate it. On they all came, shouting, after me, and every now and then 'ping' came a bullet near me. Indeed, I thought my moments were numbered, but as I neared the road at the end of the maidan a ditch presented itself. It was but a moment, I thought - dug my spurs in hard, and the mare flew over it, though she nearly fell on the other side; fortunately I recovered her, and in another moment I was leaving all behind but two sowars who followed me, and poor Hayes' horse tearing along after me. On seeing this, I put my pistol into the holster, having reserved my fire until a man was actually upon me, and took a pull at the mare, as I still had a long ride for it, and knew my riding must now stand me a good turn. So I eased the mare as much as I could, keeping those fiends about 100 yards in rear; and they, I suppose - seeing I was taking it easy and not urging my horse, but merely turning round every now and then to watch them - pulled up, after chasing me a good two miles. Never did I know a happier moment, and most fervently did I thank God for saving my life.'* He reached Mynpoorie safely but Fayrer was struck down from behind while drinking water at a well, and Barbor was cut down by troopers although he shot two and ran another through with his sword. With an act of great humanity the bodies of the three dead Englishmen were collected by some of the troopers who had refused to mutiny, were taken by them to Mainpuri and buried there.

CARMICHAEL-SMYTH, COLONEL

Colonel George Munro Carmichael-Smyth was the youngest of eight children of a physician of George III. In May 1857 he was commandant of the **3rd Light Cavalry** at **Meerut**. His bungalow still exists and is occupied, in the cantonment at Meerut. He was fifty-three years of age and had not had a particularly distinguished career, and was apparently actively disliked by both his officers and his men. It was he who ordered the parade of the skirmishers of the regiment at which they were to be given cartridges and instructed in a new drill to tear rather than bite them. *See* CARTRIDGE QUESTION. The result was a refusal by eighty-five men, court-martial, sentencing and fettering of the culprits and the outbreak of the mutiny/rebellion on 10 May 1857. The story is well-known and is repeated in essence in all general accounts of the struggle, but it is distinctly possible that Carmichael-Smyth was responsible for the ***premature*** outbreak of the mutiny, if not for the mutiny itself. 'After this I ordered each man in succession to take cartridges but with the exception of five men [three Muslims and two Hindus] they all refused to do so. None of them assigned any reason for refusing, beyond saying that they would get a bad name.' The five non-commissioned officers who had accepted the cartridges were then dismissed; and, after further explanations and orders by the Colonel had proved unavailing, the remaining eighty-five men were taken off duty and confined to their lines. 'The men of course had no real excuse for not doing what they were ordered,' commented Cornet MacNabb who thought that the men would not have disobeyed the orders 'if they had been left alone, instead of being paraded, and addressed, and all that humbug'. He added: 'Another mistake was that instead of serving out the cartridges the night before as usual, they were not given them till they got on parade, which of course made them doubt if they were the old ones . . . *But the real case is that they hate Smyth, and if almost any other officer had gone down they would have fired them off [even though] they did not want to be the first regiment that fired . . .* Craigie told me that, from what he knows of the men in his own troop, and in the Regiment, he would have guaranteed that they fired them five minutes after he had spoken to them. Craigie was told this by his men by whom he is beloved.' *See* Cadell, *The Outbreak of the Indian Mutiny, Journal of the Society for Army Historical Research, 1955; see also The Tale of Mees Dolly* in **CM**

CARRIAGE & PAIR, 'Another visitor, an English officer, gives an anecdote which is very characteristic of the barrier that obstructs the social intercourse of Europeans and Indians. On the way to Bithur, the visitor praised the equipage of his host (the **Nana Sahib**), who rejoined—"Not long ago, I had a carriage and horses very superior to these. They cost me 25,000 rupees; but I had to burn the carriage and kill the horses'. 'Why so?' 'The child of a certain sahib in Cawnpoor was very sick, and the sahib and the mem-sahib were bringing the child to Bithoor for a change of air. I sent my big carriage for them. On the road the child died; and, of course, as a dead body had been in the carriage, and as the horses had drawn that dead body in that carriage, I could never use them again.' (The reader must understand that an Indian of any rank considers it a disgrace to sell property). 'But could you not have given the horses to some friend—a Christian or a Mussulman?' 'No; had I done so, it might have come to the knowledge of the sahib, and his feelings would have been hurt at having occasioned me such a loss'. Such was the maharajah, commonly known as Nana Sahib. He appeared to be not a man of ability, nor a fool.' R. Montgomery Martin, *The Indian Empire*, London, n.d., Vol II p 249.

CARTER, JOSEPH The Peishwa Baji Rao II's widows were in constant conflict with **Nana Sahib** who had treated them with disrespect and cruelty. They apparently did their best to dissuade him from some of the excesses to which he gave his name and authority after the outbreak of the mutiny/rebellion at **Kanpur** As an example there is the story of Mrs Carter: Toll-keeper Joseph Carter of the Sheorajpur toll-bar and his expectant wife were captured near Bithur by a band of 7th Native Cavalrymen and brought to **Rao Saheb** at **Bithur**. Joseph was sent to Kanpur, where he was executed on the 10th, but when the Peishwa's widows threatened to kill themselves should any harm come to Mrs. Carter she was placed in their keeping, and delivered a baby in the Peishwa's zenana. She lived until 17 July, when **Havelock** entered Kanpur and the Nana Sahib ordered her and her baby executed.

CARTRIDGE QUESTION, THE Thought at the time by many to have been the main cause of the mutiny by the sepoys of the Bengal Army, opinion today inclines to the view that this was only the spark, or occasion, for mutiny. The latter appears to have been planned in advance. Briefly the cartridge affair was related to the introduction of a new basic weapon: at the close of the year 1856, the Government of India had decided that the old-fashioned musket ('Brown Bess') should be superseded by the Enfield Rifle. Depots for instruction in the use of the new weapon had been formed at three stations - Dum Dum, near to Calcutta; Ambala and Sialkot in Upper India. Large numbers of cartridges for the new rifle had been manufactured at Fort William in Calcutta and distributed to the various depots. A problem seems to have arisen at Dum Dum: 'a very unpleasant feeling existing among the native soldiers who are here for instruction, regarding the grease used in preparing the cartridges, some evil-disposed persons having spread a report that it consists of a mixture of the fats of pigs and cows'. Such a mixture would have been anathema to both Muslim and Hindu sepoys, particularly as the drill for the Enfield rifle required that the cartridge be torn open with the teeth. It is unlikely that the mixture was anything more significant than linseed oil and beeswax, but the cartridges provided an ideal excuse for mutiny and for rallying all to the defence of religion. On the question of the 'greased cartridges' - which so many people thought the real cause of the uprising, it is interesting to read the instructions which Government put out for their manufacture -- Orders issued in 1847 and approved by the Commander-in-Chief and the Governor General: '...the balls to be put up, five in a string, in small cloth bags with a greased patch of fine cloth, a portion carried in a ball bag attached to the girdle on the right side, the remainder in pouch. Patches to be made of calico or long cloth and issued ready greased from magazines; a portion of the greasing composition will also be issued with the patches for the purpose of renewal when required, and instructions for its preparation forwarded to magazine officers by the Military Board. (The following were those instructions)....The mode of preparing the grease and applying it to the cloth to be as follows: To 3 pints of country linseed oil add 1/4 of a pound of beeswax, which mix by melting the wax in a ladle, pouring the oil in and allowing it to remain on the fire until the composition is thoroughly melted. The cloth is then to be dipped in it until every part is saturated and held by one corner until the mixture ceases to run, after which it is to be laid out as smoothly as possible on a clean spot to cool. The above quantity of composition will answer for 3 yds of long cloth, from which 1,200 patches can be made.' These instructions were approved by the Governor General Lord Hardinge, in a letter from the Military Secretary to the Adjutant General dated 6 April 1847, and no subsequent cancelling order can be traced. The question of the Cartridges emerges time and again as the main or major contributory cause of the 'Mutiny', in that the pollution of animal fat on the cartridges was an offence to both Hindu and Muslim

sepoy, and that religion was thus being attacked by the English - some said deliberately. But was this just an excuse rather than a cause for rebellion? - 'Ironically the sepoys were found to be quite happily firing off the cartridges that were said to have started all the trouble' (From Hervey Greathed. 'Letters' etc *op.cit.*). A writer to the *Times*, countering the argument that the Sepoys' objections to the orders regarding the greased cartridges must have been bogus because the Sepoys subsequently employed Enfields in the mutiny, pointed out that military regulations had ordered the Sepoys to bite off the tip of the cartridge, whereas it was quite possible to break it off with their fingers. It was the biting and not the handling of the greased cartridge that was most objectionable; Golundazes, for instance, 'never objected to *handling* the grease applied to gun-wheels' which contained animal fat. London *Times*. 19 November, 1857, p 10. 'On May 5th cartridges OF THE OLD KIND which they and their fathers had used were served out to the 3rd Cavalry in Meerut. 85 men refused to take them.' See **FSUP** Vol I p 329 *et seq*. For a detailed account of the whole greased cartridge incident at Meerut, and the outbreak of the Mutiny there see Macnabb of Macnab Papers in the **CSAS**, Cambridge.

CASE, Lt.Col.
Commanded HM 32nd Regiment at **Lucknow** after Col. **Inglis** had taken military command of the station. Was killed in the battle of **Chinhat** 30 June 1857. He was personally brave and conscientious, but there is a strange tale attending his last moments. He was wounded, and was found lying on the ground by Captain Bassano of his regiment who immediately sought to have him removed from the field; the colonel however would have none of that, and ordered Bassano brusquely to rejoin his company, which he reluctantly did. Case was presumably killed by the sepoys as they then advanced.

CASE, Mrs Adelaide
She was the wife, subsequently widow, of Colonel Case the commanding officer of H.M.'s 32nd Regiment killed at the Battle of **Chinhat.** She was one of the besieged within the **Residency at Lucknow**; surviving, she published in 1858 the journal she had kept, under the title *Day by Day at Lucknow, see* **FQP**.

CASHIERED SUBALTERN
A subaltern (by the name of Christie, or Cox?) who had been cashiered (dismissed the service, a sentence of court-martial) for drunkenness, was resident in **Kanpur** before the outbreak and he fired a shot at a cavalry patrol who challenged him; fortunately he missed his aim, but the *sowar* naturally made a complaint in the morning. A second court-martial was convened but the officer was acquitted on the grounds that he was intoxicated at the time, and that his musket had gone off by mistake. The *sowars* of the 2nd Cavalry thereupon muttered angrily that perhaps their muskets might go off, ***by mistake***, before very long. *See* Sir George Trevelyan, *Cawnpore*, London, 1910, p 71. *See also* COX.

CASHMERE GATE
The spelling of Cashmere/Kashmir was a matter of dispute even in 1857. This was one of the major Gates of the old City of Delhi and the one by which the British burst in and assaulted the city on 14 September 1857. But to avoid repetition *see* KASHMIR GATE.

CASTE
Since the matter causes much confusion and misunderstanding to non-Hindus, and as it is of considerable significance to the story of the mutiny of the Bengal Army, it is as well to examine the meaning of this term. Possibly the best explanation, from a layman's point of view, was penned in 1974: 'Westerners are inclined to think of it (*ie* 'caste') as class, but it is very different, and the reality is obscured by a good deal of myth and many local complications. But, simplifying enormously, one may say that the Hindus were divided in the nineteenth century into more than three thousand groups who must not marry *outside* the group. The group was divided into subdivisions and usually it was not permitted to marry *within* the subdivision. The in-marrying group - in Hindi *jati* - may be compared with a battalion; the subdivision - in Hindi *gotra* - with a company. It is as though a man had to marry a girl whose father came from the same battalion as himself, but from a different company. But it is not only marriage. The group to which a man belongs affects the whole of his outward life; there are elaborate rules about eating, drinking, smoking and washing. Men from the higher groups must eat and smoke only with those of their own group and must not accept food or water from the lower, though there are almost always exceptions and some kinds of food are less subject to pollution than others. If these rules are broken, a man is polluted and may not eat, smoke or drink with his own group, still less of course with any other. He will be shunned by everyone; he is an outcast, excommunicated. He can only be cleansed, after long penance, by elaborate and expensive ceremonies. These in-marrying groups cannot be ranked exactly; to array them in an order-A is better than B, B than C, and so on- would certainly lead to controversy. There are regional variations and some groups rise or fall in general esteem; sometimes they split or coalesce. But there is some general agreement. Throughout India, the many different in-marrying groups who count as Brahman do come at the top of the tree in the dimension of caste. But they may come quite low in the *social* dimension, being often employed as cooks or office messengers. Ritual esteem does not coincide with social. This is what is so diffi-

cult for a Westerner, who is used to thinking in the one dimension of social esteem, to understand. Western observers were also often misled by a myth, in the literal sense of the word. They would be told that at the creation of mankind there were four *orders* of men: Brahmans who sprang from the head of the Creator, Rajputs who sprang from his arms, Banias from his belly and Sudras from his feet. Brahmans were priests and scholars; Rajputs were kings, barons, landowners, soldiers, Banias were traders, bankers, money lenders. These three orders were 'twice-born', wearing the sacred thread, observing far more restrictions than the others. Broadly speaking, the more abstentions and prohibitions were observed, the higher in the caste system a group was reckoned. Sudras were cultivators, messengers, clerks, artisans, certain domestic servants whose tasks did not pollute, such as those who carried a palanquin, but not scavengers, skinners, sweepers. These, with others made up a whole range of groups who were outside the caste system and who were generally called 'untouchables'. ' *See* Philip Mason. *A Matter of Honour, an account of the Indian Army, its officers and men,* Jonathan Cape, London, 1974, pp 123-4.

CASUALTIES Return of Casualties: Delhi Field Force May-Sept 1857, Killed 1,012...Wounded 2,795 (of whom many subsequently died)..Total 3,837. In HM 60th Rifles there were 389 casualties out of a total strength of 402, incl. 113 dead.

CASUALTIES 'A batch of despatches from India was printed in Tuesday's *'Gazette'*. They contain nothing new except long lists of privates killed and wounded, belonging chiefly to the Company's European regiments, and which it is impossible to find room for..' *from* the *ILN*. Now if it had been officers..........!

CATTLE, Army An army on the move in 1857 required the services of a great number of 'cattle' - the army's generic term for the animals it used to transport both men and stores and to drag heavy guns. The following gives an idea of the quantity needed: 'We are about to start an immense convoy,' an officer wrote from Delhi on 1 December (1857) 'collected here and by the Commissary at **Meerut**, for the force at **Cawnpore**. We have conjointly assembled 100 elephants, 2,000 camels, and 20,000 bullocks, a nice collection of animals to escort safely with all their attendants, nearly 10,000 men, down the road 250 miles to Cawnpore,' with an escort of only 'one European regiment reduced or weak in numbers, one troop of Horse Artillery, a squadron of Carabineers, and corps of Sikh Cavalry.' *See* London *Times*. 15 January 1858, p 8.

Army Cattle

CAUSES of the MUTINY/REBELLION There are so many causes adduced for the mutiny/rebellion that the subject is best set out in a section devoted to this alone. *See* NATURE OF THE REVOLT.

CAVE-BROWNE, John The writer of a narrative account of the measures by which the Punjab was 'saved' for government. Very one-sided (British) but readable none the less. Originally published in instalments in *Blackwoods Magazine*. The work has the merit of containing some original material as it is based on the author's own journal. *See* J.Cave-Browne, *The Punjab and Delhi in 1857,* (2 vols) Edinburgh, 1861.

CAWNPORE Modern **KANPUR**, in Uttar Pradesh, industrial city on the right bank of the Ganges. Although a settlement there was of ancient origin (it stands on the Grand Trunk Road from Calcutta to Peshawar), the modern city was almost entirely a British creation, being both an army base and staging post, and a place where industry, notably saddlery making, was established to serve the needs of the military. *See* J.F.Bruce, *The Mutiny at Cawnpore, Punjab University History Society Journal,* April 1934. 'Cawnpoor is a place of great extent, the cantonments being six miles from one extremity to the other, but of very scattered population. ...The European houses are most of them large and roomy, standing in extensive compounds, and built one story high with sloping roofs, first thatched, and then covered with tiles, a roof which is found better than any other to exclude the heat of the sun, and to possess a freedom from the many accidents to which a mere thatched roof is liable. 'The residents told him that contrary to its reputation the weather at Cawnpore was not bad. 'During the rains it was a very desirable situation, ...the cold months were remarkably dry and bracing, and ...the hot winds were not worse than in most other parts of the Dooab. The great incoveniences of the place are, as they represent it, its glare and dust, defects which are in a considerable degree removed already by the multitude of trees which they are planting in all directions.

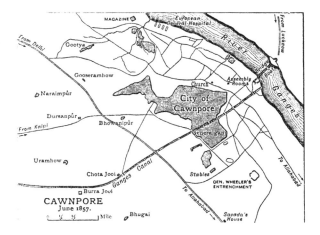

Cawnpore [Kanpur]

...On the whole, it is in many respects one of the most considerable towns which I have seen in northern India, but being of merely modern origin, it has no fine ancient buildings to shew; the European architecture is confined to works of absolute necessity only, and marked by the greatest simplicity, and few places of its size can be named where there is so absolutely nothing to see.' Heber, *Narrative*, v. II, p 38-40.

CAWNPORE, Battle of

This took place on 26 November 1857, and ranks with **Chinhat** as one of the rebels' greatest military successes. General **Windham** had been left at Kanpur by the Commander in Chief **Sir Colin Campbell**, when the latter marched to the relief of the **Lucknow Residency**. His instructions were explicit, 'not to move out to attack, unless compelled to do so by circumstances to save the bombardment of the intrenchment'. Windham considered it would be better to run the risk of meeting rather than awaiting, the approach of the combined force of the **Nana's** troops and the **Gwalior Contingent**, and so set out with 1,200 infantry, 100 *sowars* and eight guns, to face 20,000 men with forty guns. He was defeated and retired back into the (new) intrenchment in Kanpur while the rebels occupied the city and captured all the British army's baggage. Colin Campbell returned from Lucknow on 28 November, but had to wait until his wounded and the ladies *etc* from the Lucknow Residency (he rather ungallantly described them as his 'encumbrances') had been despatched by river to Allahabad before he could turn his attention to the rebels. On the 6 December there took place what might be called the second battle of Kanpur. The rebels were then divided into two distinct bodies, the Nana Sahib's men under the command of **Tatya Tope** and **Bala Sahib** having it's line of retreat on Bithur, and the Gwalior Contingent whose retreat lay towards **Kalpi**. The rebels were completely defeated, lost many men killed and thirty two of their forty guns, and fled in disorder. Kanpur was then re-occupied by the British and not again threatened for the remainder of the struggle. There are numerous factual accounts of these two battles: for the simplest *see* R.Montgomery Martin, *The Indian Empire*, Vol II pp 472-5.

CAWNPORE, before the Outbreak

'Cawnpore[Kanpur] was an important military station in 1857. Situated on the banks of the Ganges, and being fifty miles from Lucknow and a hundred miles from **Allahabad**, its stategic importance was tremendous. Ensconced between the highway to **Oudh** and the **Grand Trunk Road**, Cawnpore was highly accessible and hence the need for the British to garrison it strongly. The annexation of Oudh in 1856 made this need even more compelling. Consequently there were three Indian infantry regiments - the First, the Fifty-third, and the Fifty-sixth; the Second Light Cavalry; and the Artillery consisting of six guns manned by sixty-one Europeans and a few Indians. With the exception of three hundred (invalid) Europeans, all the three thousand soldiers were Indians. The Commanding Officer of the garrison was Major-General **Sir Hugh Wheeler**, the old war-horse, who had spent a good part of his youth in India and also had an Indian wife (But *see* WHEELER MYSTERY, **PJOT**). Wheeler was very popular amongst the sepoys and his trust in them was almost paternal, though, as future events showed, somewhat misplaced! The European soldiers lived in spacious bungalows which were more utilitarian than aesthetic and contained fond reminiscences of their moments back home. The 'native' troops lived in long rows of mud huts with thatched roofs. They were reasonably paid, but were steeped in conservatism affecting familial considerations - which often drained their pockets - and attitudes towards inter-dining.. Nepotism was rife and this enabled the development of homogeneity and camaraderie among the sepoys of similar castes and backgrounds. A trend had developed wherein extra-regimental duties were more prestigious than regimental duties. This impaired the military acumen of the Officers, their disinterest distancing them from their sepoys, and resultantly impeding army efficiency. There were some distinctive features of the armies at Cawnpore. The infantry consisted mostly of Oudh sepoys who also formed the bulk of such forces as the **Gwalior Contingent** and were predominantly Hindoo. The cavalry was dominated by Mahomedans. This dangerous preponderance of one community facilitated and extended 'combination on the part of the disaffected' (*see* **FSUP**, Vol I pp 321-2). They had left their families behind and their enforced bachelorism made it feasible for them to flow with the tide! The Mahomedans venerated the Mughal Emperor and were further piqued by the annexation. The Afghan debacle had revealed the vulnerability of the British to the sepoys.

Irrespective of community affiliations they experienced a unique fraternity in their common fear of Christianization. This deep-seated fear was exacerbated by the **rumours** from Calcutta of greased cartridges, and polluted flour in the Cawnpore markets. Cawnpore was a full-fledged district. There were about a thousand Europeans and they were employed in the army, the railways, the canal and other government departments, and trades. There were approximately one lakh Indians, forty thousand of whom resided in Cantonment bazars. There were some unsavoury elements too who had taken refuge here to evade the British law and escape to Oudh (which was outside the British Empire till 1856) if the need arose. The Cantonment Magistrate was Sir George Parker, and the Collector of the district was C.G.Hillersdon. Cawnpore had developed into a flourishing centre of commerce. The construction of the **Grand Trunk Road**, the Ganges Canal in 1854, the **Electric Telegraph**, and introduction of steam navigation on the Ganges between Calcutta and Allahabad, had reinvigorated various trades. Construction work on the railways in 1856, besides providing jobs for people, facilitated commerce. The annexation of Oudh removed the existing tariffs and duties and opened up new and exciting markets within the Company's jurisdiction. By 1857 indigo cultivation had already made the Maxwells and the Jones' families into well-entrenched landed proprietors. Billaur, Narwal and Sheorajpur were the centres of indigo cultivation. Opium was the new opiate of the traders and several opium factories were opened as trade in it became more lucrative. In fact, when profitability of indigo declined, minor planters resorted to opium smuggling! The names of W.Crump and Company and John Kirk were associated with the best wine available in Cawnpore. The most prominent merchants were the Greenway Brothers. They were bridge contractors, Lucknow Dak agents, bankers and newspaper owners. The Cawnpur Bank was under liquidation but there were private Banking houses. The other well-known firms were Bathgate, Campbell and Company, Chemists and Druggists, Brandon and Company, merchants, railway contractors and newspaper proprietors. Cotton was both cultivated and exported and leather saddles and shoes were indigenously made and sold. These merchants, though residents of long standing in India, were very conscious of being Englishmen and had created a characteristically English ethos at Cawnpore. The 'natives' followed minor trades and professions. They were mostly butchers, carpet-makers, hackerymen, blacksmiths, brickmakers, liquor manufacturers, vendors, darners, tailors, menials etc. They resented the affluence of the Englishmen, who had taken all the remunerative trade in their hands. The Indian traders and **zamindars** were particularly disaffected because there were several instances of their lands being appropriated for the construction of railways, roads and canals. The authorities felt '..The Tradesman and others only live there by sufferance with a primary regard to the wants and necessities of the cantonments.' (*see* Home, Public, 11 April 1851. Petition from Indian Merchants of Cawnpore to the Governor General). **Bithoor** [Bithur] was an important sub-division of Cawnpore. Here, **Nana Rao Dhoondoo Punt**, the adopted son of the late Maratha Peshwa, **Baji Rao II**, resided. Despite his various memorials, Nana Sahib was denied the pension, the title and the immunity of his 'jagheer', Bithoor, from the Company's Courts. As the Commissionership of Bithoor was now considered redundant, Bithoor was placed under the Collector of Cawnpore. Nana Sahib had a large entourage consisting of his family, advisers, courtiers, preachers, astrologers and several hangers-on. Though the government had permitted him control over the jagheer, he found it increasingly difficult to fend for 300 dependants and 26 widows. The significant male members of his family were **Baba Bhutt**, his elder brother, **Bala Rao** - his younger brother and adopted son of Baji Rao, Panduranga Rao - Baji Rao's grandson, Chimmaji Apa - Baji Rao's brother's grandson. The prominent ladies were his wife, Baji Rao's widows - Maina Bai and Taee Bai, and Baji Rao's daughters - Yoga Bai and Kusums Bai. His major source of strength were **Tantia Tope**, his commander and **Azimoolah**, his Chief Adviser. The powerful Courtier of Baji Rao, **Ram Chunder Subedar**, lost his considerable pension and influence with Nana. He ignored Subedar's son, Nana Narain Rao. Nana Sahib's denial of economic independence to his dependants led to acrimony which was the most pronounced in the Peshwa's widows. Chimmaji Apa went to the extent of unfairly making a legal claim on Baji Rao's property through his ignominious lawyer, **Nanak Chand**. The latter was a man of doubtful integrity and questionable credentials and discredited Nana whenever possible, in the hope of monetary gain! Even the British officials considered him untrustworthy and opportunistic. Nana Sahib felt humiliated and aggrieved, but his grievance was directed at the Governor General and the Directors. He had already corresponded with several 'native' rulers of India to persuade them to plan for the overthrow of British power. The annexation of Oudh had aroused further interest among the leaders, especially Mahomedan interest. There is evidence of Nana and Azimoolah furthering their cause abroad. In 1857 Nana, Azimoolah, Tantia and Rao Saheb visited several places. These tours were ostensibly pilgrimages or courtesy visits, but there were instances of Nana mustering financial and popular support for the revolt scheduled for 31 May 1857. (*see* **FSUP** Vol I pp 372-80). Nana's visit to Lucknow in April 1857 was looked at with scepticism by the British. It was also commonly known that Nana had withdrawn large sums of money from his London bank. Despite these suspicious movements, Nana Sahib's integrity and trustworthiness were seldom doubted by the local British officials like Wheeler, Hillersdon, Moreland and others. He entertained them frequently at Bithoor most lavishly and generously. In the period just before the outbreak at Cawnpore, there were several instances of faith reposed in Nana Sahib as an individual. Hillersdon asked Nana to protect the Treasury, which he willingly did with about 200 retainers and two guns. (After Baji Rao's death in 1851, Nana had been allowed to maintain a retinue of 500 cavalry and infantry men and three guns of small calibre at Bithoor). That these were the men who joined the revolt along with the militant sepoys of the British armies, is another story. Hillersdon had earlier toyed with the idea of sending the British wives and children to Bithoor under Nana's care. Wheeler did not doubt either Nana Sahib or his own soldiers and never expected the engagement at Cawnpore to last long. The fact was that these astute and perspicacious officials liked him and were impressed by him. Nana Sahib's eagerness to reaffirm his fondness for them was revealed when, just before the outbreak, he sent a letter to General Wheeler at the Entrenchment on 7 June 1857, warning him of the impending attack.' **Dr Kirti Narain, M.A., Ph.D. Senior Lecturer, Dept. of History. D.G.College, Kanpur.**

CAWNPORE, Description of

'The town itself is small, mean-looking, and dirty; but on the right bank of the Ganges many hundred bungalows, the barracks of the troops and the bazars, extend in a semicircle, for nearly five miles, which imparts to the whole a striking and splendid appearance. Several of the bungalows are most picturesquely situated, on the banks of the Ganges, which rise to the height of 100 feet; they are fitted up most luxuriously, and have very extensive gardens, in which tamarinds, mangoes, bananas, neemes, acacias, and fig trees, overshadow a rich carpet of flowers which charms the senses by the magnificence of its colours, and the fragrance of its perfume. In the circumjacent country there are fine fields of wheat and barley in ear, which are succeeded by crops of maize, rice, sugar-cane, cotton, gram, jowary, and indigo; potatoes, peas, and cauliflower, and many other vegetables, are grown in the gardens, but the former are insipid. This is the time of the birds of passage, some of which, especially the ortolans, are considered a dainty: they are rather larger than larks, but their taste is piquant. [Thornton reported that the ortolans flew by in such numbers in the hot winds that fifty or sixty will drop at a single shot.]' Thornton, *Gazetteer of India*. London, 1854. p 301.

CAWNPORE, EUROPEAN POPULATION

The exact number of Europeans in Kanpur in May 1857 is still a matter of minor dispute, especially as the tablets in the **Memorial Church** of All Souls, speak of the 'more than 1,000 Christian souls etc', but the most careful computation establishes the following: Artillery, 1 Company of. 59 men, and guns; Infantry, 60 men of HM's 84th, 70 men of HM's 32nd, invalids, and sick, 15 men of 1st Madras Fusiliers. [The native troops consisted of the 2nd Regiment, Light Cavalry, the 1st, 53rd, and 56th Regiments Infantry, and the Golundauze, or native gunners attached to the battery.] ...There were also nearly the whole of the soldiers' wives of HM's 32nd Regiment, which was stationed at Lucknow. 'The whole number of the European population, therefore, in Cawnpore, men, women, and children, could not have amounted to less than 750 lives.' G.W. Forrest, ed. *Selections from the Letters Despatches and Other State Papers Preserved in the Military Department of the Government of India: 1857-58*. Volume III. Calcutta, 1902. But in addition there were the European merchants and their families. Included in the 'more than 1,000' also would have been the refugees from **Fatehgarh** killed on **Nana Sahib's** orders when they drifted down on the Ganges, and an unknown number of Eurasian and native Christians.

From Kincaid's 'British Social Life in India'

CAWNPORE, MASSACRE OF

Although there were in fact three massacres carried out at this place within the space of a few weeks in the summer of 1857, it is the third or 'Massacre of the Ladies' which is best known, for it brought universal condemnation not only from the Europeans but also from prominent rebels *eg* **Hazrat Mahal the Begum of Oude[Awadh]** who believed that the rebels' cause was ruined by this one act of barbarity. Dhondu Pant, the **Nana Sahib**, smarting under his defeat on the Fatehpur road by General **Havelock**, agreed to or at least acquiesced in the murder of more than two hundred European and Anglo/Indian women and children who were being kept as hostages in a place known as the **Bibigurh**. It is said that the sepoys refused such work and that the killing was actually done by butchers from the bazar and by a member of the Nana Sahib's own guard. The bodies were thrown down a dry well. General Havelock arrived the next day, 17 July 1857, by which time the Nana Sahib and his forces had retired to **Bithur.** The first massacre concerned the opening of fire upon the Europeans from **Wheeler's intrenchment** who had surrendered their position upon promise of safe-conduct to Allahabad by river: the party was attacked from ambush as they embarked in the boats provided for them. The second massacre was of the Europeans who had floated down river from **Fatehgarh** to Bithur and were taken prisoner as they landed. *See* **SSF, CM, FSUP.** *See also* MASSACRES IN KANPUR.

CAWNPORE MERCHANTS Already in 1857 Kanpur was a thriving mercantile and manufacturing town, the basis of its prosperity being the supply of the army. There were many rich merchants like Jaganath who was a dealer in indigo. Gangraprasad Mahajan, sold tents and Sheoprasad sold coats and 'gentlemen's dresses, caps, &c.' at the Surseah Ghat. Both of the latter aided the Nana Sahib to save themselves from ruin and disgrace. Jayalkrishna we know was a well-known jeweller. For details of the European and Eurasian mercantile community *See* Zoë Yalland, *Traders & Nabobs*, Salisbury, 1987.

CAWNPORE, OUTBREAK There are many British accounts and versions of the outbreak of mutiny at **Kanpur,** which is usually set down as occurring on the evening of the 4 June 1857; but there are few statements which really give a picture from the sepoys' point of view of this timetable. Here is one: 'He [Nana] then accompanied the sepoys and they at once proclaimed him their leader. On that day (Friday 5 June) he took possession of the magazine and opened the jail. The prisoners were all released and opening the doors of the arsenal, he offered arms to those of the prisoners who would join him and told those who wished to go home, to go. On this they all joined in plundering the arsenal and after sending a good deal of property to their houses the 2nd Cavalry and 1st N.I. dropped their intentions of marching to Delhi, and of presenting the treasure and ammunition to the King of Delhi and offering their services to him and that day proceeded with the Nana five miles in the Delhi direction and encamped at Kullianpoor. At first the 53rd and 56th N.I., although mutinously inclined in consequence of the report about the cartridges, did not break out into open mutiny. The other two native corps went to their parade and having called for the native officers and sepoys of those corps had a long consultation with them on the subject of their religion, and told them if they did not join them, they would be considered as infidels, and excommunicated from caste privileges etc. At last after much talk they joined the others and agreed to go to the encamping ground at Kullianpoor where the Nana and the other regiments were. At this juncture several of the bad characters of the regiments proposed murdering their [native] officers, on which they with uncovered heads fell at the feet of the sepoys and asked for their lives to be spared. They let them go and the officers then went to the entrenchment. The sepoys of the 53rd and 56th Regiments then joined the camp of the other mutineers at Kullianpore. When the Nana saw that the three Native Infantry regiments had openly mutinied, and intended to go to Delhi, he told them that they ought not to go till they had massacred every European in Cawnpore because the *Sahibs* would take every opportunity of punishing them, and that he would be left alone'. *See* Nerput, Opium Gomashta, **FSP** quoted in **FSUP**, Vol IV, pp 503-11.

CAWNPORE: The Fate of the Leaders of the Rebellion 'Cawnpore was the epicentre of the Outbreak of 1857 and some of the leading luminaries of the rebellion were from Cawnpore and its neighbouring centres. **Nana Sahib**, the ill-fated adopted son of **Baji Rao II** was the chosen leader of the rebels. Willingly, or unwillingly, he became the central figure around whom the rebels rallied. Resultantly, the prominent men of his court at *Bithur* assumed the role of leaders, albeit under his direction. These were **Bala Rao**, his younger brother; **Baba Bhutt**, his elder brother; **Rao Saheb**, his nephew; **Tantia Tope**, his commander; **Azimoolah**, his chief adviser, and Brigadier **Jowalla Pershad**. While Rao Sahib remained at Bithur, the rest actively participated in the outbreak at Cawnpore. The massacre at **Satichaura Ghat** and **Beebeeghur** completely discredited Nana and his followers as the British held them responsible for these gruesome acts. His newly formed government at Cawnpore lasted for three weeks as the British army by 19 July occupied both Cawnpore and Bithur after defeating the rebel forces led by Bala and Jowalla Pershad. Nana hurriedly escaped from Cawnpore and Bithur with his coterie - including a wounded Bala and ladies of the family - on 17 July. Post Bithur is a long saga of the varied paths taken by these men as they battled the British in their endeavour to make the Uprising successful. Their moves and counter-moves, their movements themselves were often shrouded in such secrecy, that it became difficult for the British to formulate their plans and conclusively determine their ultimate fate. Contemporary British officers or the later historians had to depend upon *'native evidence - often second hand reports derived from interested or prejudiced sources.'* Nana Sahib, along with his compatriots, escaped to Fathpur Chaurasi. Accused of murder, rebellion and skulduggery, Nana Sahib had to be careful in implementing his plans. He remained in Awadh and directed the movements of his associates. The valiant efforts to retrieve Cawnpore failed. With Delhi lost and the British gaining ground in Lucknow, Nana Sahib left Awadh in March 1858. He proceeded towards Rohilkhand and *en route* he was joined by other revolutionary groups. In the meanwhile, the government had declared a reward of one lakh rupees to any person who would give up Nana Sahib. At **Bareilly**, he was welcomed by its ruler, **Khan Bahadur Khan**, who was heading the rebellion there. Nana helped Khan Bahadur by mustering Hindu support and organizing the rebel army. The British Commander-in-Chief, **Campbell**, attacked Bareilly in May 1858 and the rebels lost. Obscurity clouding his whereabouts made the need to apprehend him increasingly imperative. Official communication reported him to be in Awadh, while Nana was actually at **Shahjehanpore** from where, along with the Begum of Awadh, Hazratmahal, and others, he finally left for Terai in Nepal, after their conclusive defeat at Banni by Campbell. By June 1858 the condition of revolutionary armies elsewhere had deteriorated further and most of the leaders were on the run. Nana and the Begum were together and their hardships made them correspond with the Nepal leader - Jung Bahadur - for help, which was refused. Nana and the Begum reached Bahraich. With conflicting assertions as to his place of residence varying between Etawah and Gwalior - the British forces under Lord Clyde reached Bahraich, but Nana had already escaped to the Terai jungles of Nepal. Other leading revolutionaries joined him there and they still had a formidable force. There was a combat at the Rapti River and the rebels were victorious. After this the task of completely annililating them was completed by Sir Hope Grant, by April 1859. Many rebels surrendered but Grant reached Nepal to arrest the errant leaders. Jung Bahadur made it clear that Nana could not be given refuge in Nepal. The condition of the rebels was pitiable. However the ladies of Nana's family were given shelter by Jung Bahadur. Nana wrote to the British authorities (**Ishtiarnamah**) and denied complicity in the rebellion which he had joined out of helplessness. He was not a murderer, but would fight the British to the end. It was strange that the powerful British government had not been able to catch him. Nana was reported to have died of malarial fever on 24 September 1859, at Taraghurrie in the jungles of Nepal. However within a year the British started doubting the veracity of the reports of his death. The authorities caught some suspects who turned out to be fakes, rumoured to have been planted by Nana. Stories circulated of Nana donning the robes of an ascetic. He was purported to have escaped to Mecca or Turkey and had an English wife and a son. He was reported assassinated in Constantinople in 1907. Another story was that the lookalikes of Nana and Bala had killed each other in Nepal to put the British off the track. A letter supposedly written by Nana to the British in 1871 was discovered. Another unproven version was of Suraj Pratap, who claimed to be the grandson of Nana Sahib. According to him Nana returned from Nepal in 1900 and settled at Partabgarh. This theory was supported by two witnesses, Harishchandra Singh and Parmeshwar Baksh Singh who gave evidence of Nana being in Madhramau and ultimately getting drowned in a flood in 1926. Bala Rao escaped with Nana Sahib to Fathpur from Bithur. Nana sent him to Delhi and Bala later joined Tantia Tope at **Kalpi**, where the unsuccessful Third Battle of Cawnpore was fought. They crossed the Ganges at Nana Mau ki Ghat. The authorities were rather uncertain about Bala's movements and suspected him to be in Bundelkhand with Tantia. Bala had proceeded to Bahraich where he joined forces with Mahomud Husun who was supreme at Goruckpore. He escaped to Basti with Brigadier Rowcroft in pursuit. There was a skirmish at Rapti River. Bala successfully crossed the river but was driven to Tulsipur. Grant pursued Bala himself, and successfully cut him off from Goruckpore, chasing him across the border to Nepal where he took refuge in the jungles. Jung Bahadur met Nana, Bala and the Begum, but they refused to bend before the British. Bala still had two thousand fighting men, but fell sick, and sent a petition to the English on 25 April 1859, expressing his helplessness as from the beginning he had been just a pawn in the hands of Nana. He was accused of co-operation with Nana Sahib. Bala reportedly died of fever near Dhoker in Nepal somewhere close to Bhutwal. Baba Bhutt accompanied Nana to Fathpur Chaurasi and to Shahjehanpore, where he was his chief personal attendant. In the meanwhile, he had gone to Kalpi with Bala to fight the English at Cawnpore. He was present in the Nepal jungles with Nana and was ultimately accused of co-operation with Nana. He probably died there. Brigadier **Jowalla Pershad** was Nana's trusted agent. From Fathpur, he was sent by Nana to Kalpi to join the Gwalior troops and participated in the battle of Cawnpore 28 November - 6 December 1857, along with Bala Rao and Tantia Tope. Eventually, along with Nana and the Begum, he escaped to Nepal. After the reported death of Nana, he surrendered to the British. He was brought to Cawnpore fettered, was tried, and hanged on 3 May 1860 at Sati Chaura Ghat. **Azimoolah Khan** was supposedly Nana's greatest confidant. Though there were apprehensions that he might betray Nana into the hands of Clyde and pocket the reward, it is popularly believed that he went with Nana to Nepal where he died of fever - close to Nana's death - at Bhutwal in October 1859. However, Azimoolah's diary (the authenticity of which is doubtful) revealed that Nana and he were on the run and Nana's end came at Naimisharanya, Sitapur, in 1926, Azimoolah too allegedly died there in 1926. Azimoolah's indictment was the severest - conspiring with Nana to engineer the rebellion. The destinies of **Tantia Tope and Rao Saheb** were intertwined. However, Nana reposed greater confidence in Tantia. From Fathpur Chaurasi, Tantia was sent for the relief of Bithur. Defeated, he came back to Nana who later sent him to Kalpi to lead the Gwalior Contingent at Cawnpore. Rao Saheb was made the head of operations and Tantia reported to him. Nevertheless, the real authority rested with Tantia. Losing at Cawnpore, they fled to Kalpi, and while Rao Saheb remained there, Tantia gave a crushing defeat to the Raja of **Charkhari** in March 1858. In response to an appeal from the **Rani of Jhansi**, Tantia went to her help; but the rebels, defeated near Jhansi by **Sir Hugh Rose**, fled, to lose again at **Kunch**. This defeat sowed mistrust among the rebels - the infantry mistrusted the cavalry, and both were sceptical of Tantia. Rose vanquished the combined forces of Rao Saheb and the **Nawab of Banda** and occupied Kalpi which was the stronghold of the rebels. Tantia, the Rani, Rao Saheb and Nawab of Banda marched to **Gwalior**, where Scindia was defeated and his armies joined the rebels. At Gwalior Nana Sahib was declared Peshwa with Rao Saheb as the Governor and Tantia as the Dewan. Fearing the rebels would ultimately unfurl the Peshwa's banner in southern India, Hugh Rose attacked them. The Rani was killed on 17 June; and, losing considerable ammunition and men, the rebels fled. Crossing Jeypore territory and Holkar's areas, the rebels frenetically decamped, forcing armies and the populace to join them. The British seemed absolutely baffled at Tantia's escapades, and his capacity to outsmart them. But gradually the British army was closing on them and Rao Saheb was losing control over his troops. Cornered in the North, they crossed the Narbada but it was one year too late as the rebellion was petering out. Consequently, on facing the unwelcoming Southern Maratha populace, the sepoys resorted to plunder. Hounded by the British forces, they re-crossed the Narbada. With the Banda Nawab surrendering, the two fought the British alone. At Indargarh Firuz Shah joined them, and together they went to Dewasa where the English forces surprised them. How they escaped was a miracle as every route was closed to them! Tantia and Rao Saheb had been having strained relations. Tantia left Rao Saheb and his army and joined Raja Man Singh of **Narwar** in the Paron jungles. Man Singh betrayed him at the behest of Major Meade who offered him the bait of recovering his usurped lands. Tantia was captured and tried at Sipri by court-martial. He was charged with having been in rebellion and having waged war against the British government. His defence was that he had followed his master, Nana's orders and later Rao Saheb's. He had never murdered Europeans. Tantia Tope was hanged on 18 April 1859 at Sipri. The legality of his hanging was questioned as he was not a British servant and could not be considered a mutineer. Doubts were expressed by many as to whether he was the real Tantia, as Tantia was too wily to get caught. Recent claims have denied that Tantia was betrayed by Raja Man Singh, who helped his escape. Tantia allegedly died in 1912 at the age of eighty and the man hanged was Raghunath Bhagawat, a teacher. The British reported hearing he was residing at Bikaner and Kashi. As to the fate of Rao Shaib he was unable to secure honourable terms of surrender from the British. Much of his army deserted him, and disguised as a mendicant, he moved in the territories of Gwalior and Tonk. He suffered another crushing defeat at Ajmer and fled to Banswara jungles. He assumed the name of Lachman Das Pandit and wandered from place to place finally reaching Chandenee, in Jammu, with his pregnant wife. He was betrayed by a former associate, Bhim Rao, and captured in 1862. He denied being the leader of the rebels and claimed he had been a tool in the hands of Bala Nana and Tantia. He was tried and found guilty of being an instigator, and an accessory to the murder of Europeans, and of being a leader of the rebellion. None of the sixty-one deponents accused him of inciting or committing murder, but he was hanged on 20 August 1862. This historiette reveals that the ultimate fate of most of the rebel leaders of Cawnpore remained a mystery. The British themselves displayed uncertainty when in 1862 they published a list of persons who were disloyal and still at large, and, except for Rao Saheb and Jowalla Pershad, whose identity had been conclusively proved, the names of the other five rebel leaders were in it.' Principal references: **FSUP**; S.B.Chaudhury, *Civil Rebellion in the Indian Mutinies;* S.N.Sinha, *Mutiny Telegrams,* Lucknow, 1988; Motilal Bhargava, *Nana Sahib;* A.S.Misra, *Nana Sahib Peshwa and the Fight for Freedom, etc.* **Dr Kirti Narain**

CAWNPORE, THE INTRENCHMENT Many of the Europeans holed up in Wheeler's pathetic 'intrenchment', and surrounded by the forces of the Nana Sahib until their disastrous surrender at the end of June 1857, died not so much from bullets or shot and shell, as from the effects of the unremitting heat: there was no shelter from the sun, particularly after the thatched roof of a barrack was fired. Ironically, sixty years later on, an immense cool

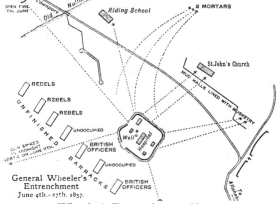

Wheeler's Entrenchment, Kanpur

underground room was discovered under one of the two barracks at the Cawnpore entrenchment; it would have afforded full protection from both sun and bullet, had the garrison known of its existence. Possibly they did know, and kept their gunpowder there, but it seems unlikely. The site has been closed up now - it was found to be infested with snakes. And then there are the tunnels leading out from the entrenchment but built many years before 1857: they are not just holes in the ground but well-built, brick lined tunnels along which men and stores could have moved easily. They are still to be seen at Kanpur. No contemporary historian mentions their existence, but are they the real reason for General Wheeler choosing this otherwise pathetically inadequate area for his entrenchment?

CAWNPORE, TREASURY Hillersdon collected some elephants from Nana Sahib with which to remove the treasury. 'Then the sepoys who were in charge of the treasury said, 'We will not allow its removal.' The opium *gomashta* Hillersdon sent told the sepoys, 'If you are all determined upon this, on Monday next a guard of Europeans will be sent down to the magazine and treasury.' The sepoys told him, 'As long as we live we will not allow this to be done because, as European guards have never been placed at the treasury or magazine before, there is no reason why they should be put in charge now.' At last after a good deal of conversation the Collector left.' Nerput, Opium Gomashta, Foreign Secret Proceedings in **FSUP**, Vol IV, pp 502-3.

CEMETERIES, On occasions the problem of dealing with the bodies of dead combatants almost overwhelmed resources. The British found this particularly difficult at Delhi and Lucknow, where casualties were severe, and where the climate demanded early action. We get some clues from the book of a later Commissioner of Delhi: 'The road running down into the old cantonment from the Flagstaff Tower leads to the Military cemetery (oldest grave there is dated 1833) alongside which a new cemetery has now been opened, that near the Kashmir Gate no longer affording resting-room. In this graveyard lie buried Sir Henry Barnard KCB, Col.Chester, Col.Yule, Capt. Fagan and many other brave officers and men, some named, but mostly nameless, who fell before Delhi, upholding the cause of their country'. ...on my recommendation Government has recently approved of the erection of a memorial cross in this cemetery. It is extraordinary and would I should hope, be impossible anywhere, except in India, that this should have been neglected for over 40 years. The English cemetery at Sebastopol and the cemetery in the Residency Lucknow, are now at least well and lovingly cared for and the same may be hoped for Delhi.' His hope was not fulfilled: true the Rajpur cemetery was looked after well until 1947, and the Memorial Cross he speaks of was erected, but their fate in recent years has been total neglect. The story is a happier one at Lucknow where in the ruins of the Residency, Government maintains the graveyard with some semblance of dignity. *See* **CM** p 91 *et seq.*

CENSORSHIP of the Press The British Government in Calcutta was made aware, somewhat belatedly, of the possibility of the spread of disaffection *via* newspapers. The Act No. 115 of 1857 resulted, and to the usual licensing and scrutiny of newspapers was added the power to prosecute and close down, in cases where the Governor-General in Council believed that press freedom had been abused. One of the first 'victims', remarkably, was the totally pro-British newspaper **'*The Friend of India*'** which offended Government by an objective and uncontroversial report on 'The Centenary of Plassey' on 25 June 1857: this was described as 'fraught with mischief' and 'calculated to spread disaffection'. Its condemnation and the official warning to the '*Friend of India*' that followed reflects not only the rising panic in Calcutta at the successes of the rebellion, but also the fact that Calcutta was completely out of touch with the realities of the situation. Readers of the '*Friend of India*' in the months following were no doubt amazed by the lack of prominence given to the rebellion: details from 'up-country' were on inside pages, very low-key, and subordinated to insignificant news from Europe. This no doubt was the '*Friend's* editorial response to the rebuke. *See* '*Friend of India*' 25 June 1857 *et seq,* and '*Hindu Patriot*' 9 July 1857.

CENTRAL INDIA AGENCY This was the British administrative organisation based on **Indore** with responsibility for co-ordinating policy among the eleven treaty states and territories of Central India. The latter with Bundelkhand comprised many states of which the principal were Dhar, Orchha, Samthar, Panna, Charkhari, Rewa, Nagode, Gwalior, Indore, Bhopal, Jaora, Rutlam and others. Colonel (later Sir Henry) **Durand** was appointed to the Central India Agency in 1857, *vice* Sir John Hamilton who proceeded on sick leave in May, and was instrumental in preventing the spread of the rebellion; held **Indore**, and reconquered Western **Malwa**. Although he successfully defended the Indore Residency he determined, when deserted by Holkar's troops, to retreat from the city, and was subsequently criticised for so doing, notably by **Sir John Kaye**. Durand evidently believed that **Holkar** was behind the attack on the Residency. Sir Robert Hamilton resumed office at Indore as Agent to the Governor General in Central India on 16 December 1857. *See* Sir George William Forrest, ed., *Selections from the Letters, Dispatches and other State Papers of the Military Department of the Government of India 1857-58*, 4 vols, Calcutta, 1893-1912, Vol IV, *and* Tapti Roy, *The Politics of a Popular Uprising: Bundelkhand in 1857,* Delhi, 1994.

CENTRAL INDIA FIELD FORCE This was the name given to the military force set up by the British to counter the rebels and restore authority in the area between the Yamuna (Jumna) and Narbada rivers, the latter being for all practical purposes the limit of territory affected by the outbreak. On 16 December 1857 Sir Robert Hamilton resumed office at Indore as Agent to the Governor General in Central India. The next day **Sir Hugh Rose** took over command of the British forces in Central India, of which the **Malwa** Field Force formed the nucleus, and which was divided into two brigades, the first at Mhow (commanded by Brigadier C.S.Stuart, Bombay Infantry), the second at **Sehore** (commanded by Brigadier C.Steuart, 14th Light Dragoons). The general plan of campaign provided for the advance of three columns, one of which, operating from Mhow under Rose would sweep the country from that place to Kalpi on the Jumna, relieving **Saugur**[Sagar] and recapturing **Jhansi**; at the same time a Madras Force under **General Whitlock**, was to cross Bundelkhand from **Jubbulpore [Jabalpur]** to **Banda**, while a Bombay Column under Major-General Roberts operated in Rajputana. It was the Central India F.F. that had the most protracted campaign of all British forces involved at this time, having to counter the well-led rebel forces of **Tatya Tope**, the **Rani of Jhansi** and **Rao Sahib** in territory largely sympathetic to the rebel cause. That the British succeeded here is in large measure due to the ability of General Rose. For details of the campaign *see Revolt in Central India 1857-59,* Simla, 1908, produced by the Intelligence Branch of Army HQ.

CERTIFICATES There were many occasions when British refugees owed their lives to the kindness and protection of Indians both high and humble. A 'certificate' was usually given by the refugee to his rescuer which might stand the recipient in good stead if he were in peril at a later date. As an example: 'Certified that we were made over to Anand Ram Chaudri for protection by the Nawab of Jhajhar during the short period we were with him. He was remarkably kind for which he deserves the highest credit. He also accompanied us into the Headquarters Camp before Delhi .' Signed by one Thos Hardy, and dated 27 July 1857.

CHALWYN, Louise   Mrs Chalwyn died in the massacre at the **Satichaura Ghat**, Kanpur on 27 June 1857. Since her name does not appear in the list of women in the **Bibigurh**, **Lt Delafosse**, a survivor, assumes she was killed in the boats. He also notes her husband Dr Edmund George Chalwyn veterinary surgeon (Bengal Army Veterinary Service 1849-57) as 'killed', which probably means that he was killed in **Wheeler's Intrenchment** before his wife died. Before her death Louise Chalwyn had written a number of letters from Calcutta, Hoshiapur and Kanpur, (those from the latter place being full of foreboding as though she knew she would not survive), and these letters may still be read as they are preserved in the Oriental and India Office collections of the British Library, see *MSS.Eur.B.344/2.*

CHAMBERS, Miss (or Mrs?)   The most horrendous story was circulated concerning this woman, and it applies, most probably, to a Mrs Chambers, wife of Captain Chambers of Meerut. It was to the effect that after being killed by the mutineers on 10 May 1857 she had been mutilated: advanced in pregnancy she had been ripped open, her child/foetus taken out 'and put round the poor lady's neck'. It was supposedly the work of the sweepers who stripped the corpses or the Muslims 'who believed they would gain entry into paradise by trying their swords on infidel corpses'. There was also a persistent rumour to the effect that her murderer was in fact a butcher from the bazar and that the sweepers, far from being the culprits, actually avenged her death: 'A butcher entered the house of Chambers the Adjutant whilst he was on parade, caught Mrs. Chambers and cut her throat. The servants caught the horrid wretch and then and there roasted him to death! Those horrors occurred at Meerut.' See the *Times* 3 and 5 April 1858. *See also* ATROCITIES.

CHAN-TOON   Mabel Mary Agnes, the author (?) or editor of a strange little book called '*Love Letters of an English Peeress to an Indian Prince*', published in London in 1912. It does not take long for the reader who is familiar with the story of **Azimullah Khan** *ie* 'Darling Azimullah', to come to the conclusion that these are the letters written to Azimullah which were discovered by **Lieutenant Frederick Sleigh Roberts** at **Bithur** in December 1857, passed round by him to a number of British army Officers' Messes, where they were greeted with hilarity and salacious gossip respecting the identity of the writer. See Roberts, *Letters Written during the Indian Mutiny,* London, 1924, reprinted Delhi 1979, *and , Forty-one Years in India,* London, 1897.

CHAND, BOOL   He was the author of '*Urdu Journalism in the Punjab*'. See Punjab University History Society Journal , April 1933.

CHANDA, Battle of   This was an engagement on 19 February 1858 between the rebels and **General Frank**'s Field Force, near Sultanpur in Oude[Awadh]. Remarkable only for the large number (over 2,500) of sepoys, as opposed to zemindari men, who took part; troops from the mutinied 20th, 28th, 48th and 71st Native Regiments were present. The rebels deserted a strongly fortified Chanda and withdrew to Lumbooah[Lambhua] but, as frequently happened, could not be pursued because the British had no cavalry.

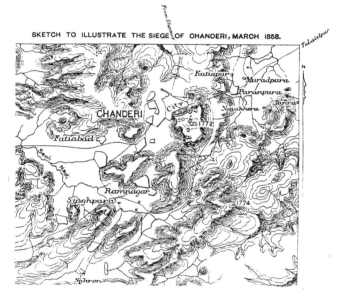

CHANDERI   This was an important town lying some 50 miles to the south of Guna and about the same distance southwest of **Jhansi**. There was a fort here of great strength. It was besieged by the British under **Brigadier C.S.Stuart** on 7 March 1858, heavy guns made a breach and the fort was taken by assault on 17 March, the rebels retreating through the town. But this was not the end of the story for when **Sir Hugh Rose** marched on **Kalpi** after the fall of **Jhansi**, the two rajas - of **Banpur** and **Shahgarh** - doubled back round his right flank, reached their own territories and the fort of Chanderi was again captured by the rebels. For details *see The Revolt in Central India,* Simla, 1908 pp.102 *et seq., and* Thomas Lowe, *Central India during the Rebellion of 1857-58,* London, 1860, and **FSUP** Vol III.

CHAPATIES   The circulation of chapaties before the outbreak of the Rebellion has always been believed to have had some significance, if only as a warning to be prepared for the signal 'Sitara Gir Parega! (A Star Shall Fall!)' which would be the watchword to rise against the British. But there was much confusion even among those who distributed the chapaties as to their meaning. Briefly a village watchman would arrive at a neighbouring village with a chapati (a flat unleavened cake) about the size of a man's palm, would hand

Distribution of Chapaties

it over with instructions to make five more like it and deliver them to neighbouring villages with similar instructions. 'For the elders; from the south to the north, from the east to the west.' The chapaties would be spread at the rate of up to 200 miles in a single night. *'Are all the chowkeydars about to strike for wages, or is anybody trying a new scheme for a parcel dawk? Is it treason or is it jest? Is there to be an explosion of feeling or only of laughter? Is the chupatti a fiery cross or only an indigestible edible, a cause of revolt or only of colic? Is the act that of a malcontent or a fool ?'*. The best description of the affair is perhaps in **Mainudin Husain**'s diary, quoted in C.T.Metcalfe's *'Two Native Narratives of the Mutiny in Delhi'*, 1898, p 39 *et seq*. See MAINODIN HASSAN.

CHAPRA (or Chupra)   An important town at the confluence of the Ganges and the Gogra, near **Patna** in Bihar. In accordance with **William Tayler's** 'ill-judged and fainthearted order', the British authorities withdrew from the town at the end of July 1857. They fled with 'a precipitation apparently injudicious and pusillanimous'. Government's embarrassment was saved (or increased, it depends on which way you look at it) by the action of a 'Mahomedan gentleman', named Qasi Ramzan Ali. It seems that this man's loyalty to government had been in doubt, but he certainly proved it now: he assumed command on the departure of the English, 'held cutcherry' in the accustomed manner, and when the civilians returned he delivered over the station to them in perfect order, courts of justice, prison, prisoners and police, the lot. In contrast to other areas at this time it does seem as though the Muslims of Bihar tended to side with the British, where the rebels/revolutionaries were mainly from the Hindu (Rajput) ranks; *see* **FSUP**, Vol IV.

CHARKARI, Raja of   Charkari, or Chirkari was a small sanad state in Bundelkhand, whose ruler Ratan Singh, assisted the British to the best of his power, but many of his own people and the countryside about him favoured the rebels. He assisted British fugitives from **Nowgong** etc, but was punished in January 1858 when his fort was beleaguered and the town plundered and burnt by the forces of **Tatya Tope**, aided by Despath. Before that the raja had attempted to conquer Jhansi but had been repulsed by the forces of the Rani and compelled to lift the investment. Tatya Tope apparently 'fined' the raja a total of 3 lakhs of rupees which were paid to the rebels in return for their leaving his territory. They did not do so, but continued to destroy or loot all the palaces and property of the raja, and compelled his surrender. This incident was the source of a controversy, not to say scandal, when afterwards it was revealed that the Commander-in-Chief, **Sir Colin Campbell**, had ordered **Sir Hugh Rose** to go to the aid of the loyal raja of Charkari but Sir Hugh, supported by **Sir Robert Hamilton** had refused to do so as this would have involved lifting his investment of Jhansi. The raja however remained loyal to the British, and his forces obtained some success against the rebels. After the restoration of British authority the raja was rewarded with a **jagir** of Rs20,000 *p.a.* in perpetuity, the privilege of adoption of an heir, a dress of honour and a salute of 11 guns. For full details of these events *see* **FSUP** Vol III; for details of Rose's refusal to obey orders, *see* T.Rice Holmes, *A History of the Indian Mutiny* pp 508-9.

CHASING NANA SAHIB   'In the *'Cornhill'* (magazine) this month Dr. Fitchet gives the following letter from General Harris: 'I knew the **Nana** quite well, having been introduced to him at **Cawnpore** as far back as '51. When the Mutiny broke out, I was with my regiment at Sabathu, and marched down to Delhi with them. The first fight was at **Badli-Ki-Serai** on the 8th June, & I was on the Ridge &c. &c. and was badly wounded and left for Sabathu again, apparently a hopeless cripple, but got all right again, and in 1858 was appointed adjutant of 18th Musbee Sikhs, and marched with them to **Lucknow**, arriving early in October. In November [1858] I was ordered with a detachment of three companies, first to Byram Ghat and then to a ford on the upper Gogra, called Chilari Ghat. A small party of seventeen Royal Engineers, under Richard Harrison, presently joined me, with orders to construct a bridge, and a regiment of Pioneers, unarmed, with a lieutenant in command, to help. I commanded the whole. Now these Musbees have never been enlisted by Goverment before. Low caste men, all or almost all expert thieves, & treacherous. I had a guard of them always at the ford, and lived myself in a tent close by. Now this ford was only about thirty miles from the **Terai**, into which the rebel forces, with the Nana, had been driven. Very shortly I found that, through my native officers, I was thoroughly posted up in all the Nana's movements. There was, as you know, a lakh of rupees reward for him, dead or alive. Two of my Subadars were always at me to allow them three or four days' leave to capture him. They kept me informed of his movements like a court circular. I always told them that I was on duty for a certain purpose, and it was impossible I could give any man leave. One Thursday Ram Singh came to me begging me still more strongly than before, saying the Nana was getting much worse, he was as you know suffering from the fever and ague and had an enlarged spleen, and he told me that the Nana had had his little finger cut off, and had burnt it as an offering to Kali, with a view of propitiating the goddess. Two days after this Ram Singh & the other Subadar came and said: 'No one will get the reward now! He died and was burnt yesterday.' And I feel quite sure it was true, for I had known for some weeks all about his movements'. Copy of a letter from General Harris, originally printed in *The Pioneer* in 1907.

CHATHAM This town on the Medway in Kent was for many years the headquarters of the Royal Engineers, a corps which played a crucial role in the suppression of the mutiny/rebellion. In mid-nineteenth century warfare it was always assumed that the battle plan, whether for the attack upon a besieged city - like Delhi - or the final breach of the enemy's defences, - as in Lucknow - would be the responsibility of the Engineers. They were therefore regarded as an elite corps, and a commission in the R.E. was much coveted. Chatham houses the Institution of Royal Engineers which holds a number of original papers concerned particularly with the work of **Colonel Richard Baird-Smith**, and **Captain Alexander Taylor** during the siege of Delhi in 1857.

The CHATTAR MANZIL photographed in 1858

CHATTAR MANZIL, The one time palace of the rulers of Awadh. Now houses the Central Drug Research Institute. The nucleus was the Kothi Farhat Baksh built by Claude Martin in 1781. It was a stronghold of the revolutionaries during the struggle of 1857. Courtesy of INTACH.

CHATTOPADHYAY, Haraprasad An eminent Indian historian of the period who argues that there would have been no popular revolt if no military mutiny first; moreover as it was regional in scope it cannot be called a national or freedom movement. *See* H.Chattopadhyay, *The Sepoy Mutiny 1857, a Social Study and Analysis,* Calcutta, 1957.

CHATTOPADHYAYA, H.B. Author of two significant articles *'Mutiny in Bihar'* and *'The Sepoy Army, its strength, composition and recruitment'*. *See Bengal Past and Present,* vols 74-5, 1955-6, and *The Calcutta Review,* May July September 1956, respectively.

CHAUBE, RAMGHARIB He collected some folk-songs from the 1857 period, and sent them for publication to William Crooke. *See The Indian Antiquary,* April and June 1911,

CHAUDHURI, Sashi Bhusan A scholar and historian of the rebellion, particularly useful for consideration of the role of non-military. *See* Bibliography. This historian emphasises the widespread nature of the disturbances, and the general passive support of the populace for the rebel forces. He is especially concerned to combat the interpretation of 1857 as the last desperate effort of the old order to maintain its privileges: the landlords who led the revolt in many areas were, he argues, unconscious tools of a nascent nationalism. *See* S.B.Chaudhuri, *Civil Rebellion in the Indian Mutinies,* Calcutta, 1957 *and, Theories of the Indian Mutiny 1857-59,* Calcutta, 1965.

CHAULAKHI KOTHI A palace in Lucknow closely associated with the name of the Begum **Hazrat Mahal**, who is said to have taken many vital decisions regarding the independence of Awadh in this kothi. So called becaue Wajid Ali Shah bought it from the royal barber Azimullah for four lakhs. Now houses two schools and a godown. Courtesy INTACH.

CHEDA This man's deposition gives a possible identification to the 'bibi' of the **Bibigurh** : 'From the day that the ladies were brought to *Mrs Batten's* house, up to the time of their massacre, I remained with them, and during this period none of the ladies ever came outside of the house, as they were not permitted to do so. I did not hear of, or see, any lady being taken away from there. Besides the four gentlemen above alluded to, no others were in confinement with the ladies.' *See Deposition of Cheda.* **Williams.**

CHHATAR SINGH A leading rebel in Bundelkhand who continued resistance long after it had ceased in other areas. For details *See The Revolt in Central India*, Simla, 1908.

CHICHESTER The County town of West Sussex. The regimental museum of the Royal Sussex Regiment (H.M.'s 35th Regiment in 1857) is situated there. It has unpublished records of the service of the 35th, including the disastrous, for the British, attack upon **Jagdishpur** which is also described in Lieutenant **Richard Parsons** book, *A Story of Jugdespore*, London, 1904. There is also the unpublished record of the 3rd Bengal European Fusiliers (later a battalion of the Royal Sussex) who were stationed at Agra in 1857, played an undistinguished part there, including the engagement at **Sussiah[Sasia]**.

CHIMNA APA Chimna Apa, who had hired **Nanak Chand** to represent him in his suit against the Nana, reported that as he and Ganesh Shastri were riding in an *ekka* the night before they came upon the **Nana** and his escort hauling artillery into the city at **Hillersdon**'s request. The Nana's men attacked them, stole a valuable sword, and ran them off into a ravine. Apa's escort was beaten, but he survived, and heard the Nana's men declare, as they rode off, that British rule would last only 'a few days longer.' Nanak Chand resolved to inform Hillersdon. From the Diary of **Nanak Chand**, quoted in Sir G.Trevelyan's *Cawnpore*, London, 1865, reprinted Delhi, 1992.

CHINHAT (or CHINHUT) In the Pargana and Tehsil of Lucknow, a village on the metalled road from Lucknow to Faizabad, at a distance of about six miles from the former: it was the scene of a decisive battle between the British led in person by **Sir Henry Lawrence**, and the mutineers, on 30 June 1857. Strictly speaking the battle actually took place at Ishmaelganj, a hamlet close to Chinhat. Lawrence himself sent a message to **Colonel Neill** at Allahabad for onward transmission to the Governor General: 'Lucknow 30th June. Went out this morning eight miles to meet the enemy, and were defeated through misconduct, chiefly of artillery and cavalry, many of whom deserted. Enemy followed us up, and we have been besieged for four hours. Shall likely be surrounded to-night. Enemy very bold and our Europeans very low...unless we are relieved in fifteen or twenty days, we shall hardly be able to maintain our ground.' *See* **FQP** p 28 *et seq. See also* Kamal-ud-Din Haidar Husaini 'Tarikh-i-Awadh or Qaisar-ut-Tawarikh', Vol II pp 212-218, quoted in **FSUP**, Vol II, p 52 *et seq. See also* Forrest, 'State Papers' Vol II pp37-8; and Gubbins 'An Account of the Mutines in Oudh etc' pp 181 *et seq*.

CHIRKI The 'home town' of **Tatya Tope**, and the place to which he retired after the defeat by the British at **Kalpi**, on 23 May 1858. It is a small town four miles from Jalaun in Central India. The rebels had been expected by the British to cross the Jumna[Yamuna] and join the others still holding out in **Oude**[Awadh], but they instead turned to the west, and it is said by the inspiration of the **Rani of Jhansi**, decided to attack **Gwalior**. For details of the Gwalior Campaign *see The Revolt in Central India*, Simla 1908 p 147 *et seq*.

CHITTAGONG A town in Eastern Bengal which was the scene of a mutiny of a detachment of the 34th BNI on 18 November 1857. They plundered the Treasury, released the prisoners from gaol, burnt down their own lines, fired the magazine, and left the station, carrying off with them three elephants, the property of Government, and the whole of the treasure they found in the collectorate, with the exception of three hundred and forty rupees in cash. They attacked none of the Europeans, and the only man who suffered at their hands was a native gaoler who protested against their proceedings. Him they killed. They then headed northwestwards, avoiding British territory. *See* Colonel G.B. Malleson, *History of the Indian Mutiny 1857-58*, Vol II London, 1879, p 419.

CHOLERA Although known in Europe where there had been a number of recent epidemics, this disease was far from understood by the British in India, among whom there were more deaths from this disease than from casualties in battle. It struck often without warning and the victim could be hale and hearty at breakfast time but dead and buried by the evening. Treatment was often bizarre: For cholera victims the British prescribed opium and calomel, or, in more extreme cases, 'alcoholic or ammoniated stimulants.' All else failing, 'acetate of lead, diluted sulphuric acid, preparations of kino or catechu, gallic acid' were prescribed. Of the one treatment that might have done any good - 'saline injections into the veins' - one leading authority declared that current research was 'conclusive against them.' Moorehead. *Researches on Disease in India*, pp 223-31.

CHOBEYPUR A Town on the Grand Trunk Road, on the right bank of the Ganges, near **Bithur**, where on 7 June 1857 a movable column of men of the 48th BNI and the 7th Bengal Cavalry mutinied and murdered their officers. Lieutenant **Boulton** escaped and rode to **Kanpur**, he was killed a few days' later. *See* John Bonham, *Oudh in 1857, some memories of the Indian Mutiny*, London, 1928.

CHOTANAGPUR in 1857 'The extensive mountainous plateau that lies between Southern Bihar, Orissa, West Bengal and the Central Provinces has been called Chotanagpur. Following Captain Wilkinson's proposals in the 1830s, the Chotanagpur Agency was consituted for four districts, namely Manbhum, Lohardagga, Hasaribagh and Singhbum, with military stations at Hasaribagh, Ranchi, Chaibasa and Purulia. This entire area, inhabited by the tribals - Santhals, Mundas, Bhuinyas, Kols, Bhogtas, among others - was drawn into the widespread insurrection of 1857. The very battalions which crushed a mighty tribal revolt like the one of the Santhals between 1855 and 1857, began to be defiant as soon as news of the rebellion of the 8th and 7th Bengal Native Infantry reached their detachments posted at Hasaribagh, on 27 July 1857. On 30 July, in a tense situation, the sepoys broke into open insurrection in their lines; they arrested, 'maltreated' and even killed loyal Indian soldiers of senior ranks - *jemadars* and *daroghas* - and broke open jails. The British officials fled the station and the sepoys continued their attacks on the treasury, the *cutchery*, the bungalows and the prison workshops. Thereafter they proceeded towards Ranchi. Meanwhile, a detachment of the **Ramgarh Regiment** under Lieutenant Graham, sent to disarm the battalions at Hasaribagh, defied orders. Under the leadership of Jemadar Madha Singh and Subedar Nadir Ali Khan, they went back to Ranchi on 1 August. On their way these men were joined by their compatriots of Hasaribagh, at Benno, about twenty miles north of Ranchi. At their approach, the British officials left the town and Jai Mangal Pandey, the son of a Brahmin priest who had been in service since 1818 welcomed the rebel army with the whole of his battalion at Doranda. The entire region thereby fell into the hands of the insurgents. Thakur Bishwanath Sahi of Burkogarh (Hatu) and Pandey Ganpat Rai of Bhowrao, two principal zamindars of Lohardagga, came forward in support of the rebels. Bishwanath Sai came to Ranchi and began to hold courts in Captain Natron's bungalow. The sepoys threatened the Grand Trunk Road, as well as the power of Sambhunath Singh, the Raja of Ramgarh, loyal to the British. The whole edifice of the Wilkinson system, based on chieftain-military alliance was now at stake. On 4 August the proclamation of padshahi Raja was circulated throughout the district. However, while Bishwanath was interested in crushing the powers of the Raja of Ramgarh and consolidating his own instead in the district, other leaders like Madho Singh insisted on joining Kuar Singh (Kunwar Singh)'s forces at Rahatgarh. Madho Singh finally persuaded most of the soldiers to move out of Ranchi, leaving Bishwanath behind. They marched through Palamau. Colonel E.T.Dalton, the Commissioner, with the help of Jagatdeo Singh of Pithaurea and the Raja of Ramgarh recaptured Hasaribagh and returned to Ranchi on 22 September. Major English intercepted the moving columns of insurgents at Chatra with a Sikh detachment on 2 October, and completely defeated and dispersed them. Seventy-seven of the insurgents were buried in one pit, the British lost fifty-six of their soldiers. The victory of the British army at Chatra, however, merely suppressed the insurrection of the indigenous infantry at Hasaribagh and Ranchi. The local politics and a deep antipathy towards the Company's rule, as well as the intransigence of the rebels, from the headman down to the cultivator, made the insurrection much more widespread in Singhbhum, Palamau and Sambalpur. In the region of Singh-bhum the local politics revolved around the rivalry between two houses, that of the Raja of Serarkella and the Raja of Parahat. On hearing of the mutiny at Ranchi and Hasaribagh, Captain Sissmore left Chaibasa; Arjun Singh, the Raja of Parahat was hesitant, while the Raja of Serarkella was openly loyal to the Company. The detachment of the Ramgarh Battalion in the meanwhile was in a state of agitation. Juggo Dewan, Raghu Deo and Baijnath Singh formed a 'war party' to propagate the urgency of the revolt against the Company's rule. However the detachment at Chaibasa revolted as late as 3 September 1857. They were surrounded and stopped on their way by the army of Arjun Singh and brought to Chakradharpur and then handed over to the authorities. Despite his profession of loyalty, however, the Commissioner asked him to appear before Birch for trial, declared Juggo Dewan as a proclaimed offender and upheld the claims of the Raja of Serarkella. Meanwhile, the *Mankis* (headmen) of the *Peers* (territorial divisions of Singhbhum) and *Killis* (clans) assembled on their own; the spies were murdered, the house and the Bazar of the thakur of Kera, an English loyalist, was attacked and plundered; an arrow, a traditional way of transmitting messages during rebellion, was said to have circulated among a large number of Kols. Arjun Singh was proclaimed a ruler. The Kols soon attacked and destroyed the Jyuntgurh police station and people inhabiting the numerous *Peers* in south Kolhan took up arms, in December 1857. All the administrative regulations like the ban on witch hunts had been systematically violated in the areas. They drove the Raja of Serarkella from Chakradharpur. A large body of Kol soldiers numbering about three to four thousand ambushed General Lushington near the Mogra River on 14 January 1858, and inflicted a defeat on the British army. The Shekhawati battalion under Colonel Foster devastated Porahat and defeated the rebels in an encounter at Srengsella Pass; but Arjun Singh evaded arrest and his chiefs often made sudden attacks on the English troops. Finally due to the intervention of the Raja of Mayuebhanj, Arjun Singh submitted in 1859; but the local chiefs continued their offensives and 'incendiarism' till as late as the middle of 1861. The civil upsurge at Palamau was as widespread as it was among the Kols in Singhbhum. The revolt began at the time of the march of the sepoys through the Palamau. Nilambar Sahi and Pitambar Sahi were the *ilaqadars* of village Chemu. Their father died as an outlaw, losing proprietary claims on his estate; since then , they had old scores to settle. They organised the tribes of the area, the Cheras and the Bhotgas. They launched a movement against the loyal Rajput jagirdars, attacked *thanas*, the offices of the coal-mines and various *ganj* (market place). Many local *ilaqadars* supported them. By the end of November, E.T.Graham himself had been besieged by the insurgents. The rebels captured the Palamau fort. In the month of January, Captain Dalton himself led the military operations, reduced the fort and burnt the home village of Nilambar and Pitambar. Finally with the withdrawal of the Chero zamindars from the insurrection, the rebellion subsided. The insurection in the Sambalpur district had an old root, almost 'a continuation of civil insurrection' of the pre-1857 period. Among the prisoners who were forcibly released from the Hasaribagh jail on 30 August 1857 were two brothers, Surendra Sai and Udwant Sai, along with their son Mitrabhamu Sai, the claimants to the throne of Sambalpur. They had earlier led a violent opposition against the British settlement in 1839 and had been arrested. During the present insurrection, they returned to Sambalpur to stake their claim. The local zamindars with their *paiks* assembled around them. Together, they stopped revenue payments, ambushed British troops and burnt down *dak* houses. The rebels also adopted long drawn out hit and run tactics. The resistance continued until 1861. Finally diplomacy, conciliatory moves as well as clever manipulations broke the ranks of the insurgents and Surendra Sai was arrested on 23 January 1864. The Santals in Manbhum were restive when sepoys of the Ramgarh Battalion met them at Purulia on 5 August. Following their usual practice of incendiarism and despoliation they marched on Ranchi; following them the Santals rebelled, and attacked the zamindars. Raja Nilmani Singh of Pachet estate had been arrested on the charge of being privy to the rebellion. The counter-insurgency operations of the British had all the aspects of diplomacy, conciliation and ruthlessness. When persons like Arjun Singh or Surenra Sahi remained elusive, diplomatic pressures as well as the tactics of offering rewards to the relatives were adopted; at the same time hardened and confirmed insurgents like Jugga Dewan and Bishwanath Sahai were publicly hanged; over two hndred sepoys were also sentenced to death, many were deported for life. The power of the State could be restored only through violence and a bloodbath. The story of the insurrection of 1857 encapsulates many narratives and many tales. In Chotnagpur, the repressive arm of the state reacted against its master; a few zamindars joined but their linkage with soldiers and the people of the surrounding areas were less strong. At the initial stage, Ranchi became the focal point for the soldiers' movement as well as for the other insurrectionists in Bihar. In areas like Singhbhum, the Kols on their own virtually arrested and disarmed the rebel soldiers. Here, the military rebellion was totally superseded by civil upsurges fuelled by local dynastic politics as well as the resistance of various Kol villages. For persons like Surendra Sai or Pitambar or Nilambar Sahi, the insurrection provided an opportunity when old wrongs in their perception could be undone; for them loyalties as well as clan linkages made mobilisation of people possible. On a much lower level, a Santhal like Rupu Majhi and a Kol like Gonoo had their tales to relate: their personal experiences, their anger against their local zamindar and the mahajun and their eagerness to demand 'justness' as well as vengeance - all these went beyond the restoration of the old order. It was through rebellion alone that many of them expressed their identity. Only at that moment of insurrection was their leadership accepted, their voices heard before they relapsed into silence for years to come.' **Dr GAUTAM BHADRA, Reader in History, Calcutta University.** *See also* Gautam Bhadra, *Four Rebels of 1857*, in *Subaltern Studies*, Vol IV. ed. Ranajit Guha, Delhi, 1985; Purushottam Kumar, *Mutinies and Rebellions in Chotanagpur, 1831-1857*, 1991; Nikil Sur, *Chotanagpur and the Rising of 1857*, Bengal Past and Present, Jan-December 1986.

CHOWKIDAR A village watchman or guard. Often the nearest thing to a policeman in the average village: paid for by the villagers themselves. *Also*, the name given to the magazine of **BACSA**, the *British Association for Cemeteries in South Asia*.

CHRISTIAN, J.G. The Commissioner of Khairabad Division in **Oude[Awadh]**, based on **Sitapur**. He thought he could contain the situation even if the 41st BNI mutinied because he had faith in two Irregular Infantry regiments and the Military Police. He was wrong. 'Christian, finding all was up, walked calmly down to the river, preceded by his wife with an infant in her arms, and her little daughter Sophy, a child of three years old, by her side. They had not gone far when Christian fell, riddled with bullets. His poor wife, heedless of her own safety, sat down beside his lifeless body, crying in her despair, ' Save my child; who will save my child?' Burns, who was one of the last to leave the house, was passing at the moment, and hearing the poor mother's cries, picked up little Sophy and carried her off. Immediately after, Mrs Christian and her infant were both killed. At the river, Burns met Sergeant-Major Morton, who was mounted, and he made the child over to him.' A succinct account of what happened may be found in John Bonham's *Oude in 1857, Some memories of the Indian Mutiny,* London 1928. *See also* CAPTIVES OF OUDE.

CHRISTIANS The majority of the troops who mutinied believed that in some way the English were threatening their religion, whether they were Hindu or Muslim. This is manifested by the bitter hatred which they showed at all times for Christians, be they Europeans or native converts. Many of the latter were pursued and killed for no other reason but that they were Christians, and likely therefore, it was thought, to throw in their lot with the English. The term was used as the greatest of insults. For example, Narain Rao (of Kanpur) and Hurdeo Baksh (talukdar of Oude[Awadh]) were both addressed, and threatened by the sepoys, as 'Christians', though neither was anything of the sort. *See also* SCRIPTURE.

CHRISTIE 'On the 1st and 2nd of June a cashiered officer called Christie came out of his bungalow (in Kanpur) in a state of intoxication at night, and fired at the cavalry patrol who had challenged him. Although he missed the sowar, yet the latter lodged a complaint in the morning, and a court-martial was assembled which acquitted Mr Christie on the ground of his having been intoxicated at the time. The sepoys became displeased at this and began to talk again of mutiny.' *Deposition of Lalla Bhudree Nath,* **Williams.**

CHUKLA A combination of villages, put together for the purpose of collecting the land revenue.

CHUKLADAR One who was appointed, or inherited, the farming of the land revenue in a district or **Chukla** of the Kingdom of Oude[Awadh].

CHUMAR A member of the skinner or shoemaker caste, which because of the material in which he dealt, was assumed to be of low social status.

CHUNDER, Bolanauth He writes in an (attractive) flowery style and includes matter relating to the struggle as, for example when he visits Kanpur he seeks to see all the sites of the Massacres etc '..But nobody could point us to the whereabouts of the well into which the unhappy Miss Wheeler had flung....and he would fain believe her to have put an end to her life, that had before it the dreary prospect of a lifelong ignominy' etc. *See* Bolanauth Chunder, *Travels of a Hindu* (2 vols) London, 1869.

CHUNDI LAL Zemindar of Moraon (Baiswara). Remained loyal to the British throughout the campaign in Oude[Awadh] and was rewarded after the final assault on Lucknow, in what was known as the 'Oude Proclamation', by the Governor General, issued on 14 March 1858: '...The first care of the Governor General will be to reward those who have been steadfast in their allegiance at a time when the authority of Government was partially overborne, and who have proved this by the support and assistance which they have given to British officers...declares that...are henceforth the sole hereditary proprietors of the lands they held when Oude came under British rule, subject only to such moderate assessment as may be imposed upon them; and that these loyal men will be further rewarded in such manner and to such an extent as upon consideration of their merits and their position, the Governor General shall determine.'

CHUPPRASSEE [Chaprassi] One in possession of the brass badge (hence the name) worn by government messengers/peons. Not usually armed, but could be used, in numbers, in support of officials.

CHURCHER, David G. An indigo planter, and one of the two European survivors, of the uprising at **Fatehgarh**, he gives a first hand account of the mutiny and siege of the Fort there. Member of a family of indigo planters of Fatehgarh, *cf* Thomas and Emery Churcher. *See* David G.Churcher, *Episodes of the Indian Mutiny,* Blackwoods Magazine May 1900.

CHURCHER, DAVID 'Testimony: I escaped after the boat catastrophe to the village of Kurbar. Not being secure I was obliged to remain in sugar cane and other fields, and I was not secure because the *zemeendars* of Kurbar were not strong enough openly to protect me from the Nuwab Raees' sowars They came to the village more than once to get the revenue money, and the Nuwab's people told the *zemeendars* of Kurbar that if (Mss. torn) and myself (Mss. torn) they would be handsomely rewarded. I have heard of no writing coming from the Nuwab's but the *zemeendars* told me *Hurkarahs* had come from the Nuwab. *Question by the Court,* At that time did...(Mss. torn) of whom did the villagers talk as the ruling power in Furruckabad ? *Answer* Of the Nuwab. *Question* Did they ever talk in such a way as to lead you to suppose that any other person shared the authority with the Nuwab? *Answer*—The *Soobahdar* of the 41st was exercising influence as well as the Nuwab: this was talked of. Re-examined on oath this 17th February- 1859. *Question* You mentioned yesterday that you escaped from the boat on the Manpoor Khutree on the (4th) fourth July 1857. Mention the circumstances attending that event. *Answer* When our boat had come opposite Manpoor it stuck on a sandbank. We left the fort at two A.M. on fourth July in three boats. We continued going on till we came opposite Manpoor. All the time we were going down, *the people on both banks continued firing at us.* The boat came aground opposite Manpoor about sixteen miles from this. We saw a boat full of armed sepoys crossing a little above Singheerampoor. This boat came on the Oudh side of the river where our boat was stuck, stopped there and at a distance of 50 yards commenced firing at us, the villagers also kept firing from the Oudh side. By this time many of our party being wounded the sepoys brought their boat closer to us. We having returned their fire. At last seeing we had no chance with them we were obliged to leave our boat. They came and joined their boat with ours. There were about forty men, women and children in the boat. In these thirteen were men. Those that were killed of course were left in the boat, the rest threw themselves overboard into the river; this was about five in the evening. Of those who jumped overboard all were drowned, except myself, Mr. Jones, Major Robertson, and Revd. Mr. Fisher. I escaped to Kurbar, I swam down the river with the assistance of an oar and got to Kurbar about midnight when the *zumeendars received me.* Major Robertson was wounded and could not go on, so I stopped there. I saw many killed. Mr. Sutherland was wounded before me, my brother was killed in the boat too, and Lieutenant Simpson also. I saw Miss Thompson cut down, she being in the water at the time. Mr. Fisher and Mr. Jones were wounded too, and Mrs. Fisher threw herself overboard. *Question* Who were the attacking party ? *Answer* There were some Mahomedans with them as well as rebel sepoys, and Singheerampoor was full of Mahomedans and sepoys. They had three guns placed there, and I saw them fire with cannon on Colonel Smith's boat. The attackers on the Oudh side were chiefly villagers. *Question* You know whom the guns belonged to? *Answer* No, and I cannot say who commanded the attacking party in the boat. I cannot say whether the attack was Nuwab's doing or not. Mungal Singh leader of the villagers on the Oudh side was wounded. *Question* Do you consider all the people in the boat with you were drowned or killed except those taken back prisoner to the Nuwab, and yourself? *Answer* Yes, I consider they were all killed or drowned with the exception of those I have named and those who were taken prisoner. *Question by the Court* What do you mean by 'those who were taken prisoners to the Nuwab'? *Answer* I heard, after I had made my escape that Mrs. Sutherland, Miss Sutherland, Mrs. Jones and her mother and a little girl of Mrs. Jones were brought away to the Nuwab, it was a common report amongst all the natives. I have never seen them since. All I know about their going back to the Nuwab is what I heard. *Cross-examined by counsel for the prisoner.* *Question* If the sepoys had remained staunch would the attack on the boats have been made? *Answer* If the 10th Regiment had remained staunch there would have been no massacre.' *Testimony of David Churcher at Trial of Nawab of Farrukhabad* in February, 1859, Farrukhabad Collectorate Mutiny Basta in **FSUP**, Vol V. pp 924-6.

CIVIL REBELLION

For details of *emeutes* in specific towns or areas *see* relevant named entries. But for an overview of the whole question of the social background to the rebellion and in particular the part played by the peasantry, the works of three historians are most significant: *see* Eric Stokes, *The Peasant Armed, The Indian Revolt of 1857,* Oxford 1986; Ranajit Guha, *Elementary Aspects of Peasant Insurgency in Colonial India,* Delhi, 1983. Reprinted, Delhi, 1992., and Sashi B.Chaudhuri, *Civil Rebellion in the Indian Mutinies,* Calcutta, 1957.

CLERICAL ADVICE

This story dated from June 1857, in Lucknow, before the storm broke there: '...In the night, on the housetop, [Major Banks] had asked me, as a Clergyman, what I should advise him to do, in case of its being certain that his wife would fall into the hands of the rebels, and that they would treat her as they had done the women at Delhi and Meerut. It was a difficult question: but I told him that, if I were certain that my wife would be so treated, I would shoot her rather than let her fall into their hands. (Colonel Inglis afterwards asked me whether I thought his wife would be justified in killing her own children, rather than let them be murdered by the natives. I said, no; for children could be but killed; whereas we had been told that at Delhi young delicate ladies had been dragged through the streets, violated by man, and then murdered.). God forgive me, if I gave the wrong advice! but I was excited; and I know that I should have killed Emmie [his wife], rather than have allowed her to be thus dishonoured and tortured by these bloodthirsty savage idolators.' Rev. Henry Polehampton. *A Memoir Letters and Diary,* London, 1858. p 270-1.

Old Fort at Muttra 1857

CLIFFORD, Richard A member of the Bengal Civil Service 1853-78. Throughout the period of the mutiny/rebellion he kept a diary at his station, **Muttra**, which includes a full narrative of events in and around that town. Two typescript versions of the diaries are preserved in the Oriental and India Office collections of the British Library, *see MSS.Eur.D.568 & 720.*

Lutkun Darwaza, The Clocktower

CLOCKTOWER

A clocktower stood just outside the **Baillie Guard Gate** at the **Lucknow Residency**. It was an unusual structure in that there was no clock in it, simply a painted clock face. It was in rebel hands during the siege, and housed a particularly effective sniper: 'One of the late King's sharpshooters, an African eunuch, had made this clocktower his favourite haunt. Like his friend, '**Bob the Nailer**', his shots were almost always fatal. So troublesome had this man become to us, and we had lost so many good soldiers by his rifle, that Capt Thomas (Madras Artillery) had been directed to shell the fellow's place of refuge, merely to enable us to get rid of this one man. The shells had been thrown with beautiful precision, and we had seen them burst just where they should have burst; but immediately afterwards a rifle bullet whizzing through the air as if in defiance of our efforts, proved that we had not succeeded in killing the marksman. Nor could we solve this riddle till after the capture of that gateway; then we learnt how he had escaped so often. The Residency was perfectly commanded by the clocktower, and the eunuch, as soon as he had observed through his telescope that we were about to shell, retired by a ladder into a sort of cavern that he had caused to be scooped out for his safety, and, returning at once to the scene of action, recommenced his firing as soon as the shell had burst. He was killed by our men at last and the telescope and rifle were found by his side.'

COGHILL PAPERS   These are held in the Centre of South Asian Studies at Cambridge University, and include material dating from the time of the mutiny/rebellion. Consists for the most part of letters to his brother and sisters in England from Major Kendal Coghill, 19th Hussars. Coghill took part in the taking of **Delhi** in September 1857, and had charge of **Bahadur Shah Zafar** after his capture. Displays a very brutal attitude to Indians, and is annoyed by the limitations set on prize money and the forbidding of personal looting. Mentions Sir **Theophilus Metcalfe**'s attitude towards the population of Delhi (exceptionally vengeful and bloodthirsty), and clearly approves of it. The papers include sixteen short stories, partly autobiographical and partly concerning the Mutiny, presumably written by Kendall Coghill, but the most valuable contents concern the actions fought at or near Delhi by the **Delhi Field Force** from 30 May to 20 September 1857 (a full list is given), including analysis of the Siege of Delhi.

COLLECTOR   The Collector-Magistrate of a District was the representative of the Government in a region of perhaps 3,000 square miles, with an average population of, say, a million of human beings. He was 'not only the head of the correctional tribunals of the whole district and of the police, he was also responsible for the collection of the dues and taxes by which it was replenished, and for the periodical transmission of the contents to headquarters. He had even to inspect the public dispensaries, to direct the rude municipal management of large towns and ...to inspect and stimulate the national schools.' see H.G.Keene. *A Servant of 'John Company,* London, 1897, pp 101-2.

COLLEGE GARDEN   This story comes from the pen of one of the officers (Colonel William Cra'ster) of the Delhi Field Force. It refers to events immediately following on the assault on Delhi (14 September 1857). 'We were then sent to the College Gardens, a large enclosure filled with trees and situated opposite to the Magazine at about 60yards distance. Here are placed an 18 pdr in the gateway to batter the magazine. The fire was very heavy both musketry and guns, and we had many men wounded. The gateway was furnished with *a massive folding gate which we opened at every shot and closed very speedily after*...On the evening of the 15th we had made a breach, which was successfully assaulted by the 61st the next morning, and the magazine was taken with slight loss...After the magazine was taken I was sent with two 5½ inch brass mortars to the Cabaul Gate to shell the Berm Bastion. On getting there I reported myself to Colonel Jones of the 61st but did not receive any very lucid instructions from him as to their disposal. *He gave me what was perhaps more to the purpose viz some very good port wine*, and told me how General Nicholson had met his death...I was aided in working the mortars by some Sikh gunners, but most of them were old men and all very raw hands, they preferred stitting down and looking on. A young Sikh belonging to an infantry regiment was much more efficient, as he cut fuses and loaded them like clockwork.' *From* Archive *MD/616* in the **Royal Artillery Institution**, Woolwich Arsenal, S.E.London.

COLLIER, Richard   An author who has attempted to recreate in dramatic language the actual events and the atmosphere of a selected number of major incidents in the course of the mutiny/rebellion, especially connected with Delhi, Kanpur and Lucknow. The result is not fiction but sometimes reads as though it were. He claims that he never intended to write a definitive history of the Indian Mutiny, since three Indian scholars, **Dr S.N.Sen**, **Dr R.C.Majumdar** and **Dr S.B.Chauduri** have all in recent years added to the earlier synoptic accounts of **Forrest**, **Kaye**, **Malleson** and **Thos Rice Holmes**. See Richard Collier, *The Sound of Fury,* London, 1963.

COLLINS, Dr Francis   An Assistant Surgeon serving with the 5th (Northumberland) Fusiliers. He reached Calcutta in July 1857 and was present with his regiment at the first and second reliefs of Lucknow. Few of the officers of this regiment survived the campaign, with the result that although the Fusiliers took an active part in the actions of this period there are few 'reminiscences' or 'memoirs' commemmorating that fact. To some extent Dr Collins fills that gap with his letters and a lengthy contribution to his regiment's magazine. The letters are preserved in the Oriental and India Office collections of the British Library, *see Photo.Eur.59. See also St George's Gazette,* Alnwick Castle, 31 May 1893 *et seq.* and letters from **Danapur and Lucknow** owned by Commander F.Collins RN of Winchester Hants.

COLONIAL TERMINOLOGY   Just as one man's 'terrorist' is another man's 'freedom fighter', so some terms in common use in the literature of the Mutiny/Rebellion have come to have automatic counter interpretations, in what **Ranajit Guha** calls the two mutually contradictory perceptions of the elite mentality and the subaltern mentality. Thus while an official document speaks of 'badmashes' as participants in rural disturbance, this can also be taken to mean 'peasants involved in a militant agrarian struggle'. A 'dacoit village' may mean that the entire population of a village is united in resistance to the armed forces of the state; similarly 'contagion' is perhaps 'the enthusiasm and solidarity generated by an uprising among various rural groups within a region'; 'fanatics' are but 'rebels inspired by some kind of revivalist or puritanical doctrine'; 'lawlessness' is no more than 'the defiance by the people of what they had come to regard as bad laws'. *See* Ranajit Guha, *Elementary Aspects of Peasant Insurgency in Colonial India,* Delhi, 1983, reprinted as paperback, Delhi, 1992. At the same time we can argue that it is altogether too far-fetched to convert 'scum of the bazars' (on the assumption that we all agree that bazars can contain some 'scum') *automatically* into 'urban freedom-fighters'. But the lesson is clear: it is desirable, indeed essential, to remember that language frequently reflects attitude, and the objective student of the mutiny/rebellion must be on guard against prejudice and prejudgement however subtly conveyed. **PJOT**

COLOURS *See* REGIMENTAL COLOURS.

COLVIN, JOHN RUSSELL (1807-57) Official in the Company's service. Lieutenant Governor of the N.W.P. from 1853. Aroused the displeasure of the government in Calcutta by his apparent dissent from official policy towards the mutineers. Died on 9 September 1857 - from natural causes - and was buried within the Fort at Agra. *See* **DNB**.

COLVIN, Sir Auckland Son of **John Russell Colvin,** whose biography he wrote, he was anxious to dispel the adverse publicity which attended the actions of the Europeans, led by J.R.Colvin the Lieutenant Governor, in **Agra** in 1857. The authorities in Agra were heavily criticised by **Lord Roberts** the subsequent Commander-in-Chief who served in **Delhi** and **Lucknow** at this time as a subaltern and won the **Victoria Cross** at **Khudaganj**. His Chapter XXI is summarised as 'Infatuation of the Authorities at Agra - a series of Mishaps - Result of indecision and incapacity.' *See* Sir A.Colvin, *Agra in 1857: a Reply to Chapter 21 of Lord Roberts 41 Years reminiscences in India.,*in *Nineteenth Century Magazine,* April 1897; *see also* FM Earl Roberts, *Forty-one Years in India,* London, 1897; *and* Sir A.Colvin, *Life of John Russell Colvin,* Oxford, 1895.

COLVIN'S PROCLAMATION This was the proclamation issued by the Lt Governor, **John Russell Colvin** on 25 May 1857 which was 'universally approved at Agra' but about which the Governor General in Council took a different view: **Canning** insisted on its withdrawal. The text of the Proclamation runs as follows: *'Soldiers engaged in the late disturbances, who are desirous of going to their own homes, and who give up their arms at the nearest government civil or military post, and retire quietly, shall be permitted to do so unmolested. Many faithful soldiers have been driven into resistance to government only because they were in the ranks and could not escape from them, and because they really thought their feelings of religion and honour injured by the measures of government. This feeling was wholly a mistake; but it acted on men's minds. A proclamation of the governor general now issued is perfectly explicit, and will remove all doubts on these points. Every evil-minded instigator in the disturbance, and those guilty of heinous crimes against private persons, shall be punished. All those who appear in arms against the government after this notification is known shall be treated as open enemies.'* For the comments of Sir Charles Trevelyan on this matter *see* The *Times,* 25 December 1857.

J.R.Colvin

COMMANDER-in-CHIEF, On the death of **Anson,** the Governor General Lord **Canning** appointed as a temporary replacement as Commander-in-Chief, Sir Patrick Grant, from Madras. This caused immense pain to Major General **Sir Hugh Massy Wheeler** who was Grant's senior in the service by many years. Thus in a letter to **Sir Henry Lawrence,** Wheeler expresses his chagrin; dated 4 June 1857, and addressed to Lawrence at Lucknow. This was written only a matter of hours before the 2nd Cavalry broke out in mutiny and two days before the Nana opened fire on the entrenchment: 'My dear Lawrence, I have kept you well informed by E.T.(Electric Telegraph) of all going on here. You will see by a letter from Ponsonby, which I enclose, the state of Benares. In fact the evil progresses, and the want of Europeans is greatly felt. Trust in any of the Native Troops is now out of the question. The 2nd Cavalry which to my knowledge has been ill disposed for a long time, is now in an almost acknowledgeable state of Mutiny and ready to start at any moment for Delhi. At one time that Regt stood alone although disaffection to some extent existed in all, and it had intended to proceed by the south of the cantonment to the Treasury and make a dash, but it is now said that the 1st N.I. is sworn to join and they now speak of going off this night or the next, doing all the mischief in their power first, this to include an attack on our position. I fear them not altho' our means are very small- under 200 men of all arms and 69 old soldiers, Railway folk and Merchants. 6- 9 ps and 12 ps......heavier of your two guns. I have heard but know not how exactly, that a small party of the Madras Fusiliers are to be in today. Of course we can offer protection to nothing out of our intrenchment. It is to be feared that the other two Native Infantry Corps may be carried off by the excitement to join the others; we will do our best. The Govn Genl. has approved M.G.Reed Provincial Commander in Chief until the arrival of Lt General Sir Pat Grant who has been sent for to Madras. The former is my Senior and I can have no objection to his nomination. The latter is long, very long my junior, and altho' if it please God to grant me life, I will continue in the due performance of my duty to the best of my ability until tranquillity is restored, I shall then take the course which I feel due to my professional Character and soldierly feelings. Gen Grant is a Friend of my own, no doubt a good soldier, but he has not the experience nor the trials that I have had; his connection with Lord Gough has carried him over me on every occasion; but as long as we were not brought into contact, I neither complained, nor envied him. I can not serve under him and it is a poor return for above 52 years of zealous, and I may proudly add, successful service, to be thus superseded. My name with the Native Army has alone preserved tranquillity here up to the present time, and the difficulties which I have had to contend with can only be known by myself. Your late lamented Secretary witnessed some of them. I have been living in a Tent in this dreadful weather until yesterday. I have performed Subaltern duty in going the Rounds at midnight, because I felt that I gave confidence, that I saw all right with my own eyes, and that doing my own duty I had a right to exact it from others. I write with a crushed spirit, for I had no right to expect this treatment. But my dear Friend, whilst I wear the Cloth I will endeavour not to disgrace it. My service has been extraordinary; with the exception of my two years in Europe; when unemployed; I have had but three months' General leave and was never absent on medical certificate. I did 41 years Regimental duty, for I had no Friends and I have had nothing from Govt that it could withhold from me and the whole has been crowned by this act, placing me under the Orders of a Regimental Lt Colonel, my junior by more than 15 years and my junior as a General officer. Believe me very sincerely yours H.M.Wheeler. Ps We got in a lac of rupees (Rs) in the usual way with a Guard and 12 sepoys from the Treasury three or four miles distant. I have had thanks *ad nauseam* but there it stops.' Wheeler to Henry Lawrence, June 4, 1857, **BM** *Add Ms 39922, Folio 10.*

COMMUNAL FRICTION

Nearly all public pronouncements of the rebels (*see* PROCLAMATIONS) emphasized the unity of the Hindus and Muslims in resisting the British, and the implication is carried that it was only the Christian faith that was inimical to the natives of India. Hinduism and Islam were spoken of as of equal significance and merit, and adherents of either were men of 'faith'. Outwardly, and publicly this was policy and practice, but there were occasional hiccups: rivalry between corps of differing religions occasionally surfaced, as in **Lucknow** when they chose a Commander for the army, and in **Kanpur** it was even more marked. Some commentators were of the opinion that had **Havelock** not arrived when he did, the Hindu and Muslim factions within the rebel camp would have fallen out to the extent not just of rivalry, but actually of making war on each other; the Hindus were under the leadership of **Nana Sahib** and the Muslims followed the **Nunne Nawab** and **Maulvi Liaquat Ali**. Nana caused the Nunne Nawab to be brought before him and coerced into helping with the attack on **Wheeler's Entrenchment**, presumably with the intention of compromising him with the British. But then the Nana committed - or permitted - a major blunder, and it concerned the age-old friction point between the two religions - the slaughter of cattle for meat. **W.J.Shepherd** reported thus: 'I am informed that the Nana had brought upon himself the contempt of the Mahomedan portion of the 2nd Cavalry from the time he interfered with the butchers in the city. On or about the 18th June, two butchers. seized by the Hindoos in the act of slaughtering cows, were brought to the Nana, by whose order their hands were cut off, and they died from loss of blood. This caused a revolt on the part of the Mahomedans, who held a consultation amongst themselves and argued thus, 'Who has made this Nana a ruler over us? Is he not a creature of our own hands; and can we not appoint anyone else we like? If he has already commenced interfering with our creed, and preventing cows being killed, which is not only lawful, but is necessary to our very existence, how much more will he not meddle with our other religious callings when he is firmly established in authority, and when our common enemy, the English, shall have been completely exterminated?' Thus arguing, they proceeded in a body to the Nana to call him to account for causing the butchers' hands to be cut off. In the meantime the Nana was informed of what was going on, and immediately ran out with bare head and bare feet in the sun to meet the troopers, and with clasped hands begged their pardon for what he had done, promising never again to interfere in this respect, and that the Mahomedans were perfectly at liberty to kill as many cows as they liked, only that they were to do it in a retired spot. The troopers, I am told, used much abusive language on that occasion to the Nana, and threatened to displace him if he did not do as they desired. From that day the 2nd Cavalry Mahomedans held the Nana's authority in contempt.' *See* W.J.Shepherd, *A Personal Narrative etc.*, Lucknow, 1886. Communal enmity was always a threat among the native ranks. A group of 33 Irregulars who had fought bravely on the British side were assigned to the 1st Punjab Cavalry to serve with Sikhs and Afghans. The Sikhs persecuted them, and 25 of them were tried for misbehaving themselves and transported for life. The remaining seven 'went boldly over to the enemy.' Muir to Beadon. William Muir, ed. *Records of the Intelligence Department of the Government of the North-West Provinces of India during the Mutiny of 1857.* Edinburgh, 1902, I: 348.

COMMUNICATIONS 'We, studying the past, have constantly to remind ourselves that we may not judge men or events in history by modern standards. It is not just that morality changes - and to commit the 'sin' of anachronism in this respect is something all historians dread - but technology has changed too. An interesting Indian comment on the events of 1857-9 recently ascribed the British victory to having had the use of the electric telegraph. If right, the inference is that the victory would have come sooner if radio communication had been invented. But that is by the way. But we do at our peril ignore the transformation that the advance of technology has made in our lives. Things regarded today as routine and run of the mill were, a hundred and fifty years' ago, impossible to carry out at all, or time-consuming and labour intensive to the point of being practically impossible. Take the sentence 'Lord Canning issued a Proclamation'. Today its contents would no doubt have been 'leaked' to the press and known to the public even before it was issued, such is our hunger for news and our sophistication in distributing it. But in March 1858, when Canning the Governor General did indeed issue a Proclamation - a very famous and significant one - there was more sophistry than sophistication around: no typewriters, photocopiers, wordprocessors or even rotary duplicators. If you wanted multiple copies you had to have an army of scribes copying in copperplate handwriting, or you committed your document to a printer who laboriously set up in hand-selected type what you wanted to say. And there was no secrecy about a printer's shop, even if your town happened to possess one. The problem was that Canning wished his proclamation to go in secret to Sir James Outram in Lucknow and thence, if agreed by him, it was to be distributed to all and sundry. For this was the proclamation that was to pacify Oude[Awadh], or so they hoped, to reconcile old adversaries, and to bring the 'rebellion' there to an end. J.W.Sherer, Magistrate of Kanpur, was instrumental. He received confidential orders that he was to prepare to lithograph 'a certain document' in an absolutely secret way. You might say that lithography, using stone in a printing process was the very latest technology in those days, allowing for the multiplication of copies of ornate text, or engravings in the most efficient manner then possible. The 'certain document' that Sherer received came from Canning via Outram who had modified it on his own responsibility. But how to get it duplicated? Sherer was lucky to find an unlikely ally in one Amarnath, a Brahmin Tehsildar. It seems that he was accustomed to the process of writing on stone, and could do it clearly and well. The next problem was to find someone in Kanpur who possessed a press, and to commandeer both his premises and his services. Sherer sent word to such an one that that night, at 9 p.m., he would come and that the press, the materials, and two of his staff - a man to work the roller and another to work the press - were to be ready and placed at his disposal, and that he would be well paid for his trouble. After dinner therefore, taking Amarnath with him in a palki gari, Sherer went to the printer's house. The printer himself was dismissed - sent to his bed in fact - and a check made that the two workmen could neither read nor write. Then Amarnath sat down with the document and wrote it upon the stone. It took a long time, and the workmen objected to staying, but were coerced into doing so. Eventually, long after midnight, all the copies required were pulled off and wrapped up and finally the stone cleansed from the writing, and all traces of the work removed. There would be no 'leak' if Sherer could help it. Then to his house, and the handover of the packet of papers given to the waiting messenger, and his task was over. Even with Outram's modifications the Proclamation was hardly an unqualified success; there were too many 'ifs' and 'buts', and dodgy bits of interpretation left to such as Colonel Barrow to explain, if he could, to would-be surrenderers. But it certainly did something to end the war, if not for the big fish, then for the small fry: the sepoys in thousands took their Rs 2 journey money, and wearily made their way home.' *from* an article by **PJOT** published in the *Statesman*.

COMMUNICATIONS Governmental communications between India and England were improved by the existence of the electric telegraph but were still very slow and unreliable. For example the first news of the outbreak of the mutiny in **Meerut** and the massacre in **Delhi** did not reach London until 26 June 1857, over six weeks after the events they described. There was no direct telegraphic link with India before 1865. The original telegram concerning the outbreak was despatched from Calcutta on 18 May 1857, and went via Bombay to Suez and then from Alexandria to Trieste by steamer, thence by landwires to London. *See* H.L.Hoskins, *British Routes to India*, London, 1966.

COMMUNICATIONS, REBEL The British certainly had problems keeping open their 'secret weapon' - the **electric telegraph,** but the rebels had even greater difficulty in communicating over long distances. Over short distances they had positive advantages: local guides who knew every step of the way between villages, supporters and sympathisers in large numbers throughout the countryside who would enthusiastically spy and report accurately. But over long distances it was another matter: frequently the rebels were deceived - or 'inacurately informed' - as to the outcome of major battles etc, and there is a distinct possibility that this lack of accurate intelligence had a significant bearing on the eventual outcome of the struggle. Take as an example two communications, both dated early December 1858: the first is a telegram, dated 6 December, that was passed between the various British officials concerned: 'Feroze Shah with 1500 sowars fled from Biswah in Oude, and was making, when last heard of, for the Ganges intending to cross near Kannouj and enter Bundelcund and join Tantia Topye. Notice has been sent to Cawnpore, Mynpoorie and Agra'. The second is a letter sent from Lucknow to Hamirpur on 8 December, augmenting the telegram: 'After the defeat of the rebels at Biswa 50 miles northwest of Lucknow on the 1st instant, a body of rebels started for the Ganges, which they reached on the 5th. They tried to cross at Nana Mhow Ghat, but finding that impossible they went up the river to Akumghat above Bilhour and below Kannouj .....the leaders of this body are Feroze Shah, a Delhi Prince, Lukkur Shah. Goolab Shah alias Peerjee. Mohsin Aly Khan of Mow Shumshabad (Furruckabad) who passed himself as an European. Fuzul Haq Moulvie, formerly *sheristadar* in the Delhi Commissioner's Office......these men were last seen at sunrise on the 6th instant at Akum Ghat, by the *Kotwal* of Sundeelah, who was taken prisoner, but contrived to escape from them, when engaged in crossing. He and two prisoners, taken by the District authorities, state that the rebels are making for Calpee (Kalpi) *en route* to Runthumbour (Ranthambore) 20 *koss* distant from Kotah in the Jyepoor (Jaipur) territory. They **hope to join Tantia Topee who, they believe, rules at Agra.'.** Tatya Tope of course never did 'rule at Agra', and if this were really the belief of Firoz Shah, it must have considerably warped his judgement.

COMPANY 'The common people never had any very intelligent comprehension of what the Company meant,' wrote Charles Raikes, Judge of the Sudder Court at Agra. 'I recollect asking an intelligent yeoman, a man who paid his annual hundred rupees or more of revenue to the State, who the Company was... ' He said, 'I don't know much about the matter, but '*muraroo hogee*' she is a female of some sort!' Charles Raikes, *Notes on the Revolt in the North-western Provinces of India*. London, 1858, pp 176-7.

CONFUSION IN KANPUR After entertaining countless reports, **Wheeler, Jack, Parker, Hillersdon** and **Wiggens** retired to bed at 1 AM on the morning of 22 May. At 6AM Hayes 'went out to have a look at the various places, and since I have been in India never witnessed so frightful a scene of confusion, fright and bad arrangement as the European barracks presented. Four guns were in position loaded, with European artillerymen in nightcaps and wide-awakes and side-arms on, hanging to the guns in groups, looking like melodramatic buccaneers. People of all kinds, of every colour, sect, and profession, were crowding into the barracks. Whilst I was there, buggies, palki-gharrees, vehicles of all sorts, drove up and discharged cargoes of writers, tradesmen, and a miscellaneous mob of every complexion, from white to tawny, all in terror of the imaginary foe; ladies sitting down at the rough mess-tables in the barracks, women suckling infants, ayahs and children in all directions, and officers too! In short, as I have written to Sir Henry, I saw quite enough to convince me that if any insurrection took or takes place, *we shall have no one to thank but ourselves,* because we have now shown to the Natives how very easily we can be frightened, and when frightened, utterly helpless. During the day (the 22nd) the shops in all the bazars were shut, four or five times, and all day the General was worried to death by people running up to report improbable stories, which in ten minutes more were contradicted by others still more monstrous.' Captain Fletcher Hayes quoted in Sir John **Kaye**,Vol. II, pp 300-1.

CONSTANTIA Popularly known as La Martinière College, in Lucknow . Played a significant part in the Revolt of 1857. Listed by INTACH. *See* MARTINIÈRE.

CONTINGENT TROOPS These were forces maintained by the rulers of native states under the terms of treaties with the British, each unit usually had three or four British officers and was equipped and drilled and trained to the English pattern. In general they were recruited from the same sources as the Bengal Army, ie a large number of the sepoys came from Oude[Awadh]. Examples were the Jodhpur Legion, the Gwalior, Malwa, Bhopal and Kotah Contingents etc. Of the corps thus maintained the Hyderabad Contingent alone remained faithful to the British, apart from the incident at **Aurangabad**.

COOPER, Frederick Henry The author of a useful account of events in the Punjab: both contemporaneous and factual. He follows events in chronological order, especially the putting down of incipient revolt; details of military movements and measures are included; the role taken by Indians who chose to be loyal to the British is also included. *But,* Cooper was a Deputy Commissioner of the Punjab, based at Amritsar and although a civilian was responsible for the 'extermination', of the disarmed 26th BNI at **Ujnalla**, in particularly inhumane and unpleasant circumstances, for which he was praised and supported by his superiors but roundly condemned by many in the House of Commons when the news reached London. He wrote this book, to vindicate his conduct which he considered 'prompt, spirited and thorough'. *See* Frederick Henry Cooper, *Crisis in the Punjab from 10th May until the Fall of Delhi,* Lahore, 1858.

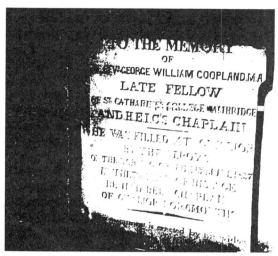

Rev. Coopland's grave at Morar

COOPLAND, Mrs Ruth M. Ruth Coopland was a survivor of the mutiny/rebellion although her husband was killed. She was the widow of the **Reverend Coopland** Chaplain at Gwalior killed by the mutineers of the **Gwalior Contingent**. His grave can still be seen at **Morar**. Her account is cold and factual and full of (understandable) animosity towards the mutineers who killed her husband, but perhaps is more interesting for her details of life in **Agra Fort** which those who were there described as besieged, although it was nothing of the kind. *See* R.M.Coopland, *A Lady's Escape from Gwalior,* London, 1859.

COOPLAND, Reverend C.W. He was the Chaplain at Gwalior on the outbreak of the mutiny/rebellion. There were of course no British troops in **Gwalior** as it was part of the princely dominions of Maharaja Scindia, but there existed a powerful army there in the form of the **Gwalior Contingent** recruited in much the same way as the Bengal army *ie* a high proportion of its sepoys came from **Oude**[Awadh], and it was officered by British Officers. It was this Contingent that mutinied. Coopland and his wife were in their bungalow when the alarm signal was given. They dressed and escaped by a bathroom, and were given shelter by an officer's servant in a neighbouring compound, but were found and he was shot dead before his wife's eyes. She eventually reached Scindia's palace safely. *See* Augier Papers in CSAS, *and* Mrs R.M.Coopland, *A Lady's Escape from Gwalior,* London, 1859.

CORNWALL's, Duke of, Light Infantry The Regimental Museum at Bodmin in Cornwall houses many memorabilia of H.M.'s 32nd Regiment which later became known as the Duke of Cornwall's Light Infantry. Historical records, private letters, maps of Lucknow by **Archdale Wilson**, the diary of Private Henry Metcalfe which gives a history of the siege of the **Lucknow Residency** are all contained in the museum's collection. The 32nd was the British regiment that was based in Lucknow on the outbreak of the mutiny/rebellion, it was involved, disastrously, in the battle of **Chinhat**, and was besieged in the **Lucknow Residency**, losing many of its number to death and disease.

COSS Or Koss: a distance equivalent to two miles or approximately 3¼ kilometres. It is said that the English use of the word 'Course' is derived from Coss.

COSSID A messenger, particularly a messenger regularly employed by Government (eg **Ungud**). They were well-paid but ran extraordiinary risks, and many lost their lives in the mutiny/rebellion. They became expert in the concealment about their persons or in the sticks they might carry of the message with which they had been entrusted. For the latter a favourite device would be to roll it tightly and insert it into a quill which would then be concealed as imagination suggested. 'Two cossids returned on the evening of the 28th with a despatch from Wheeler.' Gubbins and Captain Hawes took them aside, and the Cossids told them about the massacre. Lawrence immediately advised Neill at Allahabad not to make too hasty an approach now that Kanpur had fallen. But the spy lost the letter, and though Renaud believed his word-of-mouth, Neill did not, and ordered Renaud to proceed by slow marches to Kanpur. *See* M.R.Gubbins. *An Account of the Mutinies etc.,* London, 1858.

COSTLY BULL, a 'The intrenched people (**Wheeler's Entrenchment, Kanpur**) were so fortunate as to shoot down a Brahminee bull that came grazing within limits where his sanctity was not respected; but having floored it, how was the prey to be secured? Mrs. Glass's recipe, 'First catch your hare,' was never more appropriate. Presently a volunteer party was formed to take this bull by the horns, no trifle, since the distance from the wall was full three hundred yards, and the project involved the certainty of encountering twice three hundred bullets. But beef was scarce, and, led on by Captain Moore, eight or ten accordingly went out after the animal. They took with them a strong rope, fastened it round the hind legs and between the horns of the beast, and in the midst of the cheers from behind the mud-wall, a sharp fusillade from the rebels, diversified with one or two round shots, they accomplished their object. Two or three ugly wounds were not thought too high a price to pay for this contribution to the commissariat. The costly bull was soon made into soup, but none of it reached us in the outposts more palpably than in its irritating odour. Sometimes, however, we in the outposts had meat when there was none at head-quarters.' reported by Mowbray **Thomson**.

COTTON. LT.COL. H. The commandant of the 69th BNI was, as a result of the ineptitude of Brigadier **Polwhele** at the Battle of **Sassiah**[Sasia], on 5 July 1857, promoted to command the British troops in and around **Agra**, and given the rank of Brigadier. He had the soubriquet of 'Gun-cotton', with the implication that he was a man of considerable energy. He certainly had a strong sense of duty, incurring responsibility and unpopularity in controlling, by stringent measures, including flogging, the excesses of the troops, auxiliaries and volunteers under his command. At the battle of Agra on 10 October 1857 he distinguished himself for his personal bravery and determination, although as the enemy arrived almost at the walls of the fort without being detected his intelligence arrangements were highly defective.

Courtesans

COURTESANS The courtesans, or, more bluntly put, prostitutes of Lucknow were a numerous class in pre-annexation times ie before 1856: they were regarded with respect, and certain of them celebrated as great dancers and leaders of culture. They were still numerous in 1857 but their status had fallen. The '*Bengal Hurkaru*', April 1858, reported that their activities had contributed to the coffers of the rebel government from June 1857 to March 1858: 'Another important item of revenue was supplied by women of a profession whose designation is unmentionable. Of these there were crowds in Lucknow; and one man Mussami Abbashi had undertaken to pay to the state 60,000 rupees *p.a.* for the farm of one particular *chukla* or resort of such persons.'

COX, Mr A certain Mr Cox, formerly of the 1st Bengal Fusiliers lost both legs and died in an explosion in **Wheeler's Entrenchment** in **Kanpur** in June 1857. The same explosion, from a mortar shell, which fell in Whiting's battery, killed seven soldiers' wives and the watch-maker of Kanpur, Mr Jacobi. Now this Cox was probably the officer who had in a state of intoxication fired at a *sowar* before the outbreak at Kanpur. *See CASHIERED SUBALTERN, see also* Mowbray Thomson, *The Story of Cawnpore,* London, 1859, p 100.

CRAIGIE, CAPTAIN, H.C. Commander of the 4th Troop of the **Third Bengal Light Cavalry** stationed at **Meerut** in May 1857. If his advice had been listened to the mutiny might not have broken out as it did on 10 May 1857; instead he has given a severe reprimand for the letter he addressed to the Adjutant of the 3rd Lt.Cav., when he had heard that the Colonel, **Carmichael Smyth**, had ordered a parade at which the skirmishers were to be ordered to accept cartridges which they suspected might endanger their caste, see CARTRIDGE QUESTION. Craigie had great influence with his **sowars**, and he and his wife were protected by them when the remainder of the regiment mutinied. He was convinced that the cartridge business only required tactful handling. For further details *see* R.Montgomery Martin, *The Indian Empire* Vol II. London,1858-60, p 143 *et seq., see also* C.Hibbert, *The Great Mutiny,* London, 1978.

CRA'STER PAPER This is held in the Centre of South Asian Studies at Cambridge University, and includes a typescript copy of a Reminiscence of the Sepoy Mutiny written eight years afterwards by Colonel William R.Cra'ster, late Royal Artillery. There is the story of the siege of **Delhi**, with considerable detail of military strategy.

CRICKET Cricket was a favourite pastime among the European officers, and soon the Sepoys played it too, and they soon 'became perfect adepts in the art of bowling, batting, and fielding. At fagging they were untiring, and in catching particularly expert,' *see* Hervey, **A Soldier of the Company.** p. 103-4. Things had changed by 1944; when I 'revived' the playing of cricket at the Mahratta Light Infantry Regtl Centre, I found only the Havildar Clerks could play - but they were very keen indeed! - **PJOT**

CROOKE, WILLIAM author of '*Songs of the Mutiny*'. See *Indian Antiquary,* April and June 1911.

CRORE Equivalent to one hundred *lakhs* and set down as 1,00,00,000. Equals ten million, ie 10,000,000.

CRUELLY DESERTED In June 1857 **Sir Hugh Massy Wheeler,** Major General in the Bengal army and all the Europeans of **Kanpur** were besieged within the narrow confines of **Wheeler's Entrenchment** by the forces of the **Nana Sahib**. They had entered the entrenchment full of high hopes that they would soon be relieved, either by British troops sent up country from Calcutta by **Canning** (who certainly promised urgent relief), or by troops from Lucknow.

In the event no relief arrived, and the result was total catastrophe, *see* SATI CHAURA GHAT & BIBIGURH. Not unnaturally Wheeler felt badly let down, and the death of his favourite son, **Godfrey**, killed in the entrenchment in a particularly horrendous manner, added to his grief. The following letter explains much: 'Yesterday [23rd] morning they attempted the most formidable assault, but dared not come on, and after above 3 hours in the Trenches cheering the men, I returned to find my favourite darling son killed by a 9 pr. in the room with his Mother and sisters. He was not able to accompany me having been fearfully crippled by a severe contusion. The cannonade was tremendous, I venture to assert such a position so defended has no example, but cruel has been the Evil, Brigr. Jack, Coll. Williams, Majr. Prout, Sir G. Parker, Lieutenant Quin senr. and many ladies, also Major Lindsay died from the dreadful heat, fatigue or disease, Halliday, Reynolds, Prole, Smith, Redman, Supple, Eckford, Dempster, Jervis, Chalwin, and many more killed. Mr Hillersdon and Mr Jack killed. We have no instruments, no medicine, provisions for 8 or 10 days at farthest and no possibility of getting any as all communication with the Town is cut off. Lieutenant Bax 48th killed. If Lord Canning's promises were performed we should have 800 men, with half the number we could capture their guns and drive them before us. ***We have been cruelly deserted and left to our fate.*** We had not above 220 soldiers of all arms at first, the casualties have been numerous. Railway gents and merchants have swollen our ranks to what they are, small as that is. They have done excellent service, but neither they nor I can last for ever. We have all lost everything belonging to us, and have not even a change of linen. Surely we are not to die like rats in a cage. We know nothing of Allahabad, to which place we have sent five notes, but whether they have reached or even gone we as yet know not. The ladies, women and children have not a safe hole to lie down in and they all sleep in the trenches for safety and coolness. The barracks are perforated in every direction, and cannot long give even the miserable shelter which they now do. God bless you. Ever yours, H. M. Wheeler.' *See* Wheeler to Lawrence, 24 June, 1957, Sheo Bahadur Singh, ed. *Letters of Sir Henry Lawrence*. New Delhi, 1978.

CRUMP, C.W. An artist who claimed to have made original sketches, 'on the spot' of the Kanpur massacre. They were reproduced for sale as prints (18"x15") and are remarkable only for their subject matter, and were clearly intended to be sensationalist. All ghoulish detail is included: skeletons draped artisitically over General Wheeler's intrenchment; gore and debris liberally bespattered over the 'Chamber of Blood', and a British soldier is shown pointing out to another the fateful well. A good 'visual aid', 1858 vintage, for a highly prejudiced account of the struggle. *See* C.W.Crump, *Pictorial Record of the Cawnpore Massacre*, London, 1858.

CUNNINGHAM, M. He was the Collector of **Etawa** at the outbreak of the rebellion.

CUTCHERRY Or Kutcherry or Kacheri (*cf* **Kacheri Cemetery, Cawnpore**). A Court of justice, but used commonly to mean the office of a magistrate or civil official.

D

DABADEEN 'These people were paid the value of their boats which were burnt, amounting to Rs.4,467, through Moonshee Dabeedeen, by orders of Tantia Topee. Of this sum, a balance was left with Dabeedeen, which he took for himself. ' Dabadeen was the 'munshi' or leader of the boatmen at **Kanpur** and it was he who organised the collection of the (24?) country boats that were prepared for the Europeans to go down river to **Allahabad**. *See* SATI-CHAURA GHAT.

View of Dacca, 1857

DACCA A large town in Eastern Bengal which saw a mutiny of the detachment of the 34th Regiment BNI stationed at **Chittagong** on 18 November 1857 followed by the disarmament of the detachments of the 73rd BNI and artillery details at Dacca, which were displaying 'mutinous tendencies'. The act was performed by Lieutenant Lewis, Indian Navy. With a force of eighty-five sailors and two mountain guns, plus some 30 civilian volunteers he successfully attacked and dispersed, after a fierce struggle, some 350 sepoys: the stragglers from the rebel forces found a temporary refuge in Bhutan. For details *see* Col.G.B.Malleson, *History of the Indian Mutiny 1857-58,* Vol II London 1879 p 419.

DACOITS Lawless elements took advantage of the absence of government to make their depredations almost unchecked. They were often linked - possibly inaccurately - by British Officers attempting to restore order in former rebel-held areas (particularly in Oude[Awadh]), with the rebels: 'In the Goshaeengunje Pergunnah there are some dacoits, Anundee, Khooshal Koormee and others who amuse themselves by burning villages and committing havoc. Plunder is now the order of the day with these rebels; they lay waste a bazar and those that resist payment of blackmail they carry off to their camp and torture.' *From* the report of the District Commissioner, S. Martin, dated 10 May 1858.

Dak Runners

DAK The postal system, dependent on a network of offices combined with horse, or horse and cart, or runner services spreading throughout the NW Provinces of India. If the 'dak was running' it means that the postal service was in operation. If the 'dak is stopped' it will mean that rebels will have prevented the government mail getting through. The 'dak houses' were resthouses for the traveller, mainly government officials, to spend a night on a long journey up country.

DAK BUNGALOWS **Russell** was astonished to find that in the middle of the Mutiny, at dak bungalows still surrounded by bands of rebels, British travellers had entered complaints into the guest books. 'The kitmutgur here is uncivil,' 'There is no table cloth,' 'I could not get a napkin for dinner'. ...'We, (by which he meant the British and not the human race) are strange beings!' W.H.Russell, *My Diary in India,* London, 1860, Vol II, p 411.

DALHOUSIE, THE MARQUESS OF The tenth Earl and first Marquess was Governor General of India for a comparatively long, and certainly momentous tenure, 1847-56. Declared the Punjab a British province; began the introduction of railways and the electric telegraph; moved further against 'suttee and dacoity/thuggee'; expanded the postal system and removed internal imposts that shackled trade; annexed Lower Burma; introduced the doctrine of lapse whereby rulers who had no natural heirs were denied the right to adopt a successor, hence refused accept the adopted heir of the raja of Satara (direct descendant of Sivaji); annexed Nagpur, Jhansi and Oude[Awadh]. Fêted on his return to England, but then bitterly assailed on account of his policy of annexation and his confidence in the sepoy army - both of which were considered to have contributed to the mutiny/rebellion. *See* Sir William Hunter, *The Marquess of Dalhousie*, Oxford, 1890, *see also* Edwin Arnold, *The Marquess of Dalhousie's Administration of British India*, 2 vols, London, 1862, *and* **DNB.**

DAMODAR RAO Adopted son of **Gangadhar Rao, Maharaja of Jhansi**, whose cousin he was. Before this adoption he was called Anant Rao. After the British reconquest of Jhansi, and when he came of age he tried to claim his inheritance but it was denied most unfairly as Government had all along agreed that he could inherit his adoptive father's possessions if not his title. He was allowed a pension of Rs200 per month by Government, and was still alive living in modest circumstances in 1936, *see* **Rani of Jhansi**.

DAMOH A garrison town some fifty miles due east of Saugur[Sagar] where there were two companies of the 42nd BNI. News of the mutiny at **Saugur**[Sagar] caused great excitement here at the beginning of July 1857, and the native officers and non-commissioned officers warned the Europeans they would not be safe, so they fled on the night of 3 July the seventy miles to Narsinghpur to the south. Plundering by Bundelas from **Shahgarh** had already begun when the mutineers arrived from Saugur and demanded the government treasure; this was refused by the Subadar-Major and Havildar Ranjit Singh who also ensured that the entire detachment of the 42nd at Damoh remained faithful to government. For more details *see* **FSUP** Vol III, and *Revolt in Central India*, Simla, 1908, p 36 *et seq*.

DANAPUR *See* DINAPORE.

DARE, DAVID There is a strong rumour to the effect that **Ulrica Wheeler** (the 'Miss Wheeler' of much speculation and investigation) may have been survived by children or grandchildren. The late Mrs Zoë Yalland (*see* Bibliography) who was born in **Kanpur** and carried out meticulous research there, suspected that a talented, and apparently well connected Kanpur businessman may have been Miss Wheeler's grandson. He was adopted by a Miss Leach, sister of the Doctor who was called to Miss Wheeler's deathbed in 1902 (?). Mrs. Yalland speculated that Miss Wheeler had to have had a special reason to call for an *English* Doctor, and that the future welfare of a grandchild may have provided it.

DARIABAD In the Lucknow Division of Oude [Awadh] the only out-station of any importance was Dariabad. There was only one regiment there, the 5th Oude Irregular Infantry, under the command. of Captain Hawes. There was a large amount of treasure at the station, which some days previously Hawes was anxious to take into **Lucknow**, but finding his men would be likely to oppose its removal, the matter was not pressed at the time; but as the stations around were falling one after another he determined to make the attempt. The treasure was laden on carts and an order was given for the march. The men left the station cheering, but they had scarcely left it when an altercation arose amongst them. Some wished to proceed with the treasure, while others wished to retain it. Some firing took place, and the mutinous party prevailing, the carts were countermarched and the treasure was taken back to cantonments. When the firing began Hawes rode off and escaped miraculously in a shower of bullets. All the other officers, both civil and military, with their families, were allowed to leave unmolested and found their way without loss of life into Lucknow where they joined the other Europeans in time to be besieged in the **Residency**.

DAROGHA (or Daroga or Darogah), a superintendent or overseer. He might be an important royal official, or the overseer of a small undertaking, like a bridge of boats etc.

DAS, Manmatha Nath An expounder of the theory that interference with Indian custom contributed to the Great Rebellion, *see Western Innovations and the Rising of 1857,* in *Bengal Past and Present*, Vol 76, 1957.

DASSA BAWA Sitaram Bawa, one of Nana Sahib's emissaries, characterized **Nana Sahib** as 'always a worthless and not very clever fellow' who 'never would have been anything but for the tuition of his Guru, Dassa Bawa (said to have come from a place called Kali Dhar, beyond Kangra, this side of Jammu).' In 1855 'Nana Sahib gave the Guru, Dassa Bawa, a sunnud, granting a five-lakh jaghir and five nachatras, because Dassa Bawa had told him that he would become as powerful as the Peishwa had once been, and the sunnud was *to take effect when he came into power.*' After an astrological ceremony Nana Sahib dreamed that Hanuman had promised that he would be victorious, at which point Nana Rao presented Dassa Bawa with 25,000 rupees' worth of jewels. 'Dassa Bawa is a person who had helped and advised Nana throughout.' He was described as 'a most able man. He is **125 years old.**' See **FSUP** Vol I, pp 373-5.

DATTA, Kalikinhar K. An Indian Historian who has made many innovative contributions to the study of the Rebellion. His work is of interest also because it represents an extreme in partisan attitudes. He is anxious to discuss the extent of the civil unrest and the discontentment with foreign rule *see Some Unpublished Papers relating to the Mutiny* in *Indian Historical Quarterly,* March 1936; *also Nature of the Indian Revolt of 1867-59* in *Bengal Past and Present* Vol 73, 1954; *also Some newly discovered Records relating to the Bihar Phase of the Indian Movement,* Patna University *Journal* Vol 8, 1954; *also Popular Discontent in Bihar,* Bengal *Past and Present,* Vol 74, 1955; *also 'Some Original documents relating to the Indian Mutiny* Indian Historical Records Commission Proceedings, Vol 30, 1954; *also Contemporary Account of the Indian Movement of 1857* in *Journal of Bihar Record Society,* Vol 36, 1950.

DAVIDSON, Lt Col. Cuthbert   He was Resident at Hyderabad, Deccan at the court of the **Nizam** and was influential in helping to restore order there and at **Aurangabad** with the very active co-operation of the *Dewan* or Chief Minister, **Salar Jung**. On 17 July the Residency was attacked by a band of *Rohillas* led by Jemadar Toora Baz Khan and a *maulvi*, but the attack was beaten off. Letters from Davidson to his sister concerning the defence of the Residency, and relevant press-cuttings are preserved in the Oriental and India Office collections of the British Library, *see MSS.Eur.E.308/42.*

DE KANTZOW, C.A.   A serving officer who, like so many others, kept a journal or note of his service, and on retirement set it out for publication. He has the merit of being straightforward and readable. His book is useful for its account of events at Mainpuri which are not well chronicled elsewhere. *See* Colonel C.A. De Kantzow, *Record of Services in India, 1853-86,* Brighton, 1898.

DEATH BY ELEPHANT TRAMPLING   'At that time (11 June 1857, the Ganges at **Kanpur**) there was a strong, hot breeze blowing, and when the rebels set fire to the grass, the flames spread with the rapidity of lightning. It was heart-rending to see our Christian people, men, women and children, many of them wounded, fleeing before the devouring element! Many were overtaken by the flames and burnt to death, others were pursued by the 2nd Regiment of Sowars and taken prisoners. The men were all bound with ropes and marched like French Criminals in chain-gangs to Cawnpore. The women and children were forced along at the point of the bayonet. ...Though the Nana appropriated all the money and valuables found in the boats of the fugitives, the male prisoners were subjected to the most cruel torture for the purpose of forcing them to confess where they had buried their wealth, for some of the merchants and traders were thought to be very rich. So when the Nana's instruments of torture failed to obtain the desired information, the arch fiend put it down to stubborness and obstinacy on the part of the Englishmen and he ordered them to be bound hand and foot and trampled to death by Elephants. The Europeans were accordingly fettered and pinned to the ground. But when the Elephants, which had formerly belonged to the English Commissariat Department arrived they refused to pass over the prostrate Christians. When urged on by the mahouts, the huge beasts lifted up the prostrate bodies in their trunks and put them aside but would not trample upon them. ...The huge beasts instead of going forward retreated through the crowd, knocking down and injuring many who had come to see the hated Feringhee flattened into a pancake. After this orders were given to the men of the 2nd Cavalry to dispatch the Europeans.'

DEBI BAKSH (or Dabee Bux)   Raja of **Gonda**, one of the most determined of the **Begum of Oude's** supporters. He accompanied her and **Beni Madhoo Singh** into Nepal. It was said of him that 'he was fond of fighting and had done nothing else all his life'. He died (or was killed in a fight with **Jung Bahadur's** Gurkhas) in the Terai in November 1859, at the same time as Beni Madhoo. The Rani of Gonda and eighty nine followers then surrendered to the British.

DEBI SINGH GUNTIA   A rebel leader in the **Jabalpur** area of Central India. He apparently had as many as 1,500 followers at one time. He was captured, and hanged, on 10 November 1857. For details *see The Revolt in Central India,* Simla, 1908, p 47.

DELAFOSSE, Lieutenant Henry G.   A young officer of the 53rd N.I. Bengal Army. His regiment mutinied at **Kanpur**, and he entered **Wheeler's Intrenchments** with the other European officers and civilians. He, **Mowbray Thomson, Sullivan** and **Murphy** were the only survivors from the boats that got away at the massacre at the **Satichaura Ghat**. Described as a 'pale wiry man' he went on to have a successful military career, reaching the rank of Major General. One of the most significant parts of the account that he gave of his experiences at Kanpur is a list of those, so far as his memory served him, who had taken refuge in the entrenchment and their fate, so far as he knew it: *see* J.F.Bruce, *The Mutiny at Cawnpore,* in *Punjab University History Society Journal,* April 1934.

DELHI   The city is situated on the western or right bank of the Jumna[Yamuna] river, approximately a thousand miles up-country from Calcutta. It is of great antiquity, there being perhaps seven different cities upon the same site, and it has had a violent and turbulent history, as well as one of greatness renown for it has been an Imperial Capital for many ages. 'Dilhi bahut door hai'. ('It's a long way to Delhi'). For six centuries the city was the capital of Muslim power in Hindostan, latterly the Mughals from **Timur** and **Baber** onwards. Delhi became the focus of the whole mutiny/rebellion and when it was taken by the British the backbone of the mutiny was broken. The **Shahzadahs** were shot by **Hodson** at Humayun's tomb and the last King of Delhi tried for his life in the very Dewan-i-Khas or Audience Hall where once stood the **Peacock Throne.** *See* MUBARAK SHAH. For a fictional but highly entertaining 'history' of Delhi, based largely upon fact and sound observation, a good introduction is Khushwant Singh, *Delhi,* New Delhi, 1990. Among the monuments protected by the Delhi Administration are The **Mutiny memorial** on the Ridge; **Flagstaff Tower** near Delhi University; **Magazine**s Gate opp. GPO **Kashmere Gate;** Northern and Southern Guard House on the Ridge near Flagstaff Tower.

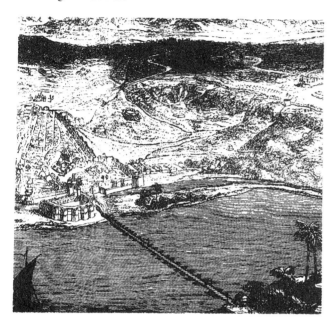

Imaginative bird's eye view of Delhi & the Ridge, 1857

DELHI, THE ASSAULT ON The British assault on Delhi, concluding the **siege** thereof, began on 14 September 1857, and six days later, on 20 September, the final rebel strongholds were taken, the King **Bahadur Shah** was captured on 21 September, and the Princes murdered by **Captain Hodson** the next day. Although he was not in command of the British troops (this was **Archdale Wilson**'s role), **John Nicholson** is the name always associated with the successful assault, although he died in its accomplishment. The assault was extremely costly in terms of killed and wounded, and it is said that these casualties were due partly to the ill-discipline and drunkenness of the British force, which was in marked contrast to the sobriety of the rebels. The British Delhi Field Force had 11,770 men on strength on 14 September 1857, by the 20th they had suffered 4,493 casualties of whom 1,254 were killed (and others died subsequently of their wounds). For details of the assault *see* **Kaye, Sen, Ball, Martin** etc; *see also* A.Llewellyn, *The Siege of Delhi,* London, 1977, Harvey Greathed, *Letters written during the Siege of Delhi,* London 1858. Charles Metcalfe, *Two Native Narratives of the Mutiny of Delhi,* London, 1890; *also* Krishan Lal, *The Sack of Delhi as witnessed by Ghalib,* in *Bengal Past & Present,* July-December 1955. For a graphic description of the city *after* the assault *see* Chapter V, *'The City of the Dead',* pp 148-171, N.K.NIGAM, *Delhi in 1857,* Delhi, 1957.

DELHI, 11 MAY 1857 'The sun rose with all its wonted glare and glitter over the domes and minarets of Delhi. In the summer heat of May the day began early...all seemed as usual; the civil surgeon, Dr Balfour, had done his round of dispensary and gaol; the cutcherries, the offices and courts, had opened for business; bargain and barter in the Chandee Chouk and the smaller bazars had begun. All spoke of peace and normality. There was no unusual bustle, no excitement, nothing to give warning of the approaching tempest... Routine was unvaried.....it was an ordinary day. Indeed early in the morning of 11 May 1857 the British had no reason in Delhi to think that the day would be any different from any other. Their thoughts would no doubt have dwelt on the excessive heat they would have to endure somehow as the sun rose higher. Even with all the help innumerable servants gave them - damping down, shading, throwing water at tatties, pulling on punkah ropes etc - and all that the (limited) technology of the time could provide to reduce discomfort, they knew they had another six weeks or so before the blessed advent of the rains that would ease their misery. Meanwhile...there was the usual routine to be endured, and for some, the same heart searching, agonising questioning of conscience or common sense: why, in the name of all that is holy, do we stay in this land, where the life expectancy of a European is well short of forty, where our children will nearly all die before they are five, where neither wealth nor good health will be our lot, where our only obvious perquisite is that we are treated like lords of creation, who must be obeyed in all things? Truly a dilemma. Yet they stayed, and there was never a shortage of recruits, male or female, to take the place of the sahib-log in India. We can only ask ourselves whether conditions for the younger sons of English and Irish gentry were so desperate at home that they had nowhere else to look for employment? One wit described India as 'outdoor relief for the upper classes', and he was not far from the truth. And for the humbler ones, were not the stews of London, Birmingham, Glasgow, or the poverty of rural Britain preferable to the earlier, and often painful, death that struck so quickly in India? Maybe the picture was distorted somehow, by chance or design. Were there still stories going round of fortunes to be made? Many a landed estate in England owed its existence to the imitators of Clive and Warren Hastings, and this would be well-known in the shires, though not often boasted about by the owners of the estates. Were the recruiting sergeants selected carefully for their powers of persuasion, or did the uniform, the drum, the Queen's shilling and the drink turn bucolic heads ? The recruits came, from Britain, in their hundreds, sometimes thousands, and more would have come if they had been allowed. Indeed there were many 'unofficial' Europeans, adventurers for the

Courtyard of the King's Palace at Delhi

most part, but respectable merchants also, who heard the strange syren call of the Orient and could not resist: they came at their own expense, and they could argue with no one if their enterprise failed. If all in Delhi was normality for the rulers on that May morning, was it equally so for the ruled? Certainly there had been weird rumours; there was evidence occasionally of discontent; there were firebrands who spoke of a restoration of the Mughals and a release for India from the stranglehold of the firenghis. Bahadur Shah II, King of Delhi, descendant of Mughal Emperors, writing sad, pessimistic, verse in his crumbling palace.....was he content with the daily round? He had the trappings of power but none of the reality, surrounded by bickering clamouring relatives for whom he sought to provide from the inadequate allowance the British paid him. Did he know that the great revolt was at last upon the way? There are those that say he did, and was party to the plan. Probably for the common folk of Delhi life that morning was as it had always been; they were anxious only to continue in peace to live and trade, to follow inherited and pre-ordained destinies, to stay out of trouble. But trouble has a way of creeping up even on the most wary, and the trouble that was about to break upon these people was not only largely undeserved but surpassed anything they had previously endured. In 1857 Europeans knew which side they were on: it was not so easy for the Indian. Even those who believed that there was a struggle to come were unsure of the outcome. Rumours abounded. The English had always seemed so invincible, would it not always be so? Was not India destined to be ruled by these foreigners, objectionable though they were in so many ways, if only because their own rulers had proved to be so inadequate? And yet....the English had been defeated recently in Afghanistan, and travellers from the North had brought stories to say that they had won, in the end, in the Crimea, but it had been touch and go. Then there was the bazar prophecy that foretold the foreigners' rule would last but a hundred years. It was now a hundred years, all but a month, since the Battle of Plassey which had set the British up. How decide? How choose? The heart said one thing, the head counselled caution.' **PJOT**.

The first of the mutineers from **Meerut** reached Delhi early on 11 May. They succeeded in persuading **Bahadur Shah** to be their leader, though it is doubtful if he was very enthusiastic at the prospect. The Europeans were murdered in some numbers, and British authority totally overthrown within hours. Details, which are not seriously in contention, of the events of this day may be found in **Ball, Sen, Kaye** etc, and an excellent, succinct, account will be found in Christopher Hibbert, *The Great Mutiny,* London, 1978, pp 91-119. *See also* MAINODIN HASSAN KHAN.

DELHI FIELD FORCE The name given to the British forces on the Ridge at Delhi which 'besieged' the city June-September 1857 before the final assault on 14 September. Originally commanded by the commander-in-chief, General **Barnard,** and then by Major-General **Archdale Wilson,** it consisted of troops from **Meerut** and **Ambala**, and the **Punjab Movable Column.** For details *see* A.Llewellyn, *The Siege of Delhi,* London, 1977.

The Tomb of Shah Alam

DELHI, KING OF   In September 1803 General Lake rescued from the Marathas the old, blind and feeble Emperor Shah Alam who was confined in a kind of imperial prison, the fort of Delhi. He was found seated upon a tattered canopy, his person emaciated by indigence and infirmity, his countenance disfigured by the loss of his eyes, and bearing marks of extreme old age, 'joined to a settled melancholy'. He died in 1806 and was succeeded in the nominal sovereignty by his eldest son Akbar Shah who enjoyed the shadow of a royal title and its annual emoluments of thirteen and a half lakhs of rupees for some thirty years, and was then succeed in his turn by his eldest son, Mirza Abu Zafar, who, on succession styled himself Mahomed Suraj-ud-din Shah Ghazi, and is known to history more familiarly as **Bahadur Shah Zafar**. When the latter's eldest son died he tried to set aside the next eldest son from the succession in favour of a younger son Jumma Bakht whose mother, his favourite Queen **Zinat Mahal**, obviously had contrived to make him declare was more worthy of the throne; the British refused to entertain this idea, although nine out of eleven of the Princes had declared their willingness to accept the King's nominee as the head of the family. *See* **Ball**. Vol I, p 69.

DELHI LANDMARKS   'Nestling in the not so green ranges of the Aravelli Mountains, within walking distance of the Delhi University Campus, stands the Monument erected to the memory of all those soldiers who laid down their lives to crush the so-called Mutiny. Unfortunately it is not in the tourists' guide books, for it would have been good to have been reminded of it once in a while, if only to recognise it as a literal landmark in the freedom struggle. It is now called **Ajith Garh**, or the place of defeat, whereas it was known formerly as the Victory Memorial, for it was the symbol of success for the Britishers. But what's in a name? What matters was that a war was fought, and one side won; the ultimate loser was not the other side: the only true victor was Death. All died for the sake of land which belongs to no one. The names of the soldiers, on the British side, who died in the struggle are engraved in stone on the monument which, in a number of respects resembles that other Delhi historical structure, the Qutub Minar: a few years ago it was posible to climb up internal steps in each of them, and see the beautiful city laid out below; now you are no longer allowed to climb, and in the case of Ajith Garh perhaps it is just as well, for the sight now would be of sky scrapers thrusting up their ugly heads, with large glazed mirrors reflecting the inverted ugliness of modern architecture. The irreverent custodian of history may attempt by change of name or hiding of statues to erase selected items from memory, but they are foolish: not only do they waste the rich inheritance of this land, but they deviate from truth: we should be grateful to the contractors who built their monuments so well; they stand, and remind us of the days so fraught with struggle. Amongst the few other landmarks which still exist around the sprawling Delhi University campus is the magnificent house, constructed in 1911 as a temporary abode, for the then Commander-in-Chief of the British Army. The trees around it are fast being felled; in the name of beautification they are being replaced by flowering plants and Mexican grass: there are more students in the gardens then in the classrooms; plumeless bipeds take the place of magnificent peacocks, shy foxes and chattering monkeys. The house has a tiled roof which leaks once in a while - is there no one left with the skill of tiling? There are twenty rooms, each large enough to accommodate a modern flatlet, each with great fireplaces to devour whole logs. Such fires would light an inner glow in our hearts and and bond those present together, so different from the new-fangled gas jets, casual and friendly on the surface but representing an on and off sort of world. Is there a moral here for our relationships? Communication breakdown between the generations might well be one of the reasons for the obnoxious graffiti on the wall of the **Flag-Staff Tower**: walls carry the scribbled messages that the writers hesitate to communicate in person. This mighty Tower, on the top of the hill originated as its name suggests, as a place from which to hoist the ruler's flag when he was in residence at Delhi: it is circular and squat, has but one door, and several barred windows, and once was a garrison post for soldiers: many fleeing Britishers found temporary refuge in it on that fateful 11 May in 1857. It is approachable by four metalled roads passing through the woods, and was once open to vehicular traffic, bur wisdom prevailed and now only pedestrians are allowed. One has to walk up the hill and look down at the world of knowledge at your feet. You can see the magnificent building which now houses the Administrative Offices of Delhi University. It has a very romantic story far away from mundane things, for it was here that Lord

Mountbatten found his love and proposed to Edwina. What a shame that architects cannot remember aesthetics when designing modern buildings, especially if they are 'competing' with something Victorian. The land shrinks and mankind expands but this should not deprive us of beautiful surroundings: so much happiness is sacrificed to utilitarian greed. No greenery, no flowing water, just a concrete jungle: perhaps we need another 'mutiny' in Delhi - to free us from modernity?' **Neelmani Bhatia, Writer and College Lecturer.**

DELHI, THE MAGAZINE On the afternoon of 11 May 1857 a party from the mutinied 38th BNI approached the Magazine in Delhi, near to the Cashmere Gate, and demanded its surrender in the name of the King of Delhi. No reply was given them so they brought scaling ladders from the King's Palace and placed them against the walls. Anticipating a fight the Indian employees of the Magazine slipped out by these scaling ladders, having first hidden the priming pouches. The insurgents then gathered in crowds on the walls; but the small group of besieged Europeans kept up an incessant fire of grape upon them. At length Conductor Buckley received a ball in his arm, and Lieutenant Forrest who had been assisting him was also struck, so further defence being thought hopeless it was decided to fire the trains of gunpowder which had been laid some hours before in readiness. Willoughby gave the order, Buckley gave the signal by raising his hat from his head and Conductor Scully fired the trains, and perished in the resultant explosion, as did also Sergeant Edwards. The other Europeans though all hurt escaped from the ruins and retreated through the sallyport on the river-face. At least five hundred Indians, some of them mutineers but the greater part being civilians, died as a result of the explosion, which took place at 3p.m. It is said that the massacre of European civilians which followed later was in revenge for these deaths. *See* **Kaye, Sen, Ball, Martin** etc *See also* A.Lllewellyn, *The Siege of Delhi*, London, 1977, Harvey Greathed, *Letters written during the Siege of Delhi*, London 1858. Charles Metcalfe, *Two Native Narratives of the Mutiny of Delhi*, London, 1898. and *MAGAZINE, DELHI.*

DELHI: MAINODIN HASSAN KHAN This man was a police thanadar on the outskirts of Delhi when the outbreak occurred.. He kept a diary, and many years later was prevailed upon to allow it to be published (after his death) in English translation. His account is one of the very few authentic contemporary Indian statements on the mutiny/rebellion, and is therefore a most valuable primary source for historians. Here are some examples/extracts from his account: ' In the month of January 1857 the house of a European gentleman and the Telegraph Office at Ranigunge were burned down. This was a concerted signal; it was calculated that the burning of a telegraph office would immediately be communicated along the line from Calcutta to the Punjab, and that those in the secret would understand on hearing of it that they too must fire houses. Information of their incendiaries was widely circulated in all directions and it is said that letters were sent from regiment to regiment inciting them to commit similar acts.'...'Among the natives it is said that late the night before (ie 10 May) a sowar had arrived from Meerut with a letter for Mr Simon Fraser the (Delhi) Commissioner, which the Jemadar took to him; he was sitting in his chair asleep after dinner, and the Jemadar had to call several times to his master before he awoke. The Jemadar then told him a sowar had brought an urgent letter from Meerut. The Commissioner however rebuffed him and taking the letter from the servant's hands mechanically put it into his pocket, falling asleep again afterwards. The servants were afraid to awaken the Commissioner again, and all that they gathered from the sowar was that he had learned from the patrol who had given him the latter, that there was a great *goolmal* at Meerut.'.....'entered the Palace on foot. I was summoned and, prostrating replied to his question that my object in seeking an audience was that plunder and butchery were going on; and all the bad characters were searching for Europeans and Christian women and children to destroy them. I begged the King to stop this, and to arrange for the restoration of order. The King replied: I am helpless; all my attendants have lost their heads and fled. I remain here alone. I have no force to obey my order; what can I do? I replied, if your Majesty will tell me the desire of your heart, possibly I may be able to carry out your commands. I described my proposed line of action. The King replied, My son, this duty will I expect from you: you have come to me in a moment of difficulty and danger; do whatever seems good to you, I command you'.....'Nawab Amin-u-din Ahmed Khan and Nawab Zia-u-din Ahmed Khan were deemed wealthy men and pressure was brought upon the King to extort money from them. Collecting their relatives and retainers they rode to the Palace with a retinue of a few armed men. On reaching the Lal Purdah Gate they were stopped by the sentries...Nawab Amin-u-din at once knocked the sentries over and pushing through the gate, passed into the Palace. On this occasion I was present. An audience was demanded of the King, who was at this time surrounded by the rebel Subahdars...the King seemed pleased to see them, but the rebels seemed to be nonplussed and scowled at them...being pressed by the King to accept the idea Amin-u-din rode with his relatives to the Mirza (Mogul)'s house. There he found large numbers of the rebels, and with difficulty obtained a seat.... one of the rebels taunted Amin-u-din with living at his ease and in safety while they were starving. This was followed by abuse...those who had accompanied A raised their carbines and covering the spokesman challenged him to repeat what he had said at the peril of his life. The rebels were for the moment awed....under the care of Bakht Khan Amin-u-din left the Prince's house and reached his own in safety. He determined to leave the city; of course he was at once suspected of de-

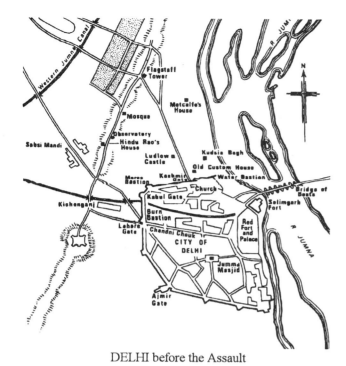

DELHI before the Assault

serting to the English. On reaching the Cashmere Gate he was stopped by the rebel guard, threatened with immediate death and detained. Unable to get out of the city he returned to his own house. ' For further extracts etc See MAINODIN HASSAN KHAN. For the full text of his diary/account, *see* Charles Theophilus Metcalfe, *Two Native Narratives of the Mutiny in Delhi*, London, 1898.

DELHI, RURAL ENVIRONS 'If rural discontent and turbulence was to be turned to the service of organised rebellion under the King of Delhi, then the only possible leaders were some Muslim *jagirdars* of whom the Nawab of **Jhajjar** was the most prominent.' See Eric Stokes, *The Peasant Armed, the Indian Revolt of 1857*, Oxford, 1986.

DELHI, THE SIEGE OF Strictly speaking Delhi never was 'besieged'; the rebels held the city and at all times the east and the south were open to them, while the British held the Ridge, and the north and the west (especially the lifeline route to the Punjab) were open to them. Nonetheless the term has persisted, and the 'Siege' lasted from 8 June 1857 when the **Battle of Badli-ki-Serai** was fought and the British occupied the Ridge, the high ground to the northwest of Delhi, until the final assault on the city on 14 September 1857. There are many accounts covering this period which saw a build-up of forces on both sides, *see* **Kaye, Sen, Ball, Martin** etc., *see also* A.Llewellyn, *The Siege of Delhi*, London, 1977, Harvey Greathed, *Letters written during the Siege of Delhi*, London 1858. Charles Metcalfe, *Two Native Narratives of the Mutiny of Delhi*, London, 1890.

MAINADIN HUSSAIN; *MAGAZINE, DELHI*; PAPERS OF A MISCELLANEOUS CHARACTER, GHALIB, SYED AHMAD KHAN, JIWAN LAL, etc.

DEOLI A cantonment town some eighty miles south of **Jaipur** in **Rajputana** and approximately the same distance north of **Kotah**. It was the depot of the **Kotah Contingent** which, although most of it later mutinied, at first gave good service to the British in suppressing rebellion in the Agra and Muttra areas. In June Deoli, in the absence of the Kotah Contingent all but a guard of 120 men, was attacked and sacked by the **Neemuch Brigade** and some of the **Malwa Contingent**, and the cantonment burnt. The European women and children had fled to **Ajmer** the day before. On 1 July 1857 Captain Forbes was sent to Deoli 'to restore confidence', and raised a force of some 800 men of the Mina tribe, not previously popular with the authorities, but now used to take the place of the Kotah Contingent.

DEVERINE, Mr He was a Telegraph employee, murdered by the rebels at Bunnee 'in the execution of his duty' in November 1857. His career, indeed the manner of his death, is obscure, but he may well have played, inadvertently a significant part. When the **Begum of Oude** was hesitating, in the last phase of the mutiny/rebellion when she was in Nepal with the remnant of her forces, as to whether or not to give herself up to the British, her decision may well have depended upon her own knowledge of events that only latterly came to light. She was promised good terms, if she had not murdered any Europeans, and it was confidently believed at first that she had not done so. Indeed witnesses had come forward to the British in Lucknow saying that the Begum had objected to the murder of prisoners which was carried on, against her wishes, by such as **Mammu Khan**. However witnesses came forward with 'strong testimony' to the effect that Deverine's head was sent to the Begum's private apartments, that she might feast her eyes on the sight, and that the bearer of the trophy was rewarded with a *killat*. As a result the Chief Commissioner concluded that she was a woman' of savage disposition who delighted in the blood of Europeans'. That she could have saved the women seems likely, but as Miss **Georgina Jackson** (one of those murdered) was the niece of the execrated former Chief Commissioner of Oude, Coverley Jackson, it seems that she may well have deliberately refused to intervene, *see* Foreign Political Consultations, 29 July 1859, No.324.

DEVIL'S WIND, The It has been asserted that this was the name by which the mutiny/rebellion was known to the inhabitants of North India for many years. It is certainly the name given to two books on the subject: a novel by M.Malgonkar and the story, by G.L.Verney, of the Naval Brigade at Lucknow, *see* Bibliography.

DHAR A town, and State, of that name in Central India, some 20 miles northwest of **Mhow**. The ruler was a minor and so was not held responsible by the British for the rebellion of his Durbar and troops. The latter bravely faced the column of the **Malwa Field Force**, commanded by **Brigadier C.S.Stuart**, in an action at Dhar on 22 October 1857, and were then besieged in Dhar Fort; on 30th they displayed a white flag and asked for terms but were told that only unconditional surrender was acceptable; they evacuated the fort in the night and fled northwards towards **Mandesore**. For further details see *The Revolt in Central India*, Simla, 1908, p 76 *et seq*. Also Khushhalial Srivastava, *The Revolt of 1857 in Central India-Malwa*, Bombay, 1966.

DHOLPUR A small state lying between Agra and Gwalior on the Chambal river. The Rana remained faithful to the British and on more than one occasion assisted Europeans fugitives to reach safety in Agra.

DHONDU PANT, *see* **NANA SAHIB**

A Dhoolie

DHOOLIE Or dooley or doolee, a kind of litter carried by two men in which the wounded (and sick) were carried in battle, and on the march. The carriers were often very brave men and suffered heavy casualties. The dhoolie itself usually resembled a stretcher fitted with a roof and side curtains made of canvas.

DHOOLIES The presence of dhoolies after these battles was evidently a cause for confused alarm back in England, where the word was unknown. .'It is said,' wrote American missionary William Taylor, 'that some editor in England, in giving his readers an account of a terrible battle in the Mutiny, made quite a sensational climax by the statement that after the engagement the 'dreadful Dhhulies came down upon Her Majesty's wounded soldiers and carried them off.' The fact was, they were thus carried off to the hospital to be treated.' Wm.Taylor, *Four Years' Campaign in India,* pp 50-1.

DHOURAIRA [Dhaurahra] A town in Kheri District of Oude[Awadh], where the Rani, as regent for her young son the Raja, attempted at first to aid the British, giving shelter to refugees from **Sitapur**, but eventually such pressure was brought upon her by neighbouring rulers that she was compelled to send the party in to the rebel court at Lucknow; they escaped but all died in the **Terai** except **Captain John Hearsey** who eventually reached Nepal and returned to the British via Naini Tal. *See* John Pemble, *The Raj the Indian Mutiny and the Kingdom of Oudh, 1801-1859,* Delhi, 1977.

DHURMA SOBHA A Brahminical Association, formed at Calcutta, for the advocacy of ancient Hindu customs, and specifically directed against European innovations (especially the recent abolition of the laws enforcing perpetual widowhood). This association, thought **General Hearsey**, was instrumental in 'tampering' with the sepoys (before the outbreak), and had been responsible for circulating if not initiating the idea that the new ammunition was part of a government plot for the destruction of the caste of the whole Bengal army.

DIARY OF MAJOR BANKS In **Lucknow** today there is a most impressive Government building known as Banks's House. You can't go in and visit it - unless of course you are on Government business, but even from the road it looks most impressive. It is hard to believe that on that spot was once the home of just one man and his family, even if he were the Commissioner of Lucknow. Major Banks was the first occupant in that capacity, immediately after the annexation of the **Kingdom of Oude[Awadh]** by the British in 1856. He has given his name to a modern Government building, but also to a little known but revealing diary which he kept during the siege of the **Residency** in Lucknow, by the 'mutineers' in 1857. It is only by chance that the diary survived at all: it was found after the siege, in the city, in the house of a pay havildar of one of the rebel regiments. He had regarded its contents as of no consequence and had written over many of the pages with his accounts. What we have recovered of the diary covers a very short period, from 2 July to 19, so why do I say it was of interest? Above all it reveals the extraordinary power struggle that took place in the Residency when the Chief Commissioner, **Sir Henry Lawrence** - loved and respected by all - was mortally wounded by a rebel roundshot at 9a.m. on 2 July. Sir Henry, whose grave is still to be seen within the Residency grounds, had been the inspiration for the besieged garrison, and his death would obviously have been much lamented by most of them: but not, perhaps, by those among them who sought to be his successor. Chief among these were the slightly sinister figure of the Financial Commissioner, **Martin Gubbins**, and Major

Banks himself. The most significant diary entry is the very first: 'It has pleased God that Sir Henry Lawrence should be very grievously wounded, it is feared mortally, at 9am this day. Sir Henry had previously notified to Government his desire that, in case of any casualty befalling himself, I (Major Banks) should fulfill the functions of Chief Commissioner, and that **Colonel Inglis** HM's 32nd Regiment, commanding all the troops, and Major Anderson, should be a military council. This morning, after being wounded, and while in the perfect possession of all his faculties, Sir Henry publicly delegated the above charge to the respective gentlemen, and these functions have now been provisionally assumed by them.' Then follow fourteen specific instructions from Sir Henry as to the defence of the Residency, including the one which we know was obeyed : 'Put on my tomb only this 'HERE LIES HENRY LAWRENCE, WHO TRIED TO DO HIS DUTY. MAY GOD HAVE MERCY ON HIM'. Colonel Inglis assumed the local rank of Brigadier forthwith. But the same day, 2 July, there was another meeting, at two o'clock in the afternoon at which Mr Gubbins, Financial Commissioner, announced that 'so long as Sir Henry lives, he would not record any objection, but would urge his claims to the chief post, should it please God to take Sir Henry.' The fact is that Gubbins could claim to be senior in the service to Banks, and as every Government officer knows, there is nothing more holy than seniority in the eyes of authority: even if mediocrity triumphs, seniority must be observed. Banks noted Gubbins's caveat but ordered that it be 'left over', with the significant words 'sufficient to the day is the evil thereof', and proceeded to take over command, as Sir Henry had instructed. For the next two days the diary is full of Banks's orders and actions, which were, apparently, carried out without demur. Then, on 4 July the worst happened: 'Our most honoured chief, Sir Henry Lawrence, KCB, has gone to his rest'. The power struggle could begin in earnest. Banks was helped by the fact that Mr Ommaney, the Judicial Commissioner, and probably a strong supporter of Gubbins, was also wounded by a rebel roundshot, and died a few days later. Now the correspondence came thick and fast, Gubbins sending three letters in quick succession demanding the 'chief civil power'. For a civil servant, in any age, to put his demands in writing is well understood, but in this case it was remarkable because Major Banks was actually living in Gubbins's House! It would seem that the two men were not talking to each other, just firing off written broadsides. Banks got on with what he considered his job. Gubbins sulked and waited. He did not have long to wait. Major Banks was shot through the head by a bullet on the roof of Gubbins's House, on 21 July, and died immediately. Unfortunately for his ambitions, Martin Gubbins had just quarrelled with Brigadier Inglis: he had much annoyed the latter by sending a totally unauthorised message to General **Havelock** who was struggling towards Lucknow with his small relieving column. What happened next is told in Gubbins's own words: 'On Major Banks' lamented death, the succession to the Chief Commissioner's duties naturally devolved upon myself, who stood next to Sir Henry Lawrence in the Civil Commission. Brigadier Inglis, however, now opposed my undertaking the office, which accordingly remained vacant. Civil authority ceased and military command remained with the Brigadier: whose correspondence was always, I am bound to say, civil and courteous.' Correspondence, the man says! - in the middle of a horrendous siege! If all he could think of was the writing of official memos, no wonder he was passed over. Poor Gubbins, he was destined to be eternally passed over, despite his seniority. It all preyed on his mind so much that he eventually committed suicide.'

PJOT

DIGNUM , PETER 'In the assault on Delhi on 14 September 1857, over 1 in 4 of the attackers became a casualty: 66 officers and 1,104 men of the British force of 3,700 were killed or wounded. With odds like that of surviving those that took part are entitled to be thought of as brave, or, as the British preferred to call it, to 'have pluck'.But there were some whose courage was suspect. This is the tale of one such: a born survivor, one of the street wise 'old soldiers' who had signed on for unlimited service, and by 1857 had already done some fifteen years in Her Majesty's 52nd Light Infantry Regiment. He was 'old' therefore in experience rather than age. His name was Peter Dignum, a private soldier and most unlikely ever to rise above that basic rank. He was, apart from anything else, a drunkard, and for many years' past had regularly been courtmartialled for 'habitual drunkenness': his sentence had been, equally regularly, twenty-five lashes with the cat-o'-nine-tails, that instrument then still used in the British Army (though banned in the Indian Army as barbaric) for flogging delinquents. In 1857, when the 'Mutiny' broke out, Pte Dignum decided that he did not fancy either active service against the mutineers or, more immediately, a hot-weather march down from the Punjab to Delhi. So he deserted. On sentry duty one fine night, when no one else on duty would be liable to challenge him, he absconded. He decamped with a girl from the bazar who had taken his fancy, and together they headed for the hills where the girl's village was sited. For a few weeks anyway he lived a life of ease, with the cool breezes of the mountains playing about his idle ears. But, the money ran out, food was short as a result and, perhaps, the bazar girl had lost her attraction. He decided to give himself up to the authorities as a deserter. He was, in the customary fashion, passed on to his regiment to be tried by general court-martial for the heinous crime of desertion. He came down country towards Delhi in the comparative luxury of a bullock-cart, drawn by two white bullocks, and with a guard for company. It was slow progress, but Dignum was in no hurry to arrive, and could reflect that his journey was considerably more comfortable than that of his comrades who had had to march every foot of the way. He arrived in the camp on the Ridge before Delhi on Saturday 12 September 1857, and was duly handed over, a prisoner to his regiment, the 52nd Light Infantry. The very next day, the General (Wilson) in charge of the British forces issued an order that invited any soldier being held prisoner for military offences to volunteer for the coming assault on Delhi. After all, they were very short of men: every trained soldier was needed. Peter Dignum duly volunteered. But...he made a bad mistake. Being anxious to save his hide he fell out as the regiment marched down to the assault on the city, complaining of being unwell. Taken into the field hospital he - as an old soldier would - deceived the doctors into believing that he was really ill. But he did not deceive his comrades: they knew he was malingering. When Delhi had finally been taken, General Wilson declared an amnesty for all those military prisoners who had taken part in the assault, no matter how heinous their crime. Dignum was of course included. His good fortune seemed limitless. In addition to a free pardon for desertion, he, because technically he had taken part in the assault, became entitled to six months' extra pay, and a share of the Delhi prize-money. But retribution was waiting. His comrades could not forgive his malingering on the day of the assault. Before long he was again on a court-martial charge of drunkenness - 'framed' this time: this time he was given the maximum sentence - fifty lashes. And instead of the sympathetic presence and support of comrades, who generally speaking hated to see their mates tied to the triangle and flogged, he had to endure the cheers and abuse of the whole regiment as punishment was administered. And what punishment! No 'going through the motions' this time: we are told that 'heavily were the lashes laid on, twenty-five by a right-handed bugler, and twenty-five by a left-handed one. That was one 'old soldier' who proved to be less clever at shirking his duty than he had thought himself to be.' **PJOT**

The Dilkusha in 1858

DILKUSHA The 'Heart's Delight' hunting lodge built in Nawabi days, to the south-east of Lucknow, and close to the **Martinière** College building. Used by **Sir Colin Campbell** as a headquarters in November 1857 when relieving the **Residency** and again in March 1858 when making the final assault upon **Lucknow**. *See* R.Llewellyn-Jones, *A Fatal Friendship,* Delhi, 1985; *see also* W.H.Russell, *My Diary in India,* London, 1860.

DINAPORE [Danapur] On the Ganges in Bihar, some ten miles from Patna which it served as a military cantonment. The Divisional Commander here was **Major General Lloyd,** an officer whose conciliatory attitude undoubtedly postponed the mutiny of the three sepoy regiments in the station which were the 7th, 8th and 40th regiments BNI. The attempt to disarm them however finally provoked the mutiny, on 25 July 1857, and they marched off to join **Kunwar Singh** and to invest **Arrah**. The events at Danapur are well documented and not in dispute, *see* **Ball, Sen, Martin, Kaye** etc.

DINAPORE, MASSACRE OF In mid-August 1857, there took place in Dinapore[Danapur] an event that was effectively hushed up at the time but which was roundly condemned by the military authorities. It has permanently stained the reputation of HM's 10th Regiment. Troops from this regiment returned from operations against **Jagdishpur** in a totally undisciplined state. General **Lloyd** who might have prevented the outrage that followed had been removed from command. The troops, drunken and insubordinate, attacked the camp of some 100 of the 40th BNI who had remained staunch and had refused to join the mutiny of their comrades. Before the massacre could be stayed at least five sepoys had been killed and twelve wounded including a woman. The Europeans tried for murder were eventually acquitted in default of legal evidence. Sir **James Outram** arrived at Dinapore on 17 August and issued a general order expressing the utmost horror and indignation at the conduct of the 10th, and relieved them of all town duties which were now given to some men of the 5th Fusiliers newly arrived from the UK; *see* **Martin**, Vol II, p 414.

DIRIG BIJYE SINGH There is a tale told of Dirbijye Singh of Morar Mhow, Oude[Awadh], who had helped the British considerably in the early days of the rebellion: Governor General proposed to reward him for his services. But Dirbijye Singh came to **Lord Clyde** and with tears in his eyes begged he would be good enough not to give him any reward, until, he said, 'you have disposed of these badmashes; for if I get anything from you, and then you leave the country, they will not only take away that which you gave me but that which I have already of my own. But once you are masters, I shall be obliged to you to be as liberal as you like.'

DIRIG BYJAI SINGH Raja of **Morar Mhow**. Remained loyal to the British throughout the campaign in

Oude[Awadh] and was rewarded after the final assault on Lucknow, in what was known as the 'Oude Proclamation', by the Governor General, issued on 14 March 1858: '...The first care of the Governor General will be to reward those who have been steadfast in their allegiance at a time when the authority of Government was partially overborne, and who have proved this by the support and assistance which they have given to British officers...' So far as the British were concerned Dirigbijai Singh's most significant assistance was in the saving of the lives of Lt.**Mowbray Thomson** and Lt **Delafosse,** two of the very few British survivors from the massacre at **Satichaura Ghat**.

DISCIPLINE 'Allahabad 12.12.57....I have no news to give excepting that there is no discipline in our army at Cawnpoor - two Europeans murdered Salik Rasi (?) the great contractor there and they have been obliged to have role (sic) called every hour to keep the troops from looting the town. This is dreadful work. The Commander in chief is not at all liked and not much thought of.....' *From* a letter of **Captain Archibald Impey**, Bengal Engineers, Collection of **PJOT**.

DISCIPLINE, REBEL After mutinying, some units of the Bengal Army took great pride in remaining well-disciplined, and smart in appearance and drill; much depended upon the influence and character of the native officers of individual corps. But the troops in **Kanpur** did not apparently maintain this high standard at all times. Many had preferred lounging in the bazar to pressing home the attack on the **Entrenchment**, while others had actually absconded and left for their homes with their share of the loot. An example of ill discipline comes from the period when news of the approach of British troops upon Kanpur from **Allahabad** was received; preparations were made to despatch troops down the **Grand Trunk Road**: 'On our way we had to pass the lines of the 2nd Light Cavalry, and here a scene, worth regarding, attracted my notice. It appears that an order had been issued by the 'authorities' to assemble five thousand troops and march them off to Futtehpoor, with nearly all the artillery available, for the purpose of attacking and repulsing the British force, said to be advancing from Allahabad to attack Cawnpore. Now, the several rebel corps belonging to the station had established themselves in their respective lines, and made themselves very comfortable in them. The 2nd Cavalry was at the time being made to assemble on its parade ground by a few troopers and sirdars of the same corps in full dress, who, duly mounted, were calling out with all their might to the men in the lines to fall in; but they appeared to feel so reluctant to leave such agreeable quarters, that neither persuasions, threats, nor imprecations seemed to have the least effect upon them. Some pretended to be saddling their horses, others packing up their luggage, &c., and a great many were seen skulking away towards the city; then it was, that the contrast between the steady regularity and ready obedience of orders under the British rule, and that of the present management, became very striking. The prisoners, however, passed on, and I did not see how the leaders managed to collect their men together. ...On entering the city, we met several sepoys, most of whom had been plundered by the villagers in the neighbouring districts while attempting to convey to their families the money they had come in possession of at Cawnpore, for immediately on the outbreak occurring, the Rajpoots and other villagers posted themselves on the roads in large bodies and well-armed, for the purpose of plundering the travellers, which work they found very profitable, as they invariably succeeded in disburthening the sepoys (who generally went singly and clandestinely in order to avoid their comrades; for where money is concerned a native will not trust even his father) of their ill-gotten booty and often treating them to a sound thrashing, send them back empty-handed. Such ill-usage, it may be supposed, caused a bitterness of feeling in the hearts of the wretched sepoys not easily to be effaced. Seeing so many prisoners, and believing them to be all villagers, their exultation was very great. The sight, as it were, added vigour to their bodies, such gestures and such menacing looks such imprecations and abuse showered on them! I can never forget the scene. I am only astonished that they were restrained from falling upon us and satisfying their revenge. They, however, contented themselves by giving vent to certain horrible sentences upon the unfortunate prisoners. One was for blowing them away from the guns, another was for cutting off both the hands and noses of all, and letting them go as living examples to others!' *See* W.J.Shepherd, *A Personal Narrative etc*, Lucknow, 1886. This activity on the part of the rebels was apparently linked to an order issued by **Nana Sahib** on 5 July 1857 to the **Kotwal, Hulas Singh**: 'It has come to our notice that some of the city people, having heard the rumours of the arrival of European soldiers at Allahabad, are deserting their houses and going into the districts; you are, therefore, directed to proclaim in each lane and street of the city that regiments of cavalry and infantry and batteries have been despatched to check the Europeans either at Allahabad or Futtehpore, that the people should therefore remain in their houses without any apprehension and engage their minds in carrying on their work.' *See* London *Times*, 29 October 1857.

DISHONOURING of EUROPEAN WOMEN

Of significance is the confidential report to Lord **Canning**, the Governor General, from **W.Muir** on this subject. Canning had asked for a confidential report, probably because of the exaggerated stories that were being circulated in London, and *possibly* because **Queen Victoria** herself asked for a confidential report. Muir was quite specific: *'Hindoos except of the lowest grade would have become outcasts had they perpetrated this offence....must be admitted there is nothing in the habits and tenets of the Musulmans population which would prevent them from taking females seized* *at the general outbreak to their homes with sinister designs........but no evidence.....the solitary exception being the story regarding **Miss Wheeler**. 30 Dec 1857. PS 5 Jan 58. There are points connected with the long detention of the Nana's victims which render Cawnpore a peculiar case... I would recommend further particular enquiry at Cawnpore. sgd W.Muir.' See Home Miscellaneous Papers, 724,725 IOL London.* 'December 1857, .. the only instance that came to my knowledge of any woman having been ill-treated in any way at Delhi was when one of the 3rd Cavalry troopers insultingly patted the cheeks of some poor creature at Daryaganj, on which her husband killed the man, and both husband and wife were immediately killed by the bystanders.' - C.B.Saunders, Offg Commissioner, Delhi, to W.Muir, 'Dishonour. It is surely most heartless to the friends of those who have perished to argue whether this is a circumstance *likely* to have occurred or not. The point is, what evidence have we? If the story of the girl [**Amelia Horne**] in Calcutta brought forward by Dr. Knighton in the *Times* is authentic, this is clearly one case. With regard poor Miss [Wheeler], if the drummer's evidence that he saw her at Futtehgurh be true, the other story of her drowning herself in a well here must be false, and *vice versa*. This case is not as yet so clear. I have not heard of any others.' J.W. Sherer, *Daily Life during the Indian Mutiny*, London, 1898.

DISRAELI, Benjamin

He was in opposition at the time of the outbreak of the rebellion and made speeches in the House of Commons critical of the Government's handling and in particular of its failure to understand the true nature of the *emeute*, which he characterised as a national revolt rather than 'a mere military mutiny'. But this is not to say that he wished the rebels well (as some historians have supposed); on the contrary he advocated a policy that would reassure Indians that Queen Victoria is a 'sovereign who will respect their laws, their usages, their customs, and above all their religion' etc. When back in power (February 1858 in Lord Derby's government) he supported the active prosecution of the war against the rebels. *See* **DNB**

DISRAELI

On the 25th Anniversary of the Royal Bucks Agricultural Association at the end of September Disraeli spoke about retribution. 'I am persuaded that our soldiers and our sailors will exact a retribution which it may, perhaps, be too terrible to pause upon. But I do without the slightest hesitation declare my humble disapprobation at persons of high authority announcing that upon the standard of England 'Vengeance' and not 'Justice,' should be inscribed. ...I for one protest against taking Nena Sahib as a model for the conduct of the British soldier. I protest against meeting atrocities by atrocities. I have heard things said and seen things written of late which would make me almost suppose that the religious opinions of the people of England had undergone some sudden change, and that instead of bowing before the name of Jesus we were preparing to revive the worship of Molech.' It was for the British to gain by the lessons of the Mutiny 'like brave and inquiring men' and establish a Government for India 'at once lasting and honourable to this country.' *See* London *Times*, 1 October, 1857, p 10.

DOST MAHOMMED KHAN, Emir of Kabul

The neutrality of Afghanistan was vital to the British cause, for the forces with which Delhi was recaptured from the rebels came from the Punjab which was thus denuded of strength. Had Dost Mahommed chosen this moment to attack from the North-West, he would have found little resistance to his advance, but instead he chose to remain honourably true to his Treaty engagements with the British government in Calcutta. *See 'Parl. Papers relating to the the Mutiny in the Punjab in 1857'. Also 'Journal of a Political Mission to Afghanistan'*, Major H.Lumsden, 1857.

DOUGLAS, BRIGADIER

He commanded a British column operating in Oude[Awadh] and Bihar. Best known for his pursuit of **Kunwar Singh** and the fatal wounding of the latter as he crossed the Ganges, *see* **Ball, Sen, Kaye etc.**

DRIGBEJAI SING

Talukdar or Raja of Mahona in Oude[Awadh]. Confusingly he has a similar name to his compatriot of **Morar Mhow** who was one of the staunchest supporters of the British. This leader was one of the last to fight against the government, but. was eventually captured and put on trial; *see Trial proceedings, Government versus Raja Drig Bijai Singh*, Lucknow Chief Court Mutiny *Basta*.

DROWNING OF NANA SAHIB

When **Havelock** occupied **Kanpur** in July 1857, the **Nana** went to **Bithur**; thence he crossed the Ganges into **Oude**[Awadh], but not before apparently convincing many of his followers that he would drown himself in mid-stream: 'To the best of my information, [the Nana] left Bithoor on the evening of the 17th. He found it impossible to get any of the soldiers to rally round him; they had thrown off restraint, and abused him and **Baba Bhut** in open terms, clamouring with threatening gestures for money, and so off, helter skelter, for Futtehgurh. That evening he embarked himself and the ladies of his family on a large boat. He had given notice that he should drown himself, I suppose as a blind to prevent pursuit, and it was understood the signal was to be, when the light was put out. The Gungapootras were watching on the shore. About mid-stream the light was extinguished, and with a yell that must have reached the boat, the mendicant Brahmins rushed up to the palace, and commenced plundering all they could lay their hands on. The crafty Nana was disembarking in the darkness on the other side; but if in so callous a heart, any bitter reflections could arise, the ingratitude of his adherents and falseness of those he had cherished, might well have induced them.' J.W. Sherer, *Daily Life during the Indian Mutiny*, London, 1898.

Drunk with Plunder

DRUNK WITH PLUNDER 'Leaning against a smiling Venus is a British soldier shot through the neck, gasping, and at every gasp bleeding to death. Here and there officers are running to and fro after their men, persuading or threatening in vain. From the broken portals issue soldiers laden with loot or plunder: shawls, rich tapestry, gold and silver brocade, caskets of jewels, arms, splendid dresses. The men are wild with fury and lust of gold - literally drunk with plunder' - extracted from the account of the looting of the **Kaiserbagh** in **Lucknow** in March 1858, contained in **W.H.Russell**'s, *My Indian Mutiny Diary*, London 1860.

DUBERLEY, Fanny   One of the (rare) English-women who, in order to stay near to their husbands, actually refuse to be parted from them. Fanny was the wife of Captain Henry Duberley, an officer of the 8th Hussars, and accompanied him to the Crimea, and is also said to have 'ridden with the 8th Hussars in India for 11½ months, and been present when the **Rani of Jhansi** was killed by a Hussar of her husband's regiment. Her immediate comment thereafter is said to have been 'Out of this damned country I will get as soon as I can' *See* Veronica Bamfield, *On the Strength; the story of the British Army Wife*, London, 1974.

DULEEP SINGH   'One afternoon, while wandering through the back streets of Thetford in Norfolk, I came across a very ancient, timber-framed house, with a sign outside saying that it was, appropriately, the 'Ancient House Museum'. Entrance free. The inside was, at first, predictable: a pictorial history of the town, artists' impressions of Stone Age men digging up flints from 'Grimes Graves' a few miles away, portraits of town worthies from past centuries etc. But then I saw something that really puzzled me. It was a notice to the effect that the Ancient House, a 15th century merchant's house in Whitehart street, had been purchased and presented to the town in 1924, for use as a museum, by one 'Prince Frederick'. Now we have a large royal family in England, but we haven't had a Frederick for 200 years, so who was this Prince I wondered? The puzzle was compounded when I saw that he had two brothers, Prince Victor and Prince Albert, and no less than five 'Princess' sisters - Catherine, Sophia, Bamba, Pauline and Irene. Quite by chance I had wandered into the little town where Maharaja Duleep Singh, last Sikh ruler of the Punjab, had settled down when he came to England, exiled effectually, in 1854. He was then only sixteen years' of age, but was accepted at Court, became a friend of Prince Albert, **Queen Victoria**'s husband, and indeed came to be known as 'Queen Victoria's Maharaja'. She is said to have been very fond of him, and anxious to be of help in his career. After spending a couple of years at **Fatehgarh**, Duleep Singh purchased a large country estate near Thetford, called 'Elvedon Hall', and proceeded to transform the interior of the house into a magnificent Indian extravaganza, which could be rivalled in England only by the Prince Regent's Royal Pavilion in Brighton. He married twice, a German lady, and an English one, which is part explanation for the English names of his children: the other explanation is that for most of his adult life he professed himself a Christian, though he reverted to Sikhism before he died. During the 'Mutiny' period, 1857-9, he made no pronouncement for or against, and was criticised for not doing so - but not by the Queen. Victoria, always passionately loyal to her friends, wrote 'His best course is to say nothing - and he must think so. It is a great mercy the poor boy is not there (ie Fatehgarh).' She was godmother to his eldest son Prince Victor, and showed great interest in his family throughout her life, although very saddened by Duleep Singh's behaviour from about 1870 onwards. By then Prince Albert, his first mentor was dead, and he had become fast friends with 'Bertie' ie the Prince of Wales, later to become Edward VII. The two men led a pretty rakish existence in London, to the despair and unhappiness of their wives, having frequent visits to Paris, and taking up and keeping mistresses. Duleep Singh's allowance from the Government was generous but inadequate to finance his extravagant life style. He petitioned for an increase, but this was not granted, and he felt bitterly betrayed. Indeed henceforth he chose to discard all things English, went to live in Paris (and Russia for a short time) and once signed himself 'Sovereign of the Sikh Nation and proud implacable foe of England'.

Prince Duleep Singh

Victoria we are told, was not amused, and commented 'He must surely be off his head!' - she thought him a kind and good man who was 'led astray' by designing intriguers. The letter to which he had put the above signature was a truly remarkable document. His first wife had just died (of a 'broken heart', some said), and a letter of condolence was sent to Duleep Singh on Bertie's instructions. Duleep Singh replied: *'The letter conveying to me Your Royal Highness's sympathy in my late bereavement has been forwarded to me from England; under other circumstances I should have felt most grateful for Y.R.H's condescension, but in the present circumstances, while your illustrious mother proclaims herself the Sovereign of a throne and an Empire, both of which have been acquired by fraud by pious Christian England, and of which YRH also hopes one day to become the Emperor, these empty conventional words addressed to me amount to an insult. For YRH's sympathy can only be expressed by one friend towards another - but which cannot ever exist between enemies. Duleep Singh, Sovereign of the Sikh nation etc.'* He died in Paris in 1893, still bitterly estranged from the British. This estrangement was not shared by his children, particularly his son Prince Frederick who brought his father's body back to Elvedon for burial, still a place of pilgrimage for Sikhs. One can only wonder what might have been the Sikhs' attitude in the Punjab had Duleep Singh rebelled against the British in 1857.   **PJOT**

Elvedon, Norfolk

DULGUNJUN SINGH, General   *See* PERSONS OF NOTE, APRIL 1859.

DUM DUM   It was always believed that it was here that the CARTRIDGE QUESTION first came to the notice of the sepoys. 'At the Presidency arsenal a low caste *clashy* asked water of a high-caste Brahmin sepoy. The latter indignantly refused. 'Oh!' rejoined the low-caste man, 'you need not be so particular, for you will all soon have no caste, when you come to put pig and bullock fat in your mouths.' The scorn of low-caste Hindus must have especially alarmed the sepoys. *See* Martin Gubbins. *Account of the Mutinies in Oudh* etc. London, 1858, p 85.

DUNBAR, CAPTAIN   An officer of HM's 10th Foot, stationed at **Danapur**; he was sent, on 29 July 1857 to relieve **Arrah** which was being attacked by **Kunwar Singh** and the mutinous regiments from Danapur. The attempt by

410 men, including 70 Sikhs, was a disastrous failure, the troops being ambushed at night only a mile or so from Arrah; the panic was so great that many of the British casualties resulted from fire from their own comrades. Dunbar and six other officers were killed with 184 men. Many also were wounded. For the failure of this expedition, Major General **Lloyd** was blamed, and subsequently dismissed from his command. *See* **Martin** Vol II, p 403.

DUNLOP, Robert H.W.

A volunteer with an irregular cavalry unit formed to deal with outbreaks of incipient revolt among the civil population. He kept a record of his activities and published the result. As it is taken from his own notes and letters, and as he was an active participant in the events he describes, the account is valuable. Useful as showing the nature of the **civil rebellion** in the Meerut District and the means used to suppress it or at least keep it in check. *See* Robert H.W.Dunlop, *Service and adventure with the Khakee Ressalah,* London, 1858.

DUPRAT

'M.Duprat was a merchant of Lucknow and Calcutta. He was French, but like other foreigners, threw in his lot with the British on the outbreak of the 'mutiny'. He had been persuaded to come to Calcutta by a 'D'Orgoni' a foreigner holding out promise of lucrative employment. But he had failed to meet him so Duprat had come to Lucknow where he succeeded modestly as a merchant. He was known to Mr **Gubbins** but more particularly to Gubbins's friend and guest, Mr Lucas. He proved himself a good and valiant soldier during the siege of the **Lucknow Residency**, taking regular duty with other officers and was a great favourite with the garrison. Duprat was a man of good family in France, formerly a soldier and then an officer of the Chasseurs d'Afrique, he had gone through all the campaigns of Lamoriciere Cavaignac Changarnier Peliddier and Canrobert in Algiers. It is said that the Nana Sahib, via Azimoolah Khan before the siege began tried to induce him to command his troops, but he replied 'no, I cannot do so. It is too late now for I have already sought the protection of the British and shall not desert them at a time like this. Besides what can there be in common between me and the assassins of women and children? Tell this to the Nana and Azimullah and be off; for if you are here half an hour hence I'll have you hanged.' In June he had fortified his house just north of the Iron Bridge over the Gomti with the Chief Commissioner's agreement. Sir Henry had availed himself of Duprat's experience as a soldier in Africa and had given him command of a hundred policemen supposedly faithful . He had powder, and ball cartridges supplied by government, two small 3-pdr guns, spare muskets, provisions and a supply of water, so if attacked and the police proved faithful (25 of them were Sikhs in whom implicit confidence was placed), it was thought he could hold his own for some weeks. Eventually Duprat had been ordered to remove all his goods into the Muchee Bhawan and even when there was doubt as to whether that Fort would be retained he was forbidden to take them away, Sir Henry saying that Government would reimburse him for any loss. Towards the end of June he moved into a house in the Residency compound and took there Mrs Hayes' library, Mr Coverley Jackson's furniture and Lady Outram's property, for which he had accepted responsibility: these items were subsequently put to a purpose that their owners could never in their wildest dreams have imagined. The splendid library of Captain Hayes, consisting of priceless Oriental mss and the standard literary and scientific works of every nation in Europe, and dictionaries of every language spoken on earth from the patois of Bretagne down to Cingalese, Malay and ancient Egyptian, were for the nonce converted into barricades, together with mahogany tables, furniture of all kinds, carriages and carts. In this humble duty they were joined by the records of all the civil service offices, large chests of stationery etc. Duprat himself behaved well in the siege, performing some heroics that 'none but a Frenchman or madman would think of'. He taunted the mutineers. He possessed a large-bored heavy rifle which he used with skill. His energy during night alarms was a source of much mirth. The rebels used to cluster round Gubbins's Garrison, especially the new bastion, shouting what appeared to be their warcry, 'Ali, Ali !', and calling on one another to advance with the words 'chalo bahadur', 'advance ye brave'. On these occasions Duprat exposing himself more than was prudent, would yell back defiance at them at the top of his voice 'Come on ye brave! ye rascals, cowards, scoundrels!' which generally provoked a discharge of musketry and matchlock balls in return. His taunts provoked more than musketry:' Cursed dog of an infidel, I know thee! Thou art Duprat the Frenchman living near the Iron Bridge. We'll yet kill you. Be sure of this. Here goes!' Duprat's House itself was an undistinguished building; it had a verandah overlooking Johannes' wall; this was now built up to a height of about six feet with mud two and a half feet thick. A sloping roof covered the verandah which besides the two feet of clear space between it and the mud wall had a number of additional loopholes made in it from which to fire. The mud wall continued outside the house leading in a straight line to the wall of the next house. It was nine feet high and a very poor affair: Anderson of the Engineers had not had time to complete it and it was thought to be a weak spot. But such as it was it protected a little yard with a well almost in the centre. It had no stockade in front of it. Duprat's house itself was a single storied one with three large rooms and a tykhana with three rooms as above plus one more under the verandah. These four rooms, underground were, at the beginning of the siege, stocked with liquors of all sorts, and other goods. Duprat with his usual generosity gave away saucissons aux truffes, hermetically sealed provisions, cigars and wine and brandy to whoever wanted any. Many took away large supplies of provisons and only signed or did not sign at all for what they took. The consequence was that poor Duprat had soon nothing left for himself; he would never have made a very successful merchant one feels. The looting by the men of the 32nd finished things off, and the thousand and one cannon balls and musket bullets which penetrated the house and in the end converted it into a heap of ruins smashed to atoms whatever was not taken away. Rees reported that on 18 July: 'My friend D and myself have been a little too liberal with our good things and instead of as usual taking brandy and water *ad lib* and smoking cheroots *ad inf* we are obliged to content ourselves with one glass of brandy and water a day and a glass of wine in the evening, and as for cigars I can only afford one or two a day now. I have at least a couple of hundred cigars left. Of pickled salmon and truffled sausages none remain now. D has given away to his friends what little he saved from being plundered.' Unfortunately, late in the siege, in August, Duprat while repelling an attack of the enemy from the roof of the building which was called Grant's Bastion, was severely wounded in the face by a musket-ball, which came in at the loophole out of which he was looking. He suffered for a long time very severely and when apparently convalescent, sank and died, heartily regretted by all. He died apparently in the greatest agony. He was buried - unwillingly it seems - by a Father Bernard the Roman Catholic clergyman who was never willing to put himself out or in danger for anyone, least of all a relapsed Catholic like Duprat. That at least is the story, though you will nowhere find a 'Father Bernard' in any list of the besieged. By mid-August Duprat's house in rear of the Cawnpore battery had been reduced nearly to ruins by continual fire. First the verandah came down then the outer wall was demolished, bringing down with it all the roof rafters'. **PJOT.**

DURAND, Sir Henry Marion(1812-1871) Major-General, KCSI. Engineers. Lt Governor of the Punjab 1870-71. Had served in the First Afghan War and Sikh Wars but is most distinguished in his civil capacity. After appointments as Political Agent at **Bhopal** and **Gwalior**, he was appointed to the Central India Agency in 1857, and was instrumental in preventing the spread of the rebellion; held **Indore**, and reconquered Western **Malwa**. Although he successfully defended the Indore Residency he determined, when deserted by Holkar's troops, to retreat from the city, and was subsequently criticised for so doing, notably by **Sir John Kaye.** Married Emily, widow of **Reverend H.S.Polehampton** who was Chaplain in the Lucknow Residency until his death during the siege. Durand died in a most bizarre and unfortunate accident: the elephant howdah in which he was seated was crushed by the gateway at Tonk through which the animal was trying to pass. Should not be confused with Sir Henry *Mortimer* Durand, best known for the Durand Line survey which established the boundary between the NW Frontier Province and Afghanistan. See **DNB,** *Burke's Peerage,* 1923 p.792. *See also Henry Mortimer Durand, Central India in 1857, being an Answer to Sir John Kaye's criticism on the Conduct of Sir H.M.Durand,* London, 1876.

DURBAR Literally this was the Court or levee of a ruling Prince, but its use is often misleading eg 'He was summoned by the Durbar' really means he was summoned by the Prince sitting in Council.

DURRIABAD See DARYABAD.

E

EARLIER MUTINIES There had been major and minor mutinies all through the history of the Indian Army, and not just of sepoys, either. European troops, about a quarter of them British, threatened to join forces with the Nawab of Bengal in 1764. Encouraged by the success of a strike among European troops, the 1st Sepoy regiment complained that they were being denied their fair share of a bonus paid to the army by the Nawab of Bengal whom they had succeeded in several bloody battles to instal on the throne of Bengal. A settlement was eventually arrived at, but emotions ran high, and flourishing a list of somewhat vague demands the 1st seized its officers, released them, and marched off to camp in a mango grove. They were brought back by sepoys of the 10th, fifty of them were selected for punishment, and 24 condemned to be blown from guns in a ceremony so inexorably ghastly that after the first four sepoys, Grenadiers who had insisted on precedence because of their rank, were lashed to artillery pieces and blown to pieces, a moan arose from among the sepoy ranks. The native officers approached their new commander, Major Hector Munro, and declared that if another sepoy were executed they could not contain their troops. Munro replied that he was merely performing his 'high and sacred duty,' and immediately ordered that the cannon be loaded with grape and directed at the native ranks, commanding that his European troops, many of whom were by now weeping, face the sepoys down. The native troops grimly dropped their muskets to the ground, and the executions proceeded; twenty men were blown away that day and four reserved for execution elsewhere as an example to the rest of the army. *See* Kaye, *A History of the Great Revolt,* V I, p 206-9. *See also* WHITE MUTINY, VELLORE.

EAST INDIA COMPANY The Honourable East India Company, formed in 1600, known by many in India simply as the Kampani or Kampani Bahadur, came to the end of its sovereign powers in India as a result of an Act of Parliament entitled 'An Act for the Better Government of India', passed on 2 August 1858, and coming into effect on 1 November of that year. On that date Queen Victoria was declared the sovereign of British India (although the title of Empress of India was not assumed for a further twenty years), and a Secretary of State appointed who henceforth, until 1947, was a member of the Cabinet. In the literal sense of toppling the Company from power it may be said that the

East India House.

Indian Rebellion of 1857 succeeded. *See* Margaret Bellasis, *Honourable Company,* London, 1952, *and* W.H.Carey, *The Good Old Days of Honourable John Company,* Calcutta, 1907.

Seal of the East India Company

EDMONSTONE, G.F. A senior member of the Bengal Civil Service; was Secretary to the Government and accompanied the Governor General wherever he went.

EDWARD VII When he was Prince of Wales visited **Kanpur**, was shown the site of **Wheeler's Entrenchment** and ordered that it be marked off by small stones 6" high. The stones may still be seen.

EDWARDES, Michael A prolific writer upon India, particularly the period of the rebellion. Member of a family associated with India for many generations. *See* Bibliography.

EDWARDES, Sir Herbert Benjamin Commissioner of the **Peshawar** Division at the outbreak of the rebellion. His was a pivotal role in keeping the Punjab from outright rebellion in 1857. *See* his papers, containing his demi-official correspondence May to September 1857, a copy of his report on the situation in his Division during the Mutiny and letters and papers of, and about, **Brigadier John Nicholson**, which are preserved in the Oriental and India Office collections of the British Library, *MSS.Eur.E.211. See also* Emma Edwardes, *Memoirs of the Life of Sir H.B.Edwardes,* 2 vols, London, 1886.

EDWARDS, Roderick Mackenzie Two letters from Roderick Edwards at Muzaffarnagar, Joint Magistrate, to the Commissioner of Meerut Division, giving some indication of events at **Muzaffarnagar** from the outbreak of the rebellion, and giving his account of his attempts to restore order, and also his Diary, 2 vols, May-September 1857, are preserved in the Oriental and India Office collections of the British Library, see *MSS.Eur.C.183. and 148.*

EDWARDS, William The author of a typical memoir of personal experience, harrowing and dangerous, but in its very danger highlighting the major problem facing so many Indians in 1857 - the conflict of loyalties, which side to join? Edwards was an English fugitive who, with a small party and with the aid of Indians loyal to government made his way from Budaon to safety in **Kanpur**. Particularly intended to give credit to the influential Indians who had assisted him when he was a fugitive, and whom he believed to have been insufficiently rewarded, notably Raja Byjenath Misr and his son Ganga Pershad. He argues that there was a well-founded plot behind the mutiny/rebellion to restore the Mughal Empire. 'I am fully satisfied that the rural classes would never have joined in rebelling with the sepoys, whom they hated, had not these causes of discontent already existed. They evinced no sympathy whatever about the cartridges, or flour said to be made of human bones, and could not then have been acted upon by any cry of their religion being in danger. It is questions involving their rights and interests in the soil and hereditary holdings, invariably termed by them as '*jan see azeez,*' 'dearer than life,' which excite them to a dangerous degree.' 'In the conversation I have had with **Hurdeo Buksh**, who is a very superior intelligent man, he has given me to understand that the native *Omlah*s, who were introduced in such shoals into Oude immediately after the annexation, were the curse of the country, and in his plain-spoken phrase, 'made our rule to stink in the nostrils of the people.' Of Christian and many other officers he spoke in terms of high commendation and respect. He never hesitated, he said, to go to **Christian,** who always treated him (as **Probyn** had, invariably, at **Fatehghar**) as a gentleman, gave him a seat, and conversed with him with affability; but with any *native* official under Government he declared he would as soon lose his wife as go.' See William Edwards, *Personal Adventures during the Indian Rebellion,* London, 1858. See also William Edwards, *Reminiscences of a Bengal Civilian,* London, 1886.

EGOTISM General Wheeler in **Kanpur** had, without doubt, considerable influence with the sepoys: he had served with them for fifty-two years and was widely respected. Yet he made the mistake of over-estimating this influence as the following letter dated 1 June 1857 and addressed to **Canning** demonstrates: 'I have this day sent eighty transport-trains bullocks in relays at four stages for the purpose of bringing up Europeans from **Allahabad**; and in a few, a very few days, I shall consider Cawnpore safe, nay, that I may aid **Lucknow**, if need be... I have left my house and am residing day and night in my tent, pitched within our entrenched position, and I purpose continuing to do so until tranquillity is restored. The heat is dreadful. I think that the fever has abated; but the excitement and distrust are such that every act, however simply or honestly intended, is open to misapprehension and misrepresentation. My difficulties have been as much from the necessity of making others act with circumspection and prudence as from any disaffection on the part of the troops. In their present state, a single injudicious step might set the whole in a blaze. It is my good fortune in the present crisis, that I am well known to the whole Native Army as one who, although strict, has ever been just and considerate to them to the best of his ability, and that in a service of fifty-two years I have ever respected their rights and their prejudices. Pardon, my Lord, this apparent egotism. I state the fact solely as accounting for my success in preserving tranquillity at a place like - Cawnpore. Indeed, the men themselves have said that my name amongst them had alone been the cause of their not following the example so excitingly set them.' Within the month he, and all but a handful of the Europeans in Kanpur were dead; *see* Sir John **Kaye**, vol. II, pp 303-4.

ELAHI BUKSH Sheikh, a musician in the 53rd BNI who remained loyal to government and entered **Wheeler's Entrenchment** with the British in Kanpur. Afterwards rewarded by being promoted in the Military Police.

ELECTRIC TELEGRAPH See TELEGRAPH, ELECTRIC.

ELEPHANTS Throughout the struggle elephants were used by both sides, and performed good service particularly in hauling heavy guns over difficult or muddy terrain. The elephant is an intelligent creature and can sometimes express its resentment at some of the tasks man gives it to do: in particular it resents loud and unexpected noises, so becomes alarmed when the guns begin to fire, and can even become panic-stricken and unmanageable. But here is one story that tells of a gruesome task given to elephants: *'All the afternoon Sepoys who had not fallen in the first attack on the Secunder Bagh were struggling to escape, but they were everywhere well watched, and I saw many of them shot down. The numbers slaughtered during the day in this building alone were estimated at 1700. In some places the bodies were piled up as many as six and eight one upon another, and this was particularly noted in the verandah. The palace itself was set on fire and many of the bodies of the dead sepoys were half burnt. The stench can therefore be imagined. I may say, before quitting the subject, that an attempt was made to bury them, and after 300 bodies had been brought out (this number was counted) it was found that it would take so much longer than it had been expected, that earth was simply thrown over the rest, and so at present they remain, and notwithstanding that 300 corpses were buried, the number remaining was so great that you could not have supposed that one had been removed. Elephants were used for the purpose of dragging them out, and it was more than curious to see the disgust they expressed at having to undertake such a disagreeable duty. A rope was attached to one or more corpses, and one each was given to the creature and he was set to pull them out. It was with some difficulty that he could be set to do this, and he showed his displeasure by trumpeting in the manner his kind are in the habit of doing when displeased, and when the opportunity offered he made further display of anger by kicking the corpse backwards and forwards like a football, and often driving it a considerable distance.'* **Dr F.Collins** in *St George's Gazette*, 31 August 1894. *See also DEATH BY ELEPHANT TRAMPLING.*

ELLENBOROUGH, Lord Edward Law, First Earl of Ellenborough, Governor General of India 1841-44; President of the Board of Control 1858. A political and personal opponent of **Canning**. Lord Ellenborough accused Canning of having contributed to the disaffection before the Mutiny by subscribing to missionary societies. But after the Mutiny Canning steered clear of missionaries, and was 'so terribly afraid of the appearance of feeling any regard even for native Christians, as to request that addresses from such expressive of sympathy and offering aid, may not be presented to him officially, though those of Hindoos and Mussulmans have been received most freely and gratefully.' *See Church Missionary Intelligencer* in B.W.Noel, *England and India*, London, 1859, p 22; *see also* **DNB**.

ELPHINSTONE, John 13th Baron Elphinstone, was Governor of Bombay 1853-60, and played an actively supportive role for the British throughout the struggle. Quantities of his correspondence with **Havelock, Outram, Sir Bartle Frere, Canning** and the President of the Board of Control in London are preserved in the Oriental and India Office collections of the British Library, *see MSS.Eur.F.87*.

EMBREE, Ainslie T. An American scholar who made an interesting attempt to present the widely differing views of certain British and Indian historians, and arrive at the true nature of the struggle of 1857-9. The views of twenty-three historians are summarised and short illustrative extracts given. No conclusion is stated, the question in the title remains unanswered, but a full range of views has been presented. The bibliography is useful but not exhaustive. *See* Ainslie T.Embree, *Mutiny or War of Independence?*, Lexington, 1963.

Elephants bringing up troops

Enfield Rifle with ramrod and bayonet

ENFIELD RIFLE, The Much was made at the time of the superiority of the British Enfield rifle over the sepoys' Brown Bess musket: greater range and accuracy and reliability. It was the Enfield's ammunition the so-called 'greased **cartridges'** that had, in some opinions, been responsible for the outbreak of the mutiny. Yet the rifle was not always superior: Sir Hugh Rose (June 1858): 'The Enfield rifles had made up a good deal for my inferiority in numbers; that advantage however no longer existed. The heat and other causes had had such an effect on the ammunition of the rifles that, their loading becoming difficult, and their fire uncertain, the men lost confidence in their arms.' An officer of the 3rd Europeans: 'The men seemed much more annoyed by their useless Enfield rifles than by the sun. No amount of force exerted by the men would drive the bullets down to the breach of their weapons.' Sergeant Forbes-Mitchell: 'Our rifles had in fact got so foul with four days' heavy work that it was almost impossible to load them, and the recoil had become so great that the shoulders of many of the men were perfectly black with bruises.' Which may explain the very frequent recourse to the use of the bayonet by British troops in this campaign.

ENGELS, Friedrich German philosopher and co-founder (with **Marx**) of modern Communism. He was contemporaneous with the mutiny/rebellion, and wrote much in the American press in support of the rebels. *See* K.Marx and F.Engels, *The First Indian War of Independence, 1857-59*, Moscow, 1959.

ENGLISH LANGUAGE **Nana Sahib** issued an order that all babus and any other person who could read or write English (presumably excepting **Azimullah**) should have their right hands and noses cut off. **Further Papers** in FSUP Vol IV, p 588.

ENGLISH, Major Frederick H.M. 53rd Regiment. Commanded a small force which defeated the mutineers at Ramghar in October 1857, accompanied by Captain J.H.Smyth, Bengal Horse Artillery. His letters and papers are preserved in the Oriental and India Office collections of the British Library, *MSS.Eur.D.787.*

ENTERTAINING THE TROOPS August 1857 **Kanpur.** 'Tomorrow a grand day comes off: every kind of amusement is got up, open to the garrison, Sikhs included. There will be mile and half-mile races, jumping in sacks and hurdle races, bobbing in treacle, putting a 24-lb. shot, running after the pig with greasy tail...'

ERSKINE, Walter Coningsby He subsequently succeeded as 12th Earl of Kellie. Served in the Bengal Army 1828-61 but was seconded for civil duties for much of the time. Was Commissioner of the Jubbulpore[Jabalpur] Division during the mutiny/rebellion and his evidence as to the state of things in Central India, including the affairs of the Rani of Jhansi, is significant. His papers and correspondence are preserved in the Oriental and India Office collections of the British Library, *see MSS.Eur.D.597,* and he was also the author of W.C.Erskine, *Chapter of the Bengal Mutiny*, Edinburgh, 1871.

EUROPEAN REBELS In 1857 there were undoubtedly Europeans fighting on the rebel side, but details are sketchy, *eg* this account by an officer of the 2nd European Regiment Bengal Fusiliers: 'Inside Delhi after the assault a European in his shirt sleeves rode up to us from towards the enemy and evidently seeing what regiment it was, exclaimed 'Where is the adjutant of the 75th?' He was told where to find the 75th, and we heard that on reaching them he asked where the adjutant of the 2nd Fusiliers was.' Eventually he was shot as a spy. Not the only instance. 'At the Battle of **Badlee ke Serai** a European was fighting against us and was disabled. On our men reaching him he begged for his life, promising to give full accounts of the enemy, but quarter was not given him. On another occasion on stable picket I distinctly saw a European among the enemy directing them where to fire.' What a pity, for the satisfaction of our curiosity, that we did not learn more of these men, from their own lips, before they died. Perhaps the matter was covered up. There is also the familiar tale of the 'Russian' at the Battle of **Chinhat** in June 1857 near Lucknow: 'Masses of the enemy cavalry had gained the road between our men retreating and the little bridge. They were apparently commanded by some European, who was seen waving his sword and attempting to make his men follow him and dash at ours. He was a handsome looking man well-built, fair about 25 years of age with light mustachios and wearing the undress uniform of a European cavalry officer, with a blue and gold-laced cap on his head. Whether he was a Russian - one suspected to be such had been seized by the authorities, confined and then re-

The Rani of Jhansi, with Damodar Rao tied to her waist. Painting on the temple wall inside Jhansi Fort.
See JHANSI

Babu Kunwar Singh 'defeats' Major Eyre at Arrah, August 1857. Wall painting at Jagdishpur.
See JAGDISHPUR and KUNWAR SINGH

The slope by which Rani Lakshmibai, mounted, leapt to safety from Jhansi Fort.
See JHANSI

Chattri, or site of the funeral pyre, of Lakshmibai, Rani of Jhansi, at Gwalior.
See JHANSI

Tatya Topi and the Rani of Jhansi. Wall paintings on the temple inside Jhansi Fort.
See JHANSI

Wall paintings on the temple inside Jhansi Fort.
See JHANSI

The Indian Heroine, Secunderbagh, Lucknow. See INDIAN HEROINES and SECUNDERBAGH

The Rani of Jhansi, statue at Gwalior. See JHANSI

The Kashmiri Gate, Delhi, today. See KASHMIRI GATE

The gateway to Jhansi Fort, from which the Europeans emerged to capture and death. See JHANSI

Babu Kunwar Singh's bust at Arrah.
See ARRAH and KUNWAR SINGH

'Portrait' of Kunwar Singh, in Maharaja College, Arrah. Most unlikely to be a true likeness. Resembles John Lang's picture of Ayodhya Prasad, q.v.

Babu Kunwar Singh in action against the British. Wall painting at Jagdishpur.
See KUNWAR SINGH

Kanwar Singh cuts off his own wounded arm. From a wall painting inside Jagdishpur.
See JAGDISHPUR and KUNWAR SINGH, DEATH OF

The billiards room in Arrah house: now the social area for girl students at the Maharaja College. A change of game!
See ARRAH

Arrah House, today.
See ARRAH HOUSE

A model of the Kurshid Manzil as it once was. Courtesy Principal of La Martiniere Girls' College, Lucknow.
See KURSHID MANZIL

Fatehgarh Fort, today. See entry of that name.

Kurshid Manzil (32nd Mess house), today.
See KURSHID MANZIL

Humayun's Tomb, Delhi, today.
See HODSON and HUMAYUN'S TOMB

The Musa Bagh today.
See MUSA BAGH

The silver mounted saddle of Nana Sahib, and to the left the sword of Tatya Topi. In the Military Heritage Museum, Lewes, Sussex.
See LEWES

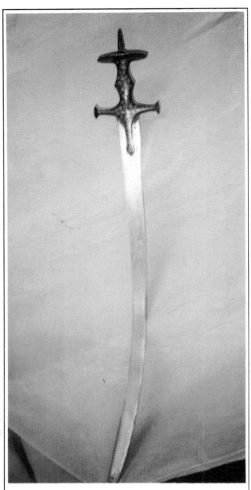

Tatya Topi's sword. On the hilt is the inscription: 'This sword was taken on Tatya Topee when captured, 8th April 1859'. Presented to Earl Canning, GCB, by Major R.J. Meade.
See LEWES

The Rani of Jhansi's gun in Jhansi Fort.
See KARHAK BIJLI

leased, - or, what is more likely, one of the renegade Christians who had changed their religion and adopted the native habits and manners and costume I cannot tell..... It is probable that their artillery was commanded by European officers, wretches for whom no punishment would be ignominious or severe enough. One of these seen several times laying a gun and giving orders apparently like one having authority. From the description given me it is not unlikely that it was either Captain----, or Captain Rotton, who had both remained in the city and during the disturbances never came near the Residency . ...Rotton was a man born in Lucknow whose daughters were married to Mussulmans and whose sons served as native officers or troopers in the late King's army. He himself commanded a portion of the artillery. Both these persons were said to have adopted the Moslem faith and neither came near to European society even before the troubles began. ...a Frenchman named Leblond, as great a villain as ever breathed, also an apostate probably likewise joined the insurgents; and a young man whose name I do not wish to mention, on account of his family, was probably the person who commanded the enemy's cavalry at Chinhut. Two of his cousins were fighting valiantly in the Residency against the rebels, another was massacred at Futteygurh, after fighting for us, a fourth was wounded in action against the Agra rebels, and a fifth also accepted a military post under government and later distinguished himself in action. The apostate had long been disowned by his relatives'. *See also* SHEIKH ABDULLAH.

EWART, Emma   Both she and her husband, Colonel J.Ewart 1st Bengal N.I., were killed at **Kanpur** but there exist three letters written by her in the months immediately preceding the tragedy at Kanpur, expressing concern about the safety of the British and Christian community there. These letters are preserved in the Oriental and India Office collections of the British Library, *see MSS.Eur.B.267.*

EWART, EMMA   'We are still free to spend our days at home, and no outbreak is at present apprehended from any of the troops here; our danger lies now in what may come from outside. The anxiety about the Commander-in-Chief's proceedings in Delhi become very intense, because we hear nothing material from him, although we hoped to have heard of the blow being struck some three days ago. ...Our European force is meanwhile increasing by arrivals from Calcutta, but, after all, we shall not have above 300 English soldiers, and they may have to cope with 3,000 natives should an attack be made. But it is useless to speculate upon what may happen. ...We are more cheerful, in spite of the great anxiety and suspense; our family party is really a charming one; we feel better able to meet difficulties and dangers for being thus associated. Dear Mrs. Hillersdon is so quiet, and gentle, and calm, never giving way to hysterical movements, nor, on the other hand, showing any want of sensibility. ...Some small items of intelligence are rather cheering. The 29th Native Infantry are said to be behaving quite loyally at Moradabad. From Lucknow we have heard of the capture of four or five emissaries of rebellion by a Sepoy of the 13th Native Infantry; so we hope some may be stanch, after all, and that the rebellion ***being so ill arranged, and there appearing no leader and no concert,*** it may die out soon and we may be at peace again. Still here we are in a most uneasy state, and we are fortifying our position and laying in provisions to provide against a siege. It is a lamentable position for the governing class of a country to be in, exceedingly humiliating and disheartening; yet we must hope for better days, and in the meantime do our duty and trust in God. Several parties of mutineers have been caught in this district with plunder upon them. Mr. Hillersdon has them confined in his gaol. Two of the 3d Cavalry are just brought in. We feel it rather dangerous work.' Ewart, Emma. Letter dated 30 May, 1857, quoted in the *Times*, 22 October, 1857.

EWART, Colonel John   General **Wheeler**, commanding at **Kanpur** received a letter from **Colvin** at **Agra** reporting that they had successfully disarmed the native regiments whose brethren had mutinied at **Muttra** three marches north. 'We have not sufficient strength of Europeans to venture such a step;' wrote Colonel Ewart at **Kanpur**, 'nor do I think **Sir H. Lawrence** will venture on it at **Lucknow.**

British reinforcements

But Sir H. Wheeler has to-day received a letter from Lord **Canning** himself, by which we learn that European troops are arriving from Madras and Ceylon, and Burmah, and that Lord Canning is pushing them up the country as rapidly as possible, some 20 men a day by dawk carriages and some 100 men a day by bullock train carts. If the journey of these detached parties is not interrupted by risings between Calcutta and this, we may hope to have our hands strengthened in a few days. The treasury here, containing some 10 or 12 lacs of rupees (£100,000 or £120,000), is situated five miles from this military cantonment. It has been hitherto thought inexpedient to bring the treasure into cantonments; but the General has now resolved on making the attempt to-morrow. Please God, he will succeed. He is an excellent officer, very determined, self-possessed in the midst of danger, fearless of responsibility, that terrible bugbear that paralyzes so many men in command. You will be glad to know that I have had the good fortune to give him entire satisfaction by my conduct and arrangements in the command of my regiment during these troubles. He has heaped praises on me. If the troops break out here it is not probable that I shall survive it. My post and that of my officers being with the colours of the regiment, in the last extremity some or all of us must needs be killed. If that should be my fate, you and all my friends will know, I trust, that I die in the execution of my duty. ...But I do not think they will venture to attack the intrenched position which is held by the European troops. So I hope in God that Emma and my child will be saved.' John Ewart's Letter dated 31 May, 1857, quoted in London *Times*, 22 October, 1857.

EWUZ KHAN Kanpur 5 June 1857. 'At gunfire in the morning the officers of the cavalry came to their lines, and took away to the entrenchments the horses, arms and furniture which had escaped plunder. A little later on the morning of the same day, the 53rd and 56th Regiments broke out.....I and Sheikh Azim-oollah, sepoy, together with **Bhowani Singh**, subadar-major, Mahboob-oolah Khan, Bukhtawur Singh, havildars, Kadirdad Khan, Abdool Rehman Khan, sowars, and Mirza Bakur Hossein, native doctor, went to the officers in the entrenchments.' *Deposition of Ewuz Khan,* **Williams**.

EXECUTION 'I was once, I now remember, present at the execution of a Sepoy taken in arms. He was ordered to be shot. The soldier told off for this purpose coolly walked up to the culprit, and put the muzzle of his rifle close to the man's head; but the cap was damp and missed fire. This occurred twice before the rebel's brains were blown out. At each failure the wretched man stood perfectly firm. He was not fettered in any way, and he merely turned his head aside as the rifle was pointed at him. I think few Englishmen could have shown more unflinching resolution, or could have met their death more bravely.' *See* J & O Wilkinson. *The Gemini Generals*. London, 1896, p 304.

EXECUTIONS 'That I may not have to return again to this disagreeable subject, I will add a word or two about executions. As a rule, those who had to die died with extraordinary, I was going to say courage, but composure is the word; the Mahomedans, with hauteur, and an angry kind of scorn; and the Hindoos, with an apparent indifference altogether astonishing. When the London steamship went down, south of the Land's End, the Captain, that noble fellow who, when offered a seat in the boat, said, 'No, thank you, I will stay with the passengers,' about noon assembled all who could come into the saloon, and gave notice that he thought the ship would keep afloat till two o'clock. One who escaped related that, in answer to this notice, an old gentleman appeared at about half past one, having arranged his dress for a journey, strapped his wrappings together, and put his money into a girdle. Even in that supreme moment the Captain could not restrain a smile. Some of the Hindoos treated death exactly as if it were a journey. One man, positively under the shadow of the fatal tree, with only three or four minutes to live, when his waist-cloth was searched for the benefit of his friends did not object to one or two articles being taken out, but demurred, peevishly enough, to giving up a few apples of the jujube tree. Of all who had to meet their end, I only remember one who died craven. He was a Mahomedan, and, whether his memory was charged with upbraiding circumstances, or whether he had never looked the subject fairly in the face, when it really come to the last scene, he was unprepared to go through it. He declared that he only nominally belonged to Islam, but was at heart a Christian; that he was prepared to eat pork and drink wine, in order to show how sincere his conversion was; and that he thought little or nothing of Mahomed. All this not availing, he grovelled on the ground, screamed; cried, and piteously entreated for life. He would betray his cause, would turn informer, would deliver hundreds, not in security and honour, to the shambles, life, only life! And the poor wretch, fainting and foaming, had to be lifted within reach of the rope.' **Sherer**

EXTREME PUNISHMENT 'An officer, one of Grant's ADCs, had his throat cut taking a stuperous nap after a nip of liquor. ...It was a horrid death for him, but he is good riddance, nobody liked him, *and he has not paid his mess bill.*' Chardin Philip Johnson, *Journal*.

EYRE, Sir Vincent (1811-1871) Major-General, KCSI, CB. Artillery. Retired September 1863. He was already, before the rebellion, well-known for his courage. He was 'the Cabool man': the officer who in the disastrous British retreat from Afghanistan had volunteered to be one of the hostages demanded by the Afghans and was then held prisoner by them. 'Such men', says Colonel **Malleson**, 'are born seldom.' Vincent Eyre was on his way from Calcutta, 'up-country', with a European battery of artillery. He reached **Danapur** on 25 July 1857, the day the three Sepoy regiments there successfully rebelled and joined **Kunwar**

Singh. Eyre offered his services to **Major-General Lloyd**, but as they were not required he moved on upstream to **Buxar**, where he learned the news that the sepoys were *en route* to **Arrah**. Leaving part of his little force to protect Ghazipur, he collected all the men he could, including a contingent of 154 men of the 5th Fusiliers under Captain L'Estrange whom he persuaded to join him, and marched to the relief of the little garrison of Arrah. Had he failed he would undoubtedly have been in considerable trouble as he had no orders to do anything but proceed up-river to **Allahabad**. The siege of **Arrah House**, and its successful relief by Major Vincent Eyre, have been described by another historian of the Great Rebellion, **I.G.Sieveking**, as 'A Turning Point in the Indian Mutiny': the point she was making, and it is a valid one, is that in the great cities and the small stations from Allahabad to Delhi where rebellion had broken out, the outnumbered British had, for the most part turned and fled, faced as they were by overwhelming odds. In Arrah alone, a handful of Europeans, aided by 45 Sikh policemen, decided that they would not run away, but would make the sepoys' task of destroying them as difficult as possible. It was sheer bloodymindedness really, they had no hope of survival, and yet, against all the dictates of common-sense, survive they did. Bluff allied to arrogance once again won the day.

Major Vincent Eyre

The fact is that if **Arrah House** had fallen, all Bihar would have gone over to Kunwar Singh, and government in Calcutta would have been effectively cut off from the scene of action in the N.W.Provinces. Major Eyre went on to **Jagdishpur** and burnt Kunwar Singh's palace. The local people still remember the old warrior-raja with pride and affection, incidentally, and have created a 'museum' out of what remains of the palace. Eyre then went on to distinguish himself in the relief of Lucknow and the pacification of Oude[Awadh] and to the British public and to writers and commentators he was a hero. But he was treated with little consideration by all in authority. Perhaps he was somehow associated in men's minds with Major General Lloyd and his ignominious conduct, or perhaps the fact that he had done what he did entirely on his own initiative annoyed his superiors. For all their failures he was made a scapegoat, and his achievements were ignored by them. Certainly he got little praise from **Sir Colin Campbell**, the new Commander-in-Chief: the latter had not arrived in India when Eyre relieved Arrah, and showed little appreciation for any military success that preceded his own campaign. In his later career Eyre was given a succession of very dull appointments, and although he did in the end achieve the rank of (Hon) Major General, he was never given the opportunity to shine. Probably his part in the Arrah House affair would have been totally forgotten had it not been for a chance meeting many years later with **Sir Hugh Rose** (aka Lord Strathnairn). Eyre was retired and on half-pay; they met in Pall Mall, London, and Sir Hugh Rose asked 'How is Lady Eyre?' 'Who is that?' was the reply. 'Why, your wife!' 'Oh, Mrs Eyre is well, thank you.' Hugh Rose went straight to the Horse Guards and got the omission rectified, in an early Gazette - a knighthood for Vincent Eyre. Author of *Military Operations at Kabul, 1843*. C.B. 1858, KCSI May 1867. *See* **DNB**; Burkes Landed Gentry; **DIB**; the *Times*, 26 & 28 September 1881.   **PJOT**

F

FAIRWEATHER PAPERS These are held in the Centre of South Asian Studies at Cambridge University, and include a copy of an unpublished typescript memoir of the mutiny/rebellion, entitled *Through the Mutiny with the 4th Punjab Infantry*, Punjab Irregular Force, by Surgeon General J.Fairweather, Indian Medical Department. The memoir concentrates on military manoeuvres, including a detailed account of the assault on **Delhi**.

FAITH, OF NANA SAHIB 'Seeing the faith perishing I have girt my loins to defend it,' Nana Sahib wrote to Raghuraj Singh of Rewa, 'and I have suffered much for it. But this is no man's doing. It is God's design; I have done my utmost; ...the faith is the faith of us all. I have endeavoured to support and defend it; all chiefs, and monarchs, Hindoo or Musulman who assist the English, the destroyers of faith, destroy their religion with their own hands.' See **FSUP** Vol III, p 367.

FAIZABAD A major city of Oude[Awadh], and former capital. Commissioner was **Colonel Goldney**. Troops of 15th Irregular Cavalry, 22nd N.I. and 6th Irregular Infantry broke out in rebellion on 8 June 1857, the two latter corps however refusing to harm their officers. Treasury was plundered and prisoners released, among whom was the **Maulvi of Faizabad**, otherwise known as Sikander Shah. 'In Fyzabad on 8th June (1857) the troops broke out into open mutiny that night. They did not go through the form of pretending a grievance, but said they were strong enough to turn us out of the country, and intended to do it. The 15th Irregular Cavalry particularly the Rissaldar in-charge tried to induce the other regiments to murder their officers, but the Artillery, 22nd Native Infantry and 6th Local Infantry not only refused to injure the Europeans but even gave them money and assisted them in procuring boats to proceed down the Ghogra.' In fact only 5 out of 21 officers reached safety in Danapur, others were murdered by the 17th Native Infantry or drowned (3) or murdered by villagers of Mahadubbur in Gorakpur. Mrs Mills with 3 children attempted to conceal herself in the city of Faizabad in the house of a havildar but he refused her food, she was obliged to disclose herself to the leader of the mutineers, who gave her money and sent her across the river into the Gorakpur district. She wandered 8-10 days, but got no assistance. Very delicate youngest child died of exposure. 'Rajah Maun Singh hearing there was an English lady in distress sent for her, provided for her wants and after a few days' rest sent her with sergeants' wives into Gorakpur. Mutineers of Fyzabad plundered 2,20,000 Rs from treasury and then followed the usual practice of releasing the prisoners from the gaol.' See Hutchinson, George, *'Narrative of Events in Oude'* 1859, pp 104-114.

FAMINE Not surprisingly in 1860 famine followed the Mutiny/rebellion, which had deprived the Northwest provinces of the resources to stave it off after the sparse harvests of 1858 and 1859. The civil servants wanted to let the market dictate supply, but 'maxims of political economy, however true in the abstract, would not suffice to keep the people quiet, or even to prolong their lives.' It eventually cost the Government 750,000 rupees to relieve the suffering of half a million people, and only 1 in 5,000 died of starvation, an improvement over past famines see A.Loveday, *The History & Economics of Indian Famines* p 45.

FANTHOME, J.F. The author, under the soubriquet 'J.F.F.' of a novel entitled 'Mariam' based upon the mutiny. The author describes himself as a 'survivor' and it is clear from the text that he has first hand knowledge of the events of 1857-8. He was of French extraction and a resident of **Agra**. According to the Preface of his book it is not just a novel, but the story of a Christian family's trials during the mutiny/rebellion. Thus: *'It is not usual to give a preface to a novel. If this were a novel, pure and simple, I should have observed the rule; but though there is an element of fiction about it, the book is meant to embody chiefly the experiences and the trials of a Christian family during the terrible political cyclone which shook the Indian Empire to its base in 1857.'* A manuscript copy of the novel is preserved in the Oriental and India Office collections of the British Library, see *MSS.Eur.D.966*.

FARNON It has often been said there were only four male European survivors of the outbreak at **Kanpur**, but this ignores a number of merchants, clerks, railway employees etc who were either European or Eurasian Chistians, who in some manner or other managed to survive. One such was a Mr Farnon: 'When the mutiny broke out in this station, I was in the employ of the railway. I had gone out to Marowlie, a place about 23 miles from here, to pay the workmen; at night the baboo woke me up, and said that he thought the cavalry would mutiny that night. I lay awake until about 4 1/2 A.M. when I heard the trampling of horses and carts, &c., going along the road: upon which I jumped out of bed, and went about two miles on the Etawah road, and then turning to my right I came to Chowbeypore on the trunk road.' See *Deposition of Thomas Ambrose Farnon*. **Williams**.

FARQUHARSON, COLONEL This officer commanded the 46th BNI at **Sialkot** when they mutinied on 9 July 1857. The sepoys were determined to mutiny but were inclined to be loyal to their own officers, which apparently did not strike them as inconsistent. When their officers rushed to the lines to try to restore order they were quite defenceless as the arms of most of them had been taken away by their servants during the night. But the regiment refused to take advantage of their condition and protected them all day with a strong guard of the steadiest men. *They even coolly offered to Farquharson, to his 'great amusement', Rs2,000 per month, and to Captain Caulfield Rs1,000 per month, if they would remain with the regiment and lead them to Delhi against the English. They also promised the two officers six months' leave in the hills every hot weather, and only to be stationed in the best cantonments once the English had been defeated. See G.Rich, The Mutiny in Sialkot,* Sialkot, 1924.

FARRUKHABAD, NAWAB of In February of 1859 the Nawab of Farrukhabad was tried at **Fatehgarh** before a panel of three judges for the 'wholesale murder of Europeans.' He was defended by a team of English barristers, who entirely blamed the 10th N.I. for the insurrection. *Testimony of David Churcher at Trial of Nawab of Farrukhabad in February, 1859,* Farrukhabad Collectorate Mutiny Basta in **FSUP** Vol V, p 926. He was convicted and sentenced to be hanged, but in May he was given the choice of exile at Mecca. He accepted, and was despatched, appearing 'indifferent to his fate' to onlookers along his way. Eventually, in August, he was let loose in the Egyptian port of Aden. *See* The *Friend of India* in **FSUP** Vol V, p 926 *et seq. Also* BONNY BYRNE.

FARRUKHABAD BREED One correspondent for the *Hindoo Patriot* regarded the British residents of Fatehgarh as especially harsh and intolerant toward Indians, and dubbed one particularly brutal and bigoted Britisher at Agra a 'gentleman of the Furruckabad breed.' The *Hindoo Patriot* quoted in **FSUP,** Vol V, p 915.

FATEHGARH A military station on the Ganges above **Kanpur**, best known for its government factory for making gun carriages. It was an unhappy station for Europeans, a posting that would be dreaded by both civil and military officers. Despite the proximity of the ostentatious and lascivious **Nawab of Farrukhabad** and the residence in exile of various discarded nobles, including the deposed boy king of the Punjab (**Duleep Singh**: now a benign converted Christian) Fatehgarh remained a mere backwater of overextended **box wallas** and disappointed officers, a begrudged annex of Kanpur Camp.

FATEHGARH FORT Near **Farrukhabad** the Company resurrected the bleak and treeless fort of Fatehgarh: ten bastions linked together by a **kankar** wall and surrounded by a 1,500 yard moat. Fatehgarh was conveniently situated to relay communications between Kanpur and the battalions posted throughout Rohilkhand, and to keep an eye

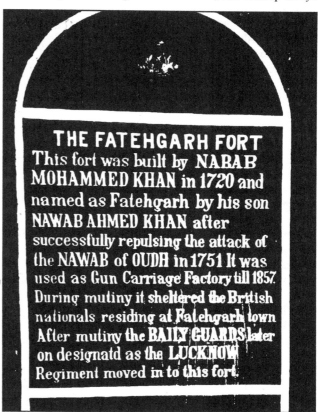

on the seasonal fords known as the Jumna Ghats. Though successive commanders fretted about the congestion of native shops and houses that crowded the Fort's gate, the cantonments were at least protected from cavalry attack by the river and a random criss-cross of dry ravines that made it approachable only by road. Fatehgarh occupied a limestone outcropping that was said to have been formed by giantesses sweeping their homes in nearby Kanauj and pitching the dirt from their doorsteps into a heap. The fort is now the Regimental Centre of the Rajput Regiment, and as such of course is not open to the public.

FATEHGARH, Gavin Jones's account Gavin Jones, an indigo planter, survived the rebellion and wrote a graphic account of it as far as his personal experiences permitted. He was not a soldier but his tale of the beleaguered British in the Fort and the attack by the rebels is professionally told: 'They commenced a mine, and worked two nights and early on the third morning sprung it. The explosion was awful; it shook the whole fort. We all thought it was over with us; but an examination proved that it had blown down only five or six yards of the wall, leaving the inner half standing. The bastion where I was, happened to be the next to that where the explosion took place; I at once ran to the spot to see what mischief was done; seeing, however, several of our party engaged in moving a gun to the breach, I returned to my own post and noticed from 100 to 150 Pathans and sepoys a long way out of range, congregating below the breach in order to attempt an escalade as soon as the dust and smoke cleared off; I at once sent notice to the others to get aid and, in the meantime, by pouring the fire of two double-barrels and eight muskets, already loaded, into them, and discharging them as they were reloaded by a native, managed to disperse them before any of my comrades came up to my aid. Somewhat later in the day a second assault was attempted, which was defeated by Mr. Fisher's shooting the leader of the party, which caused his followers to fall back. We this day lost our best gunner, Mr Ahern, who was shot through the head while laying a gun. The enemy had now brought a gun to bear upon the bungalow containing the ladies and children; the shots generally passed over, but two-or three struck the house; another gun they got to bear against the gate, and contrived to break a hinge and knocked several holes through it, but little harm was done, as we had piled up the archway with timber, which effectually stopped the shots; two of our guns were soon after disabled. The enemy then commenced another mine close to the first. The determination thus shown by them, as also the loss of three of our best men, disheartened the garrison, already worn out by fatigue from watching. It was also certain that if the second mine was completed and fired, the enemy would attack us by both breaches, which we could not possibly defend; our position became desperate; we began to Iook to the boats as our only mode of escape, the river having risen considerably by the rains.' *See Narrative of Gavin Jones* in **FSUP**,Vol V, pp 738-47.

FATEHGARH, Nana Ubbhunker's account

'There were two batteries erected [in Bithur]; one at Bundee Mata Ghat, and the other at Shookul-deo Ghat. When the boat arrived opposite the latter ghat, shots were fired at it by the Nana's people, and, from the other side of the river, the people of Jussa Singh, zemindar of Futtehpoor Chourassee, were firing. I saw all this from a distance. The Europeans in the boat also fired; after this, a gentleman put out a flag from the boat, on which firing on both sides ceased, and the people of Jussa Singh came in a boat from the other side of the river, and took away the boat containing the gentlemen to the thanah; and then, taking it to Shookuldeo Ghat, caused the gentlemen, ladies and children to leave it, and, putting them on two hackeries covered with sirkee, took them to the Rao [Nana Sahib's nephew] at the palace. It was about 9 P.M, when they reached the palace; it was raining at the time, and the Rao, seeing the Europeans, ordered them to be taken to the *burra sahib*'s house, which was done.' *See Deposition of Nana Ubbhunker*, **Williams.**

FATEHGARH, Futteh Sing's Account

The second European party to go down river from Fatehgarh in the hope of finding succour at Kanpur met the same fate as the first. The men were all killed (with the exception of three who may, or not, have tried to make some kind of bargain with the **Nana**) and the women joined the other European survivors of the **Satichaura Ghat** massacre in **Savada House**: 'On that day, (July 1857) after 12 o'clock, the gentlemen of the party of 45 above mentioned were sent for to the Nana's tent. The ladies and children were kept in the Savada house. Only the males (eighteen in number) came, and they were shot. Besides these there were three others who held a conversation with the **Nana; (Jwallah Pershad, Brigadier, Azeemoollah,** and Shah Ali, were with the Nana) The three gentlemen promised the latter they would have the Fort of Allahabad given up to them; so they were sent back to the Savada house, and the other eighteen shot by the orders of the Nana. There were many *jullads* there also, who took out in my presence gold mohurs from inside the shoes of these gentlemen. All the *jullads* were residents of this place, and, if I saw them, I think I could recognise them. The persons whom I employed could do so likewise.' *See Deposition of Futteh Singh*, **Williams.** The names of the men shot are given by Messrs **David Churcher** and **Gavin Jones,** survivors, who escaped *en route* and did not come to Kanpur, and are as follows: Basco, Mr, pensioner;. Best, Mr; Donald, Mr, senior; Donald, Mr, junior; Henderson, D., Ensign, 10th N.I.; Heathcote, T. G., Doctor, 10th N.I.; Jones, Mr, Civil Engineer; Jennings, Mr; Lowe, Captain, 10th N.I.; Maltby, J., Doctor, Civil Surgn.; Munro, R., Major, 10th N.I.; Ohern, Mr, Clothing Agency; Phillot, Major, 10th N.I.; Roach, Mr, Road Overseer; Rohan, Condr, Gun. Carr. Agency; Reid, Qr-Mr-Sergt, 10th N.I.; Sweetenham, C. W., Lt., 10th N.I.; Vibart, E. C., Capt., 2nd Cavalry; Whish, E., Lieut., 10th N. I.; Wrixen, Musician, 10th N.I. *See* W.J.Shepherd. *A Personal Narrative etc.,*Lucknow, 1886, p 113.

FATEHGARH Narrative

'When the outbreak at Meerut took place and the news came here Mr. Thornhill wrote a Persian letter to the Nawab stating that he had made arrangements for the safety of Futtehgurh and asking the Nawab to do the same for Furruckabad. Nawab wrote in reply that he would do all in his power. Several letters passed between the Nawab and Mr. Thornhill. A few days after this the 10th N.I. became mutinous. On the same day several of the Europeans got alarmed and went across the river, Colonel Smith and some others remaining in the Dak Bungalow. On the same day the *Kotwal* Meer Ukbur Aly (Mir Akbar Ali) and Wuzeer Aly the Nawab's *Naib* came to the Nawab and informed him that the Europeans had left the station with the particulars of their departure. The *Kotwal* added that in consequence of the Europeans having left, the whole city was in a turmoil and he apprehended that it would be looted and acts of violence committed as many people of bad character had come into the city from its neighbourhood. Nawab therefore directed that a proclamation should be made in the city by beat of *tom tom* to the effect that the peaceably disposed persons need be under no fear, that the Europeans would return the next day and that all wicked people attempting to make a disturbance would be apprehended and severely punished and that this was the order of the English Govt. and the proclamation made by their orders. Several respectable people from the city informed the Nawab that the *budmashes* would not listen to the proclamation as it purported to emanate from the Europeans who they said had fled and that unless other measures were adopted the city would be certainly looted. Therefore the Nawab consulting with Wuzeer Aly and others by their advice caused another proclamation to be issued to the same effect and purport only stating that it was by the orders of the Nawab. On the following day the Europeans returned to the station and Colonel Smith sent a Persian letter to the Nawab informing him that the English were all here, and asking him by what authority he had caused the proclamation to be made in his own name on the previous day. This letter was the one sent by the Collector's Sowar. In reply the Nawab sent a letter by his *Naib*, Wuzeer Aly, to Colonel Smith informing him of the reason and stating that Wuzeer Aly would give him all particulars. Wuzeer Aly went to Colonel Smith and explained all matters and brought back a letter from him stating that he was satisfied and thanking him for what (had) been done. Two days after this the Nawab received another letter from Colonel Smith requesting that 30 trustworthy men might be sent to him. The Nawab could not prevail upon any of his old servants to go to Colonel Smith. Finding this he sent for an old servant Mutloob Aly (Matlub Ali) explained the difficulty to him and requested him to try and get 30 trustworthy men for the Colonel; he said he would do so and on the following day he collected 30 men whom he thought he could rely upon and sent them with a letter by Mutloob Ali Khan to the Colonel who placed them on guard at the Puscimya Durwaza of the city. Colonel Smith wrote a reply to the Nawab thanking him for sending them. One or two days after this the Nawab received a letter from the *Seristadar* of the Collector stating that (Colonel) Smith wanted him to send him a guard of 10 men to the *cutchery daftur*. They were sent, new men being employed from the city. Five or six days after this Ahmed Yar Khan and Ushrut Khan came from Mhow with 1,500 or 2,000 Puthans and encamped in a mango tope near the Mhow-Durwaza. Three or four days after this the Collector's Sowars came in the morning to the Nawab and very shortly after 2 Companies of the 10th N.I. and the native officers with the band and colours came also. The sowars had in the meantime gone across the river to call the 41st N.I. and 10th Oudh Irregulars. The Nawab was asleep at the time the sowars came; they had him called and told him that the 10th N.I. were coming and that they were going to call the other regiments. On the 10th N.I. coming Ameer Khan *Soobahdar (Subedar)* with whom there was Ameer Khan the *Ressaldar* with some ten or a dozen troopers and [Bolee] Aly Bux, the Havildar-Major of the 10th N.I. and Emam Bux *Soobahdar* came to the Nawab. Ameer Khan *Soobahdar* was the spokesman and said, 'Up to this day we have been servants of the Company, we are now servants of the King of Delhi. Your ancestors were formerly tributary to Delhi also. So we have come to place you in your old position on the *Guddee* as Nawab under the King of Delhi.' The Nawab declined to be put on the *Guddee* in this way stating that the English (had) already put him on the *Guddee*. They replied, 'Your *Naib* Wuzeer Aly and yourself are always with the English and we believe you to be Christians, and favourable to the English.' The Nawab said, 'Leave me alone, I am content as I am. As you say you are servants of the King of Delhi you had better go to Delhi.' He said this to get rid of them. They replied, 'We are not to be got rid of in this way. If you do not do as we direct we will at once loot you and kill you.' The two Ameer Khan took him each by the hand and led him to a chair. Bhoo Aly Bux stood behind him with a drawn sword. They were all in uniform. The *Soobahdar* and *Ressaldar* then sent two servants to the Bazar for Powder and directed salute to be fired, the Nawab having told them he had none. The *Soobahdar* then requested that *sherbot*-(*sharbat*) might be served out to the sepoys as the weather was very hot. Toorab Aly (Turab Ali) was present. The *Soobahdar* asked Nawab, 'Who is your head *Karindah*'? He replied 'Toorab Aly' (who used to have charge of the Nawab's *roshakhana*). When the Nawab was forced on the *Guddee* he had on the clothes he had been sleeping in when awoke by the Collectory Sowars. The *sherbot* was served out by Toorab Aly. The officers then said they wanted to return to their Lines but as it was very hot they asked him for his carriage which he was obliged to give them, and they left and the Nawab went into the female apartments and got his food. At noon on the same day the Collector's Sowars came to the camp of the 10th N.I. accompanied by the Doobye (?) 41st N.I., Pyullal (?) 10th Oudh Irre. (Irregulars), Aga Hussein *Ressaldar* and guns and *Golundazes* and magazine. The 41st broke open the jails and let the prisoners loose and set fire to the Europeans' houses in the station, first looting them. The Collector's Sowars went and encamped with the 10th N.I. The 41st, 10th Oudh Irr. and Aga Hussein *Ressaldar* encamped near the grove in the garden named 'Laeoolah and Birra ~ In the evening of the same day the 41st, 10th Oudh Irreg. and Aga Hussein's *Ressalah* called upon the 10th N. I. to divide the treasure in their keeping among them. They refused and a dispute arose between them. The next morning the 10th N.I. began to loot the treasure and to get away with what they had individually secured. About 100 sepoys of the 10th N.I. came to the Nawab about 9 A.M. and encamping some in the Pye Bagh and some in Ukkabl Khan's house in the city, which house belongs to the Nawab's mother-in-law, some went to Kalle Bagh. Bhoo (?) Aly Bux looted a *Buggy* and went to live in the house of Hurry Naraen (Hari Narain).The 10th N.I. had ten guns, colours and magazine on the parade ground, the whole of the Regiment having left except the 100 that had gone to the city. The 41st and 10th Oudh Irr. (Irregulars) on the same day took possession of the guns and magazine left by the 10th N.I. and their colours and took them to their own Camp. Everything remained quiet that day and night. On the following day the 41st and 10th Oudh Irr. (Irregulars) went to invite the remaining portion of the 10th N.I. who were in the city to their Camp. They went with Bhoo Aly Bux at their head. Bhoo Aly Bux was then for the first time arrested by the 41st and 10th Oudh Irre: (Irregulars) who then fired upon the other sepoys and looted them The Collector's Sowars were at Ukkabl Khan's house where they had gone when the 10th N.I. left their Camp. The next day the Mhow Puthans came and encamped in the Pye Bagh. This was in the morning, and in the evening the Nawab sent Kureem (Karim) Jan to Mr. Thornhill in the fort at Futtehgurh to tell him that the Nawab had up to that time done all in his power but that he was now utterly powerless, and in the hands of the rebels. Mr. Thornhill wrote to Nawab (vide letter of 22 June). Two, or three days after, the native officers of the 41st, 10th Oudh Irreg. and Aga Hussein came to the Nawab in the evening about 5 P.M. The Nawab was going to a raised mound in his garden on which there was a seat. The native officers and Wuzeer Aly were going home, when Wuzeer Aly was fired at and the cloth on his shoulder burnt; it was not known who fired. Nawab then called Wuzeer Aly to him. He came followed by all the native officers who came with drawn swords, stating that Wuzeer Aly was a Christian, and believed that the Nawab was one also, and after threatening him, the Nawab slipped quietly away thro' (through) the crowd, which consisted of some 6 (600) or 700 persons, and got into his own house and took refuge in the female apartments and remained there 3 days without coming out. At the expiration of that time Ahmed Yar Khan came to the door of the female apartments and sent word to the Nawab requesting him to come out as he wished to speak to him. The Nawab came and Ahmed Yar Khan informed him to have no fear, that if he came out he would not be hurt. The Nawab upon this came out. The next morning all the native officers of the various regiments came to the Nawab as also the native officers of the Mhow Puthans; Aga Hussein, Gunga Singh and one of the Doobyes decided and ordered all the troops to proceed the next morning to attack the English at the Futtehgurh fort and would consider any one that did not join them as Christian, and would kill him and loot their (his) house. At the same time they had the order proclaimed in the city by *tom-tom*, saying it was the order of the officers of the force. They then went home. The Futtehgurh fort was attacked 2 days after this, about 200 of the Mhow Puthans joined the sepoys but not the general body under **Ahmed Yar Khan...**' **FPP** quoted in **FSUP**, Vol V, pp 729-733.

FATEHGARH Refugees
This is a list, not quite complete, of the refugees who were stranded near **Kanpur** on the morning of 9 June 1857 and killed on 12 June at the **Savada House.**•Mr. Alexander•Mr. and Mrs. J. Brierley and children, Misses E. and F. Brierley. Four members of the Brierly family lived within a few yards of the church at Fatehgarh He was an auctioneer, shopkeeper, tent maker and town booster.•Mr. and Mrs. Richard 'Dick Sahib' Brierley and one child. Mr. Brierly owned some of the boats in which the refugees fled. Formerly an auctioneer, Brierly was Fatehgarh's most prosperous trader, dealing primarily in what were called "Europe" goods. He commanded the party of fugitives that decided to proceed to Kanpur, and his lead boat was the largest in the flotilla. •Mr. Billington, clerk.•Ensign Reginald S. Byrne of the 10th Native Infantry. Byrne was the second son of Wale Byrne, a one-time Anglo-Indian civic leader in Calcutta, and his was a rare case of an Eurasian obtaining a commission in the regular Bengal Army. His brothers served in the 77th and 53rd Foot.

The Grave of Bonny Byrne, aged 20, in the Kacheri Cemetery, Kanpur.

Byrne married Bonny D'Fountain, *(see BONNY BYRNE)* 'a lovely young Eurasian girl' who was raised in a small bungalow in Fatehgarh by her widowed mother (there were rumours that she was the issue of a dalliance between her mother and a British major; in any case, her origin was 'discreetly hidden, the conduct of her mother no doubt making it desirable'). As Bonny entered her teens, the **Nawab of Farrukhabad** began to call on her, and so alarmed the authorities, who had already endured a scandal involving an Anglo-Indian girl and one of the Nawab's brothers, that they packed her off to Calcutta to attend the Kidderpur Girls' School. There she met Ensign Byrne, and, at the age of 14, married him and followed her new husband to his posting back in Fatehghar. Six months later, when the Mutiny broke, Reggie fought gallantly in the defence of the fort, but his teenaged bride and her mother sought refuge in the palace of the Nawab, who protected them until the approach of **Campbell's** troops. According to some commentators the scandal that attached to Bonny's retreat to the Nawab's palace contributed to the uprising at Fatehghar. Refusing safe passage to Calcutta, she returned to Kanpur and married a clerk at Maxwell's named John Chandler. She gave birth to a son, Adolphus Gennoe Chandler Byrne, and registered him as Reggie's son, though this was technically impossible. She herself died soon afterward. •Reverend and Mrs. David E. Campbell and 2 children, they had one child living in the hills who survived the Mutiny. •Post Office Inspector and Mrs W. Carr and 1 child. •Mr and Mrs William Catania and child. Mr Catania was Inspector of Post Offices. His wife Arabella was the sister of John Kew, the Postmaster.•Mr Charles Cawood, Head Clerk of the Clothing Agency, his wife and 2 children. He was the son of a pensioned clerk. •Mr Ellington •Mr and Mrs J.M.R. Elliott and 5 children of **Duleep Singh's** estate (probably students in his village schools). Elliott was a former Sergeant in the Bengal Sappers; his letters regarding Fatehgarh and Duleep Singh's vestigial estate were regarded as being of sufficient interest to be relayed to **Queen Victoria**, but they have evidently been lost •Mr and Mrs Finlay of the Clothing Agency and children. •Miss Finlay .•Mr (Pensioner) and Mrs Faulkner and children. •Rev. and Mrs J.E. Freeman. Reverend Freeman was an American Presbyterian missionary who superintended the Orphanage and tent factory. •Mr and Mrs Guise. Guise was a partner of Mr Maclean in an indigo manufacturing business. He was at one time the tutor of Maharaja Duleep Singh, who allowed him to remain on his estate after His Highness left for England in 1854. Guise and Maclean were neighbours, and lived near the Hospital Ghat. •Sergeant and Mrs Hammond of the Gun Carriage Agency and 4 children. •Surgeon Thomas Godfrey Heathcote. He was Surgeon to the 10th Native Infantry. He and his wife had at least three children back in England under the care of an Aunt named Heathcote. •Mr and Mrs J. Ives and Miss Ives. Mr Ives was a tent maker. •Reverend and Mrs Albert O. Johnson. •Mr J. Joyce, wife and children.•Mr and Mrs Kestall of the Clothing Department and 3 children. •Joint Magistrate and Mrs Robert Nisbet Lowis and two children. Mrs Lowis 'had maintained her fortitude throughout, and was indefatigable during the siege in preparing tea and refreshment for the men.' When **Gavin Jones** climbed into the boat she immediately fetched him a brandy. Probyn described Lowis as a man of 'business-like habits, sound sense, general acquaintance with his duties, and excellent temper.' •Reverend and Mrs J. McMullin. Reverend McMullin was an American Presbyterian missionary. •Mr Maclean, merchant, and Miss Maclean. Maclean was a partner with Guise in Maclean & Company, manufacturers of indigo. Another daughter was married to J.M.R. Elliott, who oversaw Duleep Singh's vacated estate. •Mr and Mrs Macklin of the Collector's Office and 8 children. Mrs Macklin had successfully held out for the unheard-of sum of Rs 1,700 when Dr Login needed her unprepossessing cottage to complete purchasing 'The Park' to prepare for the residence of Duleep Singh. •Mrs Macdonald and 3 children.•Mr and Mrs J.R. Madden of the Clothing Agency and 2 children. Mrs Madden was Elizabeth Shepherd, sister of James Shepherd the schoolmaster. •Misses Eliza and Emelia Madden. •Deputy Collector and Mrs J. Palmer and 9 children. Hingun the ayah (*See* **Williams**) was his employee. One child was supposed to have succumbed to the heat *en route* to Kanpur. •Mr R. and Miss E. Ray. •Mr John B. Kew, wife and children. Kew was the Postmaster . •Miss Kew •Reverend J. and Mrs MacMullen. MacMullen was an American Presbyterian missionary. •Mr and Mrs O'Hern. •Mr and Mrs Roach •Schoolmaster and Mrs. Shiels and 2 children. Shiels was the schoolmaster of the East India Company's school for European children, and under his direction it acquired a fine reputation and drew children from many other districts. The school was the setting for a portion of the Mutiny novel *Mariam*. •Mrs E. Shepherd and 2 children. •Miss Mary Shepherd.•Lieutenant Charles Worsley Swetenham was the son of a captain in the 10th N.I. and was born in Kanpur in 1832. He was educated at Bedford Grammar School. He was killed at Singheerampore. He died at the hands of the same regiment into which he had been born. A rebel Sepoy, seized at Ghazipur, was found riding his horse. •Mr R. Wareham. •Mr Robert Wareshaw, Head Tailor of the Clothing Agency. This list owes much to the research of **ASW.**

***FATEHGARH* SURVIVORS** Of the Europeans who first took to the boats at **Fatehgarh** and sailed down to **Kanpur** in the first week of June 1857, over one hundred were dragged before the **Nana** and ordered to be killed (some depositions say it was **Bala Rao** who insisted on their death and actually gave the order - *See Depositions* in **Forrest** or **Williams**), but there may have been two survivors. On the morning of 12 June, a day after the Fatehgarh party was killed, a dhobi carrying his wash to the river came upon an insensible five or six year-old girl lying on the bank with a deep wound to her clavicle. He handed her over to a **golundaz** who brought the child before the Nana, begging to adopt her as his own. **Teeka Singh** objected, but the Nana overruled him. She was reportedly recovering from her wounds when General **Havelock** reached Kanpur, whereupon the Golundaz took her with him. According to **Nunne Nawab** the golundaz was imprisoned by Teeka Singh for proposing to adopt the child and the little girl was probably murdered. But in 1859 **Jung Bahadur** of Nepal demanded that Bala Rao, who was then encamped in the Terai and dying of fever, return all of his European prisoners. He claimed that he no longer had any, but conceded that it was true 'that a daughter of the Judge of Furruckabad, the only European with me, who has received a bullet wound, which I caused to be cured *was* with me, but she has lately been made over to Colonel Bird.' No record of this hand-over exists, nor is it known who exactly is meant by Colonel Bird. Bala Rao's 'Judge of Farrukabad' might have been the Magistrate at Fatehgarh, Robert Thornhill, who was believed to have died with his wife Mary and, according to their memorial at All Souls Churdh, Kanpur, two children, one of whom was a three or four year-old boy, and the other an infant. A third child, whom Mary called 'My dear Georgie', evidently lived with Mary's sister in England. A possible second survivor was a woman who, 33 years later, published a memoir in *The Indian Planter's Gazette*, a transcription of which still survives in the British Library. She claimed she was the daughter of the 'Sudder Line Sahib,' or Commissary of Bazars named Sutherland who had moved to Fatehgarh in 1845 and, with his wife and children, is recorded as having been killed at Farrukabad on 23 July. Nevertheless she claimed that she had sailed to Kanpur with Brierly's party, and that while desperately fleeing the flames on the morning of 11 June, she fell into a deep hole and broke her ankle. There she lay in a stupor until morning, when a Subadar and a Sowar of the 2nd Cavalry chanced upon her while 'searching the *churs* for money and valuables which the Europeans in their flight had flung away.' The Subadar took pity on her, and sent the Sowar to fetch her a drink of water as he bandaged her ankle with a strip of his turban. 'When the Sowar had returned,' she recalled, 'he was immediately despatched for a Dhoolee in which ...I was [conveyed] to the City and placed under the care of a native nurse (dhai). ...I was now in high fever and scarcely knew or cared what should become of me.' The Subadar ordered the *dhai* to keep Miss Sutherland's presence a secret, for if Nana Sahib found out that one of his officers was protecting a European, it would mean death for all of them. Thus Miss Sutherland was kept hidden in the city, suffering from a fever that 'nearly deprived me of my reason.' She said she heard 'nothing of what was going on in and around Kanpur,' but her nurse evidently brought her fanciful bulletins from time to time, including an account of an unsuccessful attempt by Nana to have the wealthiest of the Fatehgarh party trampled by elephants.* Miss Sutherland later claimed she was left behind by the fleeing rebels. She was discovered seated in her dhoolie by officers from Havelock's expedition, who brought her with them on their entry into Kanpur on 17 July. Whatever the truth of her story, Miss Sutherland did not repay her saviour's kindness. The subadar who cared for her was eventually caught and prosecuted for mutiny at Allahabad. Miss Sutherland, by then a wife and mother, was called by the defence to testify that the subadar had risked his life to preserve hers. But she refused, the *subpoena* was withdrawn, and the subadar (possibly Ali Bux of the 2nd Cavalry) was convicted, and hanged.

FATEHGARH UPRISING There is a full account of the uprising at Fatehgarh/Futtehghur in the London *Times*. 3 November, 1857. *See also* F.R.Cosens & C.L. Wallace, *Fatehgarh and the Mutiny,* Lucknow 1933, reprint Karachi 1978.

FAVOURERS 'Favourers, here, means the small minority, chiefly civil servants, who, though advocating tremendous reprisals, are yet suggesting that it might be as well to avoid indiscriminate massacres, in which a far greater number of innocent than guilty must obviously suffer. This point of view seemed so wrong, so unmanly and unhealthy, that those who held it were denounced as agents of Nana Sahib and there were serious proposals that a number of European officials should be executed as a warning- to the others. Troops were pouring into Calcutta, but, while the ardour of the soldiers was inflamed by numerous and often untrue atrocity-stories, no one seems to have thought it necessary to tell the soldiers that the war was against certain rebels and not against the population of India. In consequence of this omission there were some regrettable incidents in the streets of the capital. A soldier had just landed when 'I seed two Moors talking in a cart. Presently I heard one of 'em say 'Cawnpore' I knowed what that meant; so I fetched Tom Walker, and he heard 'em say 'Cawnpore', and he knowed what that meant. So we polished 'em both off.' From Dennis Kincaid, *British Social Life in India,* London, 1938.

FAYRER, Sir Joseph Dr Fayrer was Civil Surgeon to the Residency at Lucknow and had overall responsibility for the newly established postal service within Oude, and he and his family were present within the Residency throughout the **Siege of Lucknow**, living in 'Fayrer's House' which was the name of one of the garrison's posts within the compound, and where **Sir Henry Lawrence** died. The ruins of this

Dr Fayrer's House, Lucknow Residency

house are preserved and may still be seen in Lucknow. Dr Fayrer was famous also as an authority on tropical diseases. *See* **FQP**; *see also* Sir Joseph Fayrer, *Recollections of my Life,* Edinburgh, 1900.

FAZIL MUHAMMAD KHAN
A **jagirdar** of **Bhopal** who took advantage of the mutiny of the **Bhopal Contingent** to rebel against the Begum. For an excellent and succinct account of these events and the uprisings at Sehore and Berasia *see* K.D.Bhargava, *Descriptive List of Mutiny Papers in the National Archives of India, Bhopal,* Vol II, New Delhi, 1963

FAZL-i-HAQ KHAIRABADI, Maulana
He gave the following as his interpretation of the causes of the struggle: 'The British Nazarenes, whose conquest of the land and cities had made their hearts wanton with rancour and nefarious designs, and who had under dread and fear brought all the self-respecting nobles to humiliation and shame to the extent, that they did not raise their head with the least resistance, started planning the conversion of the inhabitants of India, irrespective of all the classes - the rich and the poor, the great and the petty wanderers and settled ones, the men of the city as well as of the country. They had thought that with a measure as this they would not enable the people to acquire any backing and then put up resistance; it would rather compel them to surrender. All this had been staged with the avowed aim that all and sundry after becoming heathens like themselves, should converge on one and only one centre thus barring them from claiming distinction as a class among others...They cleverly started establishing educational institutions to plant the language they spoke and propogate the religion they professed. In short they did every thing to eradicate all the indigenous sciences, metaphysical beliefs and institutions of yore. Another plan they initiated was to monopolise the grain trade...Besides these they have secreted other nefarious plans *eg* dissuading the Muslims from circumcision, observance of Pardah and other religious laws of the *shariat...* the Hindu sepoys who were in great numbers, were forced to taste the fat of the cow and the Muslims that of the pig. This caused resentment in both the groups and thus disturbed their peace and tranquillity. In a state of excitement they perpetrated crimes such as killing the Christians, attacking their own general and committing robbery. They even resorted to extremes of cruelties, massacred innocent children and ladies and thus heaped upon their heads a bad name and an evil reputation. After this the rebel groups of soldiers set out from their military barracks. The administration got paralysed, the highways became unsafe and the towns and villages presented a deserted appearance'; *see* Fazl-i-Haq Khairabadi, *'Saurat-ul-Hindya',* Bijnor, 1947, pp 355-60.

FEROZEPUR
A city on the left bank of the Sutlej river in the Punjab, 175 miles from Lahore and 1,181 from Calcutta. It had once been the chief frontier station of the British, before the Sikh Wars, and contained an intrenched magazine filled with military stores scarcely inferior in amount to those of the arsenal of Fort William, and certainly rivalling the **Delhi Magazine**. There was a strong British regiment there, H.M.'s 61st, 1,000 men, and some European artillery also, together with the 10th Bengal Light Cavalry, and the 45th BNI and the 57th BNI. Brigadier Innes was in command of the station, and took some extraordinary decisions, including ordering the blowing up of the magazines of the 45th and the 57th, in an apparent attempt to safeguard the 20,000 barrels of gunpowder in the arsenal; for his defensive attitude he was summarily removed from the list of Brigadiers. Some of the native infantry mutinied, others remained loyal, the cavalry remained loyal for the time being, but, disarmed and dismounted, a part of the regiment broke into revolt on 19 August when the news reached them that their horses were to be taken away. Veterinary Surgeon Nelson was killed but the horses themselves were the greatest sufferers, some thirty-two being killed and wounded by the inept firing of a gun; *see* F.Cooper, *Crisis in the Punjab,* Lahore, 1858; Chas Raikes, *Revolt in the NWP,* London, 1858; Brigadier Innes' despatch 16 May 1857 in *Further Parliamentary Papers (No 3)* p 7, **Ball** and **Martin.**

FEROZE SHAH
See FIROZ SHAH.

FIFTH COLUMN
'Spies have a good or bad image, depending on whether they're on your side or the other fellah's. Glamorous, dangerous romantic heroic work for the cause, or sneaky dirty little tricks done on the sly by cowardly scum. Take your pick. This tale is about two spies, or fifth columnists who worked for the rebels during the siege of Delhi in 1857. The British thought them cowardly trai-

tors; we have no record of what the 'mutineers' in Delhi City thought of them: in all probability they knew nothing of their existence, or their patriotism. To the impartial observer they were men of great courage who, knowing the risk to their lives, were prepared to undertake most hazardous feats, for the sake of the cause in which they believed. Their reward can only have been satisfaction - that they were achieving something which hurt the British. No financial or material reward was possible, indeed, ironically the only pay they had came from their employers, the British Army! I can give you no names, no personal details, no families: they are anonymous martyrs, yet they were very real, and more heroic than many of the prominent rebels who strutted in the court of Bahadur Shah. The tale is simply told. On 16 August 1857 on the Ridge before Delhi, and close to the Flagstaff Tower, two men were brought before a military commission. They were charged with 'tampering with service ammunition.' At first sight this seems a strange charge, for they were both gun lascars who would have spent much of their lives handling such ammunition. They were employed by the army to help serve the artillery guns which were trained on rebel positions in and around Delhi. They did not actually fire the guns or lay them, or in any way have physical control over them, yet they had worked out a scheme which was as effective as if they had had complete control. They were used as labourers in making up the charges for the guns, and in carrying powder, cannon-balls and shells to each battery. They worked in different ways. One of them set about his task crudely, and with so little finesse that he was almost certain to be rumbled, and that quickly. He extracted some of the gunpowder from each cartridge, and substituted broken glass and gravel. The cartridge would thus weigh roughly the same as it ought to, and therefore provoke no suspicion among the gunners at first, but when a series of firing resulted in the cannon-ball merely flopping harmlessly out of the cannon's mouth, the ammunition was bound to be suspected. A cartridge was eventually opened, found to be adulterated, and the first gun lascar arrested. But the second man was altogether more astute. The British had some 5½" mortars which were used for shelling the sepoys in the suburb beyond the Subzee Mundi; and according as they advanced or retired, smaller or larger charges of gunpowder were required for the cartridges. Clearly the farther off the enemy, the more powder was needed. The powder was weighed out by a British sergeant, and poured into the bags - *ie* cartridges - through a paper funnel which was held by the lascar. The gunnery officer found after a while that his shells fell alternately short of and went beyond the mark; he sent an angry message to the sergeant. The latter could at first find no explanation, and considered perhaps that he was weighing out inferior powder, but eventually his suspicions were aroused and, while weighing out a charge he shot forth a hand and seized that of the lascar: he had indeed found the answer. By pinching the paper funnel while the powder was being poured through it, he had managed to diminish the charge in one cartridge and increase it in the next, thus explaining the erratic firing of the mortars. The young officer who was president of the military conmmission which tried the two lascars did his best to give them a fair trial, but the evidence was overwhelming, they were 'guilty as charged', and duly hanged. But no monument, no plaque, no awareness even in the rebel camp of their deeds. Unsung heroes indeed; there may have been many such.' **PJOT**

FIFTY-SECOND Regiment This BNI regiment based in and around **Jabalpur** had a somewhat strange history during the outbreak, remaining faithful for some time, and indeed helping to put down rebellion, but then it mutinied, although (with the exception of the murder of Lieutenant Macgregor) it treated its officers with kindness and courtesy. The regiment was subsequently almost totally destroyed at the **Battle of Kunch**. A full account of the mutiny, given by a sepoy of that regiment captured at **Garhakota** in February 1858, is given in Appendix III of *The Revolt in Central India*, Simla, 1908.

FIFTY-THIRD Regiment The 53rd BNI, a thousand strong, reached **Kanpur** in early 1857. Most all of them Oude[Awadh] men, averaging 5'8". Uniform British red with yellow facings. Their pay was low by British military standards, but made them aristocrats among the local people: 7 rupees a month, out of which they paid for their own food and 'a suit of summer clothing.' 'Thoroughly disciplined and martial in appearance, these native troops presented one memorable point of contrast with European forces, drunkenness was altogether unknown to them.' Mowbray Thomson, *The Story of Cawnpore*. London, 1859. p 16-17.

The Serai at the Subzee Mundi

FIRE Fire and incendiarism in general were frequent accompaniments of the mutiny/rebellion, indeed the firing of bungalows in cantonments and civil lines often preceded, by only a day or so, the full scale outbreak of mutiny. But nowhere was fire more devastating to the British than at the siege of **Wheeler's Entrenchment** in Kanpur:- ' The second barrack had from the commencement excited serious apprehension lest its thatched roof should be set on fire. An imperfect attempt had been made to cover the thatch with tiles and bricks, and any materials at hand that would preserve the roof from conflagration. But after about a week the dreaded calamity came upon us. A carcase or shell filled with burning materials settled in the thatch, and speedily the whole barrack was in a blaze. As a part of this building had been used for a hospital, it was the object of the greatest solicitude to remove the poor fellows who lay there suffering from wounds and unable to move themselves. From one portion of the barrack the women and the children were running out, from another little parties laden with some heavy burden of suffering brotherhood were seeking the adjacent building. As this fire broke out in the evening, the light of the flames made us conspicuous marks for the guns of our brutal assailants, and without regard to sex or age, or the painful and protracted toil of getting out the sufferers, they did not cease till long after midnight to pour upon us incessant volleys of musketry. By means of indomitable perseverance many a poor agonizing private was rescued from the horrible death that seemed inevitable, but though all was done that ingenuity could suggest, or courage and determination accomplish, two artillerymen unhappily perished in the flames. The livid blaze of that burning barrack lighted up many a terrible picture of silent anguish, while the yells of the advancing sepoys and the noise of their artillery filled the air with sounds that still echo in the ears of the only two survivors. ...In the burnt barrack all our medical stores were consumed; not one of the surgical instruments was saved, and from that time the agonies of the wounded became most intense, and from the utter impossibility of extracting bullets, or dressing mutilations, casualties were increased in their fatality. It was heart-breaking work to see the poor sufferers parched with thirst that could be only most scantily relieved, and sinking from fever and mortification that we had no appliances wherewith to resist.' When the ashes of the consumed barrack cooled, the men of the 32nd Regiment, who had been stationed there, raked them over with bayonets and swords, making diligent search for their lost medals. A great many of them were found, though in most instances marred by the fire. The fact that they would explore after these treasures while the sepoys were firing on them, shows the high appreciation in which the British soldier holds his decorations.' *See* Mowbray Thomson, *The Story of Cawnpore*, London, 1859.

FIROZ SHAH, Prince a cousin of the King of Delhi. Skilful commander and determined enemy. Escaped in disguise as a pilgrim to Karbela where he lived for many years. There is more detail of his movements and leadership than of many of the other leaders. His military daring, hair breadth escapes and skilful horsemanship, were spoken of with admiration on both sides of the conflict. *See* **FSUP** Vol III p 662 *et seq. See also* T.H.Kavanagh, *How I won the Victoria Cross,* London 1860, for an account of the battle of **Sandila** at which Firoz Shah was present 'riding a grey horse and being dressed in white' *See* SAHZADA SULTAN IBRAHIM.

FIROZ SHAH, PRINCE 'The delay in defeating the English has been caused by the people killing innocent children and women, without any permission from the leaders... Let all avoid such practices, and then proclaim a sacred war.' quoted by B.W.Noel, *England and India,* London, 1859, pp 72-3.

FIRST BOMBAY CAVALRY Although the Bombay Army did not (apart from an isolated incident) mutiny, it was clear that they naturally had some sympathy for their opposite numbers in the Bengal army. The most significant expression of this came at *Nasirabad* where they were ordered to charge the mutineers of the 15th and 30th BNI. They duly charged but when within a few yards of the mutineers they went 'threes about', only the officers going on. Two of the latter were killed and the other two badly wounded. It was afterwards said that there had been an understanding between the mutineers and the Cavalry; that the latter agreed to refuse to act provided their families remained unmolested. Nevertheless the 1st Bombay Cavalry afterwards obeyed orders and fought well against the rebels, notably at the battle of **Kotah-ki-Serai** and the capture of **Gwalior** by the British in June 1858.

FIRST REGIMENT This was the BNI. regiment of which **Lieutenant Godfrey Wheeler** was quartermaster. In April of 1859 the 1st N.I. camped at Bunkania, at the foot of the Nepalese hills, 8 miles from **Gonda** on the Faizabad Road. On 13 April they where defeated by Lieutenant Colonel Cormick and disappeared into oblivion. G.H.D.Gimlette, *A Postscript to the Records of the Indian Mutiny: An Attempt to Trace the Subsequent Careers and Fate of the Rebel Bengal Regiments, 1857 to 1858.* London, 1927. 'A few days (June 3 or 4) previous to the mutiny, two Mahomedans from the city (of Allahabad) came into the lines of the regiment to incite the men to mutiny; they spoke to a sepoy of the light company, and he told Gunga Pershad, 8th company, what he had heard. These sepoys seized the two men; they resisted, when drummer Peters and I aided in securing them, for which we each received a reward of 50 rupees, while the two sepoys were promoted. The Mahomedans in the corps were displeased at the men being seized, and some of them remarked to me that the sin of arresting them, especially on an *Eed* day, would rest on my head.' *Deposition of John Fitchett.* **Forrest**, G.W., ed., *Selections.* V. III, pp l-lxxii. *See also* WATKINS, QUARTER-

MASTER SERGEANT.

FITCHETT, Rev. William H, A writer of a popular history of the rebellion, which was much read when it was first published. Of significance mainly for its inclusion of a diary, verbatim, of a Mrs Soppitt, an officer's wife, which she kept throughout the siege of the Lucknow Residency. *See* Rev. William Fitchett, *The Tale of the Great Mutiny*, London, 1903.

FLAGSTAFF TOWER Delhi, on the Ridge to the north-west of the city. A strong round tower, appointed by the Brigadier as the rallying point for refugees from the mutineers on 11 May 1857. There were no troops to garrison it so it was deserted within hours and the refugees dispersed, with varying fortune. *See* **Ball**, *see also* DELHI LANDMARKS.

FLAMING ARROW **Mrs Wells** *(See* BERNERS PAPERS*)* speaks of a 'flaming arrow' fired at the **Lucknow Residency** on 1 May 1857. Her own bungalow had already been fired in the wave of incendiarism which preceded the outbreak in many areas.

FLIES 'On the 7th April 1858 the time had arrived when we were to leave Lucknow, and the change was hailed by us with delight. We were glad to get away from the captured city, with its horrible smells and still more horrible sights.....away from all the offensive odours and the myriads of flies.....some idea of the torment we suffered from these pests: when we struck tents all the flies were roosting in the roofs; when the tents were rolled up the flies got crushed and killed by bushels....after the tents were pitched again and the roofs swept down, the sweepers of each company were called to collect the dead flies and carry them out of camp. I noted down the quantity of flies carried out of my own tent. The ordinary kitchen baskets served out to the regimental cooks by the commissariat for carrying bread, rice etc, will hold about an imperial bushel, and from one tent were carried out five basketfuls of dead flies.' NB An Imperial Bushel equalled 8 gallons or 36 litres. *From* Wm Forbes-Mitchell, *Reminiscences of the Great Mutiny*, London 1910, p 243.

FOLK SONGS These, as **P.C.Joshi** points out, are one of the folk-art forms of India that have been the traditional media for approaching the masses.

Equestrian Statue of Kunwar Singh at Jagdishpur

'Oudh was annexed by the British on February 7th 1856, and its Nawab **Wajid Ali Shah** *duly exiled. The song below portrays the event. The song is very sad and shows how depressed the people became after their homeland was annexed by the British. Certain points are worth noting. Firstly Wajid Ali Shah was known to be the typical representative of a decaying feudal system - corrupt wasteful inefficient - and yet he was looked upon with sympathy, for with him went national sovereignty as well. One's own ruler, irrespective of his merits, is preferred to foreign rule. Secondly, the Oudh homeland is called 'mulk' (country). The conception of India as our common motherland had not yet emerged, one's own homeland is yet the only country they knew as their own. Thirdly, it is clearly realised that the British rely on their 'might' to 'seize' other lands. Fourthly, deep anguish was felt at the failure to resist foreign aggression and that no one 'took up arms'. It is this*

deep feeling of national humiliation, at newly lost independence, that made Oudh the storm centre of the 1857 uprising:- ***Our homeland, Hazrat, since your exile looks utterly desolate, Bereft of pomp and power is your majesty! Lost is the game; gone the Khayal. The Begums rode to a far exile; their homeland left forever. The Angrez with his might and main marched in to seize the land. Not a soul said 'nay' to him: no one took up arms against him. Laid waste the Angrez our Kaiser Bagh; for Calcutta left our King. What solace left for us, what support !'*** See P.C.Joshi ed. *Rebellion 1857. A Symposium*. Delhi, 1957. The Battle of Bibiganj (near **Arrah**) was fought on 7 August 1857 and is celebrated thus: ***'It was a night of the month of Bhado; and dark clouds covered the sky; Babu Kunwar Singh went to war at the dead of night; The Firinghis trembled with fear; the skies poured torrents of rain; And the guns showered bullets in the battle-field below; Babu Kunwar Singh's horse advanced with prancing steps; And the whole army, he kept beheading! Heads fell thumping, with the clap of each trotting hoof!'*** See also The *Indian Antiquary*, April and June 1911, containing a number of songs from 1857 collected by Ramgharib Chaube.

FOOD UNDER FIRE

Ensign Wilberforce served at the siege of **Delhi**: 'one day about luncheon time several of us were in a house which we had got into by knocking holes through walls. Question raised as to who should get us some food.. Drew lots as to who should run down the street under fire and get some food from the mess. Lot fell on a captain. He got down safely and we saw him coming back with something in his hands, when he caught sight of us watching he raised what he was holding over his head and waved it. Suddenly we saw it struck out of his hand and fall in the street. He picked it up, but when he arrived we found that the cold leg of mutton he was bringing had a bullet through it. ' ' Flies are always a nuisance in India., but on this piquet they gathered together in special numbers; it might have something to do with the fact that an elephant had died and lay, unburied, nearby. Our only amusement and that soon palled, was tying together two large bones, and seeing two adjutants (he doesn't say birds but that is what he means!) swallow one each and have a tug of war as to which should get both bones.'

FORBES-MITCHELL, William

A Sergeant in H.M.'s 93rd Regiment (later Sutherland Highlanders) who served throughout the period of the rebellion, particularly in and around Lucknow, and indeed stayed on in India after his discharge. His reminiscences, published in London in 1910, read more like a collection of anecdotes than genuine history, but make excellent reading for all that. It is possible that he has exaggerated or even concocted events so as to make a good story, but his book represents the view of the man in the ranks, who, often in war, is ignorant of the grand strategy involved, but only too well aware of the detail that affects his personal safety and comfort. *See* W.Forbes-Mitchell, *Reminiscences of the Great Mutiny 1857-9*, London 1910.

FORCES

The army of the rebel government at Lucknow: probably totalled 200,000 men; quality both of personnel and arms differed widely. Some were sepoys, particularly cavalry and artillery, and re-employed soldiers from the ex-King's army, but majority were talukdars' zamindari men, often armed only with matchlocks. The zamindari men in particular fought on occasion with great bravery and devotion particularly in attacks upon the besieged Residency. For details of these forces *see* Raikes, Chas, 'Notes on the Revolt in the NWP of India', 1858; *see also* '*Bengal Hurkaru* and *India Gazette*', 15 April, 1858, *and* STATISTICS.

FORGUES, Paul Emile Dourand

A former editor of the *Delhi Gazette* he published a number of narrative histories of the rebellion, both in French and in English. One tale that **Edward Leckey** demolished effectively was apparently the concoction of 'the former editor of the Delhi Gazette' *ie* Forgues: unfortunately it had been widely believed and had done much harm in the interim. It announced as unquestionable fact that 'Every steamer that arrived at Calcutta from the upper Provinces brought down crowds of helpless women and children, many of them without ears or lips or fingers or nose.' Leckey showed that there was not one case of such disfigurement among the Europeans. But the harm had been done: attitudes had polarised. *See* P.E.D. Forgues, *La Revolte des Cipayes*, Paris, 1861, *and* (under the soubriquet 'A Former Editor of the Delhi Gazette') *The Indian Mutiny to the Fall of Delhi*, London 1857.

FORREST, Sir George

One of the 'heavyweight' historians of the rebellion. His three volume account is one of the standard histories of the times: it was produced sufficiently long after the events to justify being regarded as objective and comprehensive. *See* Sir George Forrest, *History of the Indian Mutiny*, 3 vols, Edinburgh, 1904.

FORSYTH, Sir Douglas

A special Commissioner appointed to round up rebels. At the outbreak he was Deputy Commissioner at Ambala, and his personal friendship and influence over the Maharaja of Patiala was crucial to that Prince's attitude. He reports the remarkable fact that 'the first man not a soldier in the Punjab - and I dare say in all Upper India - who was hanged for sedition *was a Sikh.* When the news of the Mutiny at Delhi reached Rupar, a town on the Sutlej in the Umballa district, which in the first Punjab war had rendered itself notorious for its antipathy to the English rule, there was a great disturbance, and one Sikh notable went through the streets telling the people not to make any payments of revenue till they should see who was the master. On this news being reported to me, I had him seized and brought in to Umballa and there and then authority was obtained from the Chief Commissioner and a court was constituted before which he was tried and having been found guilty of sedition and inciting to rebellio he was sentenced to be hanged'. Sir Douglas Forsyth, *Autobiography and Reminiscences of*, London, 1887.

FORTY-FIRST REGIMENT The *Dooby Kee Daheena Paltan*, BNI, was raised in 1803 and served at Bhurtpore and Sobraon. They proved among the most bloodthirsty of the rebellious regiments at **Sitapur** (Seetapore). The 41st marched to **Fatehgarh** to induce the 10th to join them, and were active in the murder of the **Nawab of Farrukhabad**'s hostages. In August 1857 some proceeded to **Lucknow** to join in the siege of the **Residency** while a detachment went to join the **Nana** at **Bithur**. This second contingent held together during the Mutiny, but in January of 1858, near Fatehgarh, they were wiped out by **Sir Colin Campbell**'s **9th Lancers**. *See* G.H.D. Gimlette, *A Postscript to the Records of the Indian Mutiny: An Attempt to Trace the Subsequent Careers and Fate of the Rebel Bengal Regiments, 1857 to 1858*. London, 1927, pp 151-3.

FORTY-SECOND REGIMENT At **Saugor [Sagar]** the 42nd BNI fought against the loyal troops of the 31st and fled to **Bithur**. On 31 July they came within 10 miles of **Kanpur**, killing a thanadar and destroying a bridge. On 2 August they encamped close to Bithur, and fought **Havelock** with great courage. Crossing bayonets with the Madras Fusiliers, they were 'badly cut up,' and fled. After the Mutiny the 42nd was reconstituted as the 5th Bengal Native Infantry with a core of 50 loyal sepoys. G.H.D. Gimlette, *A Postscript to the Records of the Indian Mutiny: An Attempt to Trace the Subsequent Careers and Fate of the Rebel Bengal Regiments, 1857 to 1858*. London, 1927, pp 153-4.

FRANKS, BRIGADIER One of Sir **Colin Campbell**'s subordinate officers. He commanded a column which made a successful sweep through Oude[Awadh] in March 1858, and subsequently.

FREE SCHOOL, Kanpur After the Mutiny, the Cawnpore Free School, where **Azimullah** was educated, became Christ Church School, then College. *See* Indian National Trust for Art and Cultural Heritage (INTACH). *Preliminary Unedited Listing of Kanpur. Uttar Pradesh, Unprotected Monuments, Buildings & Structures Listed for Conservation*. Kanpur, 1992, p 106.

FREEMASONRY In 1834 the United Grand Lodge of Antient, Free and Accepted Masons of England gave a charter to 'Lodge Harmony' at Kanpur and housed it in an old and formidable building. Freemasonry had been first introduced in to the subcontinent in 1730, but it proved a continuing perplexity to the Indians. In a country permeated by religion they found it hard to believe that a society so secret and ritualistic was not religious. But then if it was religious, they wondered among themselves, how could its members also call themselves Christians? Lodge Harmony was locally known as the Bhootighar, or House of Ghosts, and most Indians would not go near it. But sometime in the early 1850s Nana Sahib actually walked through its door as an Entered Apprentice and became not only the first Indian to join Lodge Harmony but reputedly the first Indian Freemason in all of India. Nana's reasons were characteristically murky. He may have joined simply out of his desperate hunger for honours and titles, or to further ingratiate himself with the European community. But he appears to have been so conscientious a Freemason that whenever and wherever Nana Sahib died, he may very well have been cremated in his Masonic apron. It was probably due to his intercession that when the Mutiny erupted at Kanpur, Lodge Harmony was one of the few European buildings that was spared the torch. And when, on 27 June, a trunk full of Masonic paraphernalia was found in the deserted Entrenchment it was immediately given over to the Nana, who had it carried about with his dwindling effects as he fought and retreated before the British forces. A despairing coolie finally abandoned the trunk by the roadside, where it fell into the possession of a Lieutenant Sargent. Sargent's nephew eventually gave the paraphernalia to an American Mason named Hooper. At last report the gavel of the Worshipful Master of Lodge Harmony, Kanpur, resides in the Siloam Lodge in the state of Maine. *See United Grand Lodge of Antient, Free and Accepted Masons of England, The Lodge Harmony: 1836-1936.* - Kanpur, 1936, pp 27-8. **ASW**

FRENCH The English have never been good linguists apparently, but during the mutiny/rebellion they found it convenient to send messages in languages which they judged would not be understood by the rebels. Thus **Greek** was used, and on more than one occasion, French, execrable 'schoolboy' French: the following is an example, it was contained in a message sent from **Fatehgarh** to Agra, asking for help (which was not sent): 'Fort Futtehgurh, July 1st. Au magistrat de Mynpoorie, ou á un officier European attacé á une armée de soldats Europeens. Nous avons été fortement assiegés dans le Fort, de Futtehghur, par une force d'Insurgents, pendant la derniére semaine et nous avons peu d'espoir de continuer le siege si nous n'avons du secours de suite. Nous sommes en tout 100 personnes: 32 hommes, et 70 femmes et enfans. Nous vous supplions de venir de suite á notre secours, nous sommes en trés grand danger.' *See* Sir John Kaye, *History of the Sepoy War*, 3 vols, London 1864-7, Vol I.

FRÈRE, Sir Bartle Commissioner of Sinde, was apparently made aware of a number of 'prophecies' of the impending mutiny/rebellion; *see* **Martin**, Vol II, p 118-9; *see also* **DNB**.

FRIEND OF INDIA An influential English language newspaper, published in Calcutta. Valuable for contemporary comment and reports on the progress of the rebellion. Still in existence though now published under the name 'The Statesman' in Calcutta and New Delhi.

Listening for Miners

FULTON, Captain 'In the days before aircraft, if you couldn't take a stoutly defended position by direct assault - and the 'mutineers' attacking the **Lucknow Residency** in 1857 soon gave up this costly procedure - then an alternative was to dig mines and tunnels beneath those defences, blow them into the air with massive quantities of gunpowder placed in the tunnels, and then send in your attacking party through the resultant breach. It is a procedure as old as gunpowder and perhaps older, because even a tunnel alone, dug beneath the foundation of a castle, can bring down a wall or a tower, without an explosion. Hence the old name for the Engineers in the Indian Army - 'Sappers and Miners'. At the outbreak of the 'Mutiny' Captain George Fulton, Bengal Engineers, was stationed at Lucknow, and after the death of Major Anderson R.E. on 11 August 1857, he became the senior engineer officer in the beleaguered garrison: even before that date he had been the leading light in the construction of the Residency's defences. All his compatriots liked him, he was described as active, resolute and cheerful under all difficulties. A good man to have on your side. He is particularly remembered for his work in dealing with the rebels' mines. On 20 July the first mine was exploded by the rebels, at the 'Redan' battery on the edge of the Lucknow defences. It was followed by a general attack, but the latter failed because the miners had misjudged both direction and distance. But the garrison was now alerted to the fact that the rebels could, and would, dig mines, so the listening and waiting, and the digging of counter-mines began. They did not have long to wait. On 27 July, at the angle of the Sikh quarters a large mine was exploded, and killed seven British gunners: these men were experts who could be spared with difficulty. The explosion was a complete surprise, nothing had been heard of the underground diggings; it is said the nearness of the cavalry horses' stables, with constant horse movement, had drowned the sound of the miners' picks. No general attack followed, but it meant a death sentence for the horses: all were quickly converted into meat rations for the garrison, the only exception being Sir Henry Lawrence's horse, destined for his nephew. Two other mines, at the building occupied by the Martinière boys, and at Mrs Sago's Post, were exploded on 10 August, but did little permanent damage to the defences. The failure of the rebels' mining efforts can be largely attributed to Fulton, for he had commenced a vigorous programme of counter-mining, *eg* on 5 August a mine of the enemy's aimed against the guardhouse of the 'Cawnpore Battery', was destroyed by the simple process of working into the enemy's galleries, frightening away the miners - mostly Pasis - and then blowing up those galleries. Sometimes the British developed mines for offensive objectives of their own. Two were particularly effective. One started at 'Sago's Post' and went under the defences across 'no man's land', and under the enemy's guardroom: the latter was blown up with a loss to the rebels of some thirty men. More important was a mine that stretched out to, and underneath, a building called '**Johannes House'**, from which a galling and accurate fire had been kept up by enemy marksman - particularly an African eunuch formerly in the employ of the King of Awadh - which had resulted in the death of many British soldiers. This mine was exploded, resulting in the death of eighty of the rebels, and was followed by a sortie by the garrison to destroy what remained of Johannes House and one adjoining it. By the time the Lucknow Residency was relieved, the British had 15 galleries ready for countermining any further operations of the enemy. Captain Fulton was the leading spirit and the technical expert in these operations, but a laconic footnote in an official report tells us 'Captain Fulton was killed on 14 September by a round shot which struck him on the head, and the services of this distinguished and brave officer were thus lost to his country.' Not a glamorous or glorious end, but his work for the garrison had been without price'. **PJOT**

FURRUCKABAD [*Farrukhabad*] **Charles Raikes**, who was serving as Civil Commissioner under Sir **Colin Campbell**, 'rode into Furruckabad and saw in the **Kotwallie** the bodies of *two* of the Nawabs who had been implicated in the murders and robberies of our poor countrymen at **Futteygurh**. They had just been hanged by order of the Magistrate. I rode over to see their houses, and notwithstanding my indignation at the conduct of the Nawabs of Furuckabad, the sight was a sad one. A fine palace full of every luxurious appliance, mirrors, chandeliers, lustres, pictures, books, and furniture, suddenly deserted - not a human creature left, save one or two withered hags, in the Zenana; cats, parrots, pet dogs, clamorous for food. Outside, in the shady terraces and summer houses, and round the family mausoleum, wandered animals in quest of water or food, *nylghai* (blue deer), *barasingha* (twelve-horned deer), and other pet deer; on the wall a little black puppy yelping, and a dog howling piteously; in the poultry yard geese shut up, and making a frightful noise; at the stables grain for seven horses ready steeped and in separate portions, but the horses pawing, looking round, and starving, with food in their sight; monkeys, cockatoos, and an elephant, who had broken loose, and was helping himself to food formed one of the strangest yet saddest pictures I ever saw. I took care that the animals were fed. As for the princes who so lately were masters of all this luxury, nobody had ventured even to claim their bodies, and it remained for strangers to give them decent sepulture.... It seems but the other day that I, then Magistrate of Furuckabad, breakfasted in this very garden house with some of the family who, though receiving princely fortunes at our hands, have now been so ready to slay and spoil their benefactors! How odious is every scene here, once so pleasing to me.' Charles Raikes, *Notes on the Revolt in the Northwestern Provinces of India.*, London, 1858, pp 107-8.

FURRUCKABAD[Farrukhabad], Nawab of

After the re-occupation of the town by the British under **Sir Colin Campbell** there was some most unpleasant 'retribution' handed out by the civil magistrate, Power, both in the nature of the punishments awarded to so-called rebels, and in the selection of the same. Whereas the real **Nawab of Farrukhabad** was *not* executed (although tried), more than one 'Nawab' *was* executed. The strangest case of this mistaken identity concerns an unknown old man, who, if he volunteered for the deception might well be thought of as an Indian hero, since he may have saved his home town from destruction: 'An old man who had declared himself the Nawab of Farrukhabad, was given over to Sir Colin Campbell *in exchange for his not bombarding the town.* 'Poor wretch!' wrote Oliver Jones, 'he only enjoyed his usurpation for a very few hours.' In the middle of a broad avenue Jones found a tree being adapted to act as a gallows, and a large crowd of inhabitants and camp followers. 'One would have thought that, on so serious an occasion as that of an execution, especially of a person of rank, there would have been some decorum and decency of behaviour; but on the contrary, most people seemed to think very lightly of it, and were cutting their gibes and cracking their jokes.' Some country people came up with some poultry, which was seized and sold by mock auction, by an officer acting as an auctioneer; in the middle of which *good fun* the guard with the convict arrived. 'He was tied down to a charpoy, a sort of native bedstead, and carried under the fatal tree, upon which he cast an anxious look when he saw the noose suspended therefrom.' He was then stripped, flogged, and hanged. He had on a handsome shawl, which an officer took possession of on the spot, an action which requires no comment. 'The man behaved with great firmness. While the rope was being adjusted, a soldier struck him on the face; upon which he turned round with great fierceness, and said, 'Had I had a sword in my hand, you dared not have struck that blow,' his last words before he was launched into eternity.' Jones and Peel agreed that 'such a display of jesting and greediness, and the careless off-hand way with which it was done, were more likely to make the natives hate and despise us, than to inspire them with a salutary dread of our justice.' *See* Oliver Jones, *Recollections of a Winter Campaign in India in 1857-1858.* London, 1859, pp 89-90.

FUTTEHGURH   *See* Fatehgarh.

FUTTEH ISLAM

'Victory of the Mahomedan Faith'. A pamphlet, possibly written by Maulvi Ahmadullah Shah the 'Maulvi of Faizabad', undated but probably (from internal evidence) issued in mid-July 1857. It explains in detail the causes of the rebellion against the British, and the nature of that revolt, and while calling for a *Jihad* or Holy War against the infidel Christians by Muslims, it requires all Hindus also to join the struggle against the British, reminding them of their one time protection by, and loyalty and obedience to, Muslim Kings. The implication is clear that upon the successful outcome of the struggle and the departure or extermination of the British, there would again be a Muslim King in Hindustan, and there is little doubt that the Maulvi saw that role as being his. *See* English translation in *Foreign Political Proceedings,* 30 December 1859, Nos 1135-1139.

FUTTEHPORE[Fatehpur]

'It happened that for 7 months, certain villages (in Futtehpore district) had totally refused to pay any revenue (to the British), emboldened by impunity, & seeing no white faces, & encouraged by the sepoys, they had begun to attack the more loyal villagers, and murder the peaceable, and the police; whenever we came to a village of this sort, it was given over to be plundered, and burnt. This was not such a grave punishment after all, seeing that the inhabitants had generally fled with their most valuable articles, and that a Hindoo's house is built of mud, & therefore rather improved than otherwise by burning.'

FUZIL ALI

He was the leader of a gang of *dacoits* who, in the days of the former King had been very active in Awadh, but on annexation of the kingdom by the British he had tended to confine his operations to the border area in Nepal. However early in 1857 he crossed with his gang into Oude[Awadh] and began depredations in the **Bareitch** district. On his being reported to be in the neighbourhood of **Gonda** some sixteen miles to the east of Secrora, a company of the 10th Oude Irregular Infantry was sent out under Lieutenant Longueville Clarke, with orders to capture them and bring them in. They apparently had the support of the peasantry and it was some time before they were tracked down. Mr Boileau, the Deputy Commissioner of Gonda also went out in search of Fuzil Ali with a small force of mounted police; Fuzil Ali suddenly appeared and shot Mr Boileau and cutting off his head hung it in a tree. Clarke's party soon arrived but Fuzil Ali had decamped. When Clarke was reinforced a long chase took place and Fuzil Ali was eventually run to ground: in the fight that followed Fuzil Ali and most of the *dacoits* were killed, but so also were three sepoys; Clarke was wounded. Later, after the mutiny outbreak Clarke was captured by mutineers and one report has it that he was *apparently tried and condemned to death for having killed Fuzil Ali*, but what connection there was between Fuzil Ali and the mutineers cannot be established.

FUZUL HUQ (or Fazil Haq, Maulvi) An implacable foe of Government although he and his family had been in government employment for many years before the outbreak. He had been the *sheristadar* in the Delhi Commissioner's Office, and several of his relations had held high appointments under Government. He attached himself to **Firoz Shah** and left Oude[Awadh] with him in December 1858: it seems they first had intended to cross the Gogra and join the **Begum Hazrat Mahal**, but she declined their help as her funds did not permit her to accept further troops, so they turned instead to Central India where they were to join Tatya Tope. The pacification of Oude[Awadh] was greatly helped, from the British point of view, by the departure of this man and Firoz Shah. For more details of his career *see* **FSUP,** particularly Vol II.

GALL, MAJOR 'Major Gall of the **Lucknow** garrison left in disguise on 11 June (1857) for **Allahabad,** but at a serai at **Rai Bareilly** he was betrayed to a band of sepoys by a woman. Gall fired twice on the sepoys, but as they closed in he put his gun to his head and shot himself.' M.R.Gubbins. *An Account of the Mutinies in Oudh etc,* London, 1858.

Maharaja Gangadhar Rao

GANGADHAR RAO Last Maharaja of **Jhansi** and husband of **Lakshmibai**, Rani of Jhansi.

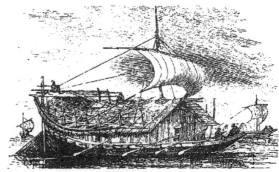

A Country Craft on the Ganges

GANGES[GANGA], River Like all other great rivers of the North West Provinces, the Ganges was both hindrance and help as a means of communication, for both sides in the conflict. In the rains the river 'ran like a sluice', was treacherous and extremely dangerous for the unskilled or unwary, and was an unreliable navigable waterway. In the dry season, although less dangerous it was still in places difficult to navigate owing to the constantly shifting sandbanks. Without experienced boatmen the refugees from **Wheeler's Intrenchment** at **Kanpur** found it practically impossible to find navigable channels for the country craft into which they crowded, hence the survivors were all (four of them) men who were able to swim away from a grounded boat. The boatmen and ferrymen on the Ganges were almost to a man supporters of the rebels, and could only be induced by threats and bribes to assist the British. Crossing the river also was hazardous, ferries being small and slow, and bridges of boats being major engineering undertakings. For General **Havelock**'s difficulties in this respect *see* J.C.Marshman, *Memoirs of Major General Sir Henry Havelock,* London,1860, p 328. *See also* STEAMERS.

GAOL FOOD At one time prisoners were allotted food to cook for themselves, thereby protecting their caste from contamination. But the system galled the British. 'Men loitered over their cooking,' wrote Kaye, 'and their eating, and made excuses to escape work.' So a new system of prison messes was devised, in which cooks were assigned to an appropriate caste group. But occasionally the wrong cook would prepare food for the wrong caste, or the wrong tray would go to the wrong prisoner and their castes would be lost, to be regained only with the most arduous, painstaking rituals and ablutions. Rumours of contamination were rampant not only in the prisons but in the bazars as well, and gave further proof that wherever the British gained complete control over Indians they would do everything they could to destroy their religion.' **Kaye,** Vol. I, p 195.

GARHA MANDLA In Central India, the power base of the ancient line of Gond kings.

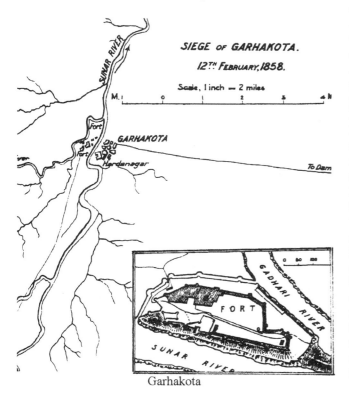
Garhakota

GARHAKOTA The town and fort of Garhakota is situated at the confluence of the rivers Sunar and Gadhari near **Saugur[Sagar]** in Central India. The fort was built of stone, was high and thick-walled: it was exceptionally strong, and **Rose** did not have sufficient strength to invest it completely. The result was that the garrison slipped out by the gate on the south side that was temporarily unguarded, and fled to the north in the direction of **Shahgarh** leaving all their cattle and stores and 13 guns loaded and ready. **Sir Robert Hamilton** reported 'I cannot but consider it most fortunate that the fort at Garhakota has been so easily obtained, for it is beyond exception the strongest and most difficult I have seen in Bundelkhand, indeed as formidable as any I have met with in India'. The British were doubly fortunate here, for the moral effect of the fall of this fortress on the rebels was considerable. For further details *see The Revolt in Central India*, Simla, 1908, *and* Thomas Lowe, *Central India during the Rebellion of 1857-58*, London, 1860, and **FSUP**, Vol III.

GARRETT, H.L.O. He was known as an intellectual opponent of **F.W.Buckler**'s theory that the **HEIC** was the 'real rebel', and for his 'Trial of Bahadur Shah II', *See Punjab University History Society Journal*, April 1932.

GAYA An important town some sixty miles south of **Patna** in Bihar, and the scene of a bizarre incident. It was a civil station but had forty European soldiers, of HM's 84th and over a hundred Sikh irregulars temporarily in the town when the order (that 'ill-judged and faint-hearted order') came from **William Tayler** the Commissioner in Patna to abandon the station and retire upon Patna. Obediently the residents did so on 31 July 1857, abandoning their houses, property, government stores and money to the tune of seven lakhs of rupees. They proceeded about three miles on the Patna road, when two of them (Alonzo Money the Collector and Mr Hollings the Opium Agent) resolved to return. They found all quiet, the police still doing their duty. They did eventually flee again, on 5 August, when news came that **Kunwar Singh** was advancing on the town, but this time took the treasure with them. *See* **Martin** and **FSUP** Vol IV. This incident illustrates the extraordinary confusion which obtained in London as to the true state of affairs in India: the Earl of Derby (Leader of the Opposition) got up in the Lords and praised Money for having opposed his superiors' orders, but in the same speech he praised Mr Tayler 'Commissioner of Arrah' (sic) for having taken a more enlarged view of affairs than the government! He obviously did not realise that it was Tayler who was Money's superior. He was attempting to make political capital out of 'governmental errors', but no one, even on the government side, knew enough about things to be able to contradict him.

GAZETTEERS These humble reference tools may be regarded (for certain basic information *eg* location, history etc) as useful sources by the student of the period. *See* particularly the *Imperial Gazetteer of India* in 7 vols, Dr.W.W. Hunter, Director of Statistics to the Government of India, London, 1881: he produced the material which was the basis for further editions over the next four decades.

GHALIB The name by which Mirza Asadullah Khan Ghalib (1797-1869) is generally known. Urdu (and Persian) poet, he is the author of a unique work dealing with the mutiny/rebellion which he called 'Dastambu' (A Posy of Flowers). He shares with **Mainodin Hassan Khan** and Munshi **Jiwan Lal** and Sir **Syed Ahmad Khan** the distinction of being the only *contemporary* Indian writers to have left written records of the struggle in the form of diaries, accounts, letters etc. His work is therefore a most valuable primary source for historians, particularly for the study of **Delhi** during and immediately after the siege and assault in 1857. On the very day that the rebellion began in Delhi (11 May 1857) Ghalib shut himself up in his house in Delhi, holding himself aloof from the Court of **Bahadur Shah** where he had been a familiar figure. To give himself something to do he began to write his experiences, adding also such news as came to him from time to time. The result is a fascinating picture of life within the walls of Delhi, written by a master of language.

His standpoint is (apparently) soon established, with such phrases as 'on that inauspicious day a handful of ill-starred soldiers from Meerut, frenzied with malice, invaded the city - every man of them shameless and turbulent, and with murderous hate for his master, thirsting for British blood.' Much later he says 'and if May drove justice out of Delhi, September (the assault on the Kashmir Gate by the British) drove out oppression and brought justice back.' Yet it would be wrong to dismiss Ghalib as a prejudiced Anglophile or sychophant: reading between the lines the anguish which the events of those months brought forth in a passionately patriotic man of letters is all too clear. He tells of the mass expulsion of the Indian population of Delhi which commenced on 20 September 1857 (although he claims that the British harmed not a hair on the head of any woman or child), of the return of the Hindu inhabitants on and after 1 January 1858, and the eventual (November 1858) return of the Muslims. We are fortunate to be able to read Ghalib in an excellent English translation, *see* Ralph Russell and Khurshidul Islam, *Ghalib 1797-1869,* Vol I : *Life and Letters,* London, 1969, reprinted as Oxford Paperback, Delhi, 1994.

GHANA GARJANA('roaring cloud')   This was the name of one of the rebels' guns during the siege of **Jhansi** . It was nicknamed 'Whistling Dick' by British soldiers.

GHARRY (or gari)   Although the direct translation

of the Hindi word is 'cart' in English, at this time it particularly meant a slow horse-drawn 'box-like vehicle in which many British officers travelled 'up-country' to the field of operations'. A man could sleep or lie in (comparative) comfort for long journeys while his luggage travelled on the roof.

GHAZI   'Sir Colin (Campbell, the British Commander-in-Chief) himself had a narrow escape. As he was riding from one company to another, his eye caught that of a Ghazi, who lay, tulwar in hand, feigning death, just before him. Guessing the ruse, he called to a soldier, 'Bayonet that man'. The Highlander made a thrust at him; but his weapon would not enter the thick cotton quilting of the Ghazi's tunic; and the impostor was just springing to his feet, when a Seik, with a 'whistling stroke of his sabre, cut off the Ghazi's head with one blow, as if it had been the bulb of a poppy!' 'They were much in evidence at the Battle of Bareilly on 5 May 1868: 'The Sikhs fell back in disorder on the advancing Highlanders, closely followed by a body of *Ghazis* - grey-bearded, elderly men, who, sword in hand, with small round bucklers on the left arm, and green cummerbunds, rushed out with bodies bent and heads low, waving their ***tulwars*** with a circular motion in the air, and uttering their war-cry - 'Bismillah Allah! deen, deen!' (Glory to God! the faith the faith!). At first they were mistaken for Sikhs whose passage had already disturbed the British ranks. But Sir Colin was close behind the 42nd, and had just time to say, 'Steady, men, steady! Close up the ranks! Bayonet them as they come!'...in a few minutes the dead bodies of the devoted band (133 in number), and some eighteen or twenty wounded on the British side, were all the tokens left of the struggle.' *From* W.H.Russell's *Diary In India* Vol II, p 14.

GHAZIPUR   A town, district headquarters and small military station, on the Ganges, between Patna and Varanasi, and some forty miles from the latter. Because it was a staging post for British troops going up country it remained relatively calm during the rebellion, although much excitement was generated by the activities of **Kunwar Singh**. *See* **FSUP,** Vol IV.

GHOLAB SINGH   A leading rebel in **Oude**[Awadh], who took part in many battles against the British including the battle of **Sandila** on 10 August 1858. For details of the battle *see* T.H.Kavanagh, *How I won the Victoria Cross,* London, 1860.

GHOUSE MOHAMED   Sepoy of the 53rd BNI who remained loyal to the British and entered the entrenchment at **Kanpur**. He was afterwards rewarded by being promoted in the Military Police, *see* **Williams**.

GHUNSERAM   'Among the prisoners with me was the murderer of a European named John Duncan, Superintendent of Roads. This murderer's name was Ghunseram, who found poor Duncan hiding himself in a village called Pewundee, about six miles east of Kanpur He treated him kindly for two or three days, but finding that the Nana had offered a reward for the heads of all Europeans or Christians, he gave notice of Mr Duncan's whereabouts and received orders to bring his head to the Nana. When he did so, his indignation was great, when, instead of receiving a large sum as he had expected, only ten rupees were offered him. A few days after this deed was done, a native woman preferred a complaint against the fiend Ghunseram, setting forth that the wretch had appropriated to himself all the money and valuables he had found upon Mr. Duncan's person. He was therefore seized and placed in confinement to be tried by the first opportunity. It was horrible to listen to his description in the jail of the manner in which he had deprived the poor man of his life. This he did with much bragging and boasting, and expressed in no measured terms his indignation at the paltry sum he had received as reward. As the fellow had really taken Mr. Duncan's valuables, he soon managed to bribe the amlah of the court and got his release. Mendes and I, however, marked him and trusted that some day or other we should be in a position to bring him to account. I am happy that on the arrival of General Havelock, this murderer was apprehended, and Mendes the drummer had the satisfaction of being present at his execution'. W.J.Shepherd., *Personal Narrative. etc.*, Lucknow, 1886, pp 100-1.

GIBBON PAPERS Held in the **Cambridge Centre of South Asian Studies** and including source material as follows: Letters to and from Captain James Gibbon, a Royal Artillery Officer who served in India from December 1857 to late 1858 and served at Lucknow and in subsequent campaigns in Oude[Awadh]; includes letters from brother officers giving their experience in many different operations.

GIMLETTE, Lt.Colonel George An army officer with a scholarly interest in the events of the rebellion of 1857-9 in India. He produced an invaluable commentary on much that had previously been taken for granted. He made an attempt to trace the career of the sepoy regiments *after* they had mutinied through the whole course of the rebellion. Here are some examples from his book: ' The 1st Oude Irregular Infantry politely mutinied after a chat with their officers, but joined the **Nana** and later took part in the massacre at **Sati Chaura.** But after mutinying large numbers of Oude Irregulars simply returned to their villages.' 'The 1st Regiment of Native Infantry was raised in 1775 by Robert Clive himself, and dressed after the European fashion in red uniforms that caused it to be known for a time as the *Lal Paltan* or Red Regiment. Later it became known as the Gilliska-Paltan, or Galliez Regiment after Captain Primrose Galliez, who was its commander for many years. The 1st served at Plassy, and for them the prophecy that the British would be run out of India on the centennial of the battle of **Plassy** must have had a special resonance. The regiment served with distinction at Korah, Delhi, Laswarrie, and in the Punjab, a chief enforcer of British hegemony. But for a year prior to the Mutiny it was cantoned with the disaffected 2nd Cavalry, whose agents worked upon its native officers, one of whom, a subadar, was among those who were forced by **Neill** to lick blood before being hanged at **Bibigurh**. Among the rebel troopers of the 1st was **Anjur Tewari,** who, after giving himself up, served as a spy and was rewarded after the Mutiny with honours, pensions, and a jagir. Despite a string of defeats, the 1st remained a cohesive regiment throughout the Mutiny. Their colours were lost, along with 120 men, at **Mangalwar,** but they remained with the **Nana Sahib** until the bitter end. In April of 1859 they camped at Bunkania, at the foot of the Nepalese hills, 8 miles from Gonda on the Fyzabad Road. On 13 April they were defeated by Lieutenant Colonel Cormick and disappeared into oblivion. 2nd Oude Irregular Infantry became the Nana's. But after mutinying large numbers of Oude Irregulars simply returned to their villages. Number 3 Field Battery under Lieutenant Ashe entered **Wheeler's Entrenchment** on 3 June, but on 5 June it was sent away and immediately joined the Nana. The 3rd Oude Irregular Infantry pursued the runaway boat after the massacre at Sati Chaura, caught up to it and took its surviving occupants prisoner. The 3rd fought against **Havelock** at **Pandu Nadee** and tried to drive off his baggage, but they were defeated. The 3rd Irregular Cavalry served at Kabul, Ferozpur, Punjab, Chillianwalla, Gujerat. They mutinied at Saugor on 1 July with the 42nd and a portion of the 31st, and though most of them joined the Nana Sahib, a few remained loyal and fought against the mutineers under Lieutenant Johnson of the 12th Irregular Cavalry in Oude and at Alambagh. The 6th Regiment had been raised in 1763, and saw action at Mysore, Bhurtpore, and Kabul. On 6 June they mutinied at Allahabad, and a detachment marched to Kanpur to join Nana Sahib; they were among the guard at Bhibigar who refused to fire on the women and children. Some of them held out to the last in 1858 on the Nepalese frontier; *see* G.H.D. Gimlette, *A Postscript to the Records of the Indian Mutiny: An Attempt to Trace the Subsequent Careers and Fate of the Rebel Bengal Regiments, 1857 to 1858*. London, 1927.

GOBIND SINGH Sepoy of the 53rd BNI who remained loyal to the British and entered **Wheeler's Entrenchment**. Later was promoted in the Military Police as a reward.

GOLAB SINGH (Or Gulab Singh) Not to be confused with **Gholab Singh of Oude[Awadh]**. He was Maharaja of Jammu and was reported, though without confirmation, to have been in correspondence with **Nana Sahib** before the outbreak of the rebellion. 'As soon as Oudh was annexed, the Nana Sahib began to receive positive replies from his fellow princes to his letters inducing them to join with him against the British. Golab Singh of Jammu was the first to send an answer.' He said that he was ready with men, money, and arms, and he sent money to Nana Sahib, through one of the Lucknow Soukars (Sahukars). The Muslim troops were expected to rise up of their own accord, but Golab Singh's money was used to entice Hindu sepoys into joining the cause.' See **FSUP** Vol I., p 374.

GOLDNEY, Lt Col., Philip, He died at **Faizabad** where he was Commissioner, although his wife, Mary Louisa Goldney and their three children escaped. Correspondence, Press cuttings and Mrs Goldney's memoirs are preserved in the Oriental and India Office collections of the British Library, see *MSS.Eur.D.279.* see also **DIB**.

GOMTI (or Gumti) The river, no wider than the Thames at London Bridge, that flows through **Lucknow**, eventually meeting the Ganges at **Ghazipur**.

GOND KINGS A pensioned descendant of the Gond Kings of Garha Mandla, one Raja Shankar Sa, lived close to **Jabalpur**, and in September 1857 planned with some sepoys of the 52nd BNI stationed there to murder the Europeans. He and his son Raghunath Sa were arrested, tried by special commission and executed by being blown away from guns on 18 September. This act sparked off the mutiny of the 52nd which till then had remained loyal, and had indeed been in action against the rebels. For further details *see* **FSUP** Vol III and *Revolt in Central India,* Simla, 1908, p 40 *et seq.*

GONDA A military station in the Division of **Baraitch** in **Oude[Awadh]**. The regiment stationed there was the 3rd Oude Irregular Infantry, commanded by Captain Mills. As late as 9 June it seemed that the regiment had determined to remain loyal to the British, but by the next morning it was clear that they intended to throw in their lot with the mutineers; all civil and military officers were allowed to proceed to **Balrampur** where the raja protected them and sent them on to Gorakhpur.

GOOD MORNING, SIR! On the day they evacuated **Wheeler's Entrenchment** in **Kanpur**, the Europeans were subjected to a mixed reception from the many thousands who congregated on the road to the **Sati Chaura Ghat**: some had come to gaze, out of idle curiosity, at Europeans so reduced, others to gloat or jeer, others to commiserate and express the hope that their old officers and their families had survived the siege. But one man took great pleasure in what he saw: 'On the Kotwal of the Bazar (a sort of manager of the native establishment of a Regiment) passing a group of officers the remark was made 'There is your old Kotwal. Doubtless he is made a General because he can speak English, though incorrectly.' They asked him 'Is the Nana coming to see us off?' upon which he replied 'Do you remember that about eighteen months or two years ago, I told you that the natives of India, would like to have the Europeans make *chapaties* for them?' *'Good morning Sir, the day is come.'*

About to leave the Entrenchment

All this time the men were employed loading up the baggage, and the women in making coarse flour cakes. They found a great difficulty in procuring fuel for fire and the time for cooking was insufficient. The former they obtained by breaking off parts of the charred frame work of the burned bungalow, considerably augmenting thereby the latter difficulty, that of time, the consequence being that (when the order was given to fall in, and the men did so, whilst the women served out what they had been at so much trouble in cooking) the cakes they served out, would have been under any other circumstances uneatable, being only half-baked flour and water. Anything, however, in the shape of food was considered a luxury and no complaints were made.' See Joseph Lee. *The Indian Mutiny, a Narrative of the events at Cawnpore,* Kanpur, 1893

GORAKHPUR An important town and district headquarters east of Faizabad and north of the Gogra river. A small detachment of the 17th BNI stationed in the town gave indications that they would follow the example of troops at Azimgarh, would mutiny and rob the Treasury. Eventually they did neither, but remained faithful to government until withdrawn from the town. The latter fell into the hands of the rebels, in particular one Mahomed Hussun (the same man who in the early days of the rebellion had protected Colonel Lennox and his family and treated them with great kindness), who set himself up as ruler, or Chukledar, of the district. Driven out by the forces of **Jung Bahadur** on 6 January 1858, he later returned (in March 1858) but in the meantime the British had apprehended, tried and hanged a large number of rebels including a leader by the name of Mushurruf Khan. For some six months the real power in this area, and Azimgarh was **Raja Kunwar Singh** of **Jagdishpur**. For details of the complex political and military developments of this time *see* **FSUP**, Vol IV.

GORDEEN The breakdown of British authority was the occasion of a great deal of wiping out of old scores. The **Nana Sahib** had been engaged in considerable litigation with the widows of his adoptive father **Baji Rao II**, and anyone who had in any way assisted these ladies (whom he persecuted abominably) was now attacked on his orders. Thus, 'On the 6th June, when the Nana's rule was proclaimed, Goordeen, Agent of Meinna Bais, and Tutee Baee, the late Peishwa's widow, together with his family were put to death by the Nana's orders, and Appa Shastree and myself, with five others, were put in confinement with irons.' Appa Shastree survived and was able to furnish considerable information to the British on the Nana's affairs. *Deposition of Appajee Luchman.* **Williams.**

GOVINDGHAR A fort adjacent to the holy city of the Sikhs, Amritsar. It was named after the great Sikh general, judge and priest, Govind Singh. The Koh-i-Noor diamond had been deposited there previous to its seizure by the British, and the possession of the fort, like that of the famous gem and also the ancient gun in **Lahore**, was looked upon as a talismanic pledge of power; *see* **Martin**, also AMRITSAR.

GRAHAM INDIAN MUTINY PAPERS

Being the diary and correspondence of the Graham family. Dr Graham was killed by the mutineers at Sialkot. His nephew James at Landour preserved and co-ordinated the correspondence. His son James committed suicide at Lucknow during the siege of the Residency which had ' a depressing effect on the whole garrison.' See **FQP**, *see also Graham Indian Mutiny Papers*, intro by A.T.Harrison, Public Record Office Northern Ireland, Belfast, 1980.

GRAHAM, DR There were coincidentally two Drs J.Graham at **Sialkot** and they were ***both*** killed by the mutineers. The elder was delayed in his drive to the fort and to safety on 9 July by the fact that his daughter was still in bed and he had to wait for her. When they did set out they were easily overtaken by two troopers one of whom shot Dr

The Death of Dr Graham

Graham; his daughter survived but is said to have been overcome by anguish, as she blamed herself entirely for the death of her father, which might not have occurred had she spent less time over her toilet that morning.. The second Dr Graham was a very popular and comparatively young man who had only just got married. He too was shot while driving to the fort, and while his young wife was unhurt, 'this sudden and terrible shock was more than Mrs Graham could endure, and she did not long survive her husband.' *See* G.Rich, *The Mutiny in Sialkot*, Sialkot, 1924.

GRAND TRUNK ROAD A well-constructed highway running from Calcutta to Peshawar via Raniganj, Benares, Allahabad, Kanpur, Aligarh, Meerut, Ambala and Lahore. The road from Bombay via Mhow, Indore, Gwalior and Agra met it at Aligarh. There was also a spur to Delhi. The building of the road preceded the coming of the British to India but they did much to maintain it and keep it free from lawlessness: 'Every two miles along the Grand Trunk Road was a police station manned by three policemen. From 10 at night to 4 in the morning two out of the three policemen patrolled the road. In addition, horse patrols were stationed and checked every night by means of a verbal order which they were obliged to convey the entire length of a district and then mail back to the originating magistrate to confirm that the road was open and they were on the job.' Charles Raikes. *Notes on the Northwest Provinces*. London, 1852, p 264. During the dry season the sands on the kutcha roads between Allahabad and Kanpur were 'halfway up the wheels occasionally,' as was the mud in places in the monsoon rains. But in the late 1830s the Grand Trunk Road advanced on Kanpur, rolling up the country from the City of Palaces like a carpet. In the stretch between Allahabad and Kanpur, the surface was 'chiefly composed of a peculiar limestone called conker, which, after being laid down for some time, well cemented by the application of water, and beaten together [became] a solid mass of extreme strength.' In fact the surface was so hard that many natives would not walk on it, preferring the soft dust along the roadside. Convict crews in gangs of a hundred laboured with the precision of soldiers. 'Letting all their battering-rams fall at the same moment' and making 'a noise like thunder,' they tamped their way through the station in the late 1830s and continued northwestward to Agra. Though the monsoons ravaged the road occasionally, creating 'deep ravines' and 'chasms of fifty or a hundred feet in length,' the British did not stop until the crews had tamped down 1500 miles of metalled road, terminating at Peshawar, just short of the Khyber Pass.

GRANT, SIR JAMES HOPE (1808-1875) A distinguished and honourable British General, best known for his activities under **Sir Colin Campbell**, in 1858-9, *ie* the campaign to pacify Oude[Awadh] and bring the great talukdars of that former kingdom to accept, though unwillingly, British rule. He could be a stern disciplinarian, and had British soldiers flogged for excesses, *eg* plundering, against villagers. He was universally respected, and because he was trusted, undoubtedly persuaded a number of the talukdars to send in their *vakils* to negotiate terms despite their general mistrust of British good faith. Details of the campaign in Oude are to be found in all the major histories of the struggle, especially **Ball, Kaye** and **FSUP**, *see also* **DNB**.

GRANT, Sir Patrick (1804-1895) Field Marshal. After a distinguished early career (Gwalior campaigns, first and second Sikh wars) became Adjutant-General of the Bengal Army in 1846. Commander-in-Chief of the Madras army 1856-61, temporarily commander-in-chief in India in 1857 before the arrival of **Colin Campbell,** *see* **DNB; DIB; Encyc.Britt.**

GRAY PAPERS These are held in the Centre of South Asian Studies at Cambridge University, and include a typescript copy of *'The Story of our Escape from Delhi in May 1857, from personal narratives by the late George Wagentreiber and Miss Haldane,'* by Miss **Wagentreiber**. Delhi Imperial Medical Hall Press, 1894.

GREASED CARTRIDGES *See* CARTRIDGE QUESTION.

Colonel E.H. Greathed

GREATHED, Harvey H. Commissioner of Meerut (though he was not actually in that place on 10 May 1857 when the 'mutiny' broke out). Proceeded to Delhi where he gave informed advice to the military. He died before the city was finally captured by the British on 20 September 1857, but his letters to his wife from the Camp on the Ridge which was occupied by the Delhi Field Force from May onwards are a commentary on events and attitudes, and provide first-hand source material particularly on the events of May 1857 in **Meerut.** *See* Harvey H.Greathed, *Letters written during the Siege of Delhi,* edited by his widow, London, 1858.

GREATHED, Colonel Edward Harris He played a significant part in the storming of Delhi in September 1857 and subsequently (somewhat controversially) in the pursuit of the rebels. His papers, and those of his brother Major General William Wilberforce Harris Greathed are held in the National Army Museum, Royal Hospital Road, Chelsea, London SW3.

GREEK The *cossids* or messengers employed by the British led a most hazardous existence: many lost their lives, or limbs, in the service of the government. They were, as a result, extremely well paid, to compensate for the risks they ran. Many devices for concealing the messages on their person were used. A common practice was for the sender to write the message - or at least the most secret part of it - in *Greek* characters (not in the Greek language), in the expectation that no one on the rebel side would then understand the message. This method was ruefully described by one British officer as the only practical advantage he had ever come across of the classical education then compulsory at all English Public Schools.

GREEN MAHUMDIE FLAG   'The Kazi and Moulvie Salamut Oollah one day having concerted together, got ready a green Mahumdee flag and set it up on the open ground near Moulvie Salamut Oollah's near the old Assembly Rooms (in **Kanpur**). The Mussulmans congregated round it from the early morning, and some measures were concerted. But in the afternoon more than 4,000 men were collected and went by way of the Filkhanah Gate to attack the **Entrenchment**s, were defeated and fled back. All the mahajuns and citizens know this, as well as the people living in the neighbourhood. *Deposition of Hulas Singh*. **Williams.**

GREENWAY   A large family of rich merchants in **Kanpur**. Although eventually killed with all other Europeans and Christians, Edward Greenway and his wife were initially imprisoned by **Nana Sahib** in the hope, it is said, of obtaining a large ransom. *See* Zoë Yalland, *Traders and Nabobs*, Salisbury, 1987.

GRIEVANCES   'During my rambles in India I have been the guest of some scores of rajahs, great and small, and I never knew one who had not a grievance. He had either been wronged by the government, or by some judge, whose decision had been against him. In the matter of the government, it was sheer love of oppression that led to the evil of which he complained; in the matter of the judge, that functionary had been bribed by the other party.' John Lang, *Wanderings in India and Other Sketches*, London, 1859, pp 111-12.

GUBBINS, Martin Richard (1812-1863)   Financial Commissioner in Oude[Awadh] 1856-7; prominent at Lucknow during the mutiny; accompanied **Sir Colin Campbell** to Kanpur, Judge of the Agra Supreme Court, 1858-62. Published *'The Mutinies in Oudh'*, 1858. Committed suicide in Leamington in 1863. A most strange and intriguing character. Government did not treat him well, yet he was clearly a most difficult man to deal with. *cf* **Earl Canning** 's remarks in a letter to **Sir Henry Lawrence**. 27 April 1857: 'I am sure that it will be necessary for you to keep a close watch and a tight hand upon that officer (*ie* Martin Gubbins). He has had, as against his late master, a triumph which it would have been unjust and mischievous to withhold from him, but *I have good reason to know that he is overmuch elevated by it.*' This presumably concerns his long-standing quarrel with the Acting Chief Commissioner of **Oude** [Awadh], **Coverley Jackson**. Gubbins appears at times to be strongly sympathetic to Indian grievance, but even in this

Martin Gubbins (Courtesy John M.Gubbins)

he is inconsistent and unreliable, and his judgement is clearly flawed. He argues at length for example in his book for the proposition that the sepoys who were recruited from Oude[Awadh] actually *gained* as a result of the annexation of the Kingdom in 1856. No other commentator, Indian or British, takes this view, which flies in the face of known facts. Much of what he says can be traced to prejudice, rumour or spite. *See* **DNB; FQP;** M.R.Gubbins, *The Mutinies in Oudh,* London , 1858.

GUDDI (Or guddee)   Literally a cushion, but taken to mean a throne of a royal or noble prince. He who sat on the guddee was taken to be the legitimate ruler.

GUERILLA WARFARE   The rebel leader whose name is mostly associated with guerilla warfare is of course **Tatya Tope** who was no doubt steeped in the Maratha tradition of campaigning light, and avoiding set piece battles when possible. But the man to whom is attributed the most cogent advice on the subject during the campaigns of 1857-9 is **Khan Bahadur Khan**. 'Do not attempt to meet the regular columns of the infidels because they are superior to you in discipline and organisation, and have big guns. But watch their movements, guard all the ghauts on the rivers, intercept their communications, stop their supplies, cut up their daks and posts; keep constantly hanging about their camps; give them no rest.' Militarily speaking this was excellent advice: where it was followed carefully the rebels were in the ascendant. *See* W.H.Russell, *My Diary in India*, 2 vols, London, 1860.

GUHA, Ranajit Indian historian. Author of *Elementary Aspects of Peasant Insurgency in Colonial India*, Delhi, 1983. Reprinted, Delhi, 1992. A scholarly and painstaking work, although it is far from easy for the layman to understand what positive conclusions are to be drawn from Dr Guha's investigations. One highly practical merit lies in his drawing attention to the dangers inherent in accepting, without question, the 'colonial terminology' in official documents of the time: the truth regarding events and attitudes can be obscured by too facile acceptance of the British use of 'lawlessness', 'dacoit', 'badmash', 'fanatical' etc. See COLONIAL TERMINOLOGY.

GUIDES, The '...When the [Punjab] Guides, under their admirable officers, plunged boldly into the thickest of the fight, shedding their blood like water for the nation who had so lately deprived them of their national liberty, every Englishman from Peshawur to Calcutta breathed more freely.' Raikes clearly regarded their composition as a kind of model for all future native regiments: one company of 'all sorts of up-countrymen... under down-country native officers;' a second company of 'fierce, relentless, and savage-looking' Pathans; a third of 'Punjabee Mahomedans' who were 'tall, sleek, good-natured, ...easily managed in the lines and ready to cut off the head of a brother Mahomedan in Hindustan when duty required;' a fourth of 'Afreedees speaking Pushtoo;' a fifth of 'Goorkhas from Nepal, brave little fellows, with high cheek bones, eagle eyes, fond of cricket, shooting, fishing, and fighting;' and a sixth company of 'tall, wiry, long-enduring' Sikhs from between the Beas and Ravee... 'In regiments thus constituted,' Raikes concluded,' there is a perpetual rivalry of company against company, or troop against troop....Thus we come back to the old lesson, 'Divide and Conquer.' Had our Bengal Army been constituted on this plan no weapon formed against us could have prospered.' Charles Raikes, *Notes on the Revolt in the North-western Provinces of India.*, London, 1858, pp 132-3.

GUJARS (or Gujoors) A tribe widespread in the Northwest Provinces who professed to be descendants of Rajputs by women of inferior castes. They were ostensibly settled and engaged in agriculture, but were also robbers and plunderers given the opportunity: the mutiny/rebellion gave them that opportunity in many areas where law and order broke down, and many 'outrages and robberies' were attributed to them. They appeared to be no more friendly to the rebels than to the British. However, having said that, it is also true that the British were inclined to categorise and stereotype when it was convenient so to do, and few officials at the time attempted to investigate the Gujars' actions or motivations. They were the 'natural scapegoats', says **Stokes**, and the British were fond of quoting the local proverbs against them. Thus: 'Kutta, Billi, do; Rangar, Gujar, do. Yih char na ho - to khule kiware so.' (The dog and the cat are two, the Rangar and Gujar are two. If it were not for these four, one could sleep with an open door) See Eric Stokes, *The Peasant Armed, the Indian Revolt of 1857*, Oxford, 1986, p127.

GUNGA SINGH, Captain Reported as Captain Gunga Singh, a subadar of the Company's service, took an active part in the rebellion at Lucknow and subsequently was at Bangur Mhow with **Prince Firoz Shah** of Delhi; see PERSONS OF NOTE, APRIL 1859.

GUNGA SINGH, General See PERSONS OF NOTE, APRIL 1859.

GUNGOO SINGH 'A certain powerful Thakoor of Mahadewa called Gungoo Singh attacked some English refugees proceeding from Futtehgurh to Kanpur in July, 1857, was seized for this offence and sentenced by me to twenty-one years' transportaton. After attacking the refugees he alienated his fellow rebels by firing a gun on a rebel police station in September.' See H.Dundas Robertson, *District Duties. etc* London, 1859, p 189. See also FATEHGARH.

Blowing from guns

GUNS, Blowing from This was a particularly unpleasant form of execution, since it caused death in a manner unacceptable to both Hindu and Muslim. It was used - largely as a deterrent - by the British, but was not invented by them: it was known to have been practised in India as early as the seventeenth century. 'A PAINLESS DEATH. He marched firmly up to the gun, a 6pdr, which was already loaded with a blank charge; his back was placed on the muzzle, and his arms tied to the two wheels. Directly this was done the Artillery officer made a signal, the gun was fired, and the wretched man had ceased to exist. The troops were marched in front of the gun to look at the body, and the parade broke off. The body was blown clean in two, the legs and stomach falling close to the gun, the head and chest being thrown forward some 3 or 4 yards, and the arms torn from the body, and lying on the ground. *It was a horrible sight; but a more painless death could hardly be devised.'* See Harvey Greathed, *Letters written during the Siege of Delhi,* 1858.

GUPTA, P.C.
A distinguished Indian Historian, particularly valuable for the story of the **Nana Sahib**, *see* Bibliography.

GURKHAS in LUCKNOW
'The next episode that I have to relate, was one that gave us considerable amusement. During the afternoon of the same day, the 17th (March 1858), we were suddenly roused by the sound of musketry and artillery fire behind, and a little to the right of the Charbagh. We could not understand what this meant, we knew that the Goorkhas (ie Jung Bahadur's soldiers) were either there or ought to be there, and we knew there was no enemy in front of them, for we had been, you may be sure, keeping a good look out to see that the enemy did not return, and we had been all over the position both in front of them and at the Alumbagh. Well, the firing increased and gradually came nearer and nearer. There was, however, no opposing fire, but we turned out ready for action, not understanding what was going on, and not wishing to be taken unawares. Suddenly, from the depths of the canal, yelling like devils, burst forth the little Goorkhas, shooting away as fast as they could load their muskets, at the air in their front, and capturing empty position after empty position, through which twenty Fusiliers had walked quietly and unmolested not six hours before. They had the good fortune to fall in with six or eight deserted guns in one of the streets, which was turned into a great achievement.' The official despatch referring to the above:- 'Jung Bahadoor was requested to move his left up the canal and take the position in reverse, from which our position at Alumbagh had been so long annoyed. This was executed very well by His Highness, and he seized the positions one after another with little loss to himself. The guns of the enemy, which the latter did not stop to take away, fell into his hands'; *see* manuscript notes of Dr Collins, Surgeon 5th Fusiliers in Museum of 5th (Northumberland) Fusiliers in **Alnwick Castle**.

GURKHAS
The Gurkhas - followers of Jung Bahadur - were not at all popular with the British; this is remarkable in view of the strong bond of affection and loyalty that was later established. Thus: '...sepoys are scattered in small parties, collecting revenue and plundering. Much of this disorder is no doubt due to the impression produced by the rapid withdrawal of the Goorkha force, which gives color (sic) to the reports industriously spread by the leaders of the enemy, that they have withdrawn in disgust, if not at enmity with the British Government; and that our successes have been owing hitherto mainly to their assistance, this last assertion being supported by the absurd vaunts of their prowess, made by the Goorkhas themselves.' 'The Bansee country is still overrun with rebels chiefly refugees who have, apparently with the connivance of the Nepalese local authorities, emerged from the Nepalese Terai.' The Gurkhas were clearly not a well-disciplined force: on their return home from Oude [Awadh], laden with booty, they seem to have broken up into small parties, each intent on getting home with their new possessions as fast as possible - the incredible result was that the looters were looted! 'An expedition guided by Mr Bird against the Doomree and Pandeepoor *Baboos* in the jungle near Goruckpore, succeeded in driving them from a stronghold they were fortifying, and in capturing all their baggage and booty including 60 cartloads captured from the Goorkhas on their return to Nepal. Mr Bird was still in pursuit....they had only just time to escape, leaving all their booty, clothes and cooking vessels behind them. Among the former were found the mess stores of the Bengal Yeomanry Cavalry, and many miscellaneous European articles. 60 hackeries were also found which they had taken from the Gorkha army on its march back to Nepaul last May'. *See Narrative of Events for Gorakhpur* w.e. 16 May and w.e. 27 June 1858, quoted in **FSUP,** Vol IV.

GWALIOR
The capital city of the Maharaja Sindhia, on the Jumna [Yamuna] river some miles to the south of Agra. Dominated by a magnificent and ancient fort. The headquarters of the **Gwalior Contingent**. The city now boasts a fine equestrian statue of the **Rani of Jhansi**: unfortunately a thief has stolen the small bronze figure of **Damodar Rao** which originally appeared tied to the Rani's back. For detailed accounts of events in Gwalior in 1857-58 *see* **Ball** or **Martin**. *Also* Khushhalial Srivastava, *The Revolt of 1857 in Central India-Malwa,* Bombay, 1966.

GWALIOR, BATTLE of
This took place on 19 June 1858, and before sunset the whole of the Lashkar area was in British hands, **Sir Hugh Rose** having exercised his usual skill outmanoeuvreing the enemy, led by Tatya Tope, in every respect. Incredibly the fort - which was of immense strength - fell on the next day to the first determined attack. For details of this action *see The Revolt in Central India,* Simla 1908, p 160 *et seq.*

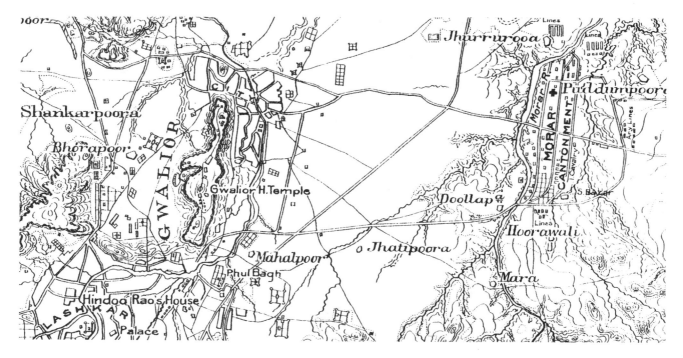

GWALIOR CONTINGENT This subsidiary force was raised as a result of a treaty with the British in 1843, when the city of **Gwalior** was restored to **Scindia**. It had British officers and was equipped, drilled and trained on the English model, and was a most formidable fighting force. It was composed of two regiments of cavalry, four field batteries, a small siege train, and seven infantry battalions, a total of 8,318 men under Brigadier Ramsay. Based mainly on the cantonment of **Morar** it also had outposts at Sipri[Shivpuri], Agra, Goona, Agar, Mehidpur, Neemuch and Asirgarh. The men of which it was composed were drawn from the same sources as the Bengal army. The Gwalior Contingent was formidable. 'A finer body of men than the 1st or Grenadier Regiment of the Contingent never stood on parade,' wrote one of its former officers in 1858, 'the front rank being composed of men not less than six feet in height, and I believe there were few men in the regiment some years ago under 5 feet 10 inches.' The infantry and artillery were primarily Hindus, and most were men from Oude. The artillery was especially efficient.' *See* London *Times*, 13 January, 1858, p 12. With great difficulty Scindia restrained the contingent, who came to regard him as their enemy, even at one stage setting up a rival to his throne, but in the end he could not prevent their marching off to **Kanpur**, under **Tatya Tope** on 15 October 1857; leaving detachments at Jalaun and **Kalpi** they reached Kanpur on 1 December. *See* **FSUP**, Vol III and *The Revolt in Central India, 1857-58*, Simla, 1908. *See also* CAWNPORE [Kanpur], BATTLE OF.

GYE PAPERS These are held in the Centre of South Asian Studies at Cambridge University, and contain duplicated copies of the *Evening Mail* dated September 1857 with articles and news of the mutiny/rebellion in India. The articles include 'The advance on **Cawnpore**', 'The Mutiny at **Fyzabad**', 'Siege of **Delhi**', 'Russian Opinions on the Indian Mutinies', 'Indian Mutinies Relief Fund'. There is also an unpublished article by Colonel A.H.Gye, written in 1970, on *'The Sepoy rebellion; a military survey of the mutiny of 1857, together with a letter from a retired officer of the Indian Army, P.Emerson, in comment.'*

H

HAIG PAPERS These are held in the Centre of South Asian Studies at Cambridge University, and contain the original pre-publication version of **Anna Madeline Jackson**'s *'A Personal narrative of the Indian Mutiny, 1857.'* written in 1880 for her children to read. She was one of the **Captives of Oude**. There is also a most interesting reminiscence of a man called *Raynor* whose grandfather survived the fall of **Delhi** in May 1857, although he was long thought to have died in the explosion of the Magazine.

HAINES, Amy *See* BENNETT, MRS AMELIA, *and* RIPON PAPERS.

HAKIM-i-WAQT This was the term applied by many to the 'temporary government' of the British in India.

HALLS, John James Assistant Surgeon at the civil station of Arrah; *see* ARRAH HOUSE; *see also* J.J.Halls, *Two Months in Arrah in 1857*, London, 1860.

HAMILTON, Sir Robert N.C. Resident at the court of **Holkar** maharaja of **Indore**, and Governor General's Agent for Central India. Influential in shaping British policy both during and after the mutiny/rebellion, *see* **DNB**.

HAMIRPUR A large town on the south or right bank of the Jumna[Yamuna] river. The Collector, Mr T.K.Lloyd acted decisively on hearing of disturbances elsewhere, seeking assistance, and getting it, from the chiefs of Charkari, Behri and Baoni, and raising a force of 500 new levies. But it did him no good for when the 53rd BNI mutinied there on 13 June 1857, his forces joined the mutineers for the most part. Of the twenty-three Europeans in Hamirpur, only two finally escaped - Lieutenant Browne, who reached Fatehpur in time to join **Havelock** on his march to **Kanpur**, and Miss Anderson who. although wounded, reached Kanpur. Hamirpur also saw the plundering of many of the townsfolk, some (as in the case of the Bengali Babus) for no other reason than that they understood English. On 18 June three boatloads of unarmed sepoys from the 44th and 67th BNI, who had been disarmed by the British at **Agra**, were passing by on the Jumna when the rebels(?) turned guns on them, killed a large number and plundered the contents of the boats. For further details *see* **FSUP** Vol III and *Revolt in Central India*, Simla, 1908, pp 29-31.

HANGING Death by hanging was the sentence for a wide variety of offences, including: 'Murder and plunder, Rioting and theft of a horse. Assisting to apprehend native Christians. Having in his possession a bag of half pice, a coin which had not yet been issued from the Treasury. Having 34 rupees the possession of which he is not able satisfactorily to account for. Conspiring and preferring false charges and being a notorious badmash.' (Sentences of Commissioners under Act xvi of 1857: Allahabad). The leaders of the rebellion could expect to be hanged if captured, unless they could prove that at some stage they had assisted Europeans to escape from the rebels, and had themselves never taken part in the murder of Europeans. Thus, **Rajah Jye Lal Singh**, the Commander-in-Chief of the Begum of Oude [Awadh] was hanged on 1 October 1859; even later **Khan Bahadur Khan** of **Bareilly** was captured and hanged. **Rao Sahib** was sentenced to death, and hanged on 8 September 1862: for some of their enemies the British reserved no mercy. Years after the events the British Press was noting the capture and sentencing of rebels, *eg* ' The murderer of the late Major Burton and his sons has been hung at Kotah, the scene of his crime,' while also reporting '**Bala Rao** Nana Sahib's nephew has escaped.,' *see* especially the *Illustrated London News*, 1859-62.

HANGING 'As for hanging, it is nothing; it is a quite common thing to have a few swung up every day; the least thing will do it. We have a provost marshall and his staff here, and they would hang a European if they found him plundering, or give him a dozen on the spot if they found him half-a-mile from his camp; but as for a native the least thing is sufficient to hang him. We have had one European hung, but they are very loth to do anything to the Europeans. The fellows missing I hear had such a lot of rupees that they could not keep up, and they were too greedy to throw it away. Cawnpore was full of all kind of liquor, from champagne to bottled beer, and our fellows used it too freely. The authorities were enraged at this.' Non-commissioned officer of the 84th. *Letter* dated 1 August, 1857, quoted in London *Times*, 29 September, 1857, p 8.

HANGMAN HUNG, The '**Bruce** was invested with full powers by the Government and with the assistance of **Baba Bhut**'s papers, and the evidence of informers, got the names of many persons who had joined the **Nana**, or, at least, had acquiesced in his rule, had paid him complimentary visits, or had sent him presents. One of the informers was a tall, stout man of the sweeper caste, and though Bruce had certainly made him no kind of promise of immunity, he had taken so prominent a prosecuting part, that he quite thought himself established as a Government agent. It came out however, quite clearly that he had jewellery in his possession which had belonged to some of the lady prisoners. I have all along supported the idea that there was no mutilation of our unfortunate countrymen and countrywomen before death, because there was no proof of it; it was not likely and it seems such gratuitous self-torture to suppose a thing which everyone would desire to be untrue. But some mutilation after death may have taken place; and in one visit to the garden in which the well stood, shortly after our arrival, I found a hand under a bush, which I took, by the slenderness of the bones, to have been that of a female. The busy ants had made all clean and white, and the hand looked like a plaster cast in an anatomical museum. It lay on the direct road to the well; and when I heard of the jewels, it at once occurred to me that it was probably severed for its rings. Bruce was pledged to no leniency, and he would not forgive this stout sweeper, who, I make no doubt, had actually joined in the massacre. He and his mates had served as hangmen in disposing of some of the rebels whom Bruce had tried. And now his own hour was come, his mates turned on him, when ordered, with a readiness that must have been very bitter to him, and led him, bound and trembling, to the scaffold on which he had himself stood so often as executioner. I was with Umurnath in the verandah of a lIttle house which he had to pass, and, seeing me, he cried at the pitch of his voice, 'Dobai Collector Sahib,' and entreated his guardians to allow him to stop and speak to me. But they were inexorable, and hurried him to his fate.' **Sherer**

HANUMAN GARHI A famous temple in Ayodhya in Faizabad District. Apparently the Mahants of the temple at first sought to restrain the mutineers and offered to give protection to Europeans, but they later changed sides; *see* **FSUP,** Vol II, p 31.

HARDA A town near Jabalpur. There were no troops there, but as in many other towns of Central India the police went over to the rebel cause, and in fact here broke into open mutiny before inducements were made to them from outside. The Deputy Magistrate, one Mazhir Jamil 'maintained a bold front' and succeeded in suppressing the outbreak, nine of the police being hanged or imprisoned.

Temple on the Ganges Kanpur

HARDEO, or HURDEO TEMPLE A Fishermen's or Boatman's Temple, on the banks of the Ganges at **Kanpur,** associated with the massacre there of some 250 British from the garrison of **Wheeler's Entrenchment** on 27 June 1857. Here is the ghat or landing place known variously as **Massacre Ghat** or **Satichaura Ghat,** although to be strictly accurate the latter is a few hundred yards up river at the village of Sati Chaura.

HARDY PAPERS These are held in the Centre of South Asian Studies at Cambridge University, and contain a detailed account of the outbreak of the mutiny at **Sitapur** (Seetapore) in June 1857, by **Captain George Hutchinson,** Bengal Engineers, Military Secretary to the Chief Commissioner of **Oude** (written in 1859).

HARRIS, Mrs Katherine Wife of the Reverend Harris, the officiating Chaplain in the Residency after the death of the Rev. **Polehampton,** she kept a diary throughout the siege. Useful for a picture of conditions within the **Lucknow Residency** during the siege, and covers the period 15 May to December 1857. Includes an account of the death of **Sir Henry Lawrence.** Mrs G.Harris, *A Lady's Diary of the Siege of Lucknow,* London, 1858.

A Trooper of the Gwalior Contingent

HATHRAS The scene of the first overt act of mutiny by the **Gwalior Contingent.** One hundred men of the 1st Cavalry of the Contingent raised the warcry of Islam and went off to Delhi on 23 May 1857.

HAVELOCK, Major General Sir Henry (1795-1857) Best known for his recapture of **Kanpur** in July 1857 and for his 'relief' (more accurately described as 'reinforcement') of the **Lucknow Residency** in September 1857. He repeatedly defeated the rebels in battle, but was unable to follow up the victory by pursuit etc because his troops never exceeded about 1,000 effectives, with limited artillery and no regular cavalry support. He co-operated with **Sir Colin Campbell** in the second relief of the Residency in November 1857 but died of dysentery on the day of the withdrawal from Lucknow. His grave is in the **Alambagh**, just to the south of Lucknow. Created KCB and a Baronet in November 1857 just before he died. Hero worshipped in England at the time, as the romantic rescuer of British women and children; eg many roads are named after him, but he was not popular with the troops he led, nor was his military judgement always sound. Quarrelled with **Colonel Neill** (whom he threatened to put under arrest), and was in danger of falling out also with **Sir James Outram,** nominally his superior who had accompanied him to Lucknow as a volunteer, nobly waiving his right to command. *See* **DNB**; NORTH-ALLERTON; Rev William Brock, *Biographical Sketch of Sir Henry Havelock,* London, 1858; L.Cooper, *Havelock,* London, 1957; J.C.Marshman, *Memoir of Major General Sir Henry Havelock,* London, 1860.

Sir Henry Havelock

HAVELOCK Havelock was 'unpopular with his soldiers to an extraordinary degree. He was a martinet very formal and precise, and seems to have maintained a rigid and perhaps somewhat sour discipline which they could not bear.' Outram was the 'life and soul of the advance' on Lucknow. Sir George Campbell, *Memories of my Indian Career,* London, 1893, pp 282-3.

HAVELOCK'S OPPONENTS Nothing reveals the nature of the conflict in Awadh more graphically than an examination of the forces opposed to **General Havelock** when he crossed the Ganges at **Kanpur** in July/August 1857. The rebel authorities in **Lucknow** had been in power for only two weeks or so, and yet they managed to summon up an army to resist invasion. The following is a significant description of the position: 'Havelock did not face a regular sepoy army when he crossed into Oude; Ubdool Hadu Khan Kandaharee was despatched with two companies from each regiment in Lucknow and 14 guns to check the General's progress, but he was defeated and his guns were captured. Tuhwar Hussein was then ordered to take command but he refused, and Mahomed Hossein who was formerly Chakladar volunteered and set out against the General...defeated but remained at **Nawabgunge**, and fell upon the British troops when the General, harassed by the men of the talooqdars Munsub Aly, Jussa Singh, Baboo Rambuksh and others, retreated into Oonao. Rebel troops mustered 10,000 strong and were reinforced by the Kanpur fugitive regiments....the General could not make head and was obliged to fall back on Cawnpore, and Mohamed Hossein with his troops encamped on the banks of the river (Ganges).' This is an extract from an undated and anonymous statement included in evidence gathered by Captain **Bruce** in 1859-60. *See* Bruce Papers, *BM Add. MSS 44,002* f10.

HAYES, Captain Fletcher He was the Military Secretary to Sir Henry Lawrence the Chief Commissioner of Oude[Awadh], but was killed before the mutiny/rebellion broke out in Lucknow. *See CAREY'S RIDE*. Towards the end of June **Monsieur Duprat** moved into a house in the Residency compound and took there Mrs Hayes (the widow of Fletcher Hayes)'s library, for which he had accepted responsibility: these items were subsequently put to a purpose that the owner could never in his wildest dreams have imagined. The splendid library of Captain Hayes, consisting of priceless Oriental mss and the standard literary and scientific works of every nation in Europe, and dictionaries of every language spoken on earth from the patois of Bretagne down to Cingalese, Malay and ancient Egyptian, were for the nonce converted into barricades, together with mahogany tables, furniture of all kinds, carriages and carts. In this humble duty they were joined by the records of all the civil service offices, large chests of stationery etc. *See* **FQP**.

HAYES, Fletcher The erudite Captain Fletcher Hayes, Military Secretary to Henry Lawrence, was sent to Kanpur to inspect Wheeler's preparations and inform him that Lawrence was sending a small force of European soldiers. 'The General,' Hayes wrote in a private letter to Secretary Edmonstone, 'was delighted to hear of the arrival of the Europeans, and soon from all sides, I heard of reports of all sorts and kinds which people kept bringing to the General until nearly one A.M., on the 22nd, when we retired to rest.'

HAZRAT MAHAL, see BEGUM OF OUDE

HEARSEY, CAPTAIN John He commanded the military police in the station of **Sitapur**, and had formerly commanded a regiment of infantry in the army of the **King of Awadh**. He was particularly popular with his men, some of whom remained loyal to him after the mutiny broke out at Sitapur on 3 June 1857. With a party that at one time included Quartermaster-sergeant and Mrs Abbot, the elder **Miss Jackson**, Mr Gonne the Deputy Commissioner of Mullapur, Captain Hastings and Messrs Brand and Carew they took to the jungle in an effort to escape from the mutineers, but were unable to keep together. Some of the party, including Georgina Jackson were captured and taken into **Lucknow** where they were put to death. Hearsey himself disguised himself as a sowar, made his way across the hills to **Naini Tal** and **Mussoorie,** ultimately reaching Lucknow in safety after its relief by the British. *See* Kaye, *see also* John Bonham, *Oude in 1857,* London, 1928, p 56 *et seq.*

HEARSEY, Sir John Bennet (1793-1865)
Lt-General; KCB; 6th Light Cavalry. A long and distinguished career found him Commanding at **Barrackpore** during the momentous events of 1857 and the incident of **Mangal Pande**. He wrote, on 11 February 1857, 'We have at Barrackpore been dwelling on a mine ready for explosion.' *See* **DIB** and The *Times* 27 October 1865.

HEARSEY, Letter General Hearsey was an old friend and confidant of General **Sir Hugh Massy Wheeler**, and it is not surprising therefore that they should exchange views on the threatening position that developed early in 1857, re the conduct of the 19th BNI etc. This letter from Wheeler to Hearsey reveals much as to the character of Wheeler in particular and is therefore quoted in full: 'Cawnpore, March 22, 1857. MY DEAR HEARSEY, My most sincere thanks are yours for your letter regarding the 19th N.I. The men are some of the finest looking in the service, but the regiment has always had the reputation of being a turbulent one. Surely the whole will be disbanded. Your advice on that head I know to be good. Some years ago, when an exception was made in favour of the native officers, the Subadar-major of my regiment, speaking of it, said to me, 'There is the mistake that Government has made. Every native officer should have been dismissed; for nothing can take place in the lines without their knowledge at least. It may not, sir, be in my power to prevent or put down a mutiny; but my commission is in your hands if, in case such a thing should ever occur, I do not acquaint you of it before it is an hour old.' They should not be allowed to be inert, but every man who cannot prove that he had actively opposed the disturbers of the peace should be, for that inertness, sent about his business. A few examples of this kind would, I feel assured, be most advantageous to the service. I think that Colonel (of the 19th N.I.) made a sad mistake in allowing himself to be dictated to instead of dictating, and his sending away the guns and cavalry before they (his men) had piled arms was most injudicious. I conceive that mutineers with arms in their hands should never be treated with or listened to. It is opposed to the first great principle of military discipline and subordination. He had the *power* to put them down, the want of which could alone justify his measures. Everything is quiet here, but from what I hear there is an unquiet feeling amongst the men, nay, *amongst the people at large* (italics by PJOT).The general tenor of all the reports is that every exertion is being made, *by the orders of Government,* to deprive the natives of their castes by making them use materials and food tainted with forbidden articles. But the way the country is left without artillery! We have guns and a European company, but no carriage for them. There should be at least a troop or battery at Cawnpore; and the *Post Guns,* which were so injudiciously taken away by Lord W. Bentinck for a miserable and paltry economy, should be forthwith restored wherever there is a wing of a corps. The two sixes are invaluable in the case of an e*meute* or disturbance; and in India you can never be certain when either will occur ....Lady Wheeler and my daughters unite in kind regards to every member of your family, with, my dear Hearsey, your old, true, and sincere friend, H. M. WHEELER.' *See* Hugh Pearse, *The Hearseys,* Edinburgh, 1905, pp 396-7.

HEARTLESS FRAUDS It would seem that even the most horrendous of situations provides someone, somewhere, with the opportunity to make money. Thus: 'A person residing at Marseilles has been practising the most heartless frauds on the relatives of those who have fallen in India, or whose fate is unknown. ...He addresses the relatives of persons whose death has been reported from India, and informs them that he has private information that the parties are not dead but in a place of concealment, and in want of money to enable them to effect their escape. He then offers on receiving a certain amount which he names to convey it to the persons in jeopardy, and thus enable them to make their way to a place of safety.' **ILN** 21 November, 1857, p 499.

HEWITT, William Henry (1790-1863) Lt-General; served Burma Multan Peshawar. Was Commanding the **Meerut** Division from January 1855 to June 1857. He was thus the G.O.C. at Meerut on the outbreak of the rebellion. He has been blamed for the weak response he made to the initial mutiny, particularly in failing to despatch British troops (of which there were many) in pursuit of the sepoys as they made their way to Delhi on the night of 10

May 1857. He was relieved of command and proceeded on furlough in 1858, and remained on furlough until his death in 1863; *see* particularly OUTBREAK OF THE MUTINY.

HIBBERT, Christopher   Eminent English historian, educated at Oriel College Oxford, and writer of one of the most readable, scholarly and objective accounts of the rebellion this century. *See* Christopher Hibbert, *The Great Mutiny India 1857,* London, 1978.

HILLERSDON, CHARLES   Collector and chief civil official at **Kanpur**. Trusted **Nana Sahib** and believed that he would aid the Europeans if the troops mutinied. Took refuge with the others in **Wheeler's Entrenchment**; was talking with his wife who had recently had a baby, when a round shot completely disembowelled him. His wife (killed by falling bricks) survived him by only a few days.

HIMAM BHARTEE   '.....yesterday a fakir came in (to **Meerut**) with a European child he had picked up on the Jumna. He had been a good deal mauled on the way; but he had made good his point. He refused any payment, but expressed a hope that a well might be made in his name, to commemmorate the act. I promised to attend to his wishes; and Himam Bhartee, of Dhunoura, will I hope long live in the memory of man'. From a *letter* written by **Greathed** the Commissioner of **Meerut**, in **Ball** p 67.

HINDI   The linqua franca of much of central and northern India, sometimes, perhaps misleadingly called 'Hindustani', and related to **Urdu**. The latter is most usually associated with Muslims, while Hindi is thought to be the language of the Hindus; the corresponding script for Hindi is called Devnagiri.

HINDOO SEPOY, The   There is a possibility - though it must be admitted it is a slim one - that had the British attempted to exploit the **communal friction** which undoubtedly obtained in **Kanpur** during the siege of **Wheeler's Entrenchment** the garrison there might have been able to escape the fate that eventually befell them. Here is one (optimistic) viewpoint: 'A promise of pardon and reward to the Hindoo portion of the mutineers would have completely gained them back to us, for they appeared to be very sorry for what they had done, and were heard to say so to the city people. In this latter belief I am the more convinced from the fact, which I forgot to mention before, of a Hindoo sepoy having, on the 21st June, after the fight I have described of that day, while we were besieged, crawled upon all-fours by a narrow drain which runs from the barracks in the intrenchment, to the road bridge near St. John's chapel. He was unarmed, and came up to about ten paces of the trenches. As soon as he appeared to view, and before the officers could prevent it, two soldiers fired upon him, and he expired instantly.This caused much regret, as it was the be

Sepoys circa 1857

lief of many in our camp that this sepoy was coming to seek pardon for himself and his comrades without reference to the rebel authorities, and would no doubt have assisted us in arranging for our departure in safety. He was no doubt sent by the Hindoo portion of the sepoys on some favourable errand to us, and his death perhaps deterred others from approaching us for he came in the day, and his comrades in all likelihood were watching him from the church compound and elsewhere to see if his attempt was attended with success.' See W.J.Shepherd, *A Personal Narrative etc.,* Lucknow, 1886

Hindu Rao's House

HINDU RAO'S HOUSE   A building which played a significant part in the period of the struggle known as the **'Siege' of Delhi** in 1857. It is situated mid way between the old city and the Ridge to northwest where the British were encamped, in what came later to be known as Civil Lines. Some parts of the building are still extant: it is now a hospital. It seems that 'Hindu Rao' was a Maratha nobleman of 'frank bluff manners and genial temperament' who had died in 1855. Before him the house had belonged to William Fraser one-time Resident at Delhi who had been murdered in 1835.

HINDUN RIVER, BATTLE of The first set-piece battle of the struggle took place on the Hindun River near Ghaziabad about fifteen miles from **Delhi**, on 30 May 1857, between the rebels commanded (nominally at least) by Mirza Mughal the King's son, and the advance division of the British force advancing from Meerut and commanded by **Archdale Wilson**. The rebels had determined to prevent the advance of the British force towards Delhi, but were beaten decisively; on returning to Delhi they were jeered and taunted, so returned to the attack the next day, but in a battle conducted mainly as an artillery duel they were again routed, but managed to preserve their guns and retreat back within the walls of Delhi. *See* **Kaye, Sen, Ball, Martin** etc *See also* A.Llewellyn, *The Siege of Delhi,* London, 1977, Harvey Greathed, *Letters written during the Siege of Delhi,* London 1858. Charles Metcalfe, *Two Native Narratives of the Mutiny of Delhi,* London, 1890.

HINDUSTANEES Here is one *British* officer's attempt to define matters: 'The inhabitants of the North-West Provinces and Oude are, both amongst themselves and by the residents of other portions of India, termed 'Hindustanees', in distinction to Bengalees, Marhattahs, Madrassees, and the various races which inhabit the peninsula. The North-West Provinces and Oude are, by the natives, looked upon as Hindustan Proper....this is the fountain-head of the lingua franca, or Hindustani, of India, as also the seat of its most venerated religious and historical associations.'
...... ' The great continent contains a good score of native populations far more distinct from each other in language, customs and religion than the nations of Europe....to talk of national insurrection, national education or national anything among a population of this description is to talk ignorantly....the utmost extent of their political cohesion is that of marbles in a bag; the sole questions open to debate are the colour and texture of the bag, or whether a marble more or less shall rattle in its interest. From the overwhelming catastrophe that such an event must have occasioned we were saved by but an accident', *see* H.Dundas Robertson, *District Duties during the Revolt,* 1859.

HINGUN The name of the Palmers' **ayah** who was with them in the **Entrenchment** and survived the massacre at the **Satichaura Ghat**, and later gave evidence, *see* **Williams.**

HINGUN LALL A Rajput of Karrakat, a large town on the left bank of the **Gomti** who befriended at considerable risk to himself, the British refugees from **Jaunpur**. 'He stated that he had a few armed men and that the enemy should cut his throat first before they reached us.' Government rewarded him with a life pension of Rs100 p.m. *See* M.A.Sherring, *The Indian Church during the Great Rebellion,* London, 1859, pp 267-276.

HISS At Lucknow 'Cawnpore,' wrote Rees, was somehow 'hissed into the ear of every one of the rebels before a thrust of the bayonet put an end to his existence.' L.E.R.Rees, *The Siege of Lucknow,* London, 1858, p 322.

HOBSON'S CHOICE 'I was led out into the verandah without another word, where sat a blacksmith amidst a heap of most formidable-looking fetters. He selected a fearfully heavy pair, and was about to clap them on my poor legs, when I told him that my sentence was three years, and therefore, in pity, he ought to give me lighter ones. *He very kindly allowed me to make my own selection*, but they were nearly all alike, and I was at a loss which to take; there was no help for it, I took up a pair, and the blacksmith put them on me. I have since weighed these fetters, which are still preserved by me, and find they are more than five pounds.' *See* W.J.Shepherd, *A Personal Narrative etc.,* Lucknow, 1886.

CAPTAIN W.S.R. HODSON

HODSON Captain Hodson, of Hodson's Horse, has always been a most controversial figure, both in his own lifetime, and since. He has inspired envy and malice, disapproval and moral condemnation, and, in some, admiration approaching idolatry. In Delhi, in September 1857 after the storming, there is a typical incident - quoting the unpublished account of a certain Capt Maisey, Judge Advocate General's office. 'It is clear that the King of Delhi gave himself up, and was not captured by Captain Hodson who was apparently obsessed with the complex to do everything for his own. Actually brought in by Rajah Ali (Meer Munshi or head of native intelligence) plus an escort. Rajah Ali met Hodson on the way back to Delhi........Next day Hodson rode out to Humayoon's tomb...not a hand was raised to save them (ie the two sons and a grandson of the King of Delhi) though there was a large crowd about the place, but the fact was the princes *wanted to come*, evidently encouraged by the news of the King's reception. They could have escaped with ease had they chosen...but they had no fear of anything more than durance vile'. Hodson himself shot the Princes: there seems no doubt of that, although his account of both the above events differs markedly from Gen. Archdale Wilson's. Hodson's account suggests that the shooting down was a sudden necessity which arose out of the fact that rescue was attempted while Hodson was conveying them under escort to Delhi: this view is fatally injured by an irrefutable scrap of evidence, that comes from the lips of Hodson's own favourite orderly, in telling the story in after years, he says: 'Prince Abul-bakr wore a talisman on his arm; so I said to Hodseyn-Sahib, Wait a bit, Huzoor! To kill him with that on will bring ill-luck. I'll take it off ere we shoot him.' No hurry there, no stress of circumstances surely, to make the immediate use of a revolver necessary. There was a general idea throughout the country that however badly the rebels might behave, nothing would induce the British to resort to extreme severity; the King of Delhi's sons, when captured by Hodson, said, with a jaunty air, 'Of course there will be a proper investigation into our conduct in the proper court.' Lucknow March 1858....'Hodson dined with us at mess. A very remarkable fine fellow - a beau sabreur, and a man of great ability. His views expressed in strong nervous language, delivered with fire and ease are very decided; but he takes a military rather than a political view of the state of our relations with India. I would like to see H at the head of his horse try a bout with the best Cossacks of the Don, or the Black Sea.' A book appeared in 1859 called HODSON, William, *Twelve Years of a Soldier's Life in India*, London, 1859. But as Captain W.S.R.Hodson was killed in Lucknow in 1858, this is a book based on his letters, edited by his brother George H. Hodson. The latter does his best to defend his brother's actions, particularly in the matter of the siege and capture of Delhi and the death of the *Shahzadas*, the Delhi Princes. See Lt General Sir George MacMunn in the *Cornhill Magazine*, 1930; see also W.H.Russell, *My Indian Mutiny Diary*, 1859; and Captain L.Trotter, *Life of Hodson of Hodson's Horse*, 1910.

HODSON'S HORSE

A regiment of irregular cavalry raised in the Punjab by **Captain W.S.R.Hodson** on 19 May 1857. Nicknamed the 'Flamingoes' on account of the scarlet turbans and sashes tied over the right shoulder. It is still a proud regiment of the Indian Army. Led by Hodson (to whom the troopers were devoted) it performed legendary feats against the rebels, but after his death, in March 1858, its reputation suffered somewhat by accusations of cowardly behaviour on more than one occasion, notably when refusing to join their officers in a charge during the campaign for the pacification of Oude[Awadh] under **Sir James Hope Grant**.

HOLIDAY The, Kanpur, 1857

'There was no enthusiasm felt for the Nana, who was soon seen to be a mere obese voluptuary with no talent for affairs and no courage in the field. But there was, among the humbler classes, a revival of the lawless element that had come down in the blood from the great anarchy of the last century, a factor never to be overlooked in the social questions of Hindustán. There was also a revival of old clan-feeling, the pleasure of foray and reprisal, unclouded by the fear of the police, or the shadow of the tax-gatherer. ***The people were out for a holiday, and enjoyed it like badly-taught school-boys.*** Lastly, the dread that the return of British power would be accompanied by the return of the auction-purchaser, an evil, perhaps necessary, and which must apparently cling to a civilisation whose resources come from land-revenue, enlisted many interests against the cause of order. Bacon long ago remarked: 'It is certain, so many overthrown estates so many votes for troubles.' See H.G.Keene, *Fifty-Seven: Some Account of the Administration of Indian Districts During the Revolt of the Bengal Army.* London, 1883, p 82.

HOLKAR

Maharaja of Indore, a treaty state in Central India. Unlike **Scindia** in **Gwalior,** this ruler appeared to be playing an ambivalent role in the struggle. Colonel **Durand,** the Officiating Agent to the Governor General in the **Central India Agency** clearly believed that Holkar had gone over to the rebels, for his note to the Commanding Officer at **Mhow** asking for help stated that Holkar was attacking the Residency. It is still not completely certain that this is not true. Holkar himself was horrified that the British whom he declared to be his allies and the Paramount Power should think this of him, and when news of the approach of two British columns from Bombay was announced he was with difficulty prevented from leaving Indore for the purpose of personally offering an explanation to Lord Elphinstone, the Governor there. His letter of explanation etc, dated 4 July 1857, is published in full in **FSUP,** Vol III and *The Revolt in Central India*, 1857-58, Simla, 1908, Appendix 1.

HOLLINGS, W.C.

Mr. Hollings, editor of the *Central Star* at **Kanpur**, was a sportsman who, after leading the **Greenway**s to his roof on the night of 8 June 1857 armed himself with double barrel shotguns and killed sixteen of his attackers. The next day (the 9th) he ran out of ammunition and offered himself as a target to the sepoys, who promptly obliged. Mortally wounded in the chest, he fell to his death from his roof. (A 'W.C. Hollings' appears on a list of merchants and tradesmen, but without any specification as to what he did for a living, in Yalland, p 347) For a brief period Hollings had been editor of the *Delhi Gazette*. *See* Zoë Yalland, *Traders and Nabobs,* Salisbury, 1987.

HOLMES, MAJOR JAMES   Commanded the 12th Irregular Cavalry at **Sagauli** (Segowlie) in Bihar about a hundred miles from **Patna**. Took it upon himself to declare martial law in the middle of June 1857 in a large tract of territory and began 'hanging right and left'. His action was declared illegal by the Lieutenant Governor, Mr Halliday, but supported by the Commissioner in Patna, **William Tayler**. On 23 July 1857 he was murdered by his own men; he and his wife were both decapitated while on an evening drive, the troopers then killing also the Surgeon and his wife and child, and the postmaster. After plundering and burning the civil station the regiment then departed, although forty of them remained loyal to government and joined **Outram's** force on its way to reinforce Havelock in September 1857. *See* **Martin**, and **FSUP**, **Vol IV**, *see also* RIPON PAPERS.

HOLMES, T.Rice   A 'heavyweight' English historian of the rebellion whose work has the merit of being distanced some forty years in time from the events he describes, which makes for an objective and historian's re-appraisal of the rebellion. The standard one-volume work. The extension to the title reads 'and of the disturbances which accompanied it among the local population' and this indicates the true scope of the book. *See* T.Rice Holmes, *History of the Indian Mutiny*, London, 1898.

Death of Adrian Hope

HOPE, Brigadier Adrian   A very popular soldier, from the 93rd Highland Regiment whose death very nearly precipitated an act of mutiny, so angry were the Highlanders with the General (**Walpole**) whom they blamed for his death: he was shot, from a tree, in the abortive and grossly mismanaged attack on the fort at **Royah** in April 1858. His dying words, to his ADC were 'They have done for me, remember me to my friends'. It was even suggested (Wm Forbes-Mitchell, *Reminiscences of the Great Mutiny*) that the man who shot him was a European renegade. Previously, in November 1857, Hope had been responsible for the almost fortuitous capture of the rebel strongpoint of **Shah Najaf** in **Lucknow,** on the occasion of **Sir Colin Campbell's** relief of the **Lucknow Residency.** As a last attempt to take the strongpoint which had resisted all attempts to storm it or batter its walls into submission, Hope had collected some fifty men and stolen cautiously through the jungle, and reached, unperceived, a portion of the wall, where he had noticed a narrow fissure: they scrambled up with difficulty and entered; pushing on, almost unopposed, they threw open the gate for the British to enter, and as they did so the garrison slipped away at the back of the 'fortress', in the gathering darkness. For an account of the near mutiny *see* W.H. Russell, *My Diary In India,* London, 1860, Vol I, p 393.

HORNE, Amy   *See* BENNETT, Mrs Amelia.

HOSHANGABAD   An important town in the **Saugur & Narbada Territories**. An uprising was expected and the place put on the defensive by the British authorities in July 1857, but in the end nothing materialised. This may well have been due to the presence of troops from the Madras army, who were not in sympathy with their opposite numbers in the Bengal army. Hoshangabad marks in effect the southernmost limit of the struggle, in 1857-59.

HOWDAH   The seat or saddle placed on the back of the elephant for dignities to sit in. Some Howdahs were most ornate and covered in precious stones and silver.

Tablet erected at Arrah House

THIS TABLET IS ERECTED TO COMMEMORATE THE LOYALTY AND DEVOTION TO DUTY OF JEMADAR HUKUM SINGH AND FORTY FIVE SIKH SEPOYS OF THE 45TH RATTRAY'S SIKHS (3RD BN 11TH SIKH REGIMENT) WHO MATERIALLY ASSISTED IN THE DEFENCE OF ARRAH HOUSE 27 JULY 3 AUGUST 1857.

HUKUM SINGH, Jemadar   Commanded the fifty Sikh policeman, from **Rattray**'s Sikhs, who aided the Europeans in the defence of *ARRAH HOUSE*. He was immensely strong, 6ft 2ins in height, and a veteran of the Sikh

Wars. His favourite expression was 'Kooch purwa nahin !' (no harm done, no matter!) which he used on every occasion when it seemed appropriate, especially when the enemy's cannon fire failed to hurt anyone, and just sent up a cloud of dust. At one time he actually stood up on the roof and threw brickbats at the enemy, not because he had no firearm, but for the sheer joy of demonstrating his strength, with roars of laughter.

HULAS SINGH Was **kotwal** of **Kanpur** under the British, and latterly under **Nana Sahib**. 'Seven or eight days after this (12-13 June 1857), I was summoned to the Nana's residence and became Kutwal of the City. I was appointed because several (two) Kutwals had been appointed in the course of the week, but they could not get on. First Kazi Waseeooddeen was appointed, and held the office for one day, then Haji Khanum's son, whose name I forget, remained in office two or three days. ...I did not seek service from the Nana. He forced me to be kutwal. Whoever did not obey his orders, he put to death. ...All my arrangements for the benefit of the troops were made from fear of losing my life....Then the mahajuns of the city, and Bidi Gunga Pershad, tent-maker; and Jugul Kishoor, jeweller; and Budri, pan-seller; and Shew Pershad, khazanchee; and others who had transactions with the Nana, recommended my appointment. On this I was sent for and when I was at the Nana's house, Ahmed Ali, tahsildar, and officer of police, gave me full instructions and a purwana of appointment. I was helpless; accepted the post and entered on its duties. I could not but carry out the *foujdaree* (criminal justice) orders which were issued by the Nana and his officers. His *cutcherry* (court house) was arranged as follows: the Nana's the highest court; 2nd, that of Deputy Collector Ramlall. In the Nana's court, Bala, and Baba Bhut, Azimoollah and Jwala Pershad, Brigadier, and others used to pass orders. Ramlall had no associate in his court, Moonshee Jwala Pershad was Ramlall's confidential adviser. These two did all the work.' *Deposition of Hulas Singh.* **Williams**, The British accepted that he had acted under duress.

HUMAYON'S TOMB The tomb of the Emperor Humayon lies to the southeast of the old city of **Delhi** on the right bank of the Jumna [Yamuna] river. It is here that **Captain Hodson** was reputed to have taken the King of Delhi, **Bahadur Shah Zafar** prisoner in September 1857 immediately after the assault on Delhi by the British, but *see HODSON.*

HUMIRPORE [HAMIRPUR] Those who took advantage of the rebellion to loot and plunder were none too fastidious. What happened at Hamirpur was typical: after the murder of the European officers the local townsfolk turned to plunder ...and were not very discriminating as to whom they robbed. Sometimes the looters or mutineers were themselves in turn robbed eg 'To shew to what extent the lust for plunder ruled, there were three boats of unarmed sepoys of the 44th and 67th Regiments, those I believe who were disarmed at Agra, passing by on the 18th June; the guns were turned on them and opened; many were killed, the boats taken and the goods found on them made over to the men of the auxiliary chiefs, the sepoys being left to get on their way as best they could', *see* **FSUP**, Vol III, p 117.

HUMILIATION, Day of On Saturday, 26 September 1857, Queen Victoria proclaimed 7 October a day of humiliation, 'so both we and our people may humble ourselves before Almighty God, in order to obtain pardon of our sins, and ...send up prayers to the Divine Majesty for imploring His blessing and assistance on our arms for the restoration of tranquillity.' London *Times,* 28 September, 1857.

HUMOUR There was not much to laugh about in 1857**,** but the British soldier could usually be relied on to find something funny in any situation, though sometimes his wit got mixed up with his ignorance: 'The places in India,' wrote Private Henry Metcalfe, 'are nearly all poors or bads.' Tuker, Francis, ed. *The Chronicle of Private Henry Metcalfe,* p 73.

HURDUT SINGH, RAJA See PERSONS OF NOTE, APRIL 1859.

Hodson 'captures' the King of Delhi at Humayun's Tomb

HURDEO BAKSH Raja of Kathiari in Oude [Awadh]; a talukdar who remained loyal to the British although threatened both from the Durbar at Lucknow and the Nawab of Farrukhabad who characterised him as a 'Christian' because he refused to join the rebellion. He was rewarded after the final assault on Lucknow, in what was known as the 'Oude Proclamation', by the Governor General, issued on 14 March 1858. *See* **FSUP,** Vol I, p 439; also

Edwards, William, *Personal Adventures during the Indian Rebellion in Rohilcund, Futtehgur and Oude,* 1858.

HURMAT KHAN

He was a man of great size and strength and a renowned swordsman, and had been employed as a professional flogger at the **Sialkot** District Gaol, but had been degraded from this position by Mr Monckton a short time before the mutiny. This man 'breathing out threatenings and slaughter' against the local authorities, and being in sympathy with his fellows in the cavalry, was a ring leader in the mutiny at the station on 9 July. First he despatched several horsemen to pursue and if possible kill Mr Monckton, who, although ill, managed to escape with the help of his servants to a place of concealment in a neighbouring village. (The Deputy Commissioner was later brought from this village into the fort, wrapped up in a sheet, as though he was a corpse being taken to burial. His Muslim servants thus twice carried him to safety under the eyes of the mutineers). Hurmut Khan then went to the 9th Cavalry lines where one squadron had mutinied and the other two were hesitating, and was mainly instrumental in causing them to break out. After this he went down to the city to murder a Court Clerk with whom he had a quarrel about a woman, which had resulted in his own degradation from his late position as flogger. Not finding the clerk at home, he cut down a servant at the door and then came to the gaol. Now the missionaries, the Hunter family, were seen driving across the plain, and others not being willing to go and kill them, Hurmat Khan went forward himself: he first shot Mr Hunter, then cut down Mrs Hunter with his sword and finally killed their baby. On the return of British authority to Sialkot, Hurmat Khan escaped to the mountains north of the town, and Captain Lawrence offered a reward of Rs1,000 for his capture; and several unsuccessful attempts were made to take him. No clue to his whereabouts was however forthcoming until July 1862 when one day a stranger from Kashmir arrived at a village some ten miles east of Sialkot. Kadir Baksh the headman of this village learnt in course of conversation that the stranger had come over the border to fetch the wife of one Fazla for a 'Hurmat Shah'. Now Kadir Baksh knew that this wife of Fazla was the very woman about whom a quarrel had existed five years before; which had resulted in Hurmat Khan's degradation from his position as flogger, and in his attempting to murder a Court Clerk who had gained this woman's affections. He therefore put two and two together and reported to the police. A spy was sent to Hurmat Khan's hut which was found to be near the city of Jammu; but the difficulty was to catch him, as he was ever on the alert and previously had always escaped to the mountains on the slightest alarm. An English gentleman at Sialkot organised a sham marriage procession. Armed men were dressed as peasants would be for a wedding and were packed in ekkas. The Englishman played the part of the dainty bride, secluded from vulgar gaze in a covered and closed ox-cart, as this Indian bride ought to be. The bells jingled merrily, the bridal party wended its way along the road leading to the murderer's hut, amid noisy talking and laughter, without exciting any susupicion. Suddenly the wedding guests were transformed into a body of armed men, who headed by the bearded 'bride', surrounded the hut. Hurmat Khan, drawing his sword, stood at the doorway. Knowing his fate if captured and owing to the desire of the police to capture him alive if possible, he was enabled to keep forty men at bay for three hours. Finally they closed in upon him in a body, and a sword-cut across the loins put an end to his life. His body was sent to Sialkot and identified on oath - in fact the whole city recognized it. It was buried in a Muslim burying ground on the south side of the city, where the city prostitutes kept a light burning on the tomb and made a pilgrimage to it once a year. **A second monument was erected to his memory by the Mohammedans of the Sadar Bazar, not far from Mr Hunter's house, and this is still visited by devotees with lights and offerings, who, not knowing facts, believe it to be the tomb of some martyr or saint.'** See G.Rich, *The Mutiny in Sialkot,* Sialkot, 1924, p 24 *et seq.*

HUSSAINEE,[HOSAINI]

'For the story of Hosaini, another legendary heroine we have again to look to **Colonel Williams**. Among other things he was trying to find out exactly what had happened to the 200 women and children (usually referred to as 'the ladies') who had survived the massacre at the **Satichaura Ghaut** on the bank of the Ganges on 30 June, or who had come down by boat, as refugees, from **Fatehgarh**. The main facts were only too clear. They had been quartered first in the old missionary building called the **Savada** (= Salvation) Kothi, or house, and had then been taken to the **Bibigurh,** or 'House of the Ladies'. They had subsequently been, literally, butchered, and their remains thrown into a dry well: they were found the next day by the relieving British force under General **Havelock**. But who had done this? Who had given the orders and who had carried them out? It has always been assumed, on the British side, that it was the **Nana Sahib** who was responsible: he was later to deny it, and others, like **Tatya Tope** his general, and many others who knew him, have said that it was not he: he was incapable of such cruelty. Colonel Williams interrogated John **Fitchett,** a Christian Drummer who survived the mutiny by becoming a Muslim: Question. 'Who attended the ladies after leaving the Savada?' Answer.' The Begum used to attend morning and evening. After the killing of the five men, the woman the Begum told the ladies they were to be killed by the Nana's orders. A lady went to Jemadar Usuf Khan of the 6th Regiment who was i/c the guard, and asked if this was true. She was told no, if the Nana had said they were to die then he, the Jemadar, would have been told. They were not to be afraid. Sepoy Kurm Alee told the Begum her orders would not be obeyed, who was she to give orders?' Who then was 'the Begum', for this was surely a nickname? Evidence was produced to show that 'The Nana had four slave girls belonging to the Peishwa (*ie* **Baji Rao,** his adoptive father, now dead), one of them named Taj Bebee, a second Chundur Khanum, a third Gatee Ufrozand the fourth Hussainee Khanum who was also called Begum. She, it is said, 'carried the order for the murder of the prisoners to the sepoy guard placed over them; and on their refusing to execute it, returned and fetched five men of the Nana's own guard, one of whom was her lover Sirdar Khan.' Amid all the conjecture facts begin to emerge. Hosaini alias the Begum was placed in charge of the ladies while they were still in Savada Kothi: she did not have a difficult task as they were in very poor physical condition and no attempt was ever made to escape through the guard of sepoys that was set permanently around them. She seems to have displayed animosity towards them from the start, and increased rather than diminished their fears. She had a staff of sweepers to assist her and to bring what little food the prisoners were allowed, unleavened dough and lentil porridge for the most part. Each day she took two of the ladies to the Nana's stables where they were set down to grind corn on a handmill for several hours. Witnesses said that they generally contrived to bring back a pocketful of flour for the children. The Nana took no interest in the prisoners at first, although his Headquarters was only 75 yards away, at a hotel owned by a Muslim. Eventually however he had to intervene because he heard that the women and children were dying in alarming numbers, perhaps as many as twenty-five a week. Both on humanitarian grounds, and, cynically speaking, because dead hostages were worth nothing to him, he gave orders that better food should be given them, and that they should be attended by a doctor: it is to the latter incidentally that we owe the complete list of the prisoners. In small numbers they were also taken outside the stifling room in which they were confined and allowed to take the air on the verandah: it has been suggested that this 'privilege', granted to the women for the sake of their health, was in fact resented by them, for on the verandah they were open to the gaze of large numbers of sightseers who came apparently to exult in their humiliation. Looking again, at the facts, the order came to the sepoy guard at the Bibigurh at about 4 p.m. on 15 July 1857, that the male prisoners should be taken out and shot. There were five of them, the two Colonels and the magistrate from Fatehgarh, and Mr Greenway, a merchant, and his 14 year old son. So far as we know they died bravely: the sepoys had no compunction in obeying these orders which almost certainly came from the Nana. Then comes the incident with the lady asking the Jemadar whether they too were to be killed, his reassurance, and the sepoy telling Hosaini Begum that 'her orders would not be obeyed': but were they her orders? - or had she brought them from someone else? The fact is that the sepoys refused to kill the ladies: some say that they poked their muskets through the windows in a token obedience of orders, and fired into the ceiling. Colonel Williams concluded that 'The evidence proves that the sepoy guard placed over the prisoners refused to murder them. The foul crime was perpetrated by five ruffians of the Nana's guard at the instigation of a courtesan. It is ungenerous as it is untrue to charge upon a nation that cruel deed.' At 5p.m. on the same day, 15 July, Hosaini Begum returned with five men, and whether or not the sepoy guards had resolved that they would not lift their weapons against the prisoners, they now did nothing to stop the five armed men from entering the Bibigurh, although it must have been obvious what they intended to do. Were they overawed by Hosaini? - a slave girl, whom one of them had roundly abused less than an hour ago? Or did she come with orders from the Nana, from his brother **Bala Rao,** from his vakil **Azimullah Khan** or from Husainee's mistress **Oula**?

The Interior of the Bibigurh, 17 July 1857

We know something of the five men. Two were Hindu peasants: one aged about 35 fair and tall with long mustachios but flat faced and wall-eyed, the other much older, short and sallow. Two were Muslim butchers from the bazar, portly strapping fellows well on in life, the larger was disfigured by small-pox. The fifth wore the red uniform of the Nana Sahib's bodyguard, and was the lover of Hosaini Begum, one Survur, or Sirdar, Khan, long in the employ of the Nana, and described as a 'vilaiti'. Survur Khan might have been an Afghan or a Pathan: the only other thing we know about him is that he had 'hair on his hands', and twice he emerged from the Bibigurh to replace his broken sword. As the five men entered, the shrieks began. They left as darkness fell and the Bibigurh was locked up. The screams had ceased but the groans continued till the morning when the five returned, and with a few sweepers dragged out the bodies and threw them into a dry well. Those who had clothes worth taking were first stripped. Some of the women were still alive, at least three could speak and implored not mercy but an end to their sufferings. Three boys were alive and ran round and round the well. All were thrown into the well, some still alive probably. It was 16 July: Havelock entered Kanpur a few hours later. We end with another statement, but this time it is not Colonel Williams and his official investigation. It is the statement of one Mahomed Ali Khan the night before he was hanged as a spy, in Lucknow, to Sergeant **Forbes-Mitchell** of the 93rd (Sutherland) Highlanders who had charge of the guard that night: 'Nana Sahib intended to have spared the women and children, but they had an enemy in his zenana in the person of a female fiend who had formerly been a slave-girl, and there were many about the Nana (Azimullah Khan for one) who wished to see him so irretrievably implicated in rebellion that there would be no possibility for him to draw back. So this woman was powerfully supported in her evil counsel, and obtained permission to have the women killed, and after the sepoys of the 6th N.I. and the Nana's own guard had refused to do the horrible work, this woman went and procured the wretches who did it. This information I have from General Tantya Tope, who quarrelled with the Nana on this same matter. What I tell you is true: the murder of the European women and children at Kanpur was a woman's crime, for there is no fiend equal to a female fiend; but what cause she had for enmity against the unfortunate ladies I don't know - I never inquired.' .... What a pity he did not! The British did not usually punish the rebels' women, even if they were known to have been active in the rebellion. Hosaini Khanum, the Begum, was never found: it was reported to Colonel Williams that 'she who distributed the food to the ladies is now (1858) with the Nana'. If this is so then she presumably died of fever with him in the **Terai**. *See* Chapter 6 of **SSF**. **PJOT**

HUSSUN RAZA He was reported as 'Hukeem Hussun Ruzah, tutor to Shurfood-Daulah, who was Prime Minister during the rebellion, this hakeem had charge of the Dewanee Adawlut under **Birjis Kudr**' *See* PERSONS OF NOTE, APRIL 1859.

HUTCHINSON, Lieutenant George He was an officer of the Bengal Engineers and, after the death of **Captain Fulton**, he directed engineering operations (particularly with regard to mining and counter-mining) during the latter part of the siege of the **Lucknow Residency**. His papers are preserved in the Oriental and India Office collections of the British Library, *see MSS.Eur.E.241*, see also George Hutchinson, *Narrative of the Mutinies in Oudh*, London, 1859.

HYDERABAD *See* NIZAM OF HYDERABAD.

HYDERABAD CONTINGENT FIELD FORCE This assembled at Edlabad in July 1857 for the specific purpose of preventing the mutiny/rebellion spreading southwards into the **Nizam's** dominions. It was commanded by Major W.A.Orr, and was composed of the 1st and 4th Cavalry, 1st, 2nd and 4th companies of artillery and two howitzers, and a wing each of the 3rd and 5th Infantry, all of the Hyderabad Contingent. In October it marched northwards to **Hoshangabad** putting down some insurgency, and then joined the force of **Brigadier C.S.Stuart** at **Dhar**.

ID (Or Eed)   A Muslim festival, in particular the one that comes at the end of the month of fasting, Ramadan (Ramazan), *ie* Id-ul-Fitr. The British were most apprehensive of rebel attacks being made to coincide with religious festivals, such as the Id. As an example *see* Letter No.2 in the *AGRA* entry: '*if they had succeeded in their night attack, the king was to have held his Eed in Camp, and the termination of our reign proclaimed.*'

ILLUSTRATED LONDON NEWS   This unique magazine was particularly given to the publishing of sensationalist reports from India at this time. In the absence of firm and accurate information most journals held back, or at least admitted that what they published might well only be speculation, but some periodicals frequently went ahead and printed what must have been known to be doubtful reports: the effect upon the families and connections of those involved can only be surmised. It took many more months of such sensationalist reporting before responsibility modified the tone. Here is an example of the earlier reporting: 'Other accounts state that the women and children of the officers, consisting of 240 persons, were taken into Kanpur and sold at auction, when, after being treated with the highest indignities, they were barbarously slaughtered by the inhabitants.' But some may have escaped this terrible fate, the *Illustrated London News* assured its readers, for it was known that 'Nana Sahib has more than a hundred European prisoners in his hands, whom he intends to hold as hostages.' Totally inaccurate. *See* **ILN** No. 877, Vol xxxi, 5 September, 1857, p 237.

Gateway Small Imambara Lucknow 1858

IMAMBARA   A religious edifice connected with the Shi'a sect of Muslims: the place where the ceremonial and processional items of the Mohurrum are kept, to do that is with Hassan and Hussein, the martyrs of Karbala. Lucknow in particular has magnificent examples of Imambaras, especially the city's Great Imambara.

IMPEY, Capt. Archibald   He was an officer of the Bengal Engineers whose recently (1993) discovered letters reveal a number of previously unknown aspects of the rebellion. He was present at the Relief of **Arrah** with **Eyre**'s Column, and then served in **Oude** under Brigadier Berkeley. (*see* T.C.Anderson, *Ubique etc*). He was in Kanpur soon after it was retaken by the British, in July 1857, and in one of his letters to his wife in Darjeeling he reveals a startling possibility: 'I can now tell you the cause of the surrender at Cawnpoor. Genl Wheeler and Whiting of our corps, (engineers) Major Moore 32nd (the virtual commander), and Dr Aske of the artillery were all for holding out, *but the European troops mutinied and after the Nana offered the terms said that if they were not accepted they would make a treaty for themselves - they would not fight any more. that and the interceptions (?) of the women caused the surrender.*'

IN MEMORIAM   In 1858 a painting entitled 'In Memoriam,' dedicated to 'the Christian heroism of the British ladies in India during the Mutiny of 1857,' caused a stir at the Royal Academy Exhibition in London. In the original version the Scottish painter Sir Joseph Noel Paton depicted 'maddened Sepoys, hot after blood' skulking toward a huddled and prayerful tableau of British women and children , breaking through the door. But so many viewers found this intolerable that after the Second (successful) Relief of Lucknow he painted them out and replaced the 'crazed rebels' by 'Highlanders arriving in the nick of time'. The prayerful tableau would appear to have been indifferent, in appearance, to foe or friend. *See* C.A. Bayly, ed. *The Raj: India and the British: 1600-1947,* p 241. and R. Montgomery Martin, *The Indian Empire*, Vol II, p 378.

INCENDIARISM Even before the outbreak of actual mutiny in the Bengal Army, the latter's general discontent and sullen spirit was manifested by widespread instances of arson: officers' bungalows, and any unoccupied military buildings or sepoy lines were mysteriously burnt down. At **Ambala** the conflagrations became so frequent and destructive that a reward of Rs1,000 was offered by government for the discovery of the incendiaries. Other places which suffered in this way included **Agra** and **Sialkot**. After the outbreak at **Meerut** on 10 May 1857, incendiarism became routine wherever mutiny occurred: bungalows being looted and then fired as a matter of course.

INDIAN ARMY ON THE MARCH 'It is a wonderful sight to see an Indian army on the march as it approaches its camping ground. The square head of the column seems enveloped by myriads of animals and men, and up-towering elephants and camels, made taller than nature by heaps of tents and baggage and furniture piled upon their backs. The elephants are provident enough to look for the sugar-cane as they march along the roadside, and generally each of these quadrupeds marches laden with a large mass of cut cane, in addition to the chairs and tables and mountains of canvas piled upon its back. The mass, dense and small in the distance, grows larger and looser all over the wide plain, as it approaches, till it seems to fill the space from horizon to horizon -- to cover the fields and permeate the forests with a shifting mass of life.' W.H.Russell, *My Diary in India.* London, 1860.

INDIA OFFICE LIBRARY & RECORDS
Situated (1996) at Blackfriars, London. This is the national repository of material on the administration of undivided British India and other Asian territories over a period of 350 years and also an international centre for South Asian Studies. The India Office Records houses the archives of the East India Company (1600-1858), the Board of Control (1784-1858), the India Office (1858-1947) and the Burma Office (1937-48). Its name is in some ways misleading and its collections include also official material dealing with those countries with which the Company or the India Office at some time had connections such as Malaysia, South Africa, Indonesia, Singapore, China and Japan, the Persian Gulf states, Aden and Afghanistan. Complementing these official archives is a growing collection of personal papers, letters, diaries and memorabilia which provides a more intimate picture of life under the British Raj. The Library was founded in 1801 by the Company for the safe custody of the Oriental manuscripts and books placed in its care. More than a hundred languages are represented in the collection of printed books which now exceeds 300,000 volumes and includes many rare items. The collection of English language newspapers published in India is the most complete in the world. Outstanding examples of the art of the miniaturist are preserved in the collection of Oriental manuscripts and the Prints and Drawings Department also includes examples of the work of many distinguished Western artists. The IOL & R has a long tradition of publishing scholarly guides and catalogues which in conjunction with a good photocopying service make the collections more accessible to those readers who are unable to visit it. It was transferred to the **British Library** in April 1982, and is now known as **'Oriental and India Office Collections of the British Library'**. It is scheduled to move, eventually, to Charing Cross, to join the main BL collection. *See* BRITISH LIBRARY.

INDIAN HEROINES In addition to the well-known major heroines of the struggle eg the Rani of **Jhansi, Lakshmibai** and the Begum of Oude, **Hazrat Mahal**, there are many stories told of individual bravery and devotion to the cause of freedom on the part of Indian women. Unfortunately although the details are often otherwise explicit the heroines remain anonymous: **A** A letter from a young officer of the 5th Fusiliers to his sister contains the following: 'We were standing in the garden of the **Secunderbagh**, *resting from killing* (sic), when a bullet came from a tree and killed a Sikh officer. We looked up, and there was a sepoy loading his musket again. We soon brought him down with our revolvers, and were astonished at the feminine regularity of his features. Some one suggested it was a woman in disguise, and so it proved. She died cursing us.' (*Narrative of the Indian Revolt*, 1858). **B** The courage shown by Indian Women was not all on the rebel side. 'All deserted Scindia in Gwalior when the rebels came, except his mother-in-law, Gujja Raja, mother of the Maharani; believing that Scindia was beleaguered at the Phoolbagh, she seized a sword, mounted her horse, and rode to the Palace, summoning all to his aid, until she found he was certainly gone.' (**FSUP** Vol III, p 456). **C** In some of the great number of personal reminiscences that found their way into print, there was evidence of women fighting, and fighting well, on the rebel side. The account of the Colonel of the 2nd European Bengal Fusiliers, during the Siege of Delhi in 1857, is typical:- 'The enemy was in position in the Subzi Mundi and Eed Ghur on the Umballah road, so a column was sent out to attack them. Nasty street fighting, then their cavalry appeared coming up the road. Broke into a half-hearted charge that did not keep up with the leader. Volleys soon brought many down. Leader's horse was shot and he was made prisoner: had led his men right gallantly and they had left him in the lurch. To our astonishment the brave cavalier turned out to be a woman. She was retained prisoner for some time: I often saw her in the Provost Marshal's tent: she always asked me for opium.' (Colonel Thos Walker, *Through the Mutiny,* 1858).

Again: 'On our first march from Delhi a comical incident, which, however, might easily have turned out rather a serious one, occurred. I was riding with the advanced files, when a young native woman, wielding with both hands a very long double-edged sword, such as is frequently used by acrobats at Indian festivities, suddenly appeared in the middle of the road and barred our way. The creature must have been mad or under the influence of bhang or some other intoxicant; for she deluged us with a torrent of abuse as she vigorously brandished the long thin blade. For a moment I was nonplussed: the situation was so entirely novel! Mad or sane, the virago evidently meant business. There was clearly no getting past her without a fight; *and that was quite out of the question. 'Shoot her, sahib,' said one of the sowars with me, little troubled with the polite consideration for the sex which the obligations of an effete civilisation inposed upon his British officer.* At that moment, as if by inspiration, a 'happy thought' flashed on my mind. 'Give her galee (abuse)' I said to the sowar; 'and give it her hot and strong, and plenty of it'. Instantly grasping the idea, the grinning sowar opened such a battery of abuse of the vilest and most comprehensive nature upon the unfortunate young person and her female relatives to the remotest degree that her own fire was promptly silenced. Encouraged by this success, the sowar redoubled his efforts; and slung such awful and shameful language with such force and precision that the rout of the enemy speedily became complete. Dropping her long sword and stuffing her fingers into her ears, she fled with a horrified shriek; and we marched triumphantly on, chuckling at the success of our tactics.' (Colonel A.R.D. Mackenzie, *Mutiny Memoirs,* 1891). **D** Here is another version of the familiar story, shown above, compiled from personal journals, and relating to the attack on the Secundra Bagh, Lucknow, in November 1857: 'I sent my A.D.C., Augustus Anson, into the garden to ascertain how the attack was succeeding when he perceived a poor native woman lying still alive amongst a crowd of dead bodies. In very moderate Hindostanee he told her to get up and make her escape through the gate. She followed his directions, but shortly afterwards he saw a Highlander, rifle in hand, as if stalking a deer, stealthily creep behind a hedge, take a deliberate aim at something in a tree, and fire. Down fell a large black object; and when the Highlander rushed up he found to his mortification that he had killed an old woman, the same poor creature whom Anson had befriended and who in her bewilderment had scrambled into a tree as the surest place of refuge. The Highlander had of course mistaken her for a sepoy, and so thus ended the poor old woman's sorrows in this world'. (General Sir Hope Grant, *Incidents of the Sepoy War,* 1873). And, again, there is another version, owing much perhaps to the above: 'Augustus Anson VC, ADC to Sir Hope Grant, was sent with a message to the Sikanderbagh, and rode into it through the big gate. There was a little desultory firing still going on between some Highlanders and the sepoys in the towers. His attention was attracted to a Highlander who at that moment was stalking some one amongst the orange trees inside the place. He saw him go down on his knees take steady aim and fire, and then heard someone fall from the tree aimed at. To his horror, and that of the Highlander also, they found it was an old woman who, Anson discovered afterwards, had been put up there a short time before by a humane officer who wished to get her out of danger. The poor Highlander was very much put out, but said he had already been fired upon several times by someone in that direction and thought the bullets had come from a man in that tree' (Related by Field Marshal Sir Garnet Wolseley). **E** in September 1858 during a sortie, 'a party of our soldiers on entering a house found among its defenders a man paralyzed with age. On seeing the hostile faces enter, far from begging for mercy which owing to his extreme age would perhaps have been granted, he took up a horsepistol and aimed at one of the men. The pistol missed fire and the soldier enraged at the opposition shown fired but before he had time to pull the trigger a woman interposed her body between the old man's and received the bullet in her breast. The man was bayonetted afterwards.' (**FSUP,** Vol IV, p 839). F Delhi....' sometimes we got into the basement of the houses before the Sepoys had time to escape. Up we used to go, batter in the door, throw ourselves flat on our faces for the volley which always came, then rush in ..bayonet etc. One time as soon as the door fell in, no volley. Looked in and saw instead of armed mutineers a quantity of beds ranged round the walls like a hosp. All tenanted by women. Called out to men: 'Turn out, we will go upstairs, the sepoys must be on the upper floor!' Was just going out when saw to my horror one of the men still in the room raise his rifle and bring it down with violence upon the bed. Ran to the man threatening dire punishment for being such a coward as to hurt a woman, when he replied (Irishman) 'Och yer honour, did ye iver see a woman with sich a prutty moustache as this?', and raised from the bed the unmistakeable face of a sepoy. Men recalled., ordered all the girls to get up - daytime so were all dressed - found the firing party of sepoys who having heard us come in through the wall sought safety under the bodies of the women. Irishman who had made the discovery explained afterwards that although the order had been given not to hurt women, no order given against kissing them, and he had merely placed his arm around the neck of the girl to give her a kiss when he became aware that his hand had come into contact with the face of the man underneath.' (R.G.Wilberforce, *An unrecorded chapter of the Indian Mutiny,* 1894).

INDIAN MILITARY LEADERS   For one assessment, *see* Burton, R.G. in *United Service Magazine* Vol 174, 1916; *see also* MILITARY DEFEAT OF THE REBELS, REASONS for.

INDIAN SOLDIER, The   'There is moreover another reason which frequently causes the British soldier to hold an entirely wrong impression of his Indian Comrade. The history of the great Mutiny of 1857, as gleaned from school books and cheap fiction, leaves one with an idea that the whole of the natives of India were up in arms against us, and that the cruelties and crimes which then occurred were the work of the ancestors of the men who fill the ranks of the Indian Army. Such is far from being a true or fair view of the case.....etc' This extract comes from a small booklet produced officially and handed to all British soldiers embarking for service in India: it shows that even sixty years after the mutiny/rebellion of 1857 opinions and attitudes were still polarised. The establishment is still unforgiving (and misunderstanding?), and seeks neither to explain or even excuse the events of the great struggle, merely pointing out that it was limited as to territory and ethnic sympathy. *See The Indian Empire,* HMSO, London, 1917.

INDISCRIMINATE SLAUGHTER   A letter defending behaviour of troops during storming of Delhi appeared in the London *Times.* Signed Iachimo. He claimed that the bodies of only two native women were found among the thousands of corpses in Delhi, and they had evidently been dead for several days. Their presence was the 'simple source' of the stories of 'the indiscriminate slaughter of women and children!' *See* London Times, 19 May, 1858, p 10.

INDORE   This was the capital of the dominions of **Holkar,** the Maratha maharaja who played an ambivalent part in the mutiny/rebellion but afterwards claimed that he had always been loyal to the British. On 1 July the Residency at Indore was besieged and after most of the men of the Bhopal Contingent had mutinied **Colonel Durand** the Officiating Agent to the Governor-General decided to retreat, which he did, with some Sikh troopers as escort, to **Sehore.** After his departure twenty-eight Christian men women and children were put to the sword. *See* **FSUP** Vol III, and *The Revolt in Central India,* Simla 1908. *Also* Khushhalial Srivastava, *The Revolt of 1857 in Central India-Malwa*, Bombay, 1966.

INFERNAL MACHINES   The name given at this time to any weapon that was unusual or non-standard. Thus the rebels would undoubtedly have regarded the banks of rockets fired by **Peel's Naval Brigade** at Lucknow in March 1858 as 'infernal machines'; similarly the rebels contrived an ingenious weapon by fixing up to forty matchlock barrels on to a frame that was mounted on wheels: numbers of these machines were captured at the end of streets when **Kotah** was assaulted by the British also in March 1858. *See The Revolt in Central India,* Simla, 1908, p 203.

INGLIS, Hon. Julia   The wife of Brigadier John Inglis, subsequently **Major-General Sir John Inglis** who succeeded to the command of the garrison of the Lucknow Residency on the death of Sir Henry Lawrence. She was a woman of considerable strength of character, and marked sense of duty who took her position seriously as senior lady within the garrison. She survived the siege and, like her friend **Mrs Adelaide Case** published the diary she had kept through the siege. The book includes the extensive notes of **Colonel F.M.Birch,** her husband's A.D.C. *See* Hon.Julia Inglis, *The Siege of Lucknow, a Diary,* London, 1892.

INGLIS, J.F.D.   Author of '*Narrative of the Outbreak of Disturbances and Restoration of Authority at **Bareilly** during the year 1857-58',* see **FSUP,** Vol I, p 428.

INGLIS, Major-General Sir John (1814-1862)
He served with H.M.'s 32nd Regiment (later the Duke of Cornwall's Light Infantry) in Canada 1837, and then in the Punjab 1848-9; commanded the regiment in Lucknow in 1857 before handing over command to **Lt.Col Case.** Succeeded **Sir Henry Lawrence** in overall command of the garrison in the Lucknow Residency on the latter's death. Made Major-General and knighted (K.C.B.) for the defence of the

Residency.. Contemporaries do not speak warmly of him; a man of limited ability and dull personality, he owed his command to seniority, though he was conscientious and person-

Brigadier Inglis

'A man universally detested', says Mrs **Wells**, wife of Dr Walter Wells (who both survived the siege). His despatches and final report after the end of the siege are unrevealing except in demonstrating his own prejudices, particularly against certain individuals with whom he had quarrelled *eg* **Martin Gubbins** who, despite his undoubted contribution to the defence he does not mention at all. Commanded in the Ionian Islands in 1860; died at Hamburg. *See* **FQP**; *see also* the Berners Papers in the **CSAS** Cambridge.

INNES, Lt General James J.McLeod
He was an engineer officer present during the operations in **Oude[Awadh]**. His most useful contribution to our knowledge of the **Lucknow Residency** siege is the series of sketches he made of the Residency buildings and the defensive position taken up by the British, from various angles. His experience is first-hand; as an engineer he is naturally most interested in that side of the operations. Some of his letters concerning the siege are owned by Mrs Christina Morley, Beaconsfield, Buckinghamshire. *See* J.J.Mcleod Innes, *Lucknow and Oudh in the Mutiny*, London, 1895, *see also* **DIB.**

INTACH
Indian National Trust for Art and Cultural Heritage (INTACH). responsible for much valuable work in advising on preservation and listing of historic buildings etc eg *Preliminary Unedited Listing of Kanpur. Uttar Pradesh, Unprotected Monuments, Buildings & Structures Listed for Conservation.* Kanpur 1992.

INTOXICANTS
The familiar picture of troops in India is to depict the British soldiers as drunken and addicted to alcohol, but to say that the Indian soldiers (with the exception of the Sikhs) were teetotallers. This is not the whole truth. Not all sepoys were abstemious. 'I have myself seen Hindoos and Muslims together cheek-by-jowl, in the arrack shop, and I have also seen them rolling drunk in the ditch,' while others 'get so intoxicated from the effects of that pernicious drug, opium, that they have not been able to tell what they were about.' But the sepoys of Hervey's Madras Army were lower caste than the sepoys of Bengal. Hervey. *A Soldier of the Company*. p 112.

INTRENCHMENT, WHEELER'S
See CAWNPORE, *see also* WHEELER'S ENTRENCHMENT.

INVESTMENT OPPORTUNITY
In August 1857, **Nana Sahib** was attempting to re-group and reform his forces: for this he would need considerable resources so he issued the following proclamation: 'Be it known to all the **Zemindars**, Chiefs, Merchants **Khundsaris** and Bankers that whoever amongst the **Zemindars** shall join me, accompanied by his men with provisions for them and ammunition, will receive credit for the price of those articles in the accounts relative to the revenue of his **Zemindaree**, and also a remission of the whole of the revenue for two years, and afterwards of 4 annas in the Rupee per annum for 8 years. That whoever amongst the **Zemindars** shall afford me aid, only in grain, bean etc as well as in balls, bullets and gun-powder, will obtain credit for the same in the accounts relating to the revenue of his **Zemindaree** and a remission of the rent for one year, and afterwards of 4 annas in the Rupee for four years; that whoever amongst the **Zemindars** from a feeling of regard for the English upon whom the wrath of God has fallen for their evil intention 'of converting Hindoos and Mahomedans to Christianity', shall hesitate to render services to the **Sirkar** or shall oppose, or desert it, or shall not procure supplies will be visited with due punishment. That, the Chief Ahmed Ali Khan alias Munwurooddowlah, who ate the salt of the Ruler of Oude, has been ruined, both in respect to this world and the next, owing to his attachment to the English. He was beaten by the **telingas** and breathed his last. His house was pillaged and he lost his honour. That the **Zemindars** will henceforward be responsible for plunder or highway robbery, committed in their **Zemindarees**. That those Chiefs, or **Khundsaris**, who shall render assistance to the **Sirkar** will obtain **Khilluts** of distinction. That after my arrival (at **Calpee**) if a **Zemindar**, merchant, or a **Khundsari** supply the English with provisions or afford them pecuniary aid, due punishment will be inflicted on him. That whoever amongst the bankers shall pay into my Treasury one Lakh of Rupees, will get interest thereon at the rate of 2 per cent (*per mensem*) until the liquidation of the principal, and whoever shall pay fifty thousand Rupees, shall receive interest at $1^{1}/_{2}$ per cent, and whoever shall deposit twenty-five thousand Rupees shall obtain interest at 1 per cent, and that no sum under twenty-five thousand Rupees will be received in the Treasury. The said Bankers

will also obtain from the King, **sunnud** of their good character and sincere attachment to the interest of the **Sirkar**. These certificates will contribute to the exaltation of their dignity and honour in the estimation of their compeers. That no demand of revenue will be made from the Bankers until debts due to them are liquidated. That whoever amongst the Chiefs or **Zemindars** shall fall into the hands of the English, will be made a Christian by them, like **Narain Rao**, son of the *Subedar*, **Ramchund Rao**. The treatise addressed to the royal troops, throws a light upon the nature of the fraudulent proceedings of the English. Whoever will offer himself a candidate for a situation, will (if his services be required) be appointed to a post; which he may be qualified to hold. Ten Regiments of *Nujeebs,* two thousand Artillery men, and 5 Regiments of Telingas will shortly be appointed'. **FPP**, 30 December 1859., Supp., Cons. No. 649, pp 68-9, **NA**.

IRREGULAR CAVALRY 'What made some troops rebel and others refuse to do so? There are some remarkable examples of decisions that seemed to turn upon no more than a word, a moment of panic, a misunderstood 'threat', the devotion or hate inspired by a particular officer. The most inexplicable cases are those where a regiment gave practical evidence of loyalty to Government - even fighting against 'mutineers' and arresting and killing their comrades - and then, by a *volte face* as swift as it was incomprehensible, suddenly became the implacable foes of the officers they had sought so recently to impress with their loyalty. Take the 5th Irregular Cavalry regiment of the Bengal Army as an example. They had declared, and sworn, fealty to Government and asked to be led against the mutineers. They were great friends of their comrades in the British army. Indeed, as late as 2 July 1857 they had an excellent opportunity to display this friendship. The British 5th Fusiliers Regiment was passing up country to join the column that was to march upon the relief of **Lucknow**. They travelled by steamer up the Ganges. On reaching Banghipur they left behind 100 men, 2 officers and a doctor to guard the line of communication, but before the main body left to travel upstream to **Benares** and **Allahabad** and beyond, they were treated to a grand tamasha with their friends of the 5th Bengal Irregulars. The regimental bands played, and the Irregulars put on an outstanding exhibition of incredible horsemanship. They practised in the large open compound of the Judge's house at Banghipur. They were wonderfully active; they threw themselves off at full gallop, ran by the side of the horses, and cut slashed and fired at imaginary enemies, and when free from assault jumped back on again without drawing rein at all. The other feats they excelled in were charging and lifting up - all done at a tearing gallop - tent pegs stuck firmly in the ground, with the point of their long spears; firing at and breaking bottles with their carbines, lying down as if dead on the backs of their horses, sitting lady's fashion, standing upright, and fighting while in that position, etc. All seemed well. The onlookers cheered and congratulated the irregulars who responded with smiles and waves. And yet.. the seed had already been sown. Soon afterwards, Sir Norman Leslie the Adjutant, Major MacDonald the Commandant and Dr Grant were sitting outside drinking tea together, when they were attacked by three men in the dusk of the evening. Leslie was cut through his shoulder and lungs with a downward stroke. MacDonald was scalped, and Grant's arm nearly cut off. Leslie died immediately; the other two beat off the assailants with chairs. A parade was ordered next morning and all men's sabres were examined; no trace of blood was found on any, and it was thought the murderers had escaped, but 'murder will out', and the full facts were brought to light. The culprits were discovered: they turned out to be two recruits and a 'good conduct man', all in the 5th Irregular Cavalry: the latter had only just received promotion as a reward for his long and uniformly good conduct. The other men of the regiment were anxious to prove that they were not at all involved with the actions of the three murderers. A Court martial was held on the spot, and the three men sentenced to be hanged. At this time the regiment was out in the jungles, a long way from any European assistance, so the Major, although wounded and weak from loss of blood, decided to carry out the sentence forthwith. He sat beside the gallows, with a loaded revolver in his hand, just in case there was any attempt at rescue. When the culprits were brought to the spot one started to address his comrades, calling upon them to rescue him and kill the Feringhee Suar - the pigs, of Englishmen, but Major MacDonald cocked his pistol and threatened if he spoke another word to shoot him through the head; whereupon, we read, 'this had the desired effect, and he suffered himself to be hanged'. *At the end of July the regiment finally mutinied, killed its officers and marched to join its comrades in Oude[Awadh].* **PJOT**

ITWARI BAHADUR Or Itbarri Bahadur. Hanged by the British in Kanpur subsequent to December 1857. 'I never forwarded on any Europeans. I furnished the executioners to murder the Europeans. I do not remember exactly how often executioners were sent for. But I recollect perfectly that I received five or six purwanahs sending for them signed by the Nana, on whose orders I sent them. Itwari, Bahadur, and his son, were the head men of the executioners. I used to send them, and they used to take others with them. Bahadur is a resident of the city.' *Deposition of Hulas Singh.* **Williams.** 'Itbarri (who has since been hanged), Bahadoor and Chota; these three-were the head **jullads** or executioners, and with them all the other jullads used to go. I only saw the above three men sent from the kotwali, but I heard that the rest used to accompany them. (They were sent) by orders of Hulas Singh, Kotwal. ...Kunkawara, sweeper of the Bransphor caste, resident of Bithoor, was employed by the Nana in cutting off heads. I know this to be true'. *Deposition of Peero.* **Williams.**

The Intrenchment at Cawnpore.

J

JACK, Brigadier Alexander An officer of the 42nd BNI. Station commander at **Kanpur**. Died of fever (in **Delafosse's** list) or 'treacherously shot' (according to the **DNB**) in **Wheeler's Entrenchment** in early June 1857. He had had a distinguished career since joining the Company's service in 1823, especially during the Sikh Wars. His position in the hierarchy within the entrenchment is obscure: although the next senior officer to **Major General Wheeler**, he does not seem to have enjoyed that officer's confidence, or it may be that he was too unwell to assume major responsibilities.

JACKSON, Anna Madeline One of the '**Captives of Oude [Awadh]**'. She was seventeen 'very beautiful' and survived the ordeal, married her cousin and in later life wrote an account of the whole affair. *See* **SSF** chapter Ten, and Madeline Jackson's MS narrative in the **Haig** Papers, **CSAS**, Cambridge. Copies of her mutiny memoirs, dated 1880 are preserved in the Oriental and India Office collections of the British Library, *see 26ff Photo.Eur.41*, and a manuscript copy of her personal narrative of the Indian Mutiny and the flight from Seetapore [Sitapur] is owned by Mrs Constance Jackson of Tenterden, Kent.

JACKSON, C. Coverley Bengal civil servant. He was Officiating Chief Commissioner of **Oude [Awadh]** in 1856. Quarrelled violently with the Financial Commissioner, **Martin Gubbins**, and the Judicial Commissioner, **M.C. Ommanney**, with the result that little real progress had been made by May 1857 in establishing British power in that province after its annexation in February 1856. Succeeded by **Sir Henry Lawrence** as substantive Chief Commissioner. Uncle to the two Jackson sisters, **Madeline** and **Georgina** and to **Sir Mountstuart Jackson**. 'Jackson was an earnest and hard-working man. He possessed a quick and brilliant brain and knew the routine of government offices backwards. But he was the very worst person to have been placed in control of Oudh at this critical juncture in its history. Jackson was utterly devoid of human sympathy. He was one of those unfortunate people who seem to spend their lives bickering with their neighbours. He bullied his subordinates and wrote unpleasant letters to his superiors, trying always to prove that he was entirely right, and that he was hampered by fools, rogues, and incompetents. ...When he was not quarrelling with his chief assistant, Gubbins, he spent most of his energies in devising petty insults and annoyances for the dethroned royal family of Oudh.' Richard Hilton, *The Indian Mutiny:* London, 1957, p 16.

JACKSON, Georgina She was nineteen and 'very beautiful'. Sister of **Anna Madeline** and Sir **Mountstuart Jackson**. Fled from **Sitapur** and took refuge with ruler of **Dhourayeah** [Dhauraha], but was rejected by the Rani, sent in to Lucknow and shot, apparently on the orders of **Hazrat Mahal**, Begum. *See* **SSF** chapter Ten.

JACKSON, Sir Mountstuart Newly arrived in India with his two sisters Madeline and Georgina, he had joined the Company's Civil Service, and was posted to **Sitapur** in Awadh under Mr Christian, the Assistant Commissioner. Was one of the 'Captives of Oude [Awadh]', dragged to Lucknow by the orders of **Loni Singh, Raja** of **Mithowlie [Mitauli]**, and murdered there along with Captain Orr. *See* **SSF** Chapter Ten and the **Haig** Papers in the **CSAS**, Cambridge.

JAGDISHPUR, or JUGDESPUR A small town, in Bihar between **Arrah** and the Ganges, which was the home of **Raja Kunwar Singh.** When Kunwar Singh was

Kunwar Singh at Home in Jagdishpur

dead and the rebellion all but over the British determined to prevent Jagdishpur ever again being used as a trap or ambush site for British troops; H.M's 35th Regiment (later the Royal Sussex Regt.) had been very nearly annihilated there. So they ordered the clearance of the Jagdishpur Jungles - the old hunting ground of Kunwar Singh - an act for which today there would be universal condemnation on ecological grounds. 'The clearance of the Jugdespore Jungles progresses satisfactorily. The work has been given to Mr Burranes of the firm of Messrs Burn & Co., railway contractors. The *'Engineers Journal'* states that 2,000 workmen are now employed in cutting and removing the trees, and that nearly 2,500 acres or 1/12th of the jungles have already been cut down. Work has been undertaken by the contractor on condition that the jungles are to be cut down at his own expense, the Govt granting him the whole of the lands rent-free for 99 years. These jungles have formed the refuge of the rebels for months past, and it is from these retreats that Koer Singh's followers have given so much trouble.' The diary of one Sgt Thomas Anderson, of E Troop Royal Horse Artillery is revealing: 'Crossed the Ganges on 28 April (1858) in a small steamer. Arrived Arrah on 5 May. Marched on Jagdishpur and saw the bones of the men of the 35th Regiment who had lost 200 men there in the jungle and two guns. Stormed their town and took their leader's palace with loss of 5 killed and 10 wounded. Enemy lost 400 killed and wounded. Acted as decoys to draw the sepoys away, then 10th and 84th Regts marched through the jungles setting fire to all the villages and towns, frequently falling on the rear of the sepoys and routing them with great slaughter. On 28 May we recaptured the two guns lost by the 35th. Pushed on and reached Buxar on 6 June ....and then joined column under Gen Sir Hope Grant following the Nana and the remnants of the sepoy army. Very near capturing the Nana but he escaped across the Raptee[Rapti] into Nepal. Then joined the Governor General's train processing through the NW Provinces.' See Illustrated London News, 29 January 1860. See also Lieutenant R.Parsons, A Story of Jugdespore, London, 1904.

JAGIR   An estate or parcel of land granted to an individual, usually in perpetuity, as a reward for services rendered. Government is the usual benefactor, and at this time granted jaghirs from the confiscated lands of those it considered had rebelled against its authority. '**Jagir** was a grant of revenue to support and maintain troops, officials, priests. Revenue, not land, became a kind of fief'. J.S.Critchely, *Feudalism*. London, 1978. pp 26-7.

JAGIRDARS of BHOPAL   Three powerful subjects of the **Begum of Bhopal** chafed under her rule: they were Waris Muhammad Khan, Fazil Muhammad Khan and **Adil Muhammad Khan**: they took advantage of the mutiny of the **Bhopal Contingent** to serve their own ends. What began as a military mutiny took on the dimensions of a general revolt and the Begum was for some time hard pressed. For an excellent and succinct account of these events and the uprisings at Sehore and Berasia, *see* K.D.Bhargava, *Descriptive List of Mutiny Papers in the National Archives of India, Bhopal*, Vol II, New Delhi, 1963.

JAI LAL SINGH, Raja (or Jey Lall Sing)   A man of great influence at the court of the **Begum of Oude [Awadh]** and **Birjis Qadr** her son. The latter became King through Jai Lal's efforts, he having great influence with the sepoys, especially the infantry. He was also given the *ilaqas* of **Daryabad** and **Azamghar**, and put in charge of the mint and the gaol as well as the army. On the restoration of British power he was at first allowed his freedom, there being no evidence against him of having instigated the murder of Europeans, but eventually he was brought to trial, and on 1 October 1859 hanged in front of the **Tarawali Koti** on the spot where he was supposed to have watched the execution of two parties of Europeans.

JAIPUR   An independent state of Rajputana whose young maharaja remained loyal to the British government throughout the struggle. He lent his 5,000 troops to the British in late May 1857 to march with the political agent, Captain W. Eden to re-establish civil government in the Muttra and Gurgaon districts: this attempt largely failed but the force did return to Jaipur as one unit, and remained neutral for the remainder of the conflict despite containing many men recruited from similar backgrounds as the Bengal army. *See* Iltudus T.Pritchard, *The Mutinies in Rajpootana, being a personal narrative of the mutiny at Nusseerabad*, London, 1864.

JANKI PERSHAD   The Depositions of many citizens and survivors of the dramatic events in **Kanpur** in May, June, July and August 1857, provide the most graphic picture still available to us, in documentary form, of what happened outside **Wheeler's Entrenchment** at that time. Thus: 'One **Azeezun**, a prostitute, who was in the service of Kulloo Mull, lived in the Lurkee Mahil in Oomrao Begum's house. She was very intimate with the men of the 2nd Cavalry and was in the habit of riding armed with the sowars. The day the flag was raised, she was on horseback in male attire decorated with medals, armed with a brace of pistols, and joined the crusade. I saw her as thousands of others did also. ..... Mahomedans raised a **Mahumdee** flag. I went with many other people and saw the flag raised two or three days after the siege commenced. When I came near the Mogul Serai, I saw a great many people collected, and a green flag raised. I heard that the flag was raised by Moulvie Salamut Oollah, and that he stood by it with a rosary in his hand, and it was rumoured that they were on the point of going to fight. There were from 2 to 3,000 armed men with him, and the Kazie of the City, Waseeooddeen, was riding about the plain with some 19 sowars of the 2nd Cavalry; great crowds were collected there. Yes, I saw Bukshee Zainolabdeen with about twenty-five or thirty chowkidars and burkundazes all armed, standing in a body at a short distance from the flag, and the rumour was that the Moulvie intended to attack the entrenchments. I stood there a short time, but as nothing occurred, soon came away. There were thousands of people from Nawab Gunge, Colonel Gunge, Orderly Bazar, and Bunsah Mow, and from the other bazars, collected there. How can I remember all their names? I heard that it was first raised near Moulvie Salamut Oollah's house, where all the Mussulmans collected and brought it on.' *Deposition of Jankee Pershad.* **Williams.**

JAORA ALIPUR   A few miles from **Gwalior** where an action took place on 21 June 1858, the rebels retreating after the **Battle of Gwalior** being pursued by Brigadier General **Napier**, were routed and dispersed, losing some 400 killed. The capture of Gwalior and the dispersal of the rebel army at Jaora-Alipur brought the regular campaign in Central India to an end, although there was a long pursuit of Tatya Tope, lasting many months, still to come. The British congratulated themselves prematurely, since the dispersal in small groups or even singly, followed by re-uniting at a given point, was always part of Tatya Tope's strategy. Sir Hugh Rose handed over his command to Napier on 29 June 1858, and left for Poona[Pune] via Mhow on that date. For details of this action, *see The Revolt in Central India*, Simla 1908, p 166 *et seq*.

JAORA, NAWAB of The chief of a small state in Central India who remained particularly loyal to the British, not only providing them with troops, but also being primarily responsible for the supplies furnished to the Malwa Field Force. Without his help it is probable that the British troops would have been able to move at only half the speed they achieved.

JATS a race of what were described as industrious and hardy cultivators, whose original seat is said to have been Ghazni, but who were then found in great numbers throughout the N.W.Provinces: in general they appeared to resent the mutiny/rebellion and to regard it as an interference with the regularity of their lives. But in some areas they actively espoused the rebel cause, *see* Eric Stokes, *The Peasant Armed*, Oxford, 1986.

JAUN MAHOMED Jaun Mohomed, a sepoy of the 8th company, 56th Native Infantry, was accused of inciting the sowars of the 2nd Cavalry, by stating that they would all be blown away from guns; this was in May 1857, between 15 and 20 May; there was then a disturbance in the 2nd Cavalry. (Jaun Mohomed was confined, but escaped from the entrenchments).'We felt sure that the 2nd Cavalry would mutiny. 'The 56th were well disposed; they told (Colonel Williams), that though all other regiments might mutiny, yet they would be true. The Colonel and officers slept amongst the men. A tent was pitched by the sepoys' pauls; (the new lines were not finished.)' *Depositions of Gobind Singh, Sheik Elahee Buksh & Ghouse Mohomed.* **Williams.**

JAUNPUR The chief place of a district of the same name, acquired by the East India Company in 1775. It stands on the river **Gomti**, 35 miles northwest from **Benares** [Varanasi] and 55 miles northeast from **Allahabad.** A mutiny took place there on 5 June 1857. It is unusual in that it involved **Sikhs** who in general remained faithful to the British, but here may have been convinced by the events at **Benares** that the British had decided to exterminate the entire Bengal Army without distinction of race or creed. The garrison consisted of 169 men of the Ludhiana Sikhs, under the command of a single European officer, Lieutenant Patrick Mara (or O'Mara?). Mara was shot by his own men, and apparently abandoned by the other Europeans while still living; Mrs Mara died of apoplexy (she was very fat), and four other Europeans or Eurasians were killed. *See* **Martin Vol II**, p 290; *see also* the *Times*, 6 August and 10 August 1857. *See also* **FSUP** Vol IV, numerous entries.

Sikhs of a Punjab (Irregular) Corps

JEEWAN LALL, Munshi *See* JIWAN LAL, MUNSHI.

JEHANGIR KHAN Those sepoys who were on leave when their regiments mutinied were faced with a most difficult dilemma. Some remained quietly in their villages and awaited the outcome; others offered their services to a local **zemindar** and often fought against the British though not in their own corps. It is said that after the suppression of the mutiny/rebellion many such men turned up in army camps and successfully applied for back-pay from the British, producing their leave certificates as 'proof' that they had not been mutineers. One man was particularly unfortunate as to arrive back from leave on the very day his corps mutinied, but at least he did survive: 'My name is Jehangeer Khan and my father is Azimutoolla Khan; I am a Pathan inhabitant of **Mooradabad** *mohullah* Fuezgunge (Faizganj) *zillah* Mooradabad and am 30 years of age. My profession is to take service. I was formerly a sowar of the 3rd troops 2nd Regt. (Regiment) L. Cavly. (Light Cavalry). I (got) enlisted in the corps 7 years before the outbreak. My Regt.(Regiment) mutinied at Cawnpore. I was with my Regt.(Regiment) so long as it was stationed at Cawnpore. I had taken leave of absence for 15 days to go home at **Futtehpore** and when my leave expired I returned to **Cawnpore** and reached it the day my Regt. (Regiment) mutinied. On the very same day I was on picquet duty near the entrenchment when the muskets were fired. The picquet guard I was in returned to the lines. Soon after I went to field officer whose name I do not know and whose tent was pitched near the picquet. The field officer asked me where the picquet was gone to. I told him in reply that it was gone to the lines. The officer asked me the reason of my standing there. I said I would stop there. The officer told me to go off lest I should be killed by men of the other Regts.(Regiments). I then returned to the lines of my Regiment. The officer who spoke to me intended to enter the entrenchment. This occurred at midnight but I forget the English month. When I reached the lines of my Regiment all of my comrades had gone off to the treasury which they wanted to enter and loot but the guard on duty consisting of 2 Companies of a Regiment whose No. I do not know would not allow us to enter and said that unless their Regiment came they would not suffer admission to any one.' *See* Lucknow Collectorate Records in **FSUP** Vol IV, pp 501-2.

JELLALABAD[Jalalabad] There appear to be two places with this name, both of which played a part in the struggle. The first is a small Fort, close to the **Alambagh** just south of Lucknow: it was attacked by **Sir Colin Campbell,** on 13 November 1857 on his way to relieve the **Residency**, then occupied and dismantled. The second is also a fort, in a village of the same name not far from **Shahjahanpur**. It was occupied by the British on 28 April 1858: the fort showed signs of recent repair, and it seems the **Maulvi of Faizabad** had intended to hold it against the troops of Campbell, but his own sepoys had fled northwards to escape the British. This Jellalabad was the scene of a disgraceful act by a Mr Money, the Civil Commissioner with the British force: he arrested and hanged the tehsildar who had acted as Deputy Collector under the Company, had continued in that function under the rebels, and had come voluntarily into camp to give himself up, having been assured of his life being spared. He met his death calmly and with great dignity. Campbell when he heard what had happened was extremely angry with Money and forbade any such action in the future; but it is probable that the harm was done, and many Indians would now believe that the word even of a British Officer could not be trusted.

JENKINS, Charles, The Joint Magistrate and Collector of **Shahjahanpur**. After the mutiny outbreak and the attack on the Europeans in the church, Jenkins was the senior officer and took charge of the survivors who first sought refuge (unsuccessfully) with the Raja of **Powain** [Pawayan] but were later massacred by the sepoys.

JENNINGS, CHAPLAIN Midgeley Died at **Delhi**. *See also* PREMONITION.

Rani of Jhansi

JHANSI A small town in Central India (now on the extreme south western edge of Uttar Pradesh) which was the scene of considerable activity during the rebellion. Originally a small independent state, ruled by a Maratha dynasty closely allied to the British, it had been recently annexed on the excuse of an absence of an 'heir of the body' to the last ruler **Gangadhar Rao**. His adopted son, **Damodar Rao**, and widow **Lakshmibai** were thus disinherited, and the latter was to become one of the leaders of the rebellion against British rule. In all there were over sixty British and Eurasians in and around Jhansi, either in the army or in official posts, and this number includes all the members of their families, men women and children. And there were a few Indian Christians also. The town is remembered for the Massacre, the Battle and the Fort. In the 'Massacre of Jhansi' sixty-six Europeans and Christians, including sixteen women and twenty children, were persuaded by false promises to come out of the Fort in which they had taken refuge, and they were then all killed, some said with shocking cruelty. The palace and the Fort were stormed by the British on 4 April 1858 after a few days' siege, but the Rani Lakshmibai was not to be found. She had left the Fort in the night, riding on her white horse, accompanied by her father Moropant and a few of her faithful family and soldiers. The little boy Damodar was fastened to her back with a silken shawl. For bibliography *see* JHANSI, Rani of..... below .

JHANSI JEWELS 'The Jhansi jewels have been sold, realising nearly £19,000 (Rs187,964). 2 necklaces ornamented with emeralds and two wristlets set with diamonds were purchased by Lord Elphinstone (Governor of Bombay) as a present for Queen Victoria. They were by far the most valuable of all, and were rated at the sum of £3,400.' This was nothing less than theft on Government's part. The Governor General Lord **Dalhousie**, while refusing to recognise the adopted son of Gangadhar Rao as ruler of **Jhansi**, did, illogically, declare that he could and should inherit much of his adoptive father's property, including his jewels and other precious items. After the campaign in Central India this was conveniently forgotten, and all the ruler's private property sequestrated. *See Illustrated London News*, 12 November 1859.

JHANSI, Rani of 'We do not know everything about the childhood of Lakshmibai, the Rani of Jhansi, but some facts are certain: she was born in Varanasi, probably in 1827 and was the daughter of Morapant Tambe, a follower of Chimaji, the brother of Baji Rao II, the last Peishwa. Her father was a learned and cultured man, and he could easily have got a post in one of the great Maratha kingdoms but he chose to be loyal to Chimaji and stayed with him on a very low salary of fifty rupees a month. His wife, Lakshmibai's mother, is said to have been very beautiful indeed and to have known all the stories of the great Hindu epics by heart: no doubt she told these tales to her little daughter. Indeed we know that Lakshmibai was, throughout her life, a devout person, often going to the Temple of Vishweshvar and other places of pilgrimage, with her parents. Even when she became a queen we know she showed great respect for Brahmins: they say she once refused to let a Brahmin priest draw up water for her from a well, saying 'You are a learned Brahmin and it is not right that you should do this for me: I will draw my own water from the well'. We think that her father took her as a young child to Bithur, to Baji Rao's palace, and that there she grew up with the ex-Peishwa's adopted son Dhondu Pant, later known as Nana Sahib, and Rao Sahib his nephew, and Tatya Tope and, very unusually for a girl, she learnt to ride and to handle weapons, and to fight in the old Maratha way, on horseback. The old Peishwa, Baji Rao, became very fond of her and admired her spirit: he called her Chhabeli, or 'sweetheart'. During her childhood she was usually known as Manu, for her given name was Manukarnika, one of the titles of the holy river Ganga. It was in the month of May 1842 that she married Gangadhar Rao, Maharaja of Jhansi, whose first wife had died childless. Henceforth she is known as Lakshmi Bai, not Manu. A son was born some nine years later but the baby died within three months. Gangadhar Rao was himself ill, and adopted a son in correct Hindu fashion, choosing a boy aged five called Anand Rao, who was a member of the royal family: henceforth the boy is known as Damodar Rao. Maharaja Gangadhar Rao made his last will and testament: he called on the British government to honour their treaty with Jhansi, to accept the young Damodar Rao as his heir, and to allow Lakshmibai to be honoured as the boy's adoptive mother and Regent, or *malika* of the Kingdom of Jhansi thoughout her lifetime. When Gangadhar died the Governor General, Dalhousie, refused to recognise the heir and declared the annexation of Jhansi by virtue of the 'doctrine of lapse'. With dignity Lakshmibai left the Fort at Jhansi and

The Rani's Palace in Jhansi

went to live in the palace in the town. There she lived quietly for three years. Still the Rani did not turn against the British, but attempted by petitions both to the Governor General and to the Directors of the H.East India Company in London, to have the decision reversed. To the viewer the Rani was most impressive: one Englishman recorded: 'In appearance she was fair and handsome, with a noble figure and a dignified and resolute, indeed stern, expression on her face which, when she was younger, had appeared so much softer and gentler. ....her dress, though that of a woman, was not the ordinary costume generally worn by females of her station in life. On her head she had a small cap of bright-coloured scarlet silk with a string of pearls and rubies encircling and laced into it, and round her neck a diamond necklace sparkled, of not less value than a lakh of rupees at least. Her bodice, freely opened in front, was drawn tightly in by a belt worked over and embroidered with gold, and it it were stuck two finely carved and silver-mounted pistols, together with a small but elegantly shaped hand-dagger, the point of which, it was whispered, had been dipped in a deadly poison, whereby a wound, even a slight one, would prove fatal. Instead of the usual petticoat, she wore a pair of loose trousers, from which protruded her small, prettily rounded bare feet.' On the outbreak of the rebellion the sepoys stationed in Jhansi hoped for the Rani's support, but she remained aloof - indeed to the extent that she was actually threatened by the sepoys. Mutiny here led to the Europeans taking refuge in the Fort from which they were treacherously persuaded to emerge on promise of safe-conduct. The point is, did the Rani know what was going to happen to the Europeans? Did she, as some said, issue the 'safe-conduct' and promise them their lives, and then break her word? She wrote a letter at once to the British Commissioner at Jubbulpore [Jabalpur] condemning the sepoys' actions, and saying she had nothing to do with the murders. In return, the Commissioner at Jubbulpore, Major **Erskine**, told her he believed her implicitly, and asked her to assume the government of Jhansi, 'till a new Superintendent arrives', because Captain Skene had been killed with all the other British. He gave her no lasting promise that Jhansi should always be hers, and as the months went by it became clear to her that the British would never give her Jhansi, and that they did not really believe her innocent - *or did not want to believe her innocent* - of the murder of the Europeans. At first she had to deal with enemies nearer home: the rulers of Orchha, Pihari and Dalia all tried to take advantage of the confusion the British had left behind, and marched on Jhansi, but the people of the little kingdom stayed loyal to their Rani, and the invaders were defeated and driven back to their own lands. It was Rani Lakshmibai's misfortune that the British army that was sent against her was led by the best General that the British had in all India. His name was Major-General Sir Hugh Rose, and he was in charge of the Central India Field Force. After defeats both at Jhansi and Kalpi (where she had joined **Tatya Tope** and **Rao Sahib** the Rani suggested to the rebels the daring exploit of seizing Gwalior, with Scindia's blessing if possible, but without him if necessary. It was near here, in battle that the Rani was killed. There was much fighting, the Rani was in the thick of it, dressed in male attire and with cropped hair: they say she fought with the horse's reins between her teeth, and with a sword *in each hand*. She was hit by a bullet from a carbine, fired by a British soldier in the 8th Hussars, who had no idea that he had hit the Rani of Jhansi. She was carried to the rear by her servants, living just long enough to order that her jewels were to be given to her soldiers, and to whisper to her companion, Ramchandra Deshmukh that he should take charge of Damodar Rao. A fine bronze equestrian monument to the memory of the Rani exists in Gwalior, spoiled only because a thief has removed the statue of Damodar which was seated behind her on the horse'. **PJOT** (From a Biography written for children, 1995). *See* **CM**; p 25 *et seq.; see also* Antonia Fraser, *Boadicea's Chariot, the Warrior Queens,* 1988; the Erskine Papers, **CSAS** Cambridge; D.V.Tahman-kar, *The Ranee of Jhansi,* 1958; Thos Lowe, *Central India during the Rebellion 1857-8,* 1860; Sir John Smyth, *The Rebellious Rani,* 1966; Dr Sinha, *The Revolt of 1857 in Bundelkhand,* 1982; R.G.Burton, *Indian Military Leaders, and Women Warriors in India,* United Services Magazine, Vols 174-5, 1916. Dr Tapti Roy, BUNDELKHAND in 1857, etc.

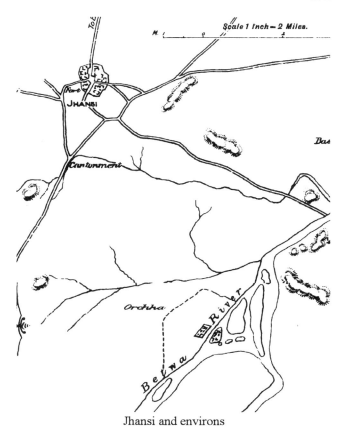

Jhansi and environs

JHANSI, SIEGE of The investment and siege of Jhansi began on 23 March 1858 after a reconnaissance by **Sir Hugh Rose,** and although interrupted by the **Battle of the Betwa River** on 31 March, it was never really relaxed, and the city was assaulted on 3 April 1858, the street fighting which followed being particularly fierce. The **Rani** escaped from the fort at night with **Damodar Rao** tied firmly on her back; she is said to have jumped on her horse down the precipitous slope: she was accompanied by some 300 *vilaities* and 25 sowars and headed towards Bhander. Some 5,000 of the garrison were killed, and British casualties were 38 killed and 181 wounded of whom 22 subsequently died. For full details of the siege and associated action *see The Revolt in Central India,* Simla, 1908, pp 106-119, *and* Thomas Lowe, *Central India during the Rebellion of 1857-58,* London, 1860, *and* **FSUP,** Vol III. *See also* Sir George William Forrest, ed., *Selections from the Letters, Dispatches and other State Papers of the Military Department of the Government of India 1857-58,* 4 vols, Calcutta, 1893-1912, Vol IV, *and* Tapti Roy, *The Politics of a Popular Uprising: Bundelkhand in 1857,* Delhi, 1994.

JHAU LAL He was a newswriter (spy), from **Khairabad** and **Sitapur**, who regularly submitted intelligence to the British (Capt A.Orr in Lucknow). He starts one of his reports thus: 'April 18th 1858. Maulvi Faizabadi, Badmash, myopic, was defeated by the English troops under Mr Right (sic) at Bari.'. See **FSUP**, Vol II, p 413.

JHELUM   A cantonment town in the north of the **Punjab**. It is situated on the east bank of the river from which it derives its name - the Hydaspes of the Greeks, which is a tributary to the Indus and the most western of the five rivers of the Punjab. *See* MARSDEN, Colonel Frederick Carleton.

JHELUM, MUTINY AT   The 14th BNI was in cantonment, and gave clear enough indications of its intention to mutiny to cause the authorities at Lahore to decide to disarm it. With this in mind three companies (approx 250 men) of H.M.'s 24th Regiment were despatched on 1 July from their quarters at **Rawalpindi**, some seventy miles distant and they arrived at Jhelum on the morning of 7 July 1857. Whether the sepoys had intended to mutiny is not completely clear, but certainly the arrival of the European soldiers precipitated things; they loaded their muskets in something approaching panic and fired at their officers (without effect) and rushed for their lines where they opened fire on the men of the 24th who were sent after them; their fire was very effective, they killed twenty-eight of the British and wounded a further fifty-one including the commanding officer Colonel Ellice. They were ejected from the lines after a fierce struggle and many took to the jungle with their arms and ammunition. Government immediately offered a reward of Rs 30 for each fugitive sepoy and many were brought in by villagers who had no sympathy with them: forty-eight were shot after drum-head court-martial on 8 July and a further twenty-five blown away from guns on the following day. The action at Jhelum typified the 'strong response' of the authorities of the Punjab to the threat of mutiny and rebellion. While it undoubtedly kept the Punjab quiet (and was therefore in British eyes fully justified) it also involved considerable harhness and injustice; for example no opportunity was given to men who did not wish to mutiny to dissociate themselves from the others. See **Ball, Sen, Kaye, Martin** etc

JHIGAN   A small village near **Banda** in **Bundelkhand**, where an action was fought on 11 April 1858, between the Madras Column commanded by Major-General **Whitlock**, and some 2,000 rebels. For details *see The Revolt in Central India*, Simla, 1908, and **FSUP**, Vol III.

JHOKUN   'The officers I left in their tents at the regimental lines, ten days before they went into the entrenchments (at Kanpur). All ladies and house servants used to go down to the lines at night and the chowkidars and syces used to remain at the houses. The Colonel used to converse with the men in the evening and do his best to allay their apprehensions regarding the cartridges and apparently with success. I don't know what took the officers into the entrenchments, as there was no active demonstration on the part of the troop to lead me to suppose that it was from fear of an immediate attack'. *Deposition of Jokhun*, **Williams**, Vol. 3.

The Jokun Bagh in 1858

JHOKUN BAGH   The name given to the garden area on the edge of **Jhansi** where the majority of the Europeans and Christians who were killed there in June 1857 were buried in a mass grave, and where subsequently certain rebels, including Mama Sahib the father of the Rani, were hanged by the British in retribution.

JIRAN, Battle of   This took place in **Rajputana** on 23 October 1857, between the Mandesar forces led by **Firoz Shah**, and a small British force (mainly Bombay army troops) led by a Captain Tucker. The latter was killed together with another officer and five other officers wounded; the affair would have been a complete victory for the rebels had it not been for the 'dash and gallantry' of these officers. As it was the town of Jiran was evacuated by Firoz Shah but he and his force remained intact. Tucker's head was carried off and placed over a gate at Mandesar. It was later replaced by the heads of two rebel leaders when Mandesar was taken by the **Malwa Field Force**; *see The Revolt in Central India*, Simla, 1908; *see also* Iltudus T.Pritchard, *The Mutinies in Rajpootana, being a personal narrative of the mutiny at Nusseerabad*, London, 1864.

JIVANLALA, RAI BAHADUR   The author of a life of Rai Jeewan Lal Bahadur, with extracts from the latter's diary pertinent to the period 1857+. *See* R.B.Jivanlala, *Short Account of the life and family of Rai Jeewan Lal Bahadur*, Delhi, 1902.

JIWAN LAL, Munshi (or Jeewan Lall)   He was an employee of the English before the outbreak, and continued to be a diarist and newswriter in their interest throughout the **'siege' of Delhi**. Munshi Jiwan Lal was an educated man, closely associated with the court life of the King of Delhi for many years both before and during the outbreak. He had been appointed Accountant of the numerous pensions paid by the British to the King's family, and became the confidential messenger between the King and the Governor General's Agent. He lived in daily contact with the King and his family, and was thoroughly familiar with everyone around the King, and with the Palace intrigues. He was resident within the walls of Delhi throughout the siege, carefully observing every event, in fear of his life but protected by the Palace. A writer by caste and profession he recorded each day's events in a diary. He was generously rewarded by the British after the resumption of power. While clearly hoping for a British victory Jiwan had none the less kept an accurate account of day to day events and his diary is a valuable primary source for historians, *eg.* 'July 20th...a number of sappers deserted from the English camp, and their officers attended the Durbar and reported that the English strength was 6,000 men. If they were attacked by the whole force in Delhi the King would probably be victorious; but if there were any delay the English would obtain reinforcements from England, and the King's forces would not be able to prevail against them. A number of cavalry troopers sought service; the King replied he had no money to pay for their services. Several unarmed soldiers applied for muskets; the King replied he had no reserve arms to give them. Muthra Dass, treasurer of Bijnor, sent the King several sepoys he had caused to be arrested. Mahommed Khan son of Nawab Najibabad presented a petition on the part of the Nawab asking for a written expression of the King's Commendation for having taken from the English Najibabad, Rampur, Bijnur, Dusnaria, Nagina. The petition was ordered to be sent to General **Moh.Bakht Khan** for reply, as the King would not commit himself....the Nawab of Jajjar (reported) that as far as lay in his power, to the extent of 3 lakhs, the King might draw upon him. The thanadar of Negumboda forwarded certain property belonging to deceased Europeans which he had discovered in the house of one Ram Gopal. The King after inspecting the property ordered it to be made over to the Begum....Fifty sepoys were enlisted and sent to assist Nawab **Walidad Khan**, the Revenue Collector. As they were passing the Calcutta Gate they were stopped by the guard who suspected them of deserting...; so they deprived them of their muskets. A Resaldar, with several troopers, wanted to get away today to Gwalior; they were also stopped by the guards at the Calcutta Gate: their weapons taken from them and they were plundered. It was rumoured today that in the recent fight with the English, a woman dressed like a sepoy had acted with great bravery; when the rebel sepoys ran away she remained opposed to a number of English, and killed one English soldier.' For the full text of Munshi Jiwan Lal's account, *see* Charles Theophilus Metcalfe, *Two Native Narratives of the Mutiny in Delhi*, London, 1898.

JODHPUR LEGION This was a force of infantry, cavalry and artillery under the command of Captain Hall, with headquarters at Erinpura which lies between Jodhpur and Udaipur in Rajputana. The majority of the Legion mutinied on 21 August 1857. Those who were recruited from the same sources as the Bengal army (*ie* mainly from high caste men of Oude [Awadh] were the mutineers, while the three companies of Bhils apparently had no sympathy with the Oude men. For a full account of the mutiny of the Legion, *see The Revolt in Central India*, Simla, 1908, p 58 *et seq.*, and *see also* I.T.Pritchard, *The Mutinies in Rajputana*, London, 1864.

JOHANNES'S HOUSE One of the houses outside the Residency compound in Lucknow had belonged to a merchant called Johannes (this man was imprisoned by the rebels and held in the jail of which **Wajid Shah Daroga** had charge: he survived the rebellion, as did his son and family who were inside the Residency during the siege). The house was close to and overlooked the defences of the south-east corner of the compound, and was, from the start, filled with rebels. The British garrison made a sortie against it on 7 July 1857 and killed a number of the rebels especially some twenty **Pasis** who were engaged in digging mines. The house was not at this time blown up and later became the favourite vantage place of **Bob the Nailer** who killed many of the garrison with his accurate rifle fire, until he too was killed; *see* M.R.Gubbins, *The Mutinies in Oude,* 1858; *see also* **FQP**.

JONES, Gavin Sibbald He was an industrialist in Kanpur. His father had had an indigo plantation near **Fatehgarh**, and he was in that town when the rebellion broke out and was one of the civilians besieged with the army officers in **Fatehgarh fort.** He was fortunate enough to escape and survive whereas most of the others perished. In later life managed a number of mills in Kanpur and became very prosperous. *See* Gavin S.Jones, *My escape from Fatehgarh,* Kanpur, 1913; *and* Gavin Sibbald Jones, *The Story of my Escape from Fatehgarh,* in *Cornhill magazine.* Jan. 1865.

JONES, Gavin 'On the 3rd of June information was received at Futtehghur [Fatehgarh] that the troops at Shahzehanpore [Shahjahanpur], and Bareilly had mutinied, and that a body of the Oude [Awadh] mutineers, consisting of an Infantry and Cavalry corps were marching into Futtehghur. This caused great anxiety as the 10th were known to be mutinously disposed, for they had given out that as soon as another corps arrived, they would rise, and murder all the Europeans, only sparing their own officers. That night a consultation was held, and it was considered absolutely necessary to send off the ladies and children to Cawnpore; and as boats had been secured, it was settled that a start should be made at once, as it had been before agreed that it was impossible to hold the fort, and it was at that time thought that the river was quite open. All was settled, when several gentlemen said that unless the Magistrate accompanied them, they would not leave the station.' Narrative of Gavin Jones quoted in **FSUP**, Vol V, pp 738-47. Of Balgobind 'He had throughout maintained that the arms of the British would ultimately triumph, and as our intimacy grew closer, I one day had the curiosity to ask him his reason for entertaining so favourable a view of our situation. The sagacious Brahmin, looking me full in the face and nodding his head significantly, replied: 'Listen; our countrymen have neither wisdom nor leaders competent to turn their advantages to account. Moreover,' he added, 'they are destitute of justice and truth, and have imbrued their hands in the blood of innocent women and children. Ram will never prosper their cause'. Gavin Jones, *The Story of my escape from Futtehgurh*, Cornhill Magazine, January 1865.

JOSHI, Puran Chandra An Indian historian specialising in the rebellion, to the history of which he made a valuable contribution. A Marxist he clearly perceives the rebellion as a revolt by the working classes, a view not apparently shared by **Karl Marx** himself. Included in his Symposium is a small collection of Folk Songs on 1857 which are most interesting; because 'folk-art forms have been the traditional media for approaching the masses.' *See* P.C.Joshi, ed, *Rebellion 1857, A Symposium,* Delhi, 1957.

JUBBULPORE [Jabalpur] A large town and the headquarters of the **Saugur & Narmada Territories** (latterly the Central Provinces), where the Commissioner and Agent to the Lieutenant Governor, **Major W.C.Erskine** was based. It was his decision probably that in the end saw the British determine that the **Rani of Jhansi** was 'guilty' of the murder of the Europeans at Jhansi, and which in turn made her rebellion against the British inevitable. At Jabalpur the 52nd BNI were thought likely to mutiny; at first they helped to establish peace in the neighbourhood, defeating large parties of Bundelas who were plundering the great Deccan Road, but afterwards they mutinied. The Gond Raja **Shankar Sa** and his son were executed here. For details *see* **FSUP** Vol III, and *Revolt in Central India*, Simla, 1908, p 36 et seq.

JUDEX Exactly who 'Judex' (possibly **William Muir**?) was is not certain, but he was a correspondent of the London Times, and his contributions, though wordy, do have the ring of both accuracy and honesty. Here is an extract: 'I thoroughly believe that the rebels have aimed at the extermination of our race, but that they have not, as a rule, especially studied our dishonour, that by far the greater part of the stories of dishonour and torture are pure inventions, and that the mutineers have generally, in their blind rage, made no distinction between men and woman in any way whatever. ...After having visited almost every place from which such stories could have come, I have not learnt one instance in which any one has survived to tell of injuries suffered. I believe there is not one mutilated, tortured, or, so far as I can gather, dishonoured person now alive. ...At **Delhi** for months the belief was that men, women, and children had been horribly and indiscriminately massacred, but no more. I never heard there a story of dishonour pretending to anything like authenticity. ...I *did* hear when I was on the Delhi side that the most horrible and too unmentionable atrocities had been committed at **Cawnpore [Kanpur]**. Well, I went to Cawnpore; what did I hear there from impartial and well-informed persons? Why, simply this, that the matter had been particularly inquired into, and that the result was the assurance that there had not been dishonours or prolonged torture, but that the women and children had been all together massacred and thrown into a well. 'But,' it was added, 'though it has not been here there is no doubt that it was so at Delhi and those places'. ...I do not assert that no unnecessary cruelty in the course of murder and no dishonour occurred, it would be opposed to human nature if, in this saturnalia of blood, it had been so, but I *do* express my strong belief that, under the known circumstances of the outbreak, there is rather to be remarked the absence, so far as we can discover, of the amount of female dishonour which might have been anticipated, than an excessive tendency that way. In short, so far as I have been able to learn, the object has throughout been... extermination rather than dishonour; and the distinction is important. ...A Caesar may be murdered; it is only a slave-driver who is tortured and deliberately dishonoured by his victims.' *See* London *Times*. 29 January, 1858, p 12.

'Our Judge', by Capt.G.F.Atkinson

JUDGES 'That is the judge of the station; a very good fellow; all judges are rather slow coaches, you know. They do the criminal business, and it is not much matter if they make mistakes, as they don't meddle with Europeans. When they can do nothing else with a fellow, in the civil service, they make him a judge.' This may have been Stewart describing **Sherer**. W.H. Russell, *My Diary in India,* London, 1860, Vol I, p 182.

JULLUNDER   A large town and cantonment in the Punjab, some seventy-five miles east of **Lahore**. It, and the neighbouring station of **Phillour**, were held at the beginning of June 1857 by H.M.'s 8th Regiment with some artillery, and a brigade from the Bengal army consisting of the 6th Light Cavalry, the 36th BNI and the 61st BNI. Incendiary fires, often the prelude to mutiny, had occurred and the decision was taken to disarm the sepoys, but the news must have reached them in advance and they mutinied on the night of 7 June, taking the road to Philhour on the Sutlej river some thirty miles away where they persuaded the 3rd BNI to join them and all four regiments (with the exception of a proportion who remained loyal and others who were on furlough) eventually joined the rebels in **Delhi,** where they are said to have distinguished themselves. Their initial mutiny was notable for the discipline with which they behaved, neither plundering nor offering violence. Brigadier Johnstone, in command at Jullunder, was severely criticised for delaying his pursuit of the mutineers but in the House of Commons was eventually 'fully and honourably acquitted of all accusations brought against him'. *See* **Ball** and **Martin**, F.Cooper, *Crisis in the Punjab,* Lahore, 1858. *See also Letter from Umballah, August 1857,* in the Times, 26 October 1857.

JUNG BAHADUR   Soi-disant MAHARAJA OF NEPAL. The fact is that Maharajah Jung Bahadur of Nepal played a double - if not treble - game from the start: None of the British - military or civilian -

who came into contact with him trusted him from the beginning, and with good cause. Nor could the rebels regard him as an ally or even a neutral: his lust for female flesh and his cupidity were the only predictable and unchanging aspects of his character. An examination of his motives/achievements in intervening in Oude [Awadh] in 1858 (to the disgust of Sir Colin Campbell the Commander-in-Chief who did not want him) reveal only an enormous appetite for loot, and an equally strong desire to share the glory of the winning side. ....And yet, if he had stood, firm and trustworthy, for the rebels, the British would have found their task very much harder. The Governor General, Earl Canning, knew this, even if the army men chose not to. But nothing could apparently be more specific than Jung Bahadur's reply to the plea of **Birjis Qadr** for help, after the defeat of the rebels in Lucknow, and when the British appeared to be winning the fight to 'pacify' Awadh. The following is the letter from H.E.Maharaja Jung Bahadur to Birjis Qadar Mirza Ramzan Ali Saheb Bahadur, of Lucknow, dated 17 July 1858: ' Your letter of the 19 May to the address of His Highness the Maharaja of Nepaul, and that of 11 May to my address , have reached their respective destinations, and their contents are fully understood. In it is written that the British are bent on the destruction of the society, religion, and faith, of both Hindoos and Mohammedans. Be it known that for upwards of a century the British have reigned in Hindostan, but up to the present moment, neither the Hindoos nor the Mohammedans have ever complained that their religion has been interfered with. *As the Hindoos and Mohammedans have been guilty of ingratitude and perfidy, neither the Nepaul government nor I can side with them.* Since the stars of faith and integrity, sincerity in words as well as acts, and the wisdom and comprehension of the British, are shining as bright as the sun in every quarter of the globe, be assured that my government will never disunite itself from the friendship of the exalted British Government, or be instigated to join with any monarch against it, be he as high as heaven. What grounds can we have for connecting ourselves with the Hindoos and Mohammedans, of Hindostan? Be it also known, that had I in any way been inclined to cultivate the friendship and intimacy of the Hindoo and Mohammedan tribes, should I have massacred nearly 5,000 or 6,000 of them on my way to Lucknow? Now as you have sent me a friendly letter, let me persuade you, that if any person, Hindoo or Mohammedan, who has not murdered a British lady, or child, goes immediately to Mr Montgomery, the Chief Commissioner of Lucknow, and surrenders his arms and makes submission, he will be permitted to retain his honour, and his crimes will be pardoned. If you still be inclined to make war on the British, no rajah or king in the world will give you an asylum, and death will be the end of it. I have written whatever has come into my plain mind and it will be proper and better for you to act in accordance with what I have said.' For Jung Bahadur's letter as above *See Foreign Secret Consultations,* 27 August 1858, Nos 97-108. Col Ramsay, the British Resident in Nepal had no illusions about Jung Bahadur. In a letter addressed to C.**Beadon,** Calcutta, dated 3 September 1860, towards the end of the final phase of the rebellion he wrote: 'Jung Bahadur from the very first had made up his mind to shield them (Nana, Bala and other relations because they were Brahmins)...had declared long beforehand that the event (death of Nana in Terai) would take place and seemed anxious that it should be taken for granted.' He quoted Jung Bahadur being surprisingly frank about the 'mutiny': 'My troops have not been shocked (*cf* Kanpur) and have nothing to avenge.' In the last few months of the struggle in 1859 there was more positive support for British requests that the rebels be turned out of Nepal. 21 December 1859 Brigadier Holdich reported complete clearance of the rebels in Nepal. 3,000 handed over by the Nepalese plus a few voluntarily surrendered. 1,000-1,500 made off to their homes. Only 300 remained - became cultivators. Begum **Hazrat Mahal**, her son, families of Nana Sahib including his wife Kashi Bai (*who was Jung Bahadur's mistress for many years*) got permanent asylum in Kathmandu. Lord Clyde, (**Sir Colin Campbell**, Commander-in-Chief) particularly disliked Jung Bahadur, whom he met only once, in camp at Lucknow before the final assault in March 1858. In a letter to Sir Patrick Grant dated 12 March 1859 he writes: 'You may recollect that when Jung Bahadoor requested that our troops should not cross the border into the Nepalese territory in pursuit of the Begum (of Oude) and the troops which had accompanied her, he voluntarily engaged to cause her and her troops and followers to leave the Nepal territory within a given time, and pointed out the ghat on the Gunduck at which he wished a detachment of our troops to be in readiness to receive the troops and followers of the Begum as they passed the river. The Resident's letter will make known to you how completely he has departed from the promises he made with respect to the troops of the Begum being disarmed and handed over to our troops preparatory to return to their homes. He has allowed them to prolong their stay in Nepal until the commencement of the hot season, and then permits their departure with their arms, and allows them to take refuge in the Teraie, in the hope evidently, that Government may be driven into giving a pardon to the rebels rather than keep our troops in the field on his frontier during the hot season. *We have been far too civil to Mr Jung Bahadoor. ....I fear the climate. It makes me very savage that by the trickery and deceit of this fellow, Jung Bahadoor, the troops may be kept out longer than we anticipated.* Most sincerely yours'.

One junior officer however was very grateful to Jung Bahadur: in a letter dated Camp Sultanpur and addressed to the Military Secretary, one Lieutenant R.H.Sankey of the Madras Engineers wrote: 'In reporting for the information of the Brigadier General the completion of the bridge over the Goomtee yesterday (19 March 1858) evening, I have much pleasure in ascribing the main portion of the design to H.H. the Maharajah Jung Bahadoor who from the outset took the most lively interest in the whole proceedings and the execution of the work, to the soldiers of the Goorkha Army whose cheerful bearing and ready adoption of the rough material at hand, was very remarkable.....His suggestion was to erect with rough piles a series of piers at intervals across the shallow portion of the river, and to place what boats we could muster in the deep channel; there were to be thus eight piers on the left portion of the stream and one abutment on the right, the boats being arranged at intervals in the centre....in consequence of the arrangements made by the Maharajah and his frequent presence on the spot encouraging the soldiers, a stream of some sixty yards in breadth was successfully spanned, *in thirteen hours!*' Although, in the event, Jung Bahadur did grant asylum to both the **Begum of Oude** and **Birjis Qadr**, he remained on good terms with the British. Finally: 'About the time (1861) this order (the 'Star of India') was instituted, Her Majesty conferred the Grand Cross of the Order of the Bath (an ancient *English* order) upon Jung Bahadour, greatly to the gratification of that native prince'. If **news-writers** are to be believed then the **Begum of Oude [Awadh]** spent considerable time and energy trying to persuade Jung Bahadur to join the rebel cause: 'The Begum has accordingly written to Jung Bahadoor, offering him Goruckpore, Azimgurh, Arrah, Chupra and the Province of Benares to unite with her. The whole energies and talent of Oudh are now devoted to attempt to buy over the Nepalese.' *See Foreign Political Consultations,* 30 December 1859 Supplement No 1374. *See* also Ramakant, *Indo-Nepalese Relations,* 1968. *See also* OLDFIELD, Dr HENRY AMBROSE.

JUNGA DOJE   Or, possibly, more accurately 'Ganga Das' was a Brigadier in the army of Maharaja **Jung Bahadur** which entered Oude [Awadh] in support of the British in February 1858.

JUWALLAH PERSHAUD   See JWALA PRASAD.

JVALA SAHAYA   The Nazim of **Bharatpur**. The author of a tribute to the services of the Rajputani Princess to Government during the period of the rebellion of 1857. *See* S.Jvala, *The Loyal Rajputani,* Allahabad, 1902.

JWALA PRASAD   Almost always referred to at this time as *Brigadier* Jwala Prasad, or Jewala Pershad. One of **Nana Sahib's** chief counsellors and military leaders. The Cawnpore [Kanpur] Depositions - *see* WILLIAMS - seem to link him strongly with the planning and execution of the **Massacre at Sati Chaura Ghat**. Jwala Prasad had been a soldier under **Baji Rao** drawing Rs.6. On his death, Nana Sahib made him a Risaldar. 'His influence was overweaning,' **Rao Sahib** said, albeit while trying to belittle his own role in the uprising. **FSUP**, Vol III, p 685. He was tall and slight with a long pock-marked face, long locks of black hair, and a fine nose, and spoke in a nasal whine. N.W.P. Political Department: January to June 1864 quoted in **FSUP**, Vol V, p 700. He was hanged by the British from a tree at the **Massacre Ghat**.

JWALA PRASAD   'Three men were sent from the hostile camp into our intrenchment to remain there the whole night as hostages for the Nana's good faith. One of them was the before-named Juwallah Pershaud; there is little doubt that this rogue was in possession of a perfect programme of the projected plans for the morrow. (*ie* the **Massacre at the Sati Chaura Ghat**). He was one of the Bithoor retainers, and had now become a very considerable personage, having floated on the tide of mutiny to high military command in the ranks of the rebel army. Juwallah condoled in most eloquent language with Sir Hugh Wheeler upon the privations he had undergone, and said that it was a sad affair at his time of life for the general to suffer so much; and that after he had commanded sepoy regiments for so many years, it was a shocking thing they should turn their arms against him. He, Juwallah, would take care that no harm should come to any of us on the morrow; and his companions used language of the same kind both for its obsequiousness and falsity.' *See* Mowbray Thomson, *The Story of Cawnpore*, London, 1859, p 155.

JYOTI SINGH   He was reported as 'Raja Joat Singh (Jyoti Singh), talukdar of Churda in the **Baraitch** Division, gave willing shelter to **Nana Rao** and his followers in his Fort for some time, fought against us in **Goruckpore** [Gorakhpur] in March and April 1858 and afterwards in November of same year in company with **Bala Rao**, made an incursion into Bunnee [Bani] committing great ravages'; *see* PERSONS OF NOTE, APRIL 1859.

K

KAISERBAGH This was the King's Palace built by **Wajid Ali Shah**, begun in 1848 and finished in 1850 at a cost it is said of eighty lakhs of rupees (= £800,000 at that time) including furniture and decorations.

Kaiserbagh Main Gate Photo 1858

The use of the word Kaiser is under dispute by differing authorities, some pointing out the obvious connection with 'Caesar' the Roman title said to have been adopted by the Kings of Oude [Awadh] and used by them on the royal seal, others saying that it means no more than 'yellow' in Hindi, and that that was the original colour of the palace. It was the scene of fierce fighting in 1857 and even fiercer looting in 1858 by British and Sikh soldiers after the capture of **Lucknow** by **Sir Colin Campbell**. For an excellent description and history of this and the other palaces etc of Lucknow see R.Llewellyn-Jones, *A Fatal Friendship*, Delhi, 1985; *see also Gazetteer of the Province of Oude*, 1877, Vol II.

KALINJAR A fort near Banda where a Lieutenant Remington held out against a rebel force sent against him by the **Nawab of Banda**; he was aided by a party of matchlockmen and guns furnished by the pro-British Raja of Panna. *See* **FSUP**, Vol III, and *Revolt in Central India*, Simla, 1908.

KALKA PERSHAD [Kalka Pershaud] 'I was in the employ of Mr. Thomas Greenway at the time of the outbreak at Cawnpore. In May 1857 we heard of the mutiny at Meerut; and about the end of the same month some Government elephants were sent to the treasury to remove the treasure kept there; but the sepoys on guard would not allow it to be taken away; this created suspicions regarding their intentions, after which the entrenchments were commenced, and the Nana came in from Bithoor.' *Deposition of Kalka Pershad of Cawnpore*. **Forrest**. Vol III.

KALLEE NADEE [Kali Nadi] Battle of There was a battle at Kallee Nuddi [Kali Nadi], before Fatehgarh on 2 January 1858, and a British officer afterwards wrote: '...eight guns fell into our hands. In the capture of one of these a striking incident occurred. It was coming along dragged by bullocks when the horsemen cut off its retreat. Deliberately turning his team towards where a squadron of the 9th Lancers stood drawn up, the sepoy driver slowly and composedly went up to them. As he approached a lancer rode out at a footspace to meet him. The sepoy, albeit well knowing his fate, came calmly on. The lancer reached him and thrust his lance through his body; heavily and without a groan he rolled down from his seat on the gun-carriage and died. No Roman ever met his fate with more stoical heroism'.

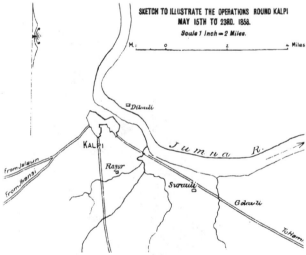

KALPI, A town on the Yamuna river south of Agra. It served more than once as a rallying point for rebel forces, particularly the **Gwalior Contingent** and remnants of the **Nana Sahib's** troops led by **Tatya Tope**. In May 1858 the rebels retired there after an unsuccessful stand at **Kunch**, and their leaders, **Rao Sahib, Tatya Tope** and the **Rani of Jhansi** seemed prepared to put up considerable resistance to defend what the Rani called their 'only arsenal', and when **Sir Hugh Rose** attacked Kalpi on 22 May, a very spirited action took place in which the rebels fought well and were ably led. The situation was afterwards described as 'critical' at one stage, but Rose captured both town and arsenal, containing fifty guns and large quantities of ammunition. For numerous references to Kalpi *see* **FSUP**, Vol III. For a detailed account of the Battle of Kalpi, *see Revolt in Central India 1857-59*, Simla 1908, p139 *et seq.*, see also Sir George William Forrest, ed., *Selections from the Letters, Dispatches and other State Papers of the Military Department of the Government of India 1857-58*, 4 vols, Calcutta, 1893-1912, Vol IV, *and* Tapti Roy, *The Politics of a Popular Uprising: Bundelkhand in 1857*, Delhi, 1994.

KAMAL-ud-Din, HAIDAR HUSAINI Author of '*Tarikh-i-Awadh* or *Qaisar-ut-Tawarikh*'. In Vol II pp 212-218 there is a detailed account of the Battle of **Chinhut** (Chinhat).

KANDOO NUDDEE [Kando Nala?] There was a battle at this place, in Oude [Awadh], on 5 March 1858 between the forces of **Maharaja Jung Bahadur** and **Mehndie Hussain**, in which the latter was defeated, largely by superior artillery fire.

KANGRA A town in the Punjab where **Montgomery** was the Commissioner. The 4th BNI was successfully disarmed on 11 July 1857 and the threat of a great revolt in the hill districts of Punjab averted.

KANKAR Kankar was a limestone mud that dried to an extraordinary toughness and was commonly used in the surfacing of walls and roads.

KANPUR (or Cawnpore) The Archaeological Survey of India is responsible for the preservation of a number of monuments in Kanpur, including 2757 Kathchahri Cemetery, 2758 Memorial Well Garden 2759 Savada Kothi and 2761 Wheeler's Entrenchment, *see* CAWNPORE.

KAPURTHALA, RAJA of Rundbir Singh, Raja of Kupoorthulla [Kaporthella, Kapurthala] played an important part in the struggle, on the British side. He was the ruler of a Punjab state that had been partly confiscated by the British on the annexation of the Jullundur Doab in 1846; he had succeeded his father in 1853, and in 1857 was 'a handsome young man of 26, with the manly bearing and address of a Seik noble, combining a general intelligence far beyond his class, and a deep sympathy with English modes of life and thought'. It was well for the British that this was so, for in May 1857 he acted decisively in support of them at **Jullundur**, furnishing them with his personal escort, and 500 men and two guns. Thereafter 'H.H. the Raja of Kupoorthulla's Contingent' served in effect as mercenaries in support of the British, and even fought in Oude[Awadh] in the Faizabad Force commanded by **General Sir Hope Grant**.

KARHAK BIJLI 'Hard Lightning'. A gun to be seen today in the Fort at Jhansi, said to have been used by the Rani of Jhansi, but this is unlikely as it was very old, and probably unserviceable, in April 1858, at the time of the siege of Jhansi.

KASHI PERSHAD Talukdar of Sissaindi south of Lucknow in Oude[Awadh]. Remained loyal to the British throughout the campaign in Oude and was rewarded after the final assault on Lucknow, in what was known as the 'Oude Proclamation', by the Governor General, issued on 14 March 1858: '...The first care of the Governor General will be to reward those who have been steadfast in their allegiance at a time when the authority of Government was partially overborne, and who have proved this by the support and assistance which they have given to British officers.'

KASHMIR GATE The gate of the city of **Delhi** nearest to the British position on the Ridge, and the obvious one by which the city was to be attacked. The assault came on 14 September 1857, and was, from the British point of view, a complete success, although once within the city walls they suffered very heavy casualties. The details of the assault on the gate are best told in the words of a report from the Chief Engineer, **Colonel Baird Smith**, to Major General **Archdale Wilson** the commander of the **Delhi Field Force**: 'The gallantry with which the explosion party, under Lieutenants Home and Salkeld, performed the desperate duty of blowing up the Cashmere gate, in broad daylight, will, I feel sure, be held to justify me in making special mention of it. The party was composed, in addition to the two officers named, of the following:- Sergeants John Smith and A.B. Carmichael, and Corporal Burgess, sappers and miners; Bugler Hawthorne, H.M.'s 52nd; fourteen native sappers and miners; ten Punjab ditto; musters covered by the fire of H.M.'s 60th Rifles.The party advanced at the double towards the Cashmere Gate. Lieutenant Home, with Sergeants Smith and Carmichael and Havildar Mahor, all the sappers leading and carrying the powder-bags, followed by Lieutenant Salkeld, Corporal Burgess, and a portion of the remainder of the party. The advanced party reached the gateway unhurt, and found that part of the drawbridge had been destroyed, but passing across the precarious footway supplied by the remaining beams they proceeded to lodge their powder-bags against the gate. The wicket was open, and through it the enemy kept up a heavy fire upon them. Sergeant Carmichael was killed wliile laying his powder-bag, Havildar Mahor being at the same time wounded. The powder being laid, the advanced party slipped down into the ditch to allow the firing party, under Lieutenant Salkeld, to perform its duty. While endeavouring to fire the charge Lieutenant Salkeld was shot through the arm and leg, and handed over the slow match to Corporal Burgess, who fell mortally wounded just as he had successfullly accomplished thc onerous duty. Havildar 'T'illuh Sing, of the Sikhs, was wounded;

and Ramloll Sepoy, of the same corps, was killed during this part of the operation. The demolition being most successful, Lieutenaut Home, happily not wounded, caused the bugler to sound the regimental call of the 52nd, as the signal for the advancing columns. Fearing that amid the noise of the assault the sound might not be heard he had the call repeated three times, when the troops advanced and carried the gateway with complete success. I feel certain that a simple statement of this devoted and glorious deed will suffice to stamp it as one of the noblest on record in military history. The perfect success contributed most materially to the brilliant result of the day, and Lieutenants Home and Salkeld, with their gallant subordinate Europeans and natives, will, I doubt not, receive the rewards which valour before the enemy so distinguished as theirs has entitled them to'. Salkeld, Home, Smith and Hawthorne were awarded the Victoria Cross. *See* **Kaye, Sen, Ball, Martin** etc *See also* A.Llewellyn, *The Siege of Delhi,* London, 1977, Harvey Greathed, *Letters written during the Siege of Delhi,* London 1858.

Bugler Hawthorne at the Kashmir Gate

The Kashmir Gate viewed in 1858

KASHMIR GATE With regard to the attack on the Kashmir Gate at Delhi on 14 September 1857 incredible stories circulated, almost within hours, of momentous events, and some continued to be put forward, embellished, for many months afterwards. Perhaps in the absence of the kind of detailed authentic news coverage we are used to today, rumour was the substitute. But who made them up in the first place? One of the oddest was that 'a naked white man was fastened to the (Kashmir) gate before it was fired. Also said that a naked white man was attached to the gun loaded with grape, just inside the gate, whose gunners were killed by the same explosion that blew in the gate.' Both stories were later denied as untrue by the third man through the gate. (Ensign Wilberforce). He continues the tale: 'We were without food since dinner the night before. Reached Church (St James) close to Cashmere (sic) gate at 11.30 a.m. Much rejoiced to see our mess servants with plenty to eat and drink. More thirsty than hungry so I and my companion ensign set to work to quench our thirst. Bottle of soda water in one hand and a long tumbler in the other, into which my companion poured some brandy. Too generous so I dreaded drinking it especially on an empty stomach. Not liking to waste it we looked around and saw a group of officers on the steps of the church, (must have been St James) engaged in animated conversation. Among them was an old man who looked as though a good peg would do him good. Drawing near to the group in order to offer the peg to the old officer we heard our Colonel say 'All I can say is that I won't retire, but will hold the walls with my regiment'. I then offered our peg to the old officer whom we afterwards knew to be General **Wilson** (the Commander of the British force, and often accused of excessive caution) : he accepted it, drank it off, and after a few minutes we heard him say *'You are quite right - to retire would be to court disaster; we will stay where we are.'* On such little matters great events depend: if English troops had left Delhi in all probability no one would be left to tell the tale.'

KAVANAGH, Thomas Henry A civilian in the uncovenanted service who was awarded the Victoria Cross for an exploit in **Lucknow**. He left the Residency in disguise and contacted the relieving force led by **Sir Colin Campbell** and assisted in guiding that force through the streets of Lucknow which he knew well. Also rewarded by government by being promoted. *See **Ball, Sen, Kaye, Malleson** etc* and Henry Kavanagh, *How I won the Victoria Cross,* London, 1860.

KAYE, Sir John William (1814-76) A much respected military historian whose work had the merit of being authentic and contemporary; educated at Eton and **Addiscombe**; Bengal Artillery 1832-41; joined staff of *Bengal Harkaru* April 1841; established the *Calcutta Review* in 1844. Entered East India Company Home Service 1856; Secretary of the India Office Political & Secret department until 1874. KCSI 1871. His major work is *History of the*

Sepoy War, 3 vols 1864-76, which, being incomplete on his death was continued by Colonel Malleson and for many is the standard and most respectable work upon the subject of the Mutiny and Rebellion. *See* Bibliography, *and* **DNB., DIB., Encyc.Britt.**

KENYON PAPERS Held in the **Cambridge Centre of South Asian Studies** and including valuable source material as follows: A large collection of long and informative letters from Captain Jonathan Fowler 8th Cavalry, Madras army. The 8th Cavalry was disbanded for unsatisfactory conduct, which appears to have been more connected with pay than with the mutiny in the Bengal army. Very ashamed that his regiment should be the only one in the Madras army proving disloyal at this juncture. Speaks of *Muslim letters intercepted urging Madras army to make a disturbance.* Also a report among the Muslim population that *Lord Elphinstone is to be King of Delhi.* Serves eventually in the north, at **Gazeepore** (sic) and **Buxhar** and **Jugdespore**. Speaks of total mismanagement by **Brigadier Douglas**.

KHAIRABAD The town in Oude [Awadh] where Harprashad was the chukledar, and where he successfully defied the British government for some months. *See* **FSUP** Vol II *and* **Rudrangshu Mukherjee**, *Awadh in Revolt 1857-58*, Delhi, 1984.

KHAKI RISALA The corps of volunteer horse raised in **Meerut** in early June 1857 by Robert Dunlop a civilian and Williams the police officer, in which clerks and refugee officers, merchants etc served, and which soon brought the district around Meerut into fair order. *See* Robert H.W.Dunlop, *Service and Adventure with the Khaki Risala*, London, 1858, reprint Allahabad, 1974.

KHALK-i-KHUDA Khalk-i-Khuda, Mulk-i-Padshah, Hukm-i-Sipah. ('God rules the world, the Emperor is on the throne, the sepoys give the orders'). Changed by the very same town-crier in the course of one day (as a British force under General Havelock marched in), to: Khalk-i-Khuda, Mulk-i-Kampani Bahadur, Hukm-i-Sahiban alishan.('God rules the world, the East India Company is on the throne, the Sahibs are giving the orders'). Kanpur, August 1857.

KHAN ALI KHAN A resident of **Shahjahanpur**, was one of the leaders in the 1st **Chinhat** affair, took a most prominent part throughout the rebellion, was wounded at **Lucknow**; *see* PERSONS OF NOTE, APRIL 1859

KHAN BAHADUR KHAN The self-styled 'viceroy of Rohilkhand' with headquarters at Bareilly. He was described as a 'Monster of cruelty', and does not seem to have been very popular with anyone. An implacable enemy of the British, and to some the representative of the most worthy of the patriotic rebels. To others perhaps typical of one species of the minor rebel leaders - no better than warlords really - who made use of the chaotic conditions obtaining in the early period of the struggle to establish themselves in an area, often by a reign of terror. It was said that the British Commander-in-Chief, **Campbell's**, victory at **Bareilly** in May 1858 was welcomed by Khan Bahadur Khan's co--religionists and the Hindus as a happy release. Eventually hanged by the British, declaring that he was sorry not to have killed more of them. 'He is, it is true, a pensioner of ours, and retired native judge, or sudder ameen; but he is also a descendant and representative of Hafiz Hushmut Khan, the chief whom we slew in the battle which led to the overthrow of his rule in Rohilcund. We conquered the province for the Nuwab Vizier of Oude, and now we have swallowed Oude and the kingdom we gave the Nawab. When he got an opportunity he grasped at what he believed to be his own, and he did so in a way which no one can approve of, for his ways were treacherous and bloody. According to the lights of his faith and civilization, the acts of which he has been guilty are not much worse than our own.' He had been appointed judge only because the British could not permit him 'to wander about before the people of India in a state of destitution.' W.H.Russell, *My Diary in India,* Vol I, p 407. 'It is true that I killed the Europeans; for this purpose I was born.' said Khan Bahadur Khan before he was executed' ...I have killed hundreds of English dogs, it was a noble act, and I triumph in having done it.' He was hanged at 7 in the morning and cut down at 8. 'Notwithstanding a large concourse of spectators, Europeans, Hindoos and Mahomedans, no disturbance or indecorum of any sort took place,' nevertheless he was buried under the jail to prevent his tomb from becoming a Muslim shrine. *See Trial of Khan Bahadur Khan* in **FSUP,** Vol V, p 615.

Nicholson attacks the Sialkot Mutineers

KHANSAMAN, the General's The khansaman, or cook, of Brigadier-General Frederick **Brind**, station commander at **Sialkot** is reported to have played a most remarkable part in the mutiny at that station. It is said that he was one of the principal figures and assisted in organizing the mutiny in the 9th Bengal Cavalry. He also arranged that the officers in the station should be left defenceless by getting their servants to steal their swords; he tampered with Brind's own pistols, removing the bullets overnight. After the mutiny had broken out he took charge of the Station Gun (a twelve-pounder) and caused it to accompany the mutineers' column.. He personally worked it 'to the last' on the island in the middle of the Ravi near **Trimmu Ghaut** where the mutineers made their last stand against **Nicholson's** column. The gun itself had an unfortunate history subsequently: it was brought back to Sialkot in triumph and resumed its function as the station gun (fired as a timekeeper), but apparently children playing around it continually filled the barrel with brickbats and stones, and caused the death of one European and the loss of a leg to two Indians. *See* G.Rich, *The Mutiny in Sialkot,* Sialkot, 1924.

KHODA BUKSH This was the name of General **Wheeler's khidmatgar,** who was with him in the **Entrenchment** and survived the massacre at the **Satichaura Ghat.** *see also* KHODA BUX.

KHODA BUX 'On the 7th of May 1857, after the target practice at the Umballa [Ambala] Depot, I came and joined my regiment at Cawnpore; I heard from people (outsiders) that the cavalry horses were to be shot, that Europeans were coming from England, and that 300 horses and 300 swords were to be taken from the cavalry regiment at Cawnpore, and given over to them. The men of the regiment were frightened because the Europeans had come, and their arms were to be given over to them. Every one talked about it. I heard from Khan Mohomed, sepoy, that all the Native force said that on the 5th of June 1857 all their arms were to be taken away from them, and they would all be called into the fort under the pretence of getting their pay; that a mine was made in the road to blow them all up: that the moment they were all assembled the mine would be fired. I immediately reported this to the Adjutant, who said it was all a lie, and told me not to believe it.' Khoda Bux or Buksh was a Jemadar in the 53rd BNI. After the suppression of the rebellion he became adjutant of Military Police, *Deposition of Khoda Buksh see* **Forrest**, Vol 3.

KHODA BUX, GENERAL Reported as Khoda Bux (Khuda Bakhsh), zemindar of Dadroe near Mohumdabad in the Daryabad District, he took an active part in the rebellion, throughout **was a Commander against the British at Nawab Ganj Bara Banki in June 1858 and at Nanpara in January1859;** *See* PERSONS OF NOTE, APRIL 1859.

KHUDAGANJ A town on the right bank of the Ganges between Cawnpore [Kanpur] and **Fatehgarh** where a battle took place on 2 January 1858 between the British force led by **Sir Colin Campbell** and some 5,000 rebels under the nominal command of the **Nawab of Farrukhabad**. The defeat of the latter led to the re-occupation of **Fatehgarh** by the British. The battle is notable from the British point of view for the award of the Victoria Cross here to Frederick Roberts, later to be Field Marshal Lord Roberts of Kandahar, and Commander-in-Chief. For details of the battle *See* Lord Roberts, *Forty-One Years in India,* London, 1921.

KHURRUK BAHADUR Was a leading commander (General) of the Nepalese force that entered Oude[Awadh] in February 1858 under **Maharaja Jung Bahadur**. He is reported to have fought well against the *nazim* **Mehndie Husain**. He 'displayed great coolness and a generous confidence in my advice' (**Captain Plowden**). His division of the army was considerable: 13 guns, 7 regiments of infantry; total 3,800 men. *See* **FSUP** Vol II.

Khurshid Manzil Photo 1858

KHURSHID MANZIL Or 'Palace of the Sun' situated near the Secundra Bagh and Residency in Lucknow. Prior to the outbreak of the rebellion it was the Officers' Mess House of H.M.'s 32nd Regiment then stationed in Lucknow, and it is as the '32nd Mess House', that it is generally known. By the gateway to its drive is a small pillar with a tablet commemorating the fact that it was at this spot that **Sir Colin Campbell** met **Outram** and **Havelock** on his relief of the **Lucknow Residency** in November 1857. Major **F.C.Maude** has added a note to his copy of a photograph of it taken in 1858, 'Sir Colin gave it a ten hours' bombardment *after the defenders had left!'* (*ie* in March 1858). The building is now the home of a Girls' School, the La Martinière Girls' College, the sister school of the Martinière College. *See WOLSELEY, FIELD MARSHAL.* Courtesy also INTACH.

KIRWI A town in **Bundelkhand**, close to **Banda**, the headquarters of Narayan Rao and Madhu or Madhava Rao. After the **Battle of Banda** General **Whitlock** marched to Kirwi on 1 June 1858. Strictly speaking if he had done what he was told he would have been assisting **Sir Hugh Rose** in his campaign against **Jhansi** and **Kalpi.** The two Raos (Madhu was a minor aged only nine) were believed by the British to be 'of the family' of the **Nana Sahib**, calling themselves 'Peshwa', and appearing to be wholeheartedly on the rebels' side; in this they were clearly influenced by Narayan Rao's *vakil* and mentor, one Radha Govind. But they were playing a double game for on the approach of Whitlock they came out and surrendered unconditionally, and their army estimated to number 15,000 melted away. Kirwi was occupied and immense booty was requisitioned by the British, afterwards awarded as prize money to the troops: 42 guns, immense quantities of shot and powder, 2,000 stand of arms, and clothing and equipment of many sepoys of mutinied regiments which showed that the Raos had recently entertained them in Kirwi. Immense wealth in specie and jewels was also found, valued at upwards of a *crore* of rupees. *See The Revolt in Central India,* Simla, 1908, pp176 and numerous entries under **Banda** and Narain Rao in **FSUP,** Vol III.

KIRWI, ACTION at   The **Battle of Banda** by no means completed the pacification of the district. On 13 August 1858 Brigadier T.D.Carpenter commanding the 1st Infantry Brigade, Saugur Field Division was faced by a rebel force of over 2,500 men including 200 sepoys in the hills near Kirwi. The rebels withdrew after losing about 100 men killed but remained as a cohesive force. *See The Revolt in Central India,* Simla, 1908, p 182.

KISHEN DIXT   In 1849 a prisoner at Bithur named Kishen Dixt sent a petition to the Commissioner in which he accused Nana Sahib of conspiring against the British. He claimed that he had been imprisoned to shut him up, and produced a document purporting to be a letter from Nana Sahib to Golab Singh, the Maharaja of Jammu and Kashmir. But the Commissioner declared the letter a forgery and supported Kishen Dixt's imprisonment, though he hoped that 'no vindictive feelings toward the prisoner should lead to his harsh or improper treatment', *see* **FSUP,** Vol I, p 12.

KOCHAK SULTAN, PRINCE   Reported as 'Prince Kochak Sultan, one of the sons of the rebel Emperor Ubdool Muzzuffer Sarajoodeen **Buhadoor Shah** of Delhie', came to Lucknow after the fall of Delhi; *See* PERSONS OF NOTE, APRIL 1859.

KOER SINGH *See* KUNWAR SINGH.

KOI HAI   Literally 'Is there any one?' - traditionally among the British the way to call for a drink, but also the name applied to those who frequently did the calling - such officers were known as 'Kwy Hies' right up to 1947. The soldiers of the Bengal army were also called 'Qui-hyes', after their officers' custom of shouting 'Koi hai?' to summon their servants. By the 1850s it had become the largest of the three armies, with 74 regiments of Native Infantry. (Madras had 52, Bombay 29.) ' Their officers were the aristocrats of the entire Company army. They came from better families with larger fortunes, descended into deeper debt, obtained more staff appointments, lived better in more extravagant cantonments than their fellow officers in the other Presidencies. But the distinctions were aesthetic and the advantages social: the 'Ducks' of Bombay and the 'Mulls' of Madras proved braver, hardier and far more loyal. And the officers of all three Presidencies felt inferior to Royal officers of the Queen's Regiments.' *See* Philip Mason. *A Matter of Honour.* Jonathan Cape, London, 1974. p 187-8.

Kolapur

KOLAPUR   This was the scene of the only mutiny to take place in the Bombay Army. Kolapur was a town and state south of Satara and north of Belgaum. On 31 July 1857 a part (140 men) of the 27th Bombay N.I. mutinied, killed three young officers, looted the bazar and the regimental quarter-guard, but were overpowered by a small force of troopers of the Southern Mahratta Horse from Satara, led by Lieutenant Kerr. The regiment was disarmed, 63 sepoys executed, 66 transported, 18 imprisoned and 14 acquitted. The mutiny may well have been linked to an attempt to restore the House of Sivaji at Satara. For a full account *see* R.Montgomery Martin, *The Indian Empire*, Vol II.

KONI PASS   Near Katangi a few miles northeast of **Jabalpur**. There was a small but decisive battle here on 26 December 1857 when the British force (consisting entirely of native troops from the Madras army, the Hyderabad Contingent and the 33rd BNI) defeated approximately 1,000 rebels. Details *see Revolt in Central India*, Simla, 1908, p 48.

KOOKRAIL   A small stream flowing in to the left bank of the River Gomti just below the city of Lucknow.

KOOLWUNT SINGH   Raja of Pudnaha. Remained loyal to the British throughout the campaign in

Oude[Awadh] and was rewarded after the final assault on Lucknow, in what was known as the 'Oude Proclamation', by the Governor General, issued on 14 March 1858.

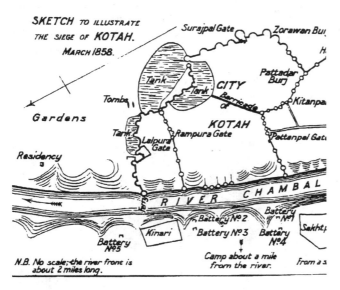

Kotah

KOTAH [KOTA] CONTINGENT Like most of the other **Contingent Forces** this one was recruited from the same areas and the same castes as the Bengal army, with which it therefore naturally sympathised. It marched from Deoli for **Agra** on 19 May 1857 and did good service for the British, in the **Muttra** district until 4 July when it mutinied. It consisted of 700 men in all, cavalry, infantry and six guns, and had apparently been quite happy in collecting revenue for the authorities, burning disaffected villages and hanging mutineers and rebels. But when it seemed likely that they would be used against the **Neemuch Brigade** they clearly decided that was enough and mutinied - all that is but a detachment of forty men under a subadar who remained loyal. They joined the Neemuch men and the **Malwa Contingent** who had mutinied at **Mehidpur** and on 5 July fought the battle of **Sussiah**[Sasia] against the British under Brigadier **Polwhele,** and drove them back into **Agra Fort**. After pausing for some ten days with the Neemuch men hoping for the co-operation of the **Gwalior Contingent**, they marched off northwards to Delhi on 18 July. For an excellent account of the Battle of Sussiah see R.Montgomery Martin, *The Indian Empire,* Vol II, p 360.

KOTAH [KOTA] REBELS This is the name given to the personal troops/levies of the Maharaja of Kotah [Kota], to distinguish them from the **Kotah Contingent**. It was a formidable force consisting of four regiments of infantry, ten troops of cavalry, some 300 artillerymen, and 3,500 armed police. The Maharaja warned the Political Agent, **Major C Burton**, that he could no longer rely on these troops, and when Burton returned to Kota in October 1857 he and his two sons were murdered. The troops were led, in their mutiny, by two leaders viz Jai Dyal, a former **vakil** of the Maharaja, and Makrab Khan, a risaldar and 'a man of character and decision', but the troops appointed the former as their commander-in-chief, see *The Revolt in Central India,* Simla, 1908, see also Iltudus T.Pritchard, *The Mutinies in Rajpootana, being a personal narrative of the mutiny at Nusseerabad,* London, 1864.

KOTAH-KI-SERAI This place is close (about twenty miles) to Gwalior and Morar in Central India and was the scene of a small but highly significant action on 17 June 1858, for it was here that the **Rani of Jhansi**, dressed as a cavalry soldier, was killed by a British Hussar (who had no idea what he had done). In her death the rebels lost their bravest and best military leader. For details of this action, see *The Revolt in Central India,* Simla, 1908, p 155 *et seq.*.

KOTHI NUR BAKSH A palace in Lucknow city owned by the son of King Mohamed Ali Shah, one Mirza Rafi-ur-Shah; he was turned out of the palace by **Havelock** who during the revolt of 1857 planned his way to the **Residency** from the top of this building. Now used as the office and residence of the Deputy Commissioner. Courtesy INTACH.

KOTRA The scene of a small but significant action in early May 1858. Kotra is a small town on the river Betwa not far from **Jhansi** in Central India. In May 1858 the British attempted to prevent the Rajas of **Banpur** and **Shahgarh** from doubling back across the Betwa and heading southwards *ie* behind British lines. The two rajas succeeded in driving back the forces of the pro-British Raja of Gursarai and fled precipitately southwards, crossing at a ford, and aided by carriage and supplies furnished them by the Raja of Jigni. For further details *see The Revolt in Central India,* Simla, 1908 , *and* Thomas Lowe, *Central India during the Rebellion of 1857-58,* London, 1860, *and* **FSUP** Vol III.

KOTWAL Usually translated as chief of police of a city, but often his functions exceeded that of a policeman, and he might be compared more accurately with a Mayor or Governor. He had a multitude of tasks of which the maintenance of law and order was only one. In times of emergency (particularly in Delhi May to September 1857) he would be used as a kind of chief procurement officer, obtaining necessary supplies by requisition, if purchase was impossible. 'Many of the papers are directives to the wretched Kotwal. He must have had a miserable life. He was expected to be a jack-of-all-trades, a supplier of anything and everything : in all probability he was rarely paid for what he provided as money was so short. I began a list of the things demanded of him, and it is formidable indeed:- he was called on to remove dead bodies - human, horses, camels etc - to send coolies, carpenters, blacksmiths, cobblers hither and thither. As on 8 August when 'all available water-carriers are to be sent at once to put out the fire in the Gunpowder factory': we can imagine that was not a popular assignment! He had to purchase and supply everything from sulphur to charpoys, fuel to spades and pulse. He was ordered to imprison 13 men accused of supplying meat to the English. We do not hear what happened to them. Sometimes the papers border on the banal. 3 July, Bundle 111, Kotwal to the King : 'Shansir Khan, Jemadar Chandni Chowk Thanah, deposes that one Dallu shot himself by accident. The deceased's father, Makhdum Bakhsh, has also made a similar statement. Natthu, a sweeper, who was examined, has said the same thing. Adds that the deceased was of unsound mind.' One wonders what Bahadur Shah made of all that! Practicalities are included. On 22 June, Paper 148, C in C to Kotwal, directs him to notify the people that no one should stay in dilapidated houses as the big gun at Salimghar is about to be fired. The same day he is directed to send without delay 10 whetters to the 6th Cavalry outside the Turkoman Gate. If you were required to find '10 whetters' in a hurry, would you manage it? On one day alone the Kotwal was required to produce hemp, carts, sweepers, provisions, ghee, flour, molasses, sugar, thatchers, hessian, oil and...torch bearers.' *See* **CM,** Chapter 10. **PJOT**

KULOO His evidence supports the theory (*see* BIBIGURH) that the British women and children were not murdered in the Bibigurh solely by the action of five hired executioners/butchers: the sepoys of the 6th BNI who were providing the guard over the women may well have fired into the prisoners as well as at the ceiling..'On the next day, leaving this place, I reached about 12 o'clock the Customs Ghat, on the banks of the Ganges at Cawnpore, and was about getting into a boat, when the man at the ghat asked for four pice, and would not allow me to get on the boat without paying this, but turned me away. I there met a cloth merchant, whose name I do not know, who had paid the fee at the ghat and received a stamp on his hand as a pass to cross over on the boat. By begging hard, I got this man to give me an impression on my arm, from the stamp-mark on his hand; by which means I crossed over and arrived about 4 o'clock near the Assembly Rooms, where I saw a crowd of about 2,000 persons, whom I questioned regarding their assemblage. They informed me that a number of ladies and children were confined in the bungalow, and were to be put to death on that day. Outside the compound of this bungalow, to the south was a Neem tree, under which I saw a person seated wearing a *pugree* covered with gold lace, who was represented to be the Nana Sahib; he was surrounded by a great number of sepoys and troopers as well as other spectators. I was told that the Nana had given orders to put to death all the ladies and children confined in the bungalow, and I heard the sepoys warning the spectators to move out of the way, so as to be clear of the bullets, as they were going to fire. On hearing this the crowd fell back, and I did the same; then I saw about 25 sepoys advance to the doors of the room which contained the ladies and children, who, firing a volley into it, retired, when another party advanced and did the same. ...I did not see any person with swords on that day...I was at this time about 50 paces from the spot and could hear the cries of the inmates of the bungalow, after the discharge of each volley. After this I left the place, and went away to my house in the Buxee Khanah, Mohulla, which was within musket range of the spot. I left and heard firing till candle light.' *See Deposition of Kuloo.* **Williams**.

KUNCH The Battle of Kunch. 7 May 1858 Among the British troops there were on this day 46 cases of sunstroke, 14 of them fatal. An officer who was present wrote:- 'The heat at day break was intense, and the mirage most remarkable. The whole of the surrounding country was dried up and covered with light brown soil, and perfectly flat, yet it appeared one beautiful lake of water, and the few trees assumed the appearance of gigantic height; and when Major Orr's force approached, so distorted was it that we could not tell whether it was friend or foe. The horses appeared twenty feet high, and riders in proportion, and the heated air ascending made them tremulous and crooked'.

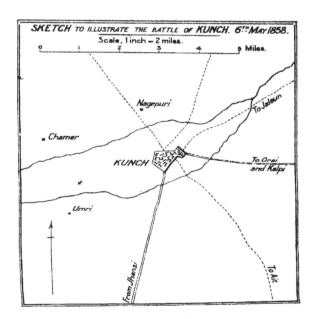

KUNCH, BATTLE of This took place on 7 May 1858, a day of excessive heat. Before any fighting took place the British suffered 46 cases of sunstroke of which 14 proved fatal. **Sir Hugh Rose** was himself incapacitated for some time. Kunch lies on the **Jhansi - Kalpi** road, and was selected by the rebels - probably by the **Rani of Jhansi** - as being the best spot to confront the advance of the British under Rose who were advancing on the great rebel stronghold of Kalpi. She, and Tatya Tope, took the field with a large army made up of sepoy regiments from the Bengal army, the **Gwalior Contingent**, the cavalry from **Kotah**, the levies of numerous rajas in revolt against the British, and the remains of the Jhansi garrison. A very severe fight developed in which the rebel infantry fought extremely well and with commendable discipline: *'in this action nothing could have been more praiseworthy than the valour displayed by the sepoys of the late Bengal Army, and nothing more disgraceful than the behaviour of the cavalry, who, in every fight I saw, distinguished themselves signally by cowardice!'* (by an officer present). The rebels were eventually beaten back and the main body retreated on the Orai road. Their defeat led to great disheartenment, also mistrust among the various factions of the rebels. It is said that **Tatya Tope** in particular was accused of abandoning the field in panic, in much the same way that he had done at the **Betwa**. For fuller details of this action in which some 600 rebels and about 20 British were killed or died of sunstroke, *see The Revolt in Central India,* Simla, 1908 pp124-129, *and* Thomas Lowe, *Central India during the Rebellion of 1857-58,* London, 1860, *and* **FSUP**, Vol III.

KUNHYE PERSHAD
'When the news of the Meerut mutiny was received, there was much excitement amongst the troops at Cawnpore [Kanpur], and reports were rife that they would follow the example of the troops at Meerut. I was on bad terms with the sowars of the 2nd Cavalry, as a year previous to the mutiny, four of the troopers had been imprisoned on a complaint made by me. I therefore feared their vengeance. I knew that many of the troopers frequented the house of a prostitute named **Azeezun,** and bought over her servant, Emam Bux. She informed me that the sowars of the 2nd Cavalry were plotting with the Nana, and that a mutiny had taken place (sic) between the parties. The facts of the case are these: the Nana had in his employ two sowars, one named Raheem Khan, resident of Bishenpore near Bithoor, and the other Muddud Ali, of Banda. The latter was discharged and commenced business as a horse dealer, but still used to visit the Nana in the way of business.' *Deposition of Kunhye Pershad.* **Williams**.

KUNHYE PERSHAD
'When the Nana was called into Cawnpore from Bithoor, these two sowars [Raheem Khan and Muddud Ali] were employed to tamper with the 2nd Cavalry; they used frequently to visit Subadar Sheeba Singh and a trooper named Shumsoodeen Khan.' *Deposition of Kunhye Pershad.* **Forrest**, Vol III

KUNWAR or KOER SINGH
A man of great qualities, respected as much by his enemies as by his friends and followers. A few more men like him and the British would not have survived the 'mutiny' of 1857-9. Because his estate was at **Jagdishpur** in Bihar, Raja Kunwar Singh is often thought only to have operated against the British in that area. This is not so, he joined the main body of rebels, and, despite his age, was indefatigable in covering great distances. He was for example in active cooperation with **Tatya Tope** at Cawnpore [Kanpur]: 'Previous to the arrival of the Gwalior mutineers at Jalaun, Kooer Singh of Jugdespore, and the 40th Native Infantry, came to Calpee via Banda on 19 October (1857). They had communication with the Gwalior mutineers, and on 3 November came in and coalesced with Kooer Singh and the 40th Native Infantry, and marched to attack Cawnpore a short time after.' And of course they there won the first battle (against **General Windham**) and caused Colin Campbell to hurry back from Lucknow to save Cawnpore [Kanpur] from being overrun - 'fine old Rajput noble and a gallant and respected gentleman, unwillingly drawn into the enemy's camp.' It is unfortunate that details of rebel successes are rarely available: government reports and despatches did not omit references to such reverses, but did not go into detail. Thus: 'In this battle (Atraulia, March 1858) the head of the rebels was Kunwar Singh with his 5 or 6 thousand men. *The Government troops by chance lost the battle and came to Koelsa. Kunwar Singh purued and followed them.'* None the less the British were generous in their praise for the old man 'Indeed they (the rebels under Kunwar Singh) would have effected the passage (of the Ganges) without any suffering, but for the timely presence of the 'Megna' gunboat, whose well-directed guns caused no small loss among the fugitives, and *above all inflicted on Koour Singh that wound which speedily rid us of the only foeman worthy of our steel, and deprived the enemy of the only leader who has displayed throughout the rebellion either skill or courage'.* Praise and respect for Kunwar Singh were mixed with a guilty conscience at the way Government had treated him (*cf* the case of the **Rani of Jhansi**). A British sergeant gives some detail on Kunwar Singh's death.' Served in the final conquest of Lucknow and Oude, was present at the death of Hodson. Was at the Martiniere. 28 March 1858 ordered to relieve the 37th Regt, besieged for some months at Azimgarh. Reached there 15 April. Ordered to fire into the town for three hours and drove the enemy out.' His troop and the cavalry were ordered to pursue them and 'we followed them for four or five miles before we came up with them. We fired on them with grape and the cavalry charged them, but they fought well, always forming square when they saw the cavalry coming. So the Horse Artillery had to gallop to the front and ply them with grapeshot. Running fight in this manner for six miles, taking three guns, a great many elephants and camels, all their baggage and that of the 37th Regt which the enemy had captured from them about two months before. At 2 am on 17 April with the cavalry and Horse Artillery in front we advanced cautiously for a few miles, until we came to a village that had been occupied by the enemy, 5,000 strong, with 6 guns, under the command of a rajah (Kunwar Singh). Villagers

told our general that the enemy had left there a few hours before, and pointed out the road they had taken. Sudden 'salute' of musketry from ambush in a clump of trees. Action front fired canister at sepoys only 200 yds away, mowed them down but they tried to turn our flank. No infantry to support us so General gave command to limber up and retire, or the sepoys would certainly have taken our guns. We followed the enemy every day until 20 April when we had another engagement with them at a place called Mannuhur, capturing four small guns. *On 21st we came up with them again at a place called Sheopore Ghat, on the banks of the Ganges where they were crossing in boats. We sank one boat in the middle of the river, drowning a great many, and we continued to fire shells among them as long as they continued in sight. It was singular that the last shell we fired wounded their chief (Kunwar Singh), of which hurt he died two days afterwards..'* See Chambers Journal No 20, 1930. See also **SSF** Chapter 9, and *Mutiny Narrative Jalaun District* quoted in **FSUP**, Vol III, p 93; and *Azimgarh Collectorate Records* quoted in **FSUP**, IV, p 104; *Mutiny Narrative, Benares Division,* pp 1-4; **FSUP**, Vol IV Chapter 4, p 404 et seq., see also K.K.Datta, *Biography of Kunwar Singh and Amar Singh,* Patna, 1957.

KUNWAR SINGH, Death of

It has always been assumed that Kunwar Singh *was in a boat*, crossing the Ganges, when he was wounded by artillery fire. But there is another suggestion. The unpublished manuscript report of a

Kunwar Singh cuts off his own wounded hand

certain Major Edward Michell, contains the following statement: ..'we overtook them in the act of crossing the Ganges. Our fire caused them a considerable loss, among the rest being that of Koor Sing himself, who received a wound from a shrapnel shell as he *was crossing on his elephant,* of which he ultimately died.' *From* the *Report to his Artillery superiors of Captain and Brevet Major J.E.Michell, R.A., dated Soraon 20th July 1858,* in archive collection MD/147 in the **Royal Artillery Institution** at Woolwich, S.E. London.

KURNAUL [KARNAL]

A town one march to the north of **Delhi**, on the **Ambala** road, to which the majority of the Delhi fugitives escaping from the rebels on 11 May 1857, made their way. Although there were no British troops in Kurnaul the fugitives were safe there because the local Nawab remained a staunch supporter of the Government.

KUTTLE

When **Wheeler** and his officers accepted the terms offered by **Nana Sahib** to surrender the **Entrenchment** at **Kanpur**, it seems that many had misgivings as to the possibility of treachery. Their concern was all too well-founded, as the massacre of **Sati Chaura Ghat** resulted, but at the time they were reassured by the report of a number of officers who were sent from the entrenchment down to the banks of the Ganges to check on the preparations for the garrison's embarkation. One of these officers however heard, but apparently failed to report to General Wheeler, what might have saved the lives of the garrison: 'A committee was next appointed, consisting of Captain Athill Turner and Lieutenants Delafosse and Goad, to go down to the river and see if the boats were in readiness for our reception. An escort of native cavalry was sent to conduct them to the ghaut. They found about forty boats moored and apparently ready for departure, some of them roofed, and others undergoing that process. These were the large up-country boats, so well known to all Indians. The committee saw also the apparent victualling of some of the boats, as in their presence a show of furnishing them with supplies was made, though before the morning there was not left in any of them a sufficient meal for a rat. Our delegates returned to us without the slightest molestation, though *I afterwards gathered that Captain Turner was made very uneasy by the repetition of the word kuttle (massacre)*, which he overheard passing from man to man by some of the 56th Native Infantry, who were present on the river's bank. ' *See* Mowbray Thomson, *The Story of Cawnpore,* London, 1859.

KWI-HYES

See KOI HAI

L

LADENDORF, J.M. A Swiss historian who produced a bibliography of the rebellion, especially useful in that it includes the work of Indian and Pakistani historians post 1947 up to the year of publication. It is annotated but is somewhat flawed by subjective remarks as to the standing of authors or their integrity. At first sight the division of the work into sections ('Scholarly Studies', 'Narrative Histories' etc) appears helpful, but it can occasionally be misleading or irritating. With nearly 1,000 entries (including duplicate and triplicate). Arranged however rather fussily under headings which are not always fully justified Decisions as to placing appear arbitrary in some cases, and it is possible to overlook valuable items as a result. Spelling mistakes are frequent. Nonetheless this remains an important reference tool for the student *see* J.M.Ladendorf, *Revolt in India*, Zug, 1966.

LADIES OF THE REGIMENT Theoretically no British regiment employed prostitutes, or 'comfort women'. Had it been known generally in England that such women existed there would have been an outcry, and Queen Victoria would certainly have been most displeased. As a result there is little reference to such persons, particularly in official correspondence, but a private letter lets the cat out of the bag: 'On July 26th (1857) we had to cross the Beas River. Engineers reported the bridge of boats fairly safe....Just then there appeared the conveyances bearing the native ladies attached to the regiment. Tommy Atkins was much agitated: would they or would they not get over? Ladies seated in ekkas, carts had tops and curtains descended hiding the dusky beauties....just as the line of ekkhas had passed the centre of the bridge a cry was heard.. It is gone!. Too true the boats which had held together during the passage of HM's soldiers, the camels and the baggage, gave way under the immense responsibility now entrusted to them, and amid despairing cries boats and ekkhas were borne away by the current of the swollen river.. which way would they be taken? ...watched by anxious eyes the boats bearing their precious burdens gradually but surely neared the shore, and as they touched, many willing hands caught and secured the wandering boats, the ekkhas were brought safely to land, and the ladies restored to their position with the regiment.'

LAHORE The city of Lahore was the capital of the province of the Punjab, newly acquired by the British as a result of the Sikh Wars and was the headquarters of the Chief Commissioner, **Sir John Lawrence**. It is 1,356 miles from Calcutta and 380 miles from **Delhi**. On 13 May 1857 all the sepoys at Lahore, amounting to 3,800, had been disarmed as a precautionary measure. On 28 September the following intelligence was published in London and caused considerable furore in the House of Commons where the action was seriously impugned: 'The 26th N.I. mutinied at Lahore on the 30th July and murdered their commanding officer, Major Spencer; but the mutineers were totally destroyed.' There is still confusion as to what exactly happened after Spencer, and a sergeant-major and a havildar had been murdered, but at all events the disarmed sepoys fled from the city, were pursued and killed in particularly brutal and inhumane manner, not by regular troops but by a mixture of irregular horse and Sikh levies, led by **Frederick Cooper** the Deputy Commissioner, based at **Amritsar**. The action was applauded by both Lawrence and Mr R.Montgomery, the Judicial Commissioner of the Punjab, and found general approval by Europeans in India, at the time but has been almost universally condemned since. *See* **Ball, Martin**, PRAKASH SINGH, COOPER, FREDERICK, MEAN MEER, UJNALLA, Well of; *see also* F.Cooper, *Crisis in the Punjab*, Lahore, 1858, p 172, and *see* Parliamentary Papers, 29 July 1859.

LAKSHMIBAI See JHANSI, Rani of.

LAL MADHO SING, Raja He was a talukdar of **Amethi**, and a friend and collaborator of Raja **Man Singh**. It was to Lal Madho's care that Man Singh sent his family for safety when his fort at **Shahganj** was attacked by the rebels led by **Mehndi Hussun**. He wished, after the fall of Lucknow in March 1858 to make his peace with the British but was deterred from doing so for fear of retaliation from the rebel talukdars. His letter to Major Barrow dated 22 June 1858 very neatly sums up the dilemma of many of his class: 'Your letter arrived. You are very kind and I have great hopes from you, but I can show that it is not from any disobedient feeling that I have refrained from sending my *Vakeel* to Mr Forbes which you call negligence. Every soul likes to preserve his life and honour, and just now that war is going on between the King's Government and the British Government. Every one can see that the King's Government army, never has, and never can overcome the British army, but all the people are getting ruined and destroyed while victory still remains undecided, for the King's army destroys all the friends of the British and the servants of the British Government destroy all who remain quiet, considering them enemies; *in fact the people are ruined in every way.*' See **FSUP**, II p 435.

LALITPUR At the beginning of the outbreak this small town in the **Chanderi** district of **Bundelkhand** had a small garrison of the **Gwalior Contingent**, but the Deputy Commissioner of Chanderi reported they were not to be relied upon, nor was the **Raja of Banpur**, despite his protestations. At the same time **Major W.C.Erskine**, the Commissioner at **Jabalpur**, was informed that the **Raja of Shahgarh** was raising levies and preparing for war. Troops were despatched from Saugur[Sagar] to Lalitpur, but got no further than **Malthone** where they mutinied. Meanwhile the Europeans at Lalitpur were confined by the raja of Banpur, who sent them to Tehri where they were well treated through the influence of the tutor of the young **Raja of Orchha,** and befriended by Muh.Ali Khan, Mukhtiar of the raja of Banpur, who did not seem to share his master's feelings towards the British. Sent on to Saugur the Europeans were again captured, this time by the Raja of Shahgarh, who imprisoned them for three months before sending them on to Saugur. For further details *see* **FSUP** Vol III and *Revolt in Central India*. Simla, 1908.

The Palace of the Raja of Orchha

LALL, BAJAY Some spectators of the attack on **Wheeler's Entrenchment at Cawnpore[Kanpur]** had a grand stand view. Bajay Lall was one such, and his evidence was used to prepare a case by the British against the **Nunne Nawab (Nane Nawab, *ie* Mohamed Ali Khan).** 'A number of Christians were killed, their houses set on fire, and all their property plundered and destroyed. I became alarmed and retired to my house from whence I could perceive all that was going on. Batteries were erected, and the fight commenced. There were four batteries erected, one near the racquet court, one near the church, another on the mall, and the fourth near the cavalry lines opposite the nullah. The Nunne Nawab was at the battery near the racquet court, Nawab Bakur Ali was at the one near the church, and on the other two, the sepoys, and a number of the Nana's men. I frequently saw the Nawab's conveyance going to and returning from the battery in question, and I often saw him seated there on a chair with a table placed before him.' *Deposition of Bajay Lall.* **Williams.**

LALLA BHUDREE NATH In May 1857, the report spread in the Cawnpore[Kanpur] Bazar that the Hindustani force at Meerut had mutinied, and committed excesses on the Europeans. This became the universal topic of conversation among the sepoys. 'One day in May, the son of a sowar whose name I do not know, came to his lessons at a school held in the same premises in which I lived; the boys were overheard by Teeka Ram, Bookseller, to say among themselves that the force here would act as that at Meerut had done. He told me this, and I at once gave information of it to General Wheeler who had long been a patron of mine. The General enjoined me always to keep him informed of anything of the kind which I might become acquainted with among the infantry or cavalry, and to be cautious and on my guard in doing so. I accordingly introduced sundry emissaries of my own disguised into the lines, and reported what I ascertained daily to the General. One day I heard that the cavalry and the 1st Native Infantry, 'Gillis' Regiment, had made up their minds to mutiny, and make off at the moment when the new cartridges were served out to them. I immediately apprised the General of this, and he gave me four cartridges, telling me to show them to the principal Native officers, and explain that there was no harm in them. I thereupon went and showed them to the Native officers of the Gillis Regiment, who said that they were quite good and that they had no objection to use them, but protested against having any given to them in the composition of which fat had been used, promising to obey the General and carry out his orders if he would give them with his own mouth a positive promise that none should be served out. I reported this to the General who replied 'I promise that within my command no other kinds of cartridges shall be served out.' *Deposition of Lalla Bhudree Nath.* **Williams**.

LANG, JOHN He was a lawyer and novelist who, having successfully represented a Meerut banker by the name of **Ajodhya Prasad** (or Jotee Pershad) in a court case against the East India Company's officials, found himself in demand to give advice to others who had quarrels with the Company, notably the **Nana Sahib** and the **Rani of Jhansi** He was the author of an interesting and relevant book '*Wanderings in India*', London, 1859.

LANG, JOHN John Lang was an Australian who had been educated at Cambridge and called to the bar. After an unpromising debut as a lawyer in Calcutta, he accepted the editorship of *The Moffusilite* in Meerut, in which capacity he became a celebrated figure, admired for his pithy style. After an endless series of stories about the attempts of a man named Gorham to gain induction into the Anglican church, Lang ran the following headline. 'The Gorham Case'....'D—n the Gorham Case.' H.G.Keene, *A Servant of 'John Company'*, London, 1897, pp 122-3. Lang was expelled from Cambridge for writing blasphemous satirical litanies, but was admitted to the Society of the Middle Temple and called to the bar in 1841. Shunned on account of his mother's convict lineage by what passed for high society in Sydney, Australia, Lang moved to India, where he wrote novels, *The Forger's Wife* (1855) about the outback, *Botany Bay* which Charles Dickens serialized in his magazine *'Household Words'* and several books on India, including *The Weatherbys* (1853) and *The Ex-Wife* (1859). He represented the Rani of Jhansi in her appeals to the East India Company, which was no doubt why Nana Sahib was so solicitous during Lang's visit to Bithur. Lang died in 1864 and was buried in Mussoorie. *See* Ruskin Bond. *Mussoorie & Landour: Days of Wine and Roses*, New Delhi, 1992.

LARKINS, Emma She was the wife of Major George Larkins, Bengal Artillery, and died at Cawnpore[Kanpur], together with her husband and children, probably, though not for certain, at the ambush at the **Satichaura Ghat** having entered one of the boats. Copies of three letters she wrote shortly before her death, expressing her great fear, particularly for her children, are preserved in the Oriental and India Office collections of the British Library, *see Phot.Eur.233*.

LAWRENCE OF LUCKNOW

'On the summit of the rising bank that connects the plain below with the little plateau on which the Residency is built, and in the shadow of the ruins of that building, Sir John Lawrence, the Viceroy held a great Durbar in November 1867, almost ten years to the day after the final lifting of the siege and the rescue of the garrison. The Viceroy sat in a chair of State and the Talukdars, the feudal chiefs, of Awadh filed past in almost total silence, affirming publicly their adherence to the British Government and to Queen Victoria. They passed slowly, with seven hundred magnificent elephants spendidly caparisoned, with bejewelled and gorgeously apparelled retainers. It was a grand spectacle. Yet there was little to gratify the heart. In that wonderful procession there were some men who, at considerable difficulty and risk, stood by the British in their hour of need; some had sheltered men and women when, maimed and wounded, they had cast themselves upon their protection. There were also not a few who had wavered and held back until they could decide on which side the hangman was. There were some too who had never been friendly, but had yielded to British power from necessity. All were perhaps thankful for their restored lands, and the hope that British protection would guarantee their permanent enjoyment. And others, some of the bravest and best and most honourable Rajput chiefs - like **Beni Madho** and Nirput Singh - fell victim to their uncompromising fidelity to the Begum **Hazrat Mahal**, and were hunted to death in the jungle like wild beasts, while a *parvenu* and time-server like Raja **Man Singh** paid state visits to the Governor General. But there was no one at the Durbar who loved the British for their own sake; not one who would not have preferred a native rule to that of Queen Victoria; not one who did not regret the unrighteous destruction of the Kingdom of Avadh. They offered their submission to the Government as a necessity with a smile, a shrug, or a scowl. Nor was there any love lost on the British side. At best there was a sincere desire to do unswerving justice to all, to protect all, curb all and to regenerate the country. At worst the memory of what was seen as the treachery and horrors of the 'mutiny' still dominated the hearts of many, and would do so for many years to come. Sir John Lawrence was sitting within a few yards of that room where his brother Sir Henry had been struck down: some fifty yards away on the other side of the Residency was his simple tomb. When the sights and sounds of the great pageant and procession had finally ended, Sir John walked round alone to his brother's grave, and stood there for many minutes by himself, wrapped in thought. His staff followed at a distance not daring to draw close. It was perhaps as well for Britain as for India, and for the cause of reconciliation, that a new Viceroy, untouched directly by the events of 1857, was soon to come to heal the festering wounds.

The impartial observer sees a military mutiny and a struggle by feudal chiefs to win back by force of arms what they had lost by force, but, at the same time, he sees the struggle of a free people to rid themselves of a hated invader: the rising in Awadh was not strictly a rebellion - for that implies simply attack upon authority - it was a fight against outsiders who had sought to take away a nation's independence. Those who fought that fight also fought for a noble cause, and for their honour.' *See* **FQP.**   **PJOT**

LAWRENCE, LORD John Laird Mair Lawrence, first Baron Lawrence (1811-1879); Viceroy of India 1863-69 when his chief concerns were sanitation, irrigation, railway extension and peace. Brother of **Sir George** and **Sir Henry** and possibly the least able of the three. He was Chief Commissioner of the Punjab 1853-57, and although not a soldier, his supportive action especially for the **Delhi Field Force** has often been quoted as the reason for the success of the eventual attack by the British on Delhi in September 1857; certainly his advice was worth following in some respects as he was familiar with Delhi and its environs. *See* **DNB** and Bibliography.

LAWRENCE, Sir George St.Patrick (1804-1884) General. Brother to **Sir Henry** and **Lord Lawrence**. Served in Afghan War and Sikh Wars. Deputy Commissioner of Peshawar, Political Agent in Mewar 1850-1857; Resident for the Rajputana States 1857-64, at Mount Abu, held chief command of the British forces there during the rebellion in 1857. Issued a proclamation on 23 May 1857 addressed to all the Chiefs of Rajputana, calling on them to preserve peace within their borders and to intercept rebel fugitives: this was promptly replied to, affirmatively, particularly by Jodhpur. Between June 1857 and January 1858 he resided alternately between **Ajmer, Beawar** and **Nasirabad,** carrying on his civil duties without interruption, and having a guard of a native officer's party of the Merwara Battalion which remained faithful to the British. He published *Forty-Three Years in India,* in 1874. *See* also **DNB**.

LAWRENCE, Sir Henry Montgomery (1806-1857) Brigadier-General, KCB, Artillery. Chief Commissioner in **Oude [Awadh]** at the time of his death within the besieged Residency at **Lucknow**. Arrived India in 1823, and had a distinguished military career, especially during the Cabul expedition and the Sikh Wars, but he was soon earmarked for civil service as his ability was recognised as outstanding. Resident in Nepal 1843-46, but most of his work in civil employ was in the Punjab where he established a considerable reputation. After three years, 1853-6, as Agent to the Governor General in Rajputana, he became Chief Commissioner and Agent to the Governor General in Oude[Awadh] in 1856, succeeding the disastrous regime of **Coverley Jackson**. At the outbreak of the mutiny/rebellion he was promoted Brigadier-General and given military command over all troops in Oude. Has been criticised particularly for his failure to concert resistance with **Sir Hugh Wheeler** at Cawnpore [Kanpur], and for the disaster for the British at the Battle of **Chinhat**. He and his wife Honoria, who predeceased him, were well-known for philanthropic works, into which they put a great deal of their own resources. He was remarkably popular with all classes and races in India, particularly with the Sikhs who gave him the credit for the successful demobilisation of the Khalsa. Founded the Lawrence Asylum for the Children of European Soldiers. Despite the jealous sensitivity of the HEIC towards its servants writing for the public, he was a voluminous contributor to the Indian (English language) newspapers. Brother of **Sir George Lawrence**, and **Lord Lawrence**. *See Burke's Peerage*, 1923, p 1349, **DNB, DIB**, Boase, *Encyc. Britt.*, and a number of biographies esp. J.L.Morison, *Lawrence of Lucknow*, London, 1934. *See also* **FQP.**

Sir Henry Lawrence

LEARNING TO SWIM There were only four male survivors of the massacre at **Satichaura Ghat**, and they escaped by swimming down river, a prodigious distance, until they reached the territory of a friendly raja, **Dirig Bijai Singh**. At least six men of the party they were with were unable to swim and so were overcome by the rebels. **Mowbray Thomson** was one who had reason to be grateful for a small investment: 'Oh! that night's rest; thankful, but weary were we; amidst many thoughts that chased each other through my distracted brain, I remember one ludicrously vivid, it was this: how excellent an investment that guinea had proved which I spent a year or two before at the baths in Holborn, learning to swim!' Mowbray Thomson, *The Story of Cawnpore*, London, 1859.

LECKEY, Edward He made a courageous attempt to refute the deliberate sensationalism of much of the earlier reporting of the mutiny/rebellion. As early as 1859, only two years after the outbreak of the 'Mutiny', and even before it was finally put down, and while tempers and feelings were still running high, this Englishman had the courage to publish, in Bombay, his little book. Let us hope for his sake that satisfaction in making truth triumphant compensated for his unpopularity in the unforgiving circles of British clubland in India. **Lord Canning** was dubbed 'Clemency Canning' because he had the good sense, as well as humanity, to call a halt to the blood-letting and the reprisals. He was the Queen's representative, as Governor General, but that did not protect him from abuse. How much more vulnerable would be the humble Mr Leckey. There were many fictions to expose, although their inventors, and their motives were often obscure. One of the first was the completely apocryphal tale of the Highlanders at Cawnpore [Kanpur] dividing the tresses of the dead Miss Wheeler. Equally ludicrous was the story of Miss Wheeler defending herself with a sword and then throwing herself down a well. As Government did its best, belatedly, to dispel such rumours, and to declare in Parliament that on investigation much alleged atrocity was found to be untrue, one must assume that it was not Government that initiated or even condoned the stories; yet it must have been convenient for the authorities to have these horrors to help whip up enthusiasm for a costly campaign to put down the rebellion, and also to have had such horrendous tales for the recruiting sergeants to put before bucolic audiences. *See* Edward Leckey, *Fictions connected with the Indian Outbreak of 1857 Exposed*, Bombay, 1859, *and* **SSF** chapter 13.

LEE, Joseph A hotel proprietor (the Railway Hotel) in **Cawnpore [Kanpur]** in the last quarter of the nineteenth century who catered for some of the tourists on early 'Cook's Tours'. He may, or may not, (opinions vary) have been a sergeant, known as 'Dobbin', in the British army at the relief of **Lucknow** by **Sir Colin Campbell**. Joe Lee was born in North Wales in 1829. and thus at the time of the Mutiny he was 28 years old. He enlisted in the 53rd Regt. of Foot in 1843, when he was only 14 years old, and embarked for India on 15 August. He remained in India from March 9 1844 to 1883, when he returned to North Wales to visit what was left of his family. 'A Non-Commissioned Officer, Regimental No. 317, Corpl. J. Lee of the old 53rd Regiment, (Shropshire) came out to India in 1844 for the Sutlej Campaign, and served throughout the same; also in the Punjab Campaign, and in the valley of Peshawar against the hill tribes, under **Sir Colin Campbell**, K.C.B., 1851, 1852. He also served throughout the Indian campaigns against the Mutineers in 1857-8 and 1859, and was present with General **Havelock** at **Cawnpore [Kanpur]** and the **Lucknow Residency,** until relieved by Lord Clyde, Commander-in-Chief, on the 24 November 1857. Present on 6 December 1857, after returning from Lucknow, when the Gwalior Contingent attacked the Camp of General **Windham**, (the Hero of the Redan) 4 days after being relieved from the Residency, on the date that General Havelock died. Present at action of Serai Ghat, Decmeber 1857: also that of Kalee Nuddee, 2 January 1858, of Shamshabad, 28 January 1858, the storming and capture of Meangunj 23 February 1858, siege of Lucknow from 2 to 19 March 1858. Present at battle of Koorsee, a village near Lucknow, 24 March 1858. Passage of Gogra, 25 March 1858. Bungaon, 3 December 1858. Tulsipore, under the Nepal Hills, 23 December 1858. Medal and two Clasps for the Indian Mutiny, also Medal and Clasps for the Sutlej and Punjab Campaigns. (Sd.) C. Bagnall. *Commanding Company.*' Lee was shot twice in the Battle of Sobraon and again in the Sutlej Campaign, and was wounded in the left arm by a sabre cut at Cawnpore [Kanpur]. He quit the army in 1865 and worked at Allahabad for a time with the railway. Around 1877 he purchased a hotel in Kanpur, and for the next thirty years was the city's resident character and chief battlefield guide. It is clear that in taking his guests around the city and its 'sights' he was both sensationalist and inaccurate, and it may be that as a result many persons were misinformed and spread that misinformation when they got home. In 1883 Joe went to America, and delivered a talk on 'The Great Indian Massacre of 1857 at Cawnpore' to an assembly at Osh Kosh, Wisconsin. 'To conclude,' Joe recalled, 'two gentlemen exhibited a magic lantern, and showed the principal scenes in India, Nana Sahib, and the notorious rebel of Cawnpore, and also scenes in connection with the late famines in India were among those that were shown. This caused much amusement and laughter, principally amongst the juveniles.' But his descriptions used to bring tears to the eyes of his nieces in North Wales. 'We could never laugh when he gave us the scenes of the Mutiny,' his niece remembered. 'It must have been horrible to witness the cruelties perpetrated by the rebels..' Among his chits garnered at his hotel is the following from a Dr. Alexander Muirhead written in 1877. 'His descriptions of the places of interest around Kanpur are marvellously good, and after hearing from his mouth what cruelties the natives of India are capable of, one wonders at the lenient policy of present Government officials.' A rumour dogged him that he lied about his part in the reliefs of Lucknow and Cawnpore, (a comrade said that he was in hospital, sick, at the time) but when Joe Lee was in his sixties he was visited by William Forbes-Mitchell, a veteran of these campaigns who, though he found some of Lee's accounts fanciful, believed him to be the man the soldiers called 'Sergeant Dobbin,' and validated all his claims in an article Lee gratefully reproduced in a pamphlet that he peddled to his guests. *See* Joseph Lee, *The Indian Mutiny: a Narrative of the Events at Cawnpore,* Cawnpore, 1893.

LENNOX, Colonel Commandant of the 22nd. N.I. at **Faizabad**. He would appear to have been popular with his troops and even with the **Maulvi of Faizabad (Ahmadullah Shah)** who protected the Colonel and his family on the outbreak of mutiny at **Faizabad** in June 1857. Colonel Lennox wrote a full account of his eventual escape to **Gorakhpur** via the lands of **Meer Mohammed Hussein Khan** who also befriended him. Lennox's statement, dated 1 July 1857, is reproduced in *Further Parliamentary Papers (No.4)* pp 46-48.

LENOX-CONYNGHAM PAPERS Held in the **Cambridge Centre of South Asian Studies** and including source material as follows: Letters and a typescript copy of 'Extracts relating to the Indian Mutiny Campaigns', 1857-8-9, by Dr A.F.Bradshaw, Assistant Surgeon, 2nd Battalion Rifle Brigade. As a non-combatant his interests transcend the purely military and he gives details (*British*) not found elsewhere, on camping, attitudes, servants, dress, amusements, shopping, looting etc. He became ill and was sent on sick leave to **Simla.** Rejoins battalion, impressed with the beauty of **Oude** [Awadh], Foot racing, horse racing, tent pegging. Repeats opinion by a Hindustani scholar that brutalities inflicted upon the European women and chilren in Cawnpore [Kanpur] done by Muslim soldiers not Hindus. Also hears discussion between **Russell** and the Chief of Police (**Bruce?**) about the massacre at Cawnpore: both thought the sepoys had been much maligned.

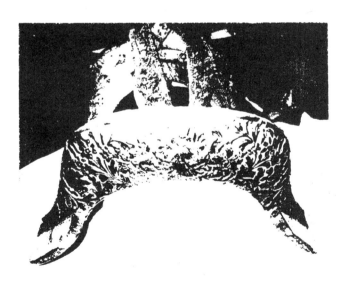

The Silver Saddle of Nana Sahib

LEWES A County town which houses the East Sussex Record Office. Here will be found Stanford's well-known and frequently reproduced plan of the City of Lucknow, illustrating the routes taken by **Havelock** and **Sir Colin Campbell** in marching to the relief of the **Residency**. Here also is a small Military Heritage Museum which contains among other memorabilia, the silver saddle of **Nana Sahib** and the sword of **Tatya Tope**.

LEWIS, Quartermaster-Sergeant Individual acts of heroism, on both sides, were common during the rebellion, but it takes much to equal the courage of a certain Quartermaster Sergeant Lewis, at the time of the Mutiny of the 17th Native Infantry at Azimghar. In addition to their British Officers each Native Infantry regiment had one or two British non-commissioned or warrant officers. Sergeant Lewis was the quartermaster sergeant of the 17th N.I.:- 'They assembled for parade and fell in by companies and in the usual manner. The only European who was present with the regiment at the time was Lewis. None of the European officers came to parade, as the mutinous state of the 17th had been known for some time and their intentions on that particular day were suspected. After waiting for some time the Sergeant assumed the command. Upon this many of the sepoys left their ranks, but fell back when ordered to do so. One of the Native officers stepped forward and suggested that the officers' (bugle) call should be sounded; to this the sergeant refused assent, and commenced to harangue the sepoys on the enormous crime of mutiny. He observed that if they rebelled they would be sure to be transported or hanged. Upon this every man of the regiment commenced to yell, and some rushing upon the sergeant cut him down, crying out 'if we are to be hanged, at all events we will kill you first'. While the Sergeant was lying wounded on the ground the sepoys discussed how they should torture him and each one proposed some torture more hideous than the rest. At last at the intervention of a native officer he was left on the ground. There he lay all night and in the morning was by a European officer of the regiment placed in a bungalow. From thence he was taken by the sepoys to a place in the Quarter Guard. Every day he was visited by the natives who, whilst amusing the wounded man used to delight in telling him for what torture he was reserved. Sergeant Lewis was rescued by Mr Venables.' One of the native officers of the Regiment concerned, one Subadar Bhoondoo Singh, was later to regret that Sergeant Lewis was not killed on that parade ground: he in fact lived to give evidence at Bhoondoo Singh's trial: 'I was quarter Master Sergeant of the 17th Regiment at Azimgarh when on the night of the 3 June 1857 just as the quarter to 9pm drum had beat, I heard the discharge of two muskets on the left of my tent. I then went out to ascertain the cause of the muskets being discharged and saw Bhoondoo Sing Subadar there, where he had no business; and when I asked him who had been guilty of discharging their muskets he was very impertinent and said it was no business of mine. I called upon the Jemadar of the guard for an explanation regarding the two shots fired. He said some budmashes in the town did it. When Bhoondoo Sing ordered me to leave the guard and said I had no business to interfere, I went to my tent and dressed in uniform and went back to the guard again and ordered the Jemadar to fall in his guard that I might inspect the muskets to discover the delinquents. The guard positively refused to fall in and the sentry fired off his piece and cried 'Company-ke-neemuck-haram!'. On the sentry firing his musket which was done by Bhoondoo Sing's orders, the whole regiment seized their arms which were piled in front of the tents. Then the bugle sounded the alarm (blown by the drum-major, a Christian), the men fell in on the parade-ground as was their wont in open column of companies. Bhoondoo Sing Subadar then ordered the bugler to sound the Officer's call but I told him not to do so and kept him by me. I then called the men to attention and reproved them, wheeled them into line and broke them into open column again, ordered arms and stood at ease. I addressed the regiment and told them they were grossly misbehaving and advised them to return to their lines which they positively refused to do. Madho Sing, orderly, who was with me and who was afterwards promoted to Adjutant by Bhoondoo Sing, called upon the regiment in the name of their religion to march on Lucknow and to follow the example set them by their brethren at Meerut and Delhi. The Havildar-Major then advised me to leave the parade ground but having received positive orders not to do so I remained; just at that moment Madho Sing stepped out and shot me through the chest, this shot was fired by Bhoondoo Sing's order. Bhoondoo Sing then in command of the regiment marched it to the kutchery for the purpose of seizing the treasure, but finding it gone marched off in pursuit, came up with it just near the Judge's house and brought it back to the cantonments. He (Bhoondoo Sing) then got all the ammunition from the magazine and brought it into the lines, marched the regiment off at about 1 o'clock am on the 4 June 1857 taking with him treasure, Government cattle, two post guns etc. Lieutenant Hutchinson 17th NI was shot at the kutchery by Madho Sing.' Bhoondoo Sing was hanged. NB the 17th NI was the regiment that took post on the left bank of the Ganges below Cawnpore [Kanpur] on Nana Sahib's orders, to destroy the Europeans should they manage to get down the river. *See Friend of India* 13 August 1857, and *Trial Proceedings,* Lucknow Collectorate Mutiny Basta.

LIAQUAT ALI, Mohammed A Maulvi or Faquir of Allahabad of considerable influence in the early days of the rebellion. Proclaimed a Jihad or Holy War against the Christians. Aroused the Muslims of **Allahabad** and the surrounding countryside and briefly ruled the city until the arrival of **Brig.General Neill** from **Benares[Varanasi].** It has been said that 'only the rabble followed him, the better quality Muslims knowing him for what he was, a fiery demagogue who saw profit and plunder for himself under the name of Holy War.' C.A.Bayly, in *Two Colonial Empires,* Lancaster, 1986 puts it in another way (p 125) 'The Maulvi of Allahabad who was influenced by Sayyid Ahmed of Rai Bareilly and a leader of the city's Muslim artisans, defied the local religious leadership there to stage a revolt in June 1857'. **Amy Horne** was his prisoner for some time. He was a native of Mogaon, a handsome young man, of the caste of

Maulvi Liaquat Ali

weavers, by profession a schoolmaster. He took possession of Allahabad on 6 June 1857 when the 6th BNI (who had just been read a letter from **Canning** thanking them for their loyalty, to which they had responded by lustily cheering) mutinied. Liaquat Ali had secured the following of the people of his village, and the Muslim **zamindars** of Chail also acknowledged him their head. He was proclaimed governor of the district in the name of the King of Delhi; his headquarters was at the Khooshro Bagh. For a week he reigned in Allahabad but there was no unanimity among the rebels, the Muslims followed Liaquat Ali, but the Pragwal Brahmins led the Hindu population, and the soldiers fought for themselves. On 11 June Neill arrived and the rebels mainly retreated. The British tried to capture the Maulvi but for fourteen years he retained his liberty, disguised as a respectable Mohamedan gentleman. He took the name, among others, of Abdul Kanem. Eventually he was betrayed, and arrested at Byculla railway station and sent to Allahabad for trial; had Rs100 in his possession, plus a hollow bamboo cane of high quality, and when a crystal knob at one end was unscrewed gold ingots to the value of Rs2,000 were found inside. He pleaded guilty but said, in mitigation, that he had protected a lady (**Amelia Bennett**) from the **Nana's** forces. Sentenced to transportation for life and despatched to Port Blair in the Andaman Islands. *See* **FSUP,** Vol I, p 445, and Vol IV pp 550-1 *Trial Depositions in Govt. versus Liaqat Ali;* also **SSF** Chapter Five, *and* the papers of Sir Theodore Cracroft Hope, Bombay Civil Service 1853-88: volume 21 comprises papers and correspondence concerned with Liaquat Ali's final arrest; they are preserved in the Oriental and India Office collections of the British Library, *see MSS.Eur.D.705* For further details of the Maulvi and even a photograph of him in captivity in the Andaman Islands *see also* H.L.Adam, *The Indian Criminal*, London, 1909, and L.P.Mathur, *Kala Pani, History of Andaman and Nicobar Islands, with a study of India's Freedom Struggle*, Delhi, 1985.

LIAQUAT ALI, MAULVI, Proclamation of

'This is a proclamation for Jehad (Mohammedan Crusade against the Christians) issued by the Imam named Leeaqut Ali of Allahabad, to both great and low men of the creed of Islam for massacring all the accursed Christians.' 'In the name of the beneficent God, the absolver of sinners. All praise to the Great and Merciful God. May the Merciful God and Preserver shine upon the Great Prophet, his Descendants and his companions, that have carried courage to the highest pitch. May the mercy of God also alight upon the Votaries of the Great Religion of Islam. After praising God and praying for blessing upon the Prophet, I, Leeaqut Alee of Allahabad a poor scholar, and a mendicant, with a view to save my readers from the trial of the Judgement Day, beg to publish some of my observations, based upon precepts extracted from the sacred Kuran and the Holy Book of our Prophet, and trust that the faithful Mussulmans will attend to my appeal, as I am only discharging a great religious duty viz. inculcating the doctrines of our sacred scriptures. Every Mussulman well knows and it is a notorious fact that the accursed Christians have been awfully tyrannizing over the whole country of Hindoostan, especially over the District of Allahabad. These Christians have been guilty of the crimes of massacring, plundering and hanging human beings, burning and destroying our houses, attacking us unawares, killing innocent learned men of our Country, and burning our sacred scriptures. Accordingly it is a duty incumbent upon every faithful Mussulman to prepare himself for *Jehad* as our gracious Prophet has said: 'Every Prophet had his duty assigned to him the duty allotted to me is that I should make *Jehad*.' ....the real Paradise lies beneath the strokes of swords. You will then obtain salvation and the honour of martyrdom, which is eternal life. You will then obtain the blessings of Paradise, and get Elysian Nymphs for your wives....God will assist you and give you speedy victory. Report this good news to the Believers.' This proclamation was found in the office of **Khan Bahadur Khan**, the rebel Nawab of **Bareilly**. For the full text *see* **FPC** Nos 749-53, NA.

LITTLEDALE, Arthur

Judge of the Saran District of Bengal, based at **Arrah**. He was one of the Europeans who determined to make a stand at **Arrah House**, and who survived both the siege and the subsequent campaign against the forces of **Kunwar Singh**. Was the senior civil officer in Arrah. *See* ARRAH HOUSE. A letter from him, dated 14 August 1857, is preserved in the Oriental and India Office collections of the British Library, *see MSS.Eur.A.72, see also* **SSF**.

LLOYD, MAJOR GENERAL

Lloyd was the divisional commander at **Danapur**; he was an old man, over seventy, and suffering severely from gout and other infirmities. But he had had an honourable career, was well liked and respected by the sepoys, and from the start of the rebellion adopted a conciliatory attitude which certainly postponed the outbreak of mutiny in Danapur and might, had he been supported by the civil authorities, have prevented it altogether. He was relieved of his command after the disastrous failure of **Captain Dunbar's** attempt to relieve **Arrah**, but it was probably as scapegoat that he was so punished, for he had no direct involvement in the fiasco. Lloyd was a critic of the policy of **Brigadier General Neill** whom he blamed for the spread of the mutiny to **Allahabad** and **Faizabad** which distracted Neill from his primary duty of relieving **Wheeler** in **Cawnpore [Kanpur]**. *See* **Martin**, Vol II, p 282 *et seq*.

LOCHIN

He was a boatman, of **Cawnpore [Kanpur]**, executed with another, Hardeo, by the British for his part in the massacre at the **Satichaura Ghat.**

LODGE HARMONY

See FREEMASONRY, & MASONIC LODGE.

LOHARI

A village and mud fort about 8 miles northwest of Punch in Central India, where a strange incident took place. It was supposed to be garrisoned by some 80 men of the Raja of Samthar who had declared for the British, but these men betrayed some of the Hyderabad Contingent cavalry to the rebels. When a party was sent to punish the garrison it was found that they were in fact sepoys of the mutinied 12th BNI. The British captured Lohari after fierce fighting, on 2 May 1858, although the *killadar*, Manowar Singh surrendered without a fight, coming out and surrendering his sword: he appeared to have no control at all over the garrison: of the latter every man was killed. *See The Revolt in Central India*, Simla, 1908, 122-4, *and* Thomas Lowe, *Central India during the Rebellion of 1857-58*, London, 1860, *and* **FSUP,** Vol III.

LONI SINGH, Raja of Mithowli[Mitauli]

He was an important Talukdar of Oude[Awadh], and one of the first to rebel against the British, though his motive might have been more mercenary than patriotic: on the suppression of the rebellion in Oude the British put him on trial 'his treason being aggravated by brutality and avarice: he having for the sum of 8,000 Rupees (£800 sterling, a prodigious sum) betrayed into the hands of the Begum of Oude the following fugitives from Seetapore who had sought his protection at Mitawlie in June 1857:- viz Captain Patrick Orr, his wife and daughter, Sir Mountstuart Jackson and sister Madeline, an orphan girl daughter of Mr Christian, Bengal Civil Service, Lieutenant G.J.H.Burnes, and Sergeant-Major A.Morton.' He was found guilty and sentenced to transportation for life, his property being confiscated to the State. *See SSF Chapter Ten; see also Trial Proceedings, Govt vs Raja Loni Singh,* Lucknow Chief Court Mutiny Basta. He, like many others caught up in the great rebellion, played an ambivalent part, seeking to please both sides, and to profit no matter who turned out to be the winner.

Looting in Lucknow

LOOTING The magistrate of Cawnpore [Kanpur], J.W.Sherer, tells us of more than one looting incident that he came across. He rode out one day in December 1857 into the city, and entering a lane found a knot of women greatly distressed, who declared to him that they had been forced to give up their nose-rings and other jewels. Moreover, they hastened to add, the culprit was in a neighbouring house. Sherer went with them and discovered a member of the new police force, still with the women's jewellery on his person. He had evidently invented an altogether delightful way of making his fortune: he found that all he had to do was show his firelock, and all parties would submit completely and immediately. The women had apparently pressed their valuables upon him. Captain **Mowbray Thomson** - one of the two British officer survivors of the **Satichaura Ghat** massacre in Cawnpore [Kanpur] - had just taken over responsibility for the local police, and was most concerned that such conduct should not be copied, so he had this particular offender flogged and dismissed from the police. Another incident a week later was even more serious. It concerned an old friend of Sherer's a tentmaker by the name of Chuni Lal, who had managed to remain loyal to the British, whom he served very well. He was a handsome, slim benevolent looking character. He was on the point of coming up to the British camp to ply his trade one day, and about to rise from the charpoy where he was resting, when he witnessed an act of barbarous looting which, though mercifully rare - perhaps because of its rarity - went largely unpunished. He saw two British soldiers enter a shop and compel the shopkeeper to hand over all his money. Chuni Lal understood and spoke English well and, showing great courage, he spoke to the two men, reminding them that they were the protectors not oppressors of the citizenry. A gentle admonition, for that is all it was, might have been ignored or brushed aside with a curse, but unfortunately Chuni Lal had added a few words that must have sounded like a threat: he said that if any officer knew what they were doing they would be punished. These words, implying, as they quite truthfully could, that Chuni Lal was on friendly terms with officers, and could speak to them in English, were enough to alarm the two thieves. They turned on him, and one of them put a musket against his body and fired it. The dead body, with still a lingering smile on its handsome face, was taken to Sherer by the murdered man's nephew, who was present when the sad event had occurred. **General Windham**, the officer commanding the troops in Kanpur at the time, was apparently most upset, and instigated vigorous enquiries into the matter, ensuring by his interest that there was no slackness in following up, and before long the two men were identified and arrested. They were subsequently hanged. For some, at least, looting did not pay. Discipline held soldiers together until the fight was won, then it was every man for himself. The usual priorities were a] personal survival, b] defeat of the enemy, c] finding liquor to drink, and d] looting.

Kaiserbagh Photo 1858

When the British army burst into the **Kaiserbagh** in Lucknow in March 1858, the priorities got a little mixed up, and the last became the first. The soldiers became 'Drunk with Plunder'. One onlooker described it as a 'strange and distressing sight', but also as 'most exciting'. In the courts and mini Palaces within the Kaiserbagh, the stuccoed gilded walls, the fresco paintings, the statues, lamps, fountains, orange groves and domes of burnished metal all gave hint and promise of wealth beyond the dreams of avarice for a poor soldier of the Queen. With loud cries the Europeans and their Indian allies burst into the Palace, pausing occasionally, almost mechanically, to fire back at windows and apertures from which still came a token resistance from determined rebels. Dead and dying sepoys were lying in the corridors throughout the maze of buildings that was the Kaiserbagh, but they were ignored by the incoming troops, wild with fury and the lust for gold. The variety of the plunder is infinite, and is matched by the ignorance of so many of the soldiers as to the true value of what they seized. Shawls, rich tapestries, gold and silver brocade, splendid dresses, curtains, priceless hangings, wrenched from their sites, were briefly assessed, discarded and thrown upon the fires that smouldered everywhere. Doors that refused to open were smashed or their locks blown out by the Enfield rifles. Inside, as like as not, for there a vast treasure at stake, the looters would find what they sought: caskets of jewels and gold, and easily portable valuables. **William Russell**, the 'Times' correspondent, witnessed the sack of the Kaiserbagh, and, by the way he describes it, implies moral condemnation *eg* he deplores the wanton destruction of precious things, jade and glass smashed, pictures ripped up, fine furniture burnt. He condemns the quarrelling, the violence, the wild menace of men 'drunk with plunder'. But...it did not stop him taking away jade bowls, and a 'present' from an Irish soldier of a nose-ring of small rubies and pearls, with a single stone diamond drop. And he tolerated, if he did not actually encourage, his servant to set up some kind of agency, weighing gold and silver plunder on the household scales and taking a percentage of the value. That was all on 14 March 1858. The plundering was stopped, by order, the very next day, but it was too late: the Kaiserbagh was an empty shell, and the victorious army, from the highest to the lowest, had taken its share of the loot.'   **PJOT**

LORD BUKSH KHAN   *See* PERSONS OF NOTE, APRIL 1859.

LORD JOWALA PERSHAUD   He was reported as 'Lord Jowala Pershaud (**Jwala Prasad**), a Kannoujeea Brahmin of Byswara (some village near Doondea Khera), was one of the four men who instigated the Cawnpore [Kanpur] rising and directed the massacre, he received the title of Resaladar from the Nawab'; *See* PERSONS OF NOTE, APRIL 1859.

LORIMER PAPERS   These are held in the Centre of South Asian Studies at Cambridge University. They are extensive and not fully arranged as yet, but include some interesting material. In addition to letters and papers connected with Major William Wyld (Assistant Military Secretary to Sir **John Lawrence** in the Punjab), and Major Laughton (Engineer Officer), there are papers concerned with the work of the latter during the siege of **Delhi.** There are photocopies of a proclamation in Urdu, dated 28 May, and four mutiny proclamations dated 25, 27 and 29 May 1857

LOVE LETTERS   *See* CHAN-TOON, Mabel Mary Agnes.

Major Maude 'lays' a gun at the Luchmantila

LUCHMANTILA   The Mosque in this part of Lucknow became the centre of a large new 'fort' - or rather fortified area - that **Sir Colin Campbell** had prepared after his assault on **Lucknow** in March 1858. It included at one angle the **Machhi Bhawan** which Sir Henry Lawrence had had partly blown up but which was now repaired. The thinking was that this should serve as a British base for the operations in Oude[Awadh] to suppress the rebellion, and also be a strongpoint if the city were attacked by the forces of the Begum. The mosque also provided quarters for British army officers of whom one, **Major F.C.Maude VC**, kept a photograph of his old quarters.

Lucknow in 1858

Lucknow Residency Today

LUCKNOW, Description of Here is one man's summary of Lucknow as it looked in 1858, from the roof of the Dilkusha Palace. 'A vision of palaces, minars, domes azure and golden, cupolas, colonnades, long facades of fair perspective in pillar and column, terraced roofs, all rising up amid a calm still ocean of the brightest verdure. Look for miles and miles away, and still the ocean spreads and the towers of the fairy city gleam in its midst. Spires of gold glitter in the sun. Turrets and gilded spheres shine like constellations. There is nothing mean or squalid to be seen. There is a city more vast than Paris as it seems, and more brilliant, lying before us. Is this a city in Oude? Is this the capital of a semi-barbarous race erected by a corrupt effete and degraded dynasty? Not Rome not Athens not Constantinople, not any city I have ever seen appears to me so striking and beautiful as this: and the more I gaze the more its beauties grow upon me......... ...sun playing on all the gilt domes and spires, the exceeding richness of the vegetation and forests and gardens which remind one somewhat of the view of the Bois de Boulogne from the hill over St Cloud....but for the thunder of the guns and the noise of balls cleaving the air, how peaceful the scene is ! Up above the gilded spires of the Kaiserbagh are to be seen many kites serenely floating in the air, and giving infinite pleasure to the gentlemen who are directing their movements.' *See* W.H.Russell, *My Diary in India*, London, 1860.

LUCKNOW, The Residency at Like the (so-called) Siege of Delhi, the Siege of the Lucknow Residency was a pivotal point in the struggle of 1857-9. After the Battle of **Chinhat** on 30 June 1857, **Sir Henry Lawrence** withdrew all his forces and dependants within the 37 acre compound of the Lucknow Residency. That same evening the Siege may be said to have begun. Eighty-seven days later, on September 25 1857, the troops of General **Havelock** entered the Residency; they did not so much relieve the garrison as reinforce it. Henceforth however the siege of the Residency is more generally regarded as a 'blockade'. This lasted until 17 November 1857 when the British Commander-in-Chief, **Sir Colin Campbell** forced his way in, and in the next few days withdrew all the garrison, starting with the women and children on 19th. This withdrawal was skilfully done, and there were no casualties, and the rebels were left in ignorance of the final withdrawal of troops, which apparently they had not been expecting. The

Residency compound was entered by the rebels at noon on 23 November, and in celebration they fired a salute of 101 guns. Lucknow, including the Residency, remained in rebel hands until 16 March 1858 when the city was reconquered and occupied by Sir Colin Campbell, and **Hazrat Mahal**, the Begum of Oude [Awadh] retreated to the north. The ruins of the Lucknow Residency are well worth a visit. They are truly awe-inspiring, a piece of India's history preserved in stark, full-sized reality. It is not as a coin on show in a museum in a glass case; it is a site, where great things happened, a matrix not an artefact. It is full of ghosts; benign ghosts, though perhaps it is as well we are not allowed to wander there alone by moonlight. It is a beautiful place and has a strength difficult to define: Lady Stokes was right, there is 'a Feeling of Quiet Power': the power comes from the honourable men on both sides who fought and died there .. The Siege of Lucknow was the subject of a large number of memoirs and reminiscences (*see* Bibliography). *See* specially **FQP**. **PJOT**

LUCKNOW, OPINION OF

One resident (**Amy Bennett, née Horne**) did not think much of Lucknow: John Hampden Cook, Amelia's stepfather, worked as an agent for the North-Western Dak Company, and had moved his family to Lucknow from the Bengal in February of 1857. The sheer strangeness of India, its interminable and inscrutable spectacle, wore down as many British as it delighted. Amelia Horne's memories of Lucknow were not fond. 'The place held out no inducements; it was so different from anything I had ever seen, the houses so strange, the streets so narrow, and the people so unlike those in the Bengal that I used to feel as if I had got into another world.' She and her family lived near a bridge, and the traffic was so dense that 'night after night, without exception, was one of merry making and rejoicing, and little sleep could we obtain. Hours have I spent staring out of the window at the richly caparisoned Elephants; the splendidly decked horses, the number of grand weddings and all the odd sights that used to distract me almost out of my senses. .... I really disliked the city very much,' Amelia concluded, 'It was a dirty one, and the people the most indecent, abusive set in the wide universe. An expensive place too, with not a particle of comfort about it, nobody boasting any modesty dared venture out, without their sight and ear being insulted with obscene language and undecorous manner, if walking out of a moonlight night it was not uncommon to have garlands of flowers thrown over you.' After Amelia was thus garlanded one night her stepfather resolved to leave Lucknow.

LUDHIANA

An important city in the **Punjab**. The Deputy Commissioner there was **G.H.Ricketts** whose correspondence is preserved in the Oriental and India Office collections of the British Library, *see MSS.Eur.D.610.*

LUSHKAR[Lashkar]

That part of **Gwalior** city in which is situated the palace of the Maharaja (Scindia). As a result sometimes there is confusion, 'the Lushkar' in such cases meaning the palace not the area. By origin the word was applied to the Camp of the Marathas, which had been established outside the boundaries of the old city; over the years it had become an integral part. When the rebels under **Tatya Tope** and **the Rani of Jhansi** occupied Gwalior in June 1858 the Lushkar Palace was the scene of a small but bitter struggle between Scindia's personal bodyguard and the rebels; the former were killed, *see* **FSUP, Vol III**, numerous entries.

LYALL, Sir Alfred Comyn

A member of the Bengal Civil Service, 1855-1887, he was serving in the **Bulandshahr** district on the outbreak of the rebellion. His observations are valuable as he was an intelligent man, and a man of courage: he took part in several campaigns as a volunteer. He was driven out of Bulandshahr (SE of Delhi) by Nawab **Walidad Khan**. His comment on the mutiny/ rebellion has often been quoted: 'The native population did not rise against the white man, but the moment they thought the white man was powerless they rose against each other, the rival castes and villages plundering and fighting in all directions: the Hindu Gujars raiding the Hindu Jats, and the Mahomedans raiding all Hindus impartially, while the English magistrate rode about from village to village followed by a few native subordinates and fighting men, vainly trying to keep order and punish the rioters...the number of natives killed by the insurgents was immeasurably greater than the number of white men.' He joined a troop of Volunteer Horse at Meerut, and came to believe that the 'Mahomedans hate us with a fanatical hate that we never suspected to exist among them, and have everywhere been the leaders in the barbarous murdering and mangling of the Christians.' Reinstated in Bulandshahr by the Movable Column after the fall of Delhi in September 1857, he was thrown on his own resources with general instructions to re-establish the authority of the British government. In this process he got the name for being a severe judge and certainly many men were hanged by him although he claimed that 'hanging was for murderers only, and mainly for the murderers of natives'. His correspondence, covering the years 1857-60 is preserved in the Oriental and India Office collections of the British Library, *See MSS.Eur.F.132/3.*

M

MACAULEY, Thomas Babington The first Baron Macauley, 1800-1859. Historian. A Commissioner of the Board of Control, 1832, Secretary 1833, member of the Supreme Council of India, 1834-38; President of the commission for composing a Criminal Code for India, 1835 (published 1837, becoming law 1860) Whig (Liberal) M.P. His most famous pronouncement on India was in 1833 in Parliament: *'That a handful of adventurers from an island in the Atlantic should have subjugated a vast country divided from the place of their birth by half the globe; a country which at no very distant period was merely the subject of fable to the nations of Europe; a country never before violated by the most renowned of Western conquerors; a country which Trajan never entered; a country lying beyond the point where the phalanx of Alexander refused to proceed: that we should govern a territory 10,000 miles from us, a territory larger and more populous than France Spain Italy and Germany put together; a territory inhabited by men differing from us in race colour language manners morals religion: these are prodigies to which the world has seen nothing similar'.* As an individual Macauley was universally disliked, but as framer of the Penal Code his influence upon India was exceptional.

MACDOWELL Mrs In Kanpur in May 1857 Major General Wheeler did his best to reassure the Europeans and urged they remain calm. He set an example by retaining his family (wife and two daughters) beside him, and this convinced others that all would be well. Most civilians had by now resolved not to flee Kanpur by boat, but a milliner named Mrs. MacDowell, who was owed a total of £500 - a prodigious sum - decided to leave and forfeited the money due to her; she safely escorted her four children to Calcutta. There is no mention of a Mr Macdowell. *See* Zoe Yalland, *Traders and Nabobs*, Salisbury, 1987, p 261.

MACGREGOR, Major General G.H. He was the Military Commissioner with Maharaja **Jung Bahadur**, when the latter led his twenty thousand Gurkhas into Oude[Awadh] to assist the British, particularly with the attack upon Lucknow in March 1858. His function was to be a liaison officer between the Maharaja and the Commander-in-Chief (**Sir Colin Campbell**), but also to advise both upon strategy and tactics. The Gurkhas were brave but undisciplined and poorly trained. A letter exists from Macgregor to the Governor General which is apparently a cover for a communication to the latter by **Jung Bahadur**: 'I have the honour to acknowledge the receipt of Your Lordship's kind letter of the 19 January (1858). It will be my earnest anxiety and effort to be always proved serviceable to Your Lordship and to the Supreme Government. I have not the good luck of being honoured with any instruction direct from his Excellency the Commander-in-Chief as yet about the movements of my force towards Oudh: but however I am in daily expectation of it. I shall be very glad to do in accordance to His Excellency's order and upon the good advices of Brigadier General MacGregor. With my best compliments to Your Lordship.' *See Foreign Political Proceedings, Suppl., Original Consultations,* 30 December 1859.

The West Gate of the Machhi Bhawan

MACHHI BHAWAN (Muchee Bhowun) Literally the 'fish-house', this was a citadel in Lucknow which gave the appearance of great strength and was deemed locally to be impregnable. It was **Sir Henry Lawrence's** first choice for a place in which the garrison should make a stand in the event of a rebellion in Lucknow, but his engineers convinced him that it would be too easy to undermine by a besieging force, so he reluctantly abandoned it. The garrison was withdrawn to the Residency by a semaphore message sent on 1 July 1857 from the Residency roof 'Spike the guns well, blow up the fort, and retire at midnight.' The commander, Colonel Palmer, managed to reach the Residency safely with 225 European troops and two 9-pdr guns. After the final assault on Lucknow by **Sir Colin Campbell** in March 1858 he had to arrange for the security of his hold on the city so he caused to be constructed a large fortified position on the south bank of the Gomti river, facing the stone bridge: this was called the Machhi Bhawan Fort, it was about half a mile each way and included at one angle the old citadel which was partly restored. *See* **FQP**.

MACHHI BHAWAN 'After the evacuation of the Muchee Bhawan in Lucknow... Our accession of strength was very necessary. We had saved all but one man who having been intoxicated and concealed in some corner, could not be found when the muster roll was called. The French say, *'Il y a un Dieu pour les ivrognes'*, and the truth of the proverb was never better exemplified than in this man's case. He had been thrown into the air, had returned unhurt to Mother Earth, continued his drunken sleep again, had awoke next morning, found the fort to his surprise a mass of deserted ruins, and quietly walked back to the Residency without being molested by a soul; and even bringing with him a pair of bullocks attached to a cart of ammunition. It is very probable that the debris of these extensive buildings must have seriously injured the adjacent houses, and many of the rebel army, thus giving the fortunate man the means of escape....our men were not a little astonished when they heard him cry, 'Arrah by Jasus, open your gates' And they let him in, convulsed with laughter.'

MACINTOSH FAMILY They were merchants in **Kanpur**. C. McIntosh was an elderly (64 years old) and well known merchant, was fluent in Hindustani and tried to disguise himself and his son Joshua (aged 25) as chowkidars, but they were recognized and, 'not knowing what to do,' hid, now disguised as Brahmins, under a bridge near Greenway Brothers' gate. After some boys pointed them out, they were hacked to pieces by sowars; their bones lay in the drain for three months before another son returned to Kanpur and buried them. In the 'Statement of Information' in the London *Times*, 16 October, 1857, p 8, he was described as 'a wealthy and respectable East Indian, and a very old resident of Cawnpore, who owned several bungalows.' Feeble Mrs. Amelia MacIntosh, aged 57 years, was discovered disguised in native clothes and hiding in her washerwoman's house. She was brought before the Nana at the Duncan Hotel and beheaded. Her trunk was laid in a ditch with her head on her breast, 'in which position it was left to decompose.' *See* W.J.Shepherd, *A Personal Narrative of the Outbreak etc.*, Lucknow, 1886 p 30.

MACKENZIE, Colonel Alfred R.D. Honorary A.D.C. to the Viceroy. The writer of a highly praised little book. On the flyleaf of a copy in the Gen.Editor's possession, Rebecca West has written 'This is a book which was much admired by Rudyard Kipling when it was published by his father's press. He spoke of it once to Beaverbrook.' It was originally printed by the 'Pioneer' of Allahabad. It includes the uprising at **Meerut**, the siege of **Delhi** and the capture of **Lucknow**. Mackenzie was a man of integrity and liberal views who deplored the severity of the reprisals against the rebels. Although he observed much he himself indulged in no looting. An English (or Scottish?) gentleman. In the Preface he writes: 'They (the reminiscences) do not pretend to any merit but that of truth. In that respect they may claim to present a record of actual events, and thus to bring before the Reader, however imperfectly, a rough sketch of the Great Indian Mutiny such as it appeared to the eyes of a young Subaltern Officer of Native Cavalry.' *See* Alfred Mackenzie, *Mutiny Memoirs*, Allahabad, 1891.

MACLAGAN PAPERS These are held in the Centre of South Asian Studies at Cambridge University. Available on microfilm are the manuscript diaries of General Robert Maclagan, 1820-94. He was educated at Edinburgh Academy and Addiscombe, and entered the service of the East India Company, in the Bengal Engineers in 1839. He served in the Sikh Wars, and was later appointed Principal of the Government Engineering College at **Roorkee**, where he was stationed throughout the mutiny/rebellion. His diaries are simple factual statements of events as they happen, with no comment or elaboration: they are therefore good primary sources of information on the happenings in Roorkee and the revolt of the Sappers and Miners there in May and June 1857.

MACNABB OF MACNAB PAPERS These are held in the Centre of South Asian Studies at Cambridge University, and include some interesting material from the Mutiny/rebellion period; there are for example long excerpts from a Mutiny letter describing the whole event of the **greased cartridges** in the **3rd Cavalry** and the punishment meted out to those who refused to accept them, in **Meerut**, May 1857. The papers are concerned with Donald Campbell Macnabb, who was Deputy Commissioner in **Jhelum** 1857-58.

MACNABB, James William *See* RAO SAHIB.

MACPHERSON, Major Samuel Charters CB. He was the Political Agent at Gwalior during the mutiny/rebellion and undoubtedly had considerable influence over **Scindia, the Maharaja of Gwalior**, although in his correspondence he often speaks of Scindia's conduct with considerable disapproval, in particular his reinstating those of his ministers and followers who had been disloyal to him. *See* William MacPherson (ed), *Memorials of Service in India from the correspondence of Major Samuel Charters MacPherson CB*, London, 1865.

MADANPUR PASS, BATTLE of This took place in **Bundelkhand** in Central India, on 3 March 1858, on the march of **Sir Hugh Rose** on **Jhansi**. The pass was defended by the **Raja of Shahgarh** and some 7,000 men, including the 52nd BNI, while the pass of **Narhat** was held by the **Raja of Banpur**. By defeating the rebels at Madanpur, Rose was able to turn the position at Narhat and take possession of the territory of Shargarh (annexed to Government), and the whole country between **Saugur[Sagar]** and **Jhansi** was in British hands with the exception of Tal Bahat. But there was another, even more significant result. The sepoys of the 52nd BNI accused their Bundela allies of running away at Madanpur, and general mistrust and panic resulted, *see The Revolt in Central India*, Simla, 1908, pp100 *et seq.*, *and* Thomas Lowe, *Central India during the Rebellion of 1857-58*, London, 1860; *and* **FSUP**, Vol III.

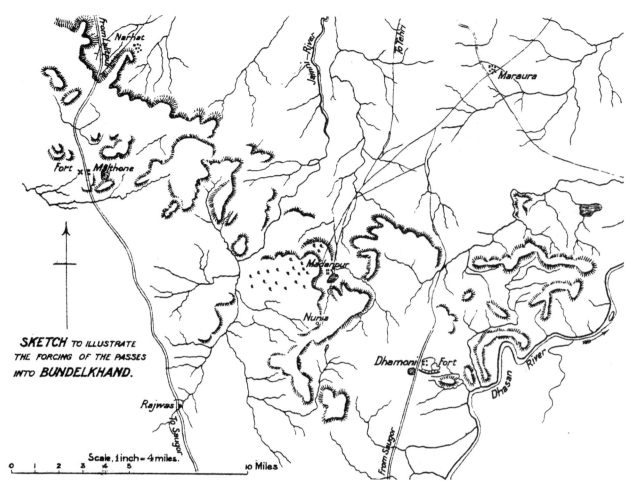

SKETCH TO ILLUSTRATE THE FORCING OF THE PASSES INTO BUNDELKHAND.

MADHU RAO A minor (aged nine) in rebellion against the British. *See* KIRWI.

MAGAZINE, DELHI To the north of the King's Palace and the Red Fort in Delhi, and close to the **Kashmir Gate,** stood the great arsenal or powder magazine of Delhi, containing a vast quantity of military stores including gunpowder in prodigious quantities. It was manned by three European subalterns, Willoughby, Forrest and Raynor, six European sergeants and conductors, and a large body of Indian lascars or coolies. Around three o'clock in the afternoon the rebels brought scaling ladders from the King who is said also to have sent a demand to Willoughby for the Magazine to be surrendered. The lascars now all used the ladders to escape from the walls, and the nine Europeans within were subjected to a harrassing fire which they returned as long as they could. Eventually however Willoughby decided to fire the train of gunpowder which had been set, and at 4p.m. the Magazine exploded killing perhaps three hundred of the rebels surrounding the walls. Three of the European sergeants were killed, but Buckley, Stewart and Shaw survived, together with the three officers although Willoughby was soon killed afterwards and Raynor had disappeared and was thought to be dead. It has been said that the death of so many of the mutineers in the explosion created so much fury among the survivors that it precipitated the murder of the European women prisoners being held then in the Palace. *See* **Ball, Kaye, Sen, Rice Holmes** etc

MAGAZINE, DELHI Many new slants upon accepted facts are provided by a study of unpublished documents, especially, letters and journals. The latter may smack rather too much of 'anecdotal evidence' sometimes, but they make interesting - and occasionally convincing - reading. Take as an example the journal of one Lt Col. W.L. Hailes, dated 8 December 1924. Talking to a brother officer in the Jat Regiment, called Raynor, he writes ....'Raynor said he had some interesting letters of his grandfather's relating to the saving of English lives by Jats during the Mutiny. It appears that his grandfather was originally in the Bengal Artillery, but getting married transferred to the Ordnance, and was the Raynor who was in the Magazine at Delhi. It appears that when the gunpowder was fired Raynor, who was the senior officer present but being Ordnance did not command (he was officially a non-combatant), Willoughby did that, and was standing with his wife who was a bride of five

Delhi Magazine, all that remains today

months, under one of the arches that did not collapse. The arch is one of those that still stands as a memorial. Neither of them was injured, except of course by shock. The terrific explosion naturally caused a panic among the attacking sepoys who disappeared, one man however approached the Raynors and befriended them. He was a Sikh 'grunthi' and an Akali. He helped and guided them down to the Yamuna River and getting them across handed them over to some Jats of Meerut district who kept them safely hidden till the fall of Delhi. Then they wished to form a band to loot the city and wanted Raynor to lead them presumably thinking that his presence would save them from the wrath of the British conquerors. When he refused they theatened to behead them both, and actually made them lay their heads across a fallen tree before laughing and saying they only wanted to test his courage. The old Akali received a pension from the Raynors of two hundred rupees a year, and our Raynor says he remembers him well coming to draw his pension, an old old man with a long white beard and a tall black pugree towering some two feet above his head, with the numerous ornaments with which a grunthi adorns himself; a very frightening figure for a small child to see.' *See* Hailes Papers, **CSAS** Cambridge.

MAGGOTTY MEAT After the mutineers had left **Sialkot** on the evening of 9 July 1857, the European refugees remained about ten days in the fort: they then returned to their (looted) bungalows. Soon after there was a false alarm and they all retired to the cavalry headquarters. It is said that during this period the butchers in the Sadar Bazar showed a 'mutinous spirit' and brought only maggotty meat to their customers; the head butcher saying it was good enough for those who had only a short time to live. **For this the man was promptly hanged**; and when his brethren showed a disposition to non-cooperate and to furnish no supplies, some of the Sikh levy 'were sent to tell them that the whole lot of them were within an ace of being hanged. This brought sufficient and uninterrupted supplies in future.' *See* G.Rich, *The Mutiny in Sialkot,* Sialkot,1924.

MAHABIRI JHUNDA A Hindu flag; a rallying point for Hindus, *cf* the green flag of Islam: the MAHMUDI JHUNDA.

MAHAJUNS Bankers or rich merchants had most to lose from the chaos that inevitably followed the rebellion. Whether it was mutineer sepoys seeking funds for their leaders, or local people looting, the mahajuns invariably suffered. 'During the 16th and 17th (April 1858) Nishan Singh with 2,000 men and 2 Horse Artillery guns... were engaged in digging for treasure in Babu Pershad Mahajun's house and are said to have found 72,000 rupees.'

MAHMUDI JHUNDA The green flag of Islam, raised, as at **Kanpur**, as a rallying point for Muslims to join the holy war against the British.

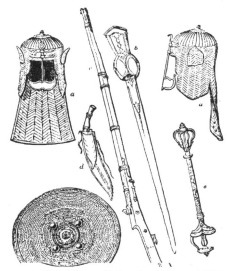

Selection of Maratha arms c.1857

MAHRATTAS [Marathas] During the Mutiny/rebellion a widespread conspiracy to restore Mahratta power was discovered at Satara, the seat of Sivaji's descendants, but it was easily suppressed. *See* Grant Duff, *A History of the Mahrattas.* 2 Vols. London, 1921. But Poona [Pune], the Peshwa's old capital, was loyal to the British in the Mutiny. J. Talboys Wheeler, *India and the Frontier States of Afghanistan, Nipal and Burma.* 2 Vols. London, 1899.

MAHRATTA [Maratha] CHIEFS 'Peshwa of Poona, Gaekwar of Gujerat, Holkar of Baroda, Sindhia of Gwalior,- these held the power. The Rajas of Satara were descendants of the great Sivaji and figureheads for the entire Mahratta confederacy, but it was the Peshwa that was head of the confederacy, although Sindhia was the most powerful.'

MAHRATTA [Maratha] NATION 'The Marathas have always formed a separate nation and still regard themselves as such,' wrote Sir Richard Temple. 'They possess plain features, short stature, a small but wiry frame. Their eyes are bright and piercing and under excitement will gleam with passion. Though not powerful physically as compared with the northern races of the Punjab and Oudh, they have much activity and an unsurpassed endurance. Born and bred in and near the western Ghat mountains and their numerous tributary ranges, they have all the qualities of mountaineers. Among their native hills they have at all times evinced desperate courage. Away from the hills they do not display remarkable valour except under the discipline supplied by able leaders of other races. They never of themselves show an aptitude for organization, but when so organized, they are reckoned among the best soldiers. After the fall of the Maratha Empire, they have betaken themselves mainly to cultivation and to the carrying business connected with agriculture. The Maratha peasantry possess manly fortitude under suffering and misfortune.

A Maratha sepoy c 1930

Though patient and good-tempered in the main, they have a latent warmth of temper and if oppressed beyond a certain endurable limit, they will fiercely turn and rend their tormentors. Cruelty is also an element in their character. Traditions of plunder have been handed down to them from early times and many of them retain the predatory instincts of their forefathers. The neighbourhood of dense forests, steep hill-sides and fastnesses hard of access, offers extraordinary facilities for the display of valour and the preservation of liberty. They work hard in the fields and possess a fund of domestic virtue. The Marathas are born equestrians and sportsmen. As a rule they are not moderate in living and are not infrequently addicted to intemperance. They often feel proud of their low origin even after attaining greatness. The Sindhias boasted of having been the slipper-bearers of the Peshwas.' Temple in Sardesai, *New History of the Marathas*. Vol I, p. 40-1

MAINODIN HASSAN KHAN 'This is the tale of Mainodin Hassan Khan, he had to make his decision - which side to join - in a hurry. More difficult even than for others he had to choose between the heart and the head. The ancestors of Mainodin were adventurers who came from the North, from distant Samarcand, to seek their fortune in the heyday of the Mughal Emperors. India had seen two thousand years of invasions : sometimes they came in their thousands over the passes of the Hindu Kush to give India a new dynasty or a new religion; sometimes they came in ones and twos. There were many such , and many fortunes were made, but it seems that Mainodin's family though brave and honourable had only limited success. At all events, in 1857 he was a Delhi policeman, in charge of the Pahargunge police station it is true, but very much a subordinate. He was an underling of the British; his masters thought well of him, especially the **Metcalfe** family, and he was known also to old **Bahadur Shah**. Mainodin was a writer too. We are lucky that as a policeman he was used to keeping a diary: from it, when the time was ripe he took facts and wove them imaginatively with a spicy touch of added fiction and produced his version of the Great Revolt. He called it *Khodang Godur*, the Mutiny Game. It was a particularly dangerous 'game' he played. This is his story. 'However the English may regard themselves', he says, 'they are trespassers in Hindustan'. This had always been his belief, yet when things started to happen with such speed and amid such confusion, he was as much taken by surprise as anyone. It was in February that a policeman had come to him and asked for orders in the matter of the chapaties.In the villages beyond the suburb of Pahargunge a strange thing was happening. A man would arrive, running, and usually in fear: he would be carrying a chapati, no bigger than a man's palm, and would hand it over to the headman of the village with the message that the village was to prepare five more and send each to another village with orders to do likewise. No reason was given, no one knew why the thing was done, but they all feared to break the chain and did what they were told. It had started 'from the North', was all they knew. There was a tradition among the Delhi folk that it had happened once before, in the time of the great Maratha power, and then it had meant 'a great disturbance would soon follow'. Mainodin tried to stop the distribution of the chapaties, indeed the magistrate Sir Theophilus Metcalfe ordered him to do so, but there was no stopping it. No village dared to be the first to ignore the message, whatever it was: ironically many headmen actually believed the British Government had started the whole thing! No one knows to this day what it all meant: disturbance soon came all right, that part at least was true. Nothing else of note happened until the morning of 11 May when Mainodin had an early visitor at his police station. Buldeo Sing, daroga, or superintendent, of the bridge of boats over the river Yamuna, that linked Delhi with Meerut to the north, came to see Mainodin. He brought incredible news. The day before, in Meerut, the Indian 3rd Cavalry had mutinied, and after setting fire to the Europeans' bungalows and killing anyone who stood in their path, had set out for Delhi to proclaim old Bahadur Shah as Emperor of India! All this despite the presence of hundreds of British troops in Meerut, who did nothing to stop them and had not even followed in pursuit. The news had actually reached Delhi the night before, in a letter addressed to the Commissioner, Mr Fraser, but he had stuffed it into his pocket unopened, and gone to sleep: none of his servants dared wake him though the messenger pleaded the urgency. So no one was alerted, no precautions were taken; more than three hundred mutinous troopers arrived at the bridge, killed an official, Namdeo, at the toll-house, and crossed the bridge into the city. Mainodin did not hesitate: it was his duty to warn the authorities. He rode hard to the courthouse in Delhi to see Mr Hutchinson, the Collector. The news had preceded him; Hutchinson was about to set out for the palace of the King of Delhi to consult Captain Douglas the commander of the palace guard, who was clearly in the front line of defence against any mutineers. ***There were no British troops in Delhi***. Hutchinson was going to his death and Mainodin told him so, but received a curt order to return to Pahargunge, outside the city walls, and there maintain order. He did so, and even paraded his little band of matchlockmen while he waited. In the still hot air, there were ominous sounds, of excited crowds perhaps on the verge of riot, and unexplained shrieks and yells. But in Pahargunge there was nothing to be seen. Away outside the city walls and far from the events that were even now being played out in the east of the city, Mainodin might have waited hours or days before anything disturbed the peace of his suburb. Coincidence however was to involve him rapidly in the drama of Delhi and cast him in a role of great significance. Around the corner, and with a clatter of hooves on the hard-baked dusty road came a European. He was riding a near exhausted horse: it would have been a poor specimen even if fresh but now, carrying the weight of a large and powerful frame, it looked close to collapse. More bizarre still was the figure that sat on it: dressed only in a torn shirt and ragged under-drawers, covered with dirt and dust and, ominously, spattered with blood, the newcomer was none other than Mainodin's patron, **Sir Theophilus Metcalfe,** Joint Magistrate of Delhi, and perhaps its best-known European citizen. Only a few words passed, only a few were needed: the mutineers were in the city and were killing every European they could find. Hutchinson was dead, and Captain Douglas, and numerous others, men and women. Metcalfe himself had ridden for the palace but had arrived too late, the mutineers were already inside. He had then ridden to the central police station and ordered out the police to guard the other gates of the city. For over fifty years the name of Metcalfe had been associated with the city, but now, suddenly, it lost its power. The City Police Chief, the Kotwal, received the order and slowly and deliberately turned and spat upon the ground. The police who stood around heard it also, and just smiled and turned away. A crowd soon gathered, Metcalfe was recognised, and the cry went up that they had got the magistrate. He had to ride for his life; his own horse had soon gone lame from the stroke of a crude sword; he had then fought for and seized the wretched mount on which he had just arrived. He asked for Mainodin's help, and got it, generously and promptly. He asked for native dress, a sword (he was given the best the police station possessed - a blade called a Jari Ghaut) and a horse. He was given all three and then declared that he would ride to his house, some three miles away to the north, and also outside the city walls. It seems he wanted to collect 13,000 rupees, mainly in gold, that he feared would otherwise be looted.

Metcalfe House in 1858

With difficulty he was dissuaded : Mainodin pointed out that Metcalfe House would be one of the mob's first targets, and that he would never get through the crowds. Instead he sent two of his own men, Kaluja Sing and Omrao Mirza, they would merge with the rioters and bring back the treasure box. It seems that Metcalfe agreed reluctantly, and although the men returned they brought him nothing, but said that they could not force their way through the crowds. Nothing more is said or heard of the treasure box, but this may be a tactful omission in Mainodin's account: certainly at a later stage when he needed to put his hand on a very substantial sum of money he had no difficulty in doing so. Mainodin and Sir Theophilus now set out together, and made their way via the Farashkhana bridge to the outer suburb of Bagh Kutapi, some three miles from the city proper. There they found what we would today call a 'safe house', the residence in fact of a man of honour called Bhura Khan Mewatti. Here the European agreed to stay for the time being, although his intention was to join any British force outside the city at the earliest opportunity. Mainodin agreed to go back to the city and to save as many Europeans as he could. He was too late, there were not many Europeans left to save. The Indian cavalry troopers from Meerut had caused a monumental panic; this, plus complete ignorance as to what was happening had paralysed the better disposed part of the native inhabitants: they could think of nothing better to do than to close up their houses and shops and set guards, if they could get them, on their property. For the ill-disposed on the other hand, the breakdown of law and order was heaven-sent. Every type of crime was perpetrated, but the murdering of Europeans and the looting of their homes turned out to be the most attractive, and lucrative. Clearly this was decision time for Mainodin but he found it impossible to decide. As he looked around he saw no native support for the Europeans anywhere. He told himself that what he had seen and heard so far was evil and was the work of evil men, and as a policeman he was pledged and committed to fight evil... and yet... if he was not for the mutineers he would be thought to be against them, and his life and that of his family would be worth nothing. Were he humble and unknown he could hope to fade into the shadows and hope for survival, if not for honour. But he was not unknown; the survival of little men, as little men always survive by keeping out of trouble, was not for him. He was well-known, respected, he could not just disappear. He decided to postpone total commitment as long as he could, to dissimulate if necessary. Had he known the English saying 'to run with the hare and hunt with the hounds', he would have approved of it at this time. He now returned to the city and went to the central police-station, believing that if law and order survived anywhere it would be there. He was wrong: he found that he had to push himself through a throng laden with loot, some of which, but by no means all, came from European houses. To his surprise he found that even the police station itself had been attacked; someone had taken a fancy to the two front doors and had carried them off. Convicts who had started the day in chains in public work gangs were now free and wreaking vengeance on any signs of authority. The Kotwal himself and his deputy were the objects of their attention: they had taken refuge on the roof of the building and would not have long survived had there not been a distraction. What happened then reads like a badly constructed play. From stage left entered two Muslim troopers in a swirl of dust crying ' For the Faith! for the Faith ! Are you all in this place for religion or against it?' Mainodin heard the Kotwal cry out from the roof 'We are all for our religion!', and in the subsequent baying of the lynch mob the Kotwal slipped away from the roof by a back way to be heard of no more. Then from stage right came two more actors: men in bright green with brilliant red turbans and mounted on camels. They thundered out as they approached: 'Hear ye, people, the drum of religion has sounded!' The crowds were terrified, Mainodin among them, and thought these men were heavenly messengers: they were never seen before or after this day. Someone then took up the cry: 'Kill the foreigners!' and the crowd then departed, heading for the Cashmere Gate beyond which lay more Europeans' houses that might not yet have been looted. What Mainodin did next appears to have been a sudden inspiration. Later he was to say that his object was to obtain some appointment to give him influence to stop the butchery of Europeans. That may be true, but perhaps also he had determined to make the best of a bad job and get what he could for himself. **What he did, no less, was to appoint himself Chief of Police, and virtual Governor of Delhi.'   PJOT**

For the remainder of this extraordinary story *see* **SSF** or Charles Metcalfe, *Two Native Narratives of the Mutiny at Delhi,* London, 1898.

MAINPURI (or Mynpoorie)   The chief town of a district of the same name, ceded by Dowlat Rao Sindia to the HEIC in 1803; some 75 miles east of **Agra**. The population was mainly Hindu (Rajput).Three companies of the 9th BNI stationed there mutinied on 23 May, the immediate cause apparently was their refusal to apprehend a compatriot who was a mutineer from **Meerut**. The news that some of their regiment had mutinied at **Alighar** precipitated events. Some of the European officers panicked and fled, but others, notably the magistrate Mr J.Power, and Lieutenant de Kantzow of the 9th BNI kept cool, and with the aid of a loyal gaol guard of about twenty men, and the significant influence of Rao Bhowani Singh, the cousin of the raja of Mainpuri, they saved the Treasury, containing three lakhs, from being looted and held on, just, to authority. But not for long: **Raja Tej Singh** of Mainpuri took control and remained in charge until defeated by Colonel Seaton in the middle of December 1857 when Mainpuri was again occupied by the British. Its strategic position was highly significant being close to the junction of the Agra and Delhi roads with that to Kanpur.

MAJUMDAR, Ramesh Chandra   An eminent Indian historian who was particularly skilful at evaluating evidence. He argues against regarding the rebellion as a national war of independence, and puts the sepoys' mutiny down to fear of loss of caste and religion. He says there was no plot, and points out the lack of a great leader. His conclusion may be summarised in his own words: 'To regard the outbreak of 1857 simply as a mutiny of sepoys is probably as great an error as to look upon it as a national war of independence.' A chapter on atrocities, committed by both sides, is particularly interesting for he strikes a balance between sensationalism and that 'drawing of the veil' syndrome which makes it impossible to come to any true conclusion. The most interesting thing to emerge is that it is well established that General Neill's indiscriminate hangings on his march to Benares[Varanasi] and Allahabad *preceded* in time the massacre of Europeans at Kanpur and cannot therefore be thought of as retribution. (However the lesser massacre of Europeans and Anglo/Indians at **Meerut** and **Delhi** did precede Neill's action). *See* R.C.Majumdar, *The Sepoy Mutiny and Revolt of 1857*, Calcutta, 1963.

MALCOLM, Neill

Under arms, the 93rd Regiment (2nd Battalion the Argyll & Sutherland Highlanders) *en route* from Mian Mir to Nowshera had arrived at Rawalpindi on 24 December 1857. They were under canvas of course. 'What a jolly time we had, and what a lot of good it had done as well! But today is Christmas Day. Needless to say there is no march; breakfast in the mess-tent takes the place of coffee-shop, and so we prepare for the serious duty of enjoying ourselves as much as possible. Unfortunately a wretched grass-cutter, too miserly to clothe himself properly, has died of cold in the night, and his caste brethren come to beg a small allowance in order that the funeral rites may be decently observed. These consist for the most part in giving themselves a feast in which they drown their sorrows - an easy task, especially for those to whom he owed nothing'. This was perhaps the kind of offensive arrogance that helped some of the English convince themselves and others that they were a superior race. It is an extract from the letter home of a young officer in the Punjab in 1857.

MALGUZARS

Those holding land who actually pay the land revenue; in contrast with those who collect or farm it.

MALLESON, George Bruce (1825-1898)

Colonel and military writer; educated at Winchester; Ensign HEIC's Bengal army 1842; Lieutenant 33rd N.I. 1847; Asst. Mil.Auditor-General 1856. A somewhat pedantic writer on military affairs including the rebellion of 1857 vide SIR JOHN KAYE; his attempt to 'reform' the spelling (in the Roman script) of Indian place and personal names was meritorious and no doubt overdue, but it makes his writing irritating and occasionally difficult to follow. Was guardian to the young Maharaja of Mysore 1869-77; CSI 1872. Some of the papers found after his death are preserved in the Oriental and India Office collections of the British Library, *see MSS.Eur. D.706:* they include papers of **Sir Robert Hamilton**, Agent General to the Governor General in Central India 1854-59, a copy of the diary of Captain William Shakespeare, Madras Army 1847-61 who served with the **Central India Force**, and letters from Major Chartres Macpherson, **Sir William Muir** and others. *See also* Bibliography *and* **DNB**.

MALTHONE

A small town on the border between the **Saugur & Narbada Territories** and **Bundelkhand**. Scene of a mutiny of sepoys of the 31st and 42nd BNI who first attacked and captured the fort of Bala Bahat under the orders of their British commandant, Major Gaussen, and then demanded that the Bundela prisoners they had taken (from the Raja of Banpur's forces) be released! In all some forty-five sepoys of the 31st BNI, mutinied, the remainder stayed aloof. The 42nd all mutinied. For details *See* **FSUP**, Vol III and *Revolt in Central India*, Simla, 1908, pp 34-5. Khushhalial Srivastava, *The Revolt of 1857 in Central India-Malwa*, Bombay, 1966.

MALWA CONTINGENT

This was recruited as were most of the **Contingent Forces** from the same areas and castes as the Bengal army. Based upon the cantonment town of **Mehidpur**. On 9 June 1857 the cavalry of the contingent mutinied some fourteen miles from **Neemuch** to which place they had been sent by Colonel **Durand**; they murdered their two officers Captain Brodie and Lieutenant Hunt and then returned to Mehidpur - eighty miles in two days, and attempted to seduce the artillery and infantry of the Contingent, but without success. They then left and joined the **Neemuch Brigade** with whom their fate was bound for the remainder of the struggle. *See The Revolt in Central India*, Simla, 1908. and I.T.Pritchard, *The Mutinies in Rajpootana*, London, 1864; *and* Khushhalial Srivastava, *The Revolt of 1857 in Central India-Malwa*, Bombay, 1966.

MALWA FIELD FORCE

Not to be confused with the **Malwa Contingent**. The Field Force was a column assembled in the Bombay Presidency to open up communications with Central India and the North-West Provinces. The column marched from Poona [Pune] on 8 June 1857 under the command of Major General Woodburn, with orders to proceed to **Mhow**, to prevent the spread of the insurrection in Malwa, and to protect the frontier of the Bombay Presidency.

Rebels kill Colonel Platt at Mhow

It consisted of five troops of the 14th Light Dragoons (just returned from service in the campaign in Persia), a battery of Horse Artillery, some of the Bombay artillery, 25th Bombay Infantry and the Pontoon Train. The column was held up at

Ahmadnagar and diverted to **Aurangabad** to cover the disaffection in the 1st Cavalry of the **Hyderabad Contingent**, and did not resume its march until 12 July. This hold-up probably had significant results in Central India in that many who were hesitating were encouraged to rebel by what seemed like a failure of support from the south for the British cause. Woodburn fell ill and was succeeded by Brigadier C.S.Stuart, who assumed command at Asirgarh while **Colonel H.M.Durand** joined the force on 22 July. For further details of the operations of this force *see Revolt in Central India*, Simla, 1908, *and* Khushhalial Srivastava, *The Revolt of 1857 in Central India-Malwa*, Bombay, 1966.

'MAMA NAROONPUNT'   The father of the third wife of the Nana Sahib, was arrested in Bombay (in 1859?) and was said to have taken 'an active part in the rebellion. His particular department was the destruction of the Telegraph Wire on the Delhi Road.'

MAMA SAHIB   His real name was Moropant Tambe (or Moro Bulwunt?), father of the **Rani of Jhansi** and one of her chief counsellors. He was eventually arrested by the British in April 1858, tried by **Sir Robert Hamilton** the Agent to the Governor General for Central India, and executed by hanging, on a tree in the **Jokun Bagh** at Jhansi. *See* **FPP**, 30 December 1859, Nos 280-88, pp 447-451, quoted in **FSUP**, Vol III.

MAMMOO [MAMMU] KHAN   He was the counsellor and paramour of the **Begum Hazrat** Mahal of Oude [Awadh], and putative father of **Birjis Qadr**. 'On the latter's crowning as King of Awadh in July 1857, Mammoo Khan was appointed *Dewan Khanah* and became very powerful although 'of low origin and illiterate'. Accompanied the Begum to Nepal on the defeat of the rebels. He was eventually dismissed by Hazrat Mahal 'for want of courage and devotion'. Gave himself up to the British and was put on trial, and hanged, *see* **SSF**; *see also* R.Montgomery Martin, *'The Indian Empire'*, Vol II, p 496 *et seq*, and *'Trial Proceedings, Govt vs Mammoo Khan'*, Lucknow Collectorate *Mutiny Basta*; and **FSUP** Vol II.

MAN SINGH (or MAUN SING) **Raja,**   A talukdar of **Oude[Awadh]** with considerable influence in that Province. Remained undecided for much of the period of the rebellion, thus afforded shelter to British women fleeing from the mutineers, but then, when General **Havelock** recrossed the Ganges to Kanpur in August believing himself to be too weak to continue with the relief of Lucknow, Man Singh clearly decided that the British were finally retiring from Oude, and leaving his headquarters at Faizabad marched in to Lucknow to join the rebels. Then, when the British star was in the ascendant again, he changed sides, and demonstrated his 'fidelity' to government by active hostilities against his fellow talukdars, the latter attacked him in his fort at Shahganj. Man Singh was not a member of the ancient aristocracy of Awadh, his grandfather having migrated, to a village near Shahganj, from his ancestral home in the district of Arrah where he was connected with the family of Raja **Kunwar Singh**. Man Singh's uncle and father both achieved wealth and status under the King of Awadh: his father was Governor of Sultanpur, and it was he who built the strong fort of Shahganj. Although the youngest of his father's three sons, Man Singh was the ablest and was soon recognised as the head of the family. He raised considerable forces ostensibly to counter Muslim threats to the sacred shrine at **Ayodhya**: a full scale communal war appeared inevitable but was averted by the intervention of the British Resident at the Court in Lucknow, and Man Singh was left triumphant. It was this action that gave him a great name among the Hindus of Awadh, who thought of him as their natural leader and a talented soldier. The British annexation of Awadh in 1856 was an embarrassment to Man Singh who lost power, privileges and position, and was left deeply in debt. He was thus, at the time of the rebellion, in a receptive mood when bribes or inducements were offered to him by government. A most informative letter dated 4 February 1858 exists (*see* **FSUP**, p 294) which very graphically sums up Man Singh's position: 'A confidential native of Mr Cook's ...has come in today with a verbal message from Maun Singh saying that he is willing to be of any assistance to us and will do whatever he is ordered. He also says that he is doing his best to get hold of the three ladies (see CAPTIVES OF OUDE) now prisoners in Lucknow to forward to us. He says it must be done very quietly and kept a secret by us otherwise their lives will be in danger. He desired the messenger to tell Brenton that **Nana Sahib** was in Lucknow getting together all the mutinous sepoys he could for the purpose of again attacking Cawnpore. He advised us to be on the alert, as this information was *puckah*; he also desired the messenger to inform Brenton that it was not his intention to fight against us, and that he has 4,000 men at Shahgunge for his own protection. He sent in Mrs Duhan and her three children yesterday who says she has received the greatest kindness from Maun Singh. If he saves the other ladies, I hope, he may escape hanging.' After the rebellion he was, despite his doubtful record, handsomely rewarded, *see* M.R.Gubbins, *The Mutinies in Oudh*, 1858; *see also Foreign Secret Consultations*, 30 October 1857, No.137. *and* **FSUP** Vol II

MAN SINGH   'Rajah Maun Singh, hitherto one of the most detested of the Amils, having come forward with his followers to the defence of the Hunnooman Gurhee, and posed as the champion of Hindooism, lost much of his unpopularity, and acquired the respect of the Rajpoots to such a degree as to enable him to act during the Mutiny as the representative and leader of the country community.' J.J.Mcleod Innes. *Lucknow & Oude in the Mutiny*, London, 1895, p 59.

MAN SINGH, RAJA of NARWAR See NARWAR.

MANDANPUR 'In January 1858 the Rahatgarh and Bhopal rebels invaded the Narsinghpur district, 4,000 strong, and attacked and burned Tendukhera. Captain Ternan with two companies 28th Madras Infantry two guns two troops 2nd Cavalry, Hyderabad Contingent, and some mounted and foot police, marched against the rebels, who retreated towards Rahatgarh. The cavalry under Captain Macintire and Lieutenant Ryall, with Dr Bradley and Captain Ternan, made a long march, and dashing into the village of Mandanpur, surprised the place, captured some insurgents, among them a son and grandson of Dilan Singh, the rebel leader of 1842, and killed others. Captain Ternan shot three, *and Dr Bradley dragged two armed men from under an earthen jar.*'

MANDESAR[Mandsaur]

This was an important town in **Scindia's** territory, about 120 miles northwest of Indore, and near to the Rajputana frontier. It was seized by some of the mutineers from **Scindia's** forces in July 1857 and shortly afterwards, on 26 August, the Shahzada Humayon, better known as **Firoz Shah** made his appearance and was greeted with great joy and excitement by the local people, especially the Muslims, many of whom were Afghans: the standard of Islam was raised. For details *see* **FSUP** Vol III p.146 and *The Revolt in Central India, 1857-58*, Simla, 1908 p.13. *Also* Khushhalial Srivastava, *The Revolt of 1857 in Central India-Malwa*, Bombay, 1966.

MANDESAR(Mundesore), Battle of

This was a significant and hard fought affair which took place on 23 November 1857 between Brigadier Stuart's Brigade of the **Malwa Field Force**, and the rebels including the mutinied **Malwa Contingent**. The result was that the rebels dispersed after sustaining heavy casualties, the main body retreated to Nangarh where 'they broke up their standards saying their gods had forsaken them'. This in turn led to the siege of **Neemuch** being lifted. The Hyderabad Contingent Field Force under Major Orr remained at Mandesar and the remainder of the Brigade marched via **Mehidpur** and Ujjain to **Indore** where **Holkar's** troops were disarmed without opposition. For further details *see The Revolt in Central India*, Simla, 1908 p81, *and* I.T.Pritchard, *The Mutinies in Rajpootana*, London,1864, *and* Thomas Lowe, *Central India during the Rebellion of 1857-58*, London, 1860., *and* **FSUP** Vol III; *and* Khushhalial Srivastava, *The Revolt of 1857 in Central India-Malwa*, Bombay, 1966.

MANGAL PANDE

A sepoy of the 34th Native Infantry, Bengal army, who has two claims to fame; his action precipitated the mutiny/rebellion of 1857, and his name gave rise to the description of all mutineers by soldiers of the British army as 'Pandies'. On Sunday 29 March 1857 he was walking up and down in front of the quarter-guard of the regiment at Barrackpore, calling for his comrades to join him in defending and dying for their religion and their caste. Of previous 'good character', it has been suggested that he was at this time under the influence of **bhang** or **opium**, or both; it is also suggested that he suffered from epilepsy, but it is significant that General **Hearsey** speaks of his motivation as 'religious frenzy', not intoxication. He fired at a British sergeant, Hewson, but missed, and fired at the Adjutant, a man called Baugh, and brought down his horse. Baugh's life was saved by a sepoy called 'Sheik Phultoo' who restrained Mangal Pande and deterred others: around 400 men were passive onlookers. The regimental commander Colonel Wheeler ordered the Jemadar of the guard to arrest Mangal Pande but he refused to do so. General Hearsey rode up and Mangal Pande turned his musket upon himself and pulled the trigger with his foot, but was not killed by the shot. He was tried, condemned and hanged on 7 April in the presence of all the troops then at Barrackpore; on 21 April the Jemadar of the guard was also hanged. Mangal Pande has had his share of praise as well as numerous detractors. A.S.Misra in his *Nana Saheb Peshwa and the Fight for Freedom* includes a picture of Mangal Pande Bazar at Meerut - a fruit and vegetable market 'which a grateful people have named after the great hero'; there is also a (possible) picture of the man himself with the controversial caption: 'For his patriotism he paid with his life but until his last breath he refused to disclose the names of those who were preparing for, and instigating, the great uprising'.

MANGALWAR

A large village on the left bank of the Ganges six miles north of **Kanpur**. The scene of **Havelock's** decision, on August 6 1857 to abandon, temporarily, his advance upon Lucknow. *See* telegram in *Further Papers (No.4) relative to the Mutinies in the East Indies*, Inclosure 56 in 2.

MANOWAR SINGH

Killadar of the fort at **Lahori** in Central India. *See The Revolt in Central India*, Simla, 1908 p.122 , *and* Thomas Lowe, *Central India during the Rebellion of 1857-58*, London, 1860, *and* **FSUP** Vol III, *and* Khushhalial Srivastava, *The Revolt of 1857 in Central India-Malwa*, Bombay, 1966.

MANSFIELD, Sir William Rose

1st Baron Sandhurst, Mansfield was Chief of Staff to Sir Colin Campbell during the campaign. He was universally regarded as very good indeed at this job, but a poor direct commander of troops. His papers are preserved in the Oriental and India Office collections of the British Library, *see MSS.Eur. D.174.*

MARDAN SINGH, Raja

See BANPUR, RAJA of.

Sir William Mansfield

MAROCHETTI'S ANGEL A beautiful figure carved in marble and placed over the **Bibigurh** well head in **Kanpur**, in February 1863. It is now in the grounds of the Memorial Church of All Saints in Kanpur. From one side the profile of the Angel is said to look stern, from the other side gentle and merciful: many who have seen it cannot see this distinction. It was the work of Baron Carlo Marochetti (1805-67), most well-known perhaps for his 'Sappho', and patronised by Queen Victoria and Prince Albert. *See* **DNB.** Lord and Lady Canning 'commissioned, at their own expense, Baron Marochetti to produce a figure in white marble.....C.B. Thornhill, then Commissioner of the Division, who had lost two brothers in the disturbances, was in charge of constructing the memorial over the well, with the assistance of Sir H. Yule, R.E., the well-known geographer and scholar. G.E. Lance, Sherer's successor laid out the surrounding plain as a memorial garden.' *See* H.G.Keene, *Handbook. etc.,* Calcutta, 1896, p 34.

MARSDEN, Colonel Frederick Carleton The Commanding officer of the 29th Bengal N.I. Must not be confused with the Colonel Frederick Marsden who was father-in-law of **Gen.Sir Hugh Massy Wheeler**. His diary, dated 1857, is of interest as it contains otherwise unrecorded detail of the mutiny/rebellion as viewed from **Jhelum** where Marsden was stationed. This diary is preserved in the Oriental and India Office collections of the British Library, *See MSS.Eur.B.139/4.*

MARTINDALE, Miss 'Just before we left **Bulandshahr,** a spy reported to me that an English lady was a prisoner in a village some twenty miles off, and that she was anxious to be rescued. As on cross-examination, however, the story did not appear to me very reliable, I told the man he must bring me some proof of the presence of the lady in the village. Accordingly on the arrival of the column at Khurja, he appeared with a piece of paper on which was written 'Miss Martindale'.This necessitated the matter being inquired into, and I obtained the Brigadier's permission to make a detour to the village in question. I started off, accompanied by Watson and Probyn, with their two squadrons of cavalry. We timed our march so as to reach our destination just before dawn; the cavalry surrounded the village and with a small escort we three proceeded up the little street to the house where the guide told us the lady was confined. Not only was the house empty, but, with the exception of a few sick and bedridden old people, there was not a soul in the village. There had evidently been a hasty retreat, which puzzled me greatly, as I had taken every precaution to ensure secrecy, for I feared that if our intention to rescue the lady became known she would be carried off. As day broke we searched the surrounding crops, and found the villagers and some soldiers hidden amongst them. They one and all denied there was the slightest truth in the story, and as it appeared a waste of time to further prosecute the fruitless search, we were on the point of starting to rejoin our camp, when there was a cry from our troopers of 'Mem-Sahib hai!', and presently an excessively dusky girl about sixteen years of age appeared, clad in native dress. We had some difficulty in getting the young women to tell us what had happened; but on assuring her that no harm should be done to those with whom she was living, she told us she was the daughter of a clerk in the Commissioner's office at **Sitapur**; that all her family had been killed when the rising took place at that station, and that she had been carried off by a sowar to his home. We asked her if she wished to come away with us. After some hesitation she declined, saying the sowar had married her (after the Mahomedan fashion), and was kind to her, and she had no friends and relations to go to. On asking her why she had sent to let us know she was there, she replied that she thought she would like to join the British force, which she heard was in the neighbourhood, but on further reflection she had come to the conclusion it was best for her to remain where she was. After talking to her for some time, and making quite sure she was not likely to change her mind, we rode away, leaving her to her sowar, with whom she was apparently quite content. I need hardly say we got unmercifully chaffed on our return to camp, when the result of our expedition leaked out.' F.S.Roberts, *Forty-one Years in India,* London, 1921. A few years afterwards Miss Martindale communicated with the civil authorities of the district, and made out such a pitiful story of ill-treatment by her husband, that she was sent to Calcutta, where 'some ladies were good enough to look after her.'

Constantia

MARTINIÈRE, La, and the RESIDENCY

Predating the Great Mutiny the connection goes back to 1845. Till the annexation of Avadh, the Resident and thereafter the Chief Commissioner, and numerous other officials residing in the Residency were members of the College Committee. These included **Sir Henry Lawrence, Sir James Outram, Sir William Sleeman**, Doctors Login and **Fayrer**, and **Ommaney** and **Major Banks**, many of whom played star roles in the making and in the growth of British power in India. Sir Henry Lawrence, a man with remarkable foresight except in the matter of the Battle of **Chinhat** (30 June 1857), had sensed trouble long before it actually came, and on 20 May decided to provision and garrison the Residency and station troops in houses from Hazratganj to La Martinière, in the Lawrence Terrace, the **Chattar Manzil** and some other places.

George Schilling, Principal of the Martinière, showed similar percipience. Immediately on receiving news of the outbreak at **Meerut** and **Delhi**, the establishment was moved into 'Constantia', the main building of the College, the bigger boys armed and assigned sentry duty, sharing it with the masters. The bridges connecting 'Constantia' with the wings were destroyed and all doors and staircases barricaded or blocked. Lawrence sent eight men and a ser-

Mr Schilling

geant of HM's 32nd Regiment to share duties as he wished to retain the Martinière as an outpost as long as possible. Provisions and water were stored in the small room on the second floor. Gingals were mounted on the bastions and the boys and masters armed with obsolete muskets. Piles of bricks and stones, for hurling at the enemy, were also laid in readiness. Two false alarms by the look-out caused wild excitement. On 17 June Lawrence ordered withdrawal to the **Residency**, and this was accomplished the next day. The smaller boys rode on elephants and, armed with muskets, the senior boys formed the rearguard. The College was accommodated in a hot, close house belonging to an Indian banker, on the extreme southern perimeter. The Martiniere was abandoned but food and clothing continued to be drawn from there till the gates were shut after Chinhat, locking out both the Martinière's flock of sheep, and the **dhobi** who had a large stock of the boys' clothing. In fact, clothes became an even greater problem than food as the siege went on and on. The hard military, domestic or hospital duty that the boys had to do soon wore out what they had on and clothing had to be purchased at the auctions of deceased officers' effects. The siege commenced on 30 June and for the next four and a half months La Martinière boys were to receive an education no other school children had ever received. Between the ages of six and sixteen, according to their capabilities the boys stood to arms; served as hospital attendants, ground wheat and corn till their weakened condition, caused by insufficient food, rendered this impossible, did domestic duties in place of the servants who had absconded, carried messages and continued their studies. A major achievement was the **semaphore** they erected on the Residency tower from instructions in a number of the Penny Encyclopaeida, which helped in maintaining contact between the Residency and the relieving force, and Outram used it to warn **Campbell** 'to give the city a wide berth' to avoid the heavy batteries of the rebels on the direct road to the Residency. For the first time Britain had called upon her schoolboys to fight for her and the Martinière boys responded magnificently. Under Schilling, six masters and the Estate Superintendent, sixty-seven boys entered the Residency and all but two came out alive, in spite of the extremely exposed position of their temporary quarters and the constant danger from bullets, cannon balls, mines and assaults, with the famous sniper **Bob the Nailer in Johannes' House,** just thirty feet across the road. Fourteen senior boys and the masters stood to arms at the Martinière Post. The close proximity to Johannes' House, a stronghold of the rebels, meant constant threat from assaults and mines. The worst nearly happened on 10 August when during the general assault a mine entirely carried away the outer room of the post, blew open the doors of the inner room and destroyed fifty feet of the defences and palisades, while the boys were at prayers. The breach was defended and the rebels driven off. No one was hurt but a soldier accompanying **Brigadier Inglis** was killed. The boys helped to dig the mine from their Post, which blew up Johannes'

House together with Bob the Nailer on 21 August. The constant threat from assaults made defence of the post night and day a most responsible one, for any breach in the perimeter defences would have been disastrous for the entire garrison. For over a month the masters and the bigger boys carried this out entirely unassisted by any regular soldiers, before being reinforced by three privates and a Corporal, later augmented to six privates and a sergeant, of the 32nd. Lawrence was mortally wounded on 2 July, dying two days later. He was attended by three Martinière boys and Roberts the senior boy, repeatedly and at imminent risk, fetched water for washing Lawrence's wound, from a well in an extremely exposed position. The boys formed a covering party for the first assault on Johannes' House as well as on every occasion when a charge had to be made from their side of the perimeter. Two boys were wounded, one while shooting at the rebels and a small one while carrying messages. Two died, both of dysentery. The rations were halved on 26 August and reduced to a quarter when Havelock and Outram arrived on 25 September, causing great hardship especially to the smaller boys who were henceforth on only two ounces. Water for drinking and washing became a major problem when a **bhistie** was hit by a cannon ball and fell into the well supplying the post. Thereafter water had to be had to be obtained at considerable risk from another well just outside the compound. The garrison was evacuated to **Dilkusha** on 19 November, but the very next day those boys who defended the post went back at dusk to defend the trench until the Residency was abandoned on 22 November. Two boys were sent back to Dilkusha on the 21st with two ponies for carrying money and other valuable Martinière property. A number of rebels were killed before the Martinière Post in the grand assault on the Residency on 22nd when the boys were compelled to retire to the basement just before the portico collapsed under the heavy cannonade, which left twenty-four cannon-balls in the Post. Then along with the rest of the garrison, they withdrew through the **Baillie Guard Gate** to Dilkusha. On 23 Mrs Schilling walked to the Martinière, Sir Colin Campbell's headquarters for the day. What she saw was appalling. Nothing remained but the bare bullet and shot-ridden walls of the Martinière. Doors and windows had vanished, the beautiful marble pavement dug up, the noble library destroyed, the superbly ornamented ceiling and walls riddled with musket balls, most of the statues smashed or irreparably damaged. The Founder's Tomb had been broken into and his bones scattered; the destruction was complete. But it was not yet over. Lucknow could not be held and through a countryside teeming with rebels and mutineers, the boys and the rest of the evacuees walked to Kanpur, taking seven days to cover fifty miles. Thence to Allahabad and Varanasi, arriving there on 15 January 1858. Studies recommenced there and the College finally returned to its own campus in 1859. Schilling and other masters received gratuities, Schilling in addition receiving a generous grant of land which he sold and returned to England, resigning his post from December 1859. All the masters and most of the boys received the Mutiny Medal. No other recognition for its unique services to England came to La Martinière till 1937 when the College was granted a Colour, in College colours of old gold and navy blue, with the Residency in one corner and the College Crest in the opposite corner, with the legend 'Defence of Lucknow, 1857'. Popularly known as 'Battle Honours', the Colours are still unique. Perhaps no other school in the world has been similarly honoured.

SRI SATISH BHATNAGAR, IAS (ret'd)

Karl Marx

MARX, Karl Often associated also with Friedrich **Engels;** German born philosopher and 'prophet' of Communism; the two men were contemporaneous with the rebellion of 1857-9, and they took a lively interest both in its nature and progress. They contributed a large number of articles to the New York Daily Tribune, much being, predictably, anti-British and anti-Colonialist, but they were also disappointed by the failure of the revolt, and even more by the reasons for that failure, which they saw as betrayal by the feudal powers in India to support the cause of self-determination, and by the absence of much evidence of true nationalist feeling in the sub-continent. *See* Marx, Engels, *The First Indian War of Independence 1857-1859*, Moscow, 1959.

MASONIC LODGE at **Kanpur** Known as Lodge Harmony this had been founded in 1834 although the building is older. It still stands, behind the Central Telegraph Office (formerly the Station **Theatre**) south of the Mall Road. This was just about the only building connected with the Europeans that survived the rebellion intact. It was a commonly held belief that **Nana Sahib** himself was a member of this Lodge and that he had given instructions that no harm should befall the building. This had been respected. Like other Masonic Lodges in India it was called *'Jadu Ghar'* and it may have been out of a sense of superstitious dread that the Masonic Lodge was spared. Pandit Moti Lal Nehru became a member of Lodge Harmony in 1885 for one year. *See* FREEMASONRY.

MASSACRE GHAT This is perhaps a more accurate designation than **Satichaura Ghat**, (and certainly the name used by those who know **Kanpur** well) for the site on the banks of the River Ganges where a massacre of some 250 of the British garrison of **Wheeler's Entrenchment** took place on 27 June 1857. Sati Chaura Ghat is in Sati Chaura village a hundred yards or so upriver, and the ghat associated with the massacre is that of the Hurdeo temple: the fishermen's or boatmen's temple, which still exists but which has been added to since 1857.

MASSACRES IN KANPUR 'Since the time of the outbreak in **Meerut**, the British in **Kanpur** had been making preparations to protect themselves. **Sir Hugh Wheeler,** the commander of the Kanpur division was convinced of the loyalty of his troops. None the less, he decided to take precautions. He decided, principally because he did not want to be too distant from the sepoy lines, not to use the Magazine adjacent to the river and which, surrounded by a strong wall, was therefore the best suited as a defensive position. Instead he chose a spot nearer the sepoy lines where there were two single storied barracks with verandahs around them and several outhouses. This site he began to entrench, to fortify with artillery and stock with provisions. As the alarm spread in the city in early June he ordered all Europeans into the entrenchment which came to be inhabited by some 900 persons. The British would remain here till 27 June. Surrounded on all sides by rebels who fired on them night and day, the British withstood the siege. Their suffering and heroism are the stock-in-trade of most popular accounts of the Mutiny. On 25 June, the British pickets saw a woman approaching the entrenchments. The woman was either Mrs **Greenway** or Mrs **Jacobi.** She carried a letter written in the hand of **Azimullah Khan** which offered terms of surrender. Negotiations began and **Nana Sahib** agreed to provide safe passage to the British down the river to **Allahabad.** On the morning of 27 June, the British left the entrenchments to proceed to **Satichaura Ghat,** where the boats were kept. According to one estimate, made after comparing different accounts, 450 persons came out of the entrenchments. As the British began to board the boats, guns opened fire from both banks and the thatched awnings of the boats were set alight. All but 130 were slaughtered; 20 of the survivors managed to escape, the rest were taken prisoner. The evidence suggests that the massacre was a stratagem in the conduct of the war. It was planned in advance. A full council met, and decided that the best way to defeat the British was to get them out of the entrenchment with the promise of a safe passage to Allahabad and then to kill them. It is significant that this council was attended not only by Nana Sahib, his men like Azimullah and rebels like **Teeka Singh,** but also by **Maulavi Liakat Ali.** A Hindu prince and a maulavi sanctioned the massacre. It also had the sanction of a qazi. According to one witness the qazi had decreed 'that to murder the Europeans having got them out of the entrenchment was lawful and proper.' The massacre could thus take on the nature of an execution, of an open and public affair. It was a spectacle watched by some 10 to 12 thousand people. And in the manner of executioners some of the rebels told a group of Englishmen, 'now repent of all your misdeeds and ask pardon of God'. The public and open character of the massacre is testified by all the eyewitness accounts that there are of the event. There was no attempt to conceal the hatred and the violence; the guns and troops were strategically placed to prevent escape. The two boats that did get away were chased and shot down. The rebels took up their positions at night and orders were given to the neighbouring zamindars and villagers to be present at the ghat. And they were present, 'armed', one witness reported, 'with swords and battle axes'. The massacre was also executed in keeping with a very definite plan. The boatmen set fire to the thatched awnings at a signal and the guns opened fire at the sound of a bugle. The rebels on horseback went into the water to slash the survivors on very definite orders from **Tatya Tope.** The operations were supervised by Teeka Singh, **Jwala Prasad,** a cavalry trooper called Nukkee and Tatya Tope, all of whom sat on a specially built platform. Everybody present was implicated in the violence, either directly or as a part of a crowd that watched and exulted. It was a spectacle of rebel power. The massacre was celebrated as a great victory; gun salutes were fired to mark the occasion. Nana Sahib took his seat on a throne and the sacred mark was put on his forehead. The city was illuminated for the victory and a proclamation announced the establishment of a new power. The massacre on the river was followed by a second massacre, the nature of which was distinct from that of the first. This involved those who had survived the slaughter at Satichaura Ghat. The men were shot in continuation of what had happened on the river. The women and children were kept as prisoners in a room called the **Bibighur.** After the rebel forces were defeated on 15 July and were forced to retreat from Kanpur, it was decided that the ladies and children imprisoned in the Bibighur were to be killed. A personal servant of Nana Sahib, named **Begum,** who was in charge of the prisoners, brought orders from the Nana for the sepoys to kill the women and children. The sepoys refused to comply and fired a few volleys aimed at the ceiling. At this four or five professional executioners were sent in armed with swords and long knives, and they cut up the prisoners. The bodies - it was said that not all were dead - were thrown into a well. The unity which had marked the first massacre had broken down. The rebels refused to obey their leader. Defeat had led to a loss of legitimacy. The leaders now had to fall back upon their personal servants and on mercenaries. The massacre at Bibighur was not an open affair. It was carried out in a closed room in secret. Sepoys who had killed on the river were unwilling to obey orders. The first massacre had been a spectacle, a show of rebel power; the second an act of retreat, the work of a leadership no longer sure of its power, its mass support and therefore of its victory.'

Dr RUDRANGSHU MUKHERJEE

MATCHLOCKMEN   The term used to describe those who fought on the rebel side but who were not from regular or irregular regiments etc of the Bengal Army, and who would be relatively poorly armed, and largely untrained, but who on occasion fought very well and bravely, particularly in Oude[Awadh]. *See also* TALUKDARS' MEN.

MATHURA,   The **Kotwal** at Mathura was ordered by the British to destroy **Nana Sahib**'s residence at Mathura, and 'to cut down all the trees of his garden' and to announce 'that anyone who has to throw filth and offal may use the garden for the purpose.' By 9 November 1857 the orders had been carried out. *See* Mathura Collectorate *Mutiny Basta* quoted in **FSUP** Vol V p. 892.

MAUDE, Major F.C. V.C., C.B.   A Royal Artillery officer who accompanied **Havelock** on the first 'relief' of the Lucknow Residency, and then accompanied **Sir Colin Campbell** in the final assault upon Lucknow in March 1858. He then remained in Lucknow for some time in the  newly fortified area of the city around the **Luchmantila** and the (repaired) **Machhi Bhawan**. His personal reminiscences are to be found in a book he wrote with his friend **J.W.Sherer**, the Magistrate of Kanpur. *See* F.C.Maude and J.W.Sherer, *Memories of the Mutiny,* 2 vols , London, 1894. Of perhaps greater significance is the collection of photographs, known as the 'Maude Album' (Album No.45 in the **Royal Artillery Institution** Library in the Old Royal Military Academy, Woolwich, S.E.London). The photographs date from 1858, some possibly from 1857, and many are clearly the work of **Felice Beato**, and are familiar to students of the mutiny/rebellion, but what makes this collection unique are the notes that accompany each photograph, from which it is clear that Maude was himself present when the photographer was at work; he also identifies figures, *eg* **Mowbray Thomson,** in a number of prints. Examples from the Maude Album follow, with Maude's comments below:-

'BAILIE GUARD GATE (CENTRE TREASURY, LEFT HOSPITAL), HELD BY THE 13TH AND 71ST N.I. THEY SAVED THE RESIDENCY'

'GREAT MOSQUE IN THE LARGE IMAMBARA. MY WAGONS ON LEFT'

'GATEWAY TO THE IMAMBARA, LUCKNOW. FUSILIERS IN FRONT.'

'BEGUM KOTHEE IN WHICH THE LATE QUEEN OF OUDE LIVED. MUCH RICH SPOIL WAS FOUND HERE AND THE 'NEST WAS WARM'. WE WERE QUARTERED HERE.'

'BREACH IN THE WALL OF THE SECUNDRA BAGH, LUCKNOW, MADE BY OUR ARTILLERY, THROUGH WHICH THE 93RD HIGHLANDERS AND THE SIKHS ENTERED.'

MAULANA ABUL KALAM AZAD   He was India's Minister for Education in 1957, the centenary of the mutiny, and commissioned a re-appraisal of the events from Dr Surendra Nath Sen, the distinguished historian. In the preface to the resultant book ('Eighteen Fifty-Seven') he wrote: *'As I read about the events of 1857 I am forced to the sad conclusion that Indian national character had sunk very low. The leaders of the revolt could never agree. They were mutually jealous and and continually intrigued against one another. They seemed to have little regard for the effects of disagreement on the common cause. In fact these personal jealousies and intrigues were largely responsible for the Indian defeat.'*

MAULVI AHMADULLAH SHAH   (or Syed Ahmadullah)   Better known as 'THE MAULVI OF FAIZABAD'. Already in trouble with the British before the outbreak for what they saw as stirring up communal violence. Known then as a 'troublesome faquir', or 'fanatic'. Called 'Sikunder Shah' in the bazar, and believed to have come originally from the far north-west beyond the Indus. Captured, by force, in Faizabad on 17 February 1857, three of his ten followers being killed. He appears to have had no support from the Hindus of the city. Imprisoned by the British, but eventually released by the mutineers and made his way to Lucknow where at one time his influence rivalled that of **Hazrat Mahal**, the Begum of Oude[Awadh]. He is said to have commanded at the Battle of Chinhat, and to have had a reputation for personal courage, and even for invincibility. He was the last leader to leave Lucknow. Escaping when it was retaken by the British in March 1858; called himself now 'King of Hindustan' and Khalifat-Ullah (God's Deputy). Rs 50,000 offered for his capture; betrayed by the Raja of **Pawayan** who claimed the reward, sending the Maulvi's head in a cloth to the Magistrate at **Shajahanpur**. The Maulvi was the author, or inspired, the document known as **'Futteh Islam'**, issued in mid-July 1857 (dated by internal evidence) which is a significant statement of the cause and nature of the struggle against the British. Also known as 'Danka Shah', from his custom of always having a drum beaten before him when he went out. *See* Capt. Geo. Hutchinson, *'Narrative of the Mutinies in Oude'*, 1859, also W.H.Carey, *'The Mahomedan Rebellion'*, 1857, *and* **FSUP** Vol II. *See also* FUTTEH ISLAM.

MAWE, Dr Thomas   A surgeon in the Bengal Medical Service 1844-57, stationed at **Nowgong** in Central India. Although he succeeded in escaping from the mutineers at Nowgong Dr Mawe died of sunstroke soon afterwards. His letters describing the early stages of the mutiny in Nowgong are preserved in the Oriental and India Office collections of the British Library, *see MSS.Eur.C.324.*

MEAN MEER   This was the name of the military cantonment five or six miles from the city of **Lahore.** In May 1857 there were 3,800 sepoys here including the 16th, 26th and 49th BNI and the 8th Light Cavalry, plus approx. 1,200 European soldiers in H.M.'s 81st regiment and Horse and Foot Artillery. Whether or not there was a conspiracy to mutiny and seize Lahore among the sepoys is open to doubt, but on 12 May a council of European officers assembled by Montgomery the Judicial Commissioner (**Sir John Lawrence** was out of Lahore at the time) decided to urge the disarmament of the sepoy regiments and Brigadier Corbett in command at Mean Meer was informed accordingly. At a remarkable parade on 13 May, ordered with skill and precision, 2,500 sepoys were disarmed without difficulty, and the remainder (on guard duty at Lahore fort and elsewhere) were subsequently disarmed on the 14th; all were then marched back, weaponless, to their lines. *See* **Ball, Martin**; *see also* F.Cooper, *Crisis in the Punjab,* Lahore 1858, p172 and *see Parliamentary Papers,* 9 February 1858, p 4. For a full account of the 'conspiracy' *see Blackwood's Magazine,* Edinburgh, January 1858, article entitled 'Poorbeah Mutiny'. For the eventual fate of the 26th BNI, *see* UJNALLA, Well of.

MEAN MEER 'But what will Lord Canning say to the proceeding at Mean Meer where the 26th Horse Artillery (No, the 26th.BNI - PJOT) after being disarmed had mutinied, murdered their commanding officer and bolted toward Delhi - their flight being by some river, a body of Punjab police and Sikhs had attacked them. Some 230 prisoners being taken were summarily executed and the residue, 500 in all, were either killed in the attack of the mounted police and villagers or drowned in their attempts to cross the river. This is said to have struck terror into the hearts of the other native regiments and they will think twice before making up their minds to go to Delhi and join their bhaees [brethren].' W.T. Seaton, *Letters.* .*See* LAHORE

MEER MEHNDEE   Reported as Meer Mehndi (Mir Menhdi), tutor to **Birjees Kadr,** superintendent of the Intelligence Department during the rebellion; *See* PERSONS OF NOTE, APRIL 1859.

MEER MOHAMMED HUSSEIN KHAN, *Nazim*. Protected Colonel Lennox and his family on their flight from **Faizabad** to **Gorakhpur,** and by all accounts treated them with great kindness. He even visited the mutineers at Faizabad in June 1857 to learn their plans, which was to march to the attack at Lucknow and then proceed to Delhi. They enquired very minutely concerning certain Europeans he had harboured. The *nazim* declared he had only fed and rested three Europeans, and then sent them on. To this the mutineers (and this must have been a surprise) replied - 'It is well; we are glad you took care of the Colonel and his family.' *See Further Parliamentary Papers* (No.4) pp 46-48.

MEERUT   A town of some size and a very important cantonment station of North India, some fifty kilometres (30 miles) northwest of Delhi. The troops stationed here in May 1857 contained a higher proportion of British soldiers than any other cantonment in India, which adds to the surprise that the outbreak occurred here: H.M.'s 6th Dragoon Guards (Carabineers), H.M.'s 60th Rifles (one battalion), a light field battery, a party of Horse Artillery, 3rd Native Light Cavalry, 11th and 20th Native Infantry, some Sappers & Miners; European troops amounted to 1,863 including 132 Commissioned Officers; Native troops numbered 2,912 including 52 Commissioned Officers. The commanding officer of the 3rd Cavalry, the man who may be said to have precipitated the mutiny, was **Colonel G.M.Carmichael Smyth**, while the Officer commanding the station was Major **General Hewitt**. Those murdered by the mutineers on that first day, 10 May 1857, were buried in the churchyard and most of their graves may still be seen.

St John's Church, Meerut, today

From the Burial register of St John's church:- '11 May 12/3 May Buried Colonel John Finnis, Capt John Taylor, Veterinary Surgeons Phillips and Dawson, Corporal Richard Mortimer, Gunner Wm Benson, Lt MacNabb, Charlotte Chambers wife of Capt Chambers, 3 unknown men, Eliza Law aged 9, John Mackenzie Chelsea Pensioner, John Marks ditto, Amelia Courtney, 2 Heatherley children, Sophia Langdale 7, Louise Macdonald, 2 unknown women, Fife-Major Murphy, Capt Macdonald, Lts Henderson and Pattle, Vincent Tregear, Eliza Dawson, 2 unknown men. On 17 May Pte Fred Kingsford and Capt Edward Fraser Sappers & Miners...' Considering that this was 'the place where it all began' surprisingly little has been written about Meerut: perhaps the fact that the action soon moved away and pictures were painted on a grander canvas led to this neglect. A magistrate at the end of the nineteenth century fortunately put on record much that would otherwise have been forgotten and lost for ever. The man who was Commissioner of Meerut in 1857, who died in Delhi and is buried beside **Nicholson** in Delhi, also left behind much useful information. *See Parliamentary Papers* 9 Februray 1858, p3; *see also* N.T.Parker, *A Memoir of Meerut,* Meerut 1904, and H.H.Greathead, *Letters written during the Siege of Delhi,* ed. by his widow, London, 1858. For a detailed account of the whole greased cartridge incident at Meerut, and the outbreak of the Mutiny there *see* Macnabb of MACNAB PAPERS in the **CSAS,** Cambridge.

MEERUT DISTRICT   For details of the rebellion in this district, *see* Eric Stokes, *The Peasant Armed,* Oxford, 1986, pp 143-175, *and also The Peasant & the Raj, Studies in Agrarian Society and Peasant rebellion in Colonial India,* Cambridge, 1978.

MEES DOLLY

'The tale of 'Mees Dolly'. That is the only name we have for her, although if I ever get to visit a certain Regimental Museum, who knows, I might find her real name. Some might call her an anti-heroine, not a romantic figure at all: it depends on your point of view. We first hear of her in a letter written in August 1857 by Captain Henry Norman (who lived to become Field Marshal Sir Henry Norman) at the so-called Siege of Delhi. He was acting as Adjutant General of the Delhi Field Force, and was thus in a good position to know most of what was happening. He speaks first of a European who had been helping the sepoys direct their guns on the British from the ramparts of Delhi. He says '....the 75th Regiment are said to have killed a European former lieutenant in the artillery, who had directed the fire on us'. This was startling enough, but his next sentence is electrifying: '...*by the way, I must mention that a European woman was hung (sic) at Meerut, being implicated in the arrangements for the first outbreak.*' This first outbreak was on 10 May 1857, and the woman hanged was Mees Dolly. Among the scholars, Indian and European, who have come to believe that there was a grand plan, a conspiracy, whether based on religion or nationalism or ambition, there is near unanimity on one point - the 'Mutiny' failed because it broke out prematurely. Mees Dolly, I believe, was responsible for the premature outbreak. Some scholars and investigators even put a starting date for the planned insurrection: in Kolapur it was said to have been 10 August; the Special Commissioner who looked into such things after the 'Mutiny' said ' I am convinced that Sunday 31 May 1857 was the day fixed..' Others believe that the soothsayers who had said that the firenghi rule would last exactly 100 years meant their prophecy to be taken literally, that the British would be overpowered by a successful rising on the hundredth anniversary of the Battle of Plassey, 23 June 1857. The aforementioned Special Commissioner, by name Mr Cracroft Wilson, placed on record his full belief in the story of a general conspiracy for a simultaneous rising. He even found evidence for there being a committee of three members in each regiment whose function it was to be in charge of 'mutiny' arrangements. If all this preparation were in hand, if all the organisation and the will existed, how come that Meerut saw a premature outbreak which, in the end, ruined all the conspirators' hopes and plans? The answer lies with the 'frail ones' of the bazar, and we can only understand this answer if we look in greater detail at the events of early May in the military cantonment of Meerut. In this large garrison town eighty-five troopers of the Light Cavalry had been court-martialled for refusing the orders of their commanding officer, Lieutenant Colonel Carmichael-Smyth, to take and use certain cartridges: they had refused, saying that they feared contamination, for the story was that the cartridges were greased with cow or pig fat, anathema to Hindu and Muslim. The eighty-five had been sentenced to ten years' hard labour, and on 9 May they had been paraded, stripped of their uniforms, and iron fetters had been fastened to their ankles. Strong men wept at the indignity, and many of the troopers cursed their officers and cried out for help to their comrades. But no move was made to save them and they were led away to gaol. Come sunset, on the same day, troopers made their way, as was their wont, to the bazar, to the 'frail ones', the courtesans who had come unofficially to be the recognized regimental women of the Light Cavalry. But the usual welcome was missing: instead taunts and sneers at their manhood. 'We have no kisses for cowards!' was the cry. Were they really men, they were asked, to allow their comrades to be fitted with anklets of iron and led off to prison? And for what? Because they would not swerve from their creed! Go and rescue them, they were told, before coming to us for kisses. Who was the first to break under the jeers? We shall never know his name. But suddenly the cry went up: 'To horse! To horse, brothers! To the gaol, to our comrades!' And the Great 'Mutiny' had begun. No consideration of caste, or religion or patriotism? .....just a taunt from a pair of painted lips! And whose lips were they? What of the European woman that Captain Norman told us was hanged at Meerut? She had, as the saying was, 'turned sour'. She was Mees Dolly. She was a European woman, pure-bred but country-born, the widow of a sergeant in the British army: in those days if such a widow did not re-marry, she had a hard time of it to survive. Mees Dolly had declined all offers of marriage, had got into trouble, probably for theft, and drifted to the bazar where she kept a house of refreshment of sorts. It seems certain that Mees Dolly was no better, as the saying is, than she should be, ran an establishment in which there were girls working for her and, enraged by the cold shoulder shown to her on all sides, was ripe for mischief and the fermentation of any kind of trouble that would let her get back at those she felt had left her friendless and deserted. She was well-known in the bazar: I cannot prove it - and nobody will ever be able to do so now - but it might well have been her house that the troopers frequented: it is fascinating to think that it might have been from her lips that the fateful taunts came! Back to facts. We know that a fortnight after the outbeak in Meerut, Mees Dolly fled from the Sudder Bazar and hid herself in a near-derelict bungalow at the government stud-farm at Hapur not far outside the city. There she was apprehended by a patrol of European cavalry, in the act of driving off in haste. She was brought into Meerut under escort. She was wanted for helping in the murder of two Eurasian girls and, significantly, for 'egging on the mutineers'. She was hanged. The logic is inescapable. If the 'mutiny' failed because it started prematurely, then the final victory of the British may be ascribed to the events of 9/10 May in Meerut. In the words of Cracroft Wilson: 'From this combined and simultaneous massacre, planned for 31 May 1857, we were, humanly speaking, saved by Lt.Col. Smyth commanding the 3rd Regiment of Bengal Light Cavalry, and the 'frail ones' of the bazaar....the mine had been prepared, the train had been laid, the spark which fell from female lips ignited it at once.' What if the 'female lips' were those of

Mees Dolly? Then she may be said to have saved, quite inadvertently, the British Raj in India: with hindsight perhaps the authorities might not have hanged her!' **PJOT**

MEHIDPUR (or Mahidpur)   This is a town and headquarters of the *zila* and *pargana* of the same name, and is on the right bank of the Sipra River some sixty miles from Mandesore[Mandsaur] and about the same distance north of **Indore**. It was the cantonment area for the **Malwa Contingent**. There was a particularly violent outbreak of mutiny here on 8 November 1857. For details *see The Revolt in Central India,* Simla, 1908, p 78.

MEHNDIE HUSSEIN, *Nazim* of **Faizabad**   He was a prominent rebel in Oude[Awadh], and an active military commander of some ability. Particularly known for his attack upon **Man Singh** at his fort of **Shahganj,** when the former finally declared for the British. He was described as a fine tall portly man, with a very agreeable face. Together with his uncle Mir Dost Ali and a number of other Oude leaders he surrendered himself under the **Queen's Proclamation** to Lord Clyde the British Commander-in-Chief in January 1859. He declared that he had been for twenty-five years in the service of the King of Oude, and could not therefore in all honour have fought against the Oude Royal family. His attitude was certainly echoed by many others in Oude. *See* Rudrangshu Mukherjee, *Awadh in Revolt 1857-1858,* Delhi, 1984.

MEMORIAL WELL GARDENS   On the site in Kanpur where the victims of the **Bibigurh** massacre were thrown into a well, the British established a beautiful garden, access to which was restricted *eg* no Indian was permitted to enter the area of the covered-over well, on which was erected a marble sculpture, known as **Marochetti's Angel**. In 1947 the latter was transferred to the churchyard of **All Saints Church** (The Memorial Church near Wheelers' entrenchment); in its place a bust of **Tatya Tope** was erected.

The cost of setting up the original gardens came from the 3 lakhs of rupees fine (=£30,000) levied upon the citizens of Kanpur for its 'too ready acceptance of the **Nana's** occupation'. It is fitting therefore that their descendants should now enjoy this very attractive open space in this modern industrial city. The Gardens were

Tatya Tope

also, in 1947, re-named **Nana Rao Park**. 'The Memorial Gardens were run by a joint committee of civil, military and religious officials. Bishop Cotton consecrated the well and its adjacent graves and monuments in the presence of Lord Elgin, Sir Hugh Rose, Mr. C. Thornhill, the Chief Commissioner of Oudh, and the civil authorities of Cawnpore, Allahabad and Lucknow, and over a thousand soldiers. The total area is near 50 acres; the iron railing for the enclosures were cast at the Ganges Canal Workshop at Roorkee. The Monument cost Rs. 67,000, and for the maintenance a grant of Rs. 50,000 is annually made by the Government of India.' H.G. Keene, *Handbook. etc.,* Calcutta 1896, p 29.

MEMORIAL, The   There was some unseemly argument as to whether the memorial should be ecumenical. The Society for Propagation of the Gospel was suspected of attempting to advance its 'extreme ecclesiastical views' by stepping forward to act as trustee. A former missionary deplored 'the unhappy *odium theologicum'* that was now afflicting the project, and sought to reassure contibutors that the Society 'recognizes no party in the Church.' Another correspondent offered to contribute two guineas of his own and raise 20 more if the church was Baptist. After all, he reasoned, Havelock 'and many of his veterans' were Baptists, and the Baptists were 'the first to send forth heralds of the cross' to Bengal. *See* London *Times,* 4 November, 1857, p 6.

MEMSAHIBS   'Before I began my research into British women travellers abroad, I knew very little about what I had never heard called other than 'The Indian Mutiny'. It was a military affair, I thought, with political overtones, and neither military nor political history interested me much. But then I chanced upon the writings of one **Katherine Bartrum** whose book, *A Widow's Reminiscences of the Siege of Lucknow,* completely changed my perception of what happened in India in 1857-8. Katherine's story encouraged me to look for other first-hand narratives by women, and soon I was utterly engrossed. What sets these contemporary women's accounts of the uprising apart from the official (*i.e.* male) ones is their unashamed immediacy. The memsahibs hardly bothered with the military news, which for all its terrible importance to both sides at the time often became measured and a little stiff in the telling by army chroniclers (even eye-witness ones). The ladies' journals and letters home are far more personal documents, allowed to be all the official ones were not: bewildered, compassionate, trivial, terrified, emotional - even downright disloyal, at times, to the British cause. And their remarkable courage does not involve the battlefield (except in so much as their menfolk might be out fighting on it): it is about enduring and coping with domestic cataclysm. The memsahibs were far less fit to cope with that, being women of their time, than the soldiers and sepoys were with the mechanics of the uprising. In the resourcefulness and sympathy they showed enduring sieges, or trying to escape (often with the devoted help of native servants) through jungle and scrub with their children and babes-in-arms, many of these women dispel the traditional image of their kind that still seems to exist. The memsahib is parodied as a ridiculous whale-boned (yet strangely stout) figure compounded of a distasteful imperial mixture of arrogance, ignorance and intolerance; perhaps she is an ex-member of the so-called 'fishing-fleet' (those cargoes of young unmarried ladies sent out to India to capture eligible bachelors) and now, several years, children, perhaps even a husband or two later, she has staled into an irritable malcontent, pining for a motherland she might never have seen and investing India with all the parochial pettiness of some parlour in the English provinces. This caricature, or one very like it, even obtained at the time of the uprising itself. In fact several commentators (British as well as Indian, *viz* **Sir John Kaye**) blamed much of the whole debacle on these harridan chatelaines of the Empire. The coming of the memsahib to India since the late seventeenth century had driven a wedge between British officers (the sahib) and Indian men (his beloved *bablogue*), irreparably spoiling what used to be a close and interdependent relationship. That much is probably true. But the policy of the **East India Company** in allowing them out in the first place was hardly the memsahibs' fault. They have been blamed as the whoresome muses - but a muse inspires in different ways, and to most horrified observers back in Britain the memsahib neatly symbolised Colonial purity. She was Britannia's virgin daughter, an angel of Albion whose sacrilegious violation at the hands of the mutineers became a metaphor for the violation of the Empire (although it should be said here that no hard evidence was ever published of the rape of British women by the rebels). British soldiers fought the Indians in blood-revenge, calling on the spirit of the slaughtered women of **Cawnpore** to guide them into battle and wearing hanks of their hair as favours - while razing Indian villages to the ground (under the orders of zealous **Brigadier-General James Neill**), along with everyone in them: man, woman or child. In the meantime, against this gusty background of rhetoric and outrage, the memsahibs who had withstood the initial outbreaks of the uprising in India, were involved in the most basic struggle of all: survival. There are some astonishing stories told amongst their number, of escapes through enemy territory (by **Mrs Mawe** and **Maria Hill**, for example); of capture and imprisonment (like **Amy Horne** or **Madeline Jackson**); of loyalty and understanding (typified in Mrs Maude, who when asked by friends whether her husband had been murdered during the Mutiny quietly told them no, he had *died*; nor would she help identify those who had robbed her as she tried to escape after he had been killed: 'they were young', she said, 'none but the young would do so'.). Something many of them have in common is the heart-breaking experience of trying to keep children alive through all the heat and privation: one poor woman, a **Mrs White**, had her husband killed, both her own elbows broken and one of her twin babies injured by a single shot during the siege of **Cawnpore**. She was seen a few days later lying on the filthy floor in **Wheeler's entrenchment** on her back, her arms quite useless, suckling a starving infant at each breast. Not all these stories are as tragic as this: there are happy endings as well as sad ones, after all, and even a welcome touch of humour in some to leaven the general grimness of many official accounts. It is impossible to be completely reverent about women like Mrs **Dunbar Douglas Muter**, who seemed to regard the entire uprising as a personal insult, or the indefatigable Mrs Wagentreiber who after organising a hair-raising escape from Delhi to Kurnaul with her family, still managed to make everyone a nice cup of tea (while all around her were collapsing with nervous exhaustion) as soon as they reached safety. Even the characters themselves found it hard on occasions to keep a straight face as they concentrated on the tricky business of hunting down one another's 'light infantry' (lice) or, towards the end of the uprising, as each was obliged to don as many clothes as she could (with various family portraits, fish-knives and forks, christening mugs and the odd musical instrument sewn in wherever possible) only to be left waddling about in the torrid heat, gently clanking, while waiting for the word to escape. What has tended to be forgotten in history's treatment of the memsahibs involved in the Mutiny/Rebellion is that they were not just symbols, nor mere statistics, but *real people*. I find their testimony to what went on in 1857-8 quite irresistible.'

Jane Robinson

MESSENGER, The Many mysteries, major and minor, remain to be solved in the long saga of the outbreak. One that has attracted perhaps more attention than it really deserves concerns the identity of the messenger sent by **Nana Sahib** on 23/24 June 1857 into **Wheeler's Entrenchment** at **Kanpur** offering to allow the garrison safe passage to Allahabad by boat on the Ganges, provided they surrendered the entrenchment and the money and guns therein. Two surviving witnesses say that it was the Eurasian Mrs Jacobie, wife (or rather widow) of the watchmaker Henry Jacobi: both witnesses would have known Mrs Jacobi well: they were **Kalka Pershad** and **Amy Horne** (herself an Eurasian). Two other surviving witnesses were equally sure that the messenger was Mrs Greenway, the aged and respectable European wife of one of Kanpur's wealthy merchants, Thomas Greenway: these were **W.J.Shepherd**, and **Mowbray Thomson**. The latter 'lifted her over the barricade in a fainting condition, when I recognised her as Mrs Greenway...the whole of the 23rd June the enemy ceased firing upon us. While the deliberations were going on Mrs Greenway stayed in my picket, though all the time eager to return to her little children (grandchildren?) whom her brutal captor had retained as hostages'. It is possible that in fact both women were involved, Mrs Greenway being the first messenger, but, in view of her age and obviously poor condition, a second messenger was sent with the follow-up missive, and that this was Mrs Jacobi. *See Deposition of Kalka Pershad,* **Williams**, *and* Amelia Bennett, *Ten Months' Cap-*

tivity etc., in *The Nineteenth Century* magazine, London, June 1913. *See also* W.J.Shepherd, *A Personal Narrative etc*, Lucknow, 1886 *and* Mowbray Thomson, *Story of Cawnpore*, London, 1859.

METCALF, Thomas R.
The author of an interesting discussion upon the effect the mutiny/rebellion had on subsequent British policy and administration. A scholarly work. He sees the sepoy revolt as 'little more than the spark which touched off a smouldering mass of combustible material'. As a result of agrarian grievances arising from British over-assessment (of land-revenue), and the passage of landed property to the moneylender, the bulk of the people in the NW Provinces gave their support to the rebel cause. *See* Thomas R.Metcalf, *The Aftermath of Revolt*, India 1857-1870, Princeton, 1965.

METCALFE, Sir CHARLES
When Resident of **Delhi** wrote (1818): 'I expect to wake some fine day and find India lost to the English Crown.'

METCALFE, CHARLES THEOPHILUS
The author of an interesting work, being translations (some reports doubt their accuracy) of accounts of the Siege of Delhi in 1857 from an Indian point of view. The diary of **Munshi Jiwan Lal** is perhaps suspect as he was a professional **newswriter** and spy for the British: he was loyal to the British Resident at Delhi, protected by the Palace and kept a careful account of events; valuable as presenting a picture of life within Delhi during the Siege. The account of **Mainudin Husain** is particularly significant. *See* C.T.Metcalfe, *Two Native Narratives of the Mutiny in Delhi*, London, 1898; *see also* Chapter Two of **SSF**.

METCALFE, SIR THEOPHILUS JOHN
(1828-1883)   Joint Magistrate of Delhi in 1857. His escape from the mutineers with the aid of Indian friends is the subject of both fact and legend. *See* **Mainudin Husain**. Also **DNB**. Also correspondence of the Metcalfe Family in the Campbell/Metcalfe Papers in the CSAS, Cambridge. *and* notes, letters, typed and manuscript extracts from printed works which are preserved in the Oriental and India Office collections of the British Library, *MSS.Eur.D.610*.

METCALFE, Sir Theophilus
From W.H. Fitchett, *Tale of the Great Mutiny*: Sir Theo Metcalfe escaped on a horse pursued by cavalry. Horse broke down. Hidden by friendly native in a cave. Pursuers came up and proposed searching the cave. 'On this my friend burst out laughing and raising his voice so that I must hear, he said 'Oh yes. search the cave. Do search it but I'll tell you what you will find, he said. You'll find a great big red devil in there, he lives up at the end of the cave. You won't be able to see him because the cave turns round at the end and the devil always stands just round the turn, and he has got a great long knife in his hand and the moment your head appears round the corner he will slice it off, and then he will pull the body in to him and eat it. Go in, do go in - the poor devil is hungry. It is three weeks since he had anything to eat and then it was only a goat. He loves men does this red devil; and if you all go in he will have such a meal.' Metcalfe guessed what he was supposed to do. Smote head off the boldest. Others fled declaring they had actually seen the red fiend. 'Why did you save my life?' 'Because you are a just man - you decided a case against me in your court, and you were right to do so'.

METCALFE'S PROPHECY
Before the outbreak Theophilus Metcalfe, the joint magistrate at Delhi, told a departing friend 'you are lucky to be going home, for we shall soon be kicked out of India, or we shall fight to the death for our existence.' Wilkinson *The Memoirs of the Gemini Generals*. London, 1896, p 30.

MHOW
This was an important military station twelve miles to the south of **Indore**. Mutiny broke out there on 1 July 1857, and British officers, including Colonel Platt the Commandant, were shot. An artillery officer however, Captain Hungerford, took command and restored order, assisted latterly by **Holkar**. For details *see* **FSUP** Vol III and *The Revolt in Central India*, Simla, 1908, pp 7-10.

MIAN MIR
See MEAN MEER.

MICHELL, Major J.E.
A Royal Artillery officer in the force of Brigadier Douglas that pursued Kunwar Singh from **Azimgarh**. *See* KUNWAR SINGH, Death of.

MILITARY DEFEAT OF THE REBELS, Reasons for
'Despite the courage displayed in any revolutionary liberation movement, whether by individual deeds or collective bravery in combat, and there was much of both in the First War of Indian Independence on the rebels' side, for ultimate success there must be adherence to the time-tested Principles of War. To assess the reasons for the rebels' military defeat in 1857/58, it is necessary to gauge to what extent, if at all, they adhered to these principles. As is well known these principles are - Selection and Maintenance of the Aim; Maintenance of Morale; Offensive Action; Concentration of Force; Economy in the use of Force; Flexibility of Action; Surprise; Security; Co-operation; and Sound Administration.   The selected aim or objective was undoubtedly to drive the British out of India, but for the maintenance of a military aim or objective, co-ordination and unity of command are necessary in order to achieve success. Despite the absence of any reliable account left behind by the rebels, it is possible to be definite that it was not an organised and methodically planned revolt, with any semblance of co-ordination or unity of command that could have ensured a modicum of success. Apart from a shared hatred of British rule, from the disparate activities of the leaders

there is hardly any indication of any unified planning on their part. Whether the **Nana Sahib** and **Maulvi Ahmadullah** of Faizabad, in their reported journeyings, were at all instrumental in inciting concerted revolt, by establishing co-ordinated links with various cantonments, is yet to be established beyond doubt. If there was such co-ordinated planning, it was virtually embryonic. The only semblance of the Maintenance of the Aim to evict the British was that the revolt became quite widespread in a little over two months, but this does not in any way suggest the co-ordination and unity of command so necessary for the rebels to have succeeded, or as **Sir Sayyid Ahmad Khan** described it in 1860 after the event, 'first here, then there, now breaking out in this place, and now in that.' Only after the eruption at **Meerut** on 10 May was it felt necessary by the rebels to establish at **Delhi** a Court or Committee of administration in order to effect unity of command, this Court or Committee consisting of six military and four civilian members. It achieved nothing, for virtually no-one accepted its authority. Prince **Mirza Mughal** who had no military experience was appointed the Commander-in-Chief on 12 May; less than two months later, it was to be Subedar **Bakht Khan,** Mirza Mughal resenting his supersession. The rebels' lack of military leadership became progressively more apparent as time went by. After the fall of Delhi when Bakht Khan moved to **Awadh** and sought to continue operations, his authority was not accepted by some of the other rebel leaders. In other centres of revolt, similar courts or committees were also established. In none did they all pull together, in order to ensure adherence to the paramount principle of war as to the Maintenance of the Aim, to drive the British out. It was not the most proficient or honest who were elected. In **Lucknow,** a ten year old prince **Birjis Qadr** was appointed the Commander in Awadh, because he was the son of **Begum Hazrat Mahal,** in preference to the seasoned Barhat Ahmed, who had defeated **Henry Lawrence** in the skirmish at **Chinhat,** on 30 June, and was now left out of the Lucknow committee or council. Most taluqdars only looked to their own interest, and some, like **Man Sing,** joined whoever appeared to be winning, changing sides on several occasions. The **Rao Sahib** ignored this principle as to the Maintenance of the Aim for having arrived at **Gwalior** in June 1858 he hung around in the region, instead of proceeding to the Deccan and there raising further insurrections amongst the Marathas, of which he was one, being the **Nana Sahib**'s nephew. As to the Maintenance of Morale of the rebels, though good initially, this did not last for much more than a year. Though the rebels received much sympathy from the people at the centres of the revolt, the country as a whole was not behind them. Most rulers, much of the intelligentsia and the majority of merchants actively supported the Government. No revolutionary movement can succeed without the ultimate support of most of the people. Militarily, by 1858, the onset of general demoralisation had commenced, contributing to the eventual defeat of the rebels. Apart from distinguished exceptions, like the **Rani of Jhansi, Kunwar Singh** and **Maulvi Ahmadullah,** the rebels were ill-served motivationally by their leaders. Most of the latter failed to provide the necessary long term motivational support to the rebels. The lack of rebel leadership after 8 June 1857 against the British on the **Ridge** was paralleled by the rebel disorganisation within Delhi. At Lucknow, there were even armed clashes between supporters of contending rebel leaders. In relation to the military necessity for consistency is Offensive Action, and the first impulse of the rebels was, by and large, always to disengage and proceed to Delhi instead of continuing to dominate offensively the locale of the various centres of revolt, eg, the **Neemuch Brigade** inflicted a reverse on **Brigadier T Polwhele** on 5 July near **Agra,** and then moved to Delhi, instead of consolidating its success at Agra itself. At the time of the revolt the aggregate number of Indian military and paramilitary forces was 313,500 (excluding those of the Indian rulers) against 38,000 British in the whole of India, that is, the British were outnumbered more than 7 to 1. Where offensive actions were undertaken, they were often piecemeal with inadequate strength, *eg* against the **Residency** at Lucknow, which was not strategically important. Where concentration of Force was achieved it was unduly protracted, *eg*, **Tantia Tope** hovering around Cawnpore right through November 1857, before defeating **Charles Windham** on 26 November, but in turn being defeated by **Campbell** on 6 December. Economy in the Use of Force was not, also, understood, *eg* Tantia Tope marching around with his aggregation of over 20,000 *en bloc,* thus, at Cawnpore on 6 December he was surprised by Campbell. This was, in fact, the end of the Nana's hopes. Co-operation of Forces was virtually non-existent, each rebel leader seemed to be fighting an independent battle instead of a co-ordinated one in order to be victorious. In order to effect Flexibility of Action, the rebels had no efficient system of communications at their disposal, and hence effective flexibility in deployment was not possible. They could not come to each other's aid in time. Tantia Tope, however, learnt from his defeat in December 1857 and resorted belatedly to more flexible guerilla tactics, but after the back of the revolt had in effect been broken at Delhi and Cawnpore. The principle of Surprise, often as not, was not put to good effect, *eg* the mutineers had the benefit of surprise on their side at **Meerut** on 10 May and yet took no pre-emptive action against the British troops there that night, even allowing for the fact that the British had artillery, but this could not be used to any effect at night. At **Peshawar** in May the British were outnumbered 5 to 1 by the rebellious Indian troops, and yet the latter were surprised. At **Najafgarh,** outside Delhi, on 25 August a larger rebel force was surprised by the sally of a smaller force led by **John Nicholson.** This rebel force thus failed to surprise the British siege train moving to Delhi. Thereafter no attempt was made to stop the siege train from reaching Delhi, due to dissension in the rebel ranks. In relation to ignoring the principle of Security, for example at Lucknow no attempt was made to fortify the north bank of the **Gomti**. Nor, at Gwalior, were any attempts made by the rebels to defend it after they had occupied it. Here the rebels were more concerned in making sweets and letting off fireworks in celebration. There was a total absence of sound Administrative Support. The rebels had no guaranteed sources of arms and ammunition and hence the appeal for aid to the French Emperor by the Nana Sahib, to which there was no response. What they had acquired from the British, could not last for ever. Despite this non-availability, for more than a year, they struggled on, latterly often only with pikes and swords, agaisnt an enemy assured of regular supplies of modern weapons. When the King was in need of money to pay the rebel forces, the rebel forces plundered their own people in Delhi. Often they plundered for themselves. Thus, to conclude this brief assessment, even if in the ultimate analysis the First War of Indian Independence was not an historical tragedy, it was undoubtedly a military tragedy for the rebels, on account of the inability to conform to the principles of War, which apply also to any war of liberation. One cannot but mention that the British were also often very inept militarily in 1857-8, but had the advantage of institutionalised leadership, thus ensuring some conformiity to the principles of war'.

Lieutenant General S.L.Menezes PVSM SC.

MILLS, Cornet William, 1st Lancers, Bombay Army, See MORAR.

MILLS, Maria, She was a survivor of the mutiny/rebellion, having escaped from **Faizabad** in **Oude [Awadh]**. Like so many others she wrote an account of her experiences but it was not published. A contemporary manuscript copy of '*A narrative of the Mutiny, a true and unvarnished tale*', is preserved in the Oriental and India Office Collections of the British Library, see *MSS.Eur.F.171*.

MINAS On 1 July 1857, Captain Forbes was sent to **Deoli** 'to restore confidence', and instructed to raise a corps of 800 Minas to replace the **Kotah Contingent**. This he successfully did and the corps took part in the campaign against the Kotah rebels under the command of Captain Macdonald. The Minas were known as a 'predatory tribe, numbering then about 500,000, and renowned as expert thieves'. They had never previously been recruited into the army by the British but proved so successful that they were afterwards made into a permanent unit known as the 42nd Deoli Regiment.

MILITARY TRAIN Literally the military train consisted of the personnel and transport which followed an army and supplied it with the necessities of war, particularly siege materials. Latterly known as the Service Corps. In 1857 the Military Train was converted to cavalry (of which the British were very short) and served particularly in **Oude[Awadh]** and around **Azimgarh**.

THE CALCUTTA GAZETTE
EXTRAORDINARY

SATURDAY, APRIL 23rd, 1859   No. 573 of 1859

NOTIFICATION

FORT WILLIAM, MILITARY DEPARTMENT

THE 22nd APRIL, 1859.

The Second Battalion Military Train is under orders for immediate Embarkation for England.

The career in India of this Corps has been short but brilliant, and eminently serviceable to its country.

Upon arriving at the Presidency it was at once converted into a Cavalry Force, and sent untrained into the Field under the late Sir Henry Havelock.

Throughout the glorious and most trying Summer Campaign of which the first Relief of Lucknow was the fruit, the Military Train bore a part which would have reflected credit upon the oldest and most experienced Cavalry Soldiers.

It has since served with distinction in various affairs under Lieutenant-General Sir James Outram, at the Capture of Lucknow, in the Operations about Azimghur, and lastly in the harassing Campaign of Shahabad.

The Military Train leaves India with the best wishes of the Viceroy and Governor-General in Council for the future Honor and Prosperity of the Battalion.

A Salute will be fired from the Guns of Fort William on the departure of the Corps.

By Order of His Excellency the Viceroy and Governor-General of India in Council.

R. J. H. BIRCH, MAJOR-GENERAL,

Secretary to the Government of India.

The Military Train

MINING The sepoys were in many places greatly in fear that the Europeans had extensively mined areas over which they might have to march or attack. Nowhere was this more in evidence than at **Lucknow,** where it was widely believed (incorrectly) that the area around the perimeter of the **Residency** had been mined and that anyone attacking over this perimeter would therefore be blown up, but also there was a similar strong rumour concerning **Wheeler's Entrenchment**. Thus: 'At last the Civil and Military Authorities, seeing that the mutinous spirit was daily spreading, dug a trench round the new hospital, paying the men employed double wages and completing the work in four days, and **undermined** the whole of it.' Nerput, Opium Gomashta, *Foreign Secret Proceedings* in **FSUP,** Vol IV, pp 502-3.

MIR MUHAMMAD HUSAIN See PERSONS

OF NOTE, APRIL 1859.

MIR WARIS ALI According to Mir Waris Ali, Tahsildar of a village in the **Kanpur** district, 'the zemindars rose up in arms and taking into their services numbers of armed men began to fight with each other taking revenge of old animosities, dispossessing those who had acquired estates by auctions, by sale, by mortgage or by gift and plundering and burning the corn stored up in principal places.' Kanpur Collectorate Records quoted in **FSUP,** Vol IV, pp 519-20.

MIRACLES 'June 6th (1858), A nice little parson dined with us who was all solicitude about a pattern for his pulpit ornaments in the new church at Delhi. He said to me, Did you observe the ball and cross on the top the church? Yes. Well the sepoys fired at them. The ball is full of bullet holes, the cross is untouched! My good friend wished to imply that something of a miraculous interposition had diverted the infidel missiles, and I did not desire to shake his faith by observing that the cross was solid while it was evident the ball was hollow.' W.H.Russell, *My Diary in India*, London, 1860.

MIRZA BOOLAKEE or BULAQI He was the son-in-law of **Bahadur Shah** King of Delhi; he went to **Lucknow** after the fall of Delhi in September 1857, and was reported as still with the rebels in Nepal in April 1859. *See* PERSONS OF NOTE, APRIL 1859.

MIRZA UBBOO or ABBAS Was a grandson of **Bahadur Shah** King of Delhi; he went to **Lucknow** after the fall of Delhi in September 1857, and was reported as still with the rebels in Nepal in April 1859. *See* PERSONS OF NOTE, APRIL 1859.

MISCELLANEOUS CORRESPONDENCE

It is the misfortune of the serious student of the events of 1857 that over 95% of the documentary evidence emanates from the British side. That is perhaps inevitable for the victor always writes the history. But there are, nonetheless, significant contemporary documents still in existence (to be fair it should be said that they were for the most part collected and preserved by the British authorities) representing correspondence and general administrative affairs of Indian leaders at this time. They are orders, *perwunnahs, hukumnamahs;* petitions, *urzees;* permits or licences, *dastaks; etc.* An excellent, indispensable source for these documents, unearthed from 'Rai Barelli Collectorate Mutiny *Basta*' or 'Lucknow Collectorate Mutiny *Basta*' etc is **FSUP** especially Vol II. Some of these documents are the source - the ***only*** source of information on developments: take for example a letter from Rana **Beni Madho Singh** addressed to **Peshwa Rao Saheb,** in which he gives Rao Sahib (Nana Sahib's nephew) the news of the fall of Lucknow, the movement of Birjis Qadr to **Bahraich**, and the collection of another army to fight the English *etc* - this is incidentally the first intimation we have that despite the fact that Nana Sahib is still alive, and active, his nephew has taken or usurped the title of 'Peshwa.'

MISS WHEELER 'The Highlanders, on coming to a body that had been barbarously exposed, and which was supposed to be that of **Sir Hugh Wheeler**'s daughter, (*see* WHEELER, Margaret Frances) cut off the tresses, and reserving a portion to be sent to their own families, sat down and counted the remainder, and swore that for every hair one of the rebels should die'. One wonders what the wife and bairns, back on the Highland croft, thought of the grisly souvenir? This is a typical contemporary tale made up to suit the mood, or inflame the lust for revenge: it was widely believed. There are other accounts - lurid journalism has been around for many years - they all oddly enough speak of 'Miss Wheeler', as though she had no Christian name, or perhaps as she was the General's daughter it was lacking in respect to mention it. One tale circulated in **Kanpur** itself for some weeks before it went out to the wider world, there to be accepted as the definitive account of 'What happened to Miss Wheeler?' for many a long year. It is said that Ulrica Wheeler was dragged from the boat at the **Satichaura Ghat** by a trooper of the mutinous 2nd Cavalry, by name Ali Khan, and taken to his house where, in the best tradition of savage warfare he had his evil way with her, but paid for it dramatically. Thus speaks the *'Friend of India'* on 3 September 1857.....' she remained with this man till night when he went out and came home drunk; so soon as he was asleep she took a sword and cut off his head, his brother's head, his wife's and two children's. She then went out and seeing other sowars said to them, 'Go in and see how nicely I have been rubbing the Resaldar's feet'. They went inside, and Miss Wheeler then jumped down a well and was killed.' For a tenderly nurtured young girl of eighteen this would have been quite a feat. The story seems to have been based on the account of a certain **Myoor Tewarree** in giving evidence at Cawnpore Camp on 15 August 1857: he speaks of the grief at 'Missy Baba's ' death, and how she was found bloated and very dead in the well the next morning by someone who thought he had seen a man jump into the well during the night. He goes on to say that none of the women were dishonoured 'except General Wheeler's younger daughter'. He also says, surprisingly, that when the sowar took her away with him to his house, she went quietly. Another version of the same tale is slightly more credible in that a revolver is substituted for the sword. She is reported to have secreted the revolver of her father and, in place of cutting everyone's head off with a sword, shot her abductor, his family and friends and then killed herself: there is a lurid illustration of Miss Wheeler 'defending her honour' in this way in Chas Ball's account of the mutiny. Oddly enough while the text tells the story of killing by the sword, the illustration is of her using a revolver. There is even one account which has her killing the trooper with his own sword, and then taking up his revolver (he would not have had one) and shooting his family with it.

This supposed fate of 'Miss Wheeler' was enacted in theatres and described in books and periodicals all over the world - the sensational accounts always included the 'fact' that she is said to have killed her captor, Ali Khan, and his entire family, although the composition of the latter varies from tale to tale.......it could well be that she heard of all these accounts and that they caused her some amusement.....In fact she married Ali Khan and lived to the early years of the twentieth century: it is said that her descendants still live in Kanpur.'
PJOT   See **SSF**

MISSIONARY ESCAPE
Butler could find no palankeens and so improvised a couple by overturning a pair of charpoys and throwing a quilt over its feet. He took with him his '*Hindustani Grammar,* two volumes of manuscript *Theological Lectures*, a couple of works on India, my Passport, my Commission, and Letter of Instructions, with my Bible, Hymn Book, and a copy of the Discipline, and sorrowfully turned away, leaving the remainder to their fate,' including Joel, a native catechist. Judge **Robertson** tried to dissuade him, telling him that **Khan Bahadur Shah**, a Rohilla noble and the Deputy Judge, had assured the station that Bareilly was secure, and had Butler not already packed he might have remained. But something told him to press on, and saying farewell to his little congregation, set forth for Naini Tal. The Butlers passed through the bazar at night but no-one molested them, only inquired who the bearers were carrying. 'The Padre Sahib,' the bearers replied. Along the way they joined up with several officers' wives who were proceeding to the hills alone with their children. They rested at Behari until the heat had subsided and then proceeded into the Terai, a belt of 'rank vegetation... reeking with malaria, and the haunt of tigers and elephants.' His palanquin bearers fell away until he had only enough to carry one, and they extorted 'heavy bucksheesh.' So Butler placed his children with his wife in the one remaining dooly and chased after a 'bullock-hackery, laden with furniture, ...about a quarter of a mile ahead, with its light fading in the distance.' He berated its drivers to take on Butler and his ayah Ann, but looking back saw that the bearers were now refusing to carry his wife and child. If they abandoned the Butlers they would take their torch with them for their own protection against tigers. Butler prayed for two minutes, at the end of which the bearers suddenly picked up the palankeen and proceeded along, and in Butler's heart 'the feeling of divine mercy and care rose above all.' [Robertson was hanged by Khan Bahadur Khan] **ASW**, from William Butler, *Land of the Veda*, New York, 1872, pp 237-8.

MISTAKEN IDENTITY
The reports of **Nana Sahib** visiting Kanpur before the outbreak may have been mistaken. **Nunne Nawab** and **Nana Narain Rao,** the son of **Subadar Ramchander Pant,** were both mistaken for him on their visits. According to Kaye he refused to enter Kanpur for the same reason his father refused: because the cantonment would not fire a salute in his honour. See **FSUP** Vol I. p 378. An even more obvious mistake was made by a certain Mr. H.F. Gibbons of Brick-court, Temple, who informed the *Times* that the Nana Sahib was not, he believed, the adopted son of the late Peshwa, 'nor does he pretend to be so, Dhoondo Punt being the adopted son of that Prince. He is the eldest son of the ex-Peishwa's Soubahdar, Ramchunder Punt, and is, as natives go, tolerably well educated. During his father's lifetime he was on ill terms with the Commissioner, Colonel Manson, and was charged by him with forgery, but the offence was not proved. He afterwards applied, through his father, the Soubbahdar, for the post of commandant of cavalry in the Gwalior service, but it was refused him. On his father's death he claimed to succeed to the bulk of the paternal wealth under a will which had been made out in Mahratta and English, but the authorities considered it a forgery, and his younger brothers, claiming equal shares, as under an intestacy, the courts held him entitled to only one-third, instead of the whole. His disaffected spirit must have been well known, and his having been left in possession of artillery and such like means of offence is one of those problems which the Government of India ought to be called upon to explain.' London *Times,* 19 August 1857.

MITAULI   See LONI SINGH

MITHOWLIE, [MITAULI]   See LONI SINGH

MOFUSSILITE
'About 1845 a clever, free-and-easy newspaper under the name of *The Mofussilite* was started at **Meerut** by **John Lang,** ...and endured for many years.'

MOHAMED ALI KHAN
, 'I then asked him if it was true that the man he had called Micky on our first acquaintance had been one of the men employed by the **Nana** to

butcher the women and children at **Cawnpore** in July? To this he replied: 'I believe it is true, but I did not know this when I employed him; he was merely recommended to me as a man on whom I could depend. If I had known then that he was a murderer of women and children, I should have had nothing to do with him, for it is he who has brought bad luck on me; it is my *kismat,* and I must suffer. Your English proverb says, 'You cannot touch pitch and escape defilement', and I must suffer; Allah is just. It is the conduct of wretches such as these that has brought the anger of Allah on our cause.' I then asked (Mohamed Ali Khan, the condemned spy) if he could give me any idea of the reason that had led the Nana to order the commission of such a cold-blooded, cowardly crime. 'Asiatics', he said, 'are weak, and their promises are not to be relied on, but that springs more from indifference to obligations than from prearranged treachery. When they make promises, they intend to keep them; but when they find them inconvenient, they choose to forget them. And so it was, I believe, with the Nana Sahib.' *See* Wm.Forbes-Mitchell. *Reminiscences of the Great Mutiny,* London, 1893.

MOHAMMED ISMAIL KHAN *Sowar,* trooper, of the 2nd Cavalry which mutinied at **Kanpur**. He abducted Amy Horne, otherwise known as **Mrs Amelia Bennett**, *see* **SSF** Chapter Five.

MOHTIMAM RASAD He was reported as Mohtimam Rasad, **bukshee** of the whole rebel forces, Nawab Naib, his family is said to be in Lucknow; *see* PERSONS OF NOTE, APRIL 1859.

MOHURRAM The Muharram: literally sacred, name of the first Muslim month; the fast held on the 10th of that month, in memory of the death of Husain the younger son of Ali, and grandson of Mahomed, who was slain on that day at Karbela in Iraq, in the 46th year of the Hegira. As the British authorities were aware of the religious background to the mutiny/rebellion they were particularly apprehensive of sacred days, such as those connected with the Mohurram or with Eid [Id].

MONCKTON PAPERS These are held in the Centre of South Asian Studies, Cambridge University and consist of a printed pamphlet entitled *'Letters from Futtehgurh',* which were printed by a 'Lady in Edinburgh quite unacquainted with the writer or her husband.' The letter writer was in fact Rose C.Monckton daughter of Thomas Taylor of Clifton and granddaughter of Sir John Cottrell Bart. She married John Rivas Monckton in 1854, a Lieutenant in the Bengal Engineers, and Superintendent of part of the **Grand Trunk Road**. The letters are dated Futtehgurh 16 and 21 May 1857, consist of the beginning of the mutiny, and a commitment of themselves (husband, wife and child) to Divine Providence; then they speak of the great alarm felt by all the Europeans, rumours and opinions on the mutiny *etc.*, and their determination not to go to the Fort, but to remain together. They did not survive.

MONEY, Gilbert Polkington He was a member of the Bengal Civil Service from 1842 to 1877, and at this period was Magistrate and Collector of **Shahjahanpur**. He survived. A copy of his diary dated 1858 and other relevant documents are preserved in the Oriental and India Office Collections of the British Library, *see Photo.Eur.151.*

MONTGOMERY, Sir Robert (1809-1887) A member of the Bengal Civil Service, appointed in 1827; he was Commissioner of the Lahore Division in 1849, and in 1857 was Judicial Commissioner of the **Punjab**, based in **Lahore.** When the fateful news arrived via the **electric telegraph** on 11 May 1857 of the uprising at Meerut, the Chief Commissioner, **Sir John Lawrence** was at **Rawalpindi**, so Montgomery was in charge, and he did not hesitate to act. He immediately called on the Brigadier and they agreed to disarm the four sepoy regiments at **Mian Meer**; this was done the next day, 12 May, long before the mutiny outbreak became common knowledge. He also took steps to warn the authorities at **Ferozepur, Multan,** and **Kangra,** and it is this promptness and decisiveness of action that earned him the reputation of 'the man who saved the Punjab'. He succeeded **Outram** as Chief Commissioner of **Oude**[Awadh] in April 1858, and from 1859-65 he was Lieutenant Governor of the Punjab. He was a devout Christian and after the suppression of the revolt he wrote: 'It was not policy that saved the Indian Empire to England and England to India. The Lord our God, He it was....'*etc. See* **DNB**; *see also* his correspondence, intelligence reports, press cuttings etc which are preserved in the Oriental and India Office Collections of the British Library, *see MSS.Eur.D.1019.*

MOOFTAHOOD-DAULAH He was reported as Mooftahood-Dowlah [Mifta-ud-Daula] grandson of the late Futteh Alli Khan, a Brahmin convert who was the King of Awadh's Treasurer; *see* PERSONS OF NOTE, APRIL 1859.

MOORE, Captain Although he was by no means the senior officer, so far as rank was concerned, in Wheeler's Entrenchment at Kanpur, Captain Moore of HM's 32nd Regiment appears to have been a kind of second in command to General Wheeler, and when the latter was utterly demoralised by the death of his son Godfrey, Moore seems to have taken over virtual command within the entrenchment; it would appear that it was he who took the fateful decision to accept Nana Sahib's terms for surrender, the mistake that led to the massacre at Satichaura Ghat. 'The whole of the activities connected with the command devolved upon Captain Moore very soon after the commencement of the attack. ...Captain Moore, who was the life and soul of our defence, was a tall, fair man, with light blue eyes, and, I believe, an Irishman by birth. He was in command of the invalid depot of the 32d Regiment when the mutiny broke out. Throughout all the harassing duties that devolved upon him, he never lost determination or energy. Though the little band of men at his direction were daily lessened by death, he was cheerful and animated to the last, and inspired all around him with a share of his wonderful endurance and vivacity. He visited every one of the pickets daily, and sometimes two or three times a day, speaking words of encouragement to every one of us. His never-say-die disposition nerved many a sinking heart to the conflict, and his affable, tender sympathy imparted fresh patience to the suffering women.' Mowbray Thomson, *The Story of Cawnpore*, London, 1859, pp 140-2.

MOORE, CAPTAIN

'On one occasion during the siege, while we were making a sortie to clear the adjacent barracks of some of our assailants, Daniell and I heard sounds of struggling in a room close at hand; rushing in together, we saw Captain Moore, our second in command, lying on the ground under the grasp of a powerful native, who was on the point of cutting the Captain's throat. A fall from his horse a few days previously, resulting in a broken collar-bone, had disabled Moore, and rendered him unequal to such a rencontre; he would certainly have been killed had not Daniell's bayonet instantly transfixed the sepoy.' Mowbray Thomson, *op.cit.*

MOORSOM, Captain William Robert

He was an officer of HM's 52nd (Light Infantry) Regiment, served as ADC to **Havelock**, and as Deputy Assistant Quartermaster General on **Sir Colin Campbell's** staff he was killed on 24 March 1858 in the final assault on **Lucknow**. But he is remembered most for the fact that, although not an Engineer Officer he was skilled in surveying, and showed great talent also for drawing. It is his survey of Lucknow that is the basis for the plans of the attack on the city and the relief of the **Residency** and most accounts of that attack carry his plans as an illustration. His letters diaries notebooks and drawings, together with six interesting photographs dated 1856, are preserved in the Oriental and India Office collections of the British Library, *see MSS.Eur.E.299.* 'A most able and promising young officer of Her Majesty's 52nd Foot, Lieutenant Moorsom, who had left us in the preceding cold weather.' Moorsom had been selected by the Commissioner of Lucknow (Jackson) to conduct a survey of the city in 1856, and on his maps all of Lawrence's defences were based. He brought his own copies with him and they helped Havelock.

MORADABAD

An important town in Rohilkhand. News of the outbreak and massacre at **Bareilly** reached here on 2 June 1857 and the troops of the 29th BNI promptly mutinied. They were disappointed with the contents of the Treasury (Rs 75,000) and threatened to blow the Treasurer away from a gun if he did not lead them to more money. Eventually however they departed, without harming the Europeans, and joined the rebel army in Delhi along with the **Bareilly Brigade**

MORAR

This was the cantonment area for **Gwalior**, where the **'Gwalior Contingent'** was stationed. Most of the Europeans who were killed at this time were buried in the cemetery at Morar, just outside Gwalior. Morar was founded in 1844 as a cantonment, the Brigadier in command and a force of all three arms being stationed here. In 1857 the most serious rising in Central India took place at this station. Lieut Stewart, Artillery, Gwalior Contingent, was the first to be murdered, on 14/15 June 1857, together with his wife and child. Also killed was Major Blake, commanding the 2nd Infantry Regiment. His murder was unusual in that he was exceptionally popular with his sepoys: great remorse was expressed at his death, and the sepoys themselves buried him.

Graves in Morar Cemetery

The Reverend Coopland, chaplain of the station, was also murdered and buried here. Also buried there is Cornet William **Mills,** 1st Lancers Bombay Army killed in action, shot through the heart in the pursuit following the charge which broke the last stand of the rebels on the parade ground at **Lushkar; Battle of Gwalior**, 19 June 1858. Also killed in Gwalior and (probably) buried at Morar were Lt Proctor, **Superintending Surgeon Kirk**, Major Sherriff, Captain Hawkins and two children, Sergeant Cronin, Quartermaster Sergeant Wild and Mr Collins. *See* D.E.Augier (Opium Dept), *List of Inscriptions on Tombs etc in Rajputana and Central India,* contained in the **Augier Papers** in the **CSAS,** *and Delhi Gazette,* 21 June 1857.

MORAR, First Battle of

Technically speaking there were perhaps two battles of Morar. The first was on 31 May 1858 when **Scindia** offered battle to the rebels. On the previous day **Tatya Tope**, the **Rani of Jhansi** and other leaders, at the head of a force of 7,000 infantry, 4,000 cav-

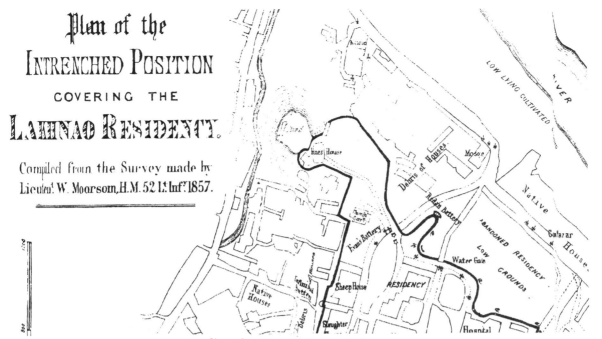

Plan of the Intrenched Position covering the Lakhnao Residency.

Compiled from the Survey made by Lieutent. W. Moorsom, H.M. 52 Lt. Infy. 1857.

From Lieutenant Moorsom's Survey

alry and 12 guns, entered Morar. At daybreak the next morning Scindia marched out with 8,000 men and 8 guns, taking up a position at Bahadurpur, two miles east of Morar and awaited the rebel attack; at 7 a.m. the latter advanced and simultaneously the whole of Scindia's army with the exception of his bodyguard, went over to the enemy; the bodyguard was attacked and put to flight after a gallant defence and Scindia fled to **Agra** with a few faithful retainers.

MORAR, Second Battle of   This took place on 16 June 1858, between the British forces commanded by **Sir Hugh Rose** and the rebels commanded by **Tatya Tope** and the **Rani of Jhansi**. The result was complete victory for Rose, and a significant moral as well as military defeat for the rebels. For details of this action *see The Revolt in Central India,* Simla 1908 p 152 *et seq.*.

MORO BULWUNT   Otherwise known as **Mama Sahib**, the father of the **Rani of Jhansi.**

MOTI MAHAL   One of the older palaces of Lucknow. It figured in the Revolt of 1857 and it was here that Dr Bartrum and Brigadier(?) Cooper were killed on 25 September 1857. The British gave the palace to Raja Dig Vijay Singh of Balrampur. Now houses a number of offices. Courtesy INTACH.

MOWBRAY THOMSON, Captain,   See THOMSON, CAPTAIN MOWBRAY.

MUBARAK SHAH   The *Kotwal* or Chief of Police in **Delhi** during the siege of that city, he left an interesting account of events and conditions in Delhi, which was translated into English by R.M. Edwards in 1859 and can be regarded therefore as a primary source. A copy of this translation is preserved in the Oriental and India Office Collections of the British Library, *see MSS.Eur.B.138.*

MUGHAL SERAI   'The Mughal Serai in **Kanpur** was built around 1830 out of mortar and bricks. **Azizun** hoisted her green Muslim flag here, and it became a rebel headquarters. It is now in a residential area.' Indian National Trust for Art and Cultural Heritage (INTACH). *Preliminary Unedited Listing of Cawnpore. Uttar Pradesh, Unprotected Monuments, Buildings & Structures Listed for Conservation.* Cawnpore, 1992, p 100.

MUGHALS, The   It is impossible, and inappropriate, to study the Rebellion of 1857 without also having some background knowledge of the Mughals, even though their effective rule in India was then a thing of the distant past. For many would say that the great struggle of that year was no more nor less than an attempt to restore the Mughals to power - in the person of their last monarch, Bahadur Shah Zafar, King of Delhi. 'It is not enough to say that Jehangir was a drunkard, that Shahjehan was a voluptuary, and to label Aurangzeb a bigot or a hypocrite. It is true that the descendants of Babur were very human and had an abundant measure of human frailty but these crude epithets do not help us understand how they created a magnificent empire out of next to nothing, how they founded a new school of art and raised buiuldings which are the wonder of the world, and how they drew together a glittering court which dazzled Frenchmen who had seen Louis le Grand at Versailles. And the mughals did even more than this - they achieved something more in the higher walks of statesmanship which nobody before or since has quite succeeded in rivalling. They cast a spell over men's hearts which is comparable only to the spell which Imperial Rome cast upon the peoples of Western Europe, and in the short space of 150 to 200 years they created among the diverse races and creeds of Northern India something approaching to a sentiment of nationality. The men who did this may have been looselivers but they must have been a great deal besides.' Sir Theo Morison, KCIE, in *Blackwoods Magazine,* 1912. 'There was a general idea throughout the country that however badly they might behave, nothing would induce us to resort to extreme severity; and I have no doubt that numbers joined the rebel cause for the sake of the plunder and the tamasha....The King of Delhi's sons, when captured by Hodson, said, with a jaunty air, 'Of course there will be a proper investigation into our conduct in the proper court!' 'The King of Delhi died at Rangoon on 11 Nov 1861 and was buried the same day. Little interest was exhibited by the Mohammedan population of Rangoon' ' Moigul Beg one of the sons of the King of Delhi who cut down Mr Fraser the Resident and afterwards took an active part in the general massacre of Christians at Delhi in May 1857 has been arrested in Scinde and sent to Delhi to stand his trial.... hanged in front of the palace of Delhi on 14 March' . *See* **ILN** 4 January 1862.

MUHAMDI [MOHAMMADI] A town of some size in the Khairabad Division of Oude[Awadh]. The Deputy Commissioner was **J.G.Thomason**. The Europeans, swelled by the refugees from Shahjahanpur were driven out of the town on 4 June 1857 by the mutineers of the Oude Irregular Force: most were massacred some fifteen miles along the road to **Arungabad**. Among the survivors was Captain Patrick Orr. *See* **Kaye**.

MUHAMMAD ALI KHAN Better known as the Nunne Nawab, Muhammad Ali Khan took a prominent part in the rebellion, in **Kanpur**, but was later determined to have been compelled to do so by the Nana Sahib. In 1861 he retired to Mecca, where he died. The rumour that the Nana Sahib died in Mecca may have had its origins in the Nunne Nawab's demise. *See* H.R.Nevill, *Cawnpore: A Gazetteer.*, Allahabad, 1929, p 119.

MUHAMMAD FAZAL AZIM KHAN A *chukledar* of southern Awadh[Oude] who remained loyal to the **Begum of Oude** long after others had defected, *see* **FSUP**, Vol II.

MUIR, Sir William A member of the Company's Civil Service, later the I.C.S., he was an influential man in this period. Held the post of Secretary to Government, North-West Provinces, at Agra, and was thus in effect No. 2 to the Lieutenant Governor, J.R.Colvin. To the scholar his significance lies in his editing *'Reports of the Intelligence Department of the N.W.P.'* 2 vols., 1902. It is possible that he was JUDEX.

MUKHERJEE, Rudrangshu Respected Indian historian who has made a particularly detailed study of the rebellion in Awadh[Oude]. *See* R.Mukherjee, *Awadh in Revolt 1857-1858,* Delhi, 1984.

MULTAN (or MOOLTAN) An area in the Punjab with a particularly violent history which could, if the tribes had risen, have given insurmountable trouble to the British. But the Khans remained supportive of government, possibly because of the personal friendships established with such English political officers as Edwardes and **Nicholson**. Indeed, in May 1857 large numbers were recruited into the 'Mooltani Horse', and these men were largely instrumental in quelling the mutiny of the Bengal regiments in the Punjab. Two regiments (67th and 69th BNI) mutinied here on 2 September 1858, being the last outbreak of the mutiny: they were overpowered and dispersed. *See* J.Royal Roseberry III, *Imoerial Rule in Punjab 1818-1881*, Delhi, 1987.

MUNDESORE *See* MANDESAR

MUNGUL PANDY *See* MANGAL PANDE.

MURRAY, BENJAMIN By no means all Christians were prepared to die for their faith, nor did the rebels/mutineers insist on that sacrifice. Many Indian Christians and Anglo/Indians became (nominal) Muslims rather than die, including probably the majority of the bandsmen of mutinied regiments. Strangely enough there appears to be no case of conversion to the Hindu faith being acceptable apostasy.*'What do you know of Benjamin Murray, the son of pensioned drum-major Murray? He was a drummer in the 6th company, 5th Regiment, on command to Ooraie, with the 2nd and 4th companies; when they mutinied, he turned Mahomedan and came back with the companies to* **Cawnpore**. *...When we were in the Free School compound, he came to us and said he was a fife-major in the Nana's employ.' Deposition of Eliza Bradshaw & Elizabeth Letts.* **Williams.**

MURRAY, Mrs Was the wife, latterly widow, of Drum-Major John Murray of the 56th Native Infantry, pensioned. She said her husband was shot in the head during the embarkation, in the boats at the **Satichaura Ghat**, at **Kanpur** in June 1857. She also says that her brother Hero was also shot in the head, and her two sons Alick and John and her grandsons Robert and Charles, ages 5 and 12, were cut down by tulwars. 'My two grandsons, Robert and Charles, aged five and 12 respectively, were cut down on the spot. My two daughters-in-law, Lewsa and Santa, were cut down, both of whom were pregnant, but the latter, being very far advanced, expected daily to be confined, was ripped open, and the child came out of her womb, which was cut on the spot.' Mrs. Murray's Account. London *Times*, 3 September 1858, p 5. None of these relatives of Mrs. Murray appear on anyone else's list of casualties. Her account is suspect in other respects, and as it had been given publication by no less than the *Times* it called forth a *riposte* from **Mowbray Thomson**, in an effort to put the record straight. *See also* MURRAY, BENJAMIN. 'Mrs. Murray, wife of the pensioned drum-major of the 56th Native Infantry, is also one of the survivors of the massacre on 27 June. She was wounded in several places, and left for dead on the bank of the river, and appears to have been tended by her only son named Benjamin Murray, who had been hiding in the city from the beginning of the outbreak, disguised as a Mahomedan, who took her away to Allahabad when General Havelock's force arrived at Cawnpore.' *See* W.J.Shepherd. *A Personal Narrative. etc.*, Lucknow, 1886.

MURPHY, Private   One of the four survivors of the **Satichaura Ghat** massacre. *See* THOMSON, CAPTAIN MOWBRAY. When the **Memorial Well Gardens** (now known as **Nana Rao Park**) were established, Private Murphy was appointed custodian.

MUSABAGH or MOOSABAGH   About a furlong beyond the fourth milestone on the Lucknow-Bareilly road, and about a mile to the right, is the Moosabagh, in which, under the spreading arms of a fine old mango-tree, will be found the solitary tomb, bearing on it the following inscription : 'Sacred to the memory of Captain F.Wale, who raised and commanded the 1st Sikh Irregular Cavalry. Killed in action at Lucknow on the 1st March 1858. This monument is erected by Captain L.B.Jones, Acting Commandant of the 1st Sikh Irregular Cavalry, as a token of regard for this officer, whom he admired both as a friend and soldier. Captain Wale lived and died a Christian soldier.' The marble plaque indicating the existence of the Moosabagh may still be seen on the main Lucknow-Bareilly road, but the track leading to it from the road is in poor condition - **PJOT.**

The ruins of the Musabagh today

It is said that it was to the Musabagh that Begum **Hazrat Mahal** and other rebel leaders first fled after the re-occupation of Lucknow by the British in March 1858. 'The Moosa Bagh having been taken this morning there is no longer an enemy in Lucknow. Mrs Orr and Miss Jackson arrived safe today..having been rescued by Capt McNeill and Lieut.Boyle of the Artillery with a small party of Goorkas; they were well protected by Meer **Wajid Alee Daroga,** and have been well treated by him.' *from* telegram sent on 20 March 1858 by **G.F. Edmonstone** to E.A.Reade, Agra. Also, previously, known as Barowen or Baronne an Indo-European house, possibly designed by **Claude Martin**. *See* R.Llewellyn-Jones, *A Fatal Friendship*, Delhi 1985

MUSIC   The struggle of 1857-9 inspired (and is still inspiring) many novels and plays: some may endure. The same cannot be said of the music (apart from **folk-songs**) that was produced at the time. The following for example have not been remembered and are never performed: 'The Relief of Lucknow Grand March,' by J. Arthur Owen. 'The Fall of Delhi New March' for the pianoforte, and 'New Song, Delhi,' by John L. Hatton.

MUSICIANS   Caste prohibitions would prevent many Hindus from becoming drummers or musicians, and the practice had therefore grown up of these roles in the army being taken by Christians, usually Eurasians. Few of them, when their regiments mutinied, were actually murdered, the majority surviving on condition that they became Muslim converts, which they seemed happy to do in order to survive.

Certainly the regimental bands appear to have remained intact, and to have played on suitable occasions (especially at Delhi and Lucknow), but perhaps less than suitable tunes: there is considerable evidence that they frequently played 'God Save the Queen!' for example. **'Musicians** The native infantry regiments in the East India Company's service usually have Christian musicians. They are native Christians of various kinds, the greater part the descendants of the old Roman Catholic converts, left without any care, and very ignorant and depraved. Many call themselves Portuguese, but scarcely differ from those before mentioned. The Portuguese in India have mixed with the low caste natives so thoroughly, that there is left scarcely any difference between them. Others of these musicians are native Protestants, who probably conducted themselves badly in their missions; and others are the descendants of the lowest East Indians, sprung from European soldiers and low native mothers. ...Some of the musicians cannot speak English at all, and the greater part of them but imperfectly.' Warren. *A Glance Backward.* p 104.

MUTER, Mrs D.Dunbar The wife of a serving officer, she, like so many others, cannot resist putting her memories into print. However she did not join the many who hastened to be published while public interest was at its height, and allowed her recollections the advantage of objectivity by delaying publication. She describes events from the beginning of the mutiny in **Meerut** in May to the fall of **Delhi** in September 1857. Included is a valuable description of life in the city and conditions obtaining there after it fell to the British. Mrs Dunbar Muter, *My Recollections of the Sepoy Revolt*, London, 1911.

MUTINY & HIS MASTER'S VOICE 'Maharajalal and Sons with its main shop in the famous Chandni Chowk bazar of Delhi, has an ancestry as old as the 'Mutiny' of 1857. The firm has a fascinating tale to tell. That is a tale of an honest public servant of the East India Company who, in spite of loyal and faithful service to the Company which came out victorious in the sepoy revolt, refused to accept any reward from his masters at the end of that revolt. Later, however, after his death his heirs were amply rewarded by the British for the services their ancestor had rendered to the British cause during the fateful days of the mutiny. It is out of that reward that the firm Maharajalal and Sons which held the agency of Gramophone company, His Master's Voice, was born. While working on records at the National Archives of India, New Delhi, about thirty years back, for my doctoral work, I chanced upon a reference to a file which purported to contain a letter written by the Principal of St. Stephen College, Delhi, in 1879, to the Home Department of the Government of India. Being a faculty member of the Hindu College which had a long history of friendly rivalry with St. Stephen College - something like Oxford-Cambridge rivalry in Britain, my curiosity was aroused and I requisitioned the said file from the record room attendant. It is from that file that I got the story which follows. The Principal, in his letter, wrote that he was about to strike off the name of a student from the College rolls for his failure to pay the college fees. But before he did so, he added, he would like to know whether the government would not like to help this student now in penurious circumstances, whose grandfather, Munshi Manoharlal, had rendered valuable service to the British during the dark days of the Mutiny and who had refused to accept any monetary or other material reward for his services after the defeat of the rebel forces. None working in the Home Department at the time the letter came, knew anything about Munshi Manoharlal and, therefore, effort was made to trace the commandants of the forces who had fought the Delhi battle against the mutineers at the northern ridge of the city. It was found that of the three commandants under whom Munshi Manoharlal had worked two were already dead. The only surviving commandant was working at the time as a political agent in the North West Frontier Agency. The file was sent to him. In reply he stated that Munshi Manoharlal was working as cashier of the British forces. In that capacity he was responsible for arranging not only cash flow and its disbursement among the troops, but also supplies of essential commodities for consumption by the fighting forces. This was a difficult and hazardous task because the city up to Kashmere Gate was under the control of the rebel forces whose authority extended right up to the southern border of the ridge. The British forces together with the Sikh troops brought from Punjab which had been annexed less than ten years earlier, on the other hand, were located on the northern side of the ridge. This site which now forms the Delhi University campus with the university and administrative offices, teaching departments and more than 12 constituent colleges, was also the locale of temporary capital complex of British India with the Viceroy and Executive Councillors residing there for two decades from 1911 to 1931. In fact, thanks to the Mutiny, the whole ridge area *ie* the ridge and the area surrounding it, has come to be invested with significance for a student of the history of modern India. Jitgarh (the victory tower) now renamed Ajit (unconquered or unconquerable) tower, the Flag staff, the Khyber pass, and a few graves of unknown soldiers killed in the mutiny battle fought on the ridge are all there besides the ancient monument, Asoka pillar, that stands opposite the Hindu Rao hospital and close to Ajitgarh. A serious student of the Indian mutiny simply cannot afford to miss looking closely at this site for clues to a number of facts about the mutiny which may have escaped notice. The name Khyber Pass which now represents a complex of derelict military barracks and a press building lying unused, was perhaps given to this area on the analogy of *the* Khyber Pass, the narrow road passing through hilly terrain that connects Peshawar (now in Pakistan) with Kabul in Afghanistan. 'Khyber Pass' in the present case meant the narrow strip of road at present running in front of the Old Secretariat building connecting it with Mall Road that marks the eastern boundary of Delhi University campus. During the Mutiny this narrow road had strategic as well as logistic importance. Its strategic importance was that it provided an effective road block to prevent any hostile forces coming from the Northwest to join the rebel Indian forces. It also served as the reception point for the Sikh troops coming from Punjab to join the British forces to fight the rebel forces. The logistic importance of Khyber pass lay in the fact that it was through this narrow path that Munshi Manoharlal arranged supplies of essential commodities for the troops from the city which was in the occupation of the rebels massed inside the Kashmere Gate. Maintaining a constant flow of supplies for the British troops even when the battle was on and Kashmere Gate was effectively controlled by rebels, was indeed a feat for which Munshi Manoharlal was justly praised and posthumously rewarded. All this was brought out in the report made by the retired military officer to whom the case was referred. In view of the meritorious services rendered by Munshi Manoharlal, the officer recommended that a landed estate, or jagir, should be given to his heirs as a grant to pull the family out of penury into which it had fallen. This was done. It was from the income of this estate that not only the college fees of the grandson of Munshi Manoharlal were paid, but also the gramophone agency firm, K. Maharajalal and Sons was established.'

Prof.B.M.BHATIA.

Ajitgarh today

MUTINY MEMORIAL aka AJIT GARH

This impressive Gothic spire dominates the Ridge above Delhi, in the University campus area. It was erected to commemorate the siege and was paid for in theory by all members of the army, regardless of rank, subscribing a day's pay. This however raised only Rs20,000, so government produced the balance and completed it at a total cost of 24,000 rupees. It is 110 feet high. It should not be confused with the granite Mutiny Memorial Cross which was erected at the entrance of the Rajpur Cemetery, also on the Ridge, where most of the dead from the **Delhi Field Force** were buried, and which now (1996) is totally enclosed within a corrugated iron structure calling itself the 'Holy Gospel Church'. See **CM** p 91 et seq.

MUTTRA A large and once wealthy ancient Hindu city some 35 miles from **Agra**. A mutiny there of two companies of the BNI on 30 May 1857 was followed by the looting of the treasury and a march of the troops to join the rebels in **Delhi**. As a result **Colvin** in Agra disarmed the 44th and 67th BNI on the 31st, sending the men to their homes on two months' furlough. Until the fall of Delhi in September 1857 the Muttra area remained disturbed, until **Greathed's** column marched south and relieved Agra.

MUZAFFARNAGAR DISTRICT For details of the rebellion here see Eric Stokes, *The Peasant Armed*, Oxford, 1986, pp 176-198, and also *The Peasant & the Raj, Studies in Agrarian Society and Peasant rebellion in Colonial India*, Cambridge 1978.

MYNPOORIE,[MAINPURI] RAJA of Raja Tej Singh was actively engaged against the British for many months; he surrendered on guarantee of his life on 10 June 1858. It was reported that although 'personally of insignificant ability' his influence over his clan was considerable, and his hereditary position made his surrender a matter of much importance. Eventually banished to **Benares [Varanasi]**, to reside in a house provided by Government, and given a pension of Rs250 *p.m*. A letter from the **Nana** was intercepted in which, after adverting to his misfortune in the rout and dispersal of the Gwalior troops, he informs the Raja of Mynpoorie that he looks to a fresh assemblage at **Calpee** [Kalpi] with the view to another attack on **Kanpur**. William Muir, ed. *Records of the Intelligence Department of the Government of the North-West Provinces of India during the Mutiny of 1857*, Edinburgh, 1902, I: p 340; see also **FSUP**, Vol V pp 856-7. See especially J.W.Kaye, *History of the Sepoy War*, Vol II, London, 1864, pp 161 *et seq*.

MYTHS**,* There are many stories told, often many years *post facto*, that may be true but are likely to be apochryphal. For example, during the so-called 'Siege' of Delhi, June-September 1857, the British sat on the Ridge to the northwest of Delhi, overlooking the old walled city where the rebels gathered in great strength around the King, Bahadur Shah Zafar. One day there occurred a truly disastrous misfortune for those in the city: the gunpowder factory - including all the accumulated stores of gunpowder - blew up with an explosion that could be heard at Meerut, 30 miles away. Until supplies could again be painstakingly brought in, the defenders of Delhi were in a really parlous state, for gunpowder was needed for muskets as well as great cannon. Fortunately perhaps for them, the British did not immediately attack. Now it has always been believed that the blowing up of the gunpowder factory in the city was purely an accident. Mainodin Hussain tells us so in his *Khodang Godur*, the correspondence of the Delhi Kotwal tells the same story, including the call to him to send firefighters to put out the resultant blaze (not a popular assignment I suspect). But some say it was **not** an accident, but was the result of the deliberate action of a spy for the English within the city walls: he is said to have directed the fire of the British guns on the Ridge straight on to the gunpowder factory. Strange how one's perception of the word 'spy' can vary: sneaky, worthless, low-minded traitor if he's on the other side; noble, selfless, brave hero if he's on yours! In the first place the 'spy' turns out to be no less than the Nawab of Loharu, who claimed, in conversation with an English official in 1866, that he had been within the walls of Delhi during the siege, and had had several secret communications with Sir Henry **Barnard**, the British Commander-in-Chief (buried in Rajpur cemetery on the Ridge), as a result of which the Nawab caused to be flown some ***kites made of red paper immediately over the quarter where the rebels had their powder manufactories, in order to direct the fire from the British guns, The result, he said, was complete success, for the guns reached their target at the first salvo. Well. maybe.

PJOT

The King's Palace at Delhi

N

NAGODE A town some forty miles west of **Rewa** in **Bundelkhand**, and the scene of the mutiny of the 50th BNI, although for many months the Europeans at that station were protected by the Raja of **Panna.** *See* numerous entries in **FSUP,** Vol IV, *and* Tapti Roy, *The Politics of a Popular Uprising, Bundelkhand in 1857*, Delhi, 1994.

NAHARGARH A town and strong fort in **Kotah**[Kota] territory, captured by a small British force on 21 January 1859. The **kotwal** was apparently the leader of the resistance, but the place is particularly remarkable for the fact that it appears to have been hostile to *all* comers: **Tatya Tope** was fired at here during his retreat only a week or two before, early in January 1859, *see The Revolt in Central India*, Simla, 1908; *see also* Iltudus T.Pritchard, *The Mutinies in Rajpootana, being a personal narrative of the mutiny at Nusseerabad*, London, 1864.

NAINI TAL 'On Monday, 1 June, fugitives from **Bareilly** began to thunder into Nynee Tal, bedraggled and exhausted, some wearing nothing but the shirt and trousers they had flung on when the sepoys opened fire. They brought news of the uprising on 31 May, and the murder of about half of Butler's little English congregation, including Brigadier Sibbald, and Lieutenant Tucker of the 68th who was shot while trying to fetch a chum up behind him on his horse. Some forty-seven Christians were murdered, among them a widower named Aspinwall who had reportedly been tied to a chair and forced to watch as his children, including a five-week baby, were slaughtered. The sepoys spared Lieutenant Gowan, the Sergeant-Major and his wife, and set them on the road out of the station, where they met up with a party of officers, women and children. Four of these officers became fed up with the pace the women and children set and decided to abandon them. But these officers were killed soon after they set off, and the women and children and the officers who had resolved to stay behind were taken in by a Hindu farmer at a village called Khaira Bajera, who hid them from the rebels for seven months. But Raikes, Judge Robertson, and the station doctor were dragged before **Khan Bahadur Khan** who, displacing Macauley's penal code with the Koran, had them hanged as infidels in front of the jail. The rebels burned almost all of the European buildings, including Butler's religious establishment and his personal library of over a thousand volumes 'which, perhaps, I loved too well.' But, as at Kanpur, they spared the Freemasons' Lodge. The officers from Bareilly were followed that same day by fugitives from **Moradabad,** who had abandoned the station just as the 29th Native Infantry broke out in rebellion. One raced up to Butler, delighted in the view of the lake and the sanitorium, and then turned to ask, 'But are we *safe* here?' Now there were eighty-seven European men to defend Nynee Tal: judges, doctors, planters, chaplains, and three generals 'grayheaded and bent with years.' Commissioner Ramsay was elected commander, and he briefly addressed his motley troops, assuring them that England would rescue them but in the meantime they must be prepared 'to fight, and, if necessary, to die to defend the ladies.' About 25 sowars had galloped up to Nynee Tal with their officers, but Ramsay could not be sure of their eternal fidelity, and assigned them to guard a nearby pass, so that only the Europeans, and a small force of aborigines hostile to the plains sowars, stood guard over the station's munitions. Khan Bahadur Khan was in no hurry to take Nynee Tal, but decided to starve them out and offered rewards for several in the garrison, including 500 rupees for Butler himself. In the rain the Europeans grieved for relatives killed on the plains below, and braced themselves for the inevitable visitation of malaria that raged soon enough among them.' **ASW.** and *see* William Butler, *Land of the Veda*, New York, 1872 pp 246-265.

NAJAFGARH 1. A town on the Ganges some sixteen miles below **Kanpur**, from which guns and infantry fired on the sole remaining boat that had escaped from the **Satichaura Ghat**. *See* **FSUP**, Vol IV. 2. The name of a small town, and canal, in the **Delhi** area, north of the city on the **Grand Trunk Road**: a battle took place here on 25 August 1857 that had a significant effect upon the outcome of the 'Siege' of Delhi.

A powerful siege-train had left **Firozpur** on 10 August, under a weak escort of British troops, and was dragging its ponderous length of five or six miles along the Grand Trunk Road that led to the Ridge above Delhi where the British were encamped. A large rebel force, with sixteen guns, from Delhi was despatched to intercept the siege-train at Najafgarh; if they had succeeded the British assault upon the walls of Delhi on September 14 would have been impossible. **Nicholson** was sent by **Archdale Wilson** to intercept the rebels. He had a large force (2,500 troops and also sixteen guns) and this indicates the importance, indeed the desperation, shown to the enterprise by Wilson. Nicholson's victory was, in some ways, the turning point in the military history of the mutiny/rebellion. For details of the battle *see* L.J.Trotter, *The Life of John Nicholson,* London, 1898, p 262 *and* F.S.Roberts, *Forty-one Years in India,* London, 1921, p 115.

NAM FORTUNATUS SUM A Latin pun, said to have been coined and despatched by an ADC of **Sir Colin Campbell** in March 1858 after the fall of the city, and roughly translated means 'I am in luck, now.' Probably produced in imitation of the more famous terse telegram of Napier some years before, after his successful campaign in Scinde - *'Peccavi'* (I have sinned).

NAMDAR KHAN The *chakledar* of **Sandila,** a chief of **Oude[Awadh]** and a leader of the rebels. Present with his followers at many battles against the British, including **Sandila** on 10 August 1858, with **Prince Firoz Shah.** He remained loyal to the **Begum's** government long after others had defected. *See* **FSUP,** Vol II, *and* T.H.Kavanagh, *How I won the Victoria Cross,* London, 1860.

NANA NARAIN RAO There is a great deal of confusion arising from the use of the common Maratha soubriquet of 'Nana'. Not all references to 'the Nana' should be assumed to refer to Dhondu Punt, the Nana Sahib. For example it is widely held that the Nana Sahib was frequently to be seen at Kanpur, riding or driving on the Mall, and mixing freely with the Europeans at that station, even playing billiards and attending Masonic Lodge. The truth is that Nana Sahib did not like Kanpur, for the same reason that his adoptive father disliked it *viz* no salute was fired when he entered the cantonment. The person known as the 'Nana' who frequently was seen in Kanpur is much more likely to have been Nana Narain Rao, eldest son of the ex-Peishwa's chief adviser, Subhadar Ramchunder Punt, who was on friendly terms with the Europeans.

NANA NARAIN RAO 'About this time, (August 1857) a resident of Bithoor, a Mahratta, named Narain Rao, wrote to me from there to say that he had always been on the English side, had been put into confinement during the supremacy of the Nana, and wished to pay his respects to the General. I showed his communication to General Havelock, and he directed a Persian answer to be prepared, stating that he must be quick about it if he did not wish his loyalty to be suspected. The title or sobriquet Nana is not uncommon amongst the Mahrattas, and this man Narain Rao was so called tbe Nana, and it was, in consequence, rather difficult to procure him a civil reception with those who could not make out who he was. He was no relation whatever to Doondoo Punt, but was the son of an adherent of the Peishwa's, called the Subahdar Sahib, whom I well remember in Mr. Thomason's time. He sometime visited Agra, and was always treated by Mr. Thomason with great respect, as having been acknowledged as a good soldier in his younger days by Sir John Malcolm and even, I believe,. by the Duke of Wellington. The son was very Mahratta-looking. I have mentioned this gentleman's name because some of the stories of the Nana's doings are based upon what he said. But though I make no question of Narain Rao's loyalty, his wish to represent himself a sufferer was so mixed up with the hope of discrediting his brothers, with whom he had a quarrel about a will, that his adventures seem to me apocryphal now that one can think of them quietly.' **Sherer**

NANA'S OFFICERS 'The following were the Nana's officers: — **Baba Bhut** and **Bala,** brothers of the Nana; Narain Dewan; Ragho Punth Aptay; Pandona Shablee; Bishnoo Punth; Patun-gir; Kesho Bhut; Anund Bhut Goottay; Narain Mohnee; Gunput Rao; and Sundhayree; and several other Mahrattas. Besides the above, the following sirdars were also present: — **Jwalla Pershad,** Brigadier; **Azeemoollah;** Ahmed Alie Khan, Vakeel; and the former Kotwal, Holass Singh, and another person, a Mussulman, whose name I do not know; he was of a tall stature and spare body. [Ahmed Ali Khan] used to visit the Nana at Bithoor occasionally.' *Deposition of Narain Kachee.* **Williams.**

NANA RAO PARK Formerly (pre-1947) known as **Memorial Well Gardens** in **Kanpur**.

NANA SAHIB aka DHONDU PANT 'When the massacre became generally known the Nana grew into a European notoriety. The French concocted his personality out of cruel instincts but with delicate and luxurious habitutes. Nana became a scented sybarite, who read Balzac, played Chopin on the piano, and, lolling on a divan, fanned by exquisite odalisques from Cashmere, had a roasted English child brought in occasionally on a pike for him to examine with his pince-nez. ....in England the desire was rather to make out the Nana to have been one of those extraordinary monsters of ferocity and slaughter who were favourite characters in the earliest dramas... ...I remember when in England in 1860 seeing a large canvas daub in a show at a fair which was said to represent the Nana, and he was really a terrific embodiment of matted hair, rolling eyes and cruel teeth. ....but the reality was extremely unlike the romance. I have heard from several who knew him especially Dr J.N.Tressider who attended him professionally that Dhoondhu Punt was an excessively uninteresting person. ....30-40, middle height, stolid features and increasing stoutness, he might well have passed for the ordinary shopkeeper of the bazaar had it not been for the Mahratta contour of his turban, of which however he did not affect a very pronounced type ...did not speak English and his habits of self indulgence had no tinge of poetry about them. He was particular about his ghee, loved the eyes of dancing girls rubbed around with lampblack and their lips rosy with the juice of betel nut, whilst his ear for music was satisfied with the rude viol and the tom-tom that accompanied their slowly revolving petticoats. ..But of any of the refinements of sensual enjoyment he was wholly ignorant. ...apparently he took pleasure to receive occasional visitors at Bithoor. Daily life at home carried on amidst surroundings of expensive discomfort. ...repose on a charpoy in a small room where a loose heifer roamed at will, watched by an attendant who caught its droppings etc. ....Nana was a heavy, dull man with a grievance. We know something of what Azeemoolah did, and Jawala Pershad, Bala Bhut, Tantia Topee and the rest, but the stolid discontented figure of the Nana himself remains in the background, rejoicing doubtless in the success of his treachery and gladly consenting probably to the cruelty, but inanimate, incapable of original ideas and more elated perhaps with the present glory of a hundred guns fired in his honour than with any distinct ideas of future dominion. ...remained so till the end. Even his death was indistinct and insignificant. ....he treated the Begums (or Princesses) his mother-in-law and her half sisters with tyranny, injustice and rapacity, shutting them up in close confinement and not allowing them to hold communication with anyone outside the zenana, which became their prison. .....only occasions on which he actually took the field were on 16 July at Kanpur and 16 August in his own palace at Bithoor. The Nana and his court possessed little or no authority over the rebel troops who, it is evident, did just as they pleased, manned the attacking batteries and joined in the assaults or not just as they deemed fit - the greater portion of them taking their ease lounging in the bazaars and on the banks of the canal, plundering provisions. ...distribution of rewards and pay occasioned much wrangling and bitter speeches against the nominal ruler whom they even threatened to replace by a Mahommedan nobleman. Two factions ...if the British had not so rapidly re-appeared upon the scene there would have been complete disruption amongst the revolters....one must doubt whether the Nana was as guilty of complicity in the murders of our women and children as he is generally believed to have been. Certain that on more than one occasion the Nana befriended the helpless creatures.'...NB the Mahommedan nobleman was Nunne Nawab. Sherer again: 'I am rather of the opinion that his hand though guilty was forced by his more bloodthirsty followers whose acts he had dared not disavow.' This far from flattering picture of the Nana is echoed again and again by commentators and scholars: 'Among those he entertained at the Palace were Sir Hugh Wheeler and his wife, the latter said to have been of the same caste as the Maharajah. He ordered his guns from Purdey and his food from Fortnum and Mason. **Wheeler** trusted him. Incidentally one **Dhondo Pant Gokhale** was the Peshwa Baji Rao II's best general who defeated Kolhapur in Oct. 1798. Nana Sahib named after him? According to Bithur lore, Baji Rao named Nana Sahib 'Dhundoo', or 'Dullard', because he had been unlucky with his natural sons, both of whom had died in infancy. To protect his adopted son from such misfortune Baji Rao gave him a humble name in order not to arouse the envy of the gods. But nobody dared call the son of the ex-Peshwa 'Dullard', and so "Punt' was added: the Maratha equivalent of Esquire. For background information on the Nana *see* J.W.Sherer, *Daily Life during the Indian Mutiny*, 1898.

Nana Sahib, from a portrait painted at Bithur by George Beechey, portrait painter to the King of Awadh

NANA SAHIB, Maharaja of Bithoor [Bithur]

Nana Sahib had a grievance. When his adoptive father Baji Rao II died the Nana inherited all his private wealth; he had hoped also to have inherited the title of Peishwa and the very large pension (or at least a substantial portion of it). Neither of these hopes was realised. The Governor General, Lord **Dalhousie** was adamant that the Nana had inherited enough: in a minute dated 15 September 1851 he ruled: 'The Lt Governor of the North Western Provinces has submitted to me a statement of the affairs of the late Peishwa Bajee Rao, and correspondence connected with it. The Acting Commissioner urges that some portion of the annual pension of 8 lacs formerly paid to the Peishwa should be continued to his heirs. The Lt Governor does not concur in this recommendation. 2. I consider that the suggestion of the Acting Commissioner is uncalled for and unreasonable. For thirty-three years the Peishwa received an annual clear stipend of Rs800,000 besides the proceeds of his Jagheer. In that time he received the enormous sum of more than two Millions and a half Sterling. He had no charges to maintain, he has left no sons of his own; and has bequeathed property to the amount of twenty-eight Lacs to his family. Those who remain have no claim whatever on the consideration of the British Government. They have no claim on its charity because the income left to them is amply sufficient for them. If it were not ample, the Peishwa out of his vast revenues ought to have made it so; and the probability is, that the property left is in reality much larger than it is avowed to be. Wherefor under any circumstances the family have no claim upon the Government; and I will by no means consent to any portion of the public revenues being conferred upon it. I request that this determination of the Government of India may be explicitly declared to the family without delay. The minor arrangements suggested by the Lt Governor may be sanctioned at once. Sd. DALHOUSIE.' There were rumours - after the event so to speak - that the Nana had been an active conspirator against the government for some years before the outbreak, and that he had tried to influence other Princes - especially those with a grievance against the Government, but that he had had no success until Oudh[Awadh] was annexed (in 1856), after which the sepoys began making their plans. Much of the 'evidence' - it is all confident assertion rather than hard proof - comes from one **Sitaram Bawa** in a statement made before H.B.Devereux, Judicial Commissioner of Mysore on 18 January 1858. While the death of Nana Sahib, in 1859, in Nepal was assumed rather than proven, a continual watch was kept for him for some years after the suppression of the revolt, and more than one 'Nana Sahib' was arrested with a view to being put on trial; correspondence dating from 1861 and 1862 between officials in Sind and the Punjab concerning captures at Karachi and Karnal is preserved in the Oriental and India Office collections of the British Library. *See* LEWES. *See MSS.Eur. D.820. See also* **FSUP** Vol I, p 372 *et seq. See also* **SSF** Chapter 3; and P.C.Gupta, *Nana Sahib at Bithur,* in *Bengal Past & Present,* Vol 77, 1958; and P.C.Gupta, *Nana Sahib & the Rising at Cawnpore,* Oxford, 1963.

NANA SAHIB'S CIRCULAR
According to a circular distributed by his agents throughout Bundelkhand in January of 1858, his object was 'not to take possession of the territories and property of the Rajahs and Chiefs of India, or to assume the supreme command of the Country, but on the other hand, it is his sole desire that after a victory shall have been obtained over the enemy, all the Chiefs may in peace enjoy the possession of the territories which they at present hold as well as those which they formerly possessed, and pass their days in the enjoyment of ease and happiness.' *See* **FSUP,** Vol I, p 459.

NANA SAHIB'S DEATH
In a letter written on 8 October, 1859, Lieutenant Colonel **Ramsay**, the Resident of Nepal, wrote **Beadon** that he had received a report from the Nana's camp in the Deokur Valley that the **Nana** had died on 24 September, according to the females of his family, 'now residing at Tara Gurrhee. But ...clear authentication of Nana's death will be very desirable, as the Nana is a Brahmin and is reputed to be very wealthy, and, in the present temper and spirit of the local authorities in the **teraie**, it is not at all unlikely that they may be conniving at his escape into the mountains.' NA. *FDP. FC.* 4 December, 1859. No. 159. The *'Englishman'* reported that Nana Sahib died on 24 September of fever at a place called 'Tara Ghurrie near to Dhang and Deokhur. He had suffered previously from repeated attacks and at one time was so ill that his attendants believed him to be dead, and the usual gifts were distributed amongst the Brahmins; he however recovered partially and did not die till 24 September. His dead body is said to have been seen by creditable witnesses, and to have been burnt in their presence with the usual Hindoo rites.' *See The Friend of India* in **FSUP,** Vol IV, p 782. 'It has once or twice, since those days, been doubted whether the Nana did die, as reported. I think the doubts were unreasonable. When Jowala Purshad was in our lock-up, as his fate was quite certain, and escape impossible, I directed a modification of his fetters, which enabled him to eat with more convenience, and he was grateful. Moreover he knew his sentence did not depend on me and so he was not afraid, and answered readily when I spoke to him. He told me, if I remember his words rightly, that he was not present when the Nana died, but that he attended when the body was burned. He spoke apparently without intention to deceive, and I fully believed him.' *See* J.W.Sherer, *Havelock's March on Cawnpore: A Civilian's Notes,* London, 1910, p 328-30.

NANA SAHIB'S DESCRIPTION
The following was the description circulated among the British stations. 'The Nana is 42 years of age. Hair black; complexion, light wheat coloured; large eyes; and fat round face; he is understood now to wear a beard; height, about 5 feet 8; he wears his hair very short (or at least did so), leaving only so much as a small skull-cap would cover; he is full in person, and of powerful frame; he has not the Mahratta hooked nose with broad nostrils, but a straight well-shaped one; he has a servant who never leaves his side, with a cut ear', William Muir, ed. *Records of the Intelligence Department of the Government of the North-West Provinces of India during the Mutiny of 1857,* Volume 1, Edinburgh, 1902.

NANA SAHIB'S 'EXPLANATION'
'When your army mutinied and proceeded to take possession of the treasury my soldiers joined them. Upon this I reflected that if I went into the entrenchments my soldiers would kill my family, and that the British would punish me for the rebellion of my soldiers; it was therefore better for me to die.' Nana Sahib in *Foreign Political Consultations* quoted in **FSUP,** Vol IV, pp 772-4.

NANA SAHIB'S PALACE
'The palace (at **Bithur**) was spacious, and though not remarkable for any architec-

tural beauty, was exquisitely furnished in European style. All the reception rooms were decorated with immense mirrors and massive chandeliers in variegated glass, and of the most recent manufacture: the floor was covered with the finest productions of the Indian looms, and all the appurtenances of eastern splendour were strewed about in prodigious abundance.

The Subedar Brothers, descendants of Ramchunder Subedar, on the site of Nana Sahib's Palace at Bithur

There were saddles of silver for both horses and camels, guns of every possible construction, shields inlaid with gold, carriages for camel-driving and the newest turn-outs from Long Acre; plate, gems, and curiosities in ivory and metal; while without in the compound might be seen the fleetest horses, the finest dogs, and rare specimens of deer, antelopes, and other animals from all parts of India. It would be quite impossible to lift the veil that must rest on the private life of this man. There were apartments in the Bithoor palace horribly unfit for any human eye; in which both European and native artists had done their utmost to gratify the corrupt master, from whom they could command any price'. See Mowbray Thomson, *The Story of Cawnpore,* London, 1859, pp 46-8. 'For downright looting commend me to the hirsute Sikh; for destructive aggression, battering, and butt-ending, the palm must be awarded to the privates of Her Britannic Majesty's Regiments. 'Look what I have found!' said a too demonstrative individual of the last-named corps, at the same time holding up a bag full of rupees for the gaze of his comrades, when an expert Sikh with a blow of his tulwar cut the canvas that held the treasure, and sent the glittering spoil flying amongst the eager spectators. A large portion of the Nana's plate was found in the wells around the palace; gold dishes, some of them as much as two feet in diameter; silver jugs; spittoons of both gold and silver, that had been used by the betel eating Brahmin, were fished up, and proved glorious prizes for somebody. Every cranny in the house was explored, floors were removed, partitions pulled down, and every square foot on the surface of the adjacent grounds pierced and dug in the search after spoil. Brazier's Sikhs have the credit of carrying off Bajee Rao's state sword, which, in consequence of its magnificent setting with jewels, is said to have been worth at least thirty thousand pounds.' Mowbray Thomson, *The Story of Cawnpore* p 49

NANA SAHIB'S PAPERS 'After the capture of **Delhi** I was appointed one of the Special Commissioners to hunt up rebels, and my chief task was to overhaul the papers of the Nana at **Cawnpore**. It was curious to see how the men who had been credited with being faithful servants of the Government hastened to ply the Nana with fulsome adulation as soon as they saw how events turned at Cawnpore. One man, writing from **Futtehpore**, reported the approach of an English steamer up the Ganges and asked for orders, on which the ***Nana told him to seize the steamer and send all the Europeans prisoners to him.*** The same man wrote another petition a day or two afterwards, saying that they were too much for him and he had, therefore, taken the precaution of running away. 'Another man, who signed himself Khuda Buksh, guard of the Government bullock-train, reported that 'there were four Europeans asleep in the bullock-van under his charge when he heard of the the Nana's victory at Cawnpore, whereupon he killed three of them, but the fourth scoundrel woke up and gave him a severe wound on the back from which he was suffering.' Across this petition the Nana had written with his own hand, 'Give the man a present of Rs. 5 and entertain him as a soldier.' 'It struck me that this man might have gone back to his old post on the Government service, and on inquiring at the Post-office, I learned that there was a man of that name, but he was a coachman and had gone towards Delhi. I telegraphed along the line and found that he had just started from **Alighar**, driving a carriage with an English officer inside. I then telegraphed to Alighar to have him stopped and examined, when the scars of the wound, just as he had described it, were found on his back. He was at once tried, convicted, and hanged. 'I made the strictest inquiries to find any trace of alleged outrages committed on European ladies and women, and satisfied myself that beyond the brutal murders of many of them, none took place. I visited the house in whch the poor women were massacred at Cawnpore, before they were thrown into the well, and saw the marks of bloody hands on the wall and near the windows, where they had evidently tried to escape.' See Douglas Forsyth, *Autobiography & Reminiscences.* Edited by his daughter. London, 1887, pp 31-3.

John Lang's 'Portrait' of Nana Sahib

NANA SAHIB'S PERWUNNAH 32 Translation of the Nana's Perwunnah 32 issued in Kanpur the night before the massacre at the Sati Chaura Ghat suggests that the massacre was planned *BUT NOT AT THE SATI CHAURA GHAT* :- The 17th Regt Native Infantry (they had mutinied at Azimgarh, see LEWIS, Quartermaster Sergeant) had now arrived on the left or Oude[Awadh] bank of the river Ganges below Kanpur; the regiment was warned to be in readiness: 'Arrangements for the destruction of the English will not be made here, but as these people will keep near the bank on the other side of the river, it is necessary that you should be prepared and make a place to kill and destroy them on that side of the river, and having obtained a victory come here. ....the English say they will go to Calcutta. ....they sought protection from the Sirkar (Nana) and said 'Allow us to get into boats and go away', therefore the Sirkar has made arrangements for their going and by ten o'clock tomorrow these people will have got into boats and started on the river.' *See Parliamentary Papers 1857,* Inclosure No.21.

NANA SAHIB'S PETITION As Nana Sahib played so important and pivotal a role in the rebellion, and as his grievance against the British Government was at the heart of his rebellion it is worth while stating at length the nature of his petition to the Court of Directors of the **HEIC**: quoting Nana Sahib's memorial of 12 December, 1852, **Kaye** represents his case fairly. 'The course pursued by the local governments is not only an unfeeling one towards the numerous family of the deceased prince, left almost entirely dependent upon the promises of the East India Company, but inconsistent with what is due to the representative of a long line of sovereigns. Your memorialist, therefore, deems it expedient at once to appeal to your Honourable Court, not merely on the ground of the faith of treaties, but of a bare regard to the advantages the East India Company have derived from the late sovereign of the Mahratta Empire. ...It would be contrary to the spirit of all treaties hitherto concluded to attach a special meaning to an article of the stipulations entered into, whilst another is interpreted and acted upon in its most liberal sense.' 'And then the memorialist proceeded to argue, that as the Peishwah, on behalf of his heirs and successors, had ceded his territories to the Company, the Company were bound to pay the price of such cession to the Peishwah and his heirs and successors. If the compact were lasting on one side, so also should it be on the other. 'Your memorialist submits that a cession of a perpetual revenue of thirty-four lakhs of rupees in consideration of an annual pension of eight lakhs establishes a *de facto* presumption that the payment of one is contingent upon the receipt of the other, and hence that, as long as those receipts continue, the payment of the pension is to follow.' 'It was then argued that the mention, in the treaty, of the 'Family' of the Peishwah indicated the hereditary character of the stipulation, on the part of the Company, as such mention would be unnecessary and unmeaning in its application to a mere life-grant, 'for a provision for the support of the prince necessarily included the maintenance of his family...' '...After this, from special arguments, the Nana Sahib turned to a general assertion of his rights as based on precedent and analogy. Your memorialist,' it was said, 'is at a loss to account for the difference between the treatment, by the Company, of the descendants of other princes and that experienced by the family of the Peishwah, represented by him. The ruler of Mysore evinced the most implacable hostility towards the Company's government; and your memorialist's father was one of the princes whose aid was invoked by the Company to crush a relentless enemy. When that chieftain fell, sword in hand, the Company, far from abandoning his progeny to their fate, have afforded an asylum and a liberal support to more than one generation of his descendants, without distinction between the legitimate and the illegitimate. With equal or even greater liberality the Company delivered the dethroned Emperor of Delhi from a dungeon, re-invested him with the insignia of sovereignty, and assigned to him a munificent revenue, which is continued to his descendants to the present day. 'Wherein is your memorialist's case different? It is true that the Peishwah, after years of amity with the British Indian Government, during which he assigned to them revenues to the amount of half a crore of rupees, was unhappily engaged in war with them, by which he perilled his throne. But as he was not reduced to extremities, and even if reduced, closed with the terms proposed to him by the British Commander, and ceded his rich domains to place himself and his family under the fostering care of the

Company, and as the Company still profit by the revenues of his hereditary possessions, on what principle are his descendants deprived of the pension included in those terms and the vestiges of sovereignty? Wherein are the claims of his family to the favour and consideration of the Company less than those of the conquered Mysorean or the captive Mogul?' 'Then the Nana Sahib began to set forth his own personal claims as founded on the adoption in his favour; he quoted the best authorities on Hindoo law to prove that the son by adoption has all the rights of the son by birth; and he cited numerous instances, drawn from the recent history of Hindostan and the Deccan, to show how such adoptions had before been recognised by the British Government. 'The same fact,' he added,' is evinced in the daily practice of the Company's Courts all over India, in decreeing to the adopted sons of princes, of zemindars, and persons of every grade, the estates of those persons to the exclusion of other heirs of the blood. Indeed, unless the British Indian Government is prepared to abrogate the Hindoo Sacred Code, and to interdict the practice of the Hindoo religion, of both of which adoption is a fundamental feature, your memorialist cannot understand with what consistency his claim to the pension of the late Peishwah can be denied, merely on the ground of his being an adopted son.' 'Another plea for refusal might be, nay, had been, based upon the fact that Badjee Rao, from the savings of his pension, had accumulated and left behind him a large amount of private property, which no one could alienate from his heirs. Upon this the Nana Sahib, with not unreasonable indignation, said: 'That if the withholding of the pension proceeded from the supposition that the late Peishwah had left a sufficient provision for his family, it would be altogether foreign to the question, and unprecedented in the annals of the History of British India. The pension of eight lakhs of rupees per annum has been agreed upon on the part of the British Government, to enable his Highness the late Badjee Rao to support himself and family; it is immaterial to the British Government what portion of that sum the late prince actually expended, nor has there been any agreement entered into to the effect that his Highness the late Badjee Rao should be compelled to expend every fraction of an annual allowance accorded to him by a special treaty, in consideration of his ceding to the British Government territories yielding an annual and perpetual revenue of thirty-four lakhs of rupees. Nobody on earth had a right to control the expenditure of that pension, and if his Highness. the late Badjee Rao had saved every fraction of it, he would have been perfectly justified in doing so. Your memorialist would venture to ask, whether the British Government ever deigned to ask in what manner the pension granted to any of its numerous retired servants is expended? or whether any of them saves a portion, or what portion, of his pension? and, furthermore, in the event of its being proved that the incumbents of such pensions had saved a large portion thereof, it would be considered a sufficient reason for withholding the pension from the children in the proportions stipulated by the covenant entered into with its servant? And yet is a native prince, the descendant of an ancient scion of Royalty, who relies upon the justice and liberality of the British Government, deserving of less consideration than its covenanted servants?' 'To disperse, however, any erroneous impression that may exist on the part of the British Government on that score, your memorialist would respectfully beg to observe that the pension of eight lakhs of rupees, stipulated for by the treaty of 1818, was not exclusively for the support of his Highness the late Badjee Rao and his family, but also for the maintenance of a large retinue of faithful adherents, who preferred following the ex-Peishwah in his voluntary exile. Their large number, fully known to the British Government, caused no inconsiderable call upon the reduced resources of his Highness; and, furthermore, if it be taken into consideration the appearance which Native princes, though rendered powerless, are still obliged to keep up to ensure respect, it may be easily imagined that the savings from a pension of eight lakhs of rupees, granted out of an annual revenue of thirty-four lakhs, could not have been large. 'But notwithstanding this heavy call upon the limited resources of the late Peishwah, his Highness husbanded his resources with much care, so as to be enabled to invest a portion of his annual income in public securities, which, at the time of his death, yielded an income of about eighty thousand rupees. Is then the foresight and the economy on the part of his Highness the late Badjee Rao to be regarded as an offence deserving to be visited with the punishment of stopping the pension for the support of his family guaranteed by a formal treaty ?' But neither the rhetoric nor the reasoning of the Nana Sahib had any effect upon the Home Government. The Court of Directors of the East India Company were hard as a rock, and by no means to be moved to compassion. They had already expressed an opinion that the savings of the Peishwah were sufficient for the maintenance of his heirs and dependents; and when the memorial came before them [on 4 May 1853], they summarily rejected it, writing out to the Government to 'inform the memorialist that the pension of his adoptive father was not hereditary, that he has no claim whatever to it, and that his application is wholly inadmissible.' See Sir John Kaye, *A History of the Great Revolt.* London, 1864-7, Vol. I, pp 104-9.

NANA SAHIB'S PROCLAMATION The following proclamation made its way from Delhi to Kanpur, in the month of June (1857): *'To all Hindoos and Mussulmans, Citizens and Servants of Hindostan, the Officers of the Army now at Delhi and Meerut send Greetings.* 'It is well known that in these days all the English have entertained these evil designs, first to destroy the religion of the whole Hindostani army and then to make the people Christians by compulsion. Therefore we, solely on account of our religion, have combined with the people, and have not spared alive one infidel, and have re-established the Delhi dynasty on these terms, and thus act in obedience to orders and receive double pay. Hundreds of guns and a large amount of treasure have fallen into our hands; therefore it is fitting that whoever of the soldiers and people dislike turning Christians should unite with one heart and act courageously, not leaving the seed of these infidels remaining. For any quantity of supplies delivered to the army the owners are to take the receipts of the officers; and they will receive double payment from the Imperial Government. Whoever shall in these times exhibit cowardice, or credulously believe the promises of those impostors the English, shall very shortly be put to shame for such a deed; and, rubbing the hands of sorrow, shall receive for their fidelity the reward the ruler of Lucknow got. It is further necessary that all Hindoos and Mussulmans unite in this struggle, and, following the instructions of some respectable people, keep themselves secure, so that good order may be maintained, the poorer classes kept contented, and they themselves be exalted to rank and dignity; also, that all, so far as it is possible, copy this proclamation, and dispatch it everywhere, so that all true Hindoos and Mussulmans may be alive and watchful, and fix in some conspicuous place (but prudently to avoid detection), and strike a blow with a sword before giving circulation to it. The first pay of the soldiers of Delhi will be 30r. per month for a trooper; and 10r. for a foot-man. Nearly 100,000 men are ready, and there are 13 flags of the English regiments and about 14 standards from different parts now raised aloft for our religion, for God, and the conqueror, and it is the intention of Cawnpore to root out the seed of the Devil. This is what we of the army here wish.' The Nana was not slow to imitate the example which had thus been set him by the Delhi people, although the specimen which he gives of the inventive faculty completely throws into the shade the tame original upon which he thus improved: 'It has been ascertained from a traveller who has lately arrived at Cawnpore from Calcutta, that previously to the distribution of the cartridges for the purpose of taking away the religion and caste of the people of Hindostan, a council was held, at which it was resolved that, as this was a matter of religion, it would be necessary to employ 7,000 or 8,000 Europeans, and to kill 50,000 Hindostanees, and then all Hindostan would be converted to Christianity.' A petition to this effect was sent to Queen Victoria and the opinion of the council was adopted. A second council was then held, to which the English merchants were admitted, and it was agreed that, to assist in carrying out the work, the same number of European soldiers should be allowed as there were Hindostanee Sepoys, lest, in the event of any great commotion arising, the former should be beaten. When this petition was perused in England, 35,000 European troops were embarked in ships with the utmost rapidity and despatched to India. Intelligence of their despatch was received in Calcutta, and the gentlemen of Calcutta issued orders for the distribution of cartridges. Their real object was to make Christians of the army under the idea that when this was done there would be no delay in Christianizing the people generally. In the cartridges the fat of swine and cows was used. This fact was ascertained from Bengalees who were employed in making the cartridges; one of these men was put to death and the rest were imprisoned. Here they were carrying out their plans. Then the Ambassador of the Sultan of Constantinople at the Court of London sent information to the Sultan that 35,000 English troops were to be despatched to India to make Christians of that country. The Sultan sent a firman to the Pasha of Egypt to the effect that he was colluding with Queen Victoria; that this was not a time for compromise; that from what his ambassador sent it appeared that 35,000 English soldiers had been despatched to India to make Christians of the people and soldiers of that country; that there was still time to put a stop to this; that if he was guilty of any neglect in the matter, what kind of a face would he be able to show to God; that that day would one day be his, since, if the English succeeded in making Christians of the people of Hindostan, they would attempt the same in his country. On the receipt of this firman of the Sultan, the Pasha, before the arrival of the English troops, made his arrangements and collected his troops at Alexandria - for that is the road to India - and on the arrival of the English army the troops of the Pasha of Egypt began firing on them with cannon from all sides, and destroyed and sank the ships so that not a single Englishman of them remained. The English at Calcutta, after issuing the order for biting the cartridges and the breaking-out of this now spreading mutiny and rebellion, were looking for assistance from the army coming from London; but God, by the exercise of His Almighty power, settled their business there. When the intelligence of the destruction of the army of London was received the Governor General felt great grief and beat his head. At the beginning of the night murder and robbery were contemplated; in the morning the body had no head, nor the head any covering! In one revolution the sky became of the same colour; neither Nadir nor Nadir's Government remained. This paper has been printed by order of Nana Sahib, 13th zeiroe, and add 1273. Higree, 8.' quoted in Mowbray Thomson, *The Story of Cawnpore*, London, 1859, pp 142-6.

NANCY DAWSON The name (taken from a popular song of the period) given to a well-sited rebel gun that kept the British pinned down at the **Alambagh** in November 1857. *'They had guns also in position at what is known as the yellow bungalow, a house situated in some gardens known as the Char Bagh, and from here they were also enabled to reach us, firing over the walls of the garden of the Alambagh. The first gun alluded to here bears the name of 'Nancy Dawson', why, I have not learnt, but suppose she is at times somewhat more than saucy. She certainly seems to be a talkative lady, but like most talkative people her bark is worse than her bite. In this instance however she, along with her neighbours, compelled us to take up ground at from quarter to half a mile to the rear of the spot first chosen and beyond her range'*. From the *St George's Gazette*,

30 April 1894.

NAND KISHOR The **Vakil** of the Maharao of **Kotah[Kota]**, blown from a gun by the Kotah rebels; his offence appears to have been no more than to have been present at a meeting between his master and **Major Burton** the Resident on 12 October 1857, at which the latter had urged that some of the principal ministers of the court be punished and dismissed. One such minister, Ratan Lal, was told of this request by Nand Kishor and promptly passed on the news to the rebels who, presumably at Ratan Lal's suggestion, seized Nand Kishor and executed him. *See The Revolt in Central India*, Simla, 1908. *See also* Iltudus T.Pritchard, *The Mutinies in Rajpootana, being a personal narrative of the mutiny at Nusseerabad*, London, 1864.

NANAK CHAND A pleader in Kanpur who not only survived the outbreak but was instrumental in supplying the British with a great deal of information. He had been on bad terms with **Nana Sahib** and with a number of other prominent local inhabitants. He claimed to have kept a detailed and meticulous diary of events throughout the period that the Nana controlled Kanpur, and this formed the basis of an exhaustive deposition to the authorities on the restoration of British power. Certainly his memory, in the matter of naming names, is quite extraordinary, but the reader is inevitably left with the impression that his evidence is influenced by his desire to pay off old scores. 'At the commencement of the outbreak, Raheem Khan, a sowar in the Nana's employ, Muddud Ali, a horse-dealer, were employed to tamper with the troops; and held consultations about the mutiny. Muddud Ali was at Cawnpore the other day. He told me he had been plundered by Sha Ali, the former Kotwal, and had therefore left the Nana's force in disgust. He could give valuable information, but he has again left Cawnpore. I will search for him.' Deposition of Nanak Chand. **Williams.**

NANAK CHAND On 17 December 1857 Nanak Chand presented his diary to **Sherer** and **Bruce**, giving each of them a copy.'Why' asks a note in the margin of a printed copy in the National Archives, written probably by **Dr. Sen**, 'did he not give it on the 17th August?' Good question, unless he spent a lot of time amplifying it or believed that the British would not have welcomed such a document any earlier. *See Nanak Chand's Diary*. NA. FDP. FCS. 30 December, 1859. 'A common informer [who] had disgusted everyone that has had anything to do with him.' G.E. Lance, Collector in Kanpur on 15 May, 1862.

NAPIER 'Colonel Napier of the Engineers was directing the blowing up of the Hindoo temples on the Cawnpore ghat, and a deputation of Hindoo priests came to him to beg that the temples might not be destroyed. 'Now listen to me', said Colonel Napier in reply, ' you were all here when our women and children were murdered, and you also well know that we are not destroying these temples for vengeance, but for military considerations connected with the safety of the bridge of boats. But if any man among you can prove to me that he did a single act of kindness to any Christian man, woman, or child, nay, if he can even prove that he uttered one word of intercession for the life of any one of them, I pledge myself to spare the temple where he worships.' I was standing in the crowd close to Colonel Napier at the time, and I thought it was bravely spoken. There was no reply, and the Brahmins slunk away. Napier gave the signal and the temples leaped into the air.'

NARAIN RAO *See* NANA NARAIN RAO

NARAYAN RAO *See* KIRWI

NARHAT A large village in the **Jhansi** district, 22 miles south of **Lalitpur.** The raja of **Banpur** defended the Pass of Narhat against **Sir Hugh Rose** in March 1858 with up to 10,000 men. *See* **FSUP,** Vol III p 303 *et seq., and* Tapti Roy, *The Politics of a Popular Uprising, Bundelkhand in 1857*, Delhi, 1994.

NARPAT SINGH (Or Nirput Singh) An old man and a cripple and one of the rebels in **Oude[Awadh]:** although he had shown little hostility to the British, he was a follower of **Beni Madho**, and is best known for his resistance when General **Walpole** attacked his jungle fort of **Royah**[Ruya], near the village of Rhodamow. It was reported of him that as 'God had taken some of his members (he was alluding to his crippled condition), he would give the rest to his country'. He is said to have been killed by the Gurkha soldiers of **Jung Bahadur**, at the same time, November 1859, as **Rana Beni Madhoo**. *See* ROYAH. *See also* **FSUP,** Vol II, p 402.

NARWAR, RAJA MAN SINGH of Scindia had refused to recognise Man Singh's right to succeed his father in the principality; the raja therefore took up arms, intimating to Brigadier Smith that he had no cause of quarrel with the British and had no connection with the rebels. But Smith was responsible for the peace of the country and was obliged to take action against Man Singh, who now allied himself with **Tatya Tope**. It is this Man Singh who eventually betrays Tatya Tope to the British. *See The Revolt in Central India 1857-59*, Army Headquarters, Simla, 1908

NASIRABAD The mutiny at Nasirabad was reported by one officer in the following terms: 'When the news from Meerut arrived every precaution was taken but it was agitators in the bazar that precipitated the crisis among the sepoys; Brigadier Macan was informed that an outbreak was imminent. On May 28th 1857 the 15th Bengal N.I. sepoys seized the guns at 4p.m. and they were soon joined by the

30th Regiment.'...'The Cavalry (1st Bombay) were ordered to charge by squadrons.

Bombay Lancer Officer

They charged, but as soon as they got within a few yards of the guns, the men went threes about, the officers going on. Major Spottiswoode fell mortally wounded; Cornet Newbury was cut to pieces among the guns; and Captain Hardy and Lieutenant Lock were badly wounded. The Europeans then assembled in rear of the cavalry lines, where the (Bombay) Lancers were drawn up but would not act. Some of the Europeans, fired at by the mutineers, escaped with difficulty: they fled at nightfall to Beawar 37 miles distant, escorted by the 1st Bombay Cavalry. Colonel Penny fell dead from his horse from excitement and exhaustion........the Adjutant of the 15th (Bengal N.I.) had his horse shot under him, but it carried him out of danger before falling dead. *It was said that he was afterwards refused compensation for the loss of his charger by the Military Auditor-General on the ground that the regiment having mutinied, he had no longer occasion to keep a charger, and therefore it was not necessary to replace it.'* Nasirabad was reoccupied by British troops on 12 June. There was a second outbreak of mutiny on 10 August but it led only to the temporary disarmament of the 12th Bombay N.I. and there were no European casualties.

NASIRABAD There was a second, potentially more dangerous to the British, outbreak here on 10 August 1857, when a trooper galloped down the front of the lines occupied by the 1st Bombay Cavalry, calling to the men to rise: the men here for the most part remained quietly in their lines. He then rode down to the 12th Bombay Infantry (latterly the 112th Infantry, later still the Mahratta Light Infantry) and did the same there. Clearly the men passively sympathised with him, did not arrest him, and refused to give him up or to proceed to the pre-arranged rendez-vous at the guns. By now Brigadier Macan had arrived, and the guns were called out. The trooper fired at the Brigadier but missed and was in turn shot by an artillery officer and later died in hospital. The 12th were then disarmed and many arrested. as it was clear that the regiment could no longer be trusted by its officers. The outcome was that on 25 August five of the ringleaders were hanged, twenty-five deserted, and all the Hindustani (as opposed to Mahratta) sepoys were discharged. The regiment afterwards performed loyal service in the field for the British, taking part in the battle of **Kotah-ki-Serai** and in the re-capture of **Gwalior**. *See* I.T.Pritchard, *The Mutinies in Rajputana,* London, 1864.

NATION The A periodical published in Ireland (then part of the United Kingdom) with marked liberal and anti-establishment views. *The Nation* was dubbed by the *Times* the 'chief organ of Sepoy sympathizers in Ireland,' and it certainly rejoiced in **Windham**'s defeat and deplored '**Neill** infamous ...whose memory will be a loathing among Christian men,' but mourned **Havelock** as the 'brave man he was, and, we believe, a good man. ...He has left behind no stain upon his character. He fought like a soldier; he did not seek to torture like a fiend.' *See* London *Times.* 11 January, 1858, p 8.

NATIONAL ARCHIVES OF INDIA New Delhi. Contains valuable primary source material. 'Bulk of the material is to be found listed in the *Press List of Mutiny Papers, 1857: being a Collection of the Correspondence of the Mutineers at Delhi, Reports of spies to English officials and other Miscellaneous Papers,* Printed in Calcutta by the Superintendent of Govt.Printing and published by the Imperial Records Dept. (which later became the National Archives of India), 201 bundles' - quotation is from Henry Scholberg , *Indian Literature of the Great Rebellion,* New Delhi, 1993. *See also* PAPERS OF A MISCELLANEOUS CHARACTER.

NATIONAL ARMY MUSEUM Formerly at Camberley in Surrey, now in Chelsea, southwest London, England, contains a wide range of 'mutiny' documents of interest but mostly of marginal significance. The collection includes plans of **Delhi** during the siege, ditto **Lucknow** and **Lucknow Residency**; maps of Delhi and Lucknow by **Lt W.R.Moorsom**; An account by **Henry Wylie Norman**, (Acting Adjutant-General 's staff), a shrewd and knowledgeable observer, of the operations in and around Delhi in 1857; the original manuscript proceedings and supporting documents of the trial of **Tatya Tope** in April 1859, etc. In addition to the archive material there are a number of photographs, watercolours, drawings and prints depicting the military campaigns of this time. A list of the fifty archive items then available is available in Janice M. LADENDORF, *The Revolt in India, 1857-58*, Zug, 1966, but a more complete account of the archive holdings may be found in volume 1 of J.D.Pearson's *Guide to Manuscripts and Documents in the British Isles Relating to South and South-East Asia*, London 1989.

NATIONAL THEATRE A number of theatrical and cinematic representations of the events of 1857-9 have been produced, notably the film ***Junoon*** in India, and the play 'H', or 'Monologues at Front of Burning Cities' by the National Theatre in London, at the Old Vic Theatre in February 1869. This was written by Charles Wood and produced by Geoffrey Reeves, the Director of the National Theatre being then Laurence Olivier. The part of General **Havelock** who was the centrepiece for the play, was played by Robert Lang. The play has not been revived.

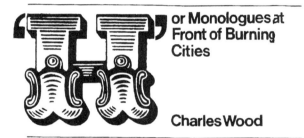

NATIVE 'Should this booklet fall into the hands of any Indian gentleman, I would like to say that, though primarily written for the European residents of Sialkot, I have tried, where possible, to avoid reproducing opinions that might be regarded as offensive. In writing of this period it is, however, difficult to avoid using words such, for example, as 'Native' instead of 'Indian', when all the records contain the former word and even the regiments were referred to as 'Native Infantry'. *From* the preface to Captain Gregory Rich, (the cantonment magistrate), *The Mutiny in Sialkot*, Sialkot, 1924.

NATIVE OFFICERS The Jemadars, Subadars and Subadar Majors of the Bengal Army were, without exception, men who had served the **HEIC** for many years; many were elderly and approaching retirement and a pension. When they were obliged to make a decision as to whether they wished to mutiny or continue to serve Government, their dilemma was extreme. Some did not hesitate and became leaders of the rebellion *eg* **Bakt Khan** at **Bareilly** who became **Bahadur Shah's** Commander-in-Chief, and **Teeka Singh** who became one of **Nana Sahib's** generals. Rarely did they actually lead or foment mutiny though they afterwards may have joined the rebels. A case in point relates to the position at **Kanpur** on 5 June 1857: 'At 9:30 about 30 of their native commissioned officers came to Wheeler to report that they were unable to dissuade their men from rebelling. As they spoke a bugle sounded and the two regiments (53rd and 56th BNI) assembled on the parade ground, only to be dispersed by a couple of rounds from the Entrenchment guns. All but a couple of native officers were directed east to the artillery hospital to persuade the most reluctant mutineers among the sepoys and native non coms to return to their British officers' command, but they were never seen again.' *see* W.J. Shepherd, *A Personal Narrative etc*, Lucknow, 1886, p16 . While the native commissioned officers and General Wheeler were speaking 'a bugle sounded, and presently afterwards we could see the two regiments drawn up in columns on their parade ground, showing a defying front; but a shot or two from our long gun immediately dispersed them, and sent them at a full gallop toward their lines, on the outside road leading to Delhi, and branching off to Nawabgunge, where their rebellious brethren were then stationed.' Shepherd's account, dated 29 August, 1857, published in the London *Times*, 7 November, 1857, p 6. *See also* BHOWANI SINGH.

NATURE OF THE CONFLICT There are many differing views as to the nature of the conflict in 1857, varying from simple mutiny by disaffected soldiers to an orchestrated patriotic rebellion designed to rid India of a hated and oppressive foreign ruler. This subject has been investigated in depth by a 'neutral' observer, A.T.Embree, *'1857 in India: Mutiny or War of Independence?'*, Boston Mass., 1963, but continues to be a matter of discussion. It might be thought that all would depend upon one's point of view, but the scholar can go deeper than that. Highly respected Indian historians like R.C.Majumdar ('*The Sepoy Mutiny & the Revolt of 1857'*, 1963) and Surendra Nath Sen ('*Eighteen Fifty Seven*', 1957) both reject the concept of a **national** war of independence. First speaks of: 'the first great and direct challenge to the British rule in India...which furnished a historical basis for the (independence) struggle,' and, 'To regard the outbreak of 1857 simply as a mutiny of sepoys is probably as great an error as to look upon it as a national war of independence.' Latter says: 'what began as a fight for religion ended as a war of independence.' Majumdar: If we turn to the other prominent leaders associated with the movement, namely Bahadur Shah, Nana Sahib, the Rani of Jhansi and Kunwar Singh, it immediately strikes us that all these four (Begum of Oude also?) were smarting under grievous injury done to them by the British and therefore bore special grudge against them... it is not easy to understand in what sense these four persons could be regarded as natural leaders.'.....' The first was a dotard and a puppet on the throne of the Mughals who inherited nothing but their name, and had little power and less knowledge of men and things. The second was an adopted child of a worthless and wicked ex-Peishwa who was mainly instrumental in ruining the Maratha power........neither the Rani of Jhansi nor Kunwar Singh, in spite of their personal ability, has any right to be called a natural leader of the country. The First was a young widow of an almost unknown ruler of a petty state, then defunct, and the second was a small talukdar in the interior of Behar, utterly impoverished beyond hopes of recovery.'

Analysis of the available historiography indicates that there are at least twenty fully supported explanations for the struggle. (For further explorations, *see* S.B.Chaudhari, *Civil Rebellion in the Indian Mutinies*, Calcutta 1957; *and Theories of the Indian Mutiny, 1857-59*, Calcutta 1965; *see also* K.K.Datta, *Nature of the Indian Revolt of 1857-9*, in *Bengal Past and Present*, Vol 73, 1954; *and Ainslie T Embree, 1857 in India: Mutiny or War of Independence*, Lexington, 1963.)   Here are some explanations, in no particular order, and selected only because they are *representative of a viewpoint, valid or invalid:* **A** 'The Bengal Army was the dupe and tool of a few disaffected aristos seeking revenge or self-aggrandizement. The Sepoys once embarked, not always willingly, on the bloody course of mutiny, their hands stained by many crimes, were committed up to the hilt. Nothing left but to fight to the death. Not generally supported by the people who only asked for a return of the peace and tranquillity of British rule. Lawlessness and anarchy were spreading. Conspirators failed perhaps narrowly to turn the mutiny into a general insurrection because the Princes and Peoples of India, on the whole, preferred the stability of the British raj to the unfettered rule of Hindu or Mussulman rajahs with the certain prospect of never-ending racial and religious tyranny' Col. Thackeray, *Chambers Journal* No20, 1930. **B** MARX & ENGELS (contemporaneous with the 'Mutiny') had every sympathy for the liberation struggle of the Indian People.....'its success depended on all round support. However there had been no such general action, for a number of historical reasons (including) the perfidy of most of the local feudal gentry who had led the revolt....' Marx calls Scindia (and others, including Jang Bahadur of Nepal) the 'English dog-man'. Holkar, Man Singh, leaders of Ghurkhas, Sikhs and many Rajputs and Mahrattas were all tainted so far as he is concerned - Karl Marx, quoted in *The First Indian War of Independence 1857-9*, Moscow, 1958. **C** Of great importance to any consideration of the struggle in Oude [Awadh], *and also by extension the entire conflict in North India 1857-9* is this significant comment by the Directors of the Honourable East India Company: *'We must admit that under these circumstances the hostilities which have been carried on in Oude have rather the character of legitimate war than than of rebellion.....the people of Oude should rather be regarded with indulgent consideration, than made the objects of a penalty, exceeding in extent and in severity, almost any which has been recorded in History as inflicted upon a subdued nation'* - Secret letter from East India House, London 19 April 1858 to the Governor General. **D** That it was just an army mutiny. A typical comment from a supporter of this idea comes from Sir John Lawrence Chief Commissioner of the Punjab in a letter quoted by R.B.Bosworth, *Life of **Lord Lawrence***, New York, 1883. *'We have been almost as much to blame for what has occurred as have the people. I have as yet neither seen nor heard anything to make me believe that any conspiracy existed beyond the army; and even in it one can scarcely say there was a conspiracy. The cartridge question was, to my mind, indubitably the immediate cause of the revolt. But the army had for a long time been in an unsatisfactory state. It had long seen and felt its power. We had gone on, year by year, adding to its numbers, without adding to our European force.'* **E** That it was a Brahmanical plot resulting from their fear of a loss of their ascendancy. This view was held by, among others, **Sir John Kaye** : *...'it was clear that a very serious peril was beginning to threaten the ascendancy of the priesthood. They saw that a reformation...once commenced, would work its way through all the strata of society...until there would scarce be a place for Hindooism to lurk unmolested. And some at least, confounding cause and effect, began to argue that all this annexation and absorption was brought about for the express purpose of overthrowing the ancient faiths of the country, and establishing a new religion in their place.....The whole hierarchy of India saw their power privileges and perquisites rapidly crumbling away from them, and they girded themselves up to arrest the devastation....Every monstrous lie exploded, every abominable practice suppressed, was a blow struck at the priesthood; for all these monstrosities and abominations had their root in Hindooism.'* **F** On the causes of the Revolt. Opinion of Major Macpherson Political Agent with Scindia , Maharajah of Gwalior (quoted in **FSUP** Vol III, p 202): 'I merely indicate their direction; that the army were predisposed to revolt, by sharing with the people of Hindoostan their feeling of dissatisfaction with our rule; they thought success certain, from the smallness of our European force, and from popular aid; and made the cartridge grievance their pretext and occasion to rise. The foremost malcontents instigated, the King of Delhi headed the revolt and all doubted deeply the stability of our power.' **G** That it was a vast political conspiracy. This is the view of, among many others, **Colonel G.B.Malleson** who says *'Indeed I may go as far as to declare that many of the actors in the drama failed to realise to their dying day that the outbreak was not merely a mutiny which they had to combat, but a vast conspiracy, the threads of which were widely spread, and which owed its origin to the conviction that a government which had, as the conspirators believed, betrayed its trust was no longer entitled to respect or allegiance.'* **H** Sir **Syed Ahmad Khan** had this summary of the nature of the revolt (written in 1858): 'As regards the rebellion of 1857 the fact is that for a long period many grievances had been rankling in the hearts of the people. In course of time, a vast store of explosive material had been collected. It wanted but the application of a match to light it, and that match was applied by the mutinous army,' *see* p116, Appendix A *The cause of the Revolt*, Hafeez Malik and Morris Dembo, transl. Sir Sayyid Ahmad Khan's *History of the Bijnor Rebellion*, East Lansing, 1972   **I** That it was an inevitable reaction to a Christian plot: there was a widespread Indian contemporary belief that there was an official government plot to rob the Hindu sepoy of his caste and to defile the Muslim, thus making them forced converts to Christianity.

J Colonel Chesney, writing in 1877, would appear to speak for many British officers in expressing surprise at the nature and progression of the struggle: 'And when the Mutiny did break out the course it took was altogether different from what was universally expected. No one who knew anything of the Bengal sepoys anticipated that the decrepid(sic) native officers would retain their places throughout the war at the head of their regiments: it was unanimously expected by the European officers who had passed their lives with them, that these old men would have to give place to younger and more adventurous spirits. Another prediction falsified by the event was that the Hindoo and Mussulman sepoys would fall out with and separate from each other. Still less was it expected that the people of the country generally would look on at the struggle with indifference and that the Princes of India, standing loyally by the Paramount Power, would take an active part against our rebellious army'. *See* Colonel G.Chesney, in *The Nineteenth Century*, August-December 1877. **K** The **Azimgarh Proclamation** (*see* Chas Ball *'History of the Indian Mutiny'* Vol 2 pp 630-32, 1860) apparently by a grandson of Bahadur Shah Zafar, issued as a manifesto or proclamation of the King at an early period of the rebellion does shed some light upon the causes of the revolt, but it is not really an affirmation of political rights or legitimacy. The British Government is condemned for its actions *ie* its treatment of certain classes, but not for seizure of power in the first place. In other words the *nature* not the *fact* of British rule is declared unacceptable. It is true that the intention of the 'Manifesto' might have been mainly to convince waverers rather than declare royal rights, but it does nonetheless read curiously. Thus: 'Natives, whether Hindoos or Mahomedans, who fall fighting against the English, are sure to go to Heaven; and those killed fighting for the English will, doubtless, go to Hell. Therefore all the natives in the British service ought to be alive to their religion and interest, and, abjuring their loyalty to the English, side with the Badshahi government, and obtain salaries of 200 or 300 rupees per month for the present, and be entitled to higher posts in the future.'   **L** An Indian summary of reasons for the failure of the rebellion from the pen of a writer of distinction ('*Travels of a Hindoo'* by Bholanauth Chunder, 1869):- 'They speak of it as a fearful epoch of unexampled atrocities on the one side, and of unparalleled retaliation on the other. There were the sepoys with the blood of murdered officers on their heads, and badmashes and bullies and cut-throats and cut-purses, all acknowledging a fraternal tie, and holding a bloody carnival. But it was impossible that twenty uncongenial parties, divided by quarrels about power, and quarrels about plunder, quarrels about caste, quarrels about religion, could long act together in an undisturbed concert. Soon as batch after batch of Englishmen arrived to re-establish the Saxon rule, they were driven like chaff before the wind.' **M** '.....the native population did not rise against the white man, but the moment they thought the white man was powerless they rose against each other, the rival castes and villages plundering and fighting in all directions: the Hindu Gujars raiding the Hindu Jats, and the Mahommedans raiding all Hindus impartially, while the English Magistrate and his young assistant rode about from village to village followed by a few native subordinates and fighting men, vainly trying to keep order and punish the rioters.' (From a letter: Alfred Lyall, Bulandshahr (50 miles from Delhi) to his mother in 1857) **N** 'Cawnpore 18 May 1857. We never can feel secure in the country - how fortunate for us that the people do not know how to combine sufficiently to subvert the Empire. You see they have no leaders, no definite causes, no real patriotism, so there is nothing likely to arise out of these disturbances but a great amount of mischief and the temporary suspension of order.' (Letter from Mrs Emma Sophia Ewart wife of Lt Col John Ewart to her sister). Most astute and accurate perhaps, but it did her no good: she died in the Massacre in Kanpur. **O** Gubbins (M.R.Gubbins, *'An Account of the Mutinies in Oudh etc'*, 1858) declared that the sepoys had gained not lost by the annexation of Oude[Awadh]. He speaks of the suspicious credulity of the people and the spreading of the wildest rumours *eg* long before the mutiny, at Simla, a report got about that the Governor General had sent orders for a certain quantity of human fat to be prepared and sent down to Calcutta, and that for this purpose the authorities were trapping hill men who were then killed and boiled down for their fat. Panic spread and all the hillmen employed around the European households fled, nor were they ever thoroughly convinced of the falsehood of the rumours. And again: Subedhar Deybee Singh told Sir Henry Lawrence that it was widely believed that Government intended to bring to India all the numerous widows of soldiers killed in the Crimean War and compel the principal zemindars of the land to marry them, and their children, not Hindus, were then to be declared heirs to the estates. And of course, the bonedust idea, the medicine bottle disturbance in the 48th NI, the cartridge question. Suggested by a Bombay army subedhar that there were too many men of the same caste and family together in the Bengal army, which rendered combination easy.'On the Bombay side I stood side by side with men of low caste, and who dared there say a word? Mahratta, Pasee, Brahmin, Chumar, Rajput and many others are there found in the same ranks. But here they are all one'. **P** That 'the sepoys had a grievance but no cause' was a widely held belief among the British. **Q** In a letter to Colvin, Raikes enumerated the causes of the disturbances: 'First... A general mutiny of the Native Army. Secondly... the violence and rapacity of the Goojhur, Mewattie, and other clans, disposed in the best times to predatory habits; also by the cruelty of the lower town-mobs and cantonment rabble. Lastly... the attempts of certain quasi-royal pensioners and landholders to revive their lost local ascendency.' Charles Raikes, *Notes on the Revolt in the North-western Provinces of India.* London, 1858 pp 123-4. **R** That it was a Russian Plot. This was the view of, among others, D.Urquhart, *The Rebellion of India*, London, 1857. Throughout the nineteenth century many people both in Britain and India were haunted by the fear of a Russian attack on India; so it was not surprising that Russia's hand was immediately 'detected' in the revolt. For Urquhart (A Member of Parliament) British troubles in India were the result of Russian machinations; and the chief agents of Russia, he insisted, were to be found within the British Cabinet! Russian involvement is hinted at in a number of Indian sources, but little evidence adduced. **S** That it was the First Indian War of Independence: such is the view expressed in **V.D.Savarkar** *The Indian War of Independence of 1857*, London, 1909: *'When, taking the searching attitude of an historian, I began to scan that instructive and glorious spectacle, I found to my great surprise the brilliance of a War of Independence shining in the 'mutiny' of 1857. The spirit of the dead seem hallowed by martyrdom, and out of the heap of ashes appeared forth sparks of a fiery inspiration.'* **T** That it was an episode in India's struggle for freedom. This is a widely held view. It was expressed succinctly by Nandalal A.Chatterjee, *A Century of India's Freedom Struggle*, Delhi, *Journal of Indian History* Vol 35 Part 2, August 1957: *'The real significance of the revolt of 1857 lies therefore in an ideological conflict which took the shape of a patriotic outburst against foreign rule...Having begun as a military outburst in various cantonment towns, the revolt assumed in various parts of Northern India the character of a popular rebellion...a study of the facts would show that the rising of 1857 failed not because it was not national or patriotic, but because it was not well-organised and well led.'* **U** That it was a Mutiny all right, but a Mutiny by the East India Company, *see* BUCKLER, F.W. **V** That it was the inevitable result of Capitalism. It is seen by **Karl Marx** as an incident in a social revolution (*'the only one ever heard of in Asia'*) : 'England in causing this social revolution was actuated only by the vilest interests, and was stated to be stupid in her manner of enforcing it: she was the unconscious tool of history in bringing about that revolution: the violence of 1857 was simply an extension of that social process'. Modern communists agree with that view: see P.C.Joshi, ed., *Rebellion:1857*, Bombay, 1957 *and* L.Hutchinson, *The Empire of the Nabobs*, London, 1937. **W** That it was the result of bad government: this was the view of Benjamin **Disraeli** among others. **X** That it was the inevitable result of foreign rule. This has been a very widespread view. As an example, Alex.Duff an influential Church of Scotland missionary in India wrote *'As regards the feelings of the great masses of the people towards the British government, the most contradictory statements have been put forth. Here, as elsewhere, extremes will be found wrong. That there ever was anything like affection or loyal attachment, in any true sense of those terms, on the part of any considerable portion of the native population towards the British power, is what no one who really knows them could honestly aver.'* **Y** That it was the inevitable result of the savage and cruel nature of Indians. This is not the serious view of any historian or commentator today, but it was certainly a widely held belief in the middle of the nineteenth century. As an example *see* W.R.Aikman, *The Bengal Mutiny, letter to Viscount Palmerston,* London, 1857.

NAVAL BRIGADE By chance HMS Shannon and HMS Pearl were in the Hoogly river at Calcutta at the beginning of the outbreak. The British took the sailors from the two ships and made them into a Naval Brigade, that is they were formed into a unit to fight on land instead of at sea. Under **Captain Sir William Peel** they did good service particularly in **Oude**[Awadh] and in the attacks upon **Lucknow** under **Sir Colin Campbell**. Large 8"guns, 65 pdrs, from the frigate Shannon were dragged across country to Lucknow, and having been used there in the assault of March 1858, were left in the grounds of the **Residency**, where they still

remain and may be seen today. For the most comprehensive story of the Naval Brigade's activities, compiled from contemporary letters and despatches, *see* Major General G.L. Verney, *The Devil's Wind*, London, 1956. *See also* Rowbotham, William, ed. *The naval brigades in the Indian Mutiny*, London, 1947. Standard history of the three brigades, Shannon and Pearl in Bengal and the little-known Pelorus on the Irrawaddy River.

NAWA'B ALI KHAN A member of the royal house of Bhopal, a pretender to the *gaddi*, he was very active as a rebel, while the Regent of Bhopal, the **Begum** remained fiercely loyal to the British. He was active in Central India throughout the latter part of 1857. For further details, *see* **FSUP**, Vol III and *Revolt in Central India,* Simla, 1908, p 45 *et seq.*

NAWAB ALI BAHADUR of Banda, *see* **Banda.**

NAWAB ALI Raja, *talukdar* of Mahmudabad (Mahomadabad) in Oude[Awadh]; possibly the first of the Oude *talukdars* to openly rebel against the British. He was joined by the **Sitapur** mutineers, and he may have been present at the Battle of **Chinhat**. *See 'Trial Proceedings, Government vs Raja Jai Lal Singh'*, Lucknow Collectorate Mutiny *Basta.*

NAWAB FAIZ MUH. KHAN One of the counsellors of the young Maharaja of **Jaipur**, by whose advice the state was kept loyal to the British throughout the struggle.

NAWABGANJ Firstly, Pargana and Tehsil; a town seventeen miles east of Lucknow and sixty-one miles west of Faizabad, on the road that joins the two. Scene of the mustering of the rebels' forces on 19 June 1857 preparatory to their march upon Lucknow and the commencement of the Siege of the Residency. Secondly, a large village in Oude[Awadh], south of Lucknow, on the Kanpur road, some five miles north of **Bashiratganj**, where **Havelock** was repeatedly held up on his drive to relieve the Residency at Lucknow. For a good account of Havelock's campaign *see* J.C.Marshman, *Memoirs of Major General Sir Henry Havelock*, 1860.

NEEMUCH BRIGADE The Neemuch Brigade which mutinied on 3 June 1857 marched in the direction of **Nasirabad**. They sacked Deoli and then marched to **Agra** where, in association with the **Kotah[Kota] Contingent** on 5 July they fought and drove back into **Agra Fort** the British garrison under Brigadier **Polwhele** at the Battle of **Sussiah**[Sassia]. They then marched for **Delhi** via **Muttra**. At Delhi they were defeated by **Nicholson** at **Najafgarh**. When Delhi fell to the British the remainder of the Neemuch Brigade joined **Khan Bahadur Khan** in **Rohilkhand.** *See The Revolt in Central India,* Simla, 1908. and I.T.Pritchard, *The Mutinies in Rajpootana,* London, 1864., *and* **FSUP**, Vols II and III.

NEEMUCH(or Nimach) One hundred and fifty miles (240 kilometres) south of **Nasirabad** and due east of Udaipur, on the borders of Rajputana. There was a foretaste of mutiny as early as 28 May among the 72nd Bengal NI and the 7th regiment of the Gwalior Contingent, but their officers calmed things down. The real mutiny occurred on 3 June. The officers were allowed to leave; but the cantonment was looted and burnt. The mutineers then marched off to **Agra**, and the ruined cantonment was re-occupied by the British. For details *see The Revolt in Central India,* Simla, 1908. and I.T.Pritchard, *The Mutinies in Rajpootana,* London, 1864.

NEEMUCH FAQUIR, The On 21 November the rebels attempted an escalade on the walls of the besieged town of **Neemuch**; they were repulsed and left their ladders and a green standard on the ground. It seems that a faquir had almost predicted this result. He, with a mirror fixed on his breast, had walked round the fort under fire, having stated that if he succeeded in completing the circle round the walls, the place would fall into the hands of the insurgents. For a considerable distance (and the Bombay sepoys within the walls were becoming anxious, for they superstitiously believed in his magic powers) the bullets of the defenders flew harmlessly round him, but at length one brought him down; a bandsman went out and brought his head and the mirror into the fort. *See The Revolt in Central India,* Simla, 1908 . *See also* Iltudus T.Pritchard, *The Mutinies in Rajpootana, being a personal narrative of the mutiny at Nusseerabad,* London, 1864.

NEEMUCH, Siege of A British force of some 800 men (Bombay army troops and Meywar Horse) was besieged in **Neemuch** in November 1857 for some fifteen days by 4,500 rebels under the command of **Firoz Shah**. The latter had also four guns but was handicapped by a shortage of ammunition. The event was notable for the passivity of the British, disagreement among its officers, and the incident of the *Neemuch faquir*. The rebels withdrew in the direction of **Mandesar**[Mandsaur], on hearing of the advance of the **Malwa Field Force** to raise the siege. After this event Neemuch and the surrounding country remained undisturbed. *See The Revolt in Central India,* Simla, 1908. *See also* Iltudus T.Pritchard, *The Mutinies in Rajpootana, being a personal narrative of the mutiny at Nusseerabad,* London, 1864

NEILL, Major A.H.S. When **Kanpur** was recaptured by the British in July 1857 many men were hanged for their part, real or alleged, in the massacres. Sometimes only the finger of suspicion was sufficient to lead to arrest. Certain men became well-known as 'informers', eg **Nanak Chand**; *see also* HANGMAN HUNG. Trials were short and informal: punishment was preceded in some cases by causing the condemned to lick a portion of the blood stained floor of the Bibigurh where the European women and children had been hacked to death. The man responsible for this procedure (which was quickly abrogated) was **Brigadier**

General James Neill who later was killed at **Lucknow**. One of his victims was Duffadar (native officer) Suffur Ali of the 2nd Cavalry which had mutinied in Kanpur: he was hanged at the **Bibigurh** on 25 July 1857. His alleged crime (which he strongly denied) was that of having cut off the head of **General Wheeler** at the **Satichaura Ghat** on 27 June. A circular, in Urdu, to the following effect was passed round the Kanpur bazaar shortly afterwards : *'Oh Mahomed Prophet! be pleased to receive into Paradise the soul of your humble servant, Suffur Ali, whose body Major Bruce's mehtar police are now defiling by lashes, forced to lick a space of the blood-stained floor of the Slaughter-House, and hereafter to be hanged, by the order of General Neill. And, O Prophet! in due time inspire my infant son Mazar Ali of Rohtuck, that he may revenge this desecration, on the General and his descendants.'*

Colonel James Neill

................On 14 March 1887, trooper Mazar Ali of the 2nd Regt.Central India Horse, was sentenced to death, and hanged, for the murder, by shooting on the parade-ground at Augur, in an apparently motiveless crime, Major.A.H.S. Neill, the son of Brigadier General James Neill. That is the familiar anecdote, to be found, among others, in Forbes-Mitchell's *Reminiscences of the Great Mutiny,* London, 1893. Unfortunately it is based upon an extraordinary coincidence, and is not true. Major A.H.S.Neill was indeed the son of Brigadier General Neill of whom Suffur Ali was a victim in Kanpur; the Urdu circular did exist and was passed round the Kanpur bazar; Suffur Ali may or may not have had an infant son Mazar Ali of Rohtuck, but most unfortunately for the anecdote the Mazar Ali who was hanged for murder at Augur in 1887 was not born until four years after the events of Kanpur and therefore cannot come into it at all. His impulse obviously was not to kill Neill in particular, but to shoot at any person who happened to be behind him, and the nearest man was Major Neill. Almost certainly an English jury would have found Mazar Ali 'guilty but insane', but at his court-martial the regimental doctor declared that he was unable to certify on oath as to the insanity. At the court martial Mazar Ali said that he had shot at a green pigeon, that a Wali (spirit) had ordered him to kill the Adjutant etc. This 'debunking' of a familiar story is contained in Major General W.A.Watson's *King George's Own Central India Horse,* see Augier Papers in the **CSAS**.

NEILL, COLONEL JAMES 'Neill did things almost more than massacre, putting to death with deliberate torture in a way that has never been proved against the natives. ...He seems to have affected a religious call to blood,' Campbell wrote after reading his letters, 'and almost gloats over the way he ordered fat subahdars and Mahomedan civil officers to be lashed till yelling, they licked the blood with their tongues, and were afterwards hanged, in all which he sees the finger of God.' It would have been 'disgusting enough' if they had been guilty of murder, 'but Neill himself does not say that they were found guilty of the murders. He executes vengeance on 'all who had taken an active part in the Mutiny,' and when we know how these things were done we may well doubt if there was any proof of that... Neill is one of those people who have been elevated into a hero on the strength of a feminine sort of violence' but Campbell found 'there was not much more in him. ...I can never forgive Neill for his very bloody work.' Sir George Campbell, *Memories of my Indian Career,* London, 1893, pp 281-2.

NERPUT An Opium *gomashta* at **Kanpur**. 'Afterwards all the Civil and Military officers, 200 European soldiers, ladies, women and children with provisions and ammunition took possession of it. They also took a lakh of Rupees inside. On that day (Tuesday 2 June) some European officers went into the magazine and spiked what guns they did not require. Some spy gave information of these proceedings to the troopers of the 2nd Cavalry, and told them that mines had been laid for blowing them up in their lines and that if they did not evacuate their lines at once they would be blown up. Accordingly the whole regiment, exasperated at this report, mounted their horses and proceeded to the parade ground. Nerput, Opium Gomashta, *Foreign Secret Proceedings* quoted in **FSUP**, Vol IV pp 503-11.

NEWS-WRITER This was a profession followed by a number of literate men at this time. The news-writer's function appears to have been to send a (preferably regular) written report to his employer *or anyone else who would* *pay for his news*, of the leading events of the area in which he lived. He would not normally serve a newspaper, but would in effect be a static spy, though he might well have denied any such function, claiming that he simply wrote down the news as it came to him, rather than seeking out military or political secret matters. At the same time it is known that spies and *hurkarus* often reported the results of their expeditions not directly to the British, but to the news-writer of the area, for onward transmission. The British employed, or paid, numerous news-writers throughout the North of India at this time, including men actually behind rebel lines, as in Delhi during the siege, or Lucknow. There must have been a temptation to enhance, or deliberately falsify local news, in order to please the ears of the recipient, and perhaps for this reason the news-writer was not always believed or his reports acted upon. Take as an example the letter of the Gonda news-writer: 'He reports the arrival at Ayodhya of Koer (Kunwar) Sing with 2,000 men of whom only 800 are sepoys. He declares *he saw him seated with **Raja Man Singh*** in the latter's *Shewala*...Raja Man Singh afterwards went down to the river and inspected the works that had been thrown up at the *ghaut* to cover the passage; quantities of wood for planking the boats had been brought over; much ammunition had also been stored.' This apparently links Man Singh firmly with the rebels, but the British chose to avoid permanently classifying him thus. *See* letter from the Commissioner of Gorakhpur to the Secretary to Government, NWP, Allahabad, dated 13 February 1858 quoted in **FSUP**, Vol III.

NEWSPAPERS Newspapers and periodicals provide a valuable resource for the historian and student of the period. Although the bias of the proprietor was as significant then as it is now in its effect on content and reporting, and although the Press in India was subject to considerable attempts by a nervous government to 'gag' it, (*see* PRESS ACT), the press reporting and comments on events have the great merit of being contemporaneous and spontaneous, the result being useful and revealing. Some of the most significant of newspapers *etc* in this category are available for consultation in many libraries and archive collections, notably the Calcutta Library, the British Library and the Bodleian Library (limited collection). Those indicated for particular study include The *Bengal Hurkaru and India Gazette;* the *Army and Navy Gazette;* the Calcutta *Englishman* and *Military Chronicle;* the *Delhi Gazette;* the *Friend of India;* the *Hindu Patriot;* the *Illustrated London News;* the *Spectator;* the *Times* (of London: this includes all **W.H.Russell's** letters and dispatches from India from February 1858 onwards), the *Lahore Chronicle; Blackwoods Magazine* etc.

NICHOLSON, Brigadier General John (1821-57) Irish, like many famous generals in the British army he was born in Dublin in relative poverty, obtained a cadetship in the Bengal Army of **HEIC** in 1839, served Afghanistan, Ghazni etc and became a legend before he was thirty, 'Nikal Seyn' being the object of worship to a mixture of Sikhs and the fakirs of Hazara. Distinguished himself during the Second Sikh War, especially at Gujerat in 1849; in civil employ at Bannu 1851-56.

His energy even more than his personal courage was the factor that made him a most powerful instrument of British policy in India. Passionately sincere, arrogantly self-confident, insubordinate without remorse when he saw cause, and 'always in the right', he provoked no ordinary emotions; he was loved, admired feared, envied and hated to a violent degree. While most Englishmen were much in awe of his unique personality, some disliked him intensely for his lack of warmth and humour and his cold calculating stance on all matters of dispute. In much the same way that Winston Churchill was the right man at the right time for Britain in 1939, so Nicholson fulfilled a similar role for the British in India during the rebellion. He was quite ruthless in his treatment of the mutineers believing that was the only way to succeed. Commanded the **Punjab Movable Column**, disarmed suspected sepoy regiments; intercepted mutineers hastening to Delhi, and destroyed them at **Trimmu Ghaut** and at the **Ravi River**; arrived at Delhi on 14 August 1857; captured thirteen guns and the camp equipment of the enemy who were manoeuvring to get at the British rear, 25 August; commanded the main storming party in the assault on Delhi 14 September 1857, was shot through the chest and died on the day Delhi finally fell to the British, 20 September; buried in 'Nicholson's Cemetery' near the Kashmir Gate, in Old Delhi where his grave is still marked. *See* **Kaye, Sen, Ball, Martin** etc *See also* A.Lllewellyn, *The Siege of Delhi*, London, 1977, Harvey Greathed, *Letters written during the Siege of*

Delhi, London 1858. Hesketh Pearson, *The Hero of Delhi,* London,1939. Captain Lionel J.Trotter, *The Life of John Nicholson,* London, 1898. *See also* **DNB**.

Nicholson at Trimmu Ghat

NIMBAHERA   A fortified town in **Meywar** in **Rajputana**, and the scene of an unfortunate incident on 18 September 1858. It was occupied by **Tonk** sepoys who were beyond the control of the Nawab who wished to remain loyal to the British; it was attacked, and the Nawab's Amil came out, unarmed, with a small party to speak with the Political Agent, Captain Showers and tender their submission in the name of the Nawab. Colonel Jackson however, who commanded the troops, drove them back at the point of the bayonet. He failed however to capture the town (which was evacuated that night) and lost 18 killed and wounded. Worst of all he alienated the Amil, 'a man of well-known respectability', who went off to Tonk, was subsequently declared a rebel and lost all his possessions (confiscated by the Nawab). The Amil had to this time been very active in supporting the British cause. Neither Showers nor Jackson came out well from the affair, indeed Colonel Jackson was obliged to retire from the service. *See The Revolt in Central India,* Simla, 1908 . *See also* Iltudus T.Pritchard, The Mutinies in Rajpootana, being a personal narrative of the mutiny at Nusseerabad, London, 1864

NINETEENTH Regiment   *See* HEARSEY, Letter.

NINTH LANCERS   For the most part the British army looked far from smart during the campaigns of 1857-9. Sometimes, as at **Lucknow** or **Kanpur** they were besieged and had no option perhaps, since many had no more than the clothes they stood up in. Other units lost morale as a result of constant heat and privation and saw little point in observing the niceties of the parade ground. But there was apparently an exception to this general slovenliness: Every man of the 9th Lancers 'was dressed in his uniform, and there were no yellow leather boots pulled up over their trousers, and all sorts of motley costume, which obtain in many regiments, and which meets with so much approbation in the present day, but which tends very much to lower the *esprit de corps* and *morale* of the regiments. ...There was no happier corps, no corps in which discipline was maintained with less punishment; and one only had to see them in the field to be satisfied that few regiments could equal, none could surpass, the gallant, the dashing, always-to-the-front 9th Lancers, every man of which, from the colonel to the bugler, looked like a gentleman.' Oliver Jones, *Recollections of a Winter Campaign in India in 1857-1858.* London, 1859.

NIZAM OF HYDERABAD   It was very fortunate for the British that the Nizam, who had been harshly treated by them in the matter of a Treaty (1853) which robbed him of much of his territory to pay for a Contingent that he did not want, died at just about the time that the outbreak occurred in **Meerut.** The annexation of Hyderabad had been openly canvassed, and it was possibly only the priority of annexing **Oude[Awadh]** that had prevented **Dalhousie** from proceeding against the Nizam. The accession of his son, Afzul-ud-Dowla, to the musnud almost certainly saved the situation for the British, for under the guidance of a shrewd minister, Salar Jung, and his uncle Shums-ul-Omrah, the new Nizam adopted a policy of staying faithful to the British, in opposition to the wishes of the great mass of the Muslim element among his subjects. His action was characterised by **Karl Marx** as a betrayal of the aspirations of the Indian people. There can be little doubt that the attitude of the Nizam and the majority of the other Princes of India was crucial to the success or failure of the rebellion. *See* the *Times* of 29 June 1857 *et seq.;* K.Marx & F.Engels, *The First Indian War of Independence,* Moscow, 1960; and *see also* **Martin**, Vol II., p55 and p 268.

NORMAN, Sir Henry Wylie (1826-1904)   Retired as Field Marshal. Joined Bengal Army 1844. Was present in a Staff capacity at **Delhi, Lucknow, Kanpur,** and the campaign in **Oude[Awadh]**. His despatches, reports, comments are particularly valuable as they are both shrewd and well-informed. Has the unusual distinction of having refused the offer to become Viceroy of India - in 1893, *see* **DNB**.

NORTH INDIAN MILITARY CULTURE, in transition, 1857, a Review   'The importance of the army for the Company, and its implications for military power, is clearly indicated by the rich archival sources located in the various state archives in India, and the **British Library** in London. This source material comprises the proceedings of the Bengal Military Consultations, and the written proceedings of Bengal Military Department, and the large number of official military records of the Company army. A valuable addition to this compodium of information are the innumerable references to the Company army in the voluminous Board of Revenue Consultations, as well as in the written proceedings of the Company's Secret, Foreign, Political an Public departments. This plethora of evidence and the range of records - from the military to the public department files - where it is scattered, testify to the vital role the army played in the economic, social and political aspects of Company rule in India. An analysis of this material explodes the myth of the Bengal Army being a monolithic force of high caste sepoys. On the contrary the consensus is that the Bengal Army accommodated and nurtured a variety of military traditions. It is true that the elite corps of the army were the high caste Rajput and Brahman sepoys from Awadh and Bihar. The loyalty of these caste sepoys was structured around the Company's promise to uphold their high ritual status. However, as the Company territorially expanded into geographically and culturally diverse regions of India the limitations of using the high caste regiments became increasingly evident. In the peripheral areas of Bengal, the Jungle Terai, the Company's recruiting agents could not locate any high caste soldiers. Those brought from the plains were found inept in the hilly terrain. In such circumstances the Company was forced to compromise with local warrior customs. The Company army recruited the hill men and accommodated them in specially created regiments called the Hill Corps. Again, in 1802 when the Company's frontier rolled up the Ganges valley into the Western Doab region, or what later came to be known as the Ceded and Conquered Provinces, it was forced to recruit local Rohilla and Afghan troopers to meet the threat of the roving bands of 'Pindari' and 'Mewati' freebooters, who obstructed the smooth collection of revenue. The mounted Afghans and Rohilla soldiers were ex-servicemen of the Mughal Empire, or else of the regional states of the late 18th century. The Company set up specially designed regular and irregular cavalry regiments to accommodate these men. These regiments were structured on the pattern of the Mughal armies. The Company's cavalry promised to restore the primarily Muslim troopers to the high status they had enjoyed in their parent armies: Mughal as well as those of the 18th century regional states of Rohilkhand and Farrukhabad. Finally, after the Gurkha war of 1815 the Company gave serious consideration to the accommodation of Gurkha recruits in its Army. This was because its own high caste regiments of the plains had faced many problems fighting in the unfamiliar hilly terrain of Nepal. The variety of military traditions of the Bengal Army ensured a careful balancing between the army, polity and society and stabilised Company rule in most parts of North India. However, from the end of the 1820s this heterogeniety of military traditions began to be threatened. This was primarily because this was an age of financial strains and the Company began to reduce its military organisations to cut down on expenditure. The Company was encouraged to go ahead with its military cuts because it felt that it had defeated its major political rivals, and there was less need for a big military establishment. But more importantly the Company officials realised that the diverse military traditions, based as they were on the sharing of power and authority with local notables and high caste sepoys, showed signs of rebounding to its disadvantage: the 1820s saw the maximum instances of desertions and mutinies in its elite caste regiments, most of which were related to infringements of the sepoys' high caste status. At the same time there were threats posed by the local notables the Company had hitherto encouraged so as to recruit its hill regiments and the cavalry regiments. In the 1830s the Company reacted to these political challenges by reducing the various components of its army: the caste regiments, the Hill Corps and the cavalry regiments. At the same time it attempted to homogenise the diverse military traditions that had constituted the Bengal Army. The high caste peasant army tradition of the Gangetic plains began to be extended as a uniform military culture all over North India. This was accompanied by an attempt to exercise a far greater interventionist control over the much reduced high caste regiments. These military reforms of the 1830s created widespread discontent in the Company army which eventually culminated in the mutiny/rebellion of 1857. The historiography of 1857 hinges around the Enfield Rifle episode, and stresses the fact that the mutiny was sparked off by the introduction of this new rifle. Locating the events of 1857 in the context of the above mentioned military culture it is evident that the 1857 events cannot be explained so simply. The mutiny was certainly not a conflict in which an antique Indian warrior tradition protested against the encroachments of a militarily modernised Company. On the contrary in 1857 Company power was under threat from its own military 'subalterns' who were incensed because the reforms of the 1830s had disturbed the power relations within which they had enjoyed financial security and a high religious and social status. From the 1840s the Company's high caste regiments had expressed their grievances quite clearly and this is reflected in their resentment over the withdrawal of an allowance - **batta** - they had hitherto obtained for service in the foreign lands of the Sindh and the Punjab. Further, in the same decade they resented permanent postings in the military outposts of Sindh and the Punjab. This separated them from their families for long periods of time - something they were not accustomed to. But more importantly, sepoy resentment was related to their apprehensions over the infringements of their high caste status, which the Company had sedulously promoted over the previous three generations. The greased cartridges scandal, in which the soldiers feared their religion was in danger because they were being made to use cartridges greased with the fat of cows and pigs ignited a religious sensitivity which the Company had kept alive ever since they joined its service. The most significant event of 1857 was the sudden displacement of the Company as the chief employer by patrons amongst the rebel leaders offering to the sepoys the material, political and ritual inducements which the Company had hitherto monopolised. In this context the actions of rebel leaders like **Kunwar Singh,** in the Shahbad district, the **Rani of Jhansi** and **Nana Sahib,** were reminiscent of the Company's efforts to project a Hindu image of the army in order to garner sepoy support. We have vivid descriptions of Nana Sahib's efforts in this directions from Vishnu Godse, a Brahman *bhikshu* from Pune who got caught in the mutiny of 1857 while on his way to participate in the *sarvuttomukh yash* (the biggest and best religious ceremony); Godse wrote in his travelogue: *'The British regiments were massacring every one they could find. Even in these dangerous times Nana Sahib was feeding Brahmans and asking them to determine his fate and actions from his horoscope. These days people will find such behaviour strange. But it was not so in those days because people had more faith in yagyas and religious cermonies rather than in the efficiency of their sword.'* (Amritlal Nagar, *Ankhon dekha gadar*, Delhi, 1986, pp 26-27). Godse also noted with appreciation the elaborate Hindu religious ceremonies he witnessed in Jhansi: according to him 'new cannons, rifles and gole barood(bombs) began to be prepared in Jhansi and public religious ceremonies for the safety of the state commenced' (ibid). Similarly, in the Ceded and Conquered Provinces the mutiny/rebellion derived its momentum from the military tradition which the Company had kept alive in the region. In this region Company's recruitment had targetted Afghan-Rohilla troopers who were ex-servicemen of the Mughal state and/or of the 18th century **Rohilkhand** and **Farrukhabad** regional polities. The Company had ensured their loyalty by promising to restore them to positions similar to those that they and their predecessors had enjoyed in earlier polities, like the Mughal Empire. This had made the Company service very popular in the region. However the large scale military retrenchments of the 1830s reduced the military establishment in the Ceded and Conquered Provinces, and uncovered the political and social tensions it had contained. As the livelihood and aspirations of the troopers came crashing down, their anti-Company mood was bound to follow. The troopers resentment was best summed up in a poem entitled *'The Old Pindaree'*, by an anonymous poet, which is published in the *Asab-i-Bhagawat-i-Hind* written in 1858 by **Sir Sayyid Ahmad Khan Bahadur**. The hero of this poem, is an old Rohilla who was reported to be of seventy years of age in 1857. He belonged to the generation of troopers who had suffered the most from the military reforms of the 1820s. The loss of prestige he experienced in his locality after the Company deprived him of his warrior profession compounded his resentment against the new Company regime: *'He has taken my old sword from me and tells me to set up a school; ...well Ram Lall the Telee mocks me, and pounded my cow last rains - He has got three greasy young urchins, and I'll see that they take pains. There comes a settlement Hakim (ie Officer) to teach me to plough and weed. I sowed the cotton he gave me, but first I boiled the seed. He likes us humble farmers and speaks so gracious and wise. As he asks of our manners and customs, I tell him a parcel of lies.'* In 1857 such disillusioned Company troopers looked to Indian patrons to restore them to the military status the Company had hitherto given them but which had now been withdrawn. At Delhi noted Company Subahdars, like Bakht Khan, furthered their military ambitions by their claims to restore the Mughal Emperor to the throne of Delhi thereby becoming popular leaders of the Company's disgruntled cavalrymen. A similar pattern could be discerned in Rohilkhand, Shahjahanpur and Bulandshahr region where local rulers restored their polities using the support of the Company's troopers who sought stable political structures to restore their lost status. If the political threat to Company power in 1857 came from the backfiring of the military tradition it had encouraged, its ability to quash the mutiny also emanated from that very military culture: the Gurkha regiments the Company had nurtured stood in its defence. In 1857 the loyal support for the Company came from the Gurkha forces which the Company had built as a military force distinct from the high caste sepoy regiments, and the cavalry regiments of the plains. The notion of 'hill Hinduism' which the Company had encouraged in its Gurkha regiments proved very handy in using the Gurkhas against the rebellious caste conscious Rajput-Brahman sepoys of the plains. The Gurkha support for the Company was most evident with the Sirmour battalion of Gurkhas. The rebellious Sappers and Miners, against whom the Sirmour Battalion of Gurkhas was despatched was reported to have attempted to lure the Gurkhas to their cause by selling them the 'religion in danger' idea which had become popular among the rebels. They tried to dissuade the Gurkhas from going to Meerut where they said, 'atta' (flour for making bread) was nothing but 'ground up bullock bones'. To this the Gurkhas were reported to have replied that the regiment was going wherever ordered and that they obeyed the bugle call (L.W.Shakespear, *History of the 2nd King Edward's own Goorkha Rifles*, Aldershot, 1912). These acts of loyalty made the Company take its Gurkha military tradition more seriously. The Gurkhas were to become the focus of British attention in the post 1857 period.' Condensed from Seema Alavi, *Sepoys and the Company: Tradition and Transition in Northern India, 1770-1830*, OUP, Delhi, 1995. **SEEMA ALAVI**

NOWGONG A military station sixty miles east of Jhansi, and headquarters of the Political Agent in Bundelkhand. When news arrived of the murder of the Europeans at the latter place, the troops (12th BNI and 14th Irreg Cavalry plus artillery) mutinied on 9/10 June 1857. The British officers and their families accordingly left the station and wandered as fugitives for upwards of a month in **Bundelkhand**, sometimes being treated well, as at Chhatarpur, at other times being attacked by the villagers who were generally hostile. Among those who died or were killed were Major Kirk, Mrs Smalley and child, Dr Mawe, Lieutenants Barber and Ewart; Captain Scott and a small party were brought as prisoners to **Banda**, where they were kindly received by the Nawab and his mother and sent in safety under escort to **Nagode** which they reached on 12 July. Others were hunted from village to village through Sihonda and Badausa, plundered and many of them killed. *See* MAWE, Dr Thomas, *and* SCOTT, Lt General PATRICK GEORGE. *See also* **FSUP** Vol III, *and* Tapti Roy, *The Politics of a Popular Uprising, Bundelkhand in 1857*, Delhi, 1994.

Nowgong Church Today

NOWSHERA A cantonment town some thirty miles east of **Peshawar**. The 55th BNI and the 10th Irreg. Cavalry mutinied there on 21 May 1857, and their action triggered the disarming of the sepoy regiments in Peshawar. For details *see* Lionel J.Trotter, *Life of John Nicholson,* London, 1898.

NUNNE NAWAB Nunne Nawab and his brothers were the sons of Aga Mir, the prime minister of Nasir-ud-din Haidar of Oude[Awadh], who retired to Kanpur in 1830 with a pension of 25,000 rupees, a portion of the interest on a loan Aga Mir had kindly persuaded his regent to extend to the Company.'When they reached near Mirza Hajee's bungalow, some six or eight troopers were despatched to take me to the Nana but I not answering their first call, again a party of about 100 troopers was sent [under a subadar of the 2nd Light Cavalry named Udhar Singh], and, effecting their entrance by forcing open the backdoor, made me their prisoner. I, of course, mounted my horse with a few of my followers, and went to the Nana, surrounded by mutineer troopers, who threatened to take my life if I should decline compliance with their wishes. 'I was first taken before Nana's younger brother, Bala Saheb, who ordered me to be disarmed, and my followers to be plundered of the silver and other valuable things they had with them. When I approached the ringleader Nana, I was commanded by his moonshee, Jawala Pershaud, to dismount my horse, which he took for himself. There were at the time about 600 arms raised at me, but Nana dissuaded them from their purpose, and ordered me to be imprisoned and placed on an elephant, as if any one is led through the streets in ignominious show, putting me in charge of two sentries. While thus confined, I received information from my men that they fired at my house six or seven guns, plundered it of all its property, amounting to ten lacs of rupees, and had my Lady not saved herself by going upstairs of a room three-storied, by means of a wooden ladder, and drawing the ladder up, they would have much ill-treated her. I heard also from creditable authorities that in this spoil the city people and most of my own servants shared. The pretence made for forcing entrance into my house was that I had concealed eight or nine Europeans therein. I was so miserably plundered that there was not a particle of my property left, one of my servants actually obliged to prepare *khitchree* for my evening meal at his own expense. 'On reaching near Mr. [Jacobi's] bungalow they began firing at the entrenchment from near the canal, butchering every European, East Indian or native Christian who unfortunately fell in with them. Now it was that I saw my two brothers, Nizam-ood-dowlah and Ameen-ood-dowlah , were violently taken to the field surrounded by troopers who led them to the Nana.' *Nunna (Nunne) Nawab's Diary* quoted in Appendix A to **Forrest**. *See also* MUHAMMAD ALI KHAN

NUR MAHOMED Was proprietor of Mahomed's Hotel, in **Kanpur** near the **Assembly Rooms,** and close also to the **Bibigurh** where the British women and children were massacred: at that time **Nana Sahib** was staying at his hotel. Nur Mahomed had the remarkable record of having run his hotel without a break from before the mutiny of the sepoys, right through the period of Nana Sahib's rule, and subsequently, without apparently suffering either investigation or punishment, under the restored British authorities: he did in fact take pleasure in conducting a tour of his establishment, pointing out where Nana had slept, where his food had been prepared etc.

NUZZER Or nuzrana, an offering or present, the significance of which lay not so much in its intrinsic value as its ceremonial purpose: it represented the acceptance of a subordinate role, and was usually accompanied, in reverse, by the presentation of the **khilat.**

NYNEE TAL *See* NAINI TAL.

OATH, The 'Our oath,' **Nana Sahib** told **Wheeler**, 'is that whoever we take by the hand, and he relies on us, we never deceive; if we do, God will judge and punish us.' *London Times,* 16 October, 1857.

OATH, on recruitment Before the mutiny/rebellion the oath administered to all sepoys on induction was as follows: 'I ...AB ... do swear to serve the Honourable Company faithfully and truly against all their enemies, while I continue to receive their pay and eat their salt. I do swear to obey all orders I may receive from my Commanders and Officers, never to forsake my post, abandon my Colours or turn my back to my enemies; that I will in all things behave myself like a good and faithful sepoy in perfect obedience at all times to the rules and customs of war...' Subsequent to the demise of the Company in 1858 the oath was made to bear allegiance to the Sovereign, *ie* Queen Victoria, and a picture of her (and latterly her successors) was brought on parade that all who swore allegiance might at least know the appearance of the one they were promising to serve!

OBSERVATORY, LUCKNOW [Tara Wali Kothi] 'From this hut my captor was obliged to remove me, as some women had discovered my existence, and, not being convinced that I was really a convert to the Moslem faith, threatened to betray the sowar to the authorities. My captor thereupon took me to a bungalow in some other part of the town, which had originally been the residence of a Mr. Simpson, and the native name of which was 'Tha-ra-walla Kotee ' (Observatory). Here I was secreted in the kitchen, which seemed a palace contrasted with the hovel I had recently occupied. In the building itself now dwelt Sha-ah Hum o Dilla, the Moulvie of Fyzabad, who had brutally murdered every European in that town. These were not his only victims, as in Lucknow he had added to the list by putting to death about fourteen women and children, together with a few native Christians. They had concealed themselves in the underground cellars of the above-named house, were dragged out by a mob, barbarously served, and then shot, after which their bodies were hacked to pieces and left to be devoured by vultures.' See Amelia Bennett, *Ten Months' Captivity After the Massacre at Cawnpore* The *Nineteenth Century.* June-July, 1913.

OBTAINING INFORMATION During the pursuit of **Tatya Tope** in Central India the rebels had the sympathy of the villagers almost entirely on their side, and so the British were severely handicapped in obtaining information, even on the field of battle itself. General Roberts was in charge of the pursuit and it is recorded that he resorted to unusual lengths to keep himself informed. Thus: 'The method which General Roberts adopted for obtaining information was to have about twenty cavalry in advance close to the rebels. They left connecting links of two or three men every two or three miles, so as to keep up the chain of communication. The advance party was *composed half of Baluch Horse, who had no sympathy with the rebels, but who could communicate very well with the villagers, and half of horsemen belonging to the Raja of Jaipur who were supposed, as Rajputs, to be on good terms and able easily to communicate with the villagers, but not to be very warm partisans of the British.'* Roberts apparently thus solved the *quis custodient custodies* dilemma, *see Blackwood's Magazine,* August 1860.

OCCUPATION OF OUDE[Awadh] 'The occupation of Oudh had been sudden and brutal; no time for upper classes to accommodate to the British, no inclination among the British to accommodate to the landholders.' Eric.Stokes, *The Peasant Armed: The Indian Revolt of 1857.* Oxford, 1986.

OLDFIELD, Henry Ambrose Dr Oldfield was a member of the Bengal Medical Service between 1846 and 1868, and in 1857 he was the Residency Surgeon in **Nepal**. His correspondence includes a recording of the views of **Jung Bahadur** on the causes of the revolt in Northern India and is preserved in the Oriental and India Office collections of the British Library, *see MSS.Eur.C.193.*

OMMANEY, M.C. Judicial Commissioner in Awadh[Oude], killed by a roundshot on 5 July 1857 during the siege of the **Lucknow Residency**. Had been 23 years in the Civil Service of the **East India Company**, and was highly thought of in official circles.

OOMRAO JAN Reported as '*Oomrao Jan* (Umrao Jan), formerly a vakeel and moonshee of Moosahibood-Dowlah (singer, who accompanied ex-King Wajid Ali Shah to exile in Calcutta), during the rebellion became one of the chief advisers of **Mammu Khan**, resident of Lucknow;' see PERSONS OF NOTE, APRIL 1859.

OOMRAO SINGH, Captain Reported as 'Captain Oomrao Singh, a subadar of the 6th Oude Irregular Force, prior to the annexation Subadar Major in Captain Barlow's regiment, resident of Kantha in the Baiswara District, but for some years had his family at Colonel Gunj Secrora, he had great influence in the Durbar of **Birjis Qadr**;' *see* PERSONS OF NOTE, APRIL 1859.

OOMRA SINGH, RANA   *See* PERSONS OF NOTE, APRIL 1859.

OPIUM   Opium, technically a Government monopoly at this time, and therefore by no means part of an illegal trade, was accepted as a fact of life in India. Its use was widespread, and strangely, although greatly desired by its devotees it does not appear to have been addictive in the way that so many 'hard' drugs are nowadays. Nonetheless it would not be high in the list of life's necessities for the educated or well-to-do, and so a reasoned defence for opium is unusual, particularly when set down by an official. It makes strange reading nowadays, but perhaps should not be taken out of its moral, social and time context. Moreover, he is talking of raw opium, not refined into the addictive drugs of today. It is worth including in this book because there are frequent references to opium and its use in the literature of the 1857 struggle: many sepoys required it before going into action (as British soldiers almost invariably drank rum in similar circumstances), some sepoys and camp followers deserted the British in the **Residency at Lucknow** not from conviction but, as their pathetic note left behind them tells us, from 'lack of opium'. 'All fakirs, devotees etc in the NorthWest Provinces drug themselves largely, and are one of the few classes who ever show that an extensive use of bungh or opium has been resorted to. All the Seikhs of the Punjaub, with the Jats, Rajpoots and other castes in the NW Provinces and Central India, consume large quantities of opium, and yet in few parts of the world is it possible to see more healthy and finer races....I constantly visited the opium shops, but neither there did I ever see, as the result of opium, anything approaching the degradation or ruin to be witnessed in enlightened Europe with its civilized drinks. Opium is in every respect, more harmless both to the recipient and his friends, particularly in a tropical climate than the wines or spirits of Europe. The claptrap philanthropists of Europe and America might possibly effect more good by initiating reform nearer their homes, than by offering England rather interested lectures, it may be, on the demoralizing effects of opium. It is an acknowledged fact, that almost all races, particularly within the tropics, seek and require a stimulant of some kind, and it may be doubted whether, when eradicating opium, it would be easy to discover another stimulant which is morally and physically less injurious.' *See* H.Dundas Robertson, *op.cit.*

ORIENTAL & INDIA OFFICE Collections of the British Library   This is the name now given to what used to be known as the India Office Library and Records. Also referred to as **OIOC** or even **OIOL**. *See* BRITISH LIBRARY.

ORR, Captain Patrick   He was one of the CAPTIVES OF OUDE[AWADH], and eventually killed at Lucknow. 'On the evening of 4 June 1857 all the Europeans at **Muhamdie,** including Captain Orr, decided to set out at once, despite the shortage of transport. They were reassured by the presence of Capt.Orr's own troops who swore loyalty - although this did not prevent them from carrying out the two almost mandatory acts of mutinous troops: they released the prisoners from the gaol and took all government money from the Treasury. The party and its escort set out with some confidence and all went well for a few miles until a halt was called between Burwun and **Arungabad**: there a trooper of the escort came to tell the Europeans to go on ahead, they would catch them up. They certainly did. Riding round in a circuitous route they set an ambush for the Europeans within half a mile of Arungabad. Captain Shiels we are told was the first to be shot. The fugitives stood very little chance: they had made no plans to deal with an attack by their own 'loyal' escort. The little group of Europeans collected forlornly under a tree, and the ladies got down from the buggy in which they were travelling. Shots were coming in from all directions. We are told that the ladies joined in praying, 'Coolly and undauntedly awaiting their fate'. We have not only an eye-witness to all this but one whose written account is still with us and can hardly be doubted. It is none other than Captain Orr! Indeed the next thing to happen is so extraordinary that it would be difficult to believe did we not have his own, written, account of what happened. He, Captain Orr, says that he *'stopped for about three minutes amongst them (ie the fugitive party), but then, thinking of my poor wife and child here, I endeavoured to save my life for their sakes. I rushed towards the insurgents, and one of my own men, Gourdeen, 6th Company, called out to me to throw down my pistol and he would save me. I did so whereupon he put himself between me and the men and several others followed his example. In ten minutes they had completed their hellish work. I was about 300 yards at the utmost. Poor Lysaght (Captain) was kneeling out in the open ground with his hands folded across his chest and although not using his firearms the cowardly wretches would not go up to him until they had shot him; then rushing foward they killed the wounded and the children, butchering them in a most cruel way'.* Captain Orr could hardly have guessed that so many years later we might read his description of the 'cowardly wretches', and wonder perhaps at this description of the men who saved his life, and wonder, too, at the strange part he played in all this. The quotation, incidentally, is from a letter that Captain Orr wrote a week or two later from the jungle outside **Mitauli**, to his brother Adolphe in **Lucknow,** by whom it was preserved for us to see. Those killed in the ambush were Captains Sneyd, Lysaght and Solomon, Lieutenants Key, Robertson, Scott, Pitt, Rutherford, Ensigns Spens, Johnston, Scott, and ten more men and women, plus four children, and 'poor good **Thomason**', and the two clerks, Mr Smith and Mr Hurst. Patrick Orr's part in the drama is further confused by the report that during the night previous to the massacre of the fugitives, which was certainly premeditated and planned, a native officer by name Lutchman came to Orr privately and hinted at what was to happen the next day, and begged him to leave the party and join his wife at Mythowlie, adding that his men had actually agreed to his going. Orr had replied that he 'could not abandon his friends' - and yet the next day, under fire, he successfully did so. Perhaps the most charitable explanation is that he thought that his presence with the fugitives might have guaranteed safety for all. Certainly his own privileged position was due to his having raised the corps the previous year in Awadh, but the way in which this 'privileged position' develops into near farce makes strange reading. It was now in fact proposed by some of the men that Captain Orr should now send for his wife and child, and marching into Sitapur should put himself at the head of the Regiment!' He did refuse this offer. For further details *see* **SSF**. **PJOT**

Artist's Impression of an Oude Fort (Naham)

OUDE, [OUDH, AVADH, AWADH] The establishment, or re-establishment, of native government in the Kingdom of Awadh took place after the Battle of **Chinhat** on 30 June 1857. There was considerable confusion at first, but then Begum **Hazrat Mahal** and her allies began to prevail and her son **Birjis Qadr** was crowned King, even though his putative father, **Wajid Ali Shah**, was still alive, exiled, in Calcutta. For an excellent and detailed account of the early days of this restored government *see* 'Begamat-i-Awadh Ke Khutut', p 42 *et seq. or* translation of same in **FSUP**, Vol II p 103 *et seq*. For an important assessment by **Rudrangshu Mukherjee** of the significance of the attack by the British on Talukdari power *see* SUMMARY SETTLEMENT of 1856, *see also* Rosie Llewellyn-Jones's account of the last days before annexation in 1856, in AWADH: THE LAST DAYS OF THE KINGDOM OF AWADH.

OUDE[AWADH], ARMY IN The troops actually present in the province at the beginning of the outbreak totalled a staggering 21,435, of whom only one regiment (800 men) and one battery of artillery were Europeans. The details of the troops are as follows: Artillery, 2 Regular batteries (12 guns) strength 500; 3 Irregular batteries (18 guns) strength 360; 1 Irregular Garrison battery strength 100; 1 European Field battery strength 75. Cavalry, 1 Regular regiment strength 600; 1 Irregular regiment(Punjab) strength 600; 3 Irregular regiments (Oude) strength 1,500; Military police equipped as cavalry strength 700. Infantry, 5 Regular regiments strength 6,000; 10 Irregular regiments strength 8,000; Military police equipped as infantry strength 2,400; 1 regiment of European infantry (HM's 32nd) strength 800. Grand total 21,435. This very large army is explained only by the British following their successful precedent in the Punjab after annexation there also, viz the raising of a large Irregular Force to ensure law and order in the newly annexed province of Oude[Awadh]. It was placed under the control of the Civil authorities, *ie* Sir Henry Lawrence, and was entirely independent of the Regular army. The men of this Irregular Force were recruited almost exclusively from the disbanded soldiery of the former King of Awadh.

OUDE[AWADH] CAMPAIGN By the end of September 1858 only about one quarter of the province had submitted to British rule. There were still many rebels in arms and the names of some of them were reported by one Khubchand Nazir, Adalat Shahjahanpur, in the following terms: 'Jani Ali Khan was at Naurangabad in the *ilaqa* of **Mohamdie**, with 7,000 horse, 2,500 *tilangas*, 5,000 *najibs*, and seventeen guns; **Raja Lonee Singh** of **Mithauli** was also at Naurangabad with 2,000 infantry, unspecified number of guns, and ammunition; **Khan Bahadur Khan** of **Bareilly** was at Haidarabad with 1,000 horse, 300 *tilangas*, 1,000 *najeebs* and eight guns; **Fazal Haq**, of **Shahjehanpur** was at Pihani, also in the *ilaqa* of Mohamdie, with 4,000 horse, 900 *najeebs* and three guns; **Firoze Shah** was at Khairabad with unspecified forces, having been sent there by the **Begum of Oude**; the Begum herself was at **Baundi** with a large force of both horse and infantry, but the latter now included few sepoys from the mutinied Bengal army: she apparently no longer trusted them.' By far and most straightforward and succinct account of the remainder of this complicated campaign will be found in Innes, *The Sepoy Revolt,* London, 1897, pp 260-71. By 7 January 1859 Lord Clyde (**Sir Colin Campbell**) was able to report to the Viceroy and Governor General, Lord Canning,

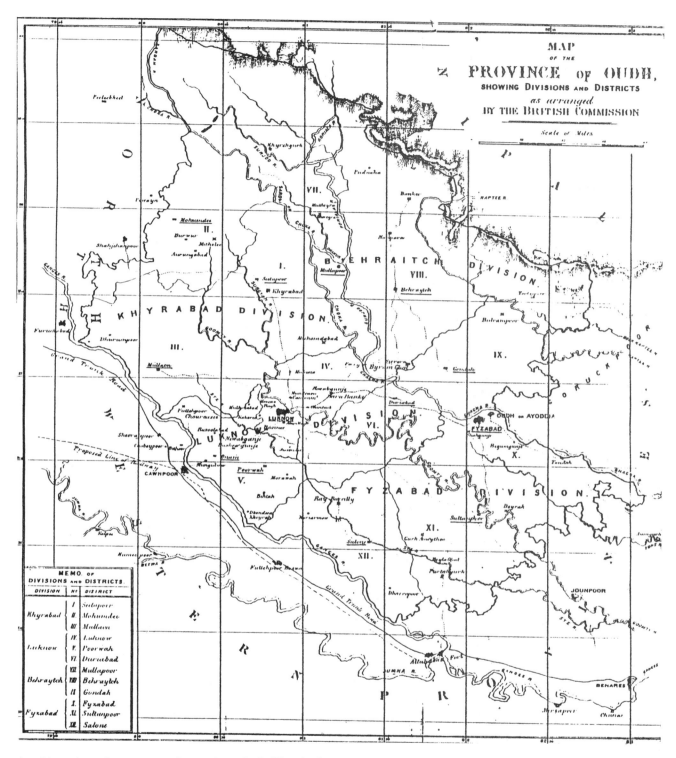

that 'there is no longer even the vestige of rebellion in the Province of Oudh', see, for a full version of this despatch which contains news of **Nana Sahib, the Begum, Bala Rao** etc, Chas Ball, *The History of the Indian Mutiny*, Vol II pp 563-66.

OUDE[AWADH] FORTS 'There are at this time in Oude 246 forts or strongholds, mounted with 476 pieces of cannon, all held by landholders of the first class, chiefly Rajpoots; not one of these landholders now feels it safe to entrust himself within the camp or cantonments of an officer of the Government... These estates are well cultivated, often in spite of all the best efforts of the contractors and collectors to prevent it, in order to reduce them to obedience. It is not at all uncommon for the landholders to have the land ploughed and the seed drilled in at night by stealth when beleaguered by the king's troops; and this accounts for the land being so much better cultivated than those of other native States in the midst of disorders that would soon make them waste in any other country or state of society.' - Colonel William Sleeman 21 June, 1849. *See* James Innes, *Lucknow & Oude in the Mutiny, London,* 1895, p 309.

OUDE[AWADH] in 1858 After the fall of **Lucknow** in March 1858 to the British, it was assumed by many of the latter that the Kingdom of Awadh had also fallen into Government hands. This was far from the case. Many of the major **Talukdars** of the kingdom continued the struggle for as long as they were able, some perhaps because they were so deeply compromised that they could hope for nothing from the British (*see* PROCLAMATION OF OUDE), but others, notably **Mehndie Hussain,** and Rana **Beni Madho Singh** on genuine patriotic grounds: having sworn allegiance to **Birjis Qadr** and his mother **Hazrat Mahal, the Begum,** they felt they had no honourable course to follow but to fight on until the end. There were still more - like Raja **Man Singh** - who had carefully sat on the fence until they had decided which side would win in the end, and then come in on the British side just in time to be able to claim that they had contributed to the British victory. As late as the end of May 1858 the civil power had made very little progress in settling the country. Where there was a column of British soldiers there was rule by government: elsewhere (and this was the greater part of the kingdom), the government writ did not run. A letter from the Deputy Commissioner (Mr S.Martin) to the Commissioner and Superintendent of the Lucknow Division (itself only a part of the territory of the Kingdom) Colonel S.Abbott, written on 17 May 1858 illustrates this point: ...'It is obvious that as long as formidable bands of armed men, consisting of our mutinous sepoys and sowars, can march about the country unmolested the Civil Power must in a great measure be in abeyance'. He then lists the very small number of tehsils and thannahs at work. The **Maulvi** of Faizabad, the **Begum** herself, and even **Nana Sahib** were active in Oude[Awadh] at this time, and although pursued by British columns were still carrying on some semblance of government. For documents illustrating this period *see* **FSUP,** Vol II Chapter 7.

'OUDE PROCLAMATION, The' Not to be confused with the **Queen's Proclamation**. This was issued in March 1858 on the authority of the Governor General, immediately following the final British assault on Lucknow. While not going so far as the Directors of the East India Company in acknowledging the right of the people of Oude[Awadh] to have fought the British, as alien invaders, it does at least recognise that resistance was almost universal. Note the slip in drafting ('their' instead of 'the') in the last sentence of this extract: 'The Army of His Excellency the Commander-in-Chief is in possession of Lucknow, and the city lies at the mercy of the British Government, whose authority it has for nine months rebelliously defied and resisted. This resistance, begun by a mutinous soldiery, has found support from the inhabitants of the city and of the province of Oude at large. Many who owed their prosperity to the British Government as well as those who believed themselves aggrieved by it have joined in this bad cause, and have ranged themselves with the enemies of the state. They have been guilty of a great crime, and have subjected themselves to a just retribution. The Capital of *their* (sic) country is now once more in the hands of the British troops.' The proclamation then went on to announce rewards for those who had remained faithful to the British, and to call on all others to submit and give their support to government if they hoped for leniency. Those whose hands 'were stained with English blood' were specifically excluded. The full text is given in The *Hindoo Patriot,* issue of 29 April 1858, and in Vol II of Chas Ball, *History of the Indian Mutiny.* The Proclamation was issued *before* the receipt of a Secret letter from the Directors in East India House, dated 24 March 1858, of which the following is an extract: 'To us it appears that, whenever open resistance shall have ceased, it would be prudent, in awarding punishment, rather *to follow the practice which prevails after the conquest of the Country which has defended itself to the last by desperate war,* than that which may perhaps be lawfully adopted after the suppression of mutiny and rebellion, such acts always being excepted from forgiveness or mitigation of punishment, as have exceeded the license(sic) of legitimate hostilities.' The Proclamation, and still more its interpretation, caused much misunderstanding and confusion, and indeed active disagreement between the Governor General, **Canning**, and the Chief Commisioner of Oude, **James Outram**: the latter felt that the terms of the Proclamation would drive many talukdars into prolonging their resistance to government. Canning felt moreover that he was not getting the support he could have expected from the home government in the matter, and threatened to resign. For the full text of these and related documents, *see* **FSUP,** Vol II p 328 *et seq.*

OUDE[AWADH], REASONS AND ARTICLES of the Draft Treaty Proposed to the Nuwab, February 1856. This was the draft treaty rejected by **Wajid Ali Shah**: 'REASONS Whereas in the year 1801, a treaty was concluded between the East India Company and his Excellency the Nuwab Vizier Saadut Ali Khan Bahadur, and whereas the 6th Article of the said treaty requires that the ruler of Oude, always advising with and acting in conformity to the counsel of the officers of the Honourable Company, shall establish in his reserved dominions such a system of administration, to be carried into effect by his own officers, as shall be conducive to the prosperity of his subjects, and be calculated to secure the lives and property of the inhabitants; and whereas the infraction of this essential engagement of the treaty by successive rulers of Oude has been continued and notorious; and whereas its long toleration of such infraction of the treaty on the part of the rulers of Oude has exposed the British Government to the reproach of having failed to fulfil the obligations it assumed towards the people of that country, and whereas it has now become the imperative duty of the British Government to take effectual measures for securing permanently to the people of Oude such a system of just and beneficent administration as the Treaty of 1801 was intended but has failed to provide. ARTICLES (summary) I. The Honourable East India Company takes over sole and exclusive administration of Oude, with full and exclusive right to its revenues, and engages to provide for the due improvement of the province. II. His Majesty and his heirs male in continual succession shall retain sovereign title of King of Oude. III. And shall be treated with corresponding honour. IV. And shall have full jurisdiction in his palace and park at Lucknow. V. Shall receive twelve (open to increase by three lakhs if desired) lakhs per annum, and also three lakhs more for his palace guard. VI. The Honourable East India Company takes upon itself maintenance of all collateral members of the Royal Family heretofore provided for by his Majesty.' James Innes, *Lucknow and Oude in the Mutiny,* London, 1895, p 313 .

OUDE[AWADH], a SPECIAL CASE   G.F. Edmonstone, Secretary to the Government of India, writing to **Outram** on 31 March, 1858, saw the people of Oude as a special case. They had been '*subjects of the British Government for little more than one year when the mutinies broke out;* **they had become so by no act of their own.** By the introduction of our rule many of the chiefs had suffered a loss of property, and all had experienced a diminution of the importance and arbitrary power which they had hitherto enjoyed; and it is no marvel that those amongst them who had thus been losers should, when they saw our authority dissolved, have hastened to shake off their new allegiance. ...That unjust decisions were come to by some of our local officers in investigating and judging the titles of landholders is, the Governor General fears, too true.' This helped explain their rebellion, and yet 'no chiefs have been more open in their rebellion than the Rajahs of **Churda, Bhinga, and Gonda**,' and none of them were hurt by the accession of the British; indeed all of their assessments had been lowered. The Governor General had to conclude that the cause of their hostility lay 'in the repugnance which they feel to suffer any restraint of their hitherto arbitrary powers over those about them, to a diminution of their importance by being brought under equal laws, and to the obligation of disbanding their armed followers, and of living a peaceful and orderly life.' Thus in almost all such cases 'the penalty of confiscation of property is no more than a just one.' *See Further Parliamentary Papers.* Volume 43 1867-58. No. 467A.

OUDE[AWADH] ZEMINDARS   A long article on the Kingdom of Awadh, based primarily on **Sleeman**'s account appeared in the London *Times*. There is a list of local zemindars and their alleged 'atrocities'. *See* London *Times,* 9 April, 1858, p 5.

OUDH   *See* OUDE[AWADH].

OULA   She was a courtesan kept by the Nana; 'this girl Oula had by her lakhs of rupees worth of property belonging to the Nana. She was also called Adala, or Sultana Adala. She was by birth and profession a courtesan, born at Mugrasa she had lived with the Nana since 1850, at first receiving Rs 300 per month, but eventually becoming a favourite and being endowed with jewels belonging to the widows of the late Peishwa, valued at Rs 50,000, a prodigious sum. Nanak Chand had been trying to fight the legal battles of these widows in the British courts before the Mutiny. It is worth remembering that Oula, and presumably her little group of slave girls had everything to gain from making sure that there was no way back into British favour for the Nana Sahib. Those who wished him to be incriminated past all redemption certainly had their way: to the British he remained the arch-fiend to the end. When he fled from his palace at **Bithur** after Havelock's arrival, Oula was sent in a boat some distance up the river, but returning to Kanpur in August 1857 was secreted in a house in the Butcher Khana, and thence via two other hiding places to her birthplace at Mugrasa, near Kanpur, under the care of a man named Maudhoo, having promised, she stated in evidence, to await one year at Kanpur for the return of the Nana to that station. Others eagerly gave evidence to Colonel Williams that the sister of Maudhoo, named Zareena, lived in the Kanpur district and that the widows' jewels were said to be secreted in her home. The fingers of accusation were pointed in many directions, and no doubt many old scores were paid off by the witnesses who came before Colonel Williams, some dragged unwillingly from obscurity, some pushing forward eagerly to tell their tale. Oula was clearly at the very height of her power during the Nana's short reign in Kanpur in June/July 1857. She is said to have accompanied him everywhere and at the Satichaura Ghaut to have been 'seated in a tent, and from behind the screen is enjoying the sight of the European ladies and gentlemen being put to death '.

OUTRAM, Sir James (1803-1863)   Lieutenant-General, Indian Army. In his early days performed great hunting exploits for which he became famous among the Bhils with whom he had a particularly strong influence. His major career was as a political officer seconded from the army, viz as Resident In **Lucknow** he had recommended annexation. Believed by many to typify the best in the British character, he was termed the 'Bayard of India' by Napier. Commanded the force which invaded Persia in January 1857, and which quickly brought the Shah to make terms (favourable to Britain's ally, **Dost Mohamed,** of Afghanistan). With **Havelock** conducted the first 'relief' of the **Lucknow Residency** in September 1857, accompanied **Campbell** on the second relief in November of that year, and remained at the **Alambagh**,

Sir James Outram

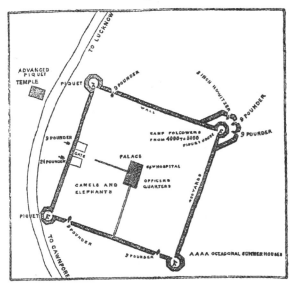

The Alambagh in 1857

in command of the garrison that the British left behind. In March 1858 on the return of the commander-in-chief and a large British force, Lucknow was finally attacked and taken and Outram took up his post as Chief Commissioner of Oude[Awadh]. An honourable man, fair and just according to his lights, he was none the less entirely blinkered in his attitude to India and Indians, being by turn stern and patronising; see **DNB**.

P

PAINTED FIGURES 'I think we remained in that village some ten or twelve days. We heard that the mutineers were making entrenchments at Bithoor. The sepoys that came every now and then from that place said that the Europeans had come in steamers, *a few of them, but that they had painted figures to make others believe there were many Europeans*. They said the British wanted to rob the Nana of his jewels. One evening, at sunset, some sepoys returned, and told us to get ready, and shortly afterwards we saw the mutineers flying; yes, many threw away their arms; we then fled to Futtehgurh', *see Deposition of John Fitchett*. **Williams.**

A Palanquin

PALANQUINS (or Palankins) These were 'oblong boxes with sliding doors and two small front windows. Underneath was a shelf with a small drawer. The compartment was stocked with a mattress, a bolster and some pillows, and an adjustable back in case the passenger wanted to sit up. There was very little room inside for much more than the passenger and a few effects and refreshments, and so baggage was carried in sacks suspended between poles carried by separate teams of bearers'. Esther Chawner, in Fanny Parks, *Wanderings*. Vol. I, p 485.

PALMER, J.A.B. The author of a comprehensive guide to the events of 10 May 1857 at **Meerut** and 11 May 1857 at **Delhi**. The authoritative and standard work? An exhaustive account based on the accessible sources containing eye-witness accounts or reminiscences or statements and reports or despatches, *see* J.A.B.Palmer, *The Mutiny Outbreak at Meerut in 1857*, Cambridge, 1966.

PALTAN A regiment of soldiers, of the regular army. Although the regiments of the Bengal Native Infantry were all given numbers, and that was their official designation, most of them were also known, by the sepoys, by the name of a former Officer, or the name of the man who had raised the regiment, *eg* 'Gillis Paltan' = 1st regiment **BNI.**

PANDIT SHIU DIN He was the private secretary to the Maharaja of **Jaipur**, by whose advice the state remained loyal to the British government throughout the struggle.

PANDU NADI *See* AONG, BATTLE OF.

PANGHATI PASS An action took place here, near **Kirwi**, on 27 August 1858 between a British Column commanded by Brigadier Carpenter and a force of rebels numbering not less than 7,500, including 500 sepoys and sowars. The rebels withdrew after losing fifty men and Kothi was occupied. The nature of this type of conflict in Bundelkhand is indicated by taking a look at the composition of the British Column: Horse artillery 6-pdrs drawn by bullocks; 2 mortars carried on an elephant; 1 brass gun on an elephant; Royal Artillery 46 men; H.M.'s 43rd Light Infantry, 160 men; 1st Madras Infantry,196 men; Sikh Horse, 30; Hyderabad sowars, 23; Levies of Native Chiefs armed with matchlock, swords and shields, 280; 10 elephants. *See The Revolt in Central India*, Simla, 1908, p183; *and* Tapti Roy, *The Politics of a Popular Uprising, Bundelkhand in 1857*, Delhi, 1994.

PANIC Early in June 1858, after **Grant**'s column had retired from a perilous position in **Oude[Awadh]**, 'a panic (among the British) took place in the night. Nobody knew exactly how it occurred but perhaps a horse got loose and ran over someone, who, thinking the enemy were coming, raised a hulla ballou and the whole line was up in a moment - such a scene as it must have been. I was out on patrol about a quarter of a mile in front and only heard the noise. It was a very dark night and the men could not see what they were doing, they began to run in every direction firing their rifles and using the butt ends freely. Rather a good story is told of the officer commanding the Artillery, a Royal! He first got knocked down in the general rush, however he managed to scramble on his knees and commenced using his revolver freely. First he shot a camel, then a Dhoolie bearer and then he fired a round through his own leg. He was then fortunately prevented from doing further damage by being knocked down with the butt of a musket by a soldier who took him for a Pazee.'

PANNA A small town/state south of Banda. The raja of Panna, although surrounded by hostile forces remained loyal to the government of the British, and indeed supplied men and guns to assist Lieutenant Remington hold out in the fort of **Kalinjar** against the **Nawab of Banda**, see Tapti Roy, *The Politics of a Popular Uprising, Bundelkhand in 1857*, Delhi, 1994.

PANWARI HEIGHTS An action took place here, near **Kirwi** in Bundelkhand on 29 December 1858, and in a way it was decisive, since the rebels, here numbering 6-7,000 including 600 sepoys, dispersed and the district was cleared of rebellion. In the action some 300 rebels were killed including **Radha Govind** and his brother, although the other leader, Ranmat Singh, escaped. For details *see The Revolt in Central India*, Simla, 1908, p185 *et seq.*, *and* Tapti Roy, *The Politics of a Popular Uprising, Bundelkhand in 1857*, Delhi, 1994.

PAPERS OF A MISCELLANEOUS CHARACTER

The following appeared in the New Delhi *'Statesman'* on 15 August 1990: 'It is about Bahadur Shah, King of Delhi, descendant of Mughal Emperors, that I wish to speak today. More particularly let us look at a remarkable offshoot, a real 'find' for historians: the papers that were collected from his palace in preparation for his trial. The trial itself, in January 1858, in the Red Fort of Delhi, was a farce. Bahadur Shah had already been promised that his life was not at stake, nor that of his Queen Zinat Mahal, or his somewhat obnoxious young son, Jumma Bakht. In fact it was more like a court of enquiry than a trial. The defendant tried to ignore both his accusers and judges: when he said anything it was obstructive or, with some justification, he would make a simple statement that he denied the court the right to try him. It is not the detail of the trial so much as the papers from the Palace that have excited me. They lay neglected in Government store, for over 60 years, neither consulted nor even catalogued. They were put into 200 bundles, not in date order, virtually unsorted, and certainly not indexed (as I found to my cost). It took me two days to read through even the synopsis, and that was speed reading. The papers were taken from Bahadur Shah's palace immediately upon its occupation by the British in September 1857: it was clearly expected that they would be of significance at his trial. But were they? The answer seems to be that they were hardly looked at let alone referred to. The fact is that where one looks for records of great political events, of regal decisions that were to decide the fate of nations and peoples, of armies and princes, of proclamations and promulgations, of decrees and judgements - one finds only hundreds, perhaps thousands, of petty petitions and grumbles, and complaints - above all of not being paid. Perhaps that is why they mouldered so long in a dusty office store. Eventually someone with a conscience, or nothing else to do, had them published: in 1921 the Imperial Record Department, in Calcutta, printed them as 'PRESS-LIST OF MUTINY PAPERS 1857'. You may be wondering why, if they were not used for the purpose for which they were collected, I am making so much fuss about them. it is because they are ***the finest direct evidence of what it was really like in Delhi May-September 1857, and, more to the point, an excellent entry to the social history of the time.*** In a short article like this it is not possible to get more than a flavour of what is contained in the papers. Some entries are dramatic and unexpected. Take for example: Paper 62, Bundle 199, 31 August 1857: King to Nawab of Jhajjar: 'I wish to migrate to Mecca and require your help.' This was the first I had heard that Bahadur Shah was so very disillusioned that he wished to leave Delhi far behind him. In other bundles we find evidence that he was contemplating suicide or wished to be taken to a local shrine where he would see out his days as a sweeper. Same bundle, paper 380, Nabi Bakhsh Khan to the King: prays that officers may be directed not to kill English women and children. Whereas Paper 269 same bundle, 11 August, the King to Ranjit Singh, Raja of Jaisalmer, directs him to kill every Englishman, or woman, wherever found in his territory. So far, fairly high-powered matters of state, but what can we make of Paper 7 in Bundle 160, 11 June?: Jagannath and other shopkeepers to the Commander-in-Chief,' requesting him to remove a man who plays English music, and thus creates disturbances near their shops.' I feel sympathy for Jagannath. I like Paper 74 Bundle 111, 3 August: a Colonel (not named) directs the Kotwal to see to it that Gopal Singh gets back his wife, who has eloped with a clerk of Paharganj thanah. Same bundle, Paper 20, 23 August, Mukhtar Saltanet directs the Kotwal to realise the rent from a dancing girl who refuses to pay, by distraining her property. Regretfully we do not know whether Gopal Singh gets his wife back or the dancing girl is made to pay up, but sometimes the Kotwal, or whoever, writes the outcome on the back of the paper. Thus in Paper 217 dated 16 August Muhammed Faiz Baksh Subahdar of the Nimach Force demands from Khuda Baksh Khan, Naib Kotwal, that he get justice from Mohd Baksh (seems like a family affair to me) and Sa'adullah who have absconded: asks him to trace them and realise from them the money they owe on account of some jewellery he had deposited with them and which was stolen while in their custody. Verso, the Kotwal records that they were traced and paid up the money demanded. Many of the papers are directives to the wretched Kotwal. He must have had a miserable life. He was expected to be a jack-of-all-trades, a supplier of anything and everything: in all probability he was rarely paid for what he provided as money was so short. I began a list of the things demanded of him, and it is formidable indeed:- he was called on to remove dead bodies - human, horses, camels etc - to send coolies, carpenters, blacksmiths, cobblers hither and thither. As on 8 August when 'all available water-carriers are to be sent at once to put out the fire in the Gunpowder factory' : we can imagine that was not a popular assignment He had to purchase and supply everything from sulphur to charpoys, fuel to spades and pulse. He was ordered to imprison 13 men accused of supplying meat to the English. We do not hear what happened to them. Sometimes the papers border on the banal. 3 July, Bundle 111, Kotwal to the King : 'Shansir Khan, Jemadar Chandni Chowk Thanah, deposes that one Dallu shot himself by accident. The deceased's father, Makhdum Bakhsh, has also made a similar statement. Natthu, a sweeper, who was examined, has said the same thing. Adds that the deceased was of unsound mind.' One wonders what Bahadur Shah made of all that! Practicalities are included. On 22 June, Paper 148, C in C to Kotwal, directs him to notify the people that no one should stay in dilapidated houses as the big gun at Salimghar is about to be fired. The same day he is directed to send without delay 10 whetters to the 6th Cavalry outside the Turkoman Gate. If you were required to find '10 whetters' in a hurry, would you manage it? On one day alone the Kotwal was required to produce hemp, carts, sweepers, provisions, ghee, flour, molasses, sugar, thatchers, hessian, oil and...torch bearers. He had also to receive, and presumably act upon, reports sent in by his subordinates that are quite extraordinary, not to say bizarre. Take paper 400 in bundle 103, dated 18 September. Khair Khan Darogha to the Kotwal. *Reports that there are daily quarrels between Dallu, weaver, and his wife.* When we remember that the British assault on the Kashmir Gate took place on 14 September, and that by the 18th they were nearing the Chandni Chowk, one is at a loss for words. There is pathos too, and many a hidden story of loss and misery. 5 August Paper 259 is typical of many: Sahu, Gardener, to the King. Reports that his brother Mamraj has been killed by the English; prays for the maintenance of the deceased's children. Included also are letters of spies, within Delhi, sent out to their English masters on the Ridge outside. The contents are noteworthy more for inaccuracy than anything else. Most spies seemed anxious only to tell the English what they wanted to hear - about bad morale among the mutineers, or shortage of money, or quarrels among the Princes and the Sepoys. All in all, a very mixed bag. We may leave the last word to Sir John Lawrence, Chief Commissioner. He commented on the documentary evidence produced at the trial: 'In brief terms, it may be said that the documentary evidence comprises the system in which the general government was conducted; the raising of loans; military arrangements; the communications with foreign powers and neighbouring chiefs; the passages in the native newspapers relating to the war between the English and the Persians. *There are also of course many papers of a miscellaneous character.'* **PJOT**

PARADE GROUND MASSACRE This was the killing, on 23 July 1857 of ten men and twenty-three women and children at **Fatehgarh** on the orders of the Nawab, but at the instigation of the Bigga Begum: the latter it is thought was trying to ensure the death of all Christians, including her rival the notorious Bonny Byrne. *See* **SSF** p.140, and F.R.Cosens and C.L.Wallace, *Fatehgarh and the Mutiny*, Lucknow, 1933. The British brought the Bigga Begum to trial but she was acquitted for lack of evidence against her.

PARSONS, Lieutenant Richard A survivor of a particularly badly organised British expedition against **Jagdishpur** in Bihar, he wrote frankly and honestly of this British defeat at the hands of the (dying) Kunwar Singh;. *see* R.Parsons, *A Story of Jugdespore*, London, 1904.

PASSIS Indigenous people of northern India, well known for their bravery, skill with bow and arrow, and loyalty when committed. The men were expert miners and were used as such by the rebels in attacking the Residency in Lucknow. They also furnished the class of village watchmen. So important were they that a specific mention is made of them in one of the **Proclamations** of **Birjis Qadr**, *see Foreign Secret Proceedings* 25 June 1858, National Archives, New Delhi.

PATNA Capital city of the province of Bihar; associated in particular at this time with the names of **Kunwar Singh**, Peer Ali Khan, and William **Tayler** the Commissioner. The city itself had a population of over 3 lakhs, a large proportion of them being Muslim. On 21 June 1857 Tayler arrested the four leading members of what he described as the Wahabi sect, and also disarmed the city; he admitted to his superiors that the only evidence against the prisoners 'was that of an untrustworthy informer who produced letters to substantiate his charge, of which only one was genuine', and that his statements *re* the distribution of money and the entertainment of fighting men 'proved incorrect'. Probably held the four leaders as hostages to ensure the cooperation of other Wahabis. No weapons were found in the Wahabis' houses. Taylor undoubtedly provoked the outbreak which came on 3 July when some 300 men broke into the premises of the Roman Catholic Mission although they stole nothing and injured no one. But a British led party of police and Sikhs was shot at, and a Dr Lyell killed. Thirty men said to have been involved in the outbreak were arrested, and fourteen, including Peer Ali Khan, a bookseller who was said to have shot Dr Lyell, were condemned and executed the same day. Tayler was attempting to connect the Wahabis with Maulvi Ali Karim in arms against the British in the Azamgarh area. He tried unsuccessfully to prosecute Lutf Ali Khan a rich banker, and even the British judge and officials complained to the Lt Governor of Tayler's 'constant indelicate and illegal interference with the cause of justice, always on the side of severity.' Tayler was removed from office, to the 'satisfaction of nearly all the residents of Patna.' The appointment of Munshi Ameer Ali as special assistant to the new Commissioner, Mr Samuells, appears to have had the effect of keeping the Muslim majority quiescent if not actively loyal to the British. Subsequent to the restoration of British authority Maulvi Ahmedulla, a leader of the Wahabis, was arrested in 1864 and tried for, among other things, 'preaching at Patna a crusade against the infidels (British)'. Banished to the Andamans, there is some suggestion that he was connected with the murder there of Lord Mayo, a Viceroy. For Tayler's version of the events *see* Wm Tayler, *The Patna Crisis,* Patna, 1882; *see also* K.K.Datta, *Unrest against British Rule in Bihar,* Patna, 1957, and *Biography of Kunwar Singh and Amar Singh,* Patna, 1957. *See also* Appendix A to Vol I of C.B.Malleson, *History of the Indian Mutiny,* London, 1878. *and* Ahmed, Qeyamuddin, *Maulavi Ali Karim, a scholar soldier of Bihar during the movement of 1857-58,* in *Indian Historical Records Commission Proceedings,* 33, 9-15, 1958.

PAYAM-e-AZADI A journal edited by Mirza Bedar Bakht, grandson of Bahadur Shah, 'under the patronage of Azimullah Khan'. 'As an organ of revolution the *Payam-e-Azadi* rendered immortal service in the cause of freedom;' *see* S.Lutfullah, *Azimullah Khan Yusufzai: the man behind the war of independence, 1857,* Karachi, 1970, p 113.

PEACOCK THRONE This was built by the Emperor Shah Jehan, at a cost of perhaps twelve millions sterling at that time; stolen from Delhi by the Persian invader Nadir Shah in 1738 and never returned.

PEARL, HMS A ship of **HM**'s navy that happened by chance to be in the Hoogly river and which provided a Naval Brigade for action against the rebels, *see* G.L.Verney, *The Devil's Wind,* London, 1956.

PEASANTS' REVOLT That the military mutiny was accompanied by what the British called 'civil disturbance' goes without question. But the nature of this disturbance has been the subject of considerable research, particularly in recent years. The names of four historians in particular are associated with this work, viz. **Eric Stokes, Ranajit Guha, C.A.Bayly** and **Thomas R. Metcalf.** The major dilemma is described by Stokes in his *The Peasant Armed*: 'Whether the upheaval of 1857 was a great movement of proto-nationalism, or the historic register of the death agonies and birth pangs accompanying the onset of acute social change, or merely a series of traditional village fights and local *jacqueries* against the moneylender, grain dealer and tax official that automatically break out on every temporary power failure of the larger state structure, any such larger conclusion turns on the assessment of peasant action.' If one recognises the rural origin of the vast majority of the sepoys of the Bengal Army, and also takes account of the sympathy (and often practical support) given in the rural areas to the rebellious sepoys, in a real sense the revolt might be described as essentially the revolt of a peasant army. Severe British revenue assessments might in part be responsible for this 'turbulence', (as the British called it) as also the misery caused to many by poverty not so much the responsibility of the British as of ecological factors. A significant source of 'turbulence' were the grazing and semi-nomadic communities of the areas most affected by the mutiny/rebellion, designated by such caste names as Gujars, Bhattis and Rangars. Certainly in the early days of the outbreak, say between 10 and 20 May 1857 the Gujars posed the most serious threat to British control of the countryside, particularly it is said in the Bulandshahr district. The arrival of the leader Nawab **Walidad Khan** from Delhi had much to do with this upsurge in violence, but it would appear that there is clear evidence that they were not just taking advantage of the breakdown of law and order in order to indulge in their ancient pastime of rapine, but were actively supporting the regime of the King of Delhi, sending in supplies etc. *See* Eric Stokes, *The Peasant Armed,* Oxford, 1986, p 162, *and also, The Peasant & the Raj, Studies in Agrarian Society and Peasant rebellion in Colonial India,* Cambridge 1978; and Robert H.W.Dunlop, *Service and Adventure with the Khakee Ressalah,* London, 1858. Reprint, Allahabad, 1974. *See also* Ranajit Guha, *Elementary Aspects of Peasant Insurgency in Colonial India,* Delhi, 1983, reprinted as paperback, 1992; *and* C.A.Bayly & D.H.A.Kolff, *Two Colonial Empires,* Lancaster, 1986, *and* Thomas R. Metcalf, *The Aftermath of Revolt,* Princeton, 1965.

PEEL, CAPT. Sir William, V.C. Commander of the frigate HMS Shannon, who, with a company of seamen from his ship, took part in the campaign in Oude[Awadh], particularly in the assault upon Lucknow in March 1858. *See* G.L.Verney, *The Devil's Wind,* London, 1956.

PEEL, DEATH of
'One circumstance which it will be impossible that I should ever forget, is a visit I paid to that hero of the naval service, Sir William Peel. After rendering invaluable assistance with his far-famed artillery, in the assault upon **Lucknow**, he was shot by a musketball in the thigh, and while suffering from his wound he caught the small-pox, and was brought into **Cawnpore** suffering severely from that fell disease. He was taken to the residence of the chaplain, the

Rev. Mr Moore, and that excellent clergyman and his lady did their utmost to alleviate his sad condition. But medical science and Christian kindness were both unavailing, and in this scion of a distinguished family, England lost one of the noblest of her sons. Upon the march from Lucknow, Sir William had suffered greatly from the paucity of medical comforts, and he entered Cawnpore in a most exhausted state, saying mournfully, 'If I were in England, the Queen would send her own physician to look after me; here I can scarcely get any attention.' One consolation remains: whether her sons sleep in the hasty grave of the battlefield, in the pathless waters of that huge sepulchre, the ocean, or in the marble magnificence of Westminster Abbey, England is always faithful to their memory, who fall nobly doing her work; and few will be longer held in the grateful recollection and high esteem of all classes in the state than the much-lamented Sir William Peel.' *See* Mowbray Thom-son, *Story of Cawnpore*, London, 1859, pp 256-8.

PEER ALI The leader of the 'Mussulman conspiracy' in **Patna**, Bihar. Also known as Peer Ali Khan; a bookseller. Had been in 'treasonable' correspondence with other Muslims for some months, apparently organising a conspiracy designed to re-establish Muslim supremacy in India, and restore the Mughal Emperors. Described by the British as 'a brutal but brave fanatic'. It was he apparently who shot Dr Lyell of the Opium Agency in Patna, 3 July 1857. Arrested and tried and condemned; his execution was delayed by the Commissioner (William Tayler) in the hope of finding out more about the conspiracy but Peer Ali refused to co-operate, and was hanged. *See Parliamentary Papers, see also* Wm Tayler, *The Patna Crisis*, Patna, 1882; K.K.Datta, *Unrest against British Rule in Bihar*, Patna, 1957, *and* Appendix A to Vol I of C.B.Malleson, *History of the Indian Mutiny*, London, 1878; *and* Ahmad, Qeyamuddin, *Maulavi Ali Karim, a scholar soldier of Bihar during the movement of 1857-58*, in *Indian Historical Records Commission Proceedings*, 33, 9-15, 1958.

PEER BUX Those who spied or carried letters for the British could expect little mercy from the rebels if they were caught: mutilation or death would follow. Even if their motive was, for the most part, greed, they were undoubtedly very brave men, and without their help the British would have been further handicapped. Peer Bux was a *cossid* of this type, and survived. 'In June 1857 (I do not remember the exact date) I was sent from **Agra** with a letter to **General Wheeler** at **Cawnpore**; when near Mutteapore, I met a number of Bengalis (some 150 in number) who had been plundered and maltreated by the residents of that village. As I approached, I saw [the villagers] drawn out prepared to stop my progress, but, being mounted on a swift camel, I avoided them. They were armed with swords, spears, bows and arrows. At Juswuntnugger, I saw the gang of dacoits under Gunga Singh plundering that village, ...which they had set on fire. On reaching **Etawah**, I found that the residents had just beaten off an attack made by some mutineers, and firing was still going on at the ghat. I saw the bodies of eight rebel sepoys, who had been killed at the outskirts of the town whilst plundering. I put up at the house of Narain Dass, a *gomashta* of Lala Joti Pershad, and, hearing that the road to Cawnpore was very unsafe, left my arms and the camel at the *gomashta's* house, disguised myself as a faqueer. Hiding the letter in the sole of my shoe, I travelled on foot by crossroads till I entered the **Grand Trunk Road** at Choteypore. The insurgent villagers were plundering each other, but I was not molested except once when I was searched on suspicion of carrying letters, and was released with a slap on the face. From Choteypore to Cawnpore I passed five police stations, at each of which there were ten sowars posted, who were the Nana's servants'. *Deposition of Peer Bux*. **Williams.**

PEISHWA The title given to the hereditary leader of the Maratha Confederacy (theoretically he was only the Minister of the Maharaja of Satara, the descendant of the great Sivaji). **Baji Rao II** was the last fully acknowledged Peishwa, although his adopted son the **Nana Sahib** aspired to the title and pension.

PENSIONER GREEN The Bridge of boats at **Cawnpore** was the only means of crossing the Ganges into **Oude[Awadh]**. The rebels broke the chain of boats and thus severed communications with Lucknow soon after the outbreak of the rebellion on June4/5 1857. Pensioner Green, Superintendent of the Bridge of Boats, lived with a native woman who managed to get him only as far as the old Native Infantry Lines, whereupon he was captured and killed on the Parade Ground. *See* W.J.Shepherd, *A Personal Narrative etc.*, Lucknow, 1886, p 30-1.

PERCUSSION CAPS The weapon with which the majority of the sepoy mutineers was armed was the muzzle-loading musket, affectionately known as 'Brown Bess'.

Percussion ignition

After loading, the weapon was fired by the explosion of a small percussion cap activated by a trigger. These caps had to be inserted after the cartridge had been rammed home down the muzzle of the musket. At **Kanpur** the rebels had enormous quantities of warlike stores available to them, left, undestroyed by the British, in the Magazine on the banks of the Ganges. But the one item of stores they lacked was a supply of percussion caps. The making of gunpowder was comparatively easy since the three major chemical constituents were available, but the making of percussion caps was virtually impossible outside specialist arsenals/factories such as **Dum Dum**. The immense and prolonged fusillade which the rebels carried out on **Wheeler's Entrenchment** early in June 1857 practically exhausted their supply of percussion caps, and so serious had the position become that they were arranging for a mass conversion of all sepoys' muskets back to the old flint-lock design. This would have been a slow and costly, and inefficient business, but they were saved the trouble by an altogether unexpected event. The British Conductor and Sergeant in charge of a flotilla of Magazine Boats blundered on shore, within four days' journey of Kanpur on 16 June; the men were seized and taken to the **Nana**, who ordered them executed immediately. The boats had taken almost a month to travel the 126 miles from Allahabad and *were filled with percussion caps,* which proved a 'godsend' to the rebels. See W.J.Shepherd, *A Personal Narrative etc,* Lucknow, 1886.

Flintlock

PERSIAN VERSE

'A few of the 2nd Cavalry troopers selected Captain Seppings, and begged, as a special favour, to save him alive, but were over-ruled by Teeka Singh and a lot of others........On every occasion, when a request of this nature was made by any one, either to spare a child or man - and many persons were desirous of getting some young European children to adopt - no sooner did they make their wishes known than the Mahomedans would get round and repeat a Persian verse as follows: *Atush kooshatun wa ukhgur goozastan, Uffaiee kooshtun wa buch-aishra neegah dashtun, Kar-i-kheerud-mundan naist.* That is to say, 'to extinguish the fire and leave the spark, to kill a snake and preserve its young, is not the wisdom of sensible men.' Besides this, whenever a Mahomedan found the lifeless body of a European or Christian lying anywhere, he immediately drew out his sword, with a "*bismillah*,' and made a gash upon the corpse, repeating the words, or some such words as 'Soonut-ool-huq-i-Kafar-un.' This act is considered by them to be equivalent to killing an infidel, and adds to their claim for entering Paradise after death...... The ladies were directed to leave the gentlemen, and when compelled to do so, they shook hands all round and separated; excepting one lady, supposed to be Mrs. Boys, (Boyes?) the wife of the Surgeon of the 2nd Light Cavalry, who with her child clung to her husband and could not be parted, she begging to be killed first. Order was then given to the sepoys to fire upon the prisoners. Captain Seppings sued for a few minutes to pray; this was allowed. They knelt down and prayed, the last prayer their mortal lips would ever utter, and now a volley of musketry opened upon them, killing a few and wounding many. The wretches then fell upon them with swords and completed the cold-blooded, cruel, slaughter'. W.J.Shepherd, *A Personal Narrative etc.,* Lucknow, 1886.

PERSONS OF NOTE, APRIL 1859

By the early months of 1859 active rebellion remained alive only in Central India, notably **Bundelkhand**, but there were still many who refused the British government's offer of clemency in return for surrender. Some no doubt feared that they would under no circumstances be pardoned as they had been actively involved in killing Europeans (this was apparently the unpardonable crime in the eyes of the authorities), but others who might well have been treated magnanimously still refused to come in, we assume from ideological conviction, or personal loyalty to a leader. By April 1859, many of these men, and women, were in Nepal, uneasy and unwanted 'guests' of **Jung Bahadur**. A list of 'Persons of Note' was compiled at this time by one Major Captain (sic) Bir Bhanjan Manjhi of the Nepalese army, and it was submitted, with comments, by T.D.Forsyth, Secretary to the Chief Commissioner, Oudh(Awadh), to the Secretary to the Government of India, Foreign Department, on 6 April 1859, as 'a list of persons who were with the Begum in the Nepaul territories' together with 'a brief description' thereof. The list is of interest as an indication of the hard core of surviving rebels, both the 'no-hopers' and the genuine patriots, though it is not necessarily easy to distinguish which is which: '*Begam of Lucknow, by name* **Huzrut Muhal**, *one of the wives of the Ex-King and mother of Berjees Kudr; took a most prominent hand in the Rebellion;* **Berjees Kudr**, *a boy of 10 or 12 years old, son of Huzrut Muhal; Nawab* **Mammoo Khan**, *formerly the darogah of the Begam, assumed the title of Nawab at Bonnree(sic), paramour of Huzrut Muhal, consequently the Chief of the rebel band; Lord Buksh Khan, supposed to be Bukht (Bakht) Khan, formerly a Subadar of artillery in the Company's army, and mutinied at* **Bareilly**, *was made rebel Commander-in-Chief at Delhie and accompanied the rebel troops from Delhie to* **Lucknow**; *Usuf (Yusuf) Khan, Commander-in-Chief, a relative of Mummoo Khan, had great power during the rebellion and plundered the city of Lucknow;* **Nawab Khan Buhadoor Khan**, *the Nawab Rais of Bans Bareilly of the family of Hafiz Ruhmut Khan, took a most active part in the insurrection in Rohilkund, was a pensioned Sudder-ool-Suddoor of Bareilly; General Abid Khan, late a resident of Kewalhar, one coss south of Mulhiabad a relative of Mummoo Khan; General* **Gunga Singh**, *formerly a subadar in the 41st NI, commanded a Division of rebel troops at Lucknow, was wounded whilst leading an attack on the Allum Bagh and is a notorious rebel. He took a principal part in the murder of British at Futtehgurh; General Ahsan Ally, a friend and relative of Mummoo Khan, fought against the Bailee Guard and Alum Bagh, in fact took a most active part throughout the insurrection; General Dulgunjun Singh, not known; General Tilak Ram Tiwari, not known; Rana Beni*

Madho Bux (**Beni Madho Bakhsh**) Talooqdar of Sunkerpore in the Baiswara District, received the title of Delair Jung from the Rebel Begum Huzrut Muhal, fought againt the Bailee Guard and took a most prominent part in the rebellion throughout; Rana Oomrao Singh (Umrao Singh), brother of the raja of Akowna (Ikauna), was in arms in the Goruckpore District, the latter part of last year; Raja Dirig Bijae Singh, Talooqdar of Mohona, resided at his fort Oomerria, took an active part in the rebellion throughout last year, killed and mutilated great numbers of our police and closed the **Cawnpore** road against all supplies coming into Lucknow; Rajah **Nurput Singh**, not known; Chowdree Nurput Singh, son of the late notorious rebel Jussa Singh of **Futtehpore** Chowrassie in Oudh, joined in the attack on the **Futtehgurh** fugitives, a firm friend of the **Nana** who resided with him when he entered Oudh; Raja Golab Singh, Karinda of Chundka Bux Talooqdar of Birwa in the **Sundeela** District, a notorious bad character, fought against us at Birwa, assumed the title of raja lately; Raja Hurdut Singh, talooqdar of Bounree, took a most prominent part in the rebellion, the Begum with her followers took refuge with him at Bounree; Thakoor Ram Gholam Singh, talooqdar of Rampore Kusseeah, fought against Brigadier Wetherall's Column in November 1858; **Raja Debi Baksh Singh**, talooqdar of **Gonda**, in the Baraitch Division, a noted rebel; Collector Mir Muhammad Husain, Nazim of **Goruckpore**, during the rebellion, and was very conspicuous throughout, pardon was offered to him by the Governor General which he would not accept; Khan Ali Khan, resident of Shahjehanpore, was one of the leaders in the 1st Chinhut affair, took a most prominent part throughout the rebellion, was wounded at Lucknow; Collector Sewsuhai (Shiv Sahay) by caste a Kait (Kayasth) or Uggurwala (Agarwal), was assistant to General Usuf Khan, resident of Lucknow, another person by name Sew Singh Collector, was a notorious bad character and was very active in plundering and creating disturbances during the insurrection; Collector Rugbur Singh (Raghubar Singh), formerly a Khasburdar or orderly of the Eunuch Deanut ood-Dowlah on a salary of 4 Rs per mensem, resident of Baiswara, a noted dacoit and rebel; Oomrao Jan (Umrao Jan), formerly a vakeel and moonshee of Moosahibood-Dowlah (singer), during the rebellion became one of the chief advisers of **Mummoo Khan**, resident of Lucknow; Bhugwan Bux (Bhagwan Bakhsh) one of the thakoors of Num, a bad character and dacoit; Mooftahood-Dowlah (Mifta-ud-Daula) grandson of the late Futteh Alli Khan, a Brahmin convert who was the King's Treasurer; Mohan Singh Chaudry, not known; Meer Mehndi (Mir Menhdi), tutor to **Birjees Kudr**, superintendent Intelligence Department during the rebellion; Raja Oudit Purgas Singh,(Udit Prakash Singh),Talooqdar of Ekona in the **Baraitch** Division; Hukeem Hussun Ruzah, tutor to Shurfood-Daulah, who was Prime Minister during the rebellion, this hakeem had charge of the Dewanee Adawlut under **Birjis Kudr**; Raja Joat Singh (Jyoti Singh), talooqdar of Churda in the **Baraitch** Division, gave willing shelter to **Nana Rao** and his followers in his Fort for some time, fought against us in **Goruckpore** in March and April 1858 and afterwards in November of same year in company with **Bala Rao**, made an incursion into Bunnee committing great ravages; Brigadier Major (sic) Gopal Singh, not known; Captain Oomrao Singh, a subadar of the 6th Oude[Awadh] Irregular Force, prior to the annexation Subadar Major in Captain Barlow's regiment, resident of Kantha in the Baiswara District, but for some years had his family at Colonel Gunj Secrora, he had great influence in the Durbar of **Birjees Kudr;** Captain Raghunath Singh, by caste a Tewarri Brahmin, prior to the annexation was in Captain Bunbury's regiment, subsequently transferred to the 2nd Oude[Awadh] Military Police under **Captain John Hearsey**, he was also a leading character during the rebellion; Captains Sangam Singh, Suraj Singh, and Ram Singh, subadars of the regular NI and were leading characters at **Lucknow** during the rebellion; Captain Ausan Singh, formerly a havildar in the artillery attached to Captain Bunbury's regiment before the annexation, subsequently a subadar in the 2nd Oude [Awadh] Military Police under Captain John Hearsey, took a prominent part in the insurrection and raised a Cavalry Corps at Lucknow; Captain Madho Singh was a pay havildar in the 2nd Oude[Awadh] Military Police; Captains Drigpal Singh, Sewdut Singh, supposed to have been promoted to their present ranks at Bounree or subsequently; Captain Gunga Singh, a subadar of the Company's service, took an active part in the rebellion at Lucknow and subsequently was at Bangur Mhow with the rebel **Prince Feroze Shah** of Delhie; Captain Nazar Ali, Not Known; Prince Kochak Sultan, one of the sons of the rebel Emperor Ubdool Muzzuffer Sarajoodeen **Buhadoor Shah** of Delhie, came to Lucknow after the fall of Delhi; Yusuf Khan, son of Azim Khan of Khyrabad where Yusuf Khan's houses are. He is a relative of **Mummoo Khan** but is supposed to hold the same position with Mummoo Khan's wife as the latter does with the Begum. Was first made Collector of Lucknow and subsequently Commander-in-Chief, he had great power during the rebellion and plundered the city of Lucknow, he is also a relative or intimate friend of Chowdree Hushmut Ally, some of Ysuf Khan's relatives live at Mullabad; Khoda Bux (Khuda Bakhsh), zemindar of Dadroe near Mohumdabad in the **Dariabad** District, he took an active part in the rebellion, throughout was a Commander against the British at Nuwab Gunj Bara Banki in June 1858 and at Nanpara in January 1859; Mohtimam Rasad, **bukshee** of the whole rebel forces, Nawab Naib, his family is said to be in Lucknow; Lord Jowala Pershaud (**Jwala Prasad**), a Kannoujeea Brahmin of Byswara (some village near Doondea Khera), was one of the four men who instigated

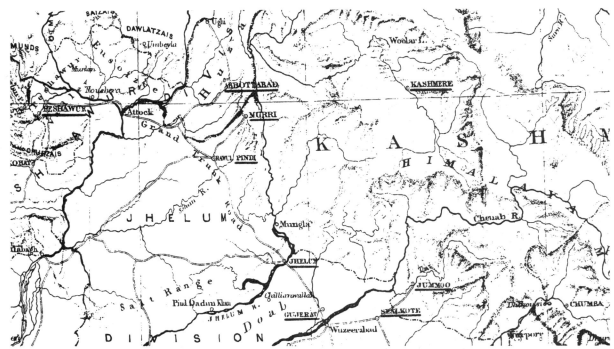

*the **Cawnpore** rising and directed the massacre, he received the title of Resaladar from the Nawab.'* See **FPC** 30 December 1859, Supp 550. **NA**. *See also* **FSUP**, Vol II, for details of individuals, where available.

PESHAWAR The city of Peshawar is situated forty miles from the River Indus, and ten from the mouth of the Khyber Pass. In 1857 the predominating characteristics of the city were Indian, but there were also many indications of Afghan life and customs. The cantonment area was one of the largest in the sub-continent, with a parade-ground estimated to be capable of containing up to 6,000 soldiers at any one time. There was a strongly Muslim population in the city and what the British called a 'reckless, restless and ruthless mob', with, in addition a host of poor and plunder-loving tribes inhabiting the surrounding hills. The political management of the area lay with Colonel **John Nicholson,** and Major Edwardes under the supervision of the Chief Commissioner of the Punjab, **Sir John Lawrence**. These men were used to taking action with decision and expedition, and the final success of the British in putting down the rebellion is due more to them than to any other factor. On 13 May 1857, only three days after the *emeute* at **Meerut**, a court-martaial met at Peshawar, consisting of Major-General **Thomas Reed,** commanding the Peshawar Division, Brigadier **Cotton**, Brigadier **Neville Chamberlain**, Colonel Edwardes, and Colonel **John Nicholson**. This meeting made the momentous decision to concentrate all troops that could be relied upon at **Jhelum**, with the intention of forming a movable column (later known as the **Punjab Movable Column** and commanded by **Nicholson**); accordingly the nucleus of this force was ordered at once to assemble, *viz* HM's 27th regiment from **Nowshera,** HM's 24th Regiment from **Rawalpindi**, one European troop of Horse Artillery from Peshawar, the Guide Corps from Murdaon, 16th Irregular Cavalry from Rawalpindi, the Kumaon Regiment from ditto, the 1st Punjab Infantry from Bunnoo, a wing of the 2nd Punjab Cavalry from Kohat, and half a company of sappers from Attock. This left Peshawar itself dangerously unguarded and Sir John Lawrence although busily raising more levies, and keen above all to send enough troops to recapture Delhi, actually contemplated handing over Peshawar to **Dost Mohamed** of Afghanistan, with a virtual undertaking that if he kept the Frontier quiet he would be rewarded by the cession of the territory permanently. Both Edwrades and Nicholson were horrified by this proposal, and were supported by the Governor General who ruled that any abandonment of territory would be deemed a sign of weakness. Meanwhile at Peshawar itself the native regiments, the 24th BNI and 27th BNI and the 51st BNI were disarmed and disbanded, (Brigadier Cotton hesitated to do so but was persuaded by Nicholson and Edwardes) while the 55th BNI at Murdaun was ordered to be disarmed, but many mutinied and fled before the action could be carried out and were captured and executed in large numbers. *See* F.H. Cooper, *Crisis in the Punjab*, Lahore, 1858, see also R.B.Smith, *Lord Lawrence*, 2 Vols, London,1885, and Lionel Trotter, *Life of John Nicholson*, London, 1898, **Kaye, Martin**, etc

PESHAWAR EXECUTIONS In August 1857 at Peshawar, 785 out of 871 captured mutineers were executed.. William Muir, ed. *Records of the Intelligence Department of the Government of the North-West Provinces of*

India during the Mutiny of 1857, Edinburgh, 1902, Vol I p 504.

Peshawar Fort

PESHAWARIE   During a tour in October 1857 of the **Bibigurh** at Kanpur, Forbes-Mitchell was told by a camp follower named Peshawarie that during the siege of **Wheeler's Entrenchment** the Nana Sahib had tried to bribe the bakers to put arsenic in the garrison's bread. They refused, and according to Peshawarie, after the massacre at the **Satichaura** ghat the Nana Sahib had the bakers thrown into their own ovens and their charred bodies fed to pigs. *See* W.Forbes-Mitchell. *Reminiscences of the Great Mutiny,* London, 1893.

PHILATELY   *See* Postage Stamps.

PHILLOUR   A town and military station in the Punjab with a fort in which 100 men of H.M.'s 8th regiment were garrisoned. The 3rd BNI were also stationed there; they mutinied and marched for Delhi on 9 June 1857. *See* JULLUNDUR, *see also* **Ball** and Martin

PHONEY WAR   In much the same way that there was a pause before the real fighting began in Europe in 1940, so there was a period after 10 May 1857 (*see* **Meerut**) when men on both sides of the conflict paused, as though waiting for a sign a signal or an excuse to break into open warfare. This situation is well summed up thus: 'The men who cared least for all this were those most in danger, our young officers in the native regiments. They rode, swam, and played at billiards with as much gaiety as though they had not nightly to sleep in the lines amongst a set of ruffians thirsting for their blood. My own position and that of my colleagues was not an enviable one. With superior opportunities of knowing the extent of the dangers which overhung not only our wives, sisters, and children, but our empire in India; we were obliged daily to take our seat on the bench, and listen to long arguments about debts and mortgages, which we suspected would soon be cleared off by the intervention of anarchy rather than law. We had to grant injunctions which nobody attended to, and to pass decrees which no man could execute. And specially irksome was this mockery to one like me, who had near and dear relations at a distant station exposed to the full fury of the mutiny. At the same time the Sudder Court had no executive power, and was helpless, or nearly so, either for good or evil.'. Charles Raikes, *Notes on the Revolt in the North-western Provinces of India,* London, 1858.

PHOTOGRAPHS   Considering that these were comparatively early days for photography, and that the Indian climate was not at all helpful to the development and processing of photographic plates, there is a remarkably large number of prints in existence showing scenes from the struggle of 1857-9. These are mainly of buildings caught up in the war, showing the damage done to them, but there are also a number of portraits of personalities and leaders involved, from **Tatya Tope** and **Zinat Mahal**, the favourite wife of **Bahadur Shah Zafar**, King of Delhi, to **Sir Colin Campbell**, the British Commander-in-Chief. Some were the work of professional photographers, like **Felice Beato**, but many were taken by amateurs *eg* Captain Robert Tytler, Assistant-Surgeon Patrick Fitzgerald, Dr John Murray, Captain Stewart *et al.* The 'Lucknow Album', as it is called, in the Oriental and India Office Collections of the British Library, contains some interesting portraits taken in 1856 in pre-rebellion Oude[Awadh] by Ahmed Ali Khan. Details of photographic portraits are given in Pauline Rohatgi, *Portraits in the India Office Library and Records,* London, 1982. *See* also MAUDE, Major F.C., V.C., C.B.

PINCKNEY, Captain   He was appointed Superintendent of **Jhansi** on the resumption of British power there. He was asked to respond to a report in the Bombay *Times* that all of the ladies massacred at Jhansi had first been raped by order of the Rani. 'The Europeans and Anglo-Indian females,' he replied, 'were not brought before the Ranee and stripped, their faces were not blackened, nor were any of them dishonoured. After the murder, the bodies were stripped, and left in Jakim Bagh, till the third day, when they were all buried in a gravel pit close to it.' Pinckney to **Erskine**, 13 April, 1858.

PINCOTT, Frederic   Author of an analytical index to Sir John **Kaye**'s *History of the Sepoy War,* and **Colonel G.B.Malleson**'s *History of the Indian Mutiny.* This is an indispensable and an invaluable guide to two major English works on the mutiny/rebellion. Indeed without Pincott much time is wasted in tracing events and references in these two major works. *See* F.Pincott, *Analytical Index to Sir John Kaye's History of the Sepoy War and Col G.B.Malleson's History of the Indian Mutiny,* London, 1880.

PLOT There are many different views as to whether there was a pre-conceived plot for the rebellion. Some - on both sides of the argument - are predictable in the sense that attitudes are ingrained and prejudices immovable, but there are many intelligent summaries that shrewdly draw attention to significant facts. One such comes from an otherwise undistinguished little book, J.Tulloch, *Volunteering in India*, London 1893 :- . 'No, says Sunker Tewarri the whole thing was a superstitious panic which led them to mutiny. There was no preconcerted organisation; all regiments fearing to initiate the move waited for each other to rise, and immediately the first successfully rose in open rebellion the rest in Indian file like imitative sheep followed as a matter of course. *JUST PANIC* did it.' There was a Plot - of sorts, according to one English woman: ' On the Queen's birthday (which was due a fortnight later) they were to parade according to their plot, with arms loaded and pouches filled with ball cartridge. It was then the custom for British regiments to stand by the side of Indian battalions and fire a *feu de joie* in favour of the event. The captain of each company of Europeans then saw the service ammo carefully removed and its place supplied by 3 rounds of blank. While the English soldier was discharging his harmless powder in the air, the bullets of his Hindoo ally were to be directed to his heart. The scheme was simple and practicable in the highest degree, and if carried out with secrecy and resolution, would have swept every European soldier in one and the same hour from the face of India....*had it been carried out, Her Majesty would have had the heartrending reflection that her birthday was the blackest in the annals of a nation whose history extends over a thousand years, and whose operations embrace the globe*' - Mrs D.D.Muter of Meerut; see *My recollections of the Sepoy Revolt,* London 1911; see also MEES DOLLY; and NATURE OF THE REVOLT.

PLOWDEN, Captain A.C. A British officer attached to the army of Gurkhas that entered Oude[Awadh] in support of the British in February 1858 under **Jung Bahadur**. He was placed in charge of the advanced division of the army, and it was he who took the major military decisions, although **Khurruck Bahadur** was nominally in command.

POLEHAMPTON, Rev. Henry He was the chaplain in the **Residency** at **Lucknow**, but did not survive the siege. His widow edited and published Rev.H.Polehampton, *A Memoir, Letters and Diary*, London, 1858. *See also* **FQP**.

POLWHELE, Colonel This officer was dismissed from his command (i/c the troops in Agra) after the fiasco of the **Battle of Sussiah**[Sasia] on 5 July 1857, near Agra. He protested, claiming somewhat pathetically that he had always done his duty etc. The reply:- *'re Colonel Polwhele, from Colonel Birch, Military Secretary, Calcutta:...it was not for any error committed as Commander of the Fort, nor yet for evincing any want of zeal for the service...that Col Polwhele was removed from command, but because he had on other occasions, more especially on that of the action on

Inside Agra Fort

the 5th July last, *evinced a total want of sufficient military talents and forethought for the very important command with which he was entrusted*...29 Sept. 1857. Fort William.'....and yet......... incredibly Thos Polwhele 1797-1885 was promoted Major-General on 1 May 1858, Lt.General 18 Mar 1870, and General 13 Dec 1876! He died, peacefully, on 23 May 1885 at Tivoli Lodge, Cheltenham (where so many old soldiers of the Raj eventually retired), aged 88. One wonders what might have happened to him if he had displayed just a trifle of military talent!

POST MORTEM **MUTILATION** Much of the mutilation that the British accepted as evidence of torture was executed *post mortem*. Dead bodies were repreatedly hacked at by passing rebels. Shepherd observed that in Kanpur, 'Whenever a Mahomedan found the lifeless body of a European or Christian lying anywhere, he immediately drew out his sword, 'with a *bismillah*,' and made a gash upon the corpse, repeating the words, or some such words as 'Soonut-ool-huq-i-Kafar-un.' This act is considered by them to be equivalent to killing an infidel, and adds to their claim for entering Paradise after death.' See W.J.Shepherd, *Personal Narrative etc.*, Lucknow, 1886, p 93; *see also* ATROCITIES.

POSTAGE STAMPS In 1957, the centenary of the mutiny/rebellion passed without undue celebration in India, although academics were encouraged or commissioned by Government to produce versions of the great struggle that would present to the world a more balanced viewpoint than the 'heavyweight' British historians like Sir John Kaye, Colonel Malleson, Charles Ball etc. Two commemorative postage stamps were issued, the 90n.p. was in mauve with nothing but a subtle pictorial reminder of the events of 1857, and the 15n.p. with an artist's impression of the **Rani of Jhansi** leading her troops into battle.

Commemmorative Postage Stamps

POWAEN [PAWAYAN, POWAIN], Raja of

He was a **Talukdar** of **Oude[Awadh]** who remained comparatively speaking neutral in the struggle between the rebels and the British. In June 1858 he attracted the wrath of **Maulvi Ahmadullah Shah**, (the Maulvi of Faizabad) because he refused help to the rebel forces. On 15 June the Maulvi attacked the raja's fort at Pawayan, mounted on a war elephant, and accompanied by some 500 sowars and the Nawab of Najebabad. On the raja refusing to open the gate of the fort, the Maulvi charged it on his elephant; in the gateway the raja's followers opened fire with matchlocks and the Maulvi and the Nawab were shot dead by the raja's nephew, Nurput Singh. The maulvi's head was promptly cut off and the raja took it into **Shahjahanpur** the next day to claim, and receive, the Rs50,000 reward. *See* **FSUP** Vol II p 418.

PRAKASH SINGH

A sepoy of the 26th BNI described by one, British, account as a 'fanatic' (= 'patriotic martyr' perhaps), said to have killed his commanding officer, Major Spencer on 30 July 1857, and to have precipitated the flight and subsequent destruction of the 26th BNI. This regiment had, in common with others, been disarmed at **Mean Meer (Lahore)** on 13 May and had since been kept in their lines exhibiting 'great sullenness'. Prakash Singh is said to have rushed out of his hut in the lines brandishing a **tulwar** and calling on his comrades to rise and kill all the **feringhees**, and to have then killed Major Spencer, while others killed the British sergeant major and a havildar. The regiment then took to flight. *See* F.Cooper, *Crisis in the Punjab*, Lahore, 1858, and Mr R.Montgomery's letter dated 29 April 1859, published in *Parliamentary Papers* dated 29 July 1859. *See also* UJNALLA, Well of.

PRAYER MEETINGS

'At this time (June 1857 in **Sialkot**), a controversy arose between the Officer commanding (Brigadier General Frederick Brind), and the Rev. Thomas Hunter, a Missionary of the Church of Scotland. The Commander opposed the organisation of the non-official English residents for self-defence and even opposed prayer-meetings, which he denounced as conventicles and used all his authority to suppress. At one time **he went so far as to threaten to hang Mr Boyle, the Chaplain**, but it seems that the prayer-meetings still went on.' *From* G.Rich. *The Mutiny in Sialkot,* Sialkot, 1924 p12.

PREMONITION

Early in 1857 there was a strange premonition of doom in St James's Church, Delhi, when **Chaplain Jennings**'s sermon, delivered in near darkness, took the theme *how unwise it is to postpone repentance.* The chaplain insisted on delivering a long sermon despite the hour. Many present, including Jennings and his family, did not long survive. **J.W.Sherer** was there doing a report on the Jumna[Yamuna] canals. Report accepted, he was posted to **Fatehpur** as Magistrate and Collector; and stayed with the Hillersdons on the way - all doomed, all were soon to be killed.

St James Church Delhi

PRESS ACT

Since the year 1835, in Sir Charles Metcalfe's day, the Press in India had been free and unrestricted, but by the middle of June 1857 the government in Calcutta became so alarmed by the tone taken by a number of newspapers that the Governor General in Council passed the Press Act, which was subsequently endorsed by the Governors of Bombay and Madras and the government in London. It came to be known in many quarters as 'The Gagging Act'. It was to last for one year only in the first instance (and indeed was *not* renewed in 1858 by **Lord Canning** the Governor General, although he was much criticised for not doing so). It is wrong to assume that the Act was intended to be used only against the vernacular press: the latter was not, as is suggested in some places, an inevitable instrument of 'sedition' and the natural ally of the mutineers and rebels:

much of the vernacular press was in fact in the hands of Christian missionaries. Strangely enough the newspaper most criticised by the Governor of Madras, Lord Harris, was the *'Examiner'* which he described as the *mouthpiece of the Roman Catholic priests*, and which he specifically accused of systematically abusing the government and creating ill-feeling between European Officers and the Madras sepoys. It was in general the English language newspapers that seemed most to offend Canning. The first to be threatened with the revocation of its licence was the **'Friend of India'**, published in Calcutta (It survives to this day as the 'Statesman'). In an article entitled *'The Centenary of Plassey'*, the paper had violently attacked the **Honourable East India Company** for its greed and for the manner in which it had acquired the sovereignty of the country. Government informed the publisher that such remarks in the existing state of affairs were dangerous *to the lives of all Europeans in the provinces not living under the close protection of British bayonets*: the editor responded by publishing government's letter and adding a variety of satirical comments which it can only be assumed were intended to provoke. Government was so offended in fact, that the paper's licence would have been withdrawn, had not the temporary Editor (who was acting for John **Marshman**), a Mr Mead, pre-empted the move by himself resigning. The licence of the *'Bengal Hurkaru'* was indeed revoked, shortly afterwards when it published a most lurid incitement to vengeance - *for every Christian church destroyed, fifty mosques should be destroyed, beginning with the Jumma Masjid in Delhi; and for every Christian man, woman, and child murdered, a thousand rebels should bleed.* For this outburst it was merely warned but ten days later it published a personal attack on **Canning** and **Beadon**, the Secretary to Government, saying that it was amazed the former had not been ordered home in irons and the latter sentenced to be tarred and feathered and ridden upon a rail etc. This was too much, government revoked the licence and restored it only on the resignation of the Editor, a Mr Blanchard. A Persian language newspaper in Calcutta, edited by one Hafiz Abdul Kadir, did lose its licence because its support for the rebellion was undisguised, thus: *'Come what may in these degenerate days, the men of Delhi must be celebrated as sons of Rustum, and very Alexanders in strength. Oh! God destroy our enemies utterly, and assist and aid our sovereign! etc.'* Other English language newspapers were warned 'for transgressing the conditions of their licences', but editors of other vernacular papers do not appear to have been over-censured: it may be that they couched their criticism of government in more subtle language. *See* VERNACULAR PRESS.

PRESS LIST OF MUTINY PAPERS

Produced in 1921 by the Imperial Record Department in Calcutta, they are of a significance that has been underestimated by historians in the past, and have been largely ignored. Yet they provide the finest direct evidence of what it was really like in Delhi in the months of May-September 1857, while the King of Delhi ruled his city and the British were confined to inaction upon the Ridge to the northwest. *See* ' *Press List of Mutiny Papers*', Imperial Record Department, Calcutta, 1921, *and* **CM** pp 67-71, *and* PAPERS of a MISCELLANEOUS CHARACTER.

PRINCES, Indian

Most of the Indian Princes supported the British, believing - probably correctly - that it was in their interest to do so. Only those in major dispute with Government, before the outbreak, actively sided with the rebels. The possible exceptions were **Kunwar Singh**, a genuine patriot, and **Bahadur Shah Zafar** the King of Delhi himself: the latter may well have been surprised and embarrassed at the position of prominence into which he was, willy nilly, thrust. And, of course the **Rani of Jhansi.** But some rulers had a most difficult time, striking the balance between the two sides.

An artist's idea of 'Scindia in his Court'

None more so than **Scindia,** Maharaja of Gwalior. The large and well-trained, and British-Officered **Gwalior Contingent**, a treaty-established force of all arms, was based in **Gwalior**, and its sepoys were recruited from the identical areas, especially Oude [Awadh], from which the sepoys of the Bengal Army came. They had sympathy, and blood brotherhood, with the government sepoys. It could only be a matter of time before they too broke out into open rebellion. In holding back the Gwalior Contingent, Scindia incurred the wrath of the rebels; in making conciliatory noises towards the same Contingent he aroused the suspicions of some of the British, suspicions which, in some quarters, were never dispelled. Probably the best and fairest summary of the position comes from the report of Major

S.C.Macpherson, the political agent in Gwalior, dated 10 February 1858 and addressed to Government (Quoted at length in **FSUP,** Vol III, p 166 *et seq.*). Macpherson had no doubt of Scindia's loyalty to Government, but recognised, shrewdly, the difficulty of his position: 'the army were predisposed to revolt, by sharing with the people of Hindoostan their feeling of dissatisfaction with our rule; thought success certain, from the smallness of our European force, and from popular aid; and made the cartridge grievance their pretext and occasion to rise. The foremost malcontents instigated, the King of Delhi headed, the revolt and all doubted deeply the stability of our power...the Princes, chiefs and best informed men generally expected that we should triumph, and took part more or less earnestly with us, or held aloof... In Gwalior the Contingent was one with the Bengal Army. Scindia's troops from our provinces shared its views. His Mahratta and Gwalior troops also shared them, but looked to his will....The Contingent rose in the usual manner, and murdered 18 men, women and children, of whom 7 were officers. *The rest escaped to the protection of Scindia or of his troops guarding the Residency.*....delayed by Scindia for two months, the Contingent, considering that Scindia had deceived them, planted their batteries against him, with the pledged support of a portion of his troops. Scindia, on 15 November 1857, was constrained to allow the Contingent to proceed with the vakeel of the Nana Sahib towards Kanpur. It was routed and dispersed there in the beginning of December, (but not before it had won a remarkable victory over the British force commanded by **General Windham** -PJOT) Should Government consider I have not misappreciated Scindia's conduct, I trust they may think it just and politic in fulfilment of my assurances to him, signally to acknowledge it.' They did.

PRIZE AGENTS Looting or plundering was, in theory, strictly forbidden in the British Army and those caught could be flogged or even hanged by the Provost Marshal. It was the task of certain officers, called Prize Agents, to receive all goods of value that had no obvious owner, including the possessions of rebel leaders, and then to sell them in open market so to speak, allocating the proceeds to be divided, on a strict scale according to rank, between all troops who had been present at the place where these goods were captured. The Prize Agents probably did their best, although there is evidence that valuables were often sold for only a fraction of their true worth, and there is evidence also of corruption, and of Government confiscating to itself all such proceeds on occasion, but the British soldier had no faith in the system, and believed, rightly, that his share of the legitimate 'profit' of his labours would be but a paltry sum. There was one exception: General Whitlock's Column was the fortunate captor of the Banda and Kirwi prize money. It became the subject of debate in Parliament in London for many years, and gigantic litigation resulted in which seventeen firms of solicitors and twenty-seven counsel were engaged.

Law expenses were over £60,000, but from the remainder *every private soldier's share was upwards of £70* - a very substantial sum indeed. The *'Madras Athenaeum'* announced that the 'Delhi prize money is to be paid immediately,' ie four and a half years after Delhi was recaptured - and the private soldier's share was just over £1 sterling (=Rs10). *See* Illustrated London News, 1 February 1862; *see also* **SSF** Chapter Five; and *see LOOTING*.

PROCLAMATION by the BEGUM of Oude[Awadh] Queen Victoria's Proclamation, issued on 1 November 1858, which the British hoped would bring the conflict to a swift end, was followed almost immediately by a counter proclamation, issued in equally regal terms by the **Begum of Oude[Awadh]**. She specifically warned her subjects against believing the deceits and deceptions of the English 'who never forgive a fault, be it great or small'. The argument is precise and to the point - what is the use of accepting the idea that the Company is dead and that now the Queen rules, when the laws, settlements, personnel, judicial procedure etc are all unchanged? She lists the occasions on which the English had deceived and betrayed its allies and broken its treaty promises; nor could the undertaking not to interfere with religion be believed in, since the present rebellion began with religion, and for it 'millions of men have been killed'. Perhaps one of the most farsighted of the suggestions of the counter proclamation comes in its last paragraph: 'In this Proclamation it is written, that when peace is restored, public works such as roads and canals will be made in order to improve the condition of the people. *It is worthy of a little reflection, that they have promised no better employment for Hindostanees than making roads and digging canals. If people cannot see clearly what this means, there is no help for them'.* For the full text of the Begum's proclamation *See Foreign Political Consultations,* 17 December 1858, Nos 250/54.

PROCLAMATION BY BIRJIS QADR 'All the Hindoos and Mahommedans are aware that four things are dear to every man; first Religion, second, Honour, third Life, fourth, Property. All these four things are safe under a native Government. No one under that Government interferes with religion. Every one is allowed to continue steadfast in his religion and persuasion, and to possess his honour according to his worth and capacity, be he a person of good descent, of any caste or denomination, Syud, Sheikh, Mughal, or Pathan, among the Mahommedans, or Brahmin, Chhuttree, Bais, or Kaith, among the Hindoos. All these retain their respectability according to their respective ranks, and all persons of a lower order such as sweeper, Chumar, Dhanook, or Passee, can claim equality with them. The life of any person of either class is not taken under that government, nor his property confiscated except for crimes and grave offences. The English are enemies of the four things above named. They wish to deprive the Hindoos and

Mahommedans of their religion, and wish them to become Christians and Nazarenes. Under their Government, thousands of people have embraced Christianity and are continuing to do so. The honour and respectability of the lower orders, nay, comparatively with the latter, they treat the former, with contempt and disrespect; and at the instance of a chamar, force the attendance of a Nawab or a Rajah, and subject him to indignity whithersoever they go. They hang the respectable people, destroy their females and children, and their troops commit acts of violence upon their females, and thus take away their honour, dig down their houses, and plunder all the property and leave them nothing. They do not kill Banyahs and Mahajans, but take away all their property and violate their women. Whithersoever they go they disarm the people, and in that state hang or shoot or blow up any one they like, and deprive any one they choose of his faith and honour. At some places they resort to the subterfuge of remitting revenue or reducing the *Juma* to the Malgozars, with the intent, that after they have established themselves and subjected them to their authority, they might do with them as they please, force them to become Christians, or hang them or dishonour them. Some foolish zemindars allow themselves to be thus imposed upon, but the shrewd avoid the snare. Therefore the Hindoos and Mahommedans are hereby warned, that whosoever among you wishes to protect his faith, honour, life and property, may come forward to fight against the English in conjunction with the forces of this Sircar (*ie* himself) and may not succomb to their strategm. This Sircar will be kind to them and contribute to their relief. And be it known to the Passees that the office of watching every town and village is their hereditary profession, but the English appoint Burkundazes in their stead and thus deprive the Passees of their livelihood. They should also, in concert with the troops of this Sircar and the zemindars, plunder the English, and their dependants, commit theft and gang robberies n their camps, and disturb their rest.' **FSC** dated 25 June 1858 Nos 68-69, **NA**.

PROCLAMATION (2nd) by BIRJIS QADR

'In the Holy Kuran the great and glorious God has said 'O ye believers! don't make friendship with Jews and Christians. He who befriends them is surely one of them that is to say, the friend of a Jew is a Jew, and the friend of a Christian is a Nazarene. God never sends his mandates to the tyrants ie infidels.' From the above it is manifest that friendship with Christians is heresy. He who has friendship with a Christian is by no means a Mussulman. Hence it is a duty of all the Mussulmans to make themselves inveterate foes of those Christians, and never enter into friendship with them; if they will not do so, they will lose their religion and become heretics. Those whose religious notions are weak, and who are given to worldly temptations say, that they fear to incur the displeasure of the Christians lest they might be involved in some misery, on the restoration of the Christian power. To them God has said 'Did ye mark those men whose minds are afflicted with faithlessness?' They are very hasty in contracting friendship with Nazarenes. They say they are afraid lest they might meet with some mishap....You should take into your consideration the outrages committed by the English at Shahjehanabad, Jhujjur, Rewaree and in the Doab. Men and women should think it their duty to extirpate the English. The futwa of the religious men of both sects (Sheeya and Soonnee) is hereby annexed...' *See* **FPC** 31 December 1858, No.1750 Enclosure of a letter from Commissioner on Special Duty, dated 27 May 1858. **NA**.

PROCLAMATION by GOVERNMENT

Immediately after the outbreak of mutiny at Meerut and Delhi, Government saw fit to issue a pompous and blustering proclamation. It is doubtful if it had any effect except perhaps to emphasise the weakness of the Europeans' position: 'Whereas it has been ascertained that in the Districts of Meerut and in, and immediately round Delhi, some short sighted Rebels have dared to raise resistance to the British Government, it is hereby declared that every Talookdar, Zemindar, or other owner of land, who may join in such resistance, will forfeit all rights of property, which will be confiscated and transferred in perpetuity to the faithful Talookdars and Zemindars of the same quarters, who may shew by their acts of obedience to the Government, and exertions for the maintenance of tranquillity that they deserve reward and favour from the State. The powerful British Government will in a marked manner recompense its friends, and punish its enemies.' *See The Mofussilite,* 15 May, 1857.

PROCLAMATIONS, by the rebels

As these constitute significant written evidence in support of the rebel cause they are most important and need to be studied in detail by students of the Great Indian Rebellion. They appear, in original and in translation in a variety of official papers, Mutiny Basta, Trial records etc, but are most conveniently found, placed together in a single section, in **FSUP**. The collection includes Proclamations by **Bahadur Shah Zafar, Nana Sahib, Birjis Qadr,** the Nawabs of **Banda** and **Farrukhabad** etc. *See* **FSUP** Vol I p 438 *et seq. See also* NANA SAHIB'S PROCLAMATION.

PROPHECY of the MUTINY

No-one had predicted the Mutiny with eerier particularity than **Henry Lawrence.** Commenting in 1843 on the Government's careless complacency, he predicted that rebels could take Delhi. 'Let this happen,' he said, 'on June 2nd, and does any sane man doubt that that twenty-four hours would swell the hundreds of rebels into thousands, and in a week every ploughshare in the Delhi States would be turned into a sword? And when a sufficient force had been mustered, which would not be effected within a month, should we not then have a more difficult game to play than Clive had at Plassey or Wellington at Assaye? We should then be literally striking for our existence at the most inclement season of the year, with the prestige of our name tarnished.' James Innes, *Lucknow & Oude in the Mutiny.* London, 1895, p 73.

PROSTITUTES   *See* COURTESANS, and LADIES OF THE REGIMENT.

PUBLIC RECORD OFFICE   This Department in London contains less than might be expected of material from this period (most relevant manuscript papers are of course in the Oriental and India Office collections of the British Library, in what was formerly known as the 'India Office Library and Records'), but there is a useful collection here of Admiralty records which include details of the activities of the **Naval Brigade**, and the ship's logs and muster books from **H.M.S. Shannon and H.M.S. Pearl**.

PUNISHMENT OFFERS   By November 1857 a committee of Western Australian 'gentlemen, landowners and others' had 'kindly volunteered' to receive transported rebels and employ them in a 'healthy' climate 'not very dissimilar to that of some of the northern provinces of Bengal.' *See* London *Times.* 19 November, 1857, p. 10. Their offer was not accepted.

PUNISHMENTS & REWARDS   When British authority was re-established there was much publicity given to the rewards and punishments handed out to those who had been loyal to government or had rebelled. The latter were harshly treated, though as the years went by, attitudes softened, and pursuit was slackened. As an example of the kind of thing that happened in many districts that had been affected by rebellion, we can look at what happened in the Kanpur district. The zemindars in the immediate vicinity of **Kanpur** joined with the mutineers and the **Nana Sahib** in attacking **Wheeler's Entrenchment**. These included the Chaudris of **Bithur**, the Raja of Thathia in **Farrukhabad**, Moti Singh of Nanamau, the Rajputs of Kakadeo and Panki Gangaganj, the Raja of Sheorajpur, the Raja of Sachendi, the Raja of Nar, the chief of the Gaurs. Without their aid perhaps the Entrenchment would not have been such a hopeless position for the British. In addition the **Tahsildars** of Narwal and Akbarpur threw in their lot with the Nana, but Afzal Ali of Ghatampur saved the treasure and records and remained at his post. Waris Ali of Derapur held out but eventually absconded, ditto Farid-uz-zaman of Rasulabad. When punishements came to be handed out the Rajas of Sheorajpur Sachendi Binaur and Nar all forfeited the whole of their estates. 61 entire villages were lost by zemindars, and some 79 others were partially lost. The rewards were equally dramatic: Ishri Prasad, the commissariat contractor who brought supplied into the Entrenchment at great risk, was given land in eight villages. Narayan Rao Nana of Bithur received village of Binaur, assessed at Rs4,500 *p.a.*, and sixteen others were similarly rewarded. For further details *See* H.R.Nevill, *Cawnpore, a Gazetteer*, Allahabad, 1929, pp 222-223.

PUNJAB   There are numerous entries dealing with the mutiny/rebellion in the Punjab (Punjaub), entries which in turn will lead to further primary and secondary sources. They include: GOVINDGARH; AMRITSAR; FEROZEPUR; JULLUNDUR; PHILLOUR; AMBALA; LAWRENCE, Sir John; COOPER, Frederick; LAHORE; MEAN MEER; PESHAWAR; DULEEP SINGH etc

PUNJAB MOVABLE COLUMN

BRIGADIER-GENERAL NEVILLE CHAMBERLAIN

This British force is often associated almost exclusively with the name of Brigadier-General John **Nicholson**, but it originated at Wazirabad, on 25 May 1857, under the command of Brigadier-General Neville Chamberlain, who was later to be appointed Adjutant General. It was formed to over-awe the Punjab when the news of the outbreak at **Meerut** was received, and in the expectation that a number of regiments of the Bengal army would follow suit and mutiny. The force consisted originally of Dawes's Troop of Horse Artillery, Bourchier's 17th Light Field Battery, a Wing of the 9th Light Cavalry, H.M.'s 52nd Light Infantry (Oxfordshire L.I.) and the 35th regiment B.N.I. The column marched for **Lahore**

Lahore

where it arrived on 2 June, moving on to **Amritsar** on the 11th. The object of these marches was apparently to over-awe the disaffected by making it known that a large body of European troops was present and was mobile. The 1st Punjab Infantry (Coke's Regiment) had joined the Column briefly but now set off for **Delhi**, on 12 June. On the 13th the

Column left for **Jullunder** where it arrived on the 19th, having crossed the river Beas by a bridge of boats. There was some doubt here expressed as to the loyalty of the 35th BNI and rumours persisted that they had intended to mutiny at the Beas and cut away the bridge of boats. On the 20th Chamberlain departed for his new appointment and on the 22nd Nicholson arrived to take command of the Column. They marched for **Phillour** where they arrived on the 24th, and here the 35th BNI and the 33rd BNI (which had just joined the Column) were disarmed as it was feared they were about to mutiny. The Column then marched in a circle, back to Sheywara, then to Jullunder and back to Amritsar on the morning of 5 July. News continued to come in of the mutiny of other regiments, the 14th BNI at **Jhelum**, the 46th BNI and a Wing of the 9th Light Cavalry at **Sialkot**, with the result that Nicholson ordered the disarming of the 59th BNI and the other Wing of the 9th Light Cavalry, whose horses were taken from them. A forced march to Goordaspur on the 11th was followed by the action at Trimmu Ghat on the river Ravi where the mutinied 46th BNI and Wing of the 9th Light Cavalry were attacked and dispersed: the Sialkot mutineers were effectually destroyed on the 16th. It is thought that the rebels mistook H.M.'s 52nd, who were dressed in 'khaki', for Sikhs or other Irregulars, as they had never before seen English Infantry in anything but red or white. Nicholson now went to Lahore to interview **Sir John Lawrence**, and returned with the news that the Column was to be sent to **Delhi** to join the Delhi Field Force encamped on the Ridge. The Column reached Phillour on 29 July and may be said to have here ended its existence as a Movable Column: it had existed for two months and had marched something in the region of 450 miles. The troops marched via **Ludhiana** (2 August), **Ambala** (5th), **Kurnaul** (8th), **Alipore** (13th), finally marching into camp at Delhi on 14 August 1857. *See* Lieut.Colonel C.K.Crosse, *Punjaub Movable Column under Nicholson,* in *United Service Magazine,* June 1896.

PURWA Was a district, or *Zillah*, lying between Lucknow and Kanpur, which, rather strangely, appears to have been one of the last to be fully subjugated by the British, despite it being so close to the line of march of British troops throughout 1857 and 1858. Indeed south east of the Kanpur-Lucknow road the land seems to have been held solidly by the rebels for many months. Although much of the discontent/rebellion in Oude[Awadh] can be traced to the land settlement the British imposed upon the annexation of the **Kingdom of Awadh**, none of the leaders in the zillah of Purwa had been **talukdars** before the British came so none had any particular grievance against the government. The three principals appear to have been Babu Devi Baksh of Purwa, who collected an army of some 3,000 men and three guns, Babu Ram Baksh of Bhagwantnagar, who had attacked the boats of the fugitives escaping from the **Satichaura Ghat** at Kanpur, and Shiv Ratan Singh of Pathun Behar. This can be taken as an indication either of the popular nature of the revolt in Oude[Awadh], or the exploitation of a state of anarchy by ruthless warlords: as so often in this struggle, much depends upon the point of view. *See Foreign Secret Consultations,* 28 May 1858, Nos 416-7.

Palanquin.

Q

QADAM-e-RASOOL This is a mosque in Lucknow city adjacent to the Shah Najaf Imambara. Built by Ghazi-ud-din Haidar it was particularly venerated as it was said to contain a stone bearing the imprint of the foot of the Prophet Mohammed. It was used as an important stronghold by the rebels in 1857-58, and is still visible today. Courtesy INTACH.

Mermaid Gateway to the Qaiserbagh

QAISARBAGH The palace of King Wajid Ali Shah, who built it 1848-1850 at a cost of Rs 80 lakhs. The quadrangle was surrounded by buildings in which the ladies of the harem used to reside. It figured in the Revolt of 1857. Courtesy INTACH. *See* KAISER BAGH

QUEEN'S BIRTHDAY A rumour had been circulated in the city, (Kanpur) that the objectionable cartridges were to be served out on 23 May, and that the artillery were to act against all who refused them. A good deal of excitement prevailed, and on 24 May, the Queen's birthday, it was not considered advisable to fire the usual salute. *See* **Sherer.**

QUEEN'S PROCLAMATION On 1 November 1858 amid much carefully orchestrated firework displays and military parades, the British played what they planned to be their trump card. A proclamation from Queen Victoria was read out in all the military stations of India declaring that the rule of the East India Company was over, and that the British Government now ruled India directly, under the Queen. Significantly, pardon was offered to all who would lay down their arms and could show they had had no hand in the murder of Europeans. Religious toleration was promised. Ancient treaties and customs would be respected. On receiving the first draft from Lord Derby the Prime Minister, the Queen asked him to revise it, 'bearing in mind that it is a female sovereign who speaks to more than a hundred millions of Eastern people on assuming government over them, *and after a bloody civil war*, giving them pledges which her future reign is to redeem, and explaining the principle of her government'. And the final text embodied all the suggestions made by H.M. The proclamation is of such major significance not only at the time but to the subsequent history of India that it should be studied in full. The Victoria Memorial in Calcutta displays the text. *See* Charles Ball, *The History of the Indian Mutiny*, Vol II pp 518-19. But *see also* the equally significant rejoinder by the **Begum of Oude**[Awadh]. *See* summary under PROCLAMATION BY THE BEGUM OF OUDE.

QUICKSANDS 'The cavalry pressed on as far as the River Raptee, the 7th Hussars capturing a gun on the other side... and cutting up a number of sowars in the middle of the river, and I am sorry to say losing Major Horne and two privates by *drowning* from their horses getting into quicksands in the river during the skirmish.' *From* Major the Hon. D.Fraser, R.A.'s Report, Camp Bankee, 3 January 1859, manuscript contained in archive *MD/147* at the **Royal Artillery Institution**, Woolwich, S.E. London.

R

RACQUETS COURT The Racquets Court still stands today, an ugly-looking brick building near the Kanpur Club. 'It was formerly the haunt of young officers playing on its echoing black marble floor before going on to the Kanpur Library to read the latest 3 month-old English papers.' The **Nunne Nawab** used the gallery of the Raquets Court as his headquarters. It was in the same vicinity that the Muslim cavalry camped, here a battery was trained on **Wheeler's Entrenchment**, and from here the shell filled with burning material was fired which carried away a part of the barracks. *See* Zoë Yalland, *A Guide to the Kacheri Cemetery*, London, 1985, p 26.

Kanpur Racquets Court today

RADHA GOVIND Was a leading rebel in Bundelkhand who continued resistance long after it had ceased in other areas. He was eventually killed by the British in the Battle of the **Panwari Heights** on 29 December 1858, together with his brother and some three hundred of his followers: the British also captured four guns, four elephants, Radha Govind's silver howdah and a quantity of arms, accoutrements and ammunition. The actual spot where Radha Govind was killed was of great sanctity, and it was said that he had determined to die there, being wearied with the constant fear of capture, and the clamours of his followers for arrears of pay and for provisions. *See The Revolt in Central India,* Simla, 1908 p.186

RAGHUBAR SINGH, Subadar A native officer of the **Kotah[Kota] Contingent** who commanded the guard of 120 men left behind at **Deoli** when the **Kotah Contingent** was sent by the British to the Agra/Muttra area to help to restore order. When the Neemuch Brigade had mutinied it set out for Agra in June 1857, and on the way burnt the cantonment at Deoli and forced Raghubar Singh and his men to accompany them. However he escaped from them a few days later and returned with some sixty of his men who also remained loyal to the British. He had also earned their gratitude by concealing the wives and children of the sergeants of the Kotah Contingent in the neighbouring village of Jehazarpur.

RAGHUNATH SA The son of the pensioned descendant of the Gond Kings of Garha Mandla. He and his father were convicted of conspiring with the 52nd BNI in Jabalpur to attack and murder the Europeans there, and were blown from guns on 18 September 1857.

RAGHUNATH SINGH reported as 'Captain Raghunath Singh, by caste a Tewarri Brahmin, prior to the annexation was in Captain Bunbury's regiment, subsequently transferred to the 2nd Oude[Awadh] Military Police under Captain John Hearsey, he was also a leading character during the rebellion'; *See* PERSONS OF NOTE, APRIL 1859.

RAGHURAJ SINGH Maharaja of **Rewa**. Possibly in league with the rebels. *See* Kameshwar Jha, *A Study of some Mutiny Letters of Sohagpur,* Indian Historical Records Commission Proceedings, 1958.

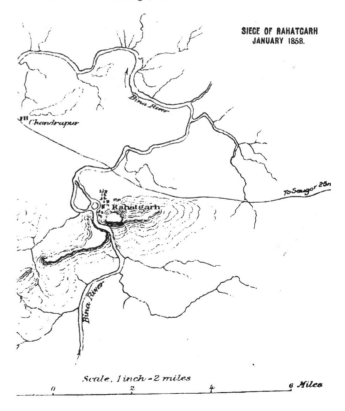

RAHATGARH A strong fortress some twenty-five miles southwest of **Saugur[Sagar]**. The territory of the latter, was, in January 1858 partially in the hands of the rebels while **Bundelkhand** was wholly in their control. Rahatgarh appeared to be the key to the whole area and was therefore selected by **Sir Hugh Rose** as his first point of attack with the **Central Indian Field Force**. It was held by Nawab **Muhammad Fazal Khan**, a man of energy and courage, related to the Regent of Bhopal; having failed in an attempt to usurp the power of the latter he had become a chief among the rebels. His garrison consisted of *vilaities* and Pathans. The fortress rivalled **Gwalior** in its strength, but was captured by the British after three days' investment; Muhammad Fazal Khan and Nawab Kamdar Khan were hanged over the gate. A detailed account of the investment and battle, critical for the restoration of British power in Central India will be found in Thomas Lowe, *Central India during the Rebellion of 1857-58,* London, 1860. *See also The Revolt in Central India*, Simla, 1908, *and* I.T.Pritchard, *The Mutinies in Rajpootana,* London,1864, *and* **FSUP** Vol III.

RAI BAREILLY (or Roy Bareilly) An important town in the south of Oude[Awadh], and long a stronghold of the rebels, particularly **Rana Beni Madhoo**. The appearance of the place at the time is described thus: 'it was long after three in the morning before we were clear of that wonderful labyrinth of deserted streets and tottering loopholed keeps, barbicans, portals, and battlemented walls, which bear witness to the former greatness of Rai Bareilly. The crenellated and turreted walls seemed, in the moonlight, of great solidity and of great height. The city is but a collection of feudal castles, old baronial forts of the nobles of Oude [Awadh] - at the base of which, and in the adjacent spaces, is a stratum of hovels, perforated by tortuous narrow paths, and surrounded by the noble old wall. Scarcely a living being came forth to look at our noisy array as it passed on. Hate and fear lived within those dark dwellings. When we first approached all the people fled.' *from* Chas Ball, *History of the Indian Mutiny,* London, n.d. (1859?) pp 537-9.

RAILWAYS In May 1857 the building of railways in India had only just begun. Many miles were planned, much engineering work had been set in motion, bridges built and embankments started: the skilled personnel were *in situ*, but the only railway actually operating ran for just 120 miles from Calcutta (Chinsurah) to Raniganj. British reinforcements mainly took this route (although a few went all the way by steamer), then changed to the tedious old bullock-cart transport as far as **Allahabad** where again there might be an alternative in the way of a steamer towing flats slowly, painfully slowly, up river against the current as far as **Kanpur**. But for the majority of troops it was the bullock-cart again from Allahabad, or shanks's pony. The European railway personnel played an active role on the British side during the struggle, notably Boyle at **Arrah**.

RAIS A nobleman. The title seems to have become attached particularly to certain characters *eg* the Nawab of Farrukhabad.

RAISING MORALE 'August 1857 in **Kanpur** was a month of great gloom and misery for the British. The rains had started with a vengeance, and to the bad news was added the inundation, mud and universal dampness and depression of the monsoon. There was much for the British to be gloomy about. General **Havelock** was in the city but news had come that, although he had done so well, with his little force, to drive the **Nana** from Kanpur, and prepare to march upon besieged **Lucknow**, he was to be superseded. True, the man so appointed was universally liked and admired - **Sir James Outram** - but it was felt, particularly among the soldiers, to be grossly unfair to take away the command from their old Bible-punching firebrand. Then there was the onset of illness among the troops: cholera, dysentery and fever raged, and death spared neither rank nor merit. A 'most accomplished and agreeable man', a Captain Young, famed for his ability as a linguist, went to **Mohammed's Hotel,** which had re-opened its doors to the British, with the same spirit of impartiality with which it had welcomed the Nana Sahib two months' previously, and had a meal of which the main constituent was some tinned provisions. The next morning he got up late, complained of feeling ill and said he would lie down. By noon he was dead, and by the evening, buried. Then there was the sad case of Brown, an army officer from a mutinied regiment, who had already survived against all odds the onslaught of the mutineers. He had been with Lloyd, a civilian, and two other Europeans at Hammespur west of **Banda**, on the Jumna[Yamuna]. The four of them stayed at their post with courage but no hope, awaiting the inevitable and overwhelming attack from the mutineers. They kept a boat on the river at the bottom of the garden for the emergency. They had not long to wait: while eating their lunch one day they heard shouts outside, the noise of muskets being fired, of a crowd bent upon murder. The four men ran down the garden and took to the boat, and floated gently down the Jumna. At night they put in to shore, but were immediately surrounded. Lloyd and one other were taken prisoner (and killed later), the other two, including Brown, somehow got across the river and took refuge in the jungle. Some days later Brown, now the sole survivor, half dead with hunger and fatigue, was taken in by a friendly zemindar. Thence he was sent in a litter, to Havelock's camp, on his way to Kanpur. His condition was terrible: he was lame and lacerated, his nerves shot to pieces, his speech incoherent, but being a naturally strong and robust young man, he made a remarkable recovery. But then - you may have guessed - he was struck down by cholera and died within twenty-four hours. They really did not know much about cholera in those days. Their instinct rightly led them to suspect water as the carrier of the disease, but the how and the when was beyond their knowledge. Coincidentally a number of men died at the same time in Kanpur, from cholera, who had all been in a certain barrack one day when the heavy rain inundated the place, and it was awash with water. An eye-witness reports, as fact, that all those who took off their shoes and stockings to paddle through the water got cholera! While those, conversely, who kept their shoes and stockings on, had no ill-effect. One hopes this piece of misleading folklore did not find its way into any medical text. Among those who had taken off his shoes and stockings, and then died, was a young officer whose fate was thought to be particularly sad. He was engaged to be married to a young lady who was then among the besieged in the **Lucknow Residency,** and it had been his dearest wish, which fate had granted, that he should be in Havelock's Column being sent to relieve the garrison of the Residency. To the grief of all those in Kanpur his death ended this romantic tale short of a happy ending. Perhaps it was as well; his fiancée, a lovely young girl, was already dead, killed by a roundshot in the Residency compound, a fact of which he died ignorant. All then was gloom and doom. This was the place of the massacres, three of them, that had been perpetrated within the last two months, there in Kanpur. As the rains came down and the earth steamed, so they thought that the very neem trees smelt of the blood of the massacred. **Delhi** was not recovered, and everywhere the rebel cause seemed to be flourishing. So what did the British do? They organised the 'Kanpur Summer Race Meeting' and, as a morale lifter if not a sporting spectacle, they say it was very successful.' **PJOT**

See also ENTERTAINING THE TROOPS.

RAJPUT A Hindu of the military tribe or order or caste, of particular nobility in the eyes of the people of Northern India. Their customs, habits and manners reflected the orders of chivalry in medieval Europe.

RAJPUTANA In 1857 this consisted of the British district of Ajmer-Merwara, and some twenty princely states. It stretched from Sind and the Punjab to the north, to the states of Central India on the south and east. The native states were, in alphabetical order: Alwar, Banswara, Bharatpur, Bikanir, Bundi, Dholpur, Dungarpur, Jaipur, Jaisalmer, Jhalawar, Jodhpur, Karauli, Kishengarh, Kotah, Lawa, Mewar or Udaipur, Partabgarh, Shahpura, Sirohi and Tonk. The British had a representative (Agent to the Governor General) with headquarters at Mount Abu, while there was a political officer at each of the courts of Jaipur, Jodhpur, Bharatpur, Kotah and Udaipur. There were no British troops in Rajputana, and although there were a few troops from the Bombay army, most of the garrisons were either from the Bengal Army or from Contingents (Kotah, Gwalior etc, Jodhpur Legion) mainly recruited from the same areas and castes as the Bengal army, and likely therefore to be sympathetic. But in 1857 the rulers and their people in general remained loyal to the British although there were exceptions. For details of the military operations in Rajputana *See The Revolt in Central India,* Simla, 1908 p 49 *et seq. See also* Iltudus T.Pritchard, *The Mutinies in Rajpootana,* being a personal narrative of the mutiny at Nusseerabad, London, 1864.

North India in 1857

RAJPUTANA FIELD FORCE This was assembled by the British in March 1858 at **Nasirabad** specifically to 'punish' the **Kotah** rebels who were considered to be the main danger to the peace of **Rajputana**. Major General H.G.Roberts commanded and the force was mainly from the Bombay army and was particularly strong in artillery, including a second-class siege train. Its fighting strength eventually reached a total of 5,500 men. Kotah[Kota] was captured on 30 March after a siege of about five days, and given over to plunder by the British troops, following presumably the age-old precedent that a city which is taken by assault at the end of a siege is open to plunder by the victor. The results in this case appear to have been more than usually unfair: 'Kotah after its capture presented a desolate appearance; the mutineers had plundered it for many months, and shot and shell had caused considerable damage. The plunder carried off by the rebels must have been great, for since the days of Zalim Shah, who resisted the Mahratta inroads, Kotah had been famed as a secure emporium for treasure, opium and valuable merchandise. The town was abandoned to plunder by the troops for five days, but the articles collected by the prize committee were of inferior value, and *hardly worth the miseries they cost the poorer classes, to whom they mostly belonged.'* See *The Revolt in Central India,* Simla, 1908 p 201 *et seq. See also* Iltudus T.Pritchard, *The Mutinies in*

Rajpootana,being a personal narrative of the mutiny at Nusseerabad, London, 1864, also **FSUP**

RAM SINGH, RAJA
Raja or maharaja of **Kotah**[Kota] in Central India. The **Kotah Contingent** mutinied at an early date, and the **Kotah Rebels** (ie the personal troops of Ram Singh) mutinied in October 1857, but the raja himself remained loyal to the British. He is described (somewhat ungratefully!) as a 'weak-minded man, of dull and apathetic temperament'. He appears however to have made up his mind at the start that the British would win, and stuck to his opinion. He eventually had to shut himself up in his palace and fort, and was practically besieged by his own soldiery, consisting mostly of Poorbeahs and Muslims, who committed many excesses and plundered the inhabitants of the town, making his attitude their excuse. *See The Revolt in Central India*, Simla, 1908 . *See also* Iltudus T.Pritchard, *The Mutinies in Rajpootana, being a personal narrative of the mutiny at Nusseerabad*, London, 1864.

RAMCHUNDER SUBADAR

The *diwan*, or chief minister of the last *peishwa*, **Baji Rao II.** The Commissioner of Kanpur found Ram Chunder's conduct 'entirely satisfactory. ...The Subadar had commanded 5,000 Horse against us during the War. He never deserted his lawful sovereign in good fortune or in bad, and at last when he saw that his master's cause was totally hopeless, he was urgent in advising him to throw himself on the mercy of the British Government, and was very instrumental in effecting his submission to Sir John Malcolm in June 1818.' The Subadar was 'uniformly most zealous in preventing political intrigue between Bajee Rao and other chiefs (several times commenced by the latter) and in regulating the Maharajah's conduct in all respects in the manner most desired by the British Government... Ram Chunder managed Baji Rao's ' internal police so efficiently since his arrival in February of 1819, and had kept the Mahrattas of Bithur under such control that the Deputy Magistrate reported 'no perceptible increase to the business of his office from the addition of the crowd of Mahrattas to the inhabitants of his district' and ' not a single crime has been substantiated against any of the Maharaja's followers.' Through Ram Chunder John Low was able to persuade Baji Rao to loan the British Rs.600,000. But Baji Rao was 'totally disappointed and annoyed by the result: no reward, no acknowledgment, no restoration of his glory. He complained bitterly that he had never received 'a single line from the Governor General on any occasion whatever. Lord Amherst's refusal to allow Baji Rao to enter his camp while his Lordship visited Kanpur annoyed him beyond measure, and was a sign that the Company had broken its faith with him. and all the while he complained he kept shooting reproachful glances in Ram Chunder's direction and later tried to remove him as his advisor, though he could never find any grounds to dismiss him.' Ramchunder Subadar's direct descendants still live in Bithur, and are highly respected locally. His son and Baji Rao's adopted son **Nana Sahib** however had fallen out before the outbreak of 1857.

RAMDEEN
The name of one of the principal boatmen in charge of the country craft that were assembled for the Europeans in **Kanpur** to travel downstream to **Allahabad**. It was he who 'fired the thatch' on a given signal, followed by the general massacre. *See* BOATS, *also* SATICHAURA GHAT, *also* Depositions in **Forrest** or Williams. Ramdeen is also the name of General **Wheeler's** bearer, who was with him in the **Entrenchment** and survived the massacre at the **Satichaura Ghat**.

RAMGHAR
See RANCHI.

RAMPORE, NAWAB of
'The Nawab of Rampore to the south of **Nynee Tal** unexpectedly offered assistance, sending rice, sugar, flour, medicine and funds, and covering the passes to the south with his own men, thus freeing up the defenders to concentrate on the passes to the east of the station. But his messengers brought alarming news that the 'King of Rohilcund' was raising an army of 20,000 men and casting cannon.' Wm Butler, *Land of the Veda*, New York, 1872. p 279-80.

RAMPORE
In addition to the large town of this name (close to **Naini Tal**), there was another Rampore, also called Singhi Rampore or Singa-rampore: a village beside the Ganges between **Fatehgarh** and **Kanpur** and the scene of a massacre on 5 July 1857 of a party of European refugees. *See* **FSUP** Vol V, *and* F.R.Cosens and C.L.Wallace, *Fatehgarh and the Mutiny*, Lucknow, 1933. *See also* E.J. Churcher, *Some reminiscences of Three-quarters of a Century in India*, London, 1909.

RAMSAY, Lt.Col. George
Was the 'Resident at Nepal', that is the British representative of the Governor General (latterly Viceroy) of India. He lived at the Residency in Khatmandu, and was the go-between for communication with **Maharaja Jung Bahadur.** It is likely that the two men neither liked nor trusted each other, but their official stance remained cordial. Much of the correspondence between the Governor General and Ramsay, dealing in particular with the closing phase of the rebellion, in 1859, is available in *Foreign and Political Proceedings* May-July 1859, quoted at length in **FSUP** Vol II.

RAMSAY, SIR JAMES Better known as the Marquess of **Dalhousie**.

RANA BENI MADHO BAKSH See PERSONS OF NOTE, APRIL 1859. *see also* BENI MADHO.

RANCHI In Chota Nagpur, a military station. Troops of the 8th BNI, together with the Ramgarh Battalion mutinied in early August 1857, and the European officers and civilians were driven out and took refuge at Hazaribagh. Government eventually restored order with the help of the Raja of Ramgarh and with some troops of the Madras army. The battle of Chattra on 2 October 1857 drove the rebels from the area. For details of the rebellion and subsequent campaign, *see* **Kaye**.

RANIGANJ The railhead, ie the end of the track from Calcutta in 1857. Many British troops disembarked here and then continued their journey up country to the theatre of war, by bullock cart. For a letter describing the organisation of this operation *See* Gibbons Papers at CSAS, James Gibbon's letter to his mother. *See also* RAILWAYS.

RANGOON Capital city of Burma, then in British hands, to which the ex-King of Delhi, **Bahadur Shah Zafar** was exiled in 1858. On 4 January 1862 the Illustrated London News reported: 'The King of Delhi died at Rangoon on 11 Nov 1861 and was buried the same day. Little interest was exhibited by the Mohammedan population of Rangoon'

RANI OF JHANSI See JHANSI, RANI OF.

RANI GHAT An old *ghat* on the earlier course of the river Ganges at Kanpur. *See* SOOKA MULL'S GHAT.

RAO HURDEO BUKSH SINGH See HURDEO BAKSH.

RAO SAHIB Nephew of Dondhu Pant, the **Nana Sahib.** He shared with **Tatya Tope** and the **Rani of Jhansi** the leadership of the forces of the Peishwa fighting the British in Bundelkhand and Central India, though his military ability was probably the least of the three. He avoided capture until 1862 when he was arrested in the hills to the north of the Punjab, and was sent to Kanpur; there he was tried and found guilty of having been concerned in the murder of Europeans, and was hanged on 20 August 1862, at the Satichaura Ghat. Documentation on his capture and trial, and *a rare photograph* are contained in the papers of James William Macnabb which are preserved in the Oriental and India Office collections of the British Library, *See MSS.Eur.F.206. See also* **FSUP** Vol IV for story of Kanpur and Bithur, *and* Tapti Roy, *The Politics of a Popular Uprising: Bundelkhand in 1857*, Delhi, 1994.

RAPE **A**. The rape of European women during the period of the 'Mutiny' is hinted at in all manner of books, periodicals and private letters, and at first the temptation is to believe that Victorian susceptibilities alone prevent the writers being more explicit. Prolonged study in depth however of the horrendous stories that were current at the time reveals a much more likely explanation for the lack of detail - *they just were not true.* Those who passed on stories of rape - or invented them - might have assumed that the sepoys would be as keen on the rape of European women as they were on acquiring European valuables, but this was not so: to the Hindus such contact would have meant loss of caste, and would not have been contemplated except with revulsion. **Mowbray Thomson**, a survivor of the Kanpur Massacres, believed that it was the physical condition of the women at the end of the siege that deterred the sepoys - perhaps as ungallant as it was inaccurate. But there were a few, a very few, tales that proved on investigation to be true. Hints and glimpses come from many quarters. Letters from Major Jonathan Fowler 8th Cavalry to his cousin Eliza Kenyon. Include interesting material on the final assault on **Jagdishpur** especially on the ineptitude of **Brigadier Douglas**. Sept-Dec 1858, but there is also mention, but with no details, of a 'EUROPEAN LADY' captured by the rebels. - *See* Kenyon Papers, CSAS. There were claims to have had evidence from one of the two women who survived the massacres (in Kanpur) but who cannot be named as 'it would be hurtful to her family': *'One of the two women who survived the Kanpur Massacre told me that when she was brought before* **Azimullah** *he said to her, 'Why are you crying? The Mughal Emperor has taken Delhi and driven the English from Northern India, when we take Kanpur and Lucknow we will march to Calcutta and be masters of Southern India and your husband (the sowar who captured her) who has been made a colonel will then be a great man and you a great woman.'* - *See* George W Forrest, *History of the Indian Mutiny*, 3 vols, Edinburgh, 1904. The very strange story of the relationship between **Maulvi Liaquat Ali** (of Allahabad fame) and **Mrs Bennett (alias Amy Horne)** depends largely on Mrs Bennett's account. But his trial and sentence to transportation for life is documented. *See* **SSF** Chapter Five, *and* **FSUP,** Vol IV, pp 643 *et seq.* **B** With regard to the possibilities of rape of Indian women by British soldiers and their Indian allies, there is much less documentation but, perversely, possibly more likelihood. The official line is that the British soldier did not make war upon women and children and that far from ill-treating Indian women he would make every effort to protect them; hence the story of the sniper in the tree in the Secunderbagh, *see INDIAN HEROINES*. After the fall of Jhansi, Sir Hugh Rose reported: 'The Commander-in-Chief will learn with pleasure that the troops under my command treated with great humanity the women and children of Jhansi; neither the desperate resistance of the rebels nor the recollections of Jhansi of last year could make them forget that in an English soldier's eyes women and children are spared; so far from hurting, the troops were seen sharing their rations with them. I gave orders also that the destitute women and children of Jhansi should be fed out of the prize grain.' If this sounds almost too good to be true, it is worth remembering that Rose was a most honourable man, and would not knowingly have been party to a deception. On the other hand the murder of wives by their husbands to prevent their dishonouring by the British was commonplace at Jhansi, as elsewhere, and many women undoubtedly committed suicide in Delhi for the same reason. In the light of insufficient evidence perhaps it is fairest to say the verdict in the case against the British soldier is 'not proven'.

RAPE QUESTION 'I asked him if he knew whether there was any truth in the report of the European women having been dishonoured before being murdered. *'Sahib'*, he replied, 'you are a stranger to this country or you would not ask such a question. Any one who knows anything of the customs of this country and the strict rules of caste, knows that all such stories are lies, invented to stir up race-hatred, as if we had not enough of that on both sides already. That the women and children were cruelly murdered I admit, but not one of them was dishonoured; and all the sentences written on the walls of the houses in Kanpur, such as, 'We are at the mercy of savages, who have ravished young and old,' and such like, which have appeared in the Indian papers and been copied from them into the English ones, are malicious forgeries, and were written on the walls after the reoccupation of Kanpur by General Outram's and Havelock's forces. Although I was not there myself, I have spoken with many who were there, and I know that what I tell you is true.' William Forbes-Mitchell. *Reminiscences of the Great Mutiny,* London, 1893, p 113.

RATTRAY, Colonel Thomas
The raiser of Rattray's Sikhs who played a significant part in the struggle. The records of the Rattray family are held in the National Army Museum, Royal Hospital Road, Chelsea, London SW3.

RAVI RIVER   See TRIMMU GHAUT.

RAWAL SHIU SINGH
A former minister of the Durbar of **Jaipur** who attempted to persuade the Jaipur Forces, in June 1857, to espouse the rebel cause, but failed. He failed also to persuade the young Maharaja to desert the British.

RAWAL
A small town north of **Indore** in Rajputana, between **Mehidpur** and **Mandesore [Mandsaur]**, memorable for a small but significant battle on 12 November 1857. The **Malwa Contingent** rebels were en route here for Mandesore; they numbered 500 men with at least two guns which they were intending to take to besiege **Neemuch**. They were overtaken by Major Orr with 337 sabres of the 1st 3rd and 4th Cavalry Hyderabad Contingent. Included in his force was **Captain Abbott**. The action began at 4p.m. and did not end till sunset when the rebels withdrew after a gallant fight, leaving seven guns in the hands of the British together with plentiful stores; they had lost 100 men killed and some 74 were taken prisoner. These prisoners were tried by Court Martial on 18 November and all were shot the same day. Had the guns and stores reached Mandesore and been sent on to Neemuch they would have added greatly to the rebel strength. For further details *see The Revolt in Central India*, Simla, 1908 p79, *and* I.T.Pritchard, *The Mutinies in Rajpootana,* London,1864, *and* Thomas Lowe, *Central India during the Rebellion of 1857-58,* London, 1860., *and* **FSUP**, Vol III.

RAWALPINDI
A large fortified town in the Punjab, forty-seven miles East South-East of Attock; it consisted mainly of flat-roofed earthern houses, with the remains of a palace built by Shah Sujah, a bazar and some mosques; it still possessed a considerable transit trade with Afghanistan. On 7 July 1857 the British decided to disarm the 58th BNI then in cantonments: they were marched on parade where there were already guns and some Europeans from HM's 14th regiment; guessing the intention of the authorities the sepoys, after a moment's hesitation, broke ranks and fled back to their lines, but being immediately surrounded they submitted and handed over their weapons. That they had intended to mutiny appeared certain when it was found that over 200 of their muskets were found loaded with ball cartridge

RAZING TO THE GROUND
'A member of the House of Commons rose... to propose a resolution that the cities of Delhi and Kanpur should be razed to the ground, in order to wipe out every vestige of the tragedies.' He received no support, not even a grudging 'hear hear!' *See* Rev. J.R. Hill. *The Story of the Kanpur Mission.* Westminster, 1909, p 51.

READE. E.A.
Commissioner of Agra, and the political co-ordinator of much of the campaign in Central India in 1858-9.

REBEL FORCES, LUCKNOW
November 1857: The strength of the mutineers at the time of Campbell's advance to relieve the Residency in November 1857 is documented in detail, but as it depends upon a spy's report, it may not be wholly reliable. The totals are given as 'Sepoys 7,950, Oude[Awadh] Regiments 5,600, Cavalry 7,720, *Talooqdars'* men 32,080: grand total available to the Begum Hazrat Mahal: 53,350.' For breakdown of these figures, in particular details of individual *talukdar's* contingents *see Foreign Secret Consultations,* 26 Feb 1858 Nos 226-228.

'LUCKNOW, March 1858: By this time, in expectation of the attack by **Sir Colin Campbell**, the Commander-in-Chief, the forces of the **Begum Hazrat Mahal** had been considerably augmented. Three lines of defence for the city had been constructed and nearly one hundred guns and mortars had been incorporated into the defences. The city contained 60,000 Sepoys of mutinied regiments and at least 50,000 volunteers and *Talooqdars'* men; in addition the population of the city - many of whom would have been armed and firmly anti-British - was estimated to have been in excess of 300,000.' *See* Charles Ball, *History of the Indian Mutiny,* Vol II, p 246.

REED, GENERAL SIR THOMAS He was Major General commanding the **Peshawar** Division on the outbreak of the mutiny/rebellion, but on 6 June was made commander of the Bengal Army pending the arrival of Sir Patrick Grant from Madras. Marched to **Delhi** but had to hand over command to **Archdale Wilson** and proceed on medical certificate. *See* **DNB**.

REGIMENTAL BANDS At Lucknow the rebels enjoyed taunting their foe with band tunes: *The Standard Bearer's March, The Girl I Left Behind Me,* and *See the Conquering Hero Comes,* among the favourites. They are even reported as playing *God Save the Queen* on a number of occasions, and *The Rogue's March* (considered most appropriate by some). The fact is of course that the bandsmen could only play what they were used to playing, and there were apparently no popular rebel songs to fall back on. *See* L.E.R. Rees, *A Personal Narrative etc.,* London, 1858 p. 341. *See also* MUSICIANS.

REGIMENTAL COLOURS The Colours of the Regiment were of great emotional, indeed, religious significance to the sepoys of the Bengal Army. The colours were forever being invoked 'as the emblem of their military honour.' At the time of the annual craft festival regiments worshipped their colours, carried them to Brahmins to be blessed. 'We put our religion in our knapsacks whenever our colours are unfurled,' an old Subadar of the Madras army said in the 1830s. At the heroic but unsuccessful siege of Bhurtpore[Bharatpur] in 1805 the colours of the 31st BNI were so torn by gunfire that it was decided to cremate them and replace them with new silks in the morning. The two standards were left to spend a night together, so that the honour of the old might be transferred to the new. But when it came time to cremate the old standard it could not be found, and would not reappear until 21 years later when the descendants of the sepoys who died at Bhurtpore returned under Lord Combermere to take the fort, and, rushing through a breach in the doomed fort's walls, drew forth pieces of the old standard they had lovingly preserved and tied them to the new colours to honour 'the fruitless valour of their fathers.' *See* BHOWANI SINGH. *See also* Philip Mason, *A Matter of Honour.* Jonathan Cape, London, 1974, pp 129-30.

REGIMENTAL NAMES The British numbered the infantry line regiments of the Bengal Army of the East India Company, but the sepoys who joined those regiments rarely used that number to identify themselves or their unit. Instead they used the *name* of the regiment (a practice which continued even after mutiny: the rebels themselves referring to a sepoy regiment by its - often British - name, rather than number). This name was usually, but not always, the surname or title of the founder or raiser or first commander of the regiment, often a British Colonel of the eighteenth century. The names of which we have record are as follows: 1st BNI: Gillis ka pultan; 2nd:Burdwan; 3rd: Sooltgen; 4th: Bailun; 5th: Grand; 7th: Burra Crawford; 8th: Nya Burdwan; 9th: Jellascar; 10th: Duffel; 11th: Runseet; 12th: Hote; 13th: Gaurud; 14th: Escotton; 15th: Doo; 16th: Hossainee; 17th: Barkur; 18th: Raja; 19th: Ung; 20th: Baillie; 21st: Neelwar; 22nd: Bole; 23rd: Chota Crawford; 24th: Dabie; 25th: Murriam; 26th: Poel; 27th: Mortdeel; 28th: Stupper; 29th: Cullenjuin; 30th: Macdoon; 31st: Broon; 32nd: Guthrie; 33rd Hilliard; 34th Bradshaw; 35th: Noke; 36th: Markum; 38th: Balanteer Tittelee; 39th: Balanteer Burral; 40th: Hamilteen; 41st: Doo bye Kee Dahena; 42nd: Balanteer

Jonsin; 43rd: Kyne ke Daheena; 44th: Kyne ke Bacon; 45th: Murrerro ke Dahena; 46th: Murrerro ke Becan; 47th: Crawin; 48th: Mutees; 49th:Royle; 50th: Christeen; 51st: Duberne; 52nd: Hindree; 53rd: Castor; 54th: Mapert; 55th: Ockterlony ka Dinah; 56th: Lambroon; 57th: Lord Moira; 58th: Bisheshwar; 64th: Harriott; 74th: Alexannder.

REGIMENTS, BRITISH This is an attempt to list the regiments and units of the British army that were engaged in the suppression of the Indian Mutiny/Rebellion:- The 5th Fusiliers; 7th Fusiliers; 8th Foot, King's Regt (Liverpool); 10th Foot; 13th Light Infantry; 19th Foot; 20th Foot (Lucknow); 23rd Fusiliers, Royal Welsh Fusiliers (Lucknow); 24th Foot; 27th Foot; 29th Foot; 32nd (Ist Battalion Duke of Cornwall's Light Infantry); 34th Foot; 35th Royal Sussex; 37th Foot; 38th South Staffs (Lucknow); 42nd Black Watch; 52nd Foot; 53rd K.S.L.I. (Lucknow); 54th Foot; 60th Rifles; 61st Foot; 64th North Staffs; 70th Foot; 73rd Foot; 75th Gordon Highlanders; 77th Foot; 78th Seaforth Highlanders; 79th Q.O.Cameron Highlanders; 80th South Staffs; 81st Foot; 82nd Prince of Wales Volunteers; 84th York and Lancs Regt; 87th Foot; 88th Foot; 90th Foot (Lucknow); 92nd Gordon Highlanders; 93rd Argyll and Sutherland Highlanders; 97th Foot; 98th Foot; Rifle Brigade 2nd Batt. (Lucknow); Rifle Brigade 3rd Batt. (Lucknow); European Bengal Fusiliers = 101st Regt - Royal Munster Fusiliers; 3rd European Regt = 2nd Bn Royal Sussex; The Military Train - RASC; Royal Artillery: Horse Art E Troop Horse Art F Troop (Lucknow) Foot Art (Lucknow only) 5th Battery; 8th Battery; 11th Battery; 12th Battery; 13th Battery. Cavalry: 2nd Dragoon Guards (Lucknow); 6th Dragoon Guards (Meerut); 7th Dragoon Guards (Sialkot); 7th Light Dragoons (Lucknow); 9th Lancers. Return of Casualties : Delhi Field Force May-September 1857, Killed 1,012...Wounded 2,795 (of whom many subsequently died)..Total 3,837. In HM 60th Rifles there were 389 casualties out of a total strength of 402, incl. 113 dead. Figures at the commencement of the Mutiny, Bengal Army : Europeans 22,698. Natives 118,663, Total 141,361. Overall proportion throughout India was 19 Europeans to 100 natives. Queen's troops in India in June 1858:- Infantry battalions...........40; Cavalry....5; +Artillery and Engineers.On 1 Jan 1857 there were about 26,000 Royal troops and 12,000 Company's European troops in India. During the ensuing fifteen months to April 1858 there were sent over 42,000 Royal troops and 5,000 Company's Europeans. This would have given a total of 85,000 British troops in India; but it was estimated that war, sickness and heat had reduced this number to 50,000 available effective men. Havelock's Force on the occasion of the first relief of Lucknow :- 5th Fusiliers; 84th Regt; Detachments 64th Regt and 1st Madras Fusiliers; 78th Regt; 90th Regt; Sikh Regt of Ferozepore; Maude's Battery; Olphert's Battery; Eyre's Battery; Volunteer Cavalry; Irregular Cavalry; 3 Engineer officers. Campbell's Force in Second relief of Lucknow in November 1857: HM's 8th; 53rd; 75th; 95th Regts; 2nd and 4th Punjab Infantry; 9th Lancers; Detachments 1st 2nd and 5th Punjab cavalry; Hodson's Horse; Bengal Sappers and Miners; Naval Brigade; Artillery: Horse and Field Batteries. Sir Hope Grant against Serai Ghat and Bithur: HM's 42nd; 93rd; 53rd; 4th Punjab; Rifles; 9th Lancers; 5th Punjab Cav; Hodson's Horse; Horse Artillery; Field Battery; Sappers. Outram's Force at the final conquest of Lucknow in March 1858, which crossed over to the left side of the Gomti: HM's 23rd and 79th; 2 Battalions Rifle Brigade; 1st Bengal Europeans; 3rd Punjab Inf; 2nd Dragoon guards; 9th Lancers; Detachments 1st 2nd 5th Punjab Cav; D'Aguilar's, Remington's and MacKinnon's troops Horse Artillery; Gibbon's and Middleton's Field Batteries. Lucknow Garrison under Grant: D'Aguilar's troop RHA; Olphert's troop Bengal HA; Gibbon's Field Battery RA; Carton's Fd Bat Bengal Art; 1 company Engineers; 3 companies Punjab and Delhi pioneers; 2nd Dragoon guards; Lahore Lt Horse; 1st Sikh cavalry; Hodson's Horse; 20th, 23rd, 38th, 90th, 97th Regts; 1st Madras Fusiliers; 5th Punjab Infantry. A clergyman in 1858, interested presumably because he was a missionary gave figures of mutineers and their associates as follows:- 60 regiments of infantry; six Light Cavalry; 10 Irregular Cavalry; 6 batteries Artillery; 9 Light Field Batteries; Malwah contingent; Gwalior Contingent = 7 Infantry regiments, 4 companies artillery, 2 regiments cavalry. Plus, he argues, 20 regiments disarmed, including the Bodyguard of the Governor General; plus also 20,000 released convicts; plus 100,000 of what he calls 'budmashes'. 'Deaths from The Indian Mutiny - the last phase '   '...temporary paralysis of British power in 1857 meant untold misery to thousands of Englishmen and, alas to a great company of Englishwomen and children: but it meant untold misery to a *FAR GREATER NUMBER OF INDIANS '.*

REH   Saline efflorescence in canals. *See* CANALS.

REHLI   A small town with a strong fort, some twenty-five miles southeast of **Saugur** in Central India. Occupied by the rebels until early October 1857 when it was recovered by one Girdhari Naik who organised the police of the place and became *killadar* until the arrival of Lieutenant Dickens at the head of 100 (loyal to Government) sepoys of the 31st BNI. The fort was then attacked by rebels and mutineers from **Garhakota** a strong place a few miles to the northeast, but Dickens and his troops and the police under Girdhari Naik beat them off with some loss, and remained in Rehli until the end of the campaign when **Sir Hugh Rose** marched through on his way to Garharkota, *see* **FSUP,** Vol III, and *Revolt in Central India,* Simla, 1908.

REILLY, or RILEY   Mr. Reilly, the Assistant Commissary of Ordnance in Kanpur, was given standing orders, by Major General **Sir Hugh Wheeler** to blow up the Magazine as soon as the rebellion began. But on his tours two armed sepoys followed him everywhere and the rest of the guard watched him so closely that he had no chance to set his charges, or 'take the least step towards blowing up the powder.'

'Who'll Serve the Queen?'

REINFORCEMENTS from ENGLAND In all a total of 1,601 Officers and 31,565 men were despatched to India specifically to put down the mutiny/rebellion. Considering the difficulty of communication and transport at the time, this was done comparatively speedily *eg* 840 Officers and 20,566 men had arrived in India by the end of November 1857. The reinforcements included the 1st 2nd 3rd and 7th Dragoon Guards, the 7th Hussars, the 8th and 17th Light Dragoons; the infantry included the 7th 18th 19th 20th 34th 38th 42nd 44th 51st 54th 56th 3rd/60th 66th 68th 69th 71st 72nd 79th 88th 97th 2nd/Rifle Brigade 3rd/Rifle Brigade 92nd 94th 98th regiments. the artillery included four troops Royal Horse Artillery, Seven Field Batteries R.A., Seventeen Companies R.A.; and Five companies of Royal Engineers. *See* F.R. Sedgwick, *The Indian Mutiny etc.*, London 1909.

REINFORCEMENTS from elsewhere The British gathered troops from wherever they could to get to India to deal with the outbreak. From the China Expeditionary Force, the 23rd 82nd 90th and 93rd Regiments were diverted; from Mauritius a Wing (approximately half the regiment) of the 4th 5th and 33rd regiments were despatched; from Ceylon a Wing of the 37th; and from Cape Colony the 13th 31st 2nd/60th 73rd 80th 89th and 95th regiments were sent. *See* F.R.Sedgwick, *The Indian Mutiny* etc., London 1909.

RESIDENCY *See* LUCKNOW, RESIDENCY.

RETRIBUTION A writer calling himself *Bona Ex Malis* suggested that the British in India were eventually going to have to stop killing captive mutineers and devise an alternative sentence for the rebel multitudes. The answer, he suggested, was transportation for life to the West Indies. 'Transportation to the high-caste East Indian is, in anticipation, as dreadful a punishment as death. ...In the British West India colonies, far from the scenes of their horrible crimes, these misguided men, dangerous in their own country, would in due course of time become useful members of the community' - and their importation would conveniently address the severe manpower shortage on British sugar plantations that had resulted from 'ultrahumane restrictions' and 'anti-slavery suspicion,' both of which would naturally be waived in the case of 'these miscreants.' *See* London *Times,* 24 September 1857, p 8.

REVELRIES 'I asked a native gentleman one day if he ever heard that our servants complained of us, or laughed at us, or tried to enter into the spirit of our revelries, and he made answer and said 'Does the sahib see those monkeys? They are playing very pleasantly. But the sahib can not say why they play, nor what they are going to do next. Well then, our poor people look upon you very much as they would on those monkeys, but that they know that you are fierce and strong, and would be angry if you were laughed at. They are afraid to laugh. But they do regard you as some great powerful creatures sent to plague them, of whose motives and actions they can comprehend nothing whatsoever.' W.H. Russell, *My Diary in India.* London, 1860.

REVENUE RECORDS In many areas the outbreak of rebellion went hand in hand with arson, and a prime objective was the destruction of *kacharis, ie* the courts and offices of local government. The loss of the collectors' records was disastrous, for they 'showed the demands, collections, and balances on account of the Government revenue for the last 30 years, and were invaluable as guidance for the remissions of the revenue when such were necessary, or for the re-settlement of certain distriicts after their term had expired....The whole of the revenue system of the North-West hinged upon their preservation and accuracy.' That 'all of these, with but few exceptions, have perished in the present rebellion' was even worse than 'if the title and deeds and leases of every landed proprietor in England were to be suddenly destroyed.' The *Friend of India* quoted in the London *Times.* 29 October, 1857, p 3.

REWA A town in Central India where the Maharaja, Raghuraj Singh appears to have played a double game and to have been in league with the rebels while ostensibly standing firm with government. *See* Kameshwar Jha, *A Study of some Mutiny Letters of Sohagpur*, Indian Historical Records Commission Proceedings, 1958. The maharaja was related by marriage to **Kunwar Singh** but he did not allow the latter to enter Rewa and rather than provoke a battle with his relative Kunwar Singh withdrew. The firmness of Lieutenant Willoughby Osborne, the Political Agent at Rewa hardened the maharaja's resolve and he determined to stand by the British. The story is a complicated one and is best looked for in the pages of **FSUP,** Vol III.

REWARD 'It is hereby notified that a reward of One Lac of Rupees will be paid to any Person who shall deliver alive at any British Military Post or Camp the Rebel **Nana Dhoondo Punt** of **Bithoor** commonly called **Nana Sahib**. It is further notified that in addition to this reward a Free Pardon will be given to any Mutineer deserter or rebel who may so deliver up the said Nana Dhoondo Punt. By order of the Right Honourable the Governor General.' Signed by Edmonstone, Secretary to the Government of India. NA. FDP. Despatch to Secretary of State, 16 July, 1859. It is said that Nana Sahib was gratified by the size of the reward offered, believing himself to be well worth it. He must have been even more gratified by the fact that so far as is known no one even attempted to betray him or claim the reward. The offer of a pardon to whoever gave up the Nana Sahib was withdrawn from the likes of the Nawabs of Farrukhabad and Banda, Tej Singh of Mainpuri, Khan Bahadur Khan of Bareilly, and Walidad Khan. *See* **FSUP,** Vol V, p 433.

RICKARDS PAPERS These are held in the Centre of South Asian Studies at Cambridge University. For the most part they consist of letters from two British officers spanning the period January 1858 to May 1860, and although there is interesting material in these first hand accounts of events in the **Ajmer, Lucknow, Udaipur, Partabghur, Gwalior** areas, little of it is new. Perhaps the most significant letter is the one dated 30 May 1860 which is in fact written by a sepoy at **Danapur**, and which describes his part, and that of Havildar Prang Singh, in the mutiny of the 6th B.N.I. at Allahabad.

RICKETTS, George Henry He was Deputy Commissioner at **Ludhiana**. Some of his correspondence is preserved in the Oriental and India Office collections of the British Library, *see MSS.Eur.D.610.* He was remarkable for an abortive but courageous attempt to cut off the mutineers from **Jullundur** on their way to **Phillour** and **Delhi**. *See* **Ball** and **Martin.**

RIDGE, The This usually refers at this time to the low ridge that dominates **Delhi** to the north west of the old city, and where the British Delhi Field Force prepared for the assault on Delhi on 14 September 1857. *See* DELHI LANDMARKS etc. The campus of Delhi University is now located there.

RIPON PAPERS The second series is concerned with material, some unique, relating to the mutiny/rebellion. They come from the estate of the 1st Marquis of Ripon (George Frederick Samuel Robinson 1827-1909) and are in the British Library in London. Included is correspondence with **Sir Bartle Frere** 1858-83; a letter from **Wheeler** to **Henry Lawrence**, 4 June 1857 (the day the mutiny broke out in Kanpur); Diaries and accounts from various eye witnesses; papers relating to General **Hearsey** commanding at Barrackpore; a letter from Major **G.Ramsay** British Resident in Nepal concerning the death of **Major James Holmes** at **Sagauli** and two accounts of experiences during the rebellion, by **Miss Amy Haines** and **Miss Sutherland.**, dated 1890. *See* Ripon Papers, British Library.

RISALA, or Rissalah A troop of horesmen - in the army.

RISSALDAR An officer of an Indian Cavalry regiment. Equivalent to **Subedar** in the Infantry. The term was apparently first used in the Irregular cavalry units and then applied to regular regiments.

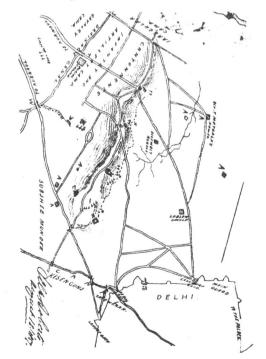

Delhi August 1857: a Sketch by Lt.Roberts

ROBERTS, Field-Marshal Frederick In 1857 Roberts was a young and very ambitious Artillery subaltern attached to the Quarter Master General's Department. He wrote many, revealing, letters at this time to his family, eg 'Camp near Lucknow. February 26th 1858. My Dearest Harriet,.....I went into the village last evening to see what damage was being done to the defences. Nearly all the houses had been burnt, except in one corner. As I approached, a very old man met me and said 'here are two cows; I have still another in the house, which I will bring out to you, but for God's sake don't burn the only property I have in the world. Yesterday morning I had five sons. See, here are three of them, the other two fled away, and I don't know whether they have shared the same fate and are dead, or whether they may have escaped. None of us ever bore arms against your government. We are all labourers. Ever since the rebellion took place, I have prayed for your success, and if all my sons are killed, I shall still pray for you, for I know under any other rule we have nothing but oppression and tyranny to expect.' ....I had not the heart to burn his house. When a prisoner is brought in I am the first to call out to have him hanged, but it does make one melancholy to come across accidents (sic!) such as I have related. Your very affect. brother, Fred Roberts'. *See F-M Lord Roberts, Letters written during the Indian Mutiny, 1924. See also Lord Roberts, Forty-one Years in India, 2 vols, London, 1897.*

ROBERTSON   He was the European Sessions Judge who, with Dr Hay and (possibly) Mr Raikes, was hanged at **Bareilly** on 31 May 1857 by **Khan Bahadur Khan**, although, at his own trial, the latter denied being concerned in their deaths. For details see **FSUP**, Vol V esp. p 355 and 609.

ROBERTSON, H. DUNDAS   He was the District Magistrate of Saharunpur and was a shrewd and brave man, and his opinions were typical of his highly influential class; the following comes from *District Duties during the Revolt,* London, 1859 :- ' At the Hurdwar Fair in May 1857 I was probably the only one in the place actually hard at work, police and magisterial duties being combined with that of allotting encampment grounds to the different Seikh and Central Indian nobles, with their large armed bands, the disarming of whose retainers before permitting an entrance into the fair, though absolutely necessary for the maintenance of order, is always a sore subject with these rough customers, whose native pride or sense of dignity induces a morbid sensibility invariably on the look out for insults.' 'During this period of uncertainty when speaking with would-be well-disposed natives, who, it was easy to observe, visited me more with the view of extracting than of furnishing information, *I was much struck with their evident satisfaction in the generally unfavourable nature of the news, and with the promise of misfortune to the English.'*... Robertson's s courage was recognised and respected by the local people - and there are no people in the world more generous in their respect for a brave man than are the people of India. He had friends in all classes, but he was - as was appropriate at that time - a snob, who believed that idealism was the exclusive privilege of the educated and well-heeled classes, and it is not surprising therefore that he relates an anecdote concerned with his friendship for the representatative of such a class. 'A native gentleman, who was fond of shooting, and had become as intimate with me as it is prudent in a civil officer in India to permit with the ever-crafty Hindustani. He was manly, loved manly sports...and was gifted neither with that cringing flattery, nor insolent arrogance... manifested by this class in the North-West. He understood how to meet you as a gentleman, *and the notorious Nana's elder brother was another of the few exceptions I have met with.* This gentleman's name was Oosman Khan, a resident of Belaspore, some fifteen miles from Saharunpore. From the commencement of the mutiny he constantly visited me...there was a mutual kindliness of feeling between us, and I was glad to see him. *At first, before perceiving that the mutiny was actually a rebellion, in which he might improve his position, his visits did, I think, originate honestly, and his advice was certainly good.* Day after day did he reiterate that the only policy which could now in our position possibly succeed was severity in the extreme; that death must now become the only punishment for the slightest opposition, and that any deviation from this course would insure the loss of the district. But when the storm was closing thicker round us, I noticed that he visited me in company with Deedar Singh, the head of the Bunjarahs. (A man powerful enough amongst his clan, of cow-herds, but totally unable to cope with Oosman Khan). Mahomedans and Bunjarahs are by nature deadly enemies, and it was easy to perceive that the wily old Mahomedan, whatever his personal attachments might be, was playing a deep game....the two men together could if necessary muster the whole population of the northern portion of the district....when to all it appeared very doubtful if we could hold out much longer Oosman Khan pressed me to take shelter in his fort...it was kindly meant...Deedar Singh, also a friend of mine, could also have protected me if necessary....but when Delhi had fallen (to the British) and no doubt remained in the Indian mind as to which side would be victorious, Oosman Khan and Deedar Singh ceased to be seen together; the natural animosities of a hostile race and creed were left to play in their full vigour.'

ROGUE'S MARCH   'On July 2nd (1857) the Bareilly mutineers marched into Delhi with their bands playing a tune supposed to be the 'ROGUES MARCH'- we thought it most appropriate.' . This comment by a British officer is rather unfair: the bands of the sepoy regiments of the Bengal Army could hardly be expected to learn new tunes overnight. But, with more justification for mirth there is another report that on solemn occasions the rebel bands did indeed play 'God Save the Queen'.

ROHILKHAND

This land lies between Oude[Awadh] and the Ganges, which river separates it from the Doab. Its headquarters was at **Bareilly**. There was much excitement among the large Muslim population in mid-May 1857 when they learned that Delhi was in the hands of the rebels and that the Mughal Emperor had been restored to his throne. Hence the area was unusual in that it appears that rebellion among the civil population actually *preceded* mutiny of the troops. Of the latter only the 66th BNI (Gurkhas) did not eventually mutiny. The military stations of Rohilkhand were as follows: **Bareilly**, 18th and 68th BNI, and the 8th Irreg.Cavalry, the 6th Company of Bengal Native Artillery; **Moradabad**, 29th BNI, some Foot Artillery; **Saharunpur**, small detachment of BNI; **Shahjahanpur**, 28th BNI, some Foot Artillery; **Budaon**, small detachment of BNI. Brigadier Sibbald commanded all troops and had responsibility also for the Kumaon district where at **Almora** the 66th BNI was stationed together with a detachment of the Sirmoor Battalion and a company of Artillery. Kumaon and Rohilkhand were included in the **Meerut** Division. See Wm Edwards, *Personal Adventures during the Indian Rebellion in Rohilcund* etc, London, 1858 (reprint Allahabad, 1974), *see also* **Ball, Martin, Kaye, Sen** etc. For an Indian (contemporary) account of the struggle in Rohilkhand, See Hafeez Malik and Morris Dembo, transl. *Sir Sayyid Ahmad Khan's History of the Bijnor Rebellion,* East Lansing, 1972

Rohilla

ROHILLAS

A people settled to the east of the Doab of the Ganges. They were originally, as the name implies, from Afghanistan, and in 1857 inhabited the districts of Bijnour, Moradabad, Bareilly and Rampur.

ROHNI

In Bihar. For details of the mutiny of the 5th Bengal Irregular Cavalry, despite its earlier loyalty to Government, *see IRREGULAR CAVALRY.*

ROHTAK

The battle of **Badli-ke-Serai** ensured that the British had an unobstructed route from the Punjab down the **Grand Trunk Road**, from Lahore via Rohtak and Alipur. Rohtak itself was a military station which, by virtue of the passage through it of the **Punjab Movable Column** and other British reinforcements down to Delhi gave no trouble to government.

ROMANCE

'There is even romance in war, but these papers have very purposely omitted as much as possible mention of woman. At Lucknow my Sikh soldiers offered to place a young and most exquisite widow, a daughter of the Royal House of Lucknow before me. ...She had herself requested my acceptance of herself, for she was utterly distracted, her young husband a rebel nawab had fallen in battle, and her servants were deserting, prize agents and plunderers of all kinds were insulting her and she entreated that I would protect her: she had her female slaves, £10,000 in cash, horses, camels, elephants and all she flung at my feet, all for peace and safety once more. ...I smiled and said by all means, I'll take care of the poor girl and will see tomorrow when she comes what your great beauty is like. That very night at an hour's notice, I had to march from Lucknow to more and more battlefields. This is a horrid abrupt end *n'est ce pas,* fair reader!' *See* George Godfrey Pearse, *Papers,* p 32.

ROONEY, Rev. JOSEPH

The Reverend Joseph Rooney of the Roman Catholic Chapel, in Kanpur, fared better than most in **Wheeler's Entrenchment**. 'The Romish priest was the only well-fed man in our party, for the Irish privates used to contribute from their scanty rations for his support: he died about the middle of the siege from sun-stroke or apoplexy.' Mowbray Thomson, *op. cit.*

ROORKEE

The headquarters of the Bengal Sappers and Miners, and the site of the Government Engineering College, *see* MACLAGAN PAPERS. On 24 May the Queen's Birthday was usually celebrated in India by the firing of an artillery (blank) salute of 21 guns. In some stations in 1857 this was dispensed with (notably **Kanpur** on **Sir Hugh Wheeler**'s orders) as potentially provocative for troops who had not yet mutinied, but in Roorkee the salute was fired deliberately 'to emphasise strength'. On 22 May the 'Roorkee Garrison Gazette' was started as a means of distributing genuine information to take the place of rumour.

ROSE, Major General Sir Hugh (1801-85).

Later (1866) created Baron Strathnairn of Strathnairn and Jhansi, and Field Marshal (1877). He was an officer with a distinguished record both as a soldier and statesman. Entering the army in 1820, he had fought in Syria twenty years' later, when he was severely wounded and captured with his own hand the Pasha of Egypt. He had succeeded Lord Stratford de Redcliffe at Constantinople prior to the outbreak of the Crimean War and played a distinguished part in the subsequent campaign. Brave, resourceful and possessed of experience in war, he was well-fitted to lead an army. It was the misfortune of the **Rani of Jhansi** (whom he much admired) that she had so able an adversary. See **DNB**. *See also* Sir George William Forrest, ed., *Selections from the Letters, Dispatches and other State Papers of the Military Department of the Government of India 1857-58,* 4 vols, Calcutta, 1893-1912, Vol IV, *and* Tapti Roy, *The Politics of a Popular Uprising: Bundelkhand in 1857,* Delhi, 1994.

Sir Hugh Rose

ROUGH JUSTICE   The elder brother of **Nana Sahib**, **Baba Bhutt**, was made Chief Commissioner of **Kanpur** by his brother, and held his court in the house of a former European resident; he chose to sit upon a billiard table when dispensing justice. Some of his sentences were both ferocious and capricious, as this example shows: 'And now came the drummer Mendes' turn, otherwise called Yaqeen Mohamed. His accuser was not present, and he, of course, denied the charge. The Baba said, 'I saw you with my own eyes taking away bales of cotton.' Now, poor Mendes had never seen such a thing since the mutiny, and answered accordingly. The reply was, '*All short men are wicked; this fellow is very short, and therefore very wicked. Give him six months' jail with fetters.*' Mendes made a salaam and came out, glad to find that nobody taxed him about being a Christian.' *See* W.J.Shepherd, *A Personal Narrative etc.*, Lucknow, 1886.

ROUNDSHOT (or Cannon-balls)   These were fearful, state-of-the-art, projectiles in 1857, but they were usually only effective when used against defences, walls, or large inanimate objects like enemy guns or emplacements; they were not regarded specifically as being effective anti-personnel weapons. Yet of course if they did hit a person, the blow would be horrendous and death would usually result. In **Wheeler's Entrenchment** at **Kanpur** a remarkably high proportion of the officers of the 1st BNI was killed in this way. Lieutenant-Colonel John Ewart, Captain Edward John Elms, Captain Athill Turner. Surgeon Arthur Wellesley Robert Newenham, Lieutenant F. Redman, Richard Murcott Satchwell, Henry Sidney Smith, and G.R. Wheeler, and Ensigns George Lindsay and J.C. Supple were the officers of the regiment. *Redman, Smith, Supple and Wheeler were all beheaded by roundshot* during the siege.

ROYAH or ROYA   Sometimes also Ruya. This was the site of a jungle fort held by a talukdar named Narpat (or Nirpat) Singh, a follower of **Beni Madho** in Oude[Awadh]. It had a two-fold significance: as the place where a very popular British Officer, **Brigadier Adrian Hope**, was killed; as the site of a battle at which British forces were temporarily repulsed. (Although the rebels evacuated the fort the next day and fled) The exaggerated account of this check on the British forces spread round Oude[Awadh] among the talukdars like wildfire and was undoubtedly responsible for a hardening of the resolve of many of them to fight on against the invaders, even though Lucknow had fallen and the cause had begun to look hopeless. The General responsible for the debâcle was **Walpole**, and, although the Commander-in-Chief **Sir Colin Campbell** remained supportive, he was very unpopular, especially with the Highland troops for whom Adrian Hope had been a great hero and inspiration. The fort was finally captured on 26 October 1858 when Narpat Singh had abandoned it and fled northwards with his guns and 500 men. For details of the attack on Royah *See* James Innes, *The Sepoy Revolt*, London,1897, p 247 *et seq*. *See also* Wm Forbes-Mitchell, *Reminiscences of the Great Mutiny*, London, 1910, p 243 *et seq*.

ROYAL ARTILLERY INSTITUTION   This is situated at the Old Royal Military Academy at Woolwich in London, and has a number of mutiny papers and diaries of interest particularly for the artillery handling. Details are set out in J.M.Ladendorf, *The Revolt in India 1857-58*, Zug, 1966 entry 976. and in J.D.Pearson, *Guide to the Manuscripts and Documents in the British Isles relating to South and South East Asia*, Vol I, London, 1989, p 250 *et seq*. Of perhaps greater significance is the collection of photographs, known as the 'Maude Album' (Album No.45 in the **Royal Artillery Institution** Library). The photographs date from 1858, some possibly from 1857, and many are clearly the work of **Felice Beato**, and are familiar to students of the mutiny/rebellion, but what makes this collection unique are the notes that accompany each photograph, from which it is clear that **Major F.C.Maude** was himself present when the photographer was at work; he also identifies figures *eg* **Mowbray Thomson,** in a number of prints: a number of these photographs are reproduced in this volume by kind permission of the Royal Artillery Historical Trust.

ROYAL UNITED SERVICES INSTITUTION, LONDON   There is a small collection of useful material here *eg* the letter from **Captain Patrick Orr** (one of the **'Captives of Oude[Awadh]'** who did not survive) dated 8 June 1857, and a sketch of **Wheeler's Int-**

renchment at Kanpur made by **Lt Moorsom**, 52nd Light Infantry and a letter from him to **Havelock** dated 18 August 1857. For a complete list of the collection *see* J.M. Ladendorf, *The Revolt in India 1857-58*, Zug, 1966 entry 978.

RUBY 'The most portable of his riches the **Nana** carried with him in his flight; the natives say that immediately before the insurrection at **Meerut** he sold out seventy lacs of government paper (£700,000.). One ruby of great size and brilliancy he is alleged to have sold recently for ten thousand rupees to a native banker; the tradition is that he carried this gem continually about his person, intending, should he be driven to extremities, to destroy himself by swallowing it; a curious mode of suicide, the efficacy of which I am not prepared either to dispute or to defend; my informant told me that the sharp edges of the ruby would cut through the vitals, and speedily destroy life'. Mowbray Thomson, *The Story of Cawnpore*, London, 1859, pp 49-50.

RUGBUR SINGH reported as 'Collector Rugbur Singh (Raghubar Singh), formerly a Khasburdar or orderly of the Eunuch Deanut-ood-Dowlah on a salary of 4 Rs *per mensem,* resident of Baiswara, a noted dacoit and rebel;' *See* PERSONS OF NOTE, APRIL 1859.

RUMOURS 1 Rumours abounded and were spread by panic and caused panic. The deposition of Mohur Singh, Deputy Collector, Meerut is typical: 'In the months of January and February 1857 it was reported that cakes had been distributed through the chowkeedars the reason for which was not known, but it was stated that they had come from the East, and that if not distributed severe punishment would be inflicted (some thought, by Government!). Previous to the outbreak rumours to the following effect very generally prevailed: 1st. That 2,000 sets of irons were being made for the sepoys. 2nd That by order of government, atta mixed with bones was to be sold. 3rd That the sepoys were to be deprived of the charge of their arms and ammunition. These reports caused the disturbance.' *See* Major G.W.Williams, *Depositions taken at Meerut*.

RUMOURS 2, 'A soubahdar of one of the Oude[Awadh] batteries informed Sir Henry Lawrence that it was believed to be the intention of Government to transport to India the numerous widows whose husbands had perished in the Crimean campaign. The principal zemindars of the country were to be compelled to marry them; and their children, who would, of course, not be Hindoos, were to be declared the heirs to the estates, so that the Hindoo proprietors of land were to be thus supplanted. ...In respect to the public gaols a perfect mania prevailed, so that burglars and murderers who could distinguish themselves as teachers were passed on from gaol to gaol to act as instructors. Reading, writing, and arithmetic were required, and sometimes geography and the planetary system were taught, as, we infer, to vagabonds and dacoits by profession. Why were we doing all this? Surely not without some hidden purpose. And the Brahmins fostered this suspicion, for they saw, in the enlightenment of the people, the certain downfall of their faith and power. ...It was whispered, and extensively believed, that the object of our Government was to destroy the religion of the Hindoos and to convert them to our own.' *London Times*. September 7, 1858, p 10.

RUMOURS 3, 'The populace (of **Fatehgarh**) embraced **Azimullah**'s theory that England was no bigger than **Farrukhabad** district, that most of its fighting population was in India, and that the British intended to corrupt their caste by issuing leather rupees silvered over to represent the ordinary coinage.'

RUMOURS 4, For another example of the kind of rumour that was current among the sepoys *See* **KHODA BUX**'s deposition.

RUN SINGH A Brigadier with the Nepalese force of **Maharaja Jung Bahadur** which entered Oude[Awadh] in February 1858 in support of the British. 'A smart and intelligent officer who led his brigade with great coolness and spirit' (**Captain Plowden**).

RUNDHIR SINGH, Raja, see Kapurthala. Raja of.

Maharaja Ranjit Singh

RUNJIT SINGH, MAHARAJA The last Sikh ruler of the Punjab, 'Lion of the Punjab'. On his death internal feuding and the ambitions of the British precipitated the two 'Sikh Wars' which ended in the annexation of the Punjab by the HEIC. Many of the sepoys who mutinied in 1857 had been employed by the Company against the Sikhs (and had conducted themselves with outstanding bravery and fortitude), and this may in part have accounted for the Sikhs great hatred for the **poorbeahs** and their willingness to help the British against the mutineers. Without Sikh help most commentators believe the British cause in North India was doomed, certainly in the first instance.

RURAL UNREST In many areas there was immense sympathy among the peasantry for the fate of their former feudal landholders dispossessed by a process inexplicable to the tenantry; this dispossession had come about in a number of ways eg reassessment of land settlements as in **Oude[Awadh]**, intervention by Government in cases of debt, as with **Kunwar Singh** in Bihar, or outright sale of land as a result of action in the British courts which might well have involved chicanery or fraud. 'To the large number of these sales during the past twelve or fifteen years, and the operation of our revenue system, which has had the result of destroying the gentry of the country and breaking up the village communities, I attribute solely the disorganization of this and the neighbouring districts in these provinces.' 'By fraud or chicanery, a vast number of the estates of families of rank and influence have been alienated, either wholly or in part, and have been purchased by new men, chiefly traders or Government officials, without character or influence over their tenantry. These men, in a vast majority of instances, were also absentees, fearing or disliking to reside on their purchases, where they were looked upon as interlopers and unwelcome intruders. The ancient proprietary of these alienated estates were again living as tenantry on the lands once theirs; by no means reconciled to their change of position, but maintaining their hereditary hold as strong as ever over the sympathies and affections of the agricultural body, who were ready and willing to join their feudal superiors in any attempt to recover their lost position and regain possession of their estates. The ancient landed proprietary body of the **Budaon** district were thus still in existence, but in the position of tenants, not proprietors. None of the men who had succeeded them as landowners were possessed of sufficient influence or power to give me any aid in maintaining the public tranquillity. On the contrary, the very first people who came in to me, imploring aid, were this new proprietary body, to whom I had a right to look for vigorous and efficient efforts in the maintenance of order. On the other hand, those who really could control the vast masses of the rural population were interested in bringing about a state of disturbance and general anarchy.' 'For more than a year previous to the outbreak, I had been publicly representing to superior authority the great abuse of the power of the civil courts, and the reckless manner in which they decreed the sale of rights and interests connected with the soil, in satisfaction of petty debts, and the dangerous dislocation of society which was in consequence being produced. I then pointed out that although the old families were being displaced fast, we could not destroy the memory of the past, or dissolve the ancient connection between them and their people; and I said distinctly, that in event of any insurrection occurring, we should find this great and influential body, through whom we can alone hope to control and keep under the millions forming the rural classes, ranged against us on the side of the enemy, with their hereditary retainers and followers rallying around them, in spite of our attempts to separate their interests. My warnings were unheeded; and I was treated as an alarmist, who, having hitherto only served in the political department of the State, and being totally inexperienced in revenue matters, could give no sound opinion on the subject. Little did I think at the time, that my fears and forebodings were so soon to be realized.' 'The leaders and promoters of the great rebellion, whoever they may have been, knew well the inflammable condition, from these causes, of the rural society in the North-Western Provinces, and they therefore sent among them the chupaties, as a kind of fiery cross, to call them to action;' see William Edwards, *Personal Adventures During the Rebellion,* London, 1858, pp 12-16. 'The main cause of rural rebellion was not the hatred of our rule but a summary adjustment by the sword of those feuds which had arisen out of the action of our Civil and Revenue laws on proprietary rights in land.- Henry Vansittart, Special Commissioner, to **Wm Muir**, 28 August 1858. But the swingeing revenue assessment was also often at the root of things. There would appear nowadays to be a concensus that the rebellion was much more than a sepoy mutiny, but much less than a national revolt. The peasantry, certainly the bulk of them, gave their support to the rebellion: there is no doubt they were deeply conservative, cherished the old ways, and were deeply distrustful of all English innovation, which, paradixically, had mostly been introduced with their welfare in mind. The country people not only joined the sepoys they occasionally even anticipated them in open revolt. 'In **Saharunpur**, in Mathura, throughout the Doab and Rohilkhand, and even into the Shahabad District of Bihar, the Mutiny assumed a distinct popular character.' *See* Thomas R.Metcalf, *The Aftermath of Revolt,* India 1857-1870, Princeton, 1965, esp. Chapter II p 46 *et seq.,* - The Mutiny and its Causes. 'The rebellion in the countryside therefore although involving extensive transfer of property was not a *jacquerie* directed against the propertied classes, but a repudiation of the British system and a return to the status quo ante,' *ibid.* 'But when British troops did appear the peasantry thoroughly sick of anarchy settled down to their peaceful avocations as if nothing had ever taken place,' - H.Dundas Robertson, *District Duties during the Revolt in the NW Provinces,* London, 1859, p.161. 'Even those among the villagers who had no property to regain welcomed the coming of the Mutiny, for it gave them an opportunity to prosecute feuds and animosities long held in check by the strong arm of the British Government. The Muslims of Rohilkhand saw in it a chance to regain their lost supremacy; the Rajputs set out to renew the martial exploits of their forefathers, while every man with a grievance to redress or an enemy to injure put the unexpected period of anarchy to good use.' See Thomas R.Metcalf, *The Aftermath of Revolt, India 1857-1870*, Princeton, 1965, esp. Chapter II .*See also* CIVIL REBELLION, and PEASANTS' REVOLT.

William Howard Russell

RUSSELL, SIR WILLIAM HOWARD

Journalist, correspondent of the London 'Times'. Already famous for his reporting of the Crimean War 1854-56, he came to India in January 1858 to cover the campaign of the Commander-in-Chief **Sir Colin Campbell** in suppressing the rebellion. Sir Colin treated him with great generosity, and made him privy to all his plans, on the condition - which Russell scrupulously observed - that he would not communicate those plans to anyone in the camp, even though he was free to write what he liked in his letters back to the 'Times'. A remarkable man, he was not only an excellent journalist he was a most shrewd observer, and was objective and fair in his observations. Russell was the 'Times' correspondent who had just returned to England with enormous prestige from covering the Crimean War as one of the world's first war correspondents. His diary is concerned with the events of 1858 not 1857, *ie* the military campaign in suppression of the revolt rather than the mutiny itself. His liberal attitude leads him to find many reasons for the Indian rejection of British rule. A very shrewd observer, his comments are invaluable to the student of the rebellion. *See 'My Diary in India, 1858-9'*, W.H.Russell, 1860. *see also* J.B.Atkins, *Life of Sir William H Russell*, 2 vols London 1911.

RUSSELL, W.H. '...I found a strongly-built man, of middle stature, with a bright eye and a merry smile, speaking with a slight Irish accent, and dressed in a frogged and braided frock coat. This was Russell, and he promised to come to dinner, and we had a most merry evening, for, in addition to other accomplishments, he sang very charmingly in a social way, and gave us, 'We will catch the whale, brave boys!' and, 'O lave us a lock of your hair!' in splendid style, the choruses being organised with great effect.' John Walter Sherer, *Daily Life During the Indian Mutiny:* London, 1898, p 152.

S

SADDLE OF NANA SAHIB The Peshwa had a magnificent saddle with crimson velour and caparisons of gold. MacNabb Family, *Papers*; *see also* LEWES.

SADHU, The At Lucknow a Captain Mayne dissuaded his men from harming a Hindu holy man they had chanced upon, seated on a leopard's skin and smeared with ashes. 'Oh don't touch him,' the Captain told his troops. 'These fellows are harmless Hindu *yogis* and won't hurt us. It is the Muhammadans that are to blame for the horrors of this Mutiny,' whereupon the sadhu pulled forth a small blunderbuss and shot the Captain in the chest. 'From that moment,' wrote Forbes-Mitchell, 'I formed the opinion that the pampered high-caste Hindu sepoys had far more to do with the Mutiny and the cowardly murders of women and children, than the Muhammadans, although the latter still bore most of the blame.' (It was important to the British that it be one or the other who was to blame, and to discount all evidence of Muslims and Hindus uniting); *see* Wm. Forbes-Mitchell, *Reminiscences of the Great Mutiny*, London, 1893, p 39.

SAGAULI, or Segowlie A place near Patna which was the scene of the murder of **Major James G.Holmes** and others. Papers concerning this massacre, together with other interesting and unique accounts are to be found in the **Ripon Papers** in the British Library.

SAHAO, ACTION at The campaign in **Bundelkhand** was fiercely fought long after the rebellion had died down elsewhere. On 5 September 1858 Brigadier J. MacDuff commanding the 2nd Brigade, about 1,000 strong, attacked a strongly posted rebel force amounting to some 3,500 men of whom 500 were sepoys: the rebels appeared determined to dispute fiercely the ground they held; the sepoys of their right wing retired with steadiness and in close order, but their left wing retreated in great disorder and this gave the British the opportunity to charge in the flank and gain the victory. In all twenty-one rebels were captured and some 250 killed; the British loss was seventeen wounded included Lieutenant Dick, severely. *See The Revolt in Central India*, Simla, 1908 p.184. *Also* Sir George William Forrest, ed., *Selections from the Letters, Dispatches and other State Papers of the Military Department of the Government of India 1857-58*, 4 vols, Calcutta, 1893-1912, Vol IV, *and* Tapti Roy, *The Politics of a Popular Uprising: Bundelkhand in 1857*, Delhi, 1994.

SAHARANPUR DISTRICT For rebellion in this district *see* Eric Stokes, *The Peasant Armed*, Oxford, 1986, pp 199-213, and also *The Peasant & the Raj, Studies in Agrarian Society and Peasant rebellion in Colonial India*, Cambridge 1978.

SAHARANPUR 'On reaching Saharunpore we found that two squadrons of the 4th Native Lancers, with several companies of the 5th Native Infantry had marched in from Umballah[Ambala]. The three European officers who accompanied this force proved our chief reinforcement, as the men under their command had already shown that they could not be trusted. On the whole the fresh arrivals did not add much to our sense of security; yet those who took an interest in the safety of the district were glad to see the 4th and 5th. *In our position they were actually reinforcements. We were playing a game of brag; the holder of the worst hand might yet be the winner. A chance existed of the native troops holding out long enough to subdue the rebellious villages, which would otherwise swamp us, and the sepoys might not be prepared to mutiny and shoot us down, till some fortunate turn in the chapter of incidents, such as the fall of Delhi or advance of Europeans from Kanpur - for still a lingering hope existed in those directions - might alter the position of both parties'*. H.Dundas Robertson. *Op.cit.*.

SAHZADAR SULTAN IBRAHIM A man calling himself 'Sahzada Sultan Ibrahim,' Prince of Delhi, had passed through Russia in the late 1860s, and in 1878 caused quite a stir in Cairo. Claiming he was none other than **Firoz Shah**, he said he was actively plotting the overthrow of the British government with the **Nana Sahib**, who, he reported, was 'living in the Black Mountains near Nepal and resides in a Hindoo temple called Sinkh. The Queen of Lucknow, whose name is Khasse Mahl, is living with him, as also one of her sons (by the King), aged about 14 years. The Nana is said to have so secured his residence that by means of machinery he could bury alive any number of men that would enter the defile. He also said that Nana Sahib is in communication with Yacoub Khan of Bokhara, by whose means and through whose territory the arms would be passed into India. The constant communication between the Nana and the Khan is principally on account of money matters as the Nana has a large quanitity of jewels concealed.' But this was nonsense, insisted the Vice Consul of Cairo, who kept the 'Prince of Delhi' in confinement for a time before dismissing him as an imposter. *See NA. FDP. FS. February, 1878. Nos. 73, 83, 106.*

SALONE A town in Oude[Awadh].'On the day after the **Sultanpur** mutiny (*ie* 9 June 1857), the regiment at Salone, to its south, threw off its allegiance; without, however, committing any particular violence. All the residents of Salone and of **Roy Bareilly** were protected and eventually escorted into safety by Hanwant Singh, the brave old Raja of Dharoopur, and by various chiefs of the great Bys clan.' J.J.Mcleod Innes. *The Mutiny in Lucknow and Oude.* London, 1895, p 85.

SALVADOR HOUSE   Another name for SAVADA KOTHI. In Kanpur.

SALYGRAM   Major W. Lindsay (killed at **Kanpur**)'s house was looted by his own servants but Salegram his head babu (clerk) collected a quantity of property and papers and stored them in his own godown where they were seen by a Committee of English officers. However when the **Gwalior Contingent** came into Kanpur after **Windham**'s retreat they burnt the godown and tried to kill the babu -- he escaped only to be shot dead in the bazar by a 'friendly' European soldier, who wanted to plunder him. The Babu's staunchness was so well known that the soldier was immediately confined and was to have been shot, but it was never carried out. The Babu was shot in a time of perfect peace.

SANKEY, Lieutenant R.H.   An officer of the Madras Engineers who was attached to the army of Maharaja **Jung Bahadur** of Nepal, which entered Oude[Awadh] in support of the British in February 1858. He had the task of bridging the **Gomti** river at **Lucknow,** and he effected this in record time thanks to the personal interest shown in the project by Jung Bahadur himself. *See* **FSUP** Vol II p.302.

'Sarvar Khan'

SARVAR KHAN or Surwur Khan   A **vilaiti** 'with hair on his hands', said to have been a long time servant of **Nana Sahib**: the evidence from a number of Depositions names him as one of the men who slaughtered the women in the **Bibigurh.** Twice his sword broke and he had to come out and go to the Nana's lodging nearby to replace it. *See* **Williams.**

SASSIAH [Sasia] *See* SUSSIAH.

SATARA   Lord Dalhousie annexed Satara or Sattarah by imposing the Doctrine of Lapse when the Raja, Sivaji's direct descendant, died without a male heir, in 1848. Sir John Kaye, *A History of the Great Revolt,* Vol I, p 71.

Satichaura Ghat

SATICHAURA GHAT   Kanpur, 27 June 1857: 'The whole rebel band had assembled on the occasion to see the English depart, and now rushed into the garrison; their number was so great that there was hardly any place to stand; some eight thousand armed men were crowding the intrenchment and occupying every inch of ground. The English were entirely in their power. The sepoys hastened them on, saying, ' Come to the boats; all is ready.' The number of the sick and wounded at this time was rather large, as I had myself seen when leaving the intrenchment; and in the confusion and hurry which ensued, and want of sufficient carriage, some twelve helpless patients were left behind, not with the intention of being abandoned, but to be sent for as soon as the doolies could be spared.'...Everything being reported ready, and carriage for the wounded having arrived, we gave over our guns, &c.; and marched out on the morning of the 27th of June, about 7 o'clock. 'We got down to the river and into the boats without being molested in the least.' Lieutenant **Delafosse** in London *Times.* 16 October, 1857, p 7. ' When a few had gone on board and others were waiting to embark on the riverside a gun opened on them with cannister (this gun and others had been masked); one boat took fire, and then another gun opened, and four boats were

British entering the boats at the Satichaura Ghat

fired; on this those who escaped the fire jumped into the water. The sepoys also fired muskets, the sowars entered the water on horseback, and cut numbers down. 15 boatloads of English were massacred; 108 women and children escaped this massacre, but many were wounded. The Nena said, 'Don't kill these; put them in prison.' London *Times*. 16 October, 1857. 'On a signal, fire is opened on boats filled with refugees. All the men taken are killed, and the women taken to Nana's camp. One boat gets away, but is brought back. James Burgess, *The Chronology of Modern India: 1494 to 1894*. 'We entered these joyfully, never for a moment expecting treachery, and were taken quite by surprise when we were fired on. The river in many places in the middle had no more than six feet of water so that most of the boats were soon aground. Some of the small ones managed to push on and even then with difficulty. The firing at first was irregular, but after a while the balls came whizzing past us as thick as hail, sinking many boats. I was on the deck of my boat seated stupefied with terror and amazement, when I was further convinced of immediate danger by seeing a party of sepoys enter the boat I was in. I was seized in an instant by the arm by one of these savages - for savages and ruffians they looked. I was asked to deliver all I possessed; money and jewels to the amount of 400 rupees, the sum I managed to take with me when I proceeded to the barracks, was now snatched from me; on replying in the negative to questions, whether I had more money and valuables by me, my person was searched rudely. My senses had very nearly forsaken me. I was in a sort of stupor. The search was made on my person while I was standing, but to speak more exactly I was made to stand while I was searched. The ruffian, as if to tantalize me, let off his gun over my head and shoulders in the most deliberate and cold-blooded manner. They afterwards shot two sweet little girls, sisters, who were between the ages of six and eight. The poor creatures were clinging to each other when they committed this diabolical act. Next they shot an Eurasian whose name was Kirkpatrick, a merchant in Cawnpore. How many others were killed by the miscreants I could not know, for I felt dizzy, and sank on the deck. For what time I remained in this state I have no idea. I returned to consciousness by feeling myself suddenly and rudely seized and thrown into the river. The next moment I was buffeting with the water. I managed with some difficulty to get to land and scrambled on shore. I crawled on my hands and knees till I reached a tree about half a mile from the banks, and hid myself as well I could. My thoughts, oh heavens! were agonizing. My sister, her husband and children had, I had not the slightest doubt, been ruthlessly murdered. I shuddered to think of their dreadful fate. My thoughts next reverted to myself. What was I to do, where could I escape, surrounded as I was on all sides by the dreadful, revengeful and blood-thirsty enemy? I had no hope of escape. I offered up a fervent prayer to God, ' Gracious and merciful Father, Thou wilt not desert me in the time of need! Oh Lord, have mercy on me!' and such like prayers burst forth from my innermost soul. I fell by degrees into a sort of drowsy fit occasioned perhaps from weariness, from which I was aroused by approaching stealthy footsteps. On an instant I sprang to my feet, but instead of the ruffians whom I expected to see, to my great relief the well-known face and form of Miss Wheeler, the General's daughter, was before me. In a few words I understood that she had been dealt with in the same way as myself *i.e.* thrown into the water by the men who perhaps thought she was not worth a bullet, that being insensible she would soon sink to the bottom of the river. Our agitation and fear, however, were so great that we had not much of consolation to offer each other. We had not been together more than an hour, I should suppose, when a party of the enemy surprised us. We were dragged in different directions, and of Miss Wheeler's fate I knew nothing till very lately. I was pushed and dragged along and subjected to every indignity. Occasionally I felt the thrust of a bayonet, and on my protesting against such treatment with uplifted hands, and appealing to their feelings as men, I was struck on my head, and was made to understand, in language too plain that I had not long to live; but before being put to death, that I would be made to feel some portion of the degradation their brethren felt at Meerut when ironed and disgraced before the troops.' *See* Amelia Bennett, *Ten Months' Captivity etc.*, The Nineteenth Century, June 1913. 'When we reached the place of embarkation, all of us, men and women, as well as the bearers of the wounded and children, had to wade knee-deep through the water, to get into the boats, as not a single plank was provided to serve for a gangway. It was 9 o'clock A.M. when the last boat received her complement. And now I have to attempt to portray one of the most brutal massacres that the history of the human race has recorded, aggravated as it was by the most reckless cruelty and monstrous cowardice.The boats were about thirty feet long and twelve feet across the thwarts, and over-crowded with their freight. They were flat down on the sandbanks, with about two feet of water rippling around them. We might and ought to have demanded an embarkation in deeper water, but in the hurry of our departure, this had been overlooked. If the rainy season had come on while we were intrenched, our mud-walls would have been entirely washed away, and grievous epidemic sickness must have been added to the long catalogue of our calamities. While the siege lasted, we were daily dreading the approach of the rains, now, alas! we mourned their absence, for the Ganges was at its lowest.…. After the men, who had not escaped in the the boats, had all been shot at the ghaut, the women and children were dragged out of the water into the presence of the Nana, who ordered them to be confined in one of the buildings opposite the Assembly rooms; the Nana himself taking up his residence in the hotel which was close at hand.' *See* Mowbray Thomson, *Story of Cawnpore*, London, 1859....'nearly 125 Europeans captured alive, consisting of twelve or fifteen males and the rest females, and confined in the subadar's kotee (= **Savada Kothi**). They were kept one whole day and night starving. There were but few among them who were uninjured. Boats were sent in pursuit of the three boats which had escaped.' *See* **Forrest**, *Diary of the Nunna Nawab*. .....'The [European males] who were taken alive were shot at and killed before the Nana, a salute of twenty-one guns were given him for the victory.' *See* **Forrest**, *Diary of the Nunna Nawab*. 'Boats were immediately provided for the conveyance of the remains of the garrison to Allahabad, and to these boats they proceeded on the morning of the 27th June. And now followed the most dastardly piece of treachery that has perhaps ever been perpetrated. Only a portion of the party had taken their places in the boats, when, by previous arrangement, the boatmen set the awnings of the boats on fire, and rushed on to the bank. A heavy fire of grape and musketry was then opened on the Europeans. Out of thirty boats, two only managed to start; one of these was shortly swamped by round shot, but its passengers were enabled to reach the leading boat. Of those on board the other twenty-eight boats, some were killed, some drowned, and the rest brought back prisoners.' This is a succinct and very brief summary of the events of 27 June 1857 at **Kanpur**. *See* J.W. Sherer, *Havelock's March on Cawnpore*, Edinburgh, 1910. For greater detail *see* the numerous Depositions of eye-witnesses and survivors in **Forrest** and **Williams**. *see also* Rudrangshu Mukherjee's summary the MASSACRES in KANPUR.

SATICHAURA SURVIVORS

There were only four male survivors of the massacre at **Satichaura Ghat, Lt Mowbray Thomson, Lt Henry Delafosse, Private Murphy, and Gunner Sullivan.** The story of their escape down river is told by both Thomson and Delafosse. Here is the latter's account: 'On the morning of the third day, the boat was no longer serviceable; we were aground on a sand bank, and had not strength sufficient to move her. Directly any of us got into the water, we were fired on by thirty or forty men at a time; there was nothing left but for us to charge and drive them away, so fourteen of us were told to go on shore and do what we could. Directly we got on shore the insurgents retired, but having followed them up too far we got cut off from the river and had to retire ourselves, as we were being surrounded; we could not make for the river, but had to go down parallel with it, and came at the river again a mile lower down, when we saw a large number of men right in front waiting for us and another lot on the other bank should we try to get across the river. On this bank, just by the force in front, was a temple; we fired a volley, and made for the temple, in which we took shelter, losing one man killed and one wounded; from the door of the temple we fired on many of the insurgents that happened to show themselves. Finding they could do nothing against us while we remained inside, they heaped wood all round, and set it on fire. When we could no longer stay on account of the smoke and heat, we threw off what clothes we had, and each taking a musket, charged through the fire.

Swimming for Life

Seven of us out of twelve got into the water, but before we had gone far two poor fellows were shot in the water. There were only five of us now left; we had to swim whilst the enemy followed us on both banks wading and firing as fast as they could. After we had gone about three miles down the stream, one of our party, an artilleryman, to rest himself, began swimming on his back, and not seeing in which direction he was swimming, floated to the shore and got killed. When we had got six miles, firing on both sides ceased, and soon after we were hailed by some natives from the Oude side, who asked us to come on shore, and they would take us to their Raja, who was friendly to the English. We gave ourselves up and were taken six miles inland to the Raja, who treated us very kindly, giving us clothes and food. We stayed with him about a month, as he would not let us leave, saying the roads were unsafe. At last he sent us off, on the 29th July (1857), to the right bank of the river to a zemindar of a village, who got us a hackery, and we took our departure on the 31st to Allahabad, but meeting a detachment of the 84th on our way, we marched up with them to Cawnpore'. *See* W.J. Shepherd. *Personal Narrative, etc.*, Lucknow, 1886, pp 91-2, *and* London *Times*, 16 October 1857. *See also* Mowbray Thomson, *The Story of Cawnpore*, London, 1859.

SAUGUR & NARBADA TERRITORIES

This was the name, in 1857, given to the administrative area that later became known as the Central Provinces, and which, post 1947 was divided between Madhya Pradesh and Maharashtra. The line of the Narbada river provides for all practical purposes the southernmost limit of the mutiny/rebellion of 1857-58: it is not without significance that it is also includes the southernmost stationing of troops of the Bengal Army. Thus Saugur[Sagar]: 3rd Irreg.Cavalry, 31st and 42nd BNI; Damoh: 42nd BNI; Jabalpur: 52nd BNI; Seoni: Madras Inf.; Narsinghpur: 28th Madras Inf.; Hoshangabad: 28th Madras Inf; Betul: 28th Madras Inf. The Territories once formed the great Kingdom of Gondwana, and latterly had been the dominions of the Nagpur Raja (Bhonsla) before being annexed or lapsing to the British government (1853).

SAUGUR[Sagar] The 3rd Irreg. Cavalry broke into open mutiny here on 1 July 1857 and were soon joined by the 42nd BNI and a few of the 31st BNI. The majority of the latter however resisted the call to mutiny and indeed asked to be led against the mutineers. This regiment was afterwards reformed as the 2nd Queen's Own Rajput Light Infantry. The Europeans took refuge in the fort and if the commanding officer at Saugur, Brigadier Sage, had shown resolution, the mutineers might have been defeated by a combination of the loyal 31st and the Company of European Artillery in the fort. But he allowed the mutineers to march out of the station towards the north some days' later. For more details of this outbreak *See* **FSUP** Vol III *and Revolt in Central India*, Simla, 1908, p 35-6, *and* Tapti Roy, *The Politics of a Popular Uprising: Bundelkhand in 1857*, Delhi, 1994.

SAUGUR, RELIEF of The fort at Saugur[Sagar] had been besieged since 29 June 1857 when it was relieved on 3 February 1858 by **Sir Hugh Rose**. It contained a considerable body of Europeans - 173 men, 67 women and 130 children. For details *see The Revolt in Central India*, Simla, 1908 pp.92-5, *and* Thomas Lowe, *Central India during the Rebellion of 1857-58,* London, 1860, *and* **FSUP** Vol III; *and* Tapti Roy, *The Politics of a Popular Uprising: Bundelkhand in 1857*, Delhi, 1994.

SAUNDERS, Charles Burslem A member of the Bengal Civil Service, he was Magistrate of Moradabad at the outbreak of the rebellion, and was appointed Commissioner of Delhi after the recapture of that city in September 1857. His correspondence, private and official, is therefore good primary source material and is preserved in the Oriental and India Office collections of the British Library, *see* MSS.Eur.C.93-4; and MSS.Eur.E.185-188.

SAVADA KOTHI (or Savada House, or Salvador House) A building in Kanpur, originally associated with missionary work hence its name. **Nana Sahib** made it his headquarters some time after the middle of June 1857 during the attack on **Wheeler's Intrenchment**. It was here that the fugitives from **Fatehgarh** were brought from Nawabganj when they were captured making their way southwards down the Ganges. This party of perhaps 130 men women and children were killed on 12 June 1857, it is said at the express order of Nana Sahib, but as there was not one survivor details are sketchy. This was the first of the three distinct massacres of Kanpur. The women who were reprieved, if not spared, at the **Satichaura Ghat** on 27 June (the second massacre) were also brought here at first and were then moved to a smaller house nearby, the infamous **Bibigurh** where they were joined on 10 July by further fugitives from Fatehgarh, making a total of about 210 who were killed on 15 July (the third and most notorious of the massacres, otherwise known as the Massacre of the Ladies).

SAVARKAR, Vinayak Damodar

His book on the rebellion was banned by the British government when it first came out, as seditious. Perhaps its greater fault is that in the anxiety to be patriotic he may have strayed from the path of true scholarship. *See* V.D.Savarkar, *The Indian War of Independence,* Calcutta, 1930 (and Bombay 1947).

SCHMIDT

This tale comes from the siege of the **Lucknow Residency**. 'In Innes' garrison was a man called Schmidt whose grandfather had been a German but who was himself born in India and who was well known as a man of dissipated habits and as having a complete disregard for truth. He had no situation, having been dismissed from some appointment under government, but he pleaded in the District Commissioner's Court at Mullapur in Oude[Awadh] in favour of Indian clients. He was however a hard-working man in the garrison, being always ready to lend a hand at anything, as fearless as anyone else, and constantly visiting one battery after another. One day he was one of a fatigue party in the Post Office, when Major Anderson saw him and asked him what business he had there, and why did he not join his garrison as an attack was expected? The reply was uncourteous enough. 'Who the devil are you to put me such a question?' 'I am an officer' was the reply. 'Damn officers' retorted Schmidt; 'we have too many of them here that are not worth their salt, and are fonder of bullying civilians, now that they can do so with impunity, than fighting the insurgents. Who are you?' 'I am Major Anderson, and will soon show you who I am'. Whereupon he called two Europeans and had him conveyed to the Quarterguard, ordering him to be detained for 48 hours. As soon as Schmidt found himself in trouble, he did his best to get out of it; sent for his friends and begged them to endeavour to get him off. They succeeded and Major Anderson ordered his release next day. Instead of being grateful for having his punishment thus remitted, Schmidt was no sooner free than he cursed Major Anderson and every officer in the garrison, swore he would return his firelock to Capt Boileau his C.O. and vowed he would no longer lift a musket in defence of 'such bullies'. He would also try to revolutionise(sic) the other garrisons where the uncovenanted were ill-treated, and if possible cause a mutiny in the entrenchments. It was in vain that he was told that neither he nor any other civilian in Lucknow fought out of love to the officers of the garrison but for his own life, for his friends, for the women and children under their common charge, and for the honour of the European name. None of these arguments had any effect. He was furious with rage at the indignity that had been offered to him, and attempted to incense the civilians in other garrisons against their superiors, assuring them that if they were all of one mind they would soon bring down the pride of these detested officers. Of course no one would listen to his nonsense and he called them all a parcel of cowards, went to Capt Boileau and threw his firelock at that officer's feet swore he would not take it up again, and begged the Brigadier to let him have a horse and have the gate opened for him and he would risk being cut up by the insurgents, but whom, if they allowed him, he would join. He then broke his sword in two and was walking away when he was ordered back, sentenced to be put in irons, and to receive 25 lashes. Before he was stripped to receive the cane, he said to the Provost, 'Remember sir I am a British subject, and a European' 'Are you indeed?' said that gentleman affecting surprise; 'well, judging from your face I did not think so, but now that I see your body I am of opinion you are;' adding, just as the culprit entertained some hope of being let off, 'However, lay on my lads.' He received his punishment stoically, went to hospital for treatment and then returned to duty as though nothing had happened. Flogging had not been made public and he always denied that the sentence had ever been executed. The following month Schmidt suddenly called out on the terrace of the Boosa-guard, 'Boys, I have some capital news to give you. Will you cheer if I tell you?' 'Yes, yes, of course' was the reply. 'Well the, (with an oath), Major Anderson is dead this afternoon.' Everyone cried out 'Shame! Shame!' and one said 'How easy it is to forgive, when one's enemy is dead; even supposing Major Anderson was your enemy, it is horrible and unmanly to speak thus of the dead.' 'Forgive him!' cried he with a sneer and an oath; 'no, never, would I forgive him', and he uttered a curse. From that day Schmidt began to ail and at last became so ill that he could no longer rise from his bed. He lingered on for weeks and just before the relief, he expired, neglected and unnoticed in the midst of rejoicing. A direct punishment from Heaven? Some thought so'. **PJOT**

SCHOLBERG, Henry

Son of an American missionary and born in India and author of a valuable source book, divided into three parts. Part 1: Published Literature in Assamese Bengali Gujarati Hindi Kannada Malyalam Marathi Oriya Persian Punjabi Sindhi Tamil Telugu and Urdu. Part 2: The Primary Sources. National Archives of India, New Delhi, Urdu Journals, Residency Vernacular Records, UP Regional Archives Bhopal State Archives, Delhi State Archives, the Western Archives, the Central Archives. Part 3: Folk Songs of the Great Rebellion, The Avadhi Songs, Songs of the Ranchi Mutiny, Bahadur Shah the Poet, the William Crooke Collection, the P.C.Joshi Collection, the

Joyce Lebra Collection *See* Henry Scholberg, *The Indian Literature of the Great Rebellion,* New Delhi, 1993.

SCINDIA, Maharaja of Gwalior
Most of the Indian Princes supported the British, believing - probably correctly - that it was in their interest to do so. Only those in major dispute with Government, before the outbreak, actively sided with the rebels. The possible exceptions were **Kunwar Singh**, a genuine patriot, and the King of Delhi himself: the latter may well have been surprised and embarrassed at the position of prominence into which he was, willy nilly, thrust. But some rulers had a most difficult time, striking the balance between the two sides. None more so than Scindia, Maharaja of Gwalior. *See* PRINCES, INDIAN.

SCOTT, Lt General Patrick George
The author of an interesting pamphlet detailing events at **Nowgong.** He was a British army officer 1842-1894. *See* Lt Gen. P.G.Scott, *A Personal Narrative of the escape from Nowgong,* which is preserved in the Oriental and India Office collections of the British Library, *see* MSS.Eur.C.324, *and* Tapti Roy, *The Politics of a Popular Uprising: Bundelkhand in 1857,* Delhi, 1994.

SEATON, Ensign William John
He was an Officer in the Madras N.I. He witnessed the arrival of **Bahadur Shah Zafar**, King of Delhi, in **Rangoon** to which place he was exiled by the British. His correspondence is preserved in the Oriental and India Office collections of the British Library, *see* MSS.Eur.A.166.

SECOND REGIMENT of LT CAVALRY
The 2nd Regiment of Light Cavalry fought at Delhi, Laswarie, Deig, Afghanistan, Ghuzni, and had received an honorary standard for its role in Lord Lake's campaign against Holkar. But at Purwandurrah in 1840 two companies of them so disgraced themselves by fleeing from a small body of **Dost Mahomed**'s horsemen that they were summarily disbanded by General Sale. Their replacement, the 11th Regiment, to which all of the 2nd's European officers were transferred, proved so brave at Mooltan[Multan], where Captain F.C. Vibart captured a standard and the regiment was awarded an extra jemadar, that in 1850 it regained its old number. The 2nd cavalry were not only the most vociferous rebels at Kanpur but the most violent. In the papers of the time a great deal was made of the fact that the regiment was primarily composed of Muslims, but they proved no more active than the 41st Native Infantry at Sitapur, which was composed primarily of Hindus. A possible explanation for the 2nd's discontent and cruelty was that even after seventeen years the wounds from its past disgrace still festered. The 2nd proved a ferocious foe, and were 'well to the fore' at the Battle of Kanpur on 17 July. In December of 1858 a remnant menaced Etawa, where, despite outnumbering the British force, they were defeated. Of the troopers in Nana Sahib's camp, they were reported to be 'the most poverty-stricken and dejected of all.' *See* G.H.D.Gimlette, *A Postscript to the Records of the Indian Mutiny: An Attempt to Trace the Subsequent Careers and Fate of the Rebel Bengal Regiments, 1857 to 1858.* London, 1927, pp 39-45.

SEGOWLIE
See SAGAULI.

SEALCOTE
See SIALKOT

SECRET ARTS OF WAR
Select Command of the East India Company to the Governor-General: Don't teach Indians the art of war, 'particularly with reference to the artillery, an arm which it ought to be our policy not to extend to the knowledge of the natives.' *See* Gupta, *Baji Rao and the East India Company,* p 104.

SECRORA
A military station in the **Baraitch** Division of **Oude[Awadh]**. The troops included Forbes' Regiment of Oude Irregular Cavalry, Boileau's 2nd Oude Irregular Infantry, and a battery of native artillery commanded by Lt.John Bonham. Until the end of May 1857 all remained quiet but by the beginning of June there were clear indications of trouble to come *eg* incendiarism. On 7 June, on the orders of Sir **Henry Lawrence**, a party of Sikhs came to Secrora, collected all the European ladies and children of the Division and took them safely into **Lucknow** on 9 June. The officers, civil and military, also survived when the troops mutinied on the 9th, although there appears to have been some disagreement or misunderstanding between them. *See* WINGFIELD. *See also* John Bonham, *Oude in 1857,* London, 1928, p 74 *et seq.*

The Attack on the Secunderbagh

SECUNDERBAGH
'Close by the spot we halted at, there was a small hole, made by one of Peel's heavy guns, in the walls of the Secunder Bagh, and an officer standing by, said, 'just peep through there', which I did, and an awful sight presented itself to me. There were hundreds of dead and dying sepoys, lying in the small court-yard, where they had been caught by our troops, and from whence there was

no exit for them to escape, and I believe not a soul got away. Some little time before we arrived, the building had been stormed by the 53rd, 93rd, and the Seikhs, and after a severe struggle, had been captured.

Secunderbagh after storming. Photo 1858

Three almost entire sepoy regiments had held this building, to say nothing of other troops, and almost all fell: no quarter was asked, and no quarter was given. The Secunder Bagh is a small palace, standing at the north extremity of a small garden, which is surrounded by high walls: at each corner of the walls there is what may be called a semi-circular bastion or tower, and in the centre of the south wall, the gate leading to the garden and palace stands; opposite to this is the palace, which stretches across the northern end of the garden... the 53rd were in readiness to dash at the gates, which the Commander-in-Chief had given directions were to be blown open by powderbags. The officer of the Engineers on whom this duty devolved told me he had arranged to have the bags ready for this purpose, but that when the moment had arrived, *the man with whom he had entrusted the powder-bags was not forthcoming.* You may imagine the poor fellow's feelings, an important duty to perform and his inability to fulfill it. Fortunately he had not too many seconds to wait, for the moment the troops entered through the breach, those of the enemy who were guarding the gate, made an attempt at escape by throwing it open, when they were met by the gallant 53rd, and a short hand-to-hand conflict ensued, which ended in the slaughter of every one here opposing them'. From the *St George's Gazette,* 31 July 1894.

SEDITIOUS LETTERS A Muslim Sepoy named Sheikh Ibrahim of the 24th Regiment at Secunderabad, rejoiced in the slaughter of the Mutiny. 'Throughout Hindustan,' he wrote his uncle, 'the infidels have been destroyed. Those who were in Bhopal, also have gone to hell. God does as it seemeth Him fit. 'The birds of the air do not move on wing without Divine permission. God it is who does justice

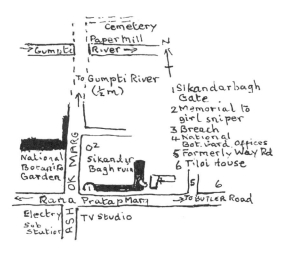

Secunderbagh today. Courtesy Satish Bhatnagar

to us, may God put into the heart of every Mussulman to destroy the infidels, and may he not leave one alive; because they have subverted the Mussulman religion, may God blacken their faces! Because they mixed bones with sugar, and distributed it and gave out cartridges made of pigs' skin. For this reason all Mussulmans have come to the front and destroyed 12,000 soldiers (English) and 4,000 English officers. All in Hindostan have sided with the King of Delhi. I am in the Nawab Secunder Begum Sahib's (Junghee) battalion.'...It is forbidden that you should eat your food with the 'accursed Christian in your bosom.' You will never get such an opportunity again! Send them all to hell.' The writer was betrayed to the British with the letter on his person, tried by general court-martial, and blown from a gun. Translated by W.R. Campbell in London *Times,* 1 March, 1858, p 10.

SEETAPORE See SITAPUR

SEHORE A garrison town in the territory of **Bhopal** some twenty miles to the west of the latter city. The Bhopal contingent troops broke into open mutiny on 7 July 1857 and the 23 Europeans of the Sehore Agency and the Contingent were forced to flee to Hoshingabad. The station and the treasury were made over to an officer of the **Begum of Bhopal** who had previously pointed out that although loyal to the British she felt their presence at Sehore caused her considerable embarrassment not to say danger as her soldiers and followers were all for rebellion. For details *see* **FSUP**, Vol III, and *The Revolt in Central India*, 1857-58, Simla, 1908. *See also* Sir George William Forrest, ed., *Selections from the Letters, Dispatches and other State Papers of the Military Department of the Government of India 1857-58*, 4 vols, Calcutta, 1893-1912, Vol IV.

SELONE 9 June 1857. Major Barrow at Selone.......'I paraded the troops of cavalry which had no European officer and in reply to my address they one and all declared they were faithful and would stand by me. The Rissaldar privately told me, out of the 85 men, he could depend only on 20. ***Subsequently all mutinied...including the Rissaldar.'***

SEMAPHORE Boy Scouts the world over are still being taught to use semaphore. It is fascinating to be able to send a message over quite a long distance just by holding out flags at different angles. But really the system had very little to recommend it: there were too many drawbacks - you had to be within sight of your correspondent, in daylight, with good visibility, no fog, mist, or gunsmoke intervening, and, because each operator had to be visible to the other, in time of war they would also of course be visible to an enemy, and an easy target for musketry. An American called Samuel Morse invented a system of dots and dashes that could send messages along wires: instantaneous communication resulted, day or night, in any weather conditions, over any distance: the **'Electric Telegraph'** had come into existence and had just been set up in India between major centres of population, when the 'Mutiny' broke out in May 1857. It was used extensively, to alert Government in Calcutta, and thence to transmit orders to mobilise and take precautionary action. Indeed one Indian historian has opined that it was the 'Electric Telegraph' which gave the British the advantage in the struggle that was to come, an advantage that resulted in victory. That may be, but it still had its major drawbacks also. The rebels could and did easily cut the telegraph wires, and cut communications instanter. In Central India guerilla forces brought this to a fine art: they would first cut the wire, then lay an ambush for the small engineer party that would be sent out to repair the damage and, wasting nothing, would use the cut wire to make musket bullets, and occasionally use the cast-iron pipes into which the telegraph poles were inserted as primitive guns. During the siege of **Lucknow Residency**, June-November 1857 something like 1,700 British and Indian soldiers and dependants were shut up in the grounds of the British Residency in Lucknow for up to five months, by a large and well-equipped army of sepoys and local levies fighting under the banner of **Birjis Qadr**, the young King of Awadh, his mother the **Begum Hazrat Mahal** and, nominally, the King of Delhi. By the end of the siege the defenders' numbers were reduced to under 1,000 by casualties, disease and desertion. They had food water and military supplies a-plenty, but what they were short of was information. Would they be relieved? Was anyone outside interested in their fate? They sent out spies but many deserted and more were killed by the besiegers. Only the scout **'Ungud'** seemed able to take out messages and bring back news, but - although he was afterwards vindicated and shown to have told the truth - he was often not believed, especially by the Indian troops within the Residency entrenchments. It was essential, to keep up morale, to get good intelligence: if the Indians in the garrison lost all hope of relief it was considered certain they would desert, and had they done so, the Residency position would not have been held for long. At the beginning of the siege in June the garrison commander, **Sir Henry Lawrence**, had determined to hold two areas: the Residency and the powerful old fort, the **Machi Bhawan**. At the last minute he decided to forsake the latter, and sent a message to its garrison to return to the Residency and to blow up the Machi Bhawan. The message was sent *by semapahore* from the roof of the Residency, not by flag fluttering signallers, but by a crude wooden semaphore 'machine'. Still, it worked. The garrisons were united. Two months' later when General **Havelock** arrived at the **Alum Bagh**, four miles to the South of the Residency, with his 'relieving' force - it didn't actually relieve anybody, it just reinforced the garrison - it was necessary to get in touch with him to co-ordinate his final advance. But how? No spy could get in or out. The semaphore? But all the engineer officers in the Residency who might be expected to understand such things were dead or incapacitated, and there weren't any Boy Scouts around then with Signaller's Badge. Incredibly, and very fortunately, the mystery of semaphore was revealed and the solution found, from a little book they came across by accident in what was left of **Martin Gubbins**'s library, in the SW corner of the Residency entrenchment.The book? A 'Manual of Military Communications'? 'The Officer's Field Service Pocket Book' ? 'An Engineers' Vade Mecum'? No, 'all necessary particulars were found under the head 'Telegraph' in the *PENNY CYCLOPAEDIA.* ' **PJOT**

SEN, Surendra Nath A distinguished historian and educator and a former Vice-Chancellor of Delhi University. He was commissioned by the Government of India, less than ten years after India won her independence, to write the 'official ' history of the mutiny/rebellion to coincide with its centennial. Although there must have been strong pressure to present the events in a strongly nationalist light, Dr Sen was judicious and free from rancour and produced a most scholarly and very readable narrative. He showed how conflicting much of the evidence is, and expresses the view that the struggle cannot be characterised as a great national uprising. A useful bibliography is included. *See* S.N.Sen, *1857,* Delhi,1957.

SEONI A town in Central India near **Jabalpur**. An outbreak of the rebellion was confidently expected here as the local population expressed sympathy with the mutineers/rebels, but the *Munsif* Ali Husain remained loyal to government and succeeded in keeping the police loyal to the British. *See* Sir George William Forrest, ed., *Selections from the Letters, Dispatches and other State Papers of the Military Department of the Government of India 1857-58*, 4 vols, Calcutta, 1893-1912, Vol IV.

Typical Sepoys 1857

SEPOY ARMY The native army had been conceived originally as an indigenous force officered by high caste natives. But in the late 1700s English commanders began self confidently to replace the Subadars, who were reduced to 'the drudgery and the dirty work.' *see* Sir John Kaye, *A History of the Great Revolt*, pp 210-11. 'Thenceforward, therefore, we dug out the materials of our army from the lower strata of society, and the gentry of the land, seeking military service, carried their ambitions beyond the red line of the British frontier, and offered their swords to the Princes of the Native States'. Kaye, *A History of the Great Revolt*, p 212. Indian officers, even of considerable experience and seniority had to serve under the most junior British officers who came to India in large numbers, both inexperienced and, in many cases, unfitted for their posts. Cadets and ensigns still in their teens commanded silver-haired old subedars and Jemadars. This was still the situation in 1947.

SETON, Rosemary The author/compiler of an invaluable guide to the source material in the India Office Library and Records (now known as the Oriental and India Office collections of the British Library). *See* Rosemary Seton, *The Indian 'Mutiny' 1857-58*, The British Library, London, 1986.

SHAFTESBURY, Earl of Anthony Ashley Cooper (1801-1885) Philanthropist and campaigner for factory reform etc. In him the disseminators of sensational stories had a most distinguished, and unlikely, ally. **Edward Leckey** devotes a whole section of his little book to 'Fictions attributed to the Earl of Shaftesbury'. Drawn along by his enthusiasm for the missionary cause in India (which in some rather obscure way sought to make capital for its policy of Christianizing India by emphazising the horrors of the mutiny), he was guilty of spreading, rather than making up, the most alarming fictions, and on occasions solemnly took the chair at meetings called to condemn 'atrocities'. One of these 'atrocities' seems to have been first reported in the Bombay Times of 31 March 1858. It concerned events at Jhansi, and reported as fact the following gruesome account:- 'Shortly after, the whole of the European community, men women and children, were forcibly brought out of their homes: in the presence of the Ranee they were stripped naked. Then commenced a scene unparalleled in historical times. She who styles herself Ranee ordered as a preliminary step, the blackening of the faces with a composition of suet (misprint for soot?) and oil, then their being tied to trees at a certain distance from each other, and having directed the innocent little children to be hacked to pieces before the eyes of their agonised parents she gave the women into the hands of the rebel sepoys to be dishonoured first by them, and then handed over to the rabble. The maltreatment was enough to kill them, and several died ere the whole of the brutal scene had transpired; but those who still lingered were put to death with the greatest cruelty, being severed limb from limb. The death the men were subjected to was by no means so intensely cruel: they fell victim to the insatiate thirst for blood of the hellish Ranee and her fiendish myrmidons.' *There is no evidence, direct or indirect, no eye-witness to corroborate any of this story.* Even the officers' servants who were present denied all knowledge of this version of events. The facts are that the adult males were first slain, then the women and children. After their death the bodies were stripped, not to dishonour them, but to get possession of their clothing which might bring in a few rupees to its possessor. Quite frankly, what happened was bad enough: there was no need to invent further horror. *See* **SSF** *and* LECKEY, Edward.

SHAH NAJAF (Or Shah Nujeef) A large and impressive building in Lucknow, built by the Nawab Ghazi-ud-din-Haidar around 1820 and intended as a mausoleum for himself. It was held by the forces of the **Begum of Oude[Awadh]** on the assault on Luck-now by **Sir Colin Campbell** in November 1857, and was identified by him as a strong point that had to be reduced before he could make further progress towards the relief of the **Lucknow Residency**. The official

Photo 1858

London Gazette of 16 January 1858 over-simplified the action: 'The Shah Nujeef was stormed in the boldest manner by the 93rd Highlanders, under **Brigadier Hope**, supported by a battalion of detachments under Major Barnston, who was, I regret to say, severely wounded; **Captain Peel** leading up his heavy guns, with extraordinary gallantry, within a few yards of the building, to batter the massive stone walls. The withering fire of the Highlanders effectually covered the Naval Brigade from great loss; but it was an action unexampled in war. Captain Peel behaved very much as if he had been laying the Shannon alongside an enemy's frigate.' In fact the artillery failed to make much impression on the building. It was taken as a result of a combination of two factors: the panic that ensued among the garrison when the British used some rockets (to cover the retreat of the artillery), this form of projectile was previously unknown to the sepoys, and the sudden appearance within the walls of some British troops: the garrison fled and evacuated the 'fortress'; see HOPE, BRIGADIER ADRIAN.

SHAHGANJ   Twelve miles south of **Faizabad Raja Man Singh** held this strong fort, in which he gave protection to English women and children at the beginning of the outbreak, and later fortified and held against the attacks of fellow talukdars (notably **Mehndi Hussan**). See **FSUP,** Vol II.

SHAHGARH, Raja of   One of the rebel chiefs of Central India who opposed **Sir Hugh Rose** until finally surrendering in August 1858, when he was sent to **Gwalior** under escort, together with his ally the **Raja of Banpur.** For details of his career see **FSUP,** Vol III, and *The Revolt in Central India, 1857-58,* Simla, 1908. *See also* Sir George William Forrest, ed., *Selections from the Letters, Dispatches and other State Papers of the Military Department of the Government of India 1857-58,* 4 vols, Calcutta, 1893-1912, Vol IV, *and* Tapti Roy, *The Politics of a Popular Uprising: Bundelkhand in 1857,* Delhi, 1994.

SHAHGUNJE[SHAHGANJ]   Not to be confused with **SHAHGANJ** above. A village on the Fatehpur Sikri road just outside Agra, between which and the village of **Sussiah[Sasia]** a battle was fought on 5 July 1857. *See* SUSSIAH, BATTLE OF.

SHAHJAHANPUR   A large town and district of **Oude**[Awadh] and the scene of a unique episode in the mutiny/rebellion. **Gilbert Money** was the Magistrate and Collector. At Shajahanpur the garrison was attacked in church by the 41st Native Infantry. The chaplain had his hand severed by a sword stroke at the door and was later murdered by villagers. The Magistrate was also struck with a sword and was killed later as he ran across the courtyard. Others were killed and wounded, but the remaining men barricaded the doors as the women and children ran to the turret. When Henry Sneyd arrived the sepoys of the 28th BNI returned to their barracks for their arms. Loyal servants hurried their masters' weapons to the church, and in the company of a small party of loyalist Sikhs the remaining men, women and children escaped, including Sneyd, his sister Anna and his brother-in-law, Captain Lysaght, only to be killed a few days later near **Arungabad.** C.Hibbert. *The Great Mutiny.* London, 1978 p 415. *See* Money's diary etc which are preserved in the Oriental and India Office collections of the British Library, Photo.Eur.151. *See also* J.Bonham, *Oudh in 1857,* London, 1928.

SHAIKH RAMZAN, Subadar   This officer of the 42nd Bengal Native Infantry raised the flag of Islam on 1 July 1857 at **Saugur[Sagar]** and called for followers, by beat of drum, and he was joined by all the men of his own regiment plus the 3rd Irregular cavalry and a few of the 31st BNI, the reminder of this latter regiment however remained loyal to the British. He assumed command in the cantonments, with the rank of General, but after a few days when threatened by the artillery from the fort and the loyal sepoys of the 31st, he led his men northwards out of Saugur. For further details *see* **FSUP,** Vol III and *Revolt in Central India,* Simla, 1908 *and* Sir George William Forrest, ed., *Selections from the Letters, Dispatches and other State Papers of the Military Department of the Government of India 1857-58,* 4 vols, Calcutta, 1893-1912, Vol IV.

SHANKAR SA   This Raja, was the descendant of the Gond Kings of Garha Mandla and in 1857 was a pensioner of government. He and his son were convicted of plotting to attack and murder the Europeans in **Jabalpur** and were blown from guns on 18 September 1857. After his death his widow seized **Ramgarh** and took part in several skirmishes. At length, being closely pursued, she dismounted from her horse and plunged a sword into her own bosom. She was taken into the English camp where she died. For further details *see* **FSUP,** Vol III, and *Revolt in Central India,* Simla, 1908.

SHANKAR SAHAI   The following verse was found in the clothes of Shankar Sahai, the rebel Raja of Jabalpur. 'O great Kali, cut up the backbiters; trample under thy feet the wicked; Ground down the enemies, the British, to dust Kill them that have remained; Destroy their common servants and children; Protect Shankar Sahae; preserve and keep the disciples; Listen to the call of the humble; Do not delay to cut off the heads of the unclean race (Malechhas) Devour them quickly, O great Kali!' His ancestors had occupied the Guddi of Gond for 1500 years until the Marathas took it from them. Sherring visited the old man in prison and watched him walk with his son to face the guns 'with great firmness.' The executioners tied his hands and feet to the wheels of a cannon, and 'a more ghastly sight I never saw before, as the blowing away of these two men. Arms and heads went high up into the air, and nearly the whole was afterwards left a prey to dogs and birds; till in the evening some remains were picked up and carried to the Ranees.' M.A.Sherring, *The Indian Church etc.,* London, 1859.

SHANNON One of H.M.'s frigates, by chance in the Hoogly river during the struggle: her crew under **Captain Sir William Peel, VC,** was used as a Naval Brigade, particularly in the assault on **Lucknow** in March 1858.

SHEEP'S MUTINY The Mutiny of the Bengal Army was apparently known for many years among some rural Indians as the 'Sheep's Mutiny.' *See* Sir John Kaye, *History of the Sepoy War*, London, 1864-7, Vol.II, p 312.

SHEIKH ABDULLAH There were, without doubt, some Europeans fighting on the rebel side. Anecdotal evidence for their presence abounds, but facts are hard to come by. In looking at relevant correspondence, the vituperation of some of those contemporaries who wrote of the events of 1857-9 leaves the modern reader gasping - with disgust, or sometimes just plain embarrassment. Serving soldiers, particularly those actively involved, are rarely the offenders in this respect: there is something about actual fighting which cleanses the emotions, and strips away hysterical prejudice. But occasionally a soldier right in the heart of things lets the side down by racist abuse or the spread of doubtful rumour. There was *some* truth in the following letter, but much conjecture also: It is from Major Kendal Coghill, 2nd European Bengal Fusiliers and is dated 'Delhi !!! Hurrah !!! 22 September.1857, to My dearest Jos' [his brother, and there then follows the story of the capture of Delhi, then on p 7]: 'Had to guard the 'King of Hindustan' and placed double sentry over him and it could not be helped I called him a pig and would have shot him dead if he had only looked up, the brute. Hodson pursued the Princes, had only 100 men and they had 10,000, but they gave up their arms, the three Princes, the Band of Christian Drummers of the 28th NI *and the English Sergeant major of the 28th NI who was formerly an artillery man, and during the siege pointed the enemy's guns on us, calling himself Shaik Abdoolah and dressing like a sepoy - the band were all killed on the spot but the three Princes were brought with the sergeant major to an open spot where the princes had commenced the slaughter and violation of our ladies themselves, and they were mercilessly killed and stripped and lay flat on the open ground till the dogs and jackals walked off with them. The sergeant major is still in our Guard in irons , and going to be blown away from a gun in presence of the force.'* The identity of this 'Sheik Abdoolah' has been the object of much investigation. There was once a 'portrait' of him in the little Museum in the Red Fort in Delhi, next to Bahadur Shah and other Indian heroes. There are other mentions of him, in print: Lt. General Macmunn, quotes an unpublished account of a certain Capt. Maisey, Judge Advocate General's office. Delhi 1857.:'I had to report on the case of a Sergeant Major Gordon of the 28th NI who escaped the Shahjehanpore massacre and was brought to Delhi by the sepoys. He'd been well treated all the time, although a prisoner and gave himself up to a cavalry patrol in the rebel camp near the Delhi jail, after the capture of the city. It was proposed to bring him to trial for aiding and abetting the rebels, as it was reported that he had manned their guns, and the feeling against him was very violent. But on quiet examination the evidence proved quite worthless, the man's story seemed indisputable, and I strongly deprecated his trial on so heinous a charge, as there was no proof even supposing him guilty, and on the other hand there was every probability that he had been a strict prisoner throughout. So the poor wretch got off trial. He was not however released but the matter was reported to the Commander-in-Chief and what the final result was I do not know.' Neither,unfortunately, do we. **PJOT**

SHEIKH HUNEEF The proprietor of the hotel sometimes referred to as 'Mohamed's Hotel' in Kanpur was a survivor. He ran the hotel before, during and after the short reign of **Nana Sahib** in 1857. Indeed he even showed the British, with pride, the rooms that Nana had occupied, and where his food had been cooked etc. 'I don't think many of the respectable people in the city joined the mutineers. All the budmashes were collected in a Rissallah called 'Shaick Panchoo,' who was a relation of Sheikh Hunneef, who kept the hotel.' Deposition of Thomas Ambrose Farnon., **Williams.**

SHEO CHURRAN DAS 'Three or four days before the troops broke out, Teeka Singh, Subadar of the 2nd Cavalry (whom the sepoys made their General after they mutinied) began to have interviews with the Nana, and said to him on one occasion, 'You have come to take charge of the magazine and treasury of the English. We all, Hindus and Mahomedans, have united for our religions, and the whole Bengal Army have become one in purpose. What do you say to it?' The Nana replied, 'I also am at the disposal of the army.' I then heard that the English told the sepoys to come unarmed into the entrenchment and receive their pay, and that the sepoys refused to enter it without their arms.' Deposition of Sheo Churran Das, **Williams.**

SHEPHERD, W.J. Anglo/Indian head assistant in the Commissariat Office in Kanpur. Took refuge with his family in **Wheeler's Intrenchment**. He volunteered to disguise himself as an Indian and act as spy and messenger for Wheeler. His offer was accepted, but as he took too much rum to fortify himself for the ordeal, he was easily picked up, rather drunk, by the rebels and imprisoned. He survived the massacres and eventually wrote an account of his experiences: much is interesting and relevant, but in part he relies upon hearsay and so can be misleading and untrustworthy. In addition to his personal experiences the book contains a complete list of all the English (and Anglo/Indian?) residents of Kanpur and their ultimate fate so far as he knew it. *See* J.W.Shepherd, *A Personal Narrative of the Outbreak and Massacre at Cawnpore,* Lucknow, 1886.

SHERER, John Walter He was a member of the Bengal Civil Service in the heady days of the 'Mutiny' of 1857. But he was much more than that. He was a writer, of power and talent, and to his accounts of his experiences in India during his career, are added a number of gentle, charming novels, and a collection or two of amusing anecdotes, commenting upon life in England as well as India. In May 1857 he was the Magistrate at **Fatehpur**, on the **Grand Trunk Road** between Allahabad and Kanpur. After the outbreak at Meerut on the 10 May, there was an uneasy peace in many stations in Northern India. Here is Sherer's own description of it: 'A Sanskrit poet describes how, in an overwhelmingly hot season, the cobra lay under the peacock's wing, and the frog, again, reclined beneath the hood of the cobra. All antipathies and antagonisms forgotten. So among the peasantry, general expectation which paralyzed activity. The thief sat down by the doorkeeper, and the bad characters sought the shelter of the miser's wall; all were waiting - waiting - they had no idea for what. Chupatties had been passed in the district, but if the affair was a signal for united action, it failed ...crime in the district ceased altogether. The courts and offices were open, but there was no business.The news grew worse. English soldiers, by twos and threes, occasionally passed through to **Cawnpore [Kanpur]** in conveyances, and these the Kotwal, or Head Constable of the town, who was really a rebel in the guise of a humourist, called the 'chorntee fouj' or ant-army.' But at last the uneasy peace came to an end, and on a cloudy and overcast afternoon, after lunch, Sherer and his guests opened the windows and venetian blinds towards the west, and sat out on the verandah, seeking the scanty breeze. They looked to the west, and saw a purple haze over the distant horizon, and then, dramatically, they distinctly heard the sound of guns, heavy guns, artillery pieces loaded with ball and grape and canister, being fired in anger not salute.They learnt from others that the firing had been going on since midday: deep rumbles of heavy ordnance. The Nana Sahib was attacking **Wheeler's Entrenchment** for the first time: the day was thus 6 June. Sherer's sub-collector, Hikmut Oollah at first professed loyalty, but when the news came that the 6th N.I. had mutinied at **Allahabad**, he showed his true colours, and with a band of armed Muslims came to Sherer's bungalow and told him in effect to depart quickly as the British government was finished. To which Sherer replied 'If I go it will only be for a month's leave'. Hikmut Oollah laughed and said, 'In that case we shall meet again!' They did: Hikmut Oollah in the dock and Sherer in the witness box. Sherer now decided, rightly, that there was no purpose in remaining in Fatehpur, so said farewell to his servants, with genuine emotion on both sides. He gave a gold seal of his father's to Nekram his bearer, and bade him take care of it. Two months' later, in Kanpur a grimy figure sank on to his verandah, undid his turban and from his coiled hair took out the family seal and returned it to its owner: he did not betray his trust. On the way to Banda they stopped at the Jumna [Yamuna] crossing at Chilatan. Two government peons came and sat ostentatiously near Sherer. 'I say', cried one, 'What would you give me for this thing?', holding up his chuprass or brazen badge. 'Four annas,' was his companion's reply, 'the brass is worth that - but the Government? ' rejoined the first speaker with a sneer. The **Banda** magistrate had sent out an escort to meet them, including a large coach belonging to the Nawab. They found Banda full of rumour but quiet. But after four days they all, led by Mayne the magistrate, withdrew, at his suggestion, to the palace of the Nawab. The latter was youngish, civil, small and slight but active, and at this point easy going and good-natured and in no way altered in his attitude to the British. He was described as morally degenerate: he certainly over-indulged in drink. He had a 'leopard' whose head was secured by magnet. This was his party piece: after a great announcement and fanfare the nawab - unfortunately drunk and hiccupping loudly - was to 'decapitate' the head with a single blow of his sword. He missed, and only with the help of two servants was able to do the mighty deed. The story of the Nawab of Banda is a complicated one, but Sherer, ever fair and ever honest, did later say that his recollection as the Europeans departed Banda for their own safety, that Mayne did say to the Nawab - as pleaded later by the Nawab as an excuse for his conduct - ' I delegate my authority to you and you must hold Banda for the British Government.' Sherer later became Magistrate in Kanpur. *See* J.W.Sherer, *Daily Life during the Indian Mutiny,* London, 1898.   **PJOT**

SHIV SAHAY He was reported as 'Collector Sewsuhai (Shiv Sahay) by caste a Kait (Kayasth) or Uggurwala (Agarwal), was assistant to General Usuf Khan, resident of Lucknow, another person by name Sew Singh Collector, was a notorious bad character and was very active in plundering and creating disturbances during the insurrection;' *see* PERSONS OF NOTE, APRIL 1859.

SHORAPUR A small state in Bundelkhand whose raja rebelled. The British, through a detachment of the Madras Column commanded by General **Whitlock**, in April 1858 forced him to dismantle his forts, dismiss his armed retainers and surrender himself as a prisoner. He was tried and condemned to be transported, but, rather than submit to such dishonour, managed to commit suicide by seizing the revolver of the British Officer who was escorting him, and blowing out his own brains. It is said to have caused immense sorrow in Shorapur where his family had ruled for thirty generations.

SHOWERS PAPERS These are held in the Centre of South Asian Studies at Cambridge University, and are particularly interesting as they contain original material. There over sixty items in all, and these include copies of narratives of occurrences in **Delhi**, written 'by a native residing within the walls of the city commencing 11 May 1857' (ie the day the mutineers arrived at and took Delhi). Native news from the city of Delhi 22 July 1857 (at the height of the siege); Dr Murray's narrative of the mutiny at **Neemuch**; correspondence with Maharaja Holkar (Showers was Officiating Political Agent, **Mewar**); an account of the proceedings and military operations which resulted in the occupation of **Nimbahera**; correspondence dated January 1860 about the supposed buried treasure at Neemuch, etc.

SHOWERS, Captain L, (later General) He was Political Agent at Mewar, and led the Meywar[Mewar] Horse during the campaign in that area. He was not perhaps an easy man to deal with and was uncooperative and obstructive with his fellow officers, not seeing the distinction as they did between military and civil command. There was great friction between him and **Brigadier General Lawrence** which is commented on, not altogether impartially, in Appendix A Volume III of Colonel Malleson's *History of the Indian Mutiny*, London, 1878-80. See *The Revolt in Central India*, Simla, 1908. See also Iltudus T.Pritchard, *The Mutinies in Rajpootana, being a personal narrative of the mutiny at Nusseerabad*, London, 1864. See also Sir George William Forrest, ed., *Selections from the Letters, Dispatches and other State Papers of the Military Department of the Government of India 1857-58*, 4 vols, Calcutta, 1893-1912, Vol IV.

SHUMSHOODEEN KHAN 'The day after this meeting, Shumshoodeen Khan was at the house of the prostitute **Azeezun**: being in liquor, he told the girl that the Peishwa's reign would soon commence, and the **Nana** in a day or two would be paramount and that he (Shumshoodeen) would fill her house with gold mohurs. Two or three days after this the troops mutinied.' Deposition of **Kunhye Pershad. Williams.**

SHUNKERPORE [Shankarpur] The town, district and fort of this name belonged to **Rana Beni Madho** who remained loyal to the end to the Royal House of Awadh. Lord Clyde arrived with his army at Kishwapur three miles from the fort on 15 November 1858, and prepared to assault the place but was forbidden by the Governor General to attack any fort in Oude[Awadh] until it was certain that the Chief who owned it had received a copy of the **Queen's Proclamation** which was issued on 1 November 1858. This worked to Beni Madho's advantage, for in the intervening period, and at dead of night he and his army slipped quietly out of the fort, without firing a shot, and making a long sweep to avoid the British pickets, got safely away. Inside the fort immense quantities of *ghee*, nuts, wheat and corn were found, also a laboratory for making gunpowder, and about 9,000 pounds of made-up powder ready for use.

SHURF-UD-DAULA He was a leading member of the **Begum of Oude[Awadh]'s** government. His death is recorded as follows: 'The Moulvie (**Ahmad Ullah Shah**) continued to stay in the city (Lucknow) after the enemy had been driven from every position but the sacred Temple of Huzrut Abbas. He had the character of being very brave and his present attitude was one of defiance....the noise and confusion was very great. A few of the houses were forced by the Seikhs and Highlanders, and from one of them I discovered a passage from the street into the Temple...we broke through two doors, and were beating at a third, when we heard sword cuts and the screams of women within the Temple. The thought occurred to us that the wretches were murdering their women. The blows at the strong door increased in force, the screams touching the heart of the generous Highlander; and, when it yielded to the hatchet, six of the 93rd, Middleton and myself, pushed through the opening. The defenders who had not gone off with the Moulvie were shot, sabred or bayoneted, and we found on the floor the warm but lifeless body of Shurf-ood-dowlah, the Prime Minister of the rebel government. He was suspected of corresponding with the English, and, refusing to buy his safety, was barbarously murdered by order of the Moulvie as he quitted the Temple; and it was this bloody deed that provoked the screams of the women.' *From* T.H.Kavanagh, *How I won the Victoria Cross*, London, 1860.

SIALKOT This is a town in the Punjab about sixty-five miles east north-east of **Lahore** with a large cantonment including, in the first place a considerable European force. The military stationed there in May 1857 were the 35th BNI commanded by Major Drake, the 46th BNI commanded by Colonel Farquharson, and the 9th Bengal Cavalry; in all about 2,200 trained sepoys. The European force was H.M.'s 52nd Light Infantry commanded by Colonel Campbell, a troop of Horse artillery under Captain Dawes, and a battery of Field Artillery commanded by Captain Bourchier, plus a musketry depot which provided an aggregate of about 900 Europeans. Brigadier-General Frederick Brind commanded the Brigade and the Station. However matters soon changed for practically all the European soldiers plus the 35th BNI and the left wing of the 9th Bengal Cavalry left Sialkot on 25 May 1857 to join the **Punjab Movable Column** assembling at Wazirabad under Brigadier-General Neville Chamberlain. **Sir John Lawrence** announced that he could not be responsible for the safety of European families who chose to remain at Sialkot, but that they would be welcome to take refuge in the Fort at Lahore; very few availed themselves of this offer, although the probability of an outbreak at Sialkot was daily increasing. Although by now the sepoys at **Mian Meer, Lahore,** had been disarmed, Brind and his officers placed complete reliance in their sepoys and refused to dishonour them by taking away their arms. Soon it was too late to even attempt to do so. When it became clear that a mutiny would soon break out, the fort of Sardar Teja Singh was designated, by the Deputy Commissioner, Monckton, as the place of refuge for all Europeans to move to when necessary, and the government treasure was quietly moved into it. At the same time on John Lawrence's instructions a Sikh Levy of about 200 was enrolled and its training begun. The 9th Bengal Light Cavalry and the 46th BNI mutinied at 4a.m. on the 9 July 1857. The officers of the Cavalry rushed to their lines and attempted to restore order but were fired on and driven off; the cavalry then rode through the station killing any European they met; the prisoners were released from the gaol, the Treasury and cantonment looted, the Courts burned and the Artillery Magazine blown up and they then left the station at 5p.m. in the direction of Gurdaspur, leaving the station in ruins. Among the Europeans who were killed were Dr J.Graham the Superintending Surgeon, and coincidentally a second Dr (J.C.) Graham the Civil Surgeon, Brigadier General **Frederick Brind**, and Captain Bishop the Brigade major, Captain Sharpe, Sergeant Nully, the Reverend Hunter and his wife and child. The two mutinous regiments were probably aware that General **Nicholson** with a 'Flying Column' was in the neighbourhood of Amritsar, hence they took the Gurdaspur road. After crossing the Ravi at **Trimmu Ghat**, they were surprised by Nicholson and around 400 killed; a further three hundred fled to an island in the middle of the river but were driven from there the next day with great loss; a further two hundred fled into the territory of the Maharaja of Jammu, but were handed over, court-martialled and the majority executed. See **Ball**, **Kaye** etc., *see also* G.Rich, *The Mutiny in Sialkot*, Sialkot, 1924, Rev.Andrew Gordon, *Our India Mission 1855-85*, New York? 1888., F.J.Montgomery, *The Mutiny at Sialkot*, 1894.

Ranbir Singh Maharaja of Jammu

SIALKOT

'It was nerve-wracking when Sikhs were called upon to execute Sikhs. In Sealcote a civil servant reported that when the Sikh commanders of the local gaol and the foot and horse police refused to escort the local Europeans and joined with the townspeople in looting and plundering, commissioners from Lahore came in to try them and sentenced them to be executed. ' 'It was a ticklish affair, as we were hanging Sikhs when we had only Sikh levies about us. The ropes broke, and the guard was ordered to shoot the half lifeless bodies; then followed three or four volleys of musketry' that alarmed nearby Europeans, who thought 'it was 'all up' and that the guard had turned against us.' Though the Sikhs had generally behaved well, 'the treachery, ingratitude, and cold-blooded cruelty we have witnessed and seen every day have more or less unsettled the minds of all, and made us mistrust everybody with a black face.' Letters dated 23 and 25 July, 1857, quoted in the London *Times*, 22 September, 1857.

SIEGE TRAIN See CAMBRIDGE COUNTY RECORD OFFICE.

SIEVEKING, Isobel Giberne Author of a book which purports to show that the siege of **Arrah House** in Bihar was the watershed of the mutiny/rebellion. Contains some letters by **John Nicholson**. *See* I.G.Sieveking, *Turning Point in the Indian Mutiny*, London, 1910.

SIKHS

As early as the second week of May 1857 it became clear that the warlike Sikh (or Seik, as the British usually spelt it) race would assist the British government in the struggle to put down the mutiny/rebellion. Apart from the occasional divergence - *see* BENARES - this was the course they followed, and their assistance was crucial, indeed many commentators believe that without Sikh help the British would not have succeeded as they did, at least not within the time span of two years; had the Sikhs actually joined the rebels en masse it is perhaps doubtful whether the British could have survived at all. The motivation of the Sikhs is complicated, and, to some extent, inexplicable in material terms. As a result of the recent Sikh Wars the British had confiscated much territory and considerably reduced the power of the Sikhs in the Punjab where they had been dominant for many years: one might have expected a war of revenge against the British conquerors, but the reverse happened. It may be that a bond of mutual respect had been established between Sikh and Englishman which was now to be paramount in their dealings with each other: certainly the influence of such men as **Nicholson** and **Sir Henry Lawrence** was keenly felt. In the course of the first eventful week of the mutiny/rebellion, it became evident that the Sikhs and Jats of the Punjab, generally, had no intention of making common cause with the Bengal army. On the contrary it appeared they had old scores of their own, which they hoped to have an opportunity of wiping off. It is said they were particularly eager to share in the capture of Delhi, because of a prophecy, that they, in conjunction with the *topee-wallahs* who should come over the sea, would lay the head of the son of the Delhi sovereign on the very same spot where that of their *Guru* had been exposed some 180 years before, by order of the Emperor Aurangzeb; and this, it is said, they actually accomplished. *See* Dolores Domin, *India in 1857-9: a study in the role of the Sikhs in the people's uprising*, Berlin, 1977; *see also* Ganda Singh, *The Indian Mutiny of 1857 and the Sikhs*, Delhi, 1969.

Sikhs

SIKHS 'Stern, wiry, dark-looking men, tall and straight of limb, their broad brows overhanging piercing black eyes their noses rather aquiline and well chiselled, their not too full lips, which, when parted, showed teeth rivalling the whitest ivory, and which were shaded by jet-black mustachios, proudly curled, and their chins covered with silky black beards, carefully parted in the middle, and combed outwards, their voluminously folded blue or red turbans, their grey tunics and bright-coloured vests, their silver-mounted fire-arms, curved scimitars, and lightly-poised lances, the gay caparisons of their well-bitted and often thorough-bred horses, the ease and grace with which they sat and managed them, their proud air and manly bearing, plainly stamped them as belonging to the aristocracy and chivalry of the northern countries of Asia; and on all occasions in this war, well and nobly have they seconded the gallantry and daring courage of their dashing leaders. They are all men of property, and their horses are their own'. Oliver Jones, *Recollections of a Winter Campaign in India in 1857-1858*, London, 1859. *See also* FORSYTH, Sir Douglas.

SIMLA [SHIMLA] Dundas Robertson felt very strongly about the 'old women' of the Hill Stations, and it is a fact that many of the British officers who found themselves in such places at the beginning of the troubles of 1857 behaved far from well. For excuse they can only point to the panic which lack of reliable information might engender. The Hill Stations were apparently safer havens, yet if the route to the coast, to the frontiers, to other European held territory were cut off by mutineers......? Many of these officers were getting on in years, were past the age of active service, and were conscious too of the heavy responsibility which the presence of large numbers of European women and children put upon their shoulders. It was not easy to stay calm in such circumstances. Nonetheless the panic which developed at one stage in Simla was a disgrace to the British name, and caused many a blush when order was restored. H.Dundas Robertson, an example of the very best type of I.C.S. officer, was struggling almost single-handed to maintain order only a few miles to the south of the hills that led up to Simla. He is scathing in his condemnation of the cowardice as he saw it of the Europeans in that place. 'The Nussooree Battalion had shown a decidedly mutinous spirit at Jutob, when ordered to march without leaving the usual guard of their own men over their women, a point on which they are extremely jealous. Jutob is some four or five miles from Simla, and separated from it by defensible ravines. At the latter place the English residents amongst whom there must have been, including officers, civilians and shopkeepers, from 150 to 200 men, instead of uniting and holding their own, which, for the protection of the ladies and children, was absolutely necessary, *actually fled and scattered in all directions*. (Mainly to Ambala: PJOT). The chief military and civil authorities residing in the place took the lead in this disgraceful evasion, and what is hardly credible, they left the ladies and children, whose husbands and fathers were fighting in the plains, to take care of themselves.

Simla in 1857

Had actual mutiny occurred at Jutob, there were few places in India, except Meerut, where the approximation in equality in the respective numbers of the English and the mutineers would have been so favourable to the former; and yet it was only here where the disgrace of unnecessary flight tarnished the English name'. Oddly enough the panic in Simla redounded to Robertson's advantage, for the Nusseeree Battalion was - unjustly - tainted with the smear of rebellion, and its destination, Delhi, was changed and instead it was sent to Saharanpur District where it helped Roberston considerably in restoring and maintaining the rule of government. *See* H.Dundas Robertson, *District Duties during the Revolt*, London, 1859.

SIRDARI SINGH   A leading rebel of Untabeg; formerly a subadar in the 1st Light Cavalry of the Bengal army; said to have been the commander of the mutineers at the battle of **Sussiah[Sasia]** near **Agra** on 5 July 1857. *See* **Ball**.

SITAPUR   A large town in **Oude**[Awadh] and the principal station in the Khairabad Division after the British annexation of the kingdom in 1856. J.G.Christian was the Commissioner, with Thornhill as his Deputy and Lieutenant Lester and Sir Mountstuart Jackson as his Assistants. There were three regiments present: the 41st BNI (commanded by Colonel Birch) destined to play a decisive part in the struggle at **Fatehgarh**; and the 9th and 10th Irregular Infantry under Captains Graves and Dorin; and a regiment of Military Police under **Captain Hearsey** who had formerly commanded one of the regiments in the King of Oude's service.

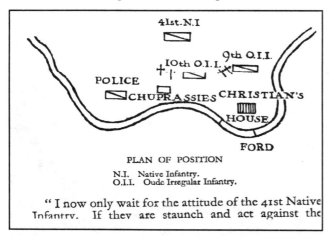

Christian was a man of considerable ability, but he did not listen to the military, and the position (*see* his own sketch, above) he took up at the beginning of June depended for its success upon the Irregulars and the Police remaining staunch even if the 41st BNI mutinied; they did not. A succinct account of what happened may be found in John Bonham's *Oude in 1857, Some memories of the Indian Mutiny*, London 1928. *See also* CAPTIVES OF OUDE.

SITAPUR (SEETAPORE)   An important centre in Oude[Awadh], the principal station in the Khairabad Division, it was the scene of a typical mutiny/uprising. A narrative of events there was written by **Captain George Hutchinson**, Bengal Engineers, Military Secretary to the Chief Commissioner Oudh[Awadh] in 1859. At first news from the outstations reached **Lucknow** effectively by the horse *dak* arrangements that Sir Henry Lawrence set up. The news that came in from Sitapur was most critical; there the 41st Bengal N.I. were stationed and although many of its officers believed the regiment would remain staunch, others believed it was biding its time. A letter from Mr J.G.Christian, the Commissioner, dated 28 May 1857 to Lieutenant (as he then was) Hutchinson out in the District as Political Officer i/c a moveable column, expresses his delight at the news that the column had been set up, and asks him to arrest **Showdree Mustafa Khan** and **Yacoob Khan**, the latter a Pathan who appeared to be raising men near Malhiabad. A letter dated 1 June, also to Hutchinson reports the mutiny at Lucknow on 30 May and reports that Col Birch with five companies of the 41st N.I. would be pursuing the mutineers towards Malhiabad. He believed the 41st would stay staunch but if not, the 9th and 10th Oudh Irregular Infantry and the Military Police would be firm. Clearly believed that it would be possible to play the regular sepoys against the irregulars. He tells the story of the Europeans attacked in church at Shahjahanpur, with the death of **Ricketts** the magistrate and with Dr Bowling and Ensign Spens badly wounded. Nine officers of the 28th N.I. and eight ladies escaped to the protection of the **Raja of Powaen[Pawayan]**. Believed **Bareilly** and **Fatehgarh** would follow. Says **Thomason** reported all was quiet at **Mohammadi** on 31 May. Says he has collected all the ladies children and women into his house for safety. House made secure. **Captain Hearsey** Officer i/c the Military Police brought in also. Believed he could rely on 250 of 9th Oudh Irregular Infantry and 330 of the 10th ditto, plus 360 Military Police, 200 chuprassees plus four guns. Advised Hutchinson on his march to avoid Bilgram, a 'Yagee Mussulman' town, also **Shahabad** and Pyhani both 'bad Mussulman towns', and on no account to go into **Farrukhabad.** Clearly Christian intended to stay put: aid from outside was impossible, so he felt it his duty to stand firm at his post and resist the rapidly rising tide of rebellion. On 27 May the vacant lines of the 10th Regt. Military Police were fired and a rise was anticipated, but none occurred and the fires wre extinguished. Men of the regiment actually showed letters to their officers which spoke of simultaneous risings by the 41st and the the 9th. On 2 June the 10th Oudh Irregular Infantry rejected some cartloads of flour sent by the kotwal and insisted that all be thrown into the river, which was done. *See* BONEDUST. The sepoys were little by little testing their strength. In a strange incident some sepoys of the same regiment plundered some fruit from the garden of the Commissioner Mr Christian. Lt Greene of the 9th and Mr Bickers, the Superintendent of the Comissioner's office went and remonstrated. The sepoys said they were very sorry, but only doing as others were doing. Christian raised many 'irregulars', from different classes trusting they would not unite to mutiny. Eighty **Passis** were recruited and placed along the banks of the nullah as guards, with their favourite weapon, the bow and arrow. Colonel Birch, the Commanding Officer of the 41st N.I. up to the last minute of his life trusted his men; a letter from him denouncing those who doubted the loyalty of the Bengal sepoy *appeared in the Calcutta press just after his murder!* On 3 June Major Apthorp of the 41st told Mr Christian his men were disaffected. Guns were loaded and the irregulars posted here and there. Colonel

Birch was shot by his own men at the Treasury, also Lt.Graves was wounded but he managed to give the alarm. The 41st quietly appropriated the treasure and then deliberately set out on the work of murdering the Europeans. Actually Christian and those in his house were murdered by the 10th Oudh Irregular Infantry. In twenty minutes all was over and a few had escaped into the jungle across the nullah. *See* CAPTIVES OF OUDE. A few like Lt.Lester and Mr Bickers and Sergeant Abbott reached Lucknow safely two days later, escorted by a small band of the 41st who remained loyal to their officers. In all twenty-four Europeans were killed at Sitapur: Mr and Mrs Christian, one child and one European nurse; Mr and Mrs Thornhill, one child and one European nurse; Lt Col Birch. Lt Salley, Sergt.Major Middleton of the 41st; Lt Graves, wife and child, Dr Hill, Sergt.Major Keogh and two children, Lt Greene of the 9th Oudh Irregular Infantry; Lt Dorin, Lt Snell wife and child of the 10th Oudh Irregular Infantry; Mr Cronenburgh, clerk. **Capt Hearsey** escaped . *See* Capt. G.Hutchinson, *Narrative of the Mutiny in Sitapur,* 1859, in the Hardy Papers in the **CSAS.**

SITARAM BAWA

Best known for his revelations- if such they be- of **Nana Sahib's** involvement in conspiracies before the outbreak of the Rebellion. He made a statement before H.B.Devereux Judicial Commissioner of Mysore on 18 January 1858, named names and gave details of the alleged conspiracy *eg* **Golab Singh Maharaja** of Jammu and Kashmir is named as a leading conspirator who had put up considerable sums of money to induce the military to rebel. **Nana Sahib** he declared to have been 'always a worthless fellow..... and could never have ordered the massacre of the women and children'. Makes interesting comment on sepoys which supports the theory that the 'mutiny' was inspired by greed alone: 'the military classes were enticed by a promise of restoring the old times of licence *and they all prefer that to a regular form of government.*' *See* **FSUP**, Vol I, p 372 *et seq*. and R.C.Majumdar, *'The Sepoy Mutiny and Revolt of 1857',* Calcutta, 1957.

SIVAJI (1627-1680)

The popular hero of the Marathas, and the founder of the dynasty which (nominally) ruled the Maratha Confederacy for two hundred years. 'Originally a freebooter who was also a dreamer, but later a soldier, statesman and administrator of unquestioned eminence.' The last Raja of Satara, a direct descendant of Sivaji died without male heir and **Dalhousie** applied his 'Doctrine of Lapse' to annex the territory to the Company's rule (along with **Jhansi** and **Nagpur**). The warcry of the Mahratta [Maratha] Light Infantry (Indian Army) has always been 'Bol! Chattrapatti Shivaji Maharaj ki jai!'.

SLEEMANABAD

A town in Central India where there was a detachment of the 52nd BNI. This regiment had remained loyal in **Jabalpur** until 18 September 1857 when it mutinied, and the news was sent at once to Lieutenants Barton and Cockburn commanding at Sleemanabad. They in turn told the native officers of the 52nd and mounted their horses. Their men expressed regret at parting with these officers and did not molest them, but said they must make the best of their way towards Delhi (they did not know it was at that very moment falling to the British). Some of them were visibly affected and shook their officers by the hand; they then robbed the *tehsili* of the little money that was there, ***the Pay Havildar handing one of the officers the balance of his month's pay***, and marched off quietly towards **Nagode**. The two officers reached **Jabalpur**, and ***their baggage was sent in by their men***. Reported in *Revolt in Central India,* Simla, 1908.

SMITH, Major Richard Baird

An Officer of the Bengal Engineers who by dint of seniority became Chief Engineer to the **Delhi Field Force** in June 1857 on the Ridge to the northwest of **Delhi**. It was his plan, nominally, that was used for the attack upon Delhi on 14 September, but as he was incapacitated for almost the entire length of the siege, the plan owes much to his subordinates, notably **Captain Alexander Taylor**, of whom it was said 'Taylor Took Delhi'. Some of Baird Smith's correspondence and official papers are preserved in the Oriental and India Office collections of the British Library, *see* MSS.Eur.B.214.

Sivaji

SMYTH, COL. G.M. CARMICHAEL

See CARMICHAEL-SMYTH.

SMYTH, Sir John

The author of a 'new' biography of **Lakshmibai**, Rani of Jhansi. This account has little new evidence to shed further light on the Rani's activities. Much remains conjecture, but at least the range of conjecture is here defined. *See* Sir John Smyth, *The Rebellious Rani*, London, 1966.

SNEYD, Elizabeth   She was in **Fatehpur** at the outbreak of the mutiny/rebellion but was one who escaped, to **Alipur**, with her family. An account of her experiences there and at other times during subsequent developments is preserved in the Oriental and India Office collections of the British Library, *see* Photo.Eur.44.

SOCIAL HISTORY OF THE REBELLION

This will be found under a variety of headings including RURAL UNREST; PEASANTS' REVOLT; CIVIL REBELLION; etc., and in the writings of Ranajit Guha, Eric Stokes, C.A.Bayly, Tapti Roy and Rudrangshu Mukherjee *etc.*; and in the contemporary accounts of Sir Sayyid Ahmed Khan, Munshi Jiwan Lal, Mainodin Hassan Khan, and Ghalib. *See* individual entries, and BIBLIOGRAPHY.

SOLDIERS of *MISFORTUNE*   It has been frequently said that the rebellion failed because of the absence of leadership among the rebels, particularly military leadership. Less has been made of the fact that on the British side also there were some senior officers who, by ill-luck in some cases, but from rank incompetence in others, seriously jeopardised the British campaign. Among these may be cited **General Hewitt** at **Meerut**, **Major General Lloyd** at **Danapur**, Colonel **Polwhele** at **Agra**, Colonel Jackson at **Nimbahera**, Major General **Whitlock** in Central India, and even General **Windham** at **Kanpur**. On the other hand, **Colin Campbell** and **Hugh Rose** were outstandingly good soldiers and leaders, but there are not many other names to place beside them. **Anson** would have been a disaster had he survived, **Archdale Wilson** at Delhi very nearly was.

SONGS OF THE MUTINY   *See* William Crooke in the Indian Antiquary, April and June 1911. 'The song of the Dancing Girl' *'Ali Kahan hy John Company, A hundred years have gone and past, When India groaned under England's rule, The avenging hour has come at last, and the Hindoo sepoy is not such a fool to eat the salt of his great tool - John Company. Ali Kahan hy John Company, Ali Kahan hy John Company oh! Subahdar of the Gillis ka Pultun - Wah wah Vajeerun! Wah wah Vajeerun! you were fond of the Captain Saheb of my company then, but now times are turned old gel, you must sing to no one but your Soobadar Vajeerun Sing again, The Topee wallas were great asses: who drew their tullubs said their masses; And after dinner filled their glasses; Then left the country to Chuprassees But now we shall be the Sepoy's Lassies, No more John Company, Ali Kahan hy John Company, Ali Kahan hy John Company oh !'* - from the Bengal Hurkaru 4 March 1858. For Indian Folk Songs of 1857 *see* P.C.Joshi ed, *Rebellion 1857, A Symposium*, Delhi, 1957, p 271 *et seq*.

The Temple at Rani Ghat today

SOOKA MULL'S GHAT   The old name for what is now known as **Rani Ghat** in Purana Kanpur. Since the Ganges has altered its course, the ghat is now some 500 metres from the river and the intervening land has been built upon or used for farming. The ghat was probably built at the same time as the old Shiva Temple which still stands there and which bears a plaque with (the boatman) Sooka Mull's name and the date 1832. The ghat is interesting as being the (reputed) place where **Nana Sahib** met the sepoys who had just mutinied and persuaded them not to go to Delhi until they had destroyed the British in **Wheeler's Entrenchment**.

SOPPITT, Mrs   *See* FITCHETT, Rev William H.

SPIES, British   Anjoor Tewarri was a famous spy, for the British, in 1857. He was a sepoy in the 1st BNI, at **Banda**, when the mutiny broke out, but having aided a European clerk to escape from the mutineers he threw in his lot with the British. Much of the information he brought in for General **Havelock's** column as it marched first on **Kanpur** and then on **Lucknow**, was of such great value that it actually influenced the General's decisions, and decided when, where and how the British should advance. He eventually built up a reputation that was so prestigious that whatever he said was taken as the literal truth; if he said that a man had been a 'traitor' to the British that man was virtually condemned forthwith. After the 'Mutiny' was suppressed he was an exceptionally powerful man, and his reports carried great weight. Some say that Anjoor abused the power that his reputation brought him - but that is another story. At first, when he was on probation so to speak as a spy, his reports were suspect; after all, the British had no particular reason for trusting him. Many intelligence reports were received which turned out to be of no use whatsoever: the spies merely reporting to their British paymasters what the latter wished to hear. Anjoor Tewarri's descriptions of the rebels' movements ahead of, or around, Havelock's Column, were always presented in the minutest detail, and though invariably proved to be very largely accurate, provoked some scepticism among the British. It sounds as though they were thoroughly ungrateful, but, as we have suggested, there were reasons for their doubts. In some ways the spy's reports were too good, too detailed, too wide-ranging and comprehensive, to be the result, as he claimed, of his own personal observation. How could he have counted the enemy numbers so carefully, overheard their plans being discussed, known in such detail what their supply position might be etc. One British officer in particular doubted Anjoor's veracity, and told him so. And as this officer was no less than **Lt.Col. Fraser Tytler** Quartermaster General to the Column and responsible therefore for intelligence gathering, Anjoor decided that he must be convinced. His opportunity came when Havelock sent part of his force to attack the rebels at **Bithur**, some 12 miles from Kanpur, and the home of the **Nana Sahib**. There were several regiments of sepoys there, with artillery and cavalry; they had thrown up excellent entrenchments and were determined it seemed to protect the Nana and his ancestral home. The details of the fight need not concern us here: Bithur was eventually taken by the British, the Nana's palace destroyed and the sepoys dispersed, but there was considerable and very hard fighting before this was achieved. The British success was in no small measure related to the accuracy of Anjoor Tewarri's intelligence report. He had, as usual, brought in the minutest description of the enemy's movements, especially those of the Nana himself, and his following. Indeed the attack was undertaken mainly on his information. But Fraser Tytler would not believe that the intelligence report was the result of personal observation on the spy's part, and asked him, cynically, where he had picked up his information. 'All right', said the spy, indignantly, 'I'll prove it to you'. As the column was in the heat of action, in the middle of the attack on Bithur itself, Fraser Tytler was as usual - for he was a brave man - in the front, calmly observing the position and movements of the enemy; suddenly he felt a tug at his foot, and Anjoor whispered to him: 'Do you see that bit of white kupra on a tree in front of you? Well, take it down quickly, and put it in your pocket!'. Almost mechanically Fraser Tytler reached up to the branch, pulled down what seemed to be a piece of cotton cloth and pocketed it. He then forgot the incident, but after the enemy had been driven out, and the action was over, Anjoor Tewarri came up again and asked Fraser Tytler if he had kept the bit of rag. 'Yes', said the QMG, and pulled it out of his pocket. 'Just see if it fits this', said the spy, and, untying the end of his loincloth, he matched the piece that had been torn from it. Thus 'The brave spy' - see how noble he becomes if he is on your side! - 'had fully proved that, on the previous night, he had taken his observations himself.' Well, probably.

PJOT

SPIES & COMMUNICATIONS

Kanpur, August 1857: 'One duty, which was by no means an unpleasant one, was the endeavour to keep open communication with **Agra**. We always found men who were willing to take the risk. They would, perhaps, not have been killed, if the letters had been found; but very probably mutilated. Later on, several of our adherents had their noses and hands cut off. The letter was written on a small piece of paper, and put into a quill; the quill, again, sealing waxed at the end, or sewn into a little case of wax cloth. The object was so small, it could be popped into the hair or held behind the teeth in an emergency. One of these tiny scrolls brought the last hand-writing of Mr. **Colvin**. Mr. (afterwards Sir William) **Muir** was the best correspondent. Sometimes **Neill** got notes direct from Delhi; these were generally in **French**. As the communication was fairly open to Calcutta, friends of those in the fort at Agra began to write overland via Cawnpore[Kanpur]. When a good number were collected, Umurnath and I set about making them as small as possible, and then got a **Kossid** to dress as a travelling beggar.

A miserable pony was procured for him. It had to be a mere bag of bones, and yet to walk to Agra. We could not give it corn, as that, doubtless, would have brought on inflammation of the stomach; and when turned loose on the fresh grass, its old teeth produced effects much like those following the mowing of a lawn by a person unacquainted with the art. On this sorry brute was bound a most moth-eaten and weather-stained pad, tied roughly with a rope. But deeply inserted in the mouldy lining were the overland letters taking comfort and refreshment to many a heart. We daubed the Kossid into a filthy and odorous fukeer; and giving his *Rosinante* a meal of some kind of thatch, the cereal it most affected, we started Her Majesty's Mail on its journey. It arrived, I am glad to add, in perfect safety.' **Sherer** See also AGRA.

SPOTTISWOODE, LT.COL. H
He was the commanding officer of the 55th BNI at **Nowshera** in the Punjab on the outbreak of the mutiny/rebellion. He believed passionately in the loyalty of the regiment, but when it was ordered from Nowshera to Murdaon across the Kabul river, the men suspected they were sent there because of mistrust in their loyalty, and taunted Spottiswoode with having brought them to a prison. The Colonel assured them to the contrary and promised to forward to headquarters any petition they might draw up. They accordingly framed one, but the most prominent grievance it contained concerned the break-up of the invalid establishment of the corps, and made no reference at all to the **cartridge question** etc. Some 120 of the regiment did remain faithful, the others absconded, many were captured and executed, see PESHAWAR. Spottiswoode was bitterly disappointed and, unable to bear the shame, committed suicide.

STANSFIELD PAPERS
These are held in the Centre of South Asian Studies at Cambridge University, and contain copies of the letters home and the diary of Ensign Henry Stansfield. The letters are vivid and full accounts of his own activities and news and comment on the conduct of the mutiny as a whole: includes **Danapur,** and **Lucknow**. The diary 23 December 1857 to March 1858 is of less interest and is concerned with troop movements, distances covered etc.

STATISTICS
In so far as they are available these are listed under appropriate entry headings. Thus: REBEL FORCES, LUCKNOW, NOVEMBER 1857 AND MARCH 1858; REGIMENTS, BRITISH; CASUALTIES; FAMILIES; REINFORCEMENTS FROM ENGLAND; REINFORCEMENTS FROM ELSEWHERE; OUDE, ARMY IN ETC.

STEAMERS
On the Ganges and other (occasionally) navigable rivers, steamers were used by the Government forces with considerable effect. Such craft were small and underpowered and rarely capable of speeds in excess of 1 knot upstream, but in the absence of a **railway** network they were still of considerable use in transporting men and stores. The steamer **'Berhampooter'** in particular was very useful: it was probably the first steamer ever to penetrate further upstream than Allahabad; in addition to carrying men and stores it was used effectively as a gunboat. It is clear that the government had plans to extend the use of these little steamers: in a letter sent from Agra to Benares in August 1857 (*see* Impey Letters in *AGRA*) there is the following: *'The armed boats from Calcutta will be of great use both on the Ganges and Jumna. If we had only one or two as a guard we might send down a fleet of boats with the women and children.'* See also J.C.Marshman, *Memoirs of Major General Sir Henry Havelock,* 1860, p 328, *and, 'The Reluctant Heroes'* in **CM** p31. Railways in India were in their infancy in 1857. True, trains started from Calcutta 'for the North West', but after a hundred miles or so came to a dead end in Bihar. So troops heading for the 'mutiny' sites of Northern India had to detrain and start marching, aided usually by an extraordinary assortment of animals and carts drawn by animals. And then..... someone thought of using steamers. Like the railways, steamers on the great rivers of India were a novelty. In August 1857 few had ever ventured up stream beyond Allahabad.

The Ganges at Allahabad

There were two reasons for this, and both were concerned with the state of the river: if the river was low then the boats risked grounding on the ever-changing sandbanks; if the river were high then the current was usually altogether too much for the feeble engines then installed - if the steamer could manage one mile an hour upstream it was thought to be doing well. But there was one steamer that performed remarkable service for the British at this time: the steamer 'Berhampooter'. There is nothing about it in despatches, nothing in the newspapers. The only mention comes in an officer's reminiscences: the skipper we know was called Dickson, and was an officer of the Honourable East India Company's Marine, and it had on board one hundred men of the 1st Madras Fusiliers (Europeans). With two 9pdr guns manned by 20 Bengal Artillery Invalid Pensioners 'reluctant heroes', it was a powerful, if slow, weapon. The Officer i/c of the troops was Captain John Spurgin, Madras Fusiliers who lived to be Lt General Sir John Spurgin KCB CSI, with Lieutenants Arnold and Bailie, both of whom were later killed. The Medical officer was Dr Rean. The

'Berhampooter' was sent upstream from Allahabad in July 1857. The original idea was that of General Neill who refused to believe that the garrrison of Kanpur had been massacred. On 3 July at Allahabad, and on the point of starting, came more authentic news, so Spurgin's orders were changed from 'proceeding at any risk and hazard' to going 'up river with caution'. Keeping abreast of Havelock's force which, moving parallel to the river was separated from the steamer by 5-8 miles, he was ordered to tow five large boats full of provisions for the Kanpur garrison (by then non-existent). They could only go at 1 knot against the strong current of the Ganges. Those on board were completely ignorant of the river bed, they had no pilot, this being the first time a steamer had gone up above Allahabad. On 6 July a large numbers of rebels on the banks opened fire. Spurgin landed (against Neill's orders?), repulsed the rebels and spiked their guns. The skirmish produced a handsome telegraphic despatch - the only one - from Sir Patrick Grant the temporary Commander in Chief, to Neill congratulating him though in fact it had nothing to do with him. This is the only official mention that I can find of the steamer's activities. As Grant soon gave way to Colin Campbell, and the latter was notorious for taking little notice of anything that had happened before he took charge *cf* the Arrah House affair - it is not surprising that the exploits of the 'Berhampooter' never reached the general public .....and yet...perhaps without this steamer the outcome of the Mutiny might have been very different. On 7 July the vessel passed a fort armed with heavy guns and had a serious engagement. Rebels followed the boat up river all day. Spurgin again landed most of his men who then attacked the rebels' camp, on which the latter fell back in confusion; while keeping up the fire of his own guns. In the dusk the Fusiliers returned safely on board. On 11th the steamer anchored, on Havelock's order and sustained a severe cannonade from the rebel artillery. It was repeatedly hulled, but there were few casualties - the troops had piled all bedding, knapsacks and kit against the bulwarks and this stopped the gunshot especially the grape, although the men's effects were reduced to rags and ribbons. The rebels had collected here on the Oudh[Awadh] bank and made preparations to cross the river to attack Havelock's flank and rear. *If this had happened the course of the rebellion might have been very different.* (*NB* in military terms General Havelock's small force did not achieve all that much but its morale-boosting effect was considerable - ditto the reverse effect on the rebels. The steamer's fire kept the rebels in check and, more important, they were able to destroy all the rebels' boats. Next day the British victory at Fatehpur opened the way to Kanpur and converted the Oudh[Awadh] rebel chief into an ally: he sent in his submission to Spurgin on 13 July. Many chiefs wavered in a similar manner, especially in Oudh *eg* Raja Man Singh. Maude calls the event a 'Brilliant page in our naval annals' - exaggeration? Even if it is not so, it is still a page in an unknown book. When Havelock met Spurgin at a ghat in Kanpur he said 'Well, I never expected to see you again!' A check was made on the medical condition of the Invalid (or Veteran more accurately called) artillerymen. It was not very sympathetically conducted as Havelock desperately needed them for the tiny column which he intended to take to Lucknow. A private letter says that General Neill had asked Havelock to send the twenty Invalids back to Allahabad. (One Invalid said that if the officer would come down into the cabin he'd show him a rupture as big as his head!) - but they were all selected none the less, and went off towards Kanpur. Havelock paraded them and proceeded in his usual pompous manner to deliver a Napoleonic oration 'My men, I have come to thank you for so nobly volunteering to assist your country in the hour of her great peril ' - and more in the same vein. Until one old fellow stepped forward, and with an apology of a salute, interrupted the General by 'Beg pardon sir, we ain't no volunteers at all : we only came 'cos we was forced to come!' Dead silence followed and the parade was dismissed immediately. Reluctant heroes indeed! Neill's Instructions for Captain Spurgin (on ½ sheet of paper) 'You are to push on as quickly as you can to Cawnpore: the object is to relieve Sir Hugh Wheeler. Land nowhere but if necesary and opposition is shown, open fire and destroy as many rebels as you can. On getting to Cawnpore to the Ghaut nearest the entrenchment camp best adapted for landing (ironically this would have been the Satichaura Ghaut), communicate with Sir Hugh, let him know what you have on board, give him all the news of Renaud's column....land your men and stores as Sir Hugh Wheeler may direct, and I hope the steamer will be made available by Sir Hugh to bring down here (Allahabad) all the ladies and children, also sick and wounded officers.'(on the same 1/2 sheet of paper) .....'3 July 57. Intelligence having been received last night that Cawnpore may have fallen, you are to proceed up the river with the greatest caution....secure the General's right flank.....Having received certain news of the state of affairs at Cawnpore, move up and relieve it if it still holds out '- Allahabad 9 July '57 telegram, from Neill to Governor General 'Keeping up the river communication is of great consquence, please impress this on the General. The Berhampooter thirty horsepower has not power enough, a light powerful steamer (sixty horse) armed with partly European crew would be invaluable for the Ganges, overawing the people taking stores etc as far as Cawnpore.' Gen Havelock 10 July sent news that he had reached one march from Fatehpur....'the steamer Berhampooter accompanied us, with 2 guns and 100 riflemen, so I hope soon to announce the recapture of Cawnpore.' Neill to Governor General 11 July . '..Heard from Renaud on 9th at Kahga, all well, from Spurgin off Newbustee, on the right, abreast of Renaud's column, getting on well. Some arrangements about coaling the steamer should be made sharp; there is none here at Benares.' 12 July at Fatehpur Havelock sends official despatch on his action. No mention of the Berhampooter. Havelock to Commander in Chief, 14 July...'General Neill is urgent with me to send back a detachment of invalid Artillery. I cannot do this without crippling my artillery force which Capt Maude so ably commands.... I have with me eleven light guns and only 72 British artillerymen including 24 invalids....' There are enigmatic references to 'Boats on the River ' and ' Mr Hannay of the Indian Navy etc'. and a further reference to the 'Berhampoota' being used to collect boats for Ganges crossing. 9 Aug 57 'The Soobahdar's son is active in our cause and was on board the steamer' *ie* the steamer trying to ascend above Kanpur....The Fifth Fusiliers were one of the first British regiments to make use of steamers to get up country towards the scene of conflict. On 18 July 1857 the headquarters of the regiment with some 170 men went on board the *Benares* and set off for Allahabad: it was slow work steaming up the river, as the current of the Ganges was strong, and as the sandbanks of the river were continually shifting, great care was necessary in navigating it. Writing of the strength of the current Dr Collins, the Assistant Surgeon of the 5th, wrote: *'The current is in many places so strong that we are constantly being washed back, and now and then the engine has not the power to force up against it. There is nothing then but out with our anchors, fix them into the bank above us, as far as the cable will reach, and then with a fatigue party at the capstan, assisted by the engines, we work the steamer, little by little, until the difficulty is overcome. It is very stiff work, but the men go at it with a will, being only anxious to reach their destination....even the pilot of the Benares turned out to be a traitor, as one evening he ran the steamer ashore and made off during the night. The consequence was that the whole of one precious day was lost before the vessel could be floated again.'* -this and further detail of the trip will be found in the *St George's Gazette* dated 31 August, 1893. *See also* **FSUP,** Vol I, p 360, *and* **FSUP,** II, p 167; *and* **CM.** *and* John Cochrane Hoseason, RN., *Remarks on the rapid transmission of troops to India and the practicability of promptly establishing effectual means of conveyance between the sea coast and the interior by the navigation of the great rivers by steam.,* London, 1858. From a collection of letters to be found in Ames in *Indian Affairs,* 2, Pamphlets 1857-60, No.8.

STEWART, Chas Edward Army officer. He was in Peshawar on the outbreak of the struggle and can give detail from personal experience of the military movements and mobilization in the Punjab which made possible the **Siege of Delhi** by the British. He also took part in the campaigns against **Lucknow** and in **Oude[Awadh]**. *See* C. E. Stewart, *Through Persia in disguise with Reminiscences of the Indian Mutiny,* London, 1911.

STIRLING In this Scottish city is the regimental depot of the Argyll and Sutherland Highlanders, known in 1857 as H.M.'s 93rd regiment which was part of **Campbell's** army in the attack on **Lucknow** and played a particularly important role in the storming of the **Secunderbagh**. An unpublished narrative account of this storming is kept at the depot. For further details *see* J.M.Ladendorf, *The Revolt in India 1857-58,* Zug, 1966 entry 981.

STOCK PAPERS These are held in the Centre of South Asian Studies at Cambridge University, and contain letters written by Catherine Ann Simons neé Stock to her family in England. She was the wife of Captain Alfred Simons of the Bengal Artillery, and like many other officers' wives, she was forced to take refuge with her two children in **Naini Tal**. Her husband was killed in the **Lucknow Residency** during the siege.

STOKES, Eric Together with C.A.Bayly, Ranajit Guha, Rudrangshu Mukherjee, Sashi Bhusan Chaudhuri etc, Eric Stokes has done much to publicise the social history of the rebellion. It was the loss of land rights, so the argument runs, that supplied the force behind the rural explosion of 1857. Unfortunately Stokes's most significant book on the subject was not finished by the author, although C.A.Bayly has produced an edited version which goes far to achieve what Prof.Stokes had intended. Bayly tells us 'Inevitably the *Peasant Armed* is an unbalanced book. The connection between the military mutiny so interestingly treated in the first two chapters, and the detailed social analysis which exists thereafter is implicit rather than explicit. This makes the book difficult for the non-specialist to follow. Clearly this is intended as a detailed treatment of the social origins of the revolt, but the lack of Eric Stokes's own conclusion is nevertheless a severe problem.' *See* Eric Stokes , *The Peasant Armed, the Indian Revolt of 1857,* Oxford, 1986.

STOKES PAPERS These are held in the Centre of South Asian Studies at Cambridge University, and are of particular interest for the study of **Gwalior** during the mutiny/rebellion. They contain a copy of an article by Profesor Eric Stokes in the St Catherine's College Society Magazine for September 1976, on the connection between the College and India, in particular reference to the life and work of the **Rev.George William Coopland**, a chaplain in the East India Company's service, 1855 in Gwalior, who went out with his wife in 1856, arriving in January 1857. He was killed in the mutiny of the **Gwalior Contingent**.

STOUT DAME From Lt Col B.W.D. Ramsay *'Recollections of Military Service and Society'* 1882: 'The King of Oude when arrested in Calcutta and taken into the Fort (15 June 57) nominated the followers who were to accompany him. A stout dame, almost 40, rushed out and abused the Europeans soundly because she was not included in the list of followers. 'Who', she said, arms akimbo, 'I should like to know, is to take care of him - to wash him, to light his pipe?' etc. There was no objection to her being added to the list, but the King laughed feebly, and had evidently left her out on purpose in the first instance.'

STRANG, Herbert   He edited a book containing nine stories of the mutiny/rebellion: the stories have in common a high charge of inflammatory emotion; the mutineers are condemned as ignorant, cruel and despicable savages with no excuse for their conduct. Perversely because it is so biassed this book is useful - it is historical propaganda *par excellence*, reminiscent of Foxe's Book of Martyrs in the degree of its prejudice. *See* H.Strang, ed., *Stories of the Indian Mutiny*, London, 1931.

STUART PAPERS   These are held in the Centre of South Asian Studies at Cambridge University, and include a manuscript copy of an unpublished *Narrative of the escape of Mrs Mill and her children from Faizabad, Oude[Awadh], during the Great Indian Mutiny in 1857.* This is the narrative of Mrs Mill copied from one written by Mrs Brown from Mrs Mill's statement, Calcutta, 4 November 1858.

STURT, John Venables   A Deputy Commissioner in Central India, he has left an interesting collection of papers describing events from 4 June to 4 October 1857 including the outbreak of the mutiny/rebellion at **Jhansi**. He was himself captured by mutineers but escaped to safety under the protection of the **Raja of Charkhari**. His papers are preserved in the Oriental and India Office collections of the British Library, *see* MSS.Eur..C.195.

SUBADAR, Manilal Bhagandas   *See* ANDERSON, George.

SUBORNING THE ENEMY   'There is a report in Lucknow, that at Cawnpoor there are three or four Regiments of Natives serving with the English. The **Begum** is in correspondence with the Native Officers of these Regiments, and it has been settled that when they engage the rebels, the Regiments are to fire blank ammunition and afterwards they will turn round upon the Europeans. A Native Officer came from Cawnpoor and arranged this and received 1,000 rupees as a present.' - from the report of a Lucknow newswriter dated 21 February 1858. Alas for the rebels, the Native Officer took the money, but did not deliver.

SULATIN   The royal family: relatives of **Bahadur Shah**, King of Delhi. Thus: 'Twenty-four of the ...Sulatin were hanged at Delhi yesterday morning. Two were brothers-in-law, two sons-in-law of the King, the remainder nephews, etc.' Muir to Sherer, 19 November, 1857. William Muir, ed., *Records of the Intelligence Department of the Government of the North-West Provinces of India during the Mutiny of 1857, I: 273*. See also SHAHZADA.

SULLIVAN, Gunner   Of the artillery. One of the four survivors of the **Satichaura Ghat** massacre. He died shortly afterwards, of cholera. *See* THOMSON, CAPTAIN MOWBRAY.

Sullivan in the 'Charge of the Thirteen' *see* Mowbray Thomson

SULTANPUR   'One party of mutineers from Azimgarh, Jaunpur, and Benares marched towards Sultanpur, on the Gomti, on the direct road to Lucknow, instead of joining their fellows at **Faizabad.** It was the one out-station, besides **Sitapur** and Faizabad, where there were any of the old Bengal troops. The Bengal regiment there was the Ist Irregular Cavalry, a gallant but very bigoted set of Mussulmans, who proved foremost in all the future operations, not only in actual fighting, but in furnishing leaders to the mutineer army. Fisher, their colonel, was an exceptionally popular and energetic officer, but he was at once shot down, though not by his own men. Other officers fell similarly under the fire of the Sepoys, but a few escaped, to find protection and escort into safety from Roostum Sah, the chief of Deyrah. Colonel Fisher had, some days before, sent off the ladies and families to the protection of the Rajah of Ameythee, who loyally sheltered them till he was able to escort them to Allahabad.' James.J. Mcleod Innes, *Lucknow and Oudh in the Mutiny,* London, 1895, p 85-6. For further details *see also* John Bonham, *Oude in 1857*, London, 1928. It was here, in northern Oude[Awadh], on 28 August 1858, that there was fought one of the last battles of what might be called the Oude Campaign. Fortunately for the British **Man Singh** now declared himself for them, and the rebels were distracted by having to besiege him in his fort of **Shahganj**. General **Hope Grant** and a powerful British force approached **Faizabad** and the forces besieging Shahganj melted away. The British Division consisted of two British regiments and one Punjab infantry regiment, three batteries of artillery, six squadrons of British cavalry and some 900 native cavalry, of whom the detachment of **Hodson's Horse** behaved badly against the **talukdari troops**. The battle was fought in the evening, the rebels came to the attack; there were said to be 14,000 men, with 15 guns. After a difficult crossing of the **Gomti** which had taken the British two days to effect, they took up a position to receive the rebel attack; the latter was checked, then driven back and finally retreated, leaving Sultanpur in Hope Grant's possession.

SUMMARY SETTLEMENT of 1856

The first task that faced the British government immediately after the annexation of Awadh in February 1856 was the systematic collection of land-revenue. The settlement operations began in May 1856 and the instructions given to the officers were precise and specific. The settlement was to be made for three years 'village by village, with the parties actually in possession'. The assessments were to be based on detailed information and statements of (i) the past five years jama (ii) the nankar grants (iii) rent free villages (iv) religious grants and (v) patwaries and chowkidaries of the old system. It was laid down as a leading principle that the settlement was to be made 'with the actual occupants of the soil', ie with village zamindars or with proprietary coparcenaries. These instructions were grounded in the belief that Awadh shared with the rest of north India the same features of agrarian relations and hence could be settled in the same way. This gave to the settlement an anti-talukadari bias from the very beginning. One settlement officer, Captain Barrow stated this very explicitly: 'The instructions for settlement made it nearly imperative on the District Officer to turn out the Talooqdar and reinstate the 'Village Zamindar', the 'Proprietor of the Soil'; no Talooqdar, middleman or farmer..was to be allowed and if long possession had given him a prescriptive right it was to be counterbalanced by a 10% Talooqdaree allowance, but this was ordered to be ignored as much as possible. If there was a village claimant he was to be put in.' Among officials who determined land-revenue policy in north India in the first half of the 19th century, men like Bird and Thomason, there was a prevalent assumption that a talukdar was somebody whose connection with land did not go back long into time. A talukdar it was assumed had established his hold over land through force, influence and fraud. For Bird and Thomason, following up on Holt Mackenzie, the village communities were the cornerstone of agrarian life in north India, and, where they did not exist, the key figure was the proprietor of every village. Hence the emphasis on making the settlement with the 'actual proprietors' and on doing away with what was seen as an imposition between proprietor and government. The execution of the instructions was fraught with difficulties. Detailed information of the kind envisaged was impossible to obtain and officers had to fall back on guesses and rough estimates based on patwaris' accounts. This made overassessment inevitable since chakladars and nazims of the nawabi regime had always rack-rented the peasantry. Over assessment in Awadh in 1856 varied from district to district; and though some districts were not overassessed on the whole, some parganas within the district were grossly overassessed. Aggregate figures for the whole province show that Awadh was assessed at a lower level than it was previous to annexation. But the authorities were aware of over assessment at the local level. In the settlement which followed the uprising of 1857 nine of the twelve districts had their assessments reduced 'owing to the discovery of over assessment in some estates'. In making a comparison of the pre-British and British revenue assessments in Awadh it is worth underlining that the British revenue demands were following on years of rack-renting. Thus while the British revenue demands could appear moderate in *relative* terms, ie compared with previous assessments, they could have been too high in *absolute* terms, ie when compared with what remained of the revenue paying capacities of the district. The most controversial aspect of the 1856 settlement was the disposal of rights. Armed with policy prescriptions to settle with village proprietors and the Thomasonian principles regarding the importance of village communities and the grasping character of the talukdar, the revenue officers set about dispossessing talukdars. Data on the losses suffered by individual talukdars is extremely scanty. Some prominent examples can be given. **Lal Madho Singh of Amethi** lost 500 out of the 800 villages he had held; **Man Singh,** the biggest talukdar of Faizabad, lost all but three of his villages; Hanwant Singh of Kalakankar and **Rana Beni Madho of Shankarpur** lost 55% and 44% of their villages respectively. Historians often underplay the extent of the upheaval caused by the Summary Settlement by only providing figures for the number of villages settled with talukdars and for those settled with others; these figures are then set beside the number of villages held by talukdars prior to annexation. A better way to underline the dispossession suffered by the talukdars is to compare the *proportion* of the total number of villages they held under the Nawab's government to the proportion of the total number of villages settled with them in the 1856 settlement. Such a comparison shows that in Nawabi times the talukdars had held about 67% of the total number of villages, by the first British revenue settlement this had been reduced to 38%. In districts like **Faizabad, Sultanpur, Partapgarh, Rae Bareli, Gonda and Bahraich** the talukdars were the hardest hit. These were also the pockets where overassessment was most noticeable. This revenue settlement had important implications. The assumption that the talukdar in Awadh was an interloper was not a valid one. They had a stake in the continuity of cultivation. The wealth of the talukdar was never directed into any kind of productive investment but it served to maintain a certain lifestyle where patronage, protection and loyalty held sway. The removal of the talukdar destroyed this world. It also meant that the surplus which the talukdar had retained and portions of which had circulated in the rural economy as salaries to retainers and through other forms of patronage were now totally extracted by the bureaucratic machinery of the raj. **RUDRANGSHU MUKHERJEE**

SUNDEELA[Sandila] A town and district headquarters in **Oude**[Awadh], long held by the rebels loyal to the Begum and the site of a battle on 10 August 1858 at which **Prince Firoz Shah** was present, although defeated.

SURVEY OF INDIA Headquarters at Dehra Dun; this is the oldest Department of Government (founded 1767). It has produced a map, scale 1:4,000,000, entitled *Sacred Places of Freedom Movement*, in which places of significance in the struggle of 1857 are marked by a crossed musket symbol etc. The very small scale however and the small amount of information conveyed limits the value of the map. The Department also produces an excellent Antique Map (reproduced from a hand-drawn original of 1868) of the Districts of Lucknow, Oonao and Roy.Bareilly.

SUSSIAH[Sasia], Battle of This was fought near Agra on 5 July 1857. Sometimes also called the Battle of Shahganj, also, even more confusingly (for there was another battle just outside Agra on 10 October 1857) the Battle of Agra. An excellent account of the battle is given in R.Montgomery Martin, *The Indian Empire*, Vol II p 360. On the rebels' side the **Kotah Contingent,** with the cavalry of the **Malwa Contingent** and the **Neemuch Brigade** fought very well against the British who were led by Brigadier **Polwhele,** subsequently dismissed for his ineptitude, and drove the British back within **Agra Fort**. The British loss was severe, more than one man in six was killed, *ie* 141, together with some thirty Europeans subsequently killed in Agra during the disgraceful panic in which the fort gates were shut and no attempt was made by the British to rescue their co-religionists and fellow-countrymen. The rebel loss was estimated as 'exceeding 500'. *See also* TOTTENHAM, Capt. John.

SUTHERLAND, Miss An account of her experiences during the mutiny/rebellion, dated 1890, is included in the **Ripon Papers** in the British Library.

SUTHERLAND, Mrs Louisa Orr One of the survivors of the **Captives of Oude**. Her obituary, giving an account of her experiences appears in the Civil & Military Gazette of 20 November 1936, and is held in the National Army Museum, Royal Hospital Road, Chelsea, London SW3.

SWEEPER POLICE In restoring law and order in **Kanpur** after the (first) restoration of British authority there in July 1857, **Major Bruce** deliberated recruited police from the lower castes, presumably in the belief that they would be more trustworthy for his purposes. This was not always the case, and **Bithur** provided an unfortunate experience in the setting up of a **thannah**: 'It had been represented that this place also required to be under military control; and Bruce had been directed to send some of what he called his sweeper police.

Contemporary Artist's Guess at

'Sweeper Police'

As far as, in some measure, disregarding caste goes, the idea was sound; but the engagement of exlusively low-caste men was, perhaps, carrying things too far. Curious adventurers turned up from time to time; and a tall, well-built Hindoo had appeared, who spoke English admirably, and had attached his fortunes to those of Bruce. Men were required to go to Bithoor, and I was sitting one afternoon with Bruce, who was enlisting volunteers. They had to give some reference, and this Hindoo questioned them with great acuteness. At length one man was brought up, and the Hindoo said: 'This fellow relies on his face, and the reference is very unsatisfactory. I had better tell him to pack.' A person capable of conducting business in so lively a manner was not to be overlooked, and when the Thana was established at Bithoor, he was sent over to preside. But he was a regular scamp; and, after day or two of business, determined to have a frolic; and so sent for wine and dancing girls, and had the Thana laid [with] carpets and lighted up, and devoted the night to music and the flowing bowl. A spy, however, sent word to the other side of the river, and a party of Sepoys and rebels got across, came quietly up, and made an attack on the revellers in the small hours. All outlook and precautions had been neglected, there was an attempt to get to arms, but of an ineffectual kind, a fight took place, several were killed. Our lively friend had taken too much to escape, or indeed to be fully aware, perhaps, of what was going on. He was murdered, and his body thrown into the street.' **Sherer**

SYED AHMAD KHAN, Sir The writer of the only Indian contemporary account of the struggle in Rohilkhand. Written in 1858 at Moradabad it does actually comprise three separate works, viz. *'Tarikh Sarkashiy-i-Dhilla Bijnor'* (History of the revolt in the District of Bijnor); *Asbab-i Baghawat-i Hind'* (Causes of the revolt of India); Prayer for Peace, at Moradabad.. At first sight it is difficult to believe in his impartiality since, for example, he does not treat the rebels of Bijnor as the equals in any sense of the East India Company, nor are they either patriots or nationalists; they had defied 'lawful British authority'. Hence Nawab Mahmud Khan the leader of the Bijnor rebels is referred to always as 'Na-Mahmud' (the Cursed One). On the other hand his work is free from the 'colonial terminology' and the consequent distortions of so much of the British literature of the period. *See* the excellent translation by Hafeez Malik and Morris Dembo, *Sir Sayyid Ahmad Khan's History of the Bijnor Rebellion,* East Lansing, 1972.

T

TABLEAU VIVANT For the most part the struggle of 1857-58 was grim and dour and there were few opportunities for an observer, however poetic, to wax lyrical about the scenes enacted. But Thomas Lowe (*Central India during the Rebellion*, London, 1860) does manage to see beauty in the crossing of the river Chambal in November 1857 just before the **Battle of Mundesore[Mandsaur]**: (apparently the passage of the river was difficult, particularly as the transport consisted mainly of bullock-carts, with a few camels etc, and it is fortunate for the British that the crossing was not defended.) 'I never saw a more animated and beautiful picture in my life than when our Brigade crossed this river. The steep, verdant, shrubby banks, covered with our various forces, elephants, camels, horses, bullocks; the deep flowing clear river, reaching on and on to the far east, to the soft blue-tinted horizon; the babble and the yelling of men, the lowing of the cattle, the grunting screams of the camels, and the trumpeting of the wary, heavily-laden elephant; the rattle of our artillery down the bank, through the river, and up the opposite side; the splashing and plunging of our cavalry through the stream - neighing and eager for the green encamping ground before them; and everybody so busy and so jovial, streaming up from the deep water to their respective grounds; and all this in the face, almost, of an enemy formed a *tableau vivant* never to be forgotten.'

TACTICS in the ENTRENCHMENT

One of the best accounts of day to day fighting in **Wheeler's Entrenchment** at Kanpur in June 1857, comes from the pen of one of the two Officer survivors - Lieutenant Mowbray Thomson: 'In order to keep us as fully acquainted as possible with their movements, I had a crow's-nest constructed twenty feet from the ground; it was made of some of the building materials lying about the place. By turns of an hour each, my men were posted up there, and through a loop-hole could overlook the movements of our troublesome neighbours, and telegraph to us beneath. As soon as any intruder quitted barrack No. 1, the signal-man fired at him. One of our party, Lieutenant Stirling, spent many hours in this elevated post, and as he was most expert with his rifle, it is quite impossible to conjecture the results, in the number of sepoys brought down by his gun.....The principal work devolving upon these outpickets was that of clearing the adjacent barracks of our assailants. They would come up from building to building, in a rabble of some hundreds, aud occasionally of thousands, as though intent upon storming our position. Their bugles sounded the advance and the charge, but no inducement could make them quit the safe side of Nos. 1 and 5; from the windows of these barracks they could pepper away upon our walls, yelling defiance, abusing us in the most hellish language, brandishing their swords, and striking up a war-dance. Some of these fanatics, under the influence of infuriating doses of **bhang**, would come out into the open and perform, but at the inevitable cost of life.

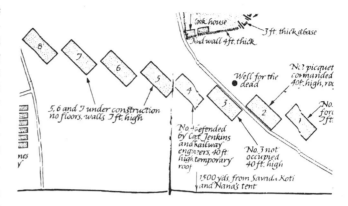

Unfinished barracks, from Yalland, *Traders & Nabobs*

Our combined pickets always swept through these barracks once, and sometimes twice a-day, in chase of the foe. They scarcely ever stood for a hand-to-hand fight, but heaps of them were left dead as the result of these sallies. As soon as we had expelled them from their covert, the musketry and artillery of the intrenched party played upon them furiously, and this process inspired them with a wholesome terror of approaching us. In some of these charges we occasionally bagged a live prisoner or two, but whether from the fatal precision of our fire, or suicide on the part of the wounded, it was strangely rare to see them otherwise than quite dead. When we did bring them in alive they expressed sorrow for their conduct, and attributed the mutiny to the *hawa*, meaning thereby an invisible influence exercised over them by the devil. It is a curious circumstance, that the Hindoos associate almost all calamity with the wind, and in not a few parts of India, the name by which the mutiny has been designated is the **devil's wind**. In the first instance, a prisoner we had taken in the barracks, who had been a private in the **1st BNI**, was sent by us into the main-guard, but he effected his escape. It was not desirable that very frequent accounts of our destitute condition should be conveyed to the rebels; so in future, to remedy this evil, all we took were despatched without reference to head-quarters.' Mowbray Thomson, *The Story of Cawnpore*, London, 1859.

TAHSIL

The office or court of a tahsildar or collector of revenue. When government attempted to restore its authority in a region which had been in rebel hands, it was the tahsil and the **thana** (police station) that they tried first to re-establish.

TAL BAHAT A strong fort near to **Jhansi** in Central India. Held by the rebels for some eight months, the *killadar* being one Jowahir Singh. It was abandoned, without a fight, on 13 March 1858 on the approach of a British force, the rebels retreating into **Jhansi**. *See The Revolt in Central India,* Simla, 1908, *and* Thomas Lowe, *Central India during the Rebellion of 1857-58,* London, 1860, *and* **FSUP**, Vol III; *and* Tapti Roy, *The Politics of a Popular Uprising: Bundelkhand in 1857,* Delhi, 1994.

TALUKDAR RESISTANCE By 17 July 1858 the principal talukdars of Faizabad Division in Oude [Awadh] had all declared in favour of the British. These included 'Rustum Sahae, **Madho Singh Amaithie**, **Man Singh Shah Gunj**, Hanuwant Singh Kalliangurh, Jugpal Singh Tiloee, Raghunath Khujoorgaon, Drig Bejay Singh Morarmou, Sudursun Singh Chundapore.' The exceptions were **Beni Madho Baksh**, Ram Buksh and Debi Singh. Submission by many others was promised to officials on the appearance of British troops in the Division. For details of the assessment of the position at this time *see* Foreign Secret Consultations, 27 August 1858, No 30 . For an unofficial and far less optimistic review *see* T.H.Kavanagh, *How I won the Victoria Cross,* London 1860; he estimates that by the middle of October 1858 less than a quarter of the Province had submitted to British rule; across the Gogra the 'rebel government reigned supreme'.

TALUKDARS' MEN This is the term given to the armed followers of the *Talukdars*, particularly in **Oude [Awadh]**. In the final struggle for Lucknow in March 1858 these men outnumbered the Sepoys in the army of the Begum **Hazrat Mahal**, and usually fought very well. Armed with matchlocks and swords (although a minority might have had flint fire-locks and muskets) they were described as a 'rabble' by Lt Colonel Strachey the Secretary to Government of the Central Provinces, and were similarly underrated by others, but they lacked organisation and training, not courage. They were composed of the men of the military classes, accustomed to the use of arms, and to constant fighting from their earliest youth, with one another (for their respective talukdars or rajas), and with the troops of the former King of Awadh, **Wajid Ali Shah** .

TALUKDARS Holders of great estates. There seems little doubt that **Oude[Awadh]** was at the very centre - in all senses - of the Mutiny/Rebellion of 1857. And the people of Oude had not appreciated or accepted the British land settlement of 1856, following annexation, that had deprived many Talukdars of their ancient rights and privileges.That many of them had usurped these privileges appears to be beside the point: the British attempt to overturn the Talukdari system in favour of the smaller landholders and for the benefit of the peasantry, was apparently a failure. (**Gubbins** the Financial Commissioner in Oude was one of the few who disagreed with this view - see M.R.Gubbins, *An Account of the Mutinies in Oudh and of the Siege of the Lucknow Residency,* London, 1858). Earl **Canning**, the Governor General (and first Viceroy) wrote on 6 October 1858...'that as the village holders did not appreciate our overturning the Talukdari system in Oudh, and turned at once to the old Talukdars when British authority was overthrown, as though they wished the talukdar to assert his former rights and resume his ancient position over them at the first opportunity....this equalled an admission that their own rights were subordinate to the Talukdar's. If they had valued their rights they would have supported government, but they have done nothing of the kind.....deserve little consideration...and therefore the Governor General has determined that a talukdari settlement shall be made'. This seems like punishing the subordinates for submitting to *force majeure*, but it had the merit of being logical. *See* Rudrangshu Mukherjee's assessment in the SUMMARY SETTLEMENT of 1856. For a statement of the conduct of individual Oude talukdars during the Revolt *See* Rudrangshu Mukherjee, *Awadh in Revolt,* Delhi, 1984, Appendix to Chapter 5, p 189.

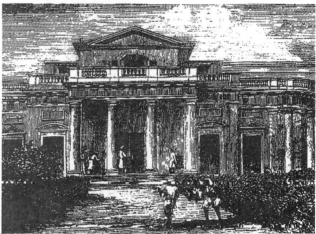

Tara Wali Kothi, Lucknow

TARA WALI KOTHI This was the Astronomical Observatory in Lucknow, now the main branch of the State Bank of India. A building in Lucknow, built as an Observatory by Nasir-ud-din Haidar, although the last King of Oude, **Wajid Ali Shah** had dismissed the establishment. The valuable astronomical instruments disappeared during the rebellion, and may have been broken up by the rebels as being of foreign manufacture and therefore suspect. The building was occupied by Maulvi Ahmadulla Shah, the Maulvi of Faizabad, and has other claims to fame/notoriety: it was the place in which **Amelia Bennett** says that she was imprisoned for some time; that part of the rebel parliament under the Maulvi's influence met there; in front of the building, and between it and the **Kaiserbagh** two separate parties of Europeans were put to death: on 24 September 1857 those sent in by the **Dhauraha** Raja *ie* **Miss Georgina Jackson**, Mrs Green, Mrs Rogers, Mr Carew, Mr J. Sullivan, and on 16 November 1857 those sent in by **Raja Loni Singh of Mitauli** *ie* **Sir Mountstuart Jackson**, Captain Orr, Lieutenant Burns and Sergeant Morton. During the Revolt of 1857 it served as the Headquarters of Begum Hazrat Mahal and Ahmad Ullah Shah. Was an important stronghold. Courtesy INTACH.

TASMA BAZ

Kanpur before the uprising was filled with *budmashes*, fugitives from justice on their way to or from **Oude**[Awadh], depending on where their crimes had been committed; in British territory they were safe from the King of Awadh's vengeance, in Oude they were safe from the long arm of British justice. There were some interesting variations of criminal activity. In the early 19th century an English private named Craegh initiated three budmashes in a gambling operation called 'Pricking the Garter,' in which a bystander was encouraged to insert a stick in a folded double strap; if, when the strap was pulled, the stick came free, he won. Craegh and his cronies dubbed the game 'Tasma-Baz' and turned it into a kind of criminal guild. They set up games in the city streets, bribed the police, and sometimes resorted to extortion, robbery and even murder. Other organized bands called Megpunnas posed as employment agents in order to lure families out onto the road, murder the parents, and sell the children as slaves. Daturias acted like Thugs but instead of strangling their victims administered poison. As a result Kanpur had an evil reputation which, to some extent, survived the mutiny/rebellion.

TATYA TOPE

Regarded by some as the most gifted of the leaders of the 1857 Rebellion. He fought in the old Maratha way, never committing his entire force to any battle, frequently retiring - even from excellent prepared positions - before a general engagement, in the hope of wearing down the enemy. His guerilla tactics were successful in prolonging the rebellion in Central India, long after it was effectually defeated in Oude[Awadh] and the North, but it can also be argued that he squandered the rebels' chief advantage, which was overwhelming numerical superiority. His one positive success was in the Battle of **Kanpur** where he led the **Gwalior Contingent** against the British forces left by the Commander-in-Chief **Sir Colin Campbell** to garrison Cawnpore[Kanpur] and block the rebels' advance into Oude[Awadh]. Eventually betrayed by **Man Singh** and hanged by the British in April 1859. Real name Ramchund Pandurang, aged about 42 when he died. Protested, with justice, at his trial that he could not be guilty of rebellion against the British as he was never subject to their government, and served only his master, Dondhu Pant, the **Nana Sahib**. Colonel C.B. Thackeray praises **Tatya Tope**, whom he calls 'a guerilla leader of genius, to be compared with De Wet in the Boer War. The only one with serious pretensions to generalship.' But **Sir Hugh Rose**, the commander of the British forces in Central India was not so sure: 'He preferred on principle to put flight before combat. He possessed no real talent. But he had a perfect genius for running away.' He had to defer for some time to the superior status of **Rao Sahib**, and even when nominally in charge of the rebel army in Central India he was not always obeyed.

Tatya Tope (from Savarkar's *The Indian War of Independence*)

A letter to the Peishwa (Nana Sahib) dated 'At Charkhari Fort 10 Feb 1858', is most revealing: 'The troops here are at variance with each other. Some are of opinion that the fort should be caused to be evacuated, while Chutta Singh and others say that 'as the Rajah solicits our protection by putting a straw between his teeth, we will not storm the fort'. The refusal of these persons to attack the fort, witholds the Rajah from siding with us. Under these circumstances, I wished the troops to return to Calpee, but they do not conform to my wish. In consequence of the different opinions of the troops, I shall be obliged first to try to make them all unanimous, and then to write to you what measures may be adopted' *See* LEWES. *See* Appendix XVII of *The Revolt in Central India 1857-59*, Simla,1908, for the full statement made by Tatya Tope after his capture in April 1859, and just before his trial and execution. *See also* **CM** p 79 *et seq., and* Sir George William Forrest, ed., *Selections from the Letters, Dispatches and other State Papers of the Military Department of the Government of India 1857-58*, 4 vols, Calcutta, 1893-1912, Vol IV, *and* Tapti Roy, *The Politics of a Popular Uprising: Bundelkhand in 1857*, Delhi, 1994.

TATYA TOPE'S CAPTURE

'Each fresh commandant who took the field fancied he could catch Tantia; prodigious marches were made, officers and men threw aside all baggage, even their tents, and accomplished upwards of forty miles daily - the rebels did fifty. The end was, all our horses were sore-backed, and the halt of a week or ten days rendered absolutely necessary.

Said to be 'Tatya Tope in camp'

Then came a new aspirant for a C.B. and Tantia's head, who brought fresh troops and camels into the field. He had perhaps not only to chase Tantia, but to keep clear of other forces commanded by a senior in rank to himself. It was wonderful the amount of energy that was thrown into the pursuit, and the hundreds of dead camels strewn over every jungle track; roads were no object, or rivers either to pursued or pursuers. On they went until dead beaten. Occasionally someone more fortunate than the rest had the luck to catch up the fugitives and cut up stragglers; but it was always in heavy jungle; they had the very best of information and never trusted themselves to the open country when any force was near. We had the very worst of information even in the territories of professedly friendly rajas. *The sympathy of the people was on their side.*' The actual capture of Tatya Tope is well documented; it owed more to the treachery of one whom he thought was his friend than to British endeavour: eg Major R Meade's account of the capture:' ..it is only necessary for me to remark that Maun Singh had been made acquainted with Sir R.Hamilton's offer to him of a yearly pension, if he came in, and of his assurance of further consideration from Government, in the event *of his performing any signal act of service, such as bringing in Tantia Topee*, the Rao etc previous to his surrender to me. The pension I was not authorised to guarantee him, but I have seen Captain Bolton's original letter to him, stating that Sir R.Hamilton would cause the *Durbar* to give him 10,000 Rs per annum, if he came in; a copy of Sir R.Hamilton's telegram promising him still further consideration as above, was however furnished to me, and I made use of its contents to urge him to the utmost, to merit the reward held out though only in general terms by some such signal service as that contemplated. *Maun Singh has now performed that service by enabling me to capture Tantia Topee*, and it appears to me for the British Government to decide what further reward shall be conferred on him for that act.....I was naturally desirous, both for myself and the officers with me, that we should share in the enterprise, but Maun Singh was so urgent that the arrangements for the capture should be left in his hands that I felt it would be hazardous to disappoint him, and of course my first duty being to secure the seizure of so great a criminal, I could suffer no personal considerations to interfere therewith: I consented therefore to place a small party of native infantry under his orders for the occasion, and having settled all that was required with him, I let him proceed to Parone about 3 p.m. to search for and get hold of Tantia Topee, and at 5 p.m. I sent after him a party of the 9th Bombay Native Infantry, under Abdoola Jeo, with orders to do what Maun Singh directed them, and to apprehend any suspicious characters that he might point out; *of course Tantia Topee's name was never mentioned, and the party had no idea of the duty on which they were proceeding.* [Note. this may have been just as well - the 9th Bombay Infantry had many Marathas in it, and Tatya of course was a Maratha, loyalty may have been strained - PJOT.] By Maun Singh's directions the *Sipahies* were placed in ambush near a hollow which he and Tantia Topee had been in the habit of frequenting and he led his unsuspecting victim there and held a long conversation with him, till after midnight when Tantia Topee fell asleep. *The Sipahies were then fetched by Maun Singh, and Tantia Topee was secured and pinioned, his arms being seized by Maun Singh himself; unfortunately during the confusion two Pundits he had with him managed to escape on horseback...*Maun Singh has requested that the money taken on Tantia Topee may be given to the men whom he despatched to track and find him out as he promised they should have it if they were successful...I have agreed to give him all but twenty-one gold *Mohurs*, which I wish to distribute amongst the party which captured Tantia Topee and I trust the General will approve of this.' He did..the blood money was distributed. Hardikar believes that Man Singh was in league with Tatya Tope and may have substituted some one else in his place: surviving relatives of Tatya Tope at **Bithur** believe he was not executed on 18 April 1859. According to his nephew, Sri Narain Lakshman Tope, Tatya Tope often visited his parents at Bithur between 1859 and 1862 in disguise, and rendered them monetary help; see Rajni Kant Gupta, *Military Traits of Tatya Tope*, Delhi, 1987, p 151. See also Intelligence Branch Chief of Staff Simla, *Revolt in Central India*, 1908; see also **FSUP,** Vol III, Chapter eight, p 555 *et seq., and* Tapti Roy, *The Politics of a Popular Uprising: Bundelkhand in 1857*, Delhi, 1994.

TATYA TOPE, PURSUIT of 'During this period (June 1858 onwards) the rebels were pursued by a Light Field Force under Lt.Colonel Holmes, and were thus kept on the move, while General Roberts followed with the main body of his force. But the troops had little chance of catching the swift-footed rebel, who carried no tents and no provisions; these he looted as required and when his horses were worn out left them on the road to die, and replaced them, sometimes from the dak stations. His light horse could hover round the British columns like shadows, and always get away from overworked irregular, or overweighted regular, cavalry. *The sympathy of the people was, moreover, with the fugitives, who obtained information and supplies without difficulty. ....when disencumbered of their guns they moved with greater celerity and secrecy, and thus more easily eluded their pursuers.'* See *The Revolt in Central India*, Simla, 1908 p 205 . See also **FSUP**, and Tapti Roy, *The Politics of a Popular Uprising: Bundelkhand in 1857*, Delhi, 1994.

TAYLER, William

Was, at the beginning of the struggle, Commissioner of **Patna**, and his prompt action in arresting the leading *Wahabi* maulvis possibly saved the town, and Bihar, for government, but he had quarrelled with Halliday, the Lieutenant Governor of Bengal; the men were incompatible, and Halliday took the excuse of Tayler's order to his subordinates to concentrate at Patna the resources in men and money of the province, to remove him from office for his 'ill-judged and faint-hearted order', which had, so Halliday averred spread alarm in every direction.

> MR. TAYLER AND THE LIEUTENANT-GOVERNOR OF BENGAL.*
>
> The names of Mr. William Tayler, Mr. Samuells, Ameer Ally, and Mr. Halliday, have become household words in Patna politics. They suffused society at one time in Calcutta, and almost monopolized the newspapers. Letters from all parts of Bengal teemed with discussions upon various points of the case. Mr. Tayler was for a time the most conspicuous person in the public eye. The position is a trying one, and nothing could have saved his case from developing into a bore but the singular popularity that he enjoyed, and the singular want of that advantage on the part of those opposed to him.
>
> But though the newspapers were full of Mr. Tayler, we have never seen in one of them a fair analysis of his case. The writers usually confined their remarks to the particular points as they arose. The vigorous and energetic measures of the Commissioner of Patna

From the Mofussilite

This was effectively the end of Tayler's official career, but he was to return to Patna to plague government for many years representing clients, mostly well-to-do Indians, in the Courts. Tayler had detractors but also many ardent supporters, of whom the most active was Colonel **Malleson**. See William Tayler, *Thirty-eight Years in India*, 2 vols, London 1881/2. See also PATNA.

TAYLOR, Captain Alexander

'An engineer officer, later promoted to a knighthood and the rank of Major General, who was the leading engineer in the British forces on the Ridge at Delhi in 1857, and responsible in the main for the battle plan which was the basis of the final assault upon the city in September 1857. His letters and papers are held in the National Army Museum, Royal Hospital Road, Chelsea, London SW3. 'There have been lots of 'Captain Taylors' in India in the last 300 years. I was, no doubt, the least distinguished, but I take pride in the name, however common it might be. At the 'siege' of Delhi in 1857 the 'mutineers' held the city and the British held the Ridge to the Northwest. Since neither protagonist controlled all sides of their respective enemy's position, it is not clear who was actually 'besieging' whom, but that is beside the point. From 1 June to 14 September 1857 the British hesitated and shrank from an all-out attack on the city. They were comparatively weak, the mutineers were strong - in numbers at least. Whereas in every engagement the British lost irreplaceable soldiers, the mutineers were constantly being reinforced by the arrival of new recruits to the army of the King of Delhi. It was this realisation that the adverse ratio of forces could only get worse as time went on that forced the British commanders to grasp the nettle and attack the city. Before they could do so they had to draw up a plan of attack. In those days that task belonged always to the Engineers. Lieut.Colonel Baird-Smith, the senior engineer with the Delhi Field Force, was instructed to draw up a battle plan, and he, in his turn (partly it must be said because he was wounded) nominated a particularly promising young man, a Captain Alexander Taylor, as the 'Director of Attack'. Why, we may ask, was a comparatively unknown and junior officer given such enormous responsibility? Sir John Lawrence, Governor of the Punjab and future Viceroy, wrote to Taylor after the siege: 'I have to congratulate you on your success at Delhi. I look on it that you and Nicholson, poor fellow (he was of course killed in the assault), are the real captors of Delhi.' Nicholson himself, just before the assault, indignant that Taylor had not had justice done to him, said 'Well, if I live through this I will let the world know who took Delhi...that Alex Taylor did it'. Perhaps it is because Nicholson did not live through it that Taylor's name has fallen into virtual oblivion, and no one today - save perhaps his humble namesake - remembers his contribution. What did he do? It is said that 'At Delhi he simply did as a duty what was laid on him, and humbly let others take the credit and honour really due to him'. He and Colonel Baird-Smith were fortunately in agreement on general principles and Taylor never had any difficulty in doing what he considered necessary, as he was invariably supported by his superior. What they both knew was that no plan of attack on the massive bastions and walls of Delhi could possibly succeed were it not the result of careful critical examination on the very ground. There could be no room for error, no doubt as to lines of fire for attacking batteries of guns etc. The ground that the British would have to use and operate from had been in the hands of the mutineers since May, and part of it was studded with large buildings and trees. What had they done with it ? Had they levelled the Custom House, the Khoodsia Bagh, Ludlow Castle? Had they fortified them, did they occupy them in force ? A project could only be prepared when these questions, and others, could be answered with certainty. It was easier said than done: the mutineers had a large picquet - perhaps the size of a whole regiment - in Ludlow Castle alone.

Ludlow Castle, Delhi. Now a Secondary School

One day however Taylor watched Ludlow Castle through a telescope from the Ridge, and came to the conclusion that although the old picquet had certainly retired, the new, relieving picquet had not yet arrived from the city. Armed only with a pistol, and taking a party of 16 men from the Guides, he slipped along from Metcalfe's House, between Ludlow Castle and the river, until he reached the Khoodsia Bagh. He explored it thoroughly, and climbed the wall next to the city: he could see the sentry on the ramparts, seemingly so close that it was difficult not to think that he must have heard the noise of his climbing the wall. Taylor later wrote: 'The Custom House, 180 yards from the sentry, lay immediately in front of me, and for more than an hour, lying on my face on the top of the wall, behind a small shrub I carefully examined it and the ground around it. I learned that it had been burnt.. but that the walls remained standing (this was eventually the site for British breaching battery No3), that the enemy had not occupied it, or the Khoodsia Bagh - the vegetation being fresh and untrampled, and that the only place in the vicinity outside the walls which was occupied in force by the mutineers was Ludlow Castle.' Further reconnaissances like this took place in July, August and September, until Taylor had built up a complete picture of the ground over which the British assault would have to be launched. On one occasion he was actually on the roof of Ludlow Castle when the mutineers sent in a new picquet, and he had to slip down quickly - 'fortunately without being observed'. In this laborious way he was able to mature a project, a plan of attack, that would bear scrutiny. Ironically, so detailed was this plan, so full of minutiae with regard to the ground between the Ridge and the City, that the General (Archdale Wilson) refused to believe that it was genuine! It was only when General Nich-olson undertook to verify things that Wilson agreed. Taylor took Nicholson at midnight into Ludlow Castle (fortunately empty at that moment) and into the Khoodsia Bagh, and got him safely back into camp. This made Nicholson a devoted supporter of Taylor, and he spent all day and every day with him in the build up to the attack, helping the young officer prepare the siege batteries and generally smoothing his way in getting men and materials, and cutting through red tape. The assault was a success - for the British - but a bloody one: they had known the risk. Alexander Taylor had a distinguished career after the 'Mutiny'. In 1861 he was sent back to his old work, the Grand Trunk Road connecting Lahore with Peshawar, known then, in the Punjab, as 'Taylor's Road'. He reached the rank of Major General and was knighted, and died in 1912 at the age of 86. He was a modest man - all Taylors are! - and never sought the glory or the reward that were rightly his.' See The Statesman and **ASR**. **PJOT**

TAYLOR, PHILIP MEADOWS (1808-76) A member of the uncovenanted Civil Service, **HEIC**, but best known as a novelist. First employment in India was with a Bombay merchant, but then entered the service of the Nizam, first as a soldier, then in a civil capacity. He was an influential correspondent for the *'Times'* 1840-53. On the outbreak of the rebellion in 1857 he kept the District of Buldana in N.Berar peaceful, and his comments upon the state of India at this time are shrewd and worthy of respect. Best known of his novels was the highly successful *'Confessions of a Thug'* (1839), reprint Oxford, 1986. See **DNB**. and Meadows Taylor, *'Letters during the Indian Rebellion'*, 1857.

TEEKA SINGH A Subadar of the 2nd Cavalry, Bengal Army, and one of the leaders of the mutiny at **Kanpur**. He soon rose to considerable power and influence with **Nana Sahib**, and appears to have become one of the most significant of Nana's advisers. He was made a Brigadier. His regiment however had many Moslem troopers in it, and whether on the grounds of religion (see COMMUNAL FRICTION) or because they envied his activities in lining his own pocket, on more than one occasion they treated him with scant respect: 'on the 20th June these lawless troopers arrested their Brigadier-General, Teeka Singh, for having sent to his own home two cart-loads of ghee and sugar seized upon the roads belonging to some Mahajun; from which he was released after much begging and entreaties both on his own part and that of the Nana. The fact is, the troopers became jealous of Teeka Singh, for he had been amassing a great deal of money derived from plunder and confiscation, and also on account of his having been presented by the Nana, as a mark of his favour, with an elephant and a pair of gold bangles, besides other khilluts'. W.J. Shepherd, *Personal Narrative etc.*, Lucknow, 1886.

TEJ SINGH, Raja See MYNPOORIE, RAJA of

TEJA SINGH, SARDAR An important minister in the government of Maharaja Ranjit Singh in the days when the Sikhs had ruled the Punjab; his fort at **Sialkot** was still standing and in good repair, and became the designated place of refuge for the Europeans at that station when mutiny appeared inevitable in June 1857. The government treasure was quietly removed thereto from the Treasury, and the fort provisioned and other preparations made by the Deputy Commissioner for a short siege.

TELEGRAPH, Electric In 1857 communications were difficult and slow. That great invention 'the electric telegraph' had recently been introduced, but in anarchic times it was very vulnerable: one snip or slash and the line was broken and dead, and all information and news cut off. In July 1857 one officer, Captain J.G. Medley, was in the Punjab, but so far as knowing what was going on he might as well have been in London; he writes: 'Beyond Delhi our news was a blank. The whole country was in the enemy's hands and our only means of communication was round by Bombay and Calcutta, where the ignorance of what was passing between Allahabad and Delhi was almost as great as our own. To give some idea of this ignorance, I may mention that it was generally rumoured in camp that Sir Hugh Wheeler with six European regiments was advancing from below to reinforce Delhi, nearly a month after he and his weak detachment had been massacred at Kanpur'. But when, as order began to be restored, the telegraph came back into use, it allowed practically instantaneous communication from the scene of action back to the government in Calcutta. As an example: Gwalior June 1858 telegraph to Governor General 'The Rani of Jhansi is killed'. Palace of Gwalior 19 June, 1858 Hugh Rose, telegraphed to Governor General 'The force under my command took Gwalior after a general action, which lasted 5 hours and a half. The enemy evacuated the fort. My Cavalry and Artillery are in pursuit.' This pursuit was in its turn aided by the entirely military use of the telegraph, thus: 'The following morning, a wounded Mahratta retainer of the Ranee was sent in to me from Captain Abbott's Flying Camp(sic). He stated that the Ranee, accompanied by 300 Vilaities and 15 sowars fled that night from the fort; that after leaving it, they had been headed back by one of the picquets where the Ranee and her party separated, she herself taking to the right with a few sowars in the direction of her intended flight to Bandiri. The observatory also telegraphed *'enemy escaping to the north-east'.* 750 miles in all of telegraph wire were totally destroyed during the rebellion, the posts being used for firewood, the wire cut up and hammered for bullets, and some of the cast-iron anti-termite tubes for the poles actually used as (extremely dangerous) guns by the rebels. See P.V.Luke, *Early History of the Telegraph in India,* Journal of the Institute of Electrical Engineers, Vol XX, 1891. See also *The Electric Telegraph,* in Calcutta Review, March 1858; *and, How the Electric Telegraph saved India,* in MacMillan's magazine 76, 401-6, October 1897.

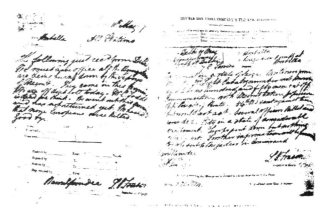

Telegrams announcing the Mutiny

TENTH REGIMENT BENGAL NATIVE INFANTRY The 10th was known as the *Duffel Ka Paltan* and was raised in 1763. It saw action at Buxar, Kora, and in the Burmese campaign. Because it had compromised caste by serving overseas (in Burma) it was jeered at by many in **Fatehgarh** as a 'Christian' regiment. At the beginning of the outbreak they appeared to be remaining staunch, but eventually succombed and joined the mutiny. Most of the 10th fled Fatehgarh with loot and were themselves looted, and some murdered, by local villagers. The survivors joined the **Nana** at **Kanpur** and remained together until 5 March, 1858, when they were defeated by Colonel Rowcroft's troops at Amora and retreated into the hills. *See* G.H.D. Gimlette, *A Postscript to the Records of the Indian Mutiny: An Attempt to Trace the Subsequent Careers and Fate of the Rebel Bengal Regiments, 1857 to 1858.* London, 1927, pp 95-7.

TERAI, The Frequent mention is made in accounts of the rebellion - especially those dealing with the campaign in **Oude[Awadh]** and with the fate of the **Nana Sahib** and other leaders - to an area known as the Terai, above all as the place of fever and sickness that was not only dangerous to British troops but appears to have been the death of numerous rebels - including Nana Sahib and **Azimullah Khan**. It is as well therefore to define what we mean by this territory:- Skirting the hot plains before they break into the first swell of the mountains, runs that continuous belt of horrendous jungle which is known locally as the Terai or 'wet land'. In some places it was thicker and more pestiferous than in others, but everywhere it was a veritable 'belt of death'. 'Rank rotting vegetation fermenting and corrupting in the oppressive atmosphere was perpetually generating the deadly miasma which hung like a grey pall about the path of the passing traveller. That it was the haunt of all sorts of dangerous wild animals from elephants and man-eating tigers to huge pythons and each smaller species of poisonous reptiles, signifies comparatively little. It was the air he was forced to inhale that was the real peril to the traveller; even taking the fatal jungle in a rapid rush he may carry away the seeds of disease to be developed later,' (NB Death of Nana Sahib, Azimoolah Khan etc of 'miasma' diseases like plague malaria typhus etc). It was reputed to furnish the best timber in the world. The sisu or sheetham the fine toon the teak the sankor the dhank and the magnificent ebony and sandal trees grew in unsurpassed splendour.

THAKOOR RAM GHOLAM SINGH *See* PERSONS OF NOTE, APRIL 1859.

THANA *See* THANNAH.

THANADAR The rural policeman - later equated with the rank of Inspector or Sub-Inspector - who was in charge of the police station known as a Thannah.

THANNAH A rural police station, the establishment and maintenance of which in any particular locality implied the dominance in that area of the party - rebel or government - that sponsored it.

THANKSGIVING, Day of 'The 1st July 1859. Proclamation. The Restoration of PEACE and TRANQUILITY to the QUEEN'S DOMINIONS in INDIA makes it the grateful Duty of the Viceroy and Governor General in Council to direct that a Day be appointed for a solemn THANKSGIVING to ALMIGHTY GOD for His signal Mercies and Protection. War is at an end; Rebellion is put down; the Noise of Arms is no longer heard where the Enemies of the State have persisted in their last Struggle; the Presence of large Forces in the Field has ceased to be necessary; Order is re-established; and peaceful Pursuits have everywhere resumed.' Names 28 July, 1859 as a Day of Thanksgiving.

THEATRE Amateur theatricals were as important for British expatriates in Kanpur as in any other station in India, but the theatre there was a particularly grand affair, and its functions were important social occasions.

In 1829 the theatre was built to a design of Lieut. Burt of the Engineers for a cost of Rs 30,000. It was a horse-shoe shape with elegant Doric columns and handsomely furnished seating arrangements and unfortunately no attention had been paid to acoustics. Performances were given once a month and followed by elaborate suppers across the road at the **Assembly Rooms**. The theatre was built partly from cantonments funds and partly from public subscription; it was restored by the Municipality which took it from the military authorities in 1868, and continued to be used as a theatre until 1905 when it was purchased for use as a Central Telegraph Office, which function it still performs. *See* Zoë Yalland. *A Guide to the Kacheri Cemetery,* p 17. Courtesy INTACH.

THIRD LIGHT CAVALRY This unit of the Bengal army was stationed at **Meerut** in May 1857, and was commanded by Colonel G.M.**Carmichael Smyth.** It was a leading regiment and had a distinguished record (Delhi, Laswaree, Deig and Bhurtpore under Lord Lake, since then

Afghanistan, Ghuznee, Aliwal, and Sobraon). It contained a high proportion of men of good family and high caste. The general weapon was the sword; but fifteen in each troop were taught to use firearms, and distinguished as carabineers or skirmishers. To these men Colonel Smyth resolved to teach the mode of tearing instead of biting the cartridges (*see* THE CARTRIDGE QUESTION). On parade eighty-five of them refused the cartridges, were court-martialled and sentenced to ten years in prison, and at a punishment parade were manacled and led off to prison. That same night, 10 May 1857, the remainder of the 3rd Light Cavalry mutinied, were joined by the other native troops in Meerut and marched for **Delhi**. The mutiny/rebellion had begun. Most general accounts of the struggle give full details of these events, *eg see* R.Montgomery Martin, *History of the Indian Empire*, Vol II, London, 1858-60, p 143 *et seq*. See also Lt.Col.G. Gimlette, *Postscript to the Records of the Indian Mutiny*, London, 1927.

THIRTY-SECOND MESS HOUSE See KURSHID MANZIL.

THOMAS, Evelyn Maude Werge (1891-1975) A writer who had intended a biography of **Brigadier General John Nicholson**. His collection of material - copies and original material - is preserved in the Oriental and India Office collections of the British Library, *see* MSS.Eur. F.171.

THOMASON J.G. British Deputy Commissioner at **Muhamdie[Mohammadi]** in **Oude[Awadh]**. Described by **Captain Patrick Orr** as 'poor good Thomason', he was killed at **Arungabad** between Mohammadi and **Sitapur** in June 1857. *See* John Bonham, *Oude in 1857*, London, 1928, p 42.

THOMPSON, Edward J. An author who sought to redress the balance in terms of the **atrocities** committed during the mutiny/rebellion, arguing that they were by no means all on the rebel side. He deals with English atrocities against Indians during the rebellion, and claims that native memories of the Mutiny are a heavy obstacle to friendship and understanding. His book is valuable, and, almost, unique in its approach. *See* E.J.Thompson, *The Other side of the Medal*, London, 1925.

THOMSON, Captain Mowbray An officer of the 53rd regiment BNI stationed in Kanpur. He was one of the only two eventual officer survivors of the Kanpur Garrison (the other was **Lt.Delafosse**). He escaped the massacre at **Satichaura Ghat** by swimming with three companions (**Lieutenant Delafosse**, Privates **Murphy** and **Sullivan** down the Ganges eventually coming ashore at Moorar Mhow, in the territory of **Raja Dirig Bijai Singh** who befriended them. He returned from Allahabad following Havelock's column and re-entered Kanpur to find all the original garrison were dead including the women who had been kept hostage for some two weeks and then slaughtered in the **Bibigurh**.

The photograph of the ruins of Wheeler's Intrenchments at this time (included in the **Maude Album**, *see* **Maude, Major F.C.**) shows Mowbray Thomson in the foreground with his arms folded, and described (no explanation is given) in Maude's notes as being 'mad'. When once asked 'When you got once more amongst all your countrymen, and the whole terrible thing was over, what on earth was the first thing you did? Did? cried he, why I went and reported myself as present and ready for duty.' Mowbray Thomson had a successful military career and retired eventually with the rank of Major General. He wrote (with the assistance of an anonymous 'ghost writer') an account of his experiences in 1857: *see* Capt. Mowbray Thomson, *The Story of Cawnpore*, London, 1859.

THORNHILL, C.B. Even a veteran like C.B. Thornhill was at a loss to explain the outbreak. 'We have seen men, who apparently had strong inducements to take part with the rebels, maintain an undeviating fidelity to the British Government; while others, who were certain to be heavy losers by a change of rulers, and had no reasonable grounds for anticipating any personal advantage from the subversion of our dominion, became our most virulent antagonists.' C.B. Thornhill, Officiating Commissioner, Allahabad Division, dated 28 April, 1859. G.W. Forrest, ed., *Selections* etc. Calcutta, 1893-1912. Vol. III, p xix.

TILAK RAM TIWARI General, *See* PERSONS OF NOTE, APRIL 1859.

TILANGAS Soldiers, by which was meant regular soldiers of the Bengal Army, in contrast to **talukdari men**; the first use of the term had been in Madras where it was applied to Madrassi sepoys.

TIMES, The London A most prestigious newspaper of the time, known as the 'Thunderer' for its willingness to speak out for truth and justice. It had just considerably enhanced its reputation during the Crimean War, in particular by drawing attention to the conditions under which British soldiers had to live, and supporting the campaign of Florence Nightingale for reform of hospitals etc. Its correspondent was the redoubtable **William Howard Russell**, but he did not arrive in India until January 1858. Previous to his arrival and before the stream of accurate and informative letters from him began to arrive in London, the *Times* was dependent on a number of local 'correspondents' who on occasions were no better than passers on of bazar rumour and gossip. Thus, 'The *Times* correspondent', in a dispatch dated the day *after* the massacre at **Satichaura Ghat**, reported that **Kanpur** had been saved by the arrival of HM's 84th and 64th Regiments, 'the latter from Bombay.' They were nearer the truth *re* **Lucknow** '...The gallant and able **Sir Henry Lawrence** [is] grimly busy in hanging the mutineers at Lucknow, with loaded field-pieces and lighted port-fires on either side of the scaffold, and I hope that there is no truth in the report that he has been forced to abandon his capital, and fall back on Kanpur', *see* the London Times, 1 August, 1857. Moreover the Editor was at first one of the most savage exponents of the doctrine of revenge: 'We claim the confidence of our readers when we tell them that we have received letters from the seat of rebellion which inform us that these merciless fiends have treated our countrymen, and, still worse, our countrywomen and their children, in such a manner that even men can scarcely hint to each other in whispers the awful details. We cannot print these narratives, they are too foul for publication. We should have to speak of families murdered in cold blood, and murder was mercy! of the violation of English ladies in the presence of their husbands, of their parents, of their children, and then, but not till then, of their assassination. ...These ruffians must be made to feel the consequences to themselves of the wrath which they have provoked. We are prepared to support our officers and soldiers in the discharge of their duty if they have retaliated upon these monsters according to the measure of their offences. ...We are very confident that we represent the feelings and opinions of the British Empire, with the exception of a small and insignificant fraction of cold-hearted theorists, when we say that the European officers and soldiers now employed in the suppression of this military Mutiny may look for the unhesitating support of their countrymen, however stern may be the measures which they think proper to employ.' *See* Editorial, London *Times*, 6 August, 1857. Fortunately for the reputation of the *Times*, this tone altered completely when more accurate and liberal-minded reports came from India (particularly of course from W.H.Russell, from February 1858 onwards), when 'cold-hearted theorists' became warmer-hearted pragmatists. **PJOT**

TIMUR
'Tamerlain or Timur, the tyrant who, from his capital at Samarkand, conquered most of what we call the Middle East, invaded India mercifully briefly, was, most significantly, the founder of the **Mughal** dynasty, [Babur was descended from Timur on his father's side and from Chingiz Khan on his mother's]. Timur apparently left Samarkand in April 1398 and by December of the same year had reached the city of **Delhi**. His army consisted almost exclusively of cavalry - 90,000 of them, and he now faced in battle the army of the King of Delhi, Mahmud. The latter had only 10,000 cavalry but had 40,000 infantry and, incredibly, 120 armour-plated elephants carrying grenadiers and archers. But the issue was not long in doubt, and a triumphant Timur entered Delhi and gave it over to a general massacre that lasted five days, and to a wholesale sack which produced an enormous quantity of gold, silver and precious stones. He swept on, in a relentless and unpitying stream of slaughter and rapine, destroying Meerut and, by Hardwar, Kangra and Jammu journeying to cross the Indus and return to his beloved Samarkand. Delhi was left as a city in which for two months 'not a bird moved wing'. Not for the last time, nor the first, its citizens were massacred at the hands of a foreign invader. But of all the Delhi massacres, Timur's was probably the most terrible. He had brought to a fine art the process of obliterating his enemies. So terrible was his name, his reputation for cruelty, his implacable hatred for all who dared to oppose him, that sometimes he had only to appear upon the battle field for the armies sent against him to melt away without a blow or a shot.' **PJOT.**

TODD 'The only Englishman resident at Bithur was a Mr Todd, who had come out in the employment of the Grand Trunk Railroad, but for some reason had exchanged his situation for that of teacher of English to the household of his Excellency Seereek Dhoondoo Punth. Mr Todd was allowed to join us in the intrenchment; when the siege began he was appointed to my picket, and was one of those who perished at the time of embarkation'. Mowbray Thomson, *The Story of Cawnpore,* London, 1859, p 51.

TOMBSTONES, and EPITAPHS Much information can be gleaned from a study of the tombstones in Christian cemeteries, dating from this period, especially in **Kanpur, Meerut, Delhi, Lucknow** (Residency), and **Morar** (the cantonment close to **Gwalior**). Although of an earlier period the following epitaph shows what may be deduced or inferred from just one tombstone (it is at Bundela):

'In loving memory of Sergt W.Brown of the Madras Artillery, died May 12th 1826, and Gunner J.Maloney of the same company, died September 27th 1828, this tomb was erected by their sorrowing widow.' See also Zoë Yalland, *Kacheri Cemetery Kanpur*, BACSA, London, 1985.

TOOLSHEPOOR, [Tulsipur] A town in the district of **Baraitch**. The young raja was one of the state prisoners held by the British in the **Residency** at **Lucknow** (he died, of natural causes, in the **Alambagh**). It is not surprising therefore that his mother, the Rani, should entertain rebel troops and show other positive acts of enmity towards the British. But when **Bala Rao** and his forces camped there in May 1859 they demanded Rs50,000 of her and as she was unable to pay she fled to the hills. *See* **FSUP,** Vol IV, numerous entries.

TOPEE-WALLAHS Literally 'hat-wearers', and generally taken to mean Europeans at this time.

TRANSPORT Apart from the rudimentary **Railways**, and the minor use of **Steamers** on some of the major rivers, the transport of men and supplies remained as it had been for hundreds of years in India. In other words it was dependent on animal power, more particularly bullocks and bullock-carts. These were rarely owned by the armies involved in the struggle (save the Artillery, which tended to possess its own animals) but were requisitioned from the countryside as and when they were required. If the local population were hostile or if the area was one over which a campaign had already been fought, then the problem of requisitioning cattle and carts became acute. This was particularly so in Oude[Awadh]. The Gurkhas under Maharaja **Jung Bahadur** entered Oude in support of the British in February 1858, but they were frequently held up (and indeed arrived 'late' for the assault on Lucknow) by lack of transport. 'On the 4th (March 1858) we commenced our march to Lucknow and made the regular marches without once halting though we had great difficulty in getting our carts along as the wretched bullocks were completely done up and the carts daily breaking down. We were obliged to abandon some for this latter cause having no spare ones nor any means of taking on the loads.' (Letter from General **MacGregor**, Military Commissioner to the Secretary to the Government with the Governor General, Allahabad, 13th March 1858). Ironically on their way out of Oude a month or so later the Gurkhas encountered similar transport difficulties - but this time their problem was to find carts for their loot, not war supplies. The British reinforcements that arrived in Calcutta by steamer or sailing ship (some old wooden warships like the 'Belleisle' were converted into transports) had generally taken some 140 days to reach India from England. From there the troops were sent up country by railway from Chinsurah to Raniganj, which was the end of the line. Thence to Benares and beyond movement was by bullock-train, with relays of bullocks some eight to ten miles apart, and six men were told off to each cart to ride and march by relief. Average advance was twenty-five to thirty miles per day, halting each day at ten o'clock for cooking, resuming the march at 4p.m. and marching on through the night for coolness: it would thus take twenty days to get to Benares[Varanasi]. On reaching Allahabad there was a further forty eight miles of railway track towards Kanpur, ending at a place called Lohunga; there were no stations built and the wagons were primitive open trucks used by the contractors for building the railway. From Lohunga the troops would revert to road marches to **Fatehpur** and **Kanpur** and beyond.

TREASURE Looting was commonplace, on both sides of the struggle. Occasionally treasure was found or captured which the soldiers would hope to share - either by helping themselves, or by the distribution of Prize Money. An example (from the Journal of Sir Hope Grant): 'Destruction of Bithoor ...Sir Colin ... directed me to proceed to Bithoor, the residence of the infamous Nana, and there to perform the work of destruction. We started on the 11th December, and on our arrival lost no time in destroying everything we could lay our hands on belonging to the low villain, blowing up his pagan temple and burning his palace. It was reported that a quantity of treasure had been concealed in a deep well, in which was 42 feet of water. With much difficulty we managed, by the aid of bullocks, to reduce this to two feet, and then we drew up a heavy log of wood. After further search we found two pewter pots. No wise disheartened, we renewed our efforts, and this time we discovered a number of gold and silver articles, which, to judge from their shape must have been of extreme antiquity. There were some curious gurrahs or pots, lamps which seemed of Jewish manufacture, and spoons of barbaric weight. All were of the purest metal, and all bore an appearance of antique magnificence....... On 24th Dec we started for Mynpoorie, leaving a small force at Bithoor to prosecute the search for treasure.' *See* Wm Forbes-Mitchell, *Reminiscences of the Great Mutiny,* 1893; *see also* **SSF,** Chapter Five, *Amy Horne & the Treasure.*

TREVELYAN, Sir George Otto Son of Sir Charles Edward Trevelyan, Governor of Madras. A writer and amateur historian who produced a very readable composite version of the events in Kanpur, based on the Depositions of 63 witnesses taken by **Colonel Williams,** Police Commissioner of NWP, plus a narrative by **Nanak Chand**, plus **Mowbray Thomson**'s account, plus the official narratives. Has been criticised for the resultant 'unscholarly account', but he is not himself to blame for the contradictory nature of the evidence he adduces. *See* Sir G.O.Trevelyan, *Cawnpore,* London 1886, reprint, Delhi, 1992.

TRIMMU GHAT The 46th BNI and the 9th Bengal Cavalry had mutinied at **Sialkot** on 9 July 1857 and after looting the cantonment had left the town the same night. General **Nicholson** caught up with them on 12 July on the banks of the river Ravi where they were endeavouring to cross at a ford known as Trimmoo Ghaut[Trimmu Ghat]. The mutineers were cut up and more than 400 killed, while many fugitives were brought in (and executed) by local villagers in the succeeding days. On the 15th it was discovered that about 300 had taken up a position on a large island in the middle of the Ravi; they were attacked and routed, many bayonetted and others drowned as they sought to cross the river. Some 200 of the original party of mutineers sought shelter over the border in the territories of the Maharaja of Jammu, but were handed over and either executed or sentenced to transportation after drum-head court-martial. The most detailed account of this 'battle' is given in **Ball** Vol I p 565.

TUCKER, H.C. Commissioner at **Benares [Varanasi]**. A man of probity and intelligence who sought to counter the actions of such avenging spirits as **Colonel Neill**. It was reported of him at the time that 'he was as remarkable for his efforts to preserve the lives of his countrymen, as some of his coadjutors were to avenge their deaths. He offered rewards for the heads of living friends rather than for those of dead foes; and his policy was decidedly the more successful of the two; for the villagers generally proved willing to hazard the vengeance of the hostile forces by saving life, but could rarely, if ever, be induced by threats or promises to earn blood-money.' *See* BLOOD MONEY; also **Martin,** Vol II, p 291.

TUCKER, Robert He was the Judge in **Fatehpur** and although he could have left the station in safety he refused to do so, putting his faith in a Deputy Collector by name Hikmat Ullah Khan. The latter went over immediately to the rebels, and Tucker was attacked in his house in which he had hurriedly barricaded himself. He was eventually overwhelmed but not before he had apparently killed about sixteen of his attackers. He was, by the inference of those who told his tale, something of an eccentric, but a brave man none the less. *See* Wm Forbes-Mitchell, *Reminiscences of the Great Mutiny,* London, 1910, *and* J.W.Sherer, *Daily Life during the Indian Mutiny,* London, 1898.

TUCKER 'Upon reaching **Futtehpore**, the first thing we did was to search for the remains of the gentleman (Tucker) who had been Commissioner (actually the Judge) of the District & who had been murdered there. He had been well known to all the natives in the locality as a good & just man, devoted to their interests & to their welfare. He was a sincerely religious man, and had erected on their main road a three-sided stone pillar with the Lord's prayer [actually the Ten Commandments] engraved in three languages. When the news of the Mutiny at **Cawnpore** reached his station, all the Englishmen there but him, had gone back to **Allahabad** (actually Banda). He would not budge, as he stoutly maintained the natives would not molest him. He was wrong. They attacked him in his house, to the flat top of which he retreated & there he sold his life, killing, it was told to us by the natives, thirteen mutineers before he ceased to breathe. We found his skull, and collected as many of his bones as we could. The only coffin we could obtain was an empty brandy case, in which we buried him with military honours. *The sole inscription upon the box that contained his bones was 'Old Cognac.'* See Captain Garnet Wolseley, *From England to Cawnpore*, Hove Reference Library, East Sussex County Council.

TUNNEL ENTRANCE The entrance to a brick-lined and well-built tunnel is still visible at Kanpur on the site of **Wheeler's Entrenchment**. There is no agreed explanation for the existence of the tunnel; local opinion believes that it existed before 1857 and may have been built by a princess who wanted an emergency escape route should her house be attacked. Perhaps **Wheeler** chose this site believing that the tunnel could be used by him in some way? - there is no evidence that he did so.

TUSSAUD, MADAME Was (and still is) a well-known wax-work show in west London and a tourist attraction. The great and the good, the famous and the infamous, are exhibited there side by side, and the artistic merit of the figures is enough to deceive many visitors into thinking they are alive. Exhibit number 325 at Madame Tussaud and Sons was 'Nena Saib' posed miserably in the 'winter costume of his country. The Indian rebellion,' read the January, 1861 catalogue, 'fruitful as it has been in atrocious crimes, showing the human beings when not civilised, to be like the ferocious tiger, that lives on gore, produced the monster Nena, who, although externally distinguished amongst his countrymen as much above them in knowledge and European usages, was at heart a savage. Instead of making war against his fellow-men, he made it against innocent women and children, whom he caused to be butchered without mercy; bringing desolation to the heroes then in India, and mourning to the homes of their relatives in England. But a retributive Providence willed it that punishment was near. Nena was totally defeated whenever he attempted to make a stand, and becoming a wanderer in his native land, died, it is said, the coward's death, despised and forsaken.' The exhibit was removed in 1878. Tussaud *et al. Madame Tussaud and Sons' Catalogue.* p 30.

TWELFTH REGIMENT BNI The 12th was called the *Hote Ka Paltan* and was raised in 1763. It served in the Carnatic, and at Laswarie, Ferozpore and Sobraon. The left wing of the 12th mutinied at **Jhansi** and fought with **Tatya Tope** in central India before they were all killed at

Sumpter Fort in April of 1858. The right wing of the 12th first offered protection to its officers and their families, but eventually joined the **Nana,** only to disappear into the **Terai**. *See* G.H.D.Gimlette, *A Postscript to the Records of the Indian Mutiny: An Attempt to Trace the Subsequent Careers and Fate of the Rebel Bengal Regiments, 1857 to 1858.* London, 1927, pp 99-105.

TWO LADIES The drummer of the 1st Regiment BNI, William Allen, with his wife and family, were **Bala Rao**'s captives in 1859. Allen died in captivity on 14 June, 1859, but his wife and sons were released into **Jung Bahadur**'s custody the following July, and reported that two ladies had also been among Bala Rao's prisoners, but that 'the rebels, being tired of carrying them about with them from place to place, murdered them.' Who these ladies were, and whether they may have been escapees from the ghat massacre, was never determined. NA. FDP. Pol. 15 July, 1859.

U

UDIT PRAKASH SINGH He was reported as Raja Oudit Purgas Singh, (Udit Prakash Singh), Talukdar of Ekona in the **Baraitch** Division; See PERSONS OF NOTE, APRIL 1859.

UJNALLA, Well of **Frederick Cooper** the Deputy Commissioner of the Punjab, based at Amritsar triumphantly boasted 'There is a well at **Cawnpore,** but there is also one at Ujnalla!' He was responsible for the pursuit of the disarmed 26th BNI which fled from **Mean Meer (near Lahore),** on 30 July 1857, and his ferocious and inhuman treatment of the survivors who apparently surrendered in the belief that they would be granted court-martial and a fair trial, was afterwards universally condemned, although at the time applauded by both **Sir John Lawrence** and **Mr Montgomery**. Within forty-eight hours of the outbreak of the mutiny (if indeed there was a mutiny) at Mean Meer, 500 men had been killed, drowned or executed. Thus 150 were shot or drowned by Montgomery's police aided by the villagers, 160 captured on an island in the middle of the Ravi river, 35 counted drowning as they attempted to get off the island, 237 summarily executed when taken (surrendered), 45 'died from fatigue' (see below), and 21 more apprehended and executed. The number of prisoners who had surrendered being too great for hanging, they were brought out ten at a time, bound together and shot; after 237 had thus been killed, Cooper was informed that the remainder 'refused to come out' of the bastion at Ujnalla. He went there, the doors were opened, and it was found that forty-five men had died, from fright, exhaustion, fatigue, heat and partial suffocation: the Black Hole of Calcutta had been re-enacted. Their bodies were thrown down a large dry well where their comrades who had been shot had also been thrown. A further forty-one sepoys were captured and sent in to **Lahore** where they were blown away from guns. See **Ball, Martin,** and F.Cooper, *Crisis in the Punjab*, Lahore, 1858.

UMBALLAH See AMBALA

UMRITSIR See AMRITSAR

UNGUD The name or pseudonym of a spy or messenger employed by the British to carry news in and out of the **Lucknow Residency.** He was referred to also as a **cossid** or messenger....'Take Ungud as an example. Was he a spy, a messenger, a dyed-in-the-wool traitor? He was an Indian, by no means young, yet he went in and out of the besieged Residency at Lucknow carrying messages from the embattled British to the relieving column led by General Havelock.

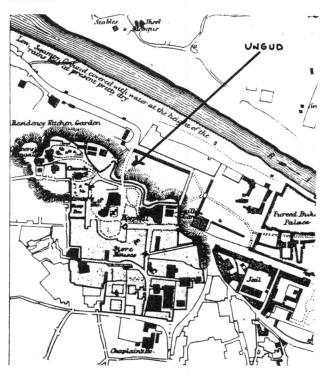

Ungud's Route into the Residency? (under the guns of the Redan Battery)

We read thus ' That night the faithful Ungud again crept into the entrenchment bringing a letter from General Havelock ' And, again, 'on 16th September Ungud the *pensioner* was again sent forth with a letter done up in a piece of quill.' The italics are mine, for in my view they explain all: Ungud was faithful to the government he had served all his life. Very creditable, you might say; an out-and-out traitor might be the reply. Whatever you believe, I think India can be very proud of Ungud, he was quite remarkably brave, and though he was, it is true, well rewarded for his labours, greed was not his motive.' See **CM** . **PJOT**

UNJUR TIWARI A spy for the British with a remarkable history. See ANJOOR TEWARI.

UNKNOWN HERO (British side) 'We all, naturally, wanted to do something, to allow that we were helping the general restoration, and as the city (Cawnpore) was getting well into our complete control, consulted the Brahmin Deputy-Collector as to whether we could get men to establish a **thana** a few miles out of the town. We found people quite willing to take service, and the young Mahomedan seemed the very fellow for the post. [He was] full of go, and anxious to bring himself forward... I asked him if he would try and form a little nucleus of British authority out in the village where it was proposed to place the thana, and he jumped at the idea. He had got a horse of his own, and he started at the head of his little band, who were all armed, and was to engage other men out there if necessity arose. Of course he was told that he was not expected to fight, and if Sepoys approached he was to fall back. But for all purposes of exercising his authority amongst the peasants he was, we thought, strong enough, and he himself quite confident. He had hardly been there two days when a large body of Sepoys, stealing across from **Calpee,** and endeavouring to get over the Ganges into **Oudh,** came suddenly upon his thana in the night. His men made some resistance, but the idea of sepoys carried a certain terror with it, and the darkness was a temptation to try to escape. The plucky fellow, notwithstanding, held out, and at last fell into his assailants' hands. They bound him, cut his throat, and hung him by his feet on a tree. There are honoured mounds above brave Englishmen all over the world; but that young hero's grave demands a leaf of laurel too. He espoused our cause; he was faithful to it to death; he fell fighting.' **Sherer.**

UNNAO Or Onao, a small town in Awadh[Oude] some nine miles to the north of Kanpur (across the Ganges) which was the scene of heavy fighting by **Havelock**'s forces advancing to relieve the garrison in Lucknow in July and August 1857. The best account of this fighting is given in J.C.Marshman's *Memoirs of Major General Sir Henry Havelock,* 1860 p 328 *et seq.*

URDU A language of India, especially North India, which in origin owes much to Persian. It is in effect a lingua franca, with its own script (writing from right to left), but is particularly associated with Muslims rather than Hindus.

USUF (YUSUF) KHAN *See* PERSONS OF NOTE, APRIL 1859.

V

VALBEZEN, E. de He was the French Consul-General in Calcutta. He describes in a book published in 1883 the dissension among government officials in Calcutta during the period of the rebllion, a matter not discussed in depth in any other publication. As a foreigner, present in India at this time, his account is valuable as being likely to be more objective than most. *See* E de Valbezen, *The English and India,* London, 1883.

VARANASI *See* also BENARES.

VARIOUS SMALL DELAYS In November 1857 **Sir Colin Campbell** relieved the **Lucknow Residency** for the second time, managing to evacuate the British garrison and the large number of women and children who had been beleaguered there. He was, naturally, anxious about their safety, and wanted them to emerge along the escape route he constructed to the **Dilkusha** as speedily as possible. Campbell 'was in a fever at the various small delays' which the ladies of the garrison 'considered necessary, and courteous as he is to women, he for once was obliged to be 'a little stern' when he found the dear creatures a little unreasonable. In order to make a proper effect most of the ladies came out in their best gowns and bonnets.' The correspondent could not say if they employed cosmetics, but considering that they had been out of communication with the 'Calcutta *modistes*' the array of fashion was said to have been 'very creditable.' But when 'the cares of the toilette prove an obstacle which an army cannot overcome, which frustrate strategic combinations, delay great sieges, and affect the fortunes of a whole campaign, it is sufficient to make Generals, at all events, wish that good Mother Eve's earlier style was now in fashion among her daughters.' *See* London *Times.* 13 April, 1858. p 10.

VELLORE In 1806, at Vellore in the Madras Presidency, 1,500 Sepoys, abetted by the princes of Mysore, suddenly mutinied, killing half of the garrison's four hundred European soldiers and all of its officers before finally being put down by a galloping troop of European dragoons from Arcot. The cause? It turned out that the Madras military authorities had not only introduced a headdress, similar to that worn by European troops, that was 'obnoxious to the Oriental' but forbidden native troops from appearing on the parade ground with the caste marks, earrings, beards or untrimmed mustaches they had been permitted to retain since 1757. The orders were deemed such gratuitous and reckless impositions on native troops that both the commander-in-chief of the Madras army, who proposed them, and the Governor of Madras, Lord William Bentinck, who approved them, were sent packing. J. Talboys Wheeler, *India and the Frontier States.* Vol. ii, p 521.

VERNACULAR PRESS 'The British had almost a hypochondriacal fear of newspapers, and resultantly the government adopted a stifling press policy. However, with Sir Charles Metcalfe announcing liberty of the press in 1835, many newspapers and journals in English and the vernacular languages cropped up. Though they grew fitfully and were often short-lived, their numbers gradually soared and their circulation increased. Vernacular journalism was most predominant in Bengal, Bombay and the North-West Provinces. In 1857, their interests were hardly political. They were scientific and literary. The Bengal press mostly discussed Hindu social practices and there was a strong educational bias in the Marathi press, while Gujrati journalism started as a commercial venture. Community backed newspapers were also introduced. For example, in 1851, Kurshedji Cama started *Rastgoftar* with Dadabhai Naoroji as the editor, and its aim was reform of the Parsis. In Delhi, a weekly *Mazhur-ul-Haq* was said to argue the Shia case, probably in answer to the *Sayyad-ul-Akhbar*'s Sunni propaganda. In the rebellion itself, the Indian press took little interest. In Bengal, Bombay and Madras, there was no indication of unrest. In the N.W.P. there were the Government supported newspapers, and it was only in Delhi and Punjab where newspapers were inclined to oppose British influence. Besides, not all of them were anti-British and emulated Raja Ram Mohan Rai who had espoused the theory that contact with British institutions was necessary for India's regeneration. In fact British innovations were reported in a laudatory way. For example, the *Buddhi Prakash* of Agra, a weekly, in Hindi, was quite awe-struck by the introduction of the first railway between Bombay and Thana in 1853 and called it an 'iron road'. However, the Indian newspapers, especially those in the vernacular languages, fanned nationalism in their own peculiar style and there were several which were definitely political in nature. The neutrality of the Indian press in 1857 has been stressed by some historians, but some say that majority of these newspapers were overtly or blatantly in favour of independence. In that age journalism and working for the country were two sides of the same coin. The main aims of vernacular newspapers were to portray the opinions of the people, shape their thinking and direct their paths. The journalists highlighted the merits and demerits of British rule. They succeeded in providing the medium for public expression of the feelings of discontent against the English which were dormant and could only have been passed on by word of mouth or ventilated from mosques and temples. The Indian press had only one idea that the Europeans mistreated them and used them for profit. A memory of the past sufferings at the hands of foreigners intensified the patriotic feelings of the Indians which had already been affected by the **cartridge** affair and the **bone dust** theory *etc*. The atmosphere that the vernacular press created forbode bad times for the British and in the Upper Indian territories there was a certain readiness to meet the forthcoming turmoil. Difficulties which the British government had to face in the international sphere were exploited by the papers for their own ends. **Bahadur Shah** was reportedly conniving with the Shah of Persia to restore him to his former glory, and in March 1857 a proclamation purported to be issued by the Shah was posted on Jama Masjid at Delhi instigating all Muslims to rise against the 'unbelievers'. The language press published this with vague and mysterious hints using ambiguous and enigmatical language 'obviously intended to be read in a contrary sense'. There were some fearless and outspoken vernacular language newspapers which reported and propagated the events of the rebellion. The *Sadiq-ul-Akhbar*, the Urdu weekly from Delhi, openly carried on an anti-British propaganda. Its copy of 7 August 1857 published in Persian the appeal issued by Khan Bahadur titled 'Victory of Religion', and was distributed in Rohilkhand. There were several others, like *Delhi Urdu Akhbar, Nur-e-Masriqui, Siraj-ul-Akhbar* in Persian from Delhi, *Akhbar-ul-Zafar* from Delhi Fort; *Agra Akhbar, Doorbeen, Sultan-ul-Akhbar, Benaras Akhbar* in Urdu, *Samachar Sudhavarshan* a Calcutta based Hindi daily, *Sarvahitkarak* in Hindi and Urdu from Agra; *Majhar-ul-Sarur* from Bharatpur, *Rajputana Akhbar* from Jaipur, *Jablabh Chitak* from Ajmer, *Gwalior Akhbar, Mufassalite etc*, which became known for arousing public opinion before and during the rebellion. Even the ones owned by Indians converted to Christianity, like *Fawaid-ul-Nazarin* taken out by a Delhi professor, Ram Chandra, and *Mahboob-i-Hind* preached ideas of self-assertion and patriotism but the newspapers of the former *genre* were more popular. There were instances of some rebels themselves setting up presses and publishing their own newspapers. For example, a press was set up at Cawnpore[Kanpur] by Azimoolah Khan, the Cawnpore rebel leader. It published a journal on 8 February 1857 in Hindi and Urdu from Delhi called *Payameazadi* and its editor was Mirza Bedar Bakht, the grandson of Bahadur Shah. It heralded the call for independence and fearlessly criticised government policies. Though short-lived, it had a great impact on the Indians and soon the British banned it and even possession of its copy was considered treasonable and the offender liable for execution. Its incendiary articles and songs, calling upon Hindu, Muslim and Sikh brethren to bow before the flag of independence, led to confiscation of its copies by the authorities. There were some newspapers in English which were owned by Indians like *The Hindu Patriot* taken out by Harish Chandra Mukerji, and the *Bengal Harkaru and India Gazette* run by Dewarkanath Tagore. These reported the events of the rebellion meticulously and were often critical of the government. These, and the papers like *The Friend of India* run by Serampore missionaries were widely read by the Englishmen. Language newspapers too covered the events of the rebellion and though they were accused of exaggeration they often published authentic news. For example, the news in *Sihare Simri*, a journal published in Lucknow in Urdu, covered **Moulvie Ahmadullah Shah's** movements with an accuracy which matched government records. The vernacular newspapers in 1857 had to face problems which hampered their independence of expression. They had practically no financial backing except the subsidy and patronage of the government, which kept them afloat. Even in Delhi and Punjab, where landholders and princes financed them, the editors were on the look-out for government grants. There was no system of advertisements which, if published, were published free of charge. Another difficulty was of finding subscribers who needed a lot of persuasion to buy a vernacular newspaper. However, simple villagers and illiterates keenly listened to the printed word when read out to them by the subscribers who were often the intelligentsia, but there were instances of newspapermen reading out the news in homes where they personally sold their papers. Hence, though circulation did not improve considerably, newspapers reached out to a large population. Lang places readership of the Bengali newspapers at ten readers to every subscriber. With the growth of education, the Indian youth started taking up journalism as a career but it was hardly remunerative. Generally, the owner was also the editor and the publisher and this put tremendous pressure on him. Besides, Indian journals had to depend on the English newspapers for news, which often meant exaggerated or delayed news. There were examples however of army or civilian officials being virtual editors of newspapers and disseminating even confidential information. They were often paid for it too. For example, Major Thomas was the virtual editor of the *Mofussilite* of Agra, flouting the government order forbidding civilians or military men corresponding with the press. There is an interesting example of a 'Moonshee' of an English press at Meerut who was also the editor of the Urdu paper *Jam-i-Jamshed* of Meerut and stole the news which passed his hands to use it on his own paper! His reporting was an example of satire disguised under effusive praise of the British. It was probably because of guarded criticism in the vernacular newspapers that the British did not see the writing on the wall. They did not give the newspapers much importance and took their contents lightly. The officials displayed a benign tolerance towards the 'native' journals which they subsidized and controlled. In fact, they felt these newspapers would educate the public and raise it from its apathy. The government did not learn the lesson palpable in 1857. It felt that censorship would not reduce the danger of labelling everyone in authority, and press should be used as a safety valve, as 'the dog that barks does not bite'. It was only in retrospect that the Judge noticed, at the trial of Bahadur Shah in 1858, the 'Mahomedan press, unscrupulously abetting rebellion'. Nevertheless, blatantly instigatory newspapers were punished and their editors prosecuted. In 1857 Munshi Ahmad Ali of *Doorbeen*, Hafiz Muhammad of *Sultan-ul-Akhbar*, Babu Sham Sunder of *Sudhabarshan* and editors of *Gulshan-i-Navbahar Mumbai, Samachar, Rastgoftar, Jami-Jamshed* and *Payameazadi* were tried and the newspapers muzzled. Canning's 'Gagging Act' of 1857 June, however, put restrictions on both the English and the vernacular press as the former was also considered provocative; though Canning realised that 'sedulously, cleverly and artfully' the 'native press had poured sedition into the minds of the people'. This act drove an irreversible wedge between the English and the Indian press.' Principal references: S.Natarajan, *A History of the Press in India*; **FSUP**; P.S.Khare, *The Growth of Press and Public Opinion in India 1857-1918*;

Dr Kirti Narain

VERNEY, Major-General G.L. The author of a book dealing with the contribution made by the naval brigade from HMS Shannon, at Lucknow in 1858, based upon the letters of his great-uncle Midshipman Edmund Verney who served with **Captain Sir William Peel VC**. The book has the merit of confining itself strictly to its chosen field - the military campaign. *See* G.L.Verney, *The Devil's Wind: the Story of the Naval Brigade at Lucknow*, London, 1956.

VIBART FAMILY Three members of this family were serving in the Bengal Army in 1857: two were killed (at **Kanpur**) viz Major Edward Vibart, Bengal Cavalry, his wife Emily and four young children, and Captain Edmund Charles Vibart, 2nd Bengal Cavalry who was killed on 27 June 1857 having escaped from **Fatehgarh** and drifted downstream to Kanpur where his party had hoped to find safety. The third, Lieutenant Edward Vibart, 54th Bengal N.I. served in Delhi, Meerut, Kanpur and Lucknow, and in his letters from April 1857 to March 1858 gives graphic detail of much that is nowhere else reported *eg* a vivid account of British reprisals in Delhi after its capture in September 1857. These letters are owned by Mrs A. Farmiloe, London, W8. Some further papers of the family are preserved in the Oriental and India Office collections of the British Library, *ee* MSS.Eur.F.135/19-23.

VICTORIA, Queen Generally speaking Queen Victoria was quite popular in India, particularly after 1858. But that was not the universal opinion. Thus 'Queen Victoria,

Victoria.Lucknow,1995. Sans orb, sans sceptre, sans throne.

that ill-starred polluted bitch'. The Queen also set a very firm sartorial standard, particularly in the matter of mourning dress. It was the women of middle class families who had most to do with the outward trappings of mourning for those they had lost in the fighting in India (their menfolk habitually dressed in black anyway), and merchants were not slow to capitalize. Hence the following advertisement: 'INDIA - FAMILY MOURNING - Skirts trimmed deeply with crape from 30s upwards to the richest quality, with Mantles and Bonnets to match. Family orders supplied on the most reasonable terms. First class Dressmaking at moderate charges - Address PETER ROBINSON, General Mourning Warehouse, 103, Oxford Street, London.' *See* FSUP, Vol II, p 160; *see also* Illustrated London News, 23 January 1859; *and* A.C.Benson ed., *Letters of Queen Victoria 1837-61*, Vol III, London 1907.

Death of Neill

VICTORIA CROSS This award for Valour had been instituted only a year or so before the outbreak, but was eagerly sought for by all ranks in the British army, but especially by the Officer cadre who saw it as a proof of their military proficiency and as a positive step in their careers. The Cross was conferred, or would have been had all survived, upon no fewer than 407 officers, warrant and petty officers, and seamen of the Royal Navy and of corresponding ranks in the Army, as well as upon one army chaplain and

three members of the Bengal Civil Service, during the struggle in India 1857-59, making a total of 411. Some were so eager to be ranked among the bravest of the brave that they exposed themselves quite recklessly to danger. One such was Col **James Neill,** who paid the penalty for his act and was killed in the streets of Lucknow.

VICTORY OF THE MAHOMEDAN FAITH
English translation of **'Futteh Islam'**.

VILAITIES Foreigners, often Europeans were meant, but sometimes it simply meant people from a great distance away. Hence mercenaries, Afghans, fighting say for the **Rani of Jhansi**, would be referred to as Vilaities. The British soldier with his genius for mishearing and misconstruing foreign languages, converted the word into 'Blighty', by which he meant far-off England. and home.

Hole in wall of Residency made by Shell that wounded Sir Henry Lawrence July 2 1857

WAGENTREIBER Family They escaped from **Delhi** on 11 May 1857, and after some adventure reached safety in **Kurnaul[Karnal]**. George was connected with the Delhi Gazette, his wife Elizabeth was the youngest daughter of one of the fourteen wives of Colonel James Skinner. Their daughter Florence wrote *The Story of our Escape from Delhi*, Delhi, 1894.

Fort William Calcutta

WAJID ALI SHAH The last King of Oude[Awadh], exiled to Calcutta by the British on the annexation of the **Kingdom of Oude[Awadh]** in 1856. Arrested on 15 June 1857 and lodged as prisoner in Fort William on suspicion of being involved with the outbreak of the rebellion. Despite considerable provocation he does not seem however to have supported the mutineers at any time although many of the sepoys were from Oude[Awadh]. He might have been expected to be a leader, for he headed a warlike and loyal nation that would have fought hard for him, but he was not. At best 'a harmless and lachrymose individual.' 'The King of Oude[Awadh] was an appalling fool, a practical joker, a profligate, and a self-styled musician who liked to show off in the streets of Lucknow, beating on a tom-tom', Sir John Kaye, *A History of the Great Revolt*, Vol I, p 132. There is evidence that after the rebellion began and when the mutineers had elected Wajid Ali's young son, Bijris Kadr, to the throne, they still possibly thought of him as acting for his father who, although by then in Calcutta and in custody in Fort William, had never actually abdicated. As Dr Rosie Llewellyn-Jones has pointed out, although the palaces of Lucknow were in the mutineers' hands in the period November 1857 to March 1858 ***they were not looted***: the implication is that the King would return to his own; the looting of the Kaiserbagh was done by British, Sikh and Gurkha soldiers in March 1858. The British chose to underplay the support given to Wajid Ali by his subjects, preferring instead to pretend that he was unpopular with them and that they would be 'grateful' therefore to be relieved of his burdensome rule. Surprisingly officials found another, unexpected, source of support, which they were quite happy to make public: '....the King has been encouraged and sustained in his resolution to adopt a course of negative opposition and passive resistance (to the treaty offered to him by Government), *by the advice I am told and believe, of Mr Brandon, a merchant at Cawnpore, whose antecedents of meddling mischievousness are well-known to his Lordship-in-Council.* This individual assures His Majesty that if deputed to England as his Agent, he will, without a doubt, obtain his restoration. The 'Central Star' of which Mr B is the proprietor, by its purchased advocacy of that person's views, has confirmed the King in his mistaken resolution.' (Major General Outram to Governor General, 7 Feb 1856. Parliamentary Papers: Oude[Awadh]). 'The ex-King when in captivity at Calcutta, has acted with a firmness which one could not have expected from a mere sensualist, as he was said to be, half-idiotic and entirely base, I am told that his conduct at the time of the annexation astonished our officers: that it was characterised by dignity and propriety. Up to the present moment he has neither consented to his deposition nor taken one farthing of the annuity which the Company settled upon him, nor has he given the least ground for believing that he has participated in the mutiny and rebellion. But Empires never make restitution; they have no consciences.' W.H. Russell, *My Diary in India,* London, 1860. See R.Llewellyn-Jones, *A Fatal Friendship*, 1985; V.T.Oldenburg, *The Making of Colonial Lucknow*, Delhi, 1984; Montgomery Martin, *The Indian Empire*, Vol II, n.d. (1859?).

WAJID ALI, DAROGA Responsible for the households of the Begums or Queens of the ex-King of Oude[Awadh], **Wajid Ali Shah**. He did not accompany the King into exile in 1856, and was in Lucknow therefore throughout the siege of the Residency etc. Assisted in the rescue of the **'Captives of Oude[Awadh]'**. See **SSF** and Veena Talwar Oldenburg, *The Making of Colonial Lucknow*, Delhi, 1989.

WAKE, Herewald Commonly known as 'Wideawake'. Magistrate, Bengal C.S. stationed in Arrah. See *ARRAH HOUSE*.

WALPOLE, Brigadier A man of seniority but little ability, he was one of **Sir Colin Campbell's** subordinate commanders. His role until the final attack on **Lucknow** in March 1858 had been without distinction, but he was then despatched by Campbell to clear the left bank of the Ganges and secure the passage of the Ramganga at **Aliganj** for the main army of the Commander-in-Chief *en route* to **Bareilly**. He had a force of upwards of 5,000 men of all arms, including some of the finest of the British infantry regiments. He attacked, ineptly, and with disastrous results, the jungle fort

The ruins of the church in the Lucknow Residency, q.v.

Lucknow Residency, q.v.

The Begum Kothi within the Residency compound, Lucknow.
See LUCKNOW RESIDENCY

The Baillie Guard Gateway.
See also LUCKNOW RESIDENCY

Memorial plaque, in St James' Church, to one of the Indian Christians killed in Delhi. Courtesy E. Layson.

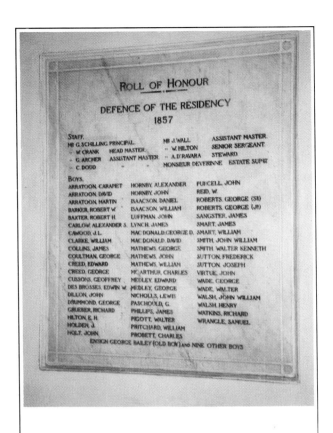

The roll of honour: staff and boys of La Martiniere College, Lucknow. Courtesy of the Principal. See MARTINIERE

Constantia, La Martiniere, Lucknow. See MARTINIERE

The name board of HMS 'Shannon' on a. 8" gun left at the Lucknow Residency can just be read. The board has since been stolen.
See SHANNON

Outdoor history lesson! Children from a Lucknow school examine HMS 'Shannon's' 65 pdr guns in the grounds of the Lucknow Residency.
See SHANNON

The site of Nana Sahib's Palace at Bithur. Nothing remains but a few well-heads, and this memorial. See NANA SAHIB

One of the seven wells of Nana Sahib's palace of Bithur. Where all his treasure was hidden when he fled? See NANA SAHIB

The Fisherman's Temple, Satichaura Ghat, Kanpur. Photo 1858, in possesion of All Souls' Church, rephotographed PJOT. See SATICHAURA GHAT

Road leading down to Massacre Ghat, Kanpur. Photo Jane Robinson. See MASSACRE GHAT OR SATICHAURA

The Fisherman's Temple, Satichaura Ghat, Kanpur, today. Courtesy Jane Robinson.
See SATICHAURA GHAT

Queen Victoria sans sceptre, sans nose, 'in store' at Lucknow.

The Granite Cross erected at Rajpur cemetery on the ridge at Delhi, now used as an altar. Courtesy E. Layson. See CEMETERIES

Mrochetti's Angel, Kanpur, q.v.

Sir Harcourt Butler 'greets' Queen Victoria among statues stored outside the State Museum in Lucknow.

All that is left today of the site of possibly the most important battle of the struggle, north of Delhi, on the Grand Trunk Road. See BADLI-KI-SERAI

Brigadier Showers and the 75th Foot storming the batteries at Badli-Ki-Serai. From the well-known sketch by Captain G.F. Atkinson, lithographed by W. Simpson. See BADLI-KI-SERAI

The watergate of the palace by which Bahadur Shah Zafar is said to have left the city of Delhi in September 1857, on the British assault.
Courtesy E. Layson.
See BAHADUR SHAH ZAFAR

The mysterious tunnels leading from Wheeler's Entrenchment, Kanpur, q.v. See also INTRENCHMENT, WHEELER'S

Antony Beevor, a descendant of Lucie Duff Gordon, and his wife Artemis, wearing Nana Sahib's dress and necklace. Courtesy Anthony Beevor. See AZIMULLAH KHAN IN ENGLAND

of **Royah** on 16 April. The British repulse there enormously encouraged many of the Oude[Awadh] Talukdars to continue or intensify their resistance. Walpole somewhat redeemed himself on 22 April when he defeated some Rohilkhand rebels at Sirsa, thus allowing him to reach the bridge of boats across the Ramganga at Tingri before it could be destroyed. It was by this bridge that **Sir Colin Campbell** and his army crossed into Rohilkhand.

WAR GAMES - with a difference

The following is illustrative of war being not much different throughout the ages *i.e.* 98% boredom and 2% terror. Colonel Thomas Walker speaks of two 'games' they played on the Ridge in the British Camp before Delhi: 'Fly Loo: each player had a lump of sugar covered with a handkerchief, and put one rupee into the pool. At a signal all hanks were taken up. Whosoever's sugar a fly landed on first scooped the pool.'...... 'We would surround a quantity of sugar with gunpowder and when well covered with flies they were blown up. These were not intellectual games but they helped to pass the hot time away'.

WAR TAX

Keeping large armies in the field costs money however the finances are organised, and the **Begum of Oude[Awadh]** had considerable difficulty at times in raising sufficient funds to pay the sepoys and others in her service. The normal land revenue was difficult in wartime to collect, and in any case would have been insufficient for her needs; windfalls like the finding of treasure in the palace of **Alli Nucky Khan** at Lucknow could not be relied upon to occur regularly, so a war tax, we are told, was levied upon the well-to-do in Lucknow and other centres: they were made to pay from 20,000 to 500 rupees per person, according to their means: anyone refusing to pay was to be dealt with by the simple expedient of being plundered. 'Four months' accounts of arrears of pay of the army have been prepared and payment is being made. This money has been raised by plundering the city' (20 February1858); *see* **FSUP**, Vol II, p 275.

WARIS MUHAMMAD KHAN

A **jagirdar** of **Bhopal** who took advantage of the mutiny of the **Bhopal Contingent** to rebel against the Begum. For an excellent and succinct account of these events and the uprisings at Sehore and Berasia, see K.D.Bhargava, *Descriptive List of Mutiny Papers in the National Archives of India, Bhopal,* Vol II, New Delhi, 1963.

WATERMAN, Capt.

From the evacuation of the **Lucknow Residency** - November 1857. 'There was one man left behind. Captain Waterman having gone to his bed in a retired corner of the Brigade Mess-house, overslept himself. He had been forgotten. At two o'clock at night he got up, and found to his horror that we had already left. He hoped against hope, and visited every outpost. All was deserted and silent. To be the only man in an open entrenchment, and 50,000 furious barbarians outside. It was horrible to contemplate. His situation frightened him. He took to his heels, and he ran, ran, ran through the Feradbuksh and the Tehree Kothee till he could scarcely breathe. Still the same silence, the same stillness, interrupted but by the occasional report of the enemy's gun or musketry. At last he came up with the retiring rear-guard, mad with excitement, breathless with fatigue. The horror of his position had been too much for his nerves and affected his intellect for some time.'

WATKINS, Quarter-master sergeant

This British non-commissioned officer must be set down as one of the most unlucky of all the combatants in the mutiny/rebellion. This is the story (relating to the mutiny at **Allahabad**) as told by **John Fitchett**, a Eurasian bandsman: 'Quartermaster-Sergeant Watkins... asked Lieutenant Hawes for a musket; he told the sergeant and myself to take arms. The sepoys, in reply to Lieutenant Hawes's question 'why the men were firing,' said, the Europeans were in the bungalows and had come to disarm them. Lieutenant Hawes assured them that there were no Europeans coming to disarm them, and that, if there were any enemies, he would lead them, and ordered the men to follow him. He had got about 100 paces from the Grenadier lines when we heard the noise of men and guns approaching; he halted his men and challenged, whether friends or foe. One of the native officers replied that they were the two companies returning from the ghat in consequence of the European officers leaving them to go to the fort. Just then Quartermaster Sergeant Watkins came up to make some report to *Lieutenant Hawes, who had his revolver in his hand, one barrel went off and shot the Quartermaster-Sergeant in the breast;* on this a sepoy of 3rd company fired and shot Lieutenant Hawes. I afterwards heard that Lieutenant Hawes was killed on the spot; but the Quartermaster-Sergeant being alive, was subsequently killed by a ball through the head. Whilst passing between the 1st and 2nd company lines, I heard some sepoys call out 'there is the one who received Rs.50 reward now flying, kill him.' I therefore passed the 3rd company lines to the 8th, and got into the house of drummer Peters. I told him to fly, and we tried to get to the parade; but Lieutenant Currie, officer of the day, who was galloping through the lines, ordered us back to our houses...We remained a few minutes, and then got to the parade ground. I saw Colonel Simpson, Captain Plunkett, Adjutant Stewart, Lieutenant Pringle, and I think four or five young officers who were doing duty with the corps. The men were between the lines and bells of arms. The two guns and companies from the ghat were a few paces in advance of the bells of arms of the Grenadier company. The Adjutant ordered Nubbee Buksh, who was the orderly bugler, to sound the 'assembly,' which was not done. As I was coming forward, a havildar pushed me back, and just then a volley was fired from the 4th and 5th companies. I do not know how many officers fell. Adjutant Stewart was wounded, ran towards the quarter-guard, and was there killed, as also Captain Plunkett before he entered the gate of his compound. I heard that a havildar was wounded. Three young officers who were doing duty got into the quarter-guard; the sentry threatened to shoot them if they did not put down their pistols; they shut the door. In the guard there were two prisoners, one a Mahomedan (I do not know his name), the other Ram Lall of the 5th company; they were got out and the sepoys then commenced firing inside; one of the young officers attempted to get away, but was immediately killed, and the other two I believe were shot inside the quarter-guard. I had followed the Adjutant, Lieutenant Stewart. A sepoy of the 3rd company, named Pirthee, came up to me, saying I must be killed, as I was one of them. Naick Mipal Singh and a sepoy of the 3rd company (I think he was a recruit, as I do not know his name) saved me, and made me sit under the tree in front of the bells of arms of the 3rd company; the sentry there would not let me move, but when he left his post, I went to drummer Peters's house. Whilst under the tree the havildar-major of the corps, Teliar Khan, came up to the sepoys assembled near the 4th and 5th companies' lines and said, ' What have you all done? But whatever was ordained has happened, it now cannot be helped. Take my advice, secure the treasure, and march for Delhi, for the Europeans from the fort will attack us at daybreak.' A number shouted, 'We will go to Delhi and serve the King.' The sepoys of the 3rd company, hearing a report that Captain Gordon was concealed in the hut of two sepoys of the Grenadier company (I forget their names), went to search for him, saying that their officers had been killed, and therefore they would not spare the officers of the Grenadier company, but they could not find him. ...I saw one or two native officers crying, and heard them say that three or four budmashes had ruined the regiment. I heard that many of the sepoys were sorry, and some of them went direct to their homes. The havildar-major was at first sorry, but said it could not be helped, and therefore persuaded the men to go to Delhi.' Deposition of John Fitchett. **Williams.** A tale of complete and utter confusion and panic.

WELLS of the DEAD Kanpur: there were in fact two such wells; the more famous contains the remains of the women massacred at the **Bibigurh**, but the other actually contains probably an even greater number (250?), of those who died or were killed in **Wheeler's Entrenchment.** 'The dead bodies of our people had to be thrown into a well outside the entrenchment, near the new unfinished barracks, and this work was generally done at the close of each day, as nobody could venture out during the day on account of the shot and shells flying in all directions like a hail storm; our intrenchment was strewed with them. ...I have seen the dead bodies of officers and tenderly brought up young ladies of rank (Colonel's and Captains' daughters) put outside in the veranda among the ruin, to await the time when the fatigue party usually went round to carry the dead to the well, ...for there was scarcely room to shelter the living.' Shepherd's *Account,* dated 29 August, 1857, published in the London *Times* 7. November, 1857. The well, covered, is still there, and visible (1995), close to **All Souls' Church**.

WELLS, Dr Walter Dr Wells was the surgeon of the 48th N.I, commanded by **Lt.Colonel Palmer** and stationed in Lucknow. His claim to notoriety is that he was insensitive enough to offend the caste prejudices of the sepoys of his regiment, before the 'mutiny' began. In April 1857 *'having occasion to visit the medicine store of the hospital, and feeling at the same time indisposed, he incautiously applied to his mouth a bottle taken from the hospital medicines containing a carminative.'* No high-caste Hindu could afterwards have partaken of the contents of the bottle: the doctor's incautious 'swig' caused an outcry, a refusal by the sepoys to take any medicine, and an addition to the legend that the English were intending to convert all the sepoys to Christianity. Wells survived the Siege of the Lucknow Residency (although wounded slightly), together with his wife and child. *See* Gubbins M.R. *'Mutinies in Oudh'*, London, 1858.

WENTWORTH-REEVE PAPERS These are held in the Centre of South Asian Studies at Cambridge University. Included is an eight-page photocopy of a MS article, undated and unsigned, giving an account of an escape from **Kanpur** by a soldier among a group of civilians. Description of troop movements and military defences. Escape by boat, ambush and final rescue by friendly raja. List of the killed and wounded, both military and women, as far as he could remember, and indicating where and when and how they were killed. There is a possibility that this was written by Lieutenant **Delafosse**.

WHEELER MYSTERY The following is the abbreviated text of a lecture given (by **PJOT**) to the Kanpur Historical Society in January 1993. 'I was hooked on the subject of the Mutiny of 1857 - or call it First War of Independence if you prefer - when I was a young man, newly commissioned into the Mahratta Lt Infantry during the Second World War. With regard to the Mutiny I thought how unfair it was that all the accounts - or nearly all - of what happened in the momentous years of 1857-9 should be written from the English point of view. As an historian I knew well the truth of the cynical old adage that 'history is written by the victor'. So I have tried hard over the many years that I have been studying those events to see the other fellah's point of view so to speak, to search actively for new and unpublished sources of information, to present alternative explanations for what is undisputed fact - in other words to present the story in as near truthful a form as can now be ascertained. An historian for whom I have a great respect is Dr Surendra Nath Sen whose book 'Eighteen Fifty Seven' was published in 1958. He had been commissioned I believe in 1957 by the Government of India, on the centenary of the great revolt, to write the official version so to speak. I suspect that his book was a disappointment to his sponsors - certainly the foreword by the then Minister of Education in New Delhi expressed surprise that Dr Sen had come to some of his conclusions. But Dr Sen was too good and too honest an historian to be willing to twist history to please his political masters. There are other historians - on both sides of the argument - who have not been so scrupulous. But all this is just by way of introduction to what I'd like to tell you about today. In reading about the Mutiny I've often been sidetracked into matters of interest that aroused my curiosity. Take for example the WHEELER FAMILY MYSTERY. The Mystery concerns the identity of the wife of Major General Sir Hugh Massy Wheeler, who was killed here at the Satichaura Ghat at the end of June 1857, and after whom the Wheeler Barracks here are named. We know that she - whoever she was - died with her husband: there is independent evidence for that. Yet there is also much apparent evidence that we cannot trust. For example, in the Memorial Church here in Kanpur, if you read the names on the marble tablets in the sanctuary, you'll find the words ' Major General Sir Hugh Massy Wheeler, Lieutenant G Wheeler, Lady Wheeler and two daughters' - they are all listed among the slain, killed by the rebels. But we know that one of those daughters, Ulrica, was not killed but was carried off by a trooper of the rebel cavalry, eventually married him and lived to old age right here in Kanpur - she may in fact have lived until 1907. Who was this Lady Wheeler? That was my main quest. A wife took the title - still does - of 'Lady' if her husband was knighted, and Wheeler had been made a KCB - Knight Commander of the Order of the Bath - some years before his death. Well the first thing to do was to look up any reference you can find to Wheeler's marriage - ten hours later at the India Office Library in London you, exhausted, find that he was married only once, to one Frances Matilda Oliver, widow, at Agra on 6 March 1842. Then where's the mystery you may ask? There's your Lady Wheeler. But it's not so simple. If you examine the records of birth and baptism of children of whom Wheeler was the acknowedged father, you'll find there are no fewer than nine, and only one was born after his marriage to Frances Matilda, that is a boy named Francis John Wheeler at Alighar on 28 May 1842 - only 2½ months after the marriage incidentally. This same boy lived to the age of 46 and died at Almora. Who then was the mother of all his other children? or, more probably, who were the mothers - as he seems at first sight to have been pretty promiscuous. Before we go any further let us introduce another complication or two. We have evidence, reliable, that some of the children were of mixed blood, particularly the two daughters, so it seems to be pointing to an Indian lady that Wheeler must have married - but not in a church or there would be a record - or taken as his beebee, which was a common practice among British Officers until the memsahibs arrived in large quantities. This view is confirmed by the story, - again from more than one source, - that the reason Wheeler did not fear an attack from the Nana Sahib after the mutiny of the troops here in Kanpur at the beginning of June 1857, *WAS THAT HIS WIFE WAS A CASTE FELLOW OF THE NANA'S, AND THAT THE NANA WOULD NEVER THEREFORE DO ANYTHING TO HARM HER.* Indeed this is adduced as a reason for Wheeler not choosing to take refuge in the very strong Magazine on the river bank but instead to construct the pathetic little entrenchment around the unfinished barracks. That there were two Lady Wheelers, two women married to Wheeler, one Indian, one British , has therefore been taken for granted from the beginning. Was he a bigamist - or a widower - when he married Frances Matilda Oliver in 1842? And which one died with him at Satichaura? That was the mystery that I set out to solve, Well, I have solved it, but it's a solution that will surprise you! Its all very complicated, and you have to go back to the year 1805 to pick up the threads of the story. Spare a glance when appropriate at the list of children I have given you. In the year 1805, two young men arrived in India, in Calcutta, in the month of March, possibly on the same ship. They were just 16 years' old, and whether or not they were together they certainly knew each other, for they both came from the same part of Ireland. They were from Anglo/Irish families of landed gentry - indeed one of them was the grandson of a peer, Lord Massy. These two young men were Hugh Massy Wheeler and Thomas Samuel Oliver. They were nominated by a director of the HEIC in the fashion of the times as Ensigns *ie* junior officers in the Bengal Army of the Company. Their fathers had both served the Company in India. The two boys turned out to be good and brave soldiers and both died in harness so to speak. More of that later. To understand what happened now you have to realise how important Kanpur was in those days to the Company's army. The British had made it a garrison town but its cantonment was far more important than that, it was a staging post and supply depot for all troops heading up country to the NW provinces of Bengal. Our two young men could expect therefore to spend some time, possibly together, in Kanpur - and so indeed they did. Young Oliver seems to have been the more precocious of the two at least where women were concerned, and we find him getting married, very young, aged 21, at Kanpur, on 21 December 1810. His bride was one Frances Matilda Marsden, the daughter of Lt Colonel Frederick Marsden, of the Company's Bengal army - who also belonged to the same class of Anglo/Irish gentry as the two young men. Frances must have been very young for she was still bearing children in 1842, thirty years later. Her first child was born in Feb 1811, just three months after her marriage to Oliver, and this must have caused much scandal. Indeed I have come to the conclusion that this was a shotgun marriage both because the girl was pregnant and because she was so young. Her outraged father - who was in Kanpur at this time - must have insisted on the marriage. There may have been neither love nor willing acceptance of it on either side - I am tending to the possibility of rape having been involved. Frances does not appear to have had another child by her husband. He fathered one in 1823 when he returned from a spell in Java and another in Ceylon, but the mother is not mentioned, as she would have been had it been his wife. But this same Frances Matilda Oliver does appear to have had many more children - she is named in the baptismal record for some, and the name with which she is linked is the other young man of our story, Hugh Massy Wheeler. The years go by, the scandal must have been considerable, even in those pre-Victorian days - an officer fathering children by the wife of a brother officer must have been thought outrageous. Mind you, as you can see, Oliver was not exactly idle either. Eventually fate took a hand, and Oliver was killed in action, honourably, fighting Afghans near Kabul, on 23 Nov 1841. Four months or so later Wheeler married his lover, Frances Matilda, and attempted rather late to legitimise their relationship and their children. But how could Frances Matilda be the mother of all Wheeler's children when we know that two of them - at least - were of mixed blood? I seemed to be getting nowhere, yet the truth was not far away. That the scandal was well-known to the succeeding generations of the Wheeler family seems obvious, for they closed ranks and gave little information away. I can only find that one of the children, George, came back to Europe: all the others lived and died in India, though distinguished in their careers. One grandson was killed in the first world war, having won the VC, but I can as yet find no details of him. It still seemed to me that there were two questions really that needed answering: Firstly, who was the Lady Wheeler who died here at Satichaura? And secondly who was the Indian Lady Wheeler? The first question was much easier to answer than the second. And the vital clue came when I re-read the account of Lieutenant Thomson, one of the two survivors of Wheeler's Entrenchment. He tells the story of Lt Godfrey Wheeler, the General's beloved son, who was killed during the siege of the entrenchment in June 1857. Apparently the young man had been wounded, was lying on a charpoy being fanned by one of his sisters when a round shot (a cannon ball) came through the wall and took off his head - in the presence of his sisters and his father *and his mother*. Now we know - see list - that Godfrey's mother was Frances Matilda, and we know therefore that the Lady Wheeler who was in the entrenchment and died with her husband was undoubtedly Frances Matilda Wheeler, formerly Oliver, née Marsden. *QED*. Big sigh of relief. All done, no more records to sift through, satisfaction at last. But is it all done? What of the second question? Who, where, when was the Indian Wife, the mother of Wheeler's two mixed blood daughters (and possibly some of his sons) - those same sisters who were fanning their wounded brother when a cannonball ended his life ? It seemed I had got nowhere, but quite by chance I found the answer and it was so simple I could have kicked myself. I had a stroke of luck. In trying to find some record of the birth of Frances Matilda Marsden, in order to see how old she was when she married Oliver in 1810, I found myself looking up the Marsden family, and, by chance, I came across a little privately published book, written by a clergyman by the name of Marsden in 1912, and called 'The History of the Marsden Family'. I don't know if you have people who anxiously compile their family tree - it is a common practice for men in retirement. With great respect the fruit of their labours is rarely of interest to anyone but themselves. Well the Rev Marsden devotes a lot of time to Colonel Frederick Marsden's brother, for William Marsden was a noted numismatist and orientalist. But he does just mention Frederick, Frances Matilda's father. He says that he married *once*, to the daughter of the Rev Henry of Waterford, Ireland, and they are recorded as *dsp* - now that is shorthand for died *sine prole*, died without issue, *ie* childless! Surely there must be a mistake! What of Frances Matilda? Ah but then, you see that Fred Marsden's marriage was in 1802, and even if they had had a daughter she would only have been 7-8 years old in 1810 when we know that Frances Matilda married Oliver. I always thought she was young but not that young! particularly as her first child is born three months later! So....a solution dawned on me. Frances Matilda was the daughter of the Irish colonel Frederick Marsden, *but her mother was not a European*. Marsden must have taken a beebee, as so many did, with a longlasting and loving relationship to follow. He acknowledged their daughter as his own and gave her his surname. The Indian mother might well have been a castefellow of the Nana, for Marsden had certainly served down Bombay and Pune way for some time. I'd like some help from my Maratha friends on the plausibility of this. *Hence Frances Matilda was of mixed blood* and that would have shown through on some if not all of her children by Wheeler. Suddenly everything falls into place. *There never was an Indian Lady Wheeler!*, General Wheeler, far from being the philanderer we thought him was true and faithful to the one love of his life, throughout his life. Colonel Oliver was a philanderer - as he had so early demonstrated by his seduction of Frances Matilda, and the latter, after her unhappy marriage to Oliver had in effect finished, was a true faithful and loving partner to Wheeler throughout her life, dying with him at Satichaura in Kanpur. 'NB Last (1932) 'informed' reference I can find to the lady is :- 'Lady Wheeler Sir Hugh's wife was an Indian and belonged to the same caste at the Nana Sahib' ! **PJOT**

WHEELER, Godfrey Richard   Born at Nimach[Neemuch] 28 November 1826, natural son of **Frances Matilda Oliver** and **Hugh Massy Wheeler**, educated at Mr John Mackinnon's Seminary at Mussoorie: Killed by a roundshot in the intrenchment at Kanpur, June 1857. Lieutenant and quartermaster 1st Bengal Native Infantry.

WHEELER, Lady Frances Matilda   Wife of **Sir Hugh Massy Wheeler** with whom she died at the Sati Chaura Ghat on the Ganges at Kanpur in June 1857. She was the natural daughter of Colonel Frederick Marsden of the Hon E.I.Coy's service, by an Indian lady, unknown. Married first Thomas Oliver, Lieutenant, Bengal Army, in 1810 by whom she had a child. Then lived with, scandalously, Hugh Massy Wheeler for many years, giving birth to eight children by him; married him on the death of her husband in the Afghan War 1842. 'the General has a half-caste wife who formerly misbehaved & lived for some years with him before they were married, she being the wife of another.' (Lindsay Family, *Papers*.) *See* also Rev. B.A.Marsden, *Memoirs of Family of Marsden*, 1914, p 107.

WHEELER, Major General Sir Hugh Massy

KCB. Born Ballywire, Ireland, arrived in India 17 Mar 1805. Son of Capt Hugh Wheeler, Honourable East India Company Service.(One account born Clonbeg Co Tipperary; but no real dispute as Clonbeg is just down the road from Ballywire). Born 30 June 1789. His mother was Margaret (died 1838), 2nd daughter of Hugh 1st Lord Massy, and widow of Godfrey Evan Baker (who was still alive in 1783.) *NB* His mother and father may well have been cousins, because Elizabeth, 3rd daughter of Col Hugh Massy, and younger sister of Hugh the 1st Baron Massy, married a certain Francis Wheeler of Ballywire. Married Frances Matilda at Agra 6 March 1842. She had previously married Thos Sam.Oliver at Kanpur 21 Dec 1810. Best known for his part in the events in Cawnpore (modern Kanpur) in 1857. He was the Major General commanding troops in the Kanpur Area at the outbreak of the Sepoy Mutiny. Hoped by his influence with the sepoys with whom he was very popular, and his relationship with Dondhu Pant the Nana Sahib, at Bithur, 12 miles away, that he would avoid any excesses even if the sepoys under his command did rebel. Withdrew all Europeans and Anglo/Indians and native Christians into a poorly constructed 'intrenchment' which offered little resistance to a determined enemy. Surrendered the intrenchment to Nana Sahib on promise of liberty and safe conduct by river to Allahabad. Died in the resulting massacre at the Sati Chaura Ghat on the Ganges, 27 June 1857, together with his wife and elder daughter. *See Burke's Peerage* 1923 p.1525 s.n.Massy B; Boase - *Dict. of Modern English Biography*; **DNB**; *Gentleman's Magazine 1857*, ii, 460. *See* also **SSF**.

WHEELER, Margaret Frances

Baptised at Allygurh[Aligarh] on 6 March 1842 the same day as her parents (**Sir Hugh and Lady Frances Wheeler**), married. She was their youngest child. Her pet or family name was Ulrica. She was abducted (or rescued?) by a rebel trooper at the Sati Chaura Ghat in June 1857 and saved from the general massacre that followed. Subsequently she married the trooper, but never returned to the surviving members of her family who presumed her dead. Died 1907, Kanpur. *See 'Traders & Nabobs'*, Zoë Yalland 1989; and **SSF**.

WHEELER, S.G.

Colonel commanding the 34th Regiment, Native Infantry, Bengal Army, Barrackpur (near Calcutta). Of significance only in that he was one of the few army officers who attempted to convert his soldiers to Christianity. Reprimanded by the Governor General for 'distributing tracts'. His activities provided ammunition to the opponents of British rule as he was held to be representative of an 'official' attempt to convert India to Christianity.

WHEELER, Ulrica

See WHEELER, Margaret Frances.

WHEELER'S ENTRENCHMENT

Major General Wheeler, the commander in Kanpur was subsequently criticised for his choice, in May 1857, of a place in which to make defence against possible sepoy attack.

The Entrenchment, Photo 1858

He rejected the Magazine (which with hindsight, all say would have been his best choice) and instead chose to create an entrenchment (often intrenchment) around some unfinished barracks in the area of the cantonment. The Entrenchment consisted of two barracks formerly used by the European Dragoons as a hospital. The longest barrack had a thatched roof and according to **Shepherd**'s map measured 50 feet wide by almost 400 feet long. The smaller barrack had a pucca roof and measured about 40 feet by 200 feet.

Each consisted of a central arcade of apartments lined on both sides with verandahs 'made of beams and solid masonry.' The brick walls of the barracks were 2 feet thick. (Shepherd, p 18) Wheeler's plan was to dig a roughly rectangular trench around the barracks, measuring about 500 by 800 feet. The Magazine 'could have been held for an indefinite time against almost any force, and contained abundant supplies of every description. It has been urged that in that year the water supply was defectve, but there is no record of any such plea: and in fact it would seem that Wheeler not only depended implicitly on the **Nana** but had made up his mind to hold on to the line of communication with **Allahabad,** confident that succour would speedily arrive on the **Grand Trunk Road.**' **Delafosse** told Mrs. Germon that Wheeler 'did not make the entrenchment at the Magazine because he had no idea there was any ammunition in it whereas a great deal of what has been fired at us [at Lucknow] came from there, in addition to what was expended at Cawnpore [Kanpur].' Wheeler positioned his Entrenchment as 'a place of refuge' on the Calcutta side of the station, so he could meet the reinforcements he expected from Calcutta 'at the nearest possible point, without being exposed to the perils of a struggle through an unknown city filled with lawless plunderers and escaped convicts, and bad characters of all sorts.' General MacMunn sums up the dilemma facing Wheeler thus: 'The localities possible for a place of refuge were two.

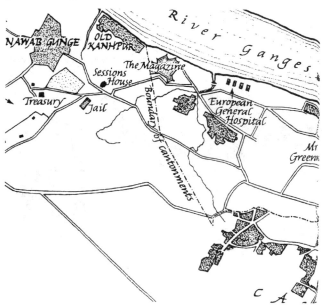

One was the magazine some miles north of the military station, defensible, but hard to get at, depending largely on the river for water, with the actual running water some distance away across the sandbank until the river rose. This magazine or ordnance depot was full of military stores, many guns of various sizes and a store of ammunition. It was guarded by the sepoy guard, and had the usual staff of ordnance officers and warrant officers with *lascars*. There was thus this position, none too well placed to receive several hundred women and children with their goods and chattels. It was... hard to get to in times of excitement and some way from the trunk road and the road to **Lucknow.** The other possible site, was two large barracks built for European troops, one too a hospital fitted with those sanitary out-houses and conveniences which would be so essential a feature of a place of refuge in hot weather, with also a good inexhaustible well. The buildings were out in the open, and close to the road from the south. Believing that the soldiery would march away north and on balance of argument as it appeared to him, Wheeler chose the latter. It is difficult to say on the data before him that he was wrong.' Sir George MacMunn, *The Indian Mutiny in Perspective*. London, 1931, p 100. *See* H.R.Nevill, *Kanpur: A Gazetteer*, Allahabad, 1929, p 212., *see also* W.J.Shepherd, *A Personal Narrative of the Outbreak and Massacre at Kanpur*, Lucknow, 1886.

WHITING, CAPTAIN FRANCIS Captain Francis Whiting, Engineer in the Canal Department and Postmaster Roach were the civilians who, with Captain Moore, were authorized by **Wheeler** to negotiate with the **Nana Sahib**. Whiting commanded the northwestern end of the **Entrenchment** and was 'a man of stout heart and clear brain.' Whiting, with **Moore**, had argued for accepting the Nana's terms. He was the son of a merchant and was born in London in 1822. He was educated at **Addiscombe**, joined the service in 1842, and saw action in the Sutlej campaign. A plaque in his honour at **All Souls Church** (the Memorial Church at Kanpur) says he 'was shot by Mutinous Sepoys on the 28th June 1857 in the 35th year of his age. One of the devoted band who defended the Kanpur entrenchments under Sir H. Wheeler, he was in command of the boat which escaped about 30 miles down the river and fell while pushing her off from a sand bank on which she had grounded. *In God have I put my trust. I will not fear what flesh can do unto me. Ps. lvi. 4*'.

WHITLOCK, Major-General He commanded the Madras Column that was supposed to co-operate with and supplement the Central India Force commanded by **Sir Hugh Rose**. Whitlock was appointed in November 1857, his troops were to be concentrated at **Jabalpur** for the advance on **Banda**. His column was made up from the **Saugur** Field Division, and the **Nagpore**[Nagpur] Moveable Column, and consisted mainly of troops from the Madras Army, untainted by mutiny or rebellion. He was a poor commander, slow, obstinate, overcautious not to say timid. He did not leave Jabalpur until 17 February, after a quite unnecessary delay, and as he marched northwards he refused all requests from the Commissioner, **Major Erskine**, to turn aside and dismantle rebel forts etc, with the result that the country was not pacified by his march, and the rebels would part in his front and form up again behind his column. He is best known for his capture of the town of **Banda** but even here his reputation is sullied, as there is a strong suspicion that he was more interested in capturing booty, for the Prize money involved, than in defeating the rebels. A good account of his conduct is given in the official account of these military operations, *see The Revolt in Central India,* Simla, 1908, and **FSUP**, Vol III.

WIGGENS, COLONEL L.M.

Although described as 'Major' Wiggens by **Gubbins**; he was Deputy Judge Advocate General at **Kanpur** in 1857, and possibly the fourth senior officer in the station commanded by **Major General Sir Hugh Massy Wheeler**. During the siege of **Wheeler's Entrenchment** he received and despatched correspondence with **Lucknow**. Gubbins, in an Appendix, quotes extracts from a private letter Wiggens sent to Colonel Halford, 71st BNI, dated Cawnpoor, 24 June 1857: 'I was agreeably surprised to receive your most welcome letter of the 21st, the messenger of which managed cleverly to find his way here; but that surprise was exceeded by the astonishment felt by us all, at the total want of knowledge you seem to be in regarding our position and prospects; while we have been, since the 6th of the month, equally in the dark respecting the doings of the world around us. Your loss at Lucknow is frightful, in common with that of us all; for since the date referred to, every one here has been reduced to ruin. On that date they commenced their attack, and fearfully have they continued now for eighteen days and nights; while the condition of misery experienced by all is utterly beyond description in this place. Death and mutilation, in all their forms of horror, have been daily before us. The numerical amount of casualties has been frightful, caused both by sickness and the implements of war, the latter having been fully employed against our devoted garrison by the villainous insurgents, who have, unluckily, been enabled to furnish themselves therewith from the repository which contained them. We await the arrival of succour with the most anxious expectation, after all our endurance and sufferings; for that, Sir Henry Lawrence has been applied to by Sir Hugh, and we hope earnestly it will be afforded, and that *immediately,* to avert further evil. If he will answer that appeal with 'deux cents soldats Britanniques,' we shall be doubtless at once enabled to improve our position in a vital manner: and *we deserve* that the appeal should be so answered forthwith. You will be grieved to learn that among our casualties from sickness, my poor dear wife and infant have been numbered. The former sank on the 12th, and the latter on the 19th. I am writing this on the floor, and in the midst of the greatest dirt, noise, and confusion. Pray urge our reinforcement to the Chief Commissioner [Lawrence]..'Yours, &c.,' (Signed) L.M. Wiggens.' **Mowbray Thomson** writing from recollection of matters within the entrenchment recorded the following: 'I deeply regret, however, to have to record the fact that there was one officer of high rank, and in the prime of life, who never showed himself outside the walls of the barrack, nor took even the slightest part in the military operations. This craven-hearted man, whose name I withhold out of consideration for the feelings of his surviving relatives, seemed not to possess a thought beyond that of preserving his own worthless life. Throughout three weeks of skulking, while women and children were daily dying about him, and the little band of combatants was being constantly thinned by wounds and death, not even the perils of his own life could rouse this man to exertion; and when at length we had embarked at the close of the siege, while our little craft was stuck upon a sandbank, no expostulation could make him quit the shelter of her bulwarks, though we were adopting every possible expedient to lighten her burden. It was positively a relief to us when we found that his cowardice was unavailing; and a bullet through the boat's side that despatched him caused the only death that we regarded with complacency.' Knowing, as we do, the fate of all the other senior British officers within the entrenchment, the process of elimination appears inexorably to point to Colonel Wiggens as Thomson's cowardly officer; but as he was not actually named the identification must remain speculative. See M.R.Gubbins, *An Account of the Mutinies in Oudh etc.,* London, 1858, *and* Mowbray Thomson, *The Story of Cawnpore,* London, 1859.

WIGS ON THE GREEN

This unusual metaphor comes in a letter written on 5 June by a married engineer in **Kanpur** to his friend Bews in **Fatehpur**, 'we are quite prepared, and if the fellows break out there will be wigs on the green.' Two months later, all Bews would ever find of his friend and his family was a miniature of the engineer's wife at an abandoned Sowar outpost across the river from Kanpur. *See* J.W.Sherer, *Daily Life During the Indian Mutiny: Personal Experiences of 1857*, p 70.

WILBERFORCE, Lt. Colonel

Wilberforce, in Delhi in 1857 during the so-called 'siege' was a man of spirit and courage, yet his patronising attitude to all who were not true-blue English gentlemen can leave us today gasping with embarrassment. Take this as an example of his attitude : 'So highly did we esteem the little priest indeed, that he dined at our Mess an honoured guest - the first, and I should think the only instance of an Italian Jesuit priest dining at the mess-table of a Queen's Regt. I saw a good deal of the good little man, and I once asked him how he could, in action, discriminate between the faithful and the heretic. His characteristic answer was 'Ah my friend, in Rome the saints are good and the Virgin Mary is very good, but here where the cholera is doing its deadly work, and where the bullets are flying around, the saints are no good - the Blessed Virgin even is no good. All I do is : I hold this (shewing his crucifix) before the eyes of the dying man and say, Look at the figure of Jesus! Jesus Christ died for you! Believe on him, and you are saved !'

WILLIAMS, G.W., Major

Superintendent of Cantonment Police, NWP. Latterly Lt.Colonel, Commissioner of Military Police, NWP. Originally from Meerut. He took the depositions of witnesses both in Meerut and Kanpur, as a supplement to the 'Narrative of Events'. These depositions are of great importance to scholars, but at the same time can be very frustrating, since they leave unanswered as many questions as they solve. Moreover the witnesses are on occasion clearly lying since their evidence is contradictory, and they are doing their best to incriminate others while exculpating themselves. *See Depositions taken at Cawnpore[Kanpur] under the direction of Lt.Colonel G.W.Williams, Commissioner of Military Police, NWP*, no imprint, 1858. Colonel Williams and a staff drawn from his own office and the Cantonment Magistrate's took evidence. A Mr. Longdens translated the depositions, but he died during the preparation of the report and was replaced by Mr. Blanchett, Mr Stewart Reid and Lieutenant Hennessy. NA. FDP. FCS. 30 December, 1859. No. 662-80.

WILSON, Sir Archdale

KCB and First Baronet. Born 1803. Lieutenant General. Best known as the very cautious General i/c **Delhi Field Force** whose indecision almost led to disaster for the British. **John Nicholson** when mortally wounded in the assault rose on one elbow and swore he had enough strength left to shoot Wilson if - as he was threatening to do - he ordered the troops to retire. But as the final assault in September 1857 was successful (though costly), he was rewarded with a KCB and later a Baronetcy as Sir Archdale Wilson of Delhi. Later commanded the artillery at the assault on Lucknow in March 1858. It is a curious fact that most of the guns employed on either side, both in attacking and defending **Delhi**, had been cast by him when holding the appointment of superintendent of the Calcutta Foundry. Died 1874. *See* **DNB**

WILSON PAPERS

General Sir Archdale Wilson

These are held in the Centre of South Asian Studies at Cambridge University. They consist of a typescript copy of a short biography and notes on **Major General Sir Archdale Wilson's** career in India, and the letters written to his wife during the Indian mutiny/rebellion from 2 May 1857 to 18 March 1858. They contain a day to day account of the fighting and conditions generally and his own feelings. In an Appendix is the general order which was read at the head of each division of the army just before the assault on **Delhi**, 6 September 1857.

WILSON, J.Cracroft.

Commissioner on Special Duties, and the author of *'Narrative of Events attending the Outbreak of Disturbances and the Restoration of Authority in the District of Moradabad in 1857-8'*. One of his conclusions is highly significant: he decided that there was indeed a conspiracy to rebel, and that plans were carefully laid, and that the plan miscarried only because the spark (at **Meerut**) was ignited prematurely. *See* the *'Sad Tale of Mees Dolly'* in **CM**.

General Windham

WINDHAM, Sir Charles Ash

KCB. Born 1810. Established a good reputation in the Crimean War, but was defeated in November 1857 at Kanpur by **Tatya Tope** and the **Gwalior Contingent**. He was Major-General left behind i/c troops at Kanpur in November 1857 when Sir Colin Campbell marched to the relief of the garrison of the Lucknow Residency. Before dawn on 26 November, Windham advanced with 1,200 bayonets and eight guns. 'The enemy were found in considerable numbers, and the clouds of dust betrayed the movement of troops. They could be distinguished dressed in their scarlet uniforms' The British troops advanced 'cheerfully' until six heavy rebel guns opened fire with grape. Beating their way through fields 'encumbered with high standing corn, topes of trees, walls, &c.' the troops encountered some 3,000 rebel soldiers. They did not run at the first sight of British troops but fought hard to hold their ground, and retreated grudgingly through a nearby village, eventually killing 16 men and wounding 78 more. Windham halted to rest his troops and savour his success, but from a rise in the ground he was distressed to find that the main body of Contingent troops were nearby and advancing in good order and enormous numbers, undismayed by the defeat of their advance guard. Windham drew his troops back to their positions around the city and found, on his return, that a letter had been received announcing that Lucknow had been taken and the expedition was on its way back to Kanpur. The effect of Windham's precipitate action was that the rebels were enabled to temporarily re-occupy Kanpur, and, to the disgust of the troops that had marched to Lucknow, capture all the British army's baggage. Promoted Lieutenant-General, 1863. Commanded forces in Canada 1867. Died 1870. *See* **DNB**; *see also* Lt.Col. John Adye, *Defence of Cawnpore by the Troops under the orders of Major General Charles A Windham in November 1857*, London 1858. *See also* KANPUR, BATTLE of.

WINGFIELD, Mr He was Commissioner of the **Baraitch** Division of **Oude[Awadh]**. When the troops mutinied he escaped, via **Gonda**, but in circumstances that leave a question-mark over his conduct: 'Everything was still going on as usual in the Battery, and after evening stable duties I returned to the Commissioner's house and found Boileau, Hale and the Doctor, sitting outside waiting for dinner.We noticed the absence of our host, Mr Wingfield, and on making inquiries we were told he had gone out for his usual evening ride and had not yet returned. We waited some time for him, and as it was getting dark, and long after the usual time for his return, we began to feel very anxious about him. At last we went in to dinner, and we had scarcely been seated when a note was received, explaining his absence. *It was merely a few lines in pencil, to the effect that he did not like the state of affairs at Secrora, and seeing that Boileau's regiment was on the verge of mutiny, he had gone out as if for his usual evening ride, and was then on his way to Gonda. He had left without giving us a word of warning, leaving everything standing in his house*. After this, Boileau gave it as his opinion that, now the Commissioner had left, we should be quite justified in following his example and making our escape. In support of this he produced a circular letter from Sir Henry Lawrence, sent to all out-stations, saying: 'Should the mutiny of the troops be deemed inevitable, the officers are at liberty to leave their men and consult their own safety,' and he considered the time had come to take advantage of this permission. When the question was put to me, I said my men had behaved well and I could not think of deserting them; that I was determined to stay and make every effort to save the guns, and that if he and the others left, I should march the Battery to Lucknow on the following morning.' *from* John Bonham, *Oude in 1857*, London, 1928, p 77.

WOLSELEY, Field Marshal Field Marshal Viscount Wolseley (a Captain in 1858 at Lucknow) in *Stories of a Soldier's Life*, 1903, was in no doubt as to the nature of the relationship between British and Indian: '*In an army of aliens, acting in their own country in the midst of their friends and relations, indeed of a whole population who loathed and abhorred the religion and daily habits of their officers, there must be no delay, no hesitation to nip in the bud all incipient mutiny. ...We won India by the sword, and whilst humanity and a Christian spirit incites us at all times to do what we can to make the sepoy and the people generally happy, prosperous and contented, that sword must always be kept sharp and ready for use at any moment.*'

WOMEN, attitudes towards The British claimed to have preserved an exemplary attitude towards the rebels' women, even if they were found with weapons in their hands, and it is true that there is no case on record of court-martial for rape by any British soldier. Nevertheless there is an uncomfortable feeling left by the so oft-repeated assertions of blamelessness in this matter, that the cynical might think the protestations overdone. Take this as an example of one form of embarrassment: 'Good bit of news.. the infamous Ranee of Jhansie has been slain in a fight sustained with Sir Hugh Rose's troops. This last piece of intelligence gave universal satisfaction: for, had she fallen into our hands alive, *we might have had some trouble disposing of her*. She had been a great friend of the Nana, and had tortured and killed great numbers of our women and children, evidently taking a delight in the sufferings of her victims.' (I.S.A.Herford, *Stirring Times under Canvas*, 1862). 'An emissary came out from Zeenut Muhul, the favourite wife of the king, a great political personage, offering to exercise her influence with the king, to bring about some arrangement. I sent word, we wished her personally all happiness, and had no quarrel with women and children, but could hold no communications with anyone belonging to the palace. I have probably told you this before'. After the assault, ...'.there has been great slaughter among the pandies. The women are well treated, and seek refuge in our lines with perfect confidence.' 18 Sept (day before he died): 'If the king wishes to have the lives of his family and his own spared, he had better surrender the Palace, and I should be glad to save that slaughter. Great numbers of women have thrown themselves on our mercy, and have been safely passed on. One meets mournful processions of these unfortunates, many of them evidently quite unaccustomed to walk, with children and sometimes old men.' 'A Joan of Arc was made prisoner yesterday; she is said to have shot one of our men, and to have fought desperately. She is a 'Jehadin', a religious fanatic, and sports a green turban, and was probably thought to be inspired. She is to be sent prisoner to Umballa' Camp Delhi 19 July (1857) from Hervey Greathed, *Letters Written during the Siege of Delhi*, 1858). '.....Many will be glad to learn that women and children are suffered to go unmolested. This is a stretch of mercy I should not have been prepared to make had I a voice in the matter. It ought to be remembered that many of these very women (or fiends in female form) were foremost in inflicting cruelty upon our own women and children; and it must be fresh in your memory that when the mutineers came out of the city for a grand attack upon our camp, while Nicholson's force was at Nufjugurh, they were followed by crowds of these very women, whose sole object on venturing out was to loot our camp when the mutineers took possession of it.....these coolie women were with the men who looted all the European houses in Delhi, and they are, therefore, to my thinking, equally deserving of punishment. However it is the General's hoorum (hukm?) that they should be spared; and I hope he won't rue it. I wonder if one of these women would have spared one of our women if she had the chance of murdering her? Mercy to such wretches is a mistake; they are not human beings, or at best wild beasts deserving only the death of dogs.' (*See* Illustrated London News, 16 January 1858). It is certain that the rebel soldiers themselves had no faith whatsoever in the British soldier's refusal to make war on women and children. Thus: 'The bodyguards of the Ranee (of Jhansi) were conspicuous for their desperate defence. Maddened with opium they defended their stables until they were burnt out, and then rushing on their assailants with their clothes in a blaze, they attempted not to cut their way through but to sell their lives as dearly as possible...one *velaitie*, after an unsuccessful endeavour to blow himself and his wife up attempted to hew her in pieces so that she might not fall into our hands. *And yet Sir Hugh Rose with just and honest pride can point to the humanity of British troops, who on every occasion shewed the most tender solicitude for honour and safety of (rebel) women and children of the conquered town (Jhansi).* See Friend of India, 10 June 1858.

WRITINGS ON THE WALL The graffiti artist seems to date from 1857. Examples include the 'writing on the walls' in Kanpur: the so-called last messages, the calls for vengeance, the graphic descriptions of dishonouring of British womanhood and the demands that their sufferings should be avenged. The **Bibigurh** the site of the massacre of the 200 women and children, the most terrible single act of the entire rebellion, was found to be covered, particularly behind the door, with these 'messages' from the frantic women in their last hour. *They were all forgeries.* **Sir William Russell**, the *'Times'* correspondent, had no doubts. In his *'My diary in India'*, he reports: 'One fact is clearly established; that the writing behind the door on the walls of the slaughterhouse, on which so much store was laid in Calcutta, did not exist when Havelock entered the place, and therefore was not the work of any of the poor victims. It has excited many men to fury, the cry has gone all over India. It has been scratched on the wall of Wheeler's entrenchment, and on the walls of many bungalows. God knows the horrors and atrocity of the pitiless slaughter needed no aggravation. Soldiers in the heat of action need little excitement to vengeance. The words that thrilled all our hearts were written in cold blood by 'someone' who came in after the massacre and the arrival of Havelock in Cawnpore. It has been proved most irrefragably (what a lovely word! - PJOT) and completely that the writing did not exist on the wall of the building when it was first visited by our officers and the civilians who accompanied the force.' *See* **SSF** Chapter 13.

Graffiti in the Bibigurh

XYZ

YOUNG, Captain 'One of the earliest casualties after our arrival was the death of Captain Young, who had served under **Havelock** in Persia, had followed him to **Cawnpore [Kanpur]** as a volunteer, and was now occupied in raising police at **Futtehpore[Fatehpur]**, a most hazardous service, as he was alone in the midst of an excited multitude of natives. He dined with General **Neill**, went to sleep in Colonel Olphert's tent, and died of cholera the next morning. This officer was, as well as a thorough soldier, a most accomplished linguist, and was famous for that rare attainment amongst Europeans, his most exquisite Persian writing.' Mowbray Thomson, *The Story of Cawnpore*, London, 1859, pp 204-7.

YOUNGHUSBAND, Lieutenant George An officer of the 13th Bombay N.I. who served at **Delhi** and in **Kanpur** and **Bithur** but was then killed at Fatehgarh in January 1858. His nephew was Sir Francis Younghusband and the latter's papers which include letters of his uncle, are preserved in the Oriental and India Office collections of the British Library, see MSS.Eur.F.197.

YULE, Sir George He was Commissioner of Bhagalpur and a man of eccentric and remarkable talent and habits. He personally organised 'a little campaign' for the purpose of protecting the Purnea District from the attentions of the 11th Irregular Cavalry and the **Dacca** mutineers. A detachment of the 5th Fusiliers had been left behind under Lieutenant Chapman for the protection of Monghyr; Mr Yule mounted the detachment on his own elephants, and followed the mutineers up. On one occasion the enemy made a very plucky charge, compelling the British to form square to resist cavalry. Yule was not quick enough in getting inside the square, and had to throw himself under the bayonets. But there is one other anecdote about him which almost defies belief: *'Mr Yule was, as described by the Times correspondent, a mighty hunter, and more, those who knew him think, a grand fellow. He commenced his campaign with a day's pig-sticking, highly characteristic of the man, and ended it, this is really almost too good, but he wished to make others enjoy what he enjoyed, by taking Chapman and his detachment a day's tiger-shooting in the jungles, mounted on elephants, in which, for very good reasons they were unfortunately unsuccessful. He finished his campaign by offering in his extreme generosity, to pay the entire expenses out of his own pocket'.* This tale comes the pen of Dr Francis Collins who was an Assistant Surgeon with the Fifth Fusiliers in 1858, and is printed in the *St George's Gazette,* dated 30 June 1893.

YULE, Sir George He was Commissioner of Bhagalpur from 1856 to 1861. His correspondence (and some press cuttings) concerning the suppression of the mutiny in **Bhagalpur** are preserved in the Oriental and India Office collections of the British Library, see MSS. Eur.e. 257/26.

YUSUF KHAN He was reported as 'Yusuf Khan, son of Azim Khan of Khairabad where Yusuf Khan's houses are. He is a relative of **Mammu Khan** but is supposed to hold the same position with Mammu Khan's wife as the latter does with the Begum. Was first made Collector of Lucknow and subsequently Commander-in-Chief, he had great power during the rebellion and plundered the city of Lucknow, he is also a relative or intimate friend of Chowdree Hushmut Ally, some of Ysuf Khan's relatives live at Mullabad'; see PERSONS OF NOTE, APRIL 1859.

ZAR, ZAMIN, ZAN A Persian proverb. All trouble, it was said, comes from these three Zs - meaning Gold; Land; Women.

ZEMINDAR Or Zamindar; a term not easy to translate. A landlord who *held* the land but did not *own* it. A small zemindar might hold a single village, a large zemindar might control a large number of villages: he had the right to collect the rent or land revenue, and was a man of great influence.

ZEMINDARI COURAGE At the battle of Nawabganj, on the Faizabad road 18 miles from Lucknow in June 1858.....'....four companies of the Rifles, the troop of Horse Artillery and some cavalry, now crossed the stream, followed in time by the main body; and then we found we had struck at the centre of the enemy who, having been thus surprised, had as yet been unable to concentrate. Their forces appeared to be divided into four parts each commanded by its separate leader, and of course acting without any unanimity. Still their attacks were rigorous, if unsuccessful, and we had much ado to repel them.......a large body of fine Zemindaree men brought two guns into the open and attacked us in rear. *I have seen many battles in India, and many brave fellows fighting with a determination to conquer or die, but I never witnessed anything more magificent than the conduct of these Zemindarees.'* It is interesting that this tribute is being paid to 'zemindaree men', *ie talukdars'* retainers, not regular army sepoys. It ties in with the notion that a man fights best when he believes wholeheartedly in his cause. These men of Oude[Awadh] felt they had every right to defend their land against the foreign usurper: among those who agreed with this viewpoint were the Directors of the East India Company itself! The tribute goes on: *' In the first instance they attacked Hodson's Horse who would not face them, and by their unsteadiness placed in great jeopardy two guns which had been attached to the regiment. Fearing that they might be captured I ordered up the 7th Hussars and the other four guns belonging to the battery to within a distance of 500 yards from the enemy, opened a fire of grape which mowed them down with terrible effect, like thistles before the scythe. Their chief, a big fellow with a goitre on his neck, nothing daunted, caused two green standards to be planted close to the guns, and used them as a rallying point, but our grape fire was so destructive that whenever they attempted to serve their pieces they were struck down. Two squadrons of the 7th Hussars under Sir William Russell and two companies of the 60th Rifles now came up, and forced the survivors to retire, waving their swords and spears at us, and defiantly calling out to us to come on. The gallant 7th Hussars charged through them twice, and killed the greater part of them. Around the two guns alone there were 125 corpses....after three hours' fighting the day was ours; we took 6 guns and killed about 600 of the enemy. Our loss in killed and wounded was 67....in addition 33 more died of sunstroke, and 250 men were taken into hospital.'* See Sir Hope Grant, *The Sepoy War*, 1873.

Queen Zinat Mahal

ZINAT MAHAL The youngest and favourite wife of the King of Delhi, **Bahadur Shah Zafar**, and the mother of his son Mirza Jiwan Bakht: they both accompanied him into exile in Rangoon in October 1858. It is said that she went with him most unwillingly, being 'quite tired of him - a troublesome, nasty, cross old fellow.'

ZUBR SINGH Zemindar of Gopal Kher in Oude[Awadh]. Remained loyal to the British throughout the campaign in Oude and was rewarded after the final assault on Lucknow, in what was known as the 'Oude Proclamation', by the Governor General, issued on 14 March 1858.

ZUHOOR-OOL-HUSSUN [Zahur-Ul-Hasan]

The British Government's treatment of those who had fought against them and been captured or who had surrendered as a result of the **Queen's Proclamation**, differed widely according to the social status of the rebel and *whether or he had been guilty of the murder of Europeans*: for such there was no mercy either in the heat of battle nor for many years after the rebellion was finally put down. There is, incidentally, no evidence to suggest that there was any policy of regarding a man of Oude[Awadh] as being 'entitled' to fight against the British as a foreign power, despite the Court of Directors' misgivings on this subject. As an example of implacable pursuit of certain offenders, take the case of Zuhoor-ool-Hussun. [Zahur-ul-Hasan] ....'Now upon the scene comes one Zuhoor-ool Hussun. Exactly what he had done to earn the gratitude of a friend of **Captain Patrick Orr**'s is not clear, but the said friend in **Bareilly**, had, before the annexation of **Oude**, asked Orr to do his best for this man, whose major accomplishment seems to have been that he was a good Persian scholar. Orr had tried very hard to oblige his friend, and had eventually got him the appointment of *vakil* to **Loni Singh**, Rajah of **Mitauli**. It was an important post and Orr might have expected Zuhoor to be grateful. Indeed he seemed to be so, and was loud in his protestations of gratitude and considered himself to be most deeply indebted to his benefactor. His power and significance grew as the months went by: he had frequent meetings with Orr in his official position of Assistant Commissioner for the district. But by early July 1857 the Durbar, or Court, in Lucknow, was now the effective government in Oude [Awadh]. From thence came orders to all zemindars to come to Lucknow, or send their vakils, with armed retainers to help the cause. Loni Singh sent Zuhoor-ool Hussun, and he left Mitauli at the end of July with 300 armed followers. Despite his previous protestations of gratitude never once had Zuhoor attempted to help Orr and his companions, (these people the **'Captives of Oude'** were prisoners of Loni Singh) and now he asked the raja whether he was to mention in Lucknow the existence of the Europeans. Not unless, replied Loni Singh, some advantage could be gained thereby, in which case he would willingly forward them all to Lucknow. We know all this because the conversation was faithfully reported to Orr by another, lesser, agent of Loni Singh, one Mehndi Hussun who had been present when Zuhoor got his instructions. Mehndi had given a number of proofs of kindness towards the captives, and this warning of what was afoot did at least keep Orr fully informed of treachery, even though there was little that he could do about it. On 6 August Loni Singh sent word to Orr that he had heard troops were on the way to Mitauli from Lucknow with orders to demand the persons of the Europeans and that for their safety therefore they should leave the fort of Kachiani at once and take refuge again in the jungle. Perhaps this was a ruse to hedge his bets should the British, against all probability, win the final battle; perhaps he thought the Europeans would be taken more easily by the troops in open country. A Passi guard, friendly to Orr, went out and brought back news that there were indeed 250 soldiers, from five different regiments, within a day's march of Mitauli. The 'five different regiments' detail is interesting: it is a phenomenon often seen among the mutineers: they did not trust each other and feared betrayal to the British by some corps who might only be paying lip-service to rebellion. So, frequently, they sent out parties, mixed up from different units, so that none might betray the cause. In this case the mixture of regiments had a bizarre side-effect. On the night of 6 August therefore, with heavy hearts, the party went once more into the jungle. The monsoon rains had come and conditions were atrocious. Moreover a servant reported that foul play was being contemplated. The raja was in fact urging the 250 soldiers to go into the jungle to take the Europeans and offered guides and support. But the mixture of regiments seems to have reduced the courage of the soldiers to the lowest common denominator, or perhaps they were all too intent on watching each other: the effect anyway was that they refused to attack the pathetic handful of Europeans, and returned to Lucknow. It would be interesting to know how they explained their failure to the Begum and the Durbar: unfortunately there is no record. The fugitives grew sick; it was impossible to protect themselves from the rain. They were to remain in these conditions until 20 October. Before that date at least two letters came to Orr. One was from Captain Gordon at Cawnpore[Kanpur]: it told of the massacres, but it also told of the advance of Havelock's column. The other was from General Outram who accompanied Havelock and was to be Sir Henry Lawrence's successor as Commissioner in Oude: it pledged all help and support and was accompanied by an order to Loni Singh instructing him to remain faithful, and threatening punishment if he did not. What of Zuhoor-ool Hussun? It seems he was disappointed not to see his victims dragged into Lucknow by the troops, and started for Mitauli in person, promising the Durbar to bring them in himself. He asked, as an afterthought, whether they were required alive, or would their heads do? He was told to bring them in alive. On arrival at Mitauli Zuhoor hesitated for a day or two: we do not know the reason: it may be that he had to convince the raja that there was no danger in what he wanted to do, or it may be that he too was frightened of the handful of fugitives. He need not have worried : they were weak with fever, clad only in rags without shoes or hats. They could not resist. Zuhoor took no chances however: he went with no fewer than 300 armed men into the jungle; Orr's pistol and gun were taken from him and the fugitives were told to follow the raja's men: they were not told where they were being taken. Sergeant-major Morton asked if he could take with him a piece of cloth that served him as a carpet: no, was the reply. Annie, Mrs Orr attempted to carry away a sheet with which to cover her head and that of her daughter: this also was refused, and with a blow to the head that sent her sprawling to the ground. At the edge of the jungle two rough country carts awaited them, and in the first village they came to, a mile and a half away, a blacksmith appeared with heavy fetters for the men. Burnes's mind went, and Sergeant-major Morton had a convulsive fit from which he recovered with great difficulty. The men were all duly fettered: Orr with the heaviest irons: even the string by which to hold them up when walking was refused. The march to Lucknow took six days, from 20-26 October. Each day the procession started at 8 a.m. and went on till evening. 150 men of the escort marched in front, with a piece of artillery loaded at all times, the other 150 came behind: all this to guard 4 men, 2 women and 2 little children. The sun was terrible: they had a raging thirst; they were given food, but it was literally thrown at them. Their sufferings must have been as much mental as physical. In each village they were exposed to the gaze of the villagers, abused in tones of brutal contempt, and loaded with every indignity. Zuhoor decided that only by inhuman treatment could he wipe out the debt that he owed to this man: he must have burned with hate for Captain Orr - how dared this foreigner place him under an obligation, there can be no other explanation. After the rebellion was put down Zuhoor must have felt, by taking on a new identity, that he was safe, but his eventual fate was recorded in an English periodical: 'ZAHOOR OOL HUSSEIN. The accompanying portrait is a likeness of one of the leaders in the Indian Mutiny of 1857 from a photo taken by Lt Col Stewart. Rebel so long evaded capture - about to be forwarded to Oude for trial. ....stated to have delivered up in 1857-8 to the Begum of Lucknow a party of nine Europeans including Major Orr and Mr Hart (?) by which act they met their death...employed under the assumed name of Sufraz Hoosein in service of a talookdar of His Highness the Nizam in district of Copal... Capt Torin Thatcher, Superintendent of Police Ahmedabad obtained permission and proceeded to the Deccan in August (1862) accompanied by some men able to identify the rebel. Further information collected and net closed in but Zahoor obtained wind of the intention and fled a few hours before. A confidant and a Mussulmanee woman living with him were however captured and he himself was arrested within the limits of the Dharwar collectorate. 18 Aug taken to Lingsagoor and thence escorted by Nizam's contingent to Sholapore to which place Captain Thatcher repaired to establish his identity. At once recognised and he himself admitted he was living under an assumed name. Then brought to Poona and now in the gaol there with 4 other persons lately in his service. One of these (Ghareebun his mistress) has been living with him since 1856, a second, Iklak Hoosein, a Syed, described as having one eye, a mendicant by profession, and 2 others Alli Buksh and Shad Ahmed

Khan, inferior servants. *Police Commissioner reports this as a remarkable instance of the retribution which is slowly but surely overtaking all the principal criminals and leaders of the rebellion. The house and family of Zahoor ool Hussein are in the NW but he attempted to evade pursuit by abandoning both for the extreme SW of the Nizam's dominions. Information obtained in the distant province of Guzerat and by means of this information an offender is brought to justice for crimes committed five years ago in the Province of Oude, at the other extremity of India.'* He was subsequently hanged. **PJOT**

See **SR**; see also Illustrated London News, January 1862.

ZUHOOREE, MOONSHEE, [Munshi Zahori]

'At this time the 1st Regiment of Infantry and 2nd Cavalry stationed here (Kanpur) seemed to be in an excited state, when a sepoy named Jan Mahomed went to the cavalry and held a consultation, which proceeding was brought to the notice of the General by Subadar Bhowanee Singh, and led to the confinement of Jan Mahomed.' Deposition of Moonshee Zuhooree. **Williams**

ZULFIQUAR, SYED

The grave of this Muslim saint, who was killed during the mutiny/rebellion is well-preserved at Atrauli, district Aligarh. Courtesy: INTACH.

Zemindar, Hindoo landowner.

THE 'INDIAN MUTINY' 1857-9

SUMMARY OF MAJOR EVENTS

1857

January 22	The Sepoys at **DumDum** become uneasy about the new cartridges.
Jan 26	Treaty with **Dost Mohamed** at **Peshawar** agreeing to assist him against Persia
Jan 27	Generals **Outram** and **Havelock** at Bushir. Persian campaign swiftly over.
Jan-March	Unrest among the sepoys on the greased **cartridge** question; outbreaks at **Berhampore** and **Barrackpore**. The **chapaties** pass from village to village.
April 24	**Meerut.** 3rd Bengal Light Cavalry; mutinous conduct followed by court-martial.
May 3	Mutinous conduct by 7th (Oude) Infantry at **Lucknow**. Disarmed and leaders imprisoned
May 10	**Meerut.** 3rd Light Cavalry, 11th Native Infantry, 20th Native Infantry mutiny. **THE MUTINY BEGINS.** Mutineers kill officers and other Europeans and burn bungalows, and then set off for **Delhi.**
May 11	Mutineers arrive Delhi; proclaim **Bahadur Shah Zafar** as Emperor, murder all Europeans they can find. Delhi troops 5th Light Fd Battery Foot Artillery, 3rd Bn 2nd Company, 38th N.I. and 54th N.I. all mutiny. **Magazine** blown up by Lieutenant Willoughby and handful of British. European survivors retreat first to **Flagstaff Tower** on the Ridge, and then to **Karnal.** City of Delhi and surrounding area completely in the control of the rebels. Sappers and Miners march from Roorki to **Meerut Telegram** to Ambala alerts the Punjab
May 12	News of **Meerut** outbreak reaches Muzaffanagar, Moradabad, Saharanpur, Etawa, Aligarh etc. **Bahadur Shah** sits in Dewan-i-Khas; his three sons apply for army commands.
May 13	The 45th and 57th BNI at **Ferozepur** mutiny and try to capture magazine, but are checked by the Europeans; men of the 45th scattered and many killed. The Sepoy regiments at **Lahore** disarmed.
May 16	Fifty European prisoners, men women and children are massacred in Delhi. Sappers and Miners mutiny at Roorki.

May 20	At **Aligarh** a portion of the 9th N.I. mutiny, followed at **Mainpuri** and **Bulandshahr** on the 22nd and **Etawa** on the 23rd, by the remainder of the regiment. Chiefs of Bikaner, Alwar, Jaipur and Datia invited by King of Delhi.
May 21,22	Sepoys at **Nowshera** and **Mardan** mutiny. Four regiments at **Peshawar** disarmed
May 25	**Nowshera**, the 55th N.I. are pursued by **Nicholson** and many killed Etawa re-occupied by the British.
May 26	Disaffection noticed in the **Gwalior Contingent**.
May 27	General George **Anson** dies at Karnal. Sir Henry **Barnard** takes command. Martial law declared at Etawa.
May 28	**Nasirabad.** 15th N.I., 30th N.I. 6th Light Fd Battery mutiny. and march off to Delhi
May 29	Mutinous conduct at **Lucknow** in **Oude[Awadh]** and other smaller stations. Europeans hunted and killed in many areas of the NW Provinces.
May 30	Mutiny at **Lucknow** of 7th Light Cavalry, and (portions of) the 13th N.I., 48th N.I., and 71st N.I. Outbreak at Mathura; rebels march to **Delhi**. **Battle of the Hindun River**.
May 31	**Bareilly**, 18th N.I., 68th N.I., 15th Light Fd Battery, and 8th Irreg.Cavalry mutiny. **Khan Bahadur Khan,** a government pensioner takes the lead, and is proclaimed ruler under the King of **Delhi.**. At **Shahjahanpur** the 28th N.I. mutinies and attacks the Christians in the church. At **Agra** the native regiments are disarmed. Date believed by many to have been fixed for the uprising of the Bengal Army. Moradabad in open revolt.
June 1	At **Moradabad** the 29th N.I. mutinies. Sepoys at **Mathura** shoot officers and march for Delhi. Outbreak at Budaon.
June 2	The 5th N.I. mutinies at **Saharanpur**.
June 3	At **Neemuch** the entire force mutinies viz 1st Light Cavalry, 4th Troop 1st Brigade horse artillery, 15th Light Fd Battery, 72nd N.I.. At **Azamgarh** the 17th N.I. mutinies. At **Sitapur** the 41st N.I. mutinies with particular ferocity against the Europeans both military officers and civilians. Abbas Ali proclaims himself ruler in Moradabad.
June 4	Partial mutiny at **Benares[Varanasi]** by 37th N.I. General **Neill** arrives and disarms some troops. Mutiny at **Cawnpore[Kanpur]** by 2nd Light Cavalry, 1st N.I., 74th N.I., and 56th N.I. First party of Europeans leaves **Fatehgarh**.
June 5	53rd N.I. joins the mutiny at **Cawnpore**, and General Wheeler calls all Europeans and Christians into his 'entrenchments'. At **Jaunpur** the regiment of Ludhiana mutinies.
June 6	Siege of **Wheeler's Intrenchment** at **Cawnpore** begins. Nana Sahib goes over to the mutineers and is proclaimed **Peishwa.** 12th N.I. at **Jhansi** mutiny, and the massacre of the Europeans follows; 4th Company 9th Bn Artillery at **Azamgarh** mutiny; at **Allahabad** (part) 6th N.I. mutiny. **Nawab of Rampur** visits Moradabad to take charge.
June 7	At **Jullundur** the 6th Light Cavalry, 36th N.I. and 61st N.I. mutiny. then march to Delhi.

	Liaqat Ali moves to **Allahabad** and appoints officers.
June 8	The **Battle of Badli-ki-Serai** on the Grand Trunk Road just northwest of Delhi: as a result British troops occupy the **Ridge**, and the 'siege' of Delhi begins. At **Phillour** the 3rd N.I. mutiny. At **Faizabad** the 22nd N.I., and the 13th Lt Fd Battery 5th Bn 2nd company mutiny.
June 9	Mutiny at **Nowgong**. Mutiny of **Malwa Contingent** cavalry.
June 10	At **Rohtak** the 60th N.I. mutiny; at **Nowgong** the 4th Coy 9th Bn artillery mutiny. **Fatehgarh** refugees reach **Bithur** but are fired on by **Nana's** troops.
June 11	Brigadier General **James Neill** arrives at **Allahabad**. Outbreak at **Jalaun**. Jhansi rebels leave for **Delhi**.
June 12	Massacre by **Nana Sahib** of 130 European fugitives from **Fatehgarh** at **Bithur**. Mutiny at **Lalitpur**.
June 13	Partial mutiny of the 1st regiment of Irregular Cavalry of the Nizam's Contingent at **Aurangabad**. Mutiny at **Hamirpur**. Disarmament of the 47th BNI at **Mirzapur** intended but not carried out and regiment assists in pacification of the district.
June 14	**Gwalior Contingent** mutiny and murder officers and families. Sepoys at **Barrackpore** are disarmed. Outbreak at **Banda:** Nawab proclaims his rule and issues proclamation. Mutiny at **Sipri**. **Allahabad** in total disorder.
June 15	**Wajid Ali Shah** ex-King of **Oude** [Awadh] and chief councillors imprisoned in Fort William in Calcutta. **Bareilly Brigade** reaches Moradabad en route for **Delhi**; Walidad Khan at Malagarh asks King of Delhi they be given to him.
June 16	**Maulvi Liaqat Ali** flies from **Allahabad**. Nawab's rule proclaimed at **Fatehgarh**.
June 18	At **Fatehgarh** the 10th N.I. (previously loyal) eventually break into mutiny. Mewati and Pathan villages in **Allahabad** district attacked and destroyed by the British.
June 19	**William Tayler** Commissioner at **Patna** arrests four Maulvis and disarms citizens.
June 25	At **Sialkot** the 35th N.I. mutiny. At **Kanpur** the **Nana Sahib** offers terms to Wheeler in the entrenchment. Mutiny at **Malthone**. Sack of **Deoli**.
June 27	Massacre at **Satichaura Ghat**, Cawnpore[Kanpur].
June 28	**Fatehgarh Fort** besieged by mutineers. **Nagode** gaol broken into.
June 30	British defeated at **Chinhat**; **Lucknow Residency** surrounded and besieged. Bulandshahr captured by Walidad Khan.
July 1	At **Saugur**[Sagar] the 42nd N.I. and the 3rd Irreg.Cavalry mutiny; at **Mhow** the 1st Light Cavalry and the 23rd N.I. mutiny. **Holkar's** troops mutiny at **Indore**, and attack the Residency. **Nana Sahib's** rule proclaimed at **Hamirpur**. **Neemuch** rebels reported at Fatehpur Sikri.

July 2	At **Mhow** the 2nd Coy 6th Bn artillery mutiny. **Bakht Khan** arrives at **Delhi** with the **Bareilly Brigade**. Pearson's Battery and Cavalry revolt at **Agra**.
July 3	Mutiny at **Damoh**. Report (10 weeks' premature) of British re-occupation of **Delhi**.
July 4	Massacre at **Rampore** on Ganges of European fugitives from **Fatehgarh/Farrukhabad**. **Sir Henry Lawrence**, Chief Commissioner of Oude, dies of his wounds in the **Lucknow Residency**. The **Kotah Contingent** mutinies at **Agra**.
July 5	Death (from cholera) of **General Barnard**, the British Commander-in-Chief. Battle of **Shahganj** or **Sussiah[Sasia]**, just outside **Agra**: the Kotah Contingent is attacked by Brigadier **Polwhele**, but the British are beaten back and forced to take refuge in **Agra Fort**.
July 6	**Nana Sahib** issues Proclamation.
July 7	At **Jhelum** the 14th N.I. mutiny. **General Havelock** leaves **Allahabad** en route for **Kanpur**.
July 9	At **Sialkot** the 9th Light Cavalry and the 6th Infantry mutiny. The British abandon **Sehore**.
July 12	At **Ambala** the 5th N.I. mutiny. **Fatehpur (Futtehpore)** occupied by **Havelock**
July 15	Action at **Aong** (**Renaud** killed), and at **Pandu Nadi**. 'Massacre of the Ladies' ie **Bibigurh** massacre at **Cawnpore[Kanpur]**.
July 16	**Nana Sahib** defeated by **Havelock** at **Fatehpur**, and first **Battle of Cawnpore[Kanpur]**. **Nana Sahib** orders all Bengali babus to be apprehended.
July 17	**Havelock** enters Cawnpore and the Nana Sahib retreats to Bithur. General Reed hands over command in **Delhi** to **Archdale Wilson**. March of the Nagpore[Nagpur] Movable Column.(Rebel) **Neemuch Brigade** arrive in **Delhi**.
July 18	Action at Benaika.
July 22	42nd BNI from **Saugur[Sagar]** arrive at **Kalpi**.
July 25	The **Danapur** Brigade mutiny, ie the 7th N.I., 8th N.I.,40th N.I.
July 27	Siege of **Arrah House** begins.
July 28	'**Parade Ground Massacre**' at **Farruckhabad** (Fatehgarh). **Azimgarh** evacuated by the British
July 29	Havelock defeats rebels at **Oonao[Unnao]**
July 30	At **Lahore** the 26th N.I. mutiny; ditto the 8th N.I. at **Hazaribagh**.
July 31	The large rebel force from **Mhow** and **Indore** arrives at **Gwalior**, leaving **Scindia** helpless to prevent their onward march towards **Agra**.
August 2	At **Ranchi** the **Ramgarh Battalion** mutiny. **Mhow** relieved by the British
August 3	Arrah House relieved by **Major Eyre**.
August 5	Havelock defeats the rebels at **Bashiratganj**. **Kunwar Singh** reported to have proclaimed himself 'King of Shahabad.'

August 8	**Khan Bahadur Khan's** forces from **Bareilly** reported as advancing to attack **Naini Tal**.
August 8	Rebels' gunpowder factory in **Delhi** blown up - probably an accident.
August 10	Second outbreak at **Nasirabad**.
August 11	King of Delhi requires chiefs to contribute money for the cause.
August 11	**Eyre** burns **Kunwar Singh's** palace at **Jagdishpur**.
August 12	Second outbreak at **Neemuch**. **Havelock** defeats Oude[Awadh] rebels in the third battle of **Bashiratganj**.
August 13	**Havelock** forced, by weakness of his force, to withdraw back to **Kanpur**. British abandon **Gorakhpur**.
August 14	5th Irreg.Cavalry mutiny at **Bhagalpur**. Movable Column under **John Nicholson** arrives at Delhi. **Kunwar Singh** arrives at Sassaram.
August 16	**Havelock** defeats the forces of **Nana Sahib** at **Bithur**.
August 17	Sir **Colin Campbell** takes over duties as British Commander-in-Chief.
August 19	At **Ferozepur** the 10th Light Cavalry mutiny. **Kunwar Singh** at Akbarpur. **Amar Singh** threatens to burn Dehri.
August 20	**Kunwar Singh** at Rohtasgarh.
August 21	Attack on **Mount Abu**.
August 23	Mutiny at Erinpura.
August 25	**Firoz Shah**, Shahzada, placed on the musnud at **Mandsaur**.
August 26	Outbreak at **Mandsaur**
August 29	At **Peshawar** the disarmed 51st N.I. mutiny, and many are slaughtered. **Kunwar Singh** arrives at Ramgarh, and plunders Ghorawal.
August 31	Dhar seized by the rebels
September 7	**Kunwar Singh** closes road to **Rewa**.
Sept. 8	Rebels attack **Neemuch**. Defeat of **Jodhpur** army by rebels at **Awah**. **Kunwar Singh** marches through Mirzapur, and reported to have arrived in **Rewa**. **Jung Bahadur's** Gurkhas arrive at **Jaunpur**.
Sept.14	British begin the assault on **Delhi**.
Sept.16	At **Nagode** the 50th N.I. mutiny.
Sept.17	Rebels destroyed at Narsingarh.
Sept.18	At **Jabalpur** the 52nd N.I. mutiny. **Shankar Shah** and his son executed. Abortive attack on rebels at Awah. Mutiny at **Nagode**.
Sept.19	At **Deoghar** and **Rampurhat** the 32nd N.I. mutiny. **Havelock** and **Outram** set out from Kanpur for **Lucknow**.
Sept.20	After fierce fighting **Delhi** finally conquered by the British.

Sept.21	**Captain Hodson** claims to have captured the King of Delhi, **Bahadur Shah Zafar** at **Humayun's Tomb.** British abandon **Damoh.** Muhammad Hasan enters **Gorakhpur.**
Sept.22	**Hodson** murders the Mughal Princes.
Sept.25	**Lucknow Residency** 'relieved' (reinforced) by Havelock. Neill and Tytler killed, together with 600 other British casualties.
Sept.27	Action at Katangi.
Sept.28	Thana of **Bithur** attacked by 42nd BNI aided by Raja of Sheorajpur.
Sept.29	**Kunwar Singh** reaches **Banda**. Said to be accompanied by 1,800 of **Danapore** rebels.
October 2	Capture of **Rehli**.. Muslim governor still rules **Gorakhpur** for the King of Delhi.
Oct.3	Rebels from **Delhi** arrive at Hathras.
Oct.5	Walidad Khan reaches **Bareilly** with 500 followers. **Greathed's** Column meets opposition at **Aligarh**.
Oct.8	**Nawab of Banda** attacks fortress of Ajaigarh.
Oct.10	**Greathed's** Column defeats the mutineers (Indore rebels) at **Agra.** Rebel attack on Bhopawar and Sirdarpur. **Mehndi Hasan** reported as enlisting men at Hasanpur.
Oct.12	Troops of the Madras army at Mirzapur move out on to the road to **Rewa**.
Oct.13	Two Shahzadas, Mirza Bakhtawar Shah and Mirza Mendu, sons of the **King of Delhi**, tried by the British and shot beside the Jumna.
Oct 15	**Gwalior Contingent** finally joins the rebels. Outbreak at **Kotah**.
Oct.17	Rebels from **Delhi** reported as within 20 miles of **Kanpur**.
Oct.19	**Kunwar Singh** with the 40th BNI reaches **Kalpi** via **Banda**.
Oct.20	Nawab of Jhajjhar brought into Delhi in British custody.
Oct.21	Sack of Patan.
Oct.22	Action at Dhar.
Oct.23	British repulsed at **Jiran**
Oct.29	**Tatya Tope** with **Gwalior Contingent** arrives at **Jalaun**.
Oct.31	Dhar captured by the British.
November 1	Severe action at Khajuha near **Fatehpur** between the **Banda** rebels and British column headed by Colonel Powell (killed) and Captain Peel and his Naval Brigade. The rebels were beaten but there were casualties on both sides.
Nov. 7	Gwalior Contingent and 40th BNI (with **Kunwar Singh**) join up and begin advance on **Kanpur.**
Nov.8	Mutiny at **Mehidpur**.
Nov.11	Rebels reported to have entered the district of **Allahabad** in some numbers.

Nov.17	**Sir Colin Campbell** relieves and evacuates the **Lucknow Residency**, leaving **Outram** at the **Alambagh**; rebels see his withdrawal from Oude as a great victory. Defeat of the **Jodhpur Legion**.
Nov 18	Sepoys of 34th BNI at **Chittagong** desert, carrying off the treasure.
Nov.22	At **Dacca** the 73rd N.I. and the 4th Coy 9th Bn artillery mutiny.
Nov.23	Battle of **Mandsaur**. Rebels on the Narbada defeated.
Nov.24	Havelock dies of dysentery and is buried in the Alambagh.
Nov.27	**Gwalior Contingent** attacks Nawabganj (Kanpur) and forces the British to retire into entrenchments.
Nov.28	**General Windham** is defeated by the **Gwalior Contingent** in the second battle of Cawnpore[Kanpur]. The city occupied by the rebels who take possession of the baggage of all British troops with **Sir Colin Campbell.** Muhammad Hasan still holds **Gorakhpur**.
December 3	**Campbell** despatches all the ladies and sick, ex Lucknow Residency, to **Allahabad.** Nazim of the Nawab of **Fatehgarh** with a large force reaches Etawa.
Dec. 6	**Campbell** defeats **Tatya Tope** in the third battle of Cawnpore[Kanpur].
Dec.14	British enter **Indore.**
Dec 15	Three of **Holkar's** regiments are disarmed at **Indore**
Dec 19	**Mainpuri** recaptured by the British.
Dec.26	Fight at **Koni Pass**.
Dec.27	Raja Tej Singh of **Mainpuri** defeated in Battle of Mainpuri and flees towards **Lucknow**

1858

January 6	Campbell re-occupies **Fatehgarh.** **Sir Hugh Rose** begins Central India campaign.
Jan.7	Trial of Bahadur Shah Zafar begins in Red Fort Delhi.(Ends March 9th)
Jan.14	**Rani of Jhansi** issues a proclamation against the British - 'Victory of Religion'.
Jan.19	**Awah** captured by the British
Jan.23	Rebels mustering again at **Kalpi**.
Jan.27	Battle of Shamsabad near **Fatehgarh**.
Jan 28	Fortress of **Rahatgarh** taken by Rose
Jan.31	Action at Barodia.
February 3	**Sir Hugh Rose** relieves **Saugur[Sagar]**
Feb.5	Action at **Ayodhya** between rebels and **Jung Bahadur's** Gurkhas.
Feb 7	**General Whitlock** with Madras troops arrives **Jabalpur.**
Feb.10	**Nana Sahib** at Naubatganj.
Feb.11	Fall of **Garharkota**.
Feb.12	**Kunwar Singh** reported as ariving at Ayodhya.

Feb.18	Proclamation issued by **Firoz Shah** at **Bareilly**.
Feb.25	**Nana Sahib** reported to be at **Kalpi.**
Feb.26	**Whitlock** relieves **Damoh**.
Feb.27th	**Rose** advances from **Saugur[Sagar]**.
March 1	Tehri troops beaten by **Rani's** forces. **Charkhari** captured.
March 2	Populace of **Saharanpur** disarmed by British: 5,000 weapons collected.
March 2-21	Campbell retakes Lucknow and drives out the rebels led by the **Maulvi of Faizabad** and the **Begum of Oude. Lucknow** itself stormed on 14th.
March 3	**Raja of Banpur** occupies the Narat Pass against Rose who attacks the **Raja of Shahgarh** holding the **Madanpur Pass** and drives them out.
March 5	Raja of Rooroo, discovered in 'treasonable correspondence' commits suicide.
March 16	**Nana Sahib** reported at **Shahjahanpur.**
March 17	Brigadier Stuart takes **Chanderi** by storm. Battle of Atraulia, victory for **Kunwar Singh**.
March 18	**Nawabs of Farrukhabad** and **Banda**, Raja Tej Singh of **Mainpuri, Khan Bahadur Khan** of **Bareilly, Walidad Khan** specifically excluded from the benefit (ie free pardon etc) of British Proclamation offering One Lakh of rupees reward to any rebel betraying **Nana Sahib**.
March 21	Rose arrives at **Jhansi.**
March 22	Millman besieged in **Azimgarh** by **Kunwar Singh**
March 23	Investment of Jhansi begins.
March 24	Explosion in **Khan Bahadur Khan's** gunpowder factory at **Bareilly**. **Nana Sahib** arrives at **Bareilly**.
March 26	**Kunwar Singh** occupies **Azamgarh**.
March 30	**Kotah** captured by **Rajputana Field Force**
April 1	**Battle of Betwa River.** Rose defeats **Tatya Tope**.
April 3	**Jhansi** captured and sacked.
April 5	**Jhansi fort** taken by the British. **Rani**, with her step-son, reaches **Kunch**.
April 6	Lord Mark Kerr relieves **Azimgarh.**
April 11	Action at **Jhigan**.
April 15	General Walpole defeated at **Royah**; Brigadier **Adrian Hope** killed.
April 16	**Kunwar Singh** driven from **Azamgarh** by General Lugard.
April 17	**Rao Sahib** issues proclamation to Chiefs of **Bundelkhand**. **Kunwar Singh** attacked by Brigadier Douglas near Azmatgarh.
April 18	Battle of **Banda.**
April 21	**Kunwar Singh** crosses the Ganges at Sheopur Ghat and is mortally wounded while doing so.

April 22	**Prince Firoz Shah** reaches Moradabad.
April 24	Forces of Nawab of **Rampur** drive **Firoz Shah** from Moradabad.
April 23	**Kunwar Singh** defeats the British under Le Grand at **Jagdishpur**. Rose captures **Kalpi**.
April 26	Death of **Kunwar Singh**. **Jung Bahadur** leaves **Gorakhpur** for Butwal.
April 28	**Rao Sahib** encamped at Jalalpur to oppose **Whitlock**. Gurkha army reaches **Faizabad** and **Ayodhya**
May 2	Capture of Lohari. Moradabad district made over to the Nawab of Rampore.
May 3	Maulvi Ahmadullah Shah with large force comes from **Mohammadi** to **Shahjahanpur**.
May 6	**Battle of Bareilly**, included the famous charge of the **Ghazis** of whom 133 were bayonetted. Bareilly taken but rebel leaders escaped.
May 8	**Battle of Kunch**, Rose defeats Tatya Tope
May 10	General Lugard occupies **Jagdishpur**.
May 11	**Amar Singh** defeated by Corfield near Piru but manages to escape. Battle of **Shahjahanpur**.
May 13	Rebels menace British position and camp at **Jagdishpur**.
May 15	Rebels leaders including **Nana Sahib, Khan Bahadur Khan** and **Maulvi Ahmadullah Shah** all in vicinity of **Shahjahanpur**.
May 22	**Battle of Kalpi**, Rose captures the arsenal of Kalpi.
May 24	Second Battle of **Shahjahanpur**: rebels defeated by **Campbell** and driven back to **Mohammadi** but followed up and British occupy Mohammadi. Nizam Ali Khan reported to be in the Shahi pargana.
May 25	**Hamirpur** occupied by the British.
May 27	**Chanderi** and its fort evacuated by the rebels.
May 28	Dara Shikoh and Haji Shikoh, princes of the **Delhi** royal family arrested at Hasanpur in the Moradabad district.
June 1	**Rani of Jhansi, Rao Sahib and Tatya Tope** capture **Gwalior. Nawab of Banda** also present. **Lushkar** and **Gwalior fort** occupied. **Scindia** leaves for Dholpur. **Madho Singh** captured by **Rose**.
June 3	Engagement at Rath between **Charkari** raja and Martand Rao Tatya: latter killed.
June 6	**Kirwi** captured by **Whitlock** and enormous booty taken. News from Etawa that one Ganga Singh has been appointed Nazim for the **Gwalior** area.
June 11	**Tej Sing**, Raja of Mainpuri, surrenders.
June 12	Battle of **Nawabganj**; **Sir Hope Grant** defeats 16,000 rebels in final decisive battle in Oude[Awadh]. **Amar Singh** returns to Buxar. **Khan Bahadur Khan** attacks **Shahjahanpur**.
June 15	**Maulvi Ahmadullah Shah** attacks Pawayan, and is killed.
June 16	Battle of **Morar**. Raja of Pawayan brings head of **Mailvi Ahmadullah Shah** to the British at **Shahjahanpur** to claim the reward.

June 17	Battle of **Kotah-ki-Serai**; death of **Rani of Jhansi**. Mahbub Khan hanged at Aligarh.
June 19	Battle of **Gwalior**. City is occupied
June 20	**Gwalior Fort** captured by Rose. Battle of **Jaora-Alipur**. **Scindia** returns to Gwalior.
July 5	**Banpur** raja surrenders to the British.
July 9	**Tonk** occupied by rebels; **Firoz Shah** reported to be with them.
July 10	**Shahgarh** raja surrenders. Rebel force reaches Rampur.
July 11	Ramnagar occupied by the rebels.
July 23	Kishan Singh hanged.
July 26	Pawayan and **Shahjahanpur** reported as 'seriously threatened ' by Oude rebels.
July 27	Force besieging **Shahganj** breaks up.
July 31	Sir Hope Grant relieves **Raja Man Singh** besieged by rebels.
Aug 2	'By **Act 21 and 22 Victoria, c.106,** all the East India Company's territories are vested in Her Majesty, and all its powers excercised in her name.' To take effect on November 1st 1858.
Sept 2	Mutiny at **Multan** by two regiments of Bengal Infantry; overpowered, fled, cut off.
Sept.5	Battle of Bijapur near Guna. Rebels heavily defeated.
October 8	Pawayan attacked by large rebel force but latter had to retire. Raja of Pawayan now actively supporting the British.
Nov. 1	**Queen Victoria's Proclamation** abolishing the rule of the Honourable East India Company in India and instituting her own.
Nov 12	Battle of **Shankarpur**, **Beni Madho** driven northwards, reported as joining up with **Bala Rao**..
Nov.24	**Amar Singh** in the Palamau district
Nov 25	**Raja of Gonda** defeated by Sir Patrick Grant, and Gonda occupied
December 5	**Nana Sahib** was reported as crossing the Ganges between **Fatehgarh** and **Kanpur**.
Dec.6	**Firoz Shah** and Walidad Khan reported at Aroul.
Dec.17	**Firoz Shah** has brief encounter with British force under Napier.
Dec 23	**Bala Rao** driven from **Tulsipur** and retreats into Nepal.
Dec 24	Engagement at Partabgarh
Dec 29	Engagement of Zirapur. Rebels defeated. Pursuit of **Tatya Tope** continues
December	Bihar rebels finally dispersed.

1859

January 2	Muhammad Hasan reported at Tulsipur
Jan 7	The rebellion in Oude[Awadh] officially declared at an end.
Jan 9	Prince Firoz Shah cuts his way across Oude and the Ganges and joins Tatya Tope temporarily
Jan.29	**Nawab of Farrukhabad** arrives at **Fatehgarh** under arrest.
March 28	**Nana Sahib** and **Begum of Oude** reported to be at Butwal.

April 7	**Khan Bahadur Khan, Begum of Oude, Nana Sahib** and **Birjis Qadr** reported as in the fort of Niacote in Nepal.
April 8	**Tatya Tope** betrayed and captured.
April 18	Tatya Tope hanged.
April 20	**Nana Sahib** sends Ishtiharnama to **Queen Victoria**. Petition to the English from **Bala Rao**.
May-June	European troops of HEIC object to transfer to the Crown and a state of mutiny results
July 8	State of peace officially declared throughout India
July 31	Still some rebel activity in Central India, notably under **Barjur Singh**.
August 2	500 rebels under **Firoz Shah** enter Bala Behat jungle.
September 24	The death of the **Nana Sahib** reported.
December	**Amar Singh** captured in the **Terai** by **Jung Bahadur's** troops and handed over to the British.
Dec.9	Capture of **Khan Bahadur Khan** by **Jung Bahadur** reported.
Dec.17	**Khan Bahadur Khan** and **Mammu Khan** lodged in **Lucknow** Gaol.

1860

February 5	**Amar Singh** dies in **Gorakhpur** gaol hospital while awaiting trial.
March 24	**Khan Bahadur Khan** of **Bareilly** hanged on the spot where he had raised the flag of revolt.
May 3	**Jwala Prasad** hanged at the **Satichaura Ghat**.
	Firoz Shah, Shahzada reported to be in Kandahar

1861

	Firoz Shah reported in Bokhara

1862

	Firoz Shah reported in Teheran
April 9	**Rao Sahib** reported captured by the British. Tried and hanged on 20 August

1863

April 4	Calcutta Times reports the names of 'the Lucknow Begum' and her son **Birjis Kudr** as having been expunged from the list of proscribed outlaws.

1877

December 17	**Firoz Shah** reported died at Mecca.

Bn = Battalion; Coy = Company; Fd = Field; Irreg = Irregular; Lt = Light; N.I. = Native Infantry;

GLOSSARY

Abkari	Revenue from liquor, drugs opium etc
Achar	Custom *eg* daily worship, religious rites.
Adawlut [Adalat],	A Court of Justice.
Amil,	An Official of Government, particularly a Revenue Collector under the Nawabs.
Amla,	Government Officials, connected with Courts of Justice.
Arzi,	Petition or address.
Ashraf,	Respectable, well-born.
Atta,	Flour: any grain ground to powder.
Aurang,	Place of manufacture and wholesale disposal.
Babu,	A title of respect, particularly in Bengal.
Badmash,	Disreputable person.
Badshah,	King or Emperor.
Bagee/Baghi,	An Arabic word meaning a mutineer or rebel.
Bangar,	Upland, as distinct from Khadir or riverain land.
Banias,	Or bunniyas. Shopkeepers, particularly of foodstuffs. 'Merchants' or even bankers.
Bankar,	Income from the produce of forest land.
Batta,	Discount or commission; special payment to troops in respect of active service.
Bazar,	A Market.
Bhaichara,	Land tenure founded on brotherhood.
Bhisti,	Water-carrier.
Bhumiyawal,	A general Plundering.
Bhusa,	Or Boosa: chaff, or straw.
Bhut,	A Sanskrit word meaning a demon or goblin.
Bigha,	⅝ ths of an Acre.
Biradari,	Brotherhood.
Bobachee,	Or Bawarchi, a Cook.
Budgerow,	A travelling boat of a large size.
Budmashes,	Or Badmashes, persons of low character, suspected criminal elements.
Bukshee,	Paymaster.
Burkundaze,	An armed auxiliary used in rural areas to support the local *thanah* or *tehsildari*.
Chowkee,	A guard-post, or village lock-up, Anglicised to 'Chokey' *ie* Prison.
Chowkidar,	Watchman.
Chukla,	A combination of Villages, put together for the purpose of collecting Land Revenue.
Chukladar,	One appointed to, or inheriting, the farming of the land revenue in a Chukla of the Kingdom of Oude[Awadh].
Chumar,	A member of the skinner or shoemaker caste, assumed to be of low social status.
Chunari,	A kind of dyed cloth.
Chupprassee,	[Chaprassi]. One in possession of the brass badge (hence the name) worn by government messengers/peons.
Crore,	One hundred lakhs, *ie* 1,00,00,000. Equals ten million, *ie* 10,000,000.

Cutcherry,	Or kutcherry or kacheri. A court of justice, but commonly means the office of a magistrate or civil official.
Dak,	The Postal System.
Dalal,	A broker.
Darogha,	Or daroga or darogah, a superintendent or overseer.
Dastaks,	Permit or licence or summons to appear, a *sub poena*.
Debi,	Goddess.
Deen,	Or Din, faith or religion.
Devnagiri,	The script in which Hindi is normally written if it is not 'romanised'. It reads from left to right.
Dewan,	Or diwan, a minister of a royal court.
Dharam,	Moral duty, religion.
Dhoolie,	Or Dooley or doolee, a kind of litter carried by two men in which the wounded were carried in battle, and on the march.
Dufter,	Or Dafter, an office, or building in which there were offices sited.
Durwasa,	Or Darwasa, a gate or gateway, particularly of a city or palace.
Elaqua,	Or Ilaka: an Estate.
Eed,	Or Id; a Muslim festival, particularly that which celebrates the end of the fasting month of Ramadan (Ramazan), the Eed-ul-Fitr.
Fakeer,	Or Faquir; A holy man usually a mendicant..
Farman,	A mandate, order or command.
Fasli,	The agricultural year - of significance in revenue collection.
Feringhee,	Literally 'Frank'; a foreigner, usually a European.
Foujdari Adawlut,	A Criminal Court.
Ganj,	A frequently used ending to the name of a locality, signifying village or market place.
Garhi,	A measure of time, equal to 24 Minutes, also a small fort.
Ghee,	Clarified butter fat.
Gharry, Or Gari,	Although the direct translation of the Hindi word is 'cart' in English, at this time it paricularly meant a slow horse-drawn box-like vehicle.
Ghat,	Or ghaut, a landing place on a river bank, consequently used also for a ford or ferry.
Ghazi,	Or ghazee; a heroic soldier, Muslim, prepared for martyrdom in fighting unbelievers. Used contemporaneously by the British to mean fanatic.
Golundaz,	A man trained in the use of primitive cannon.
Gomashta,	Agent.
Goonda,	Hooligan or ruffian.
Guddi,	Or Guddee; literally a cushion, but taken to mean a throne of a royal or noble prince.
Gujar,	A semi-nomadic, semi-agricultural caste of western U.P.
Hakim,	A difficult word to translate exactly in English, since its meaning depends on pronunciation; a superior officer of a ruler's court, a man attracting respect; or a doctor.
Havildar,	A 'non-commissioned' officer of the army or police, roughly equating to the rank of sergeant.
Hookumnamah,	An order, from a superior to a follower.
Howdah,	A 'saddle' on an elephant or camel.
Hundi,	A bill of exchange.
Hurkara,	A messenger, or carrier of letters and news.

Id,	Or Eed. A Muslim festival, in particular the one that comes at the end of the month of fasting, Ramadan (Ramazan), *ie* Id-ul-Fitr.
Ilaka,	Or Ilaqa; An Estate.
Ishtahars,	Proclamations, especially by the Rebels.
Jaghir,	Or Jagir. The revenue on an estate or parcel of land granted to an individual, usually in perpetuity, as a reward.
Jagirdar,	The holder of a jagir, or jaghir.
Jalkar,	Produce from lakes streams *etc*.
Jat,	Standard agricultural caste of Western U.P.
Jemadar,	A native infantry officer, a platoon commander, *ie* equivalent to a lieutenant. Or, the controller of a household, a superior servant.
Jheel,	A lake, including a temporary inundation at the time of the monsoon rains.
Jihad,	Or jehad; a holy war, proclaimed usually against unbelievers.
Juar,	Large millet.
Jullad,	An executioner.
Kabar,	A grave or tomb.
Kaffirs,	A term of abuse, used by both Hindus and Muslims, meaning 'infidels', and applied mainly to Europeans and Indian Christians.
Kakun,	Inferior grain.
Kanungo,	Village revenue official.
Karinda,	An agent; one who acts for another. *cf* vakil.
Kashtkar,	Cultivator.
Khadir,	Low, alluvial land
Khandsari,	Dealer in sugar.
Kharif, Or Khareef,	A farm crop that is harvested in the autumn; of significance in revenue collection.
Khasburdar,	An orderly.
Khillat,	A robe of honour granted by a prince to an underling as a sign of recognition of his worth.
Khitmadgar,	A servant.
Khundsaree,	A dealer in sugar *etc*. Generally a rich merchant.
Killadar,	The commander of a fort.
Kisan,	A peasant.
Kist,	An instalment particularly of the revenue.
Koran,	Or Quran, the sacred book of the Muslims.
Kothi,	A mansion or important dwelling house, *eg* the Begum Kothi in Lucknow; sometimes just a warehouse.
Kotwal	Chief of police, a mayor.
Kukri,	The curved short sword, or large dagger, of a Gurkha soldier.
Kunkawara,	A sweeper employed by the Nana Sahib to cut off heads.
Lakh,	Equal to One Hundred Thousand, Written as 1,00,000.
Lotah,	Or lota, a metal pot, usually brass, particularly personal to the owner if of high caste; used for drawing and dispensing of water.
Mahajun,	A banker or money-lender; sometimes the term is used simply to describe a rich man.
Mahal,	A palace. Whereas in the plural, as 'mahals', the usual meaning is wives esp. of the ex-king of Oude,[Awadh] Wajid Ali Shah.

Mahant,	The head of a monastery or leader of a sect; a chief priest.
Mahaut,	Elephant driver.
Maulavi,	A Muslim scholar or Divine.
Meer Munshi,	A chief scribe .
Mofussil,	The suburbs or rural areas, in contrast with the city.
Moonsurrim,	A head clerk, especially of a court of justice.
Morcha,	A fortification or entrenchment; particularly a gun emplacement or battery.
Mundee,	Or mundi, a market-place.
Munshi,	A scribe or clerk, or, more particularly, a language teacher or translator.
Munsif,	Civil Judge of the lowest rank.
Musahibs,	Courtiers.
Musjid,	Or masjid. A mosque, hence jumma musjid or friday mosque.
Musnud,	A large cushion; cf guddi. Throne.
Mutsuddee,	A clerk or scribe, particularly concerned with accounts?
Naib,	A deputy: can be applied to a wide range of appointments.
Naik,	A 'non-commissioned ' officer in the army. The first rung on the ladder of promotion. Equivalent to corporal.
Najib,	Or nujeeb; one who has been recruited as a soldier but is untrained; armed only with a matchlock and tulwar. A militiaman.
Nawab,	Or nuwab, an arabic word meaning viceroy or more literally viceroys, since it is the plural of naib, a vicegerent. A nabob.
Nazim,	A man of considerable influence; appointed as the head of a large district or block of territory with very wide powers, to rule on behalf of an overlord.
Nuddee,	A river.
Nullah,	A water-course, the channel of a torrent; most generally a *dry* watercourse.
Nuzzer,	Or nuzrana, an offering or present.
Oubhree,	A small room, in many village houses, in hich grain was kept. Because it was secure it also was used sometimes as a 'dungeon' or temporary prison.
Pahar,	A measure of time equivalent to three hours.
Panchayat,	Council of village elders.
Pandit,	A learned Hindu Brahmin.
Pargana,	An administrative area, consisting of a number of villages grouped together usually under the name of the most significant.
Parwana,	An order or warrant or licence.
Patwari,	Village accountant.
Peon,	A foot-messenger, usually in attendance at an official's court or office.
Perwannah,	Or parwana. A command or royal order.
Pir,	A Muslim saint.
Poorbeah,	A sepoy possibly from Oude[Awadh], literally an easterner.
Puja,	Worship, prayers.
Pultun,	A regiment, probably a corruption of battalion.
Purdah,	A veil screen or curtain.
Qaba,	A long gown worn by men that fits closely to the body.
Qasid,	*See* COSSID (main entry).

Quran,	Or Koran; the sacred book of the Muslims.
Rabi	Spring harvest.
Rais,	A nobleman, notable or man of position.
Raiyat,	A cultivator or peasant.
Rajcumar,	A prince of a Hindu ruling family, in particular the heir to the throne or 'crown prince'.
Rajput,	A Hindu of the military tribe or order or caste.
Rangar,	A nomadic herdsman caste of western U.P. and Haryana.
Risala,	A troop of horse.
Roomal,	A waistband or kerchief: especially the cloth used by *thugs* for garotting their victims.
Rubee,	The name given to the crop harvested in the spring as opposed to the autumn season: of particular use in revenue collection terms.
Ryots,	The name given to the humblest people, normally the cultivators of the land. Raiyats.
Sahib,	Technically just means 'master' or 'lord', but was the term used most frequently to apply to Europeans, hence sahib-log - Europeans.
Sadr	Chief seat of government.
Salar,	A general or chief.
Sanad,	A grant or charter.
Sarkar,	Or Sirkar, means Government.
Serai,	Secure resting place for travellers.
Shahzada,	A prince of the royal family, that is of the Mughal royal house.
Sharista,	A Court of Justice, an official office.
Sheristadar,	An official of a court of justice, a headman or superintendent.
Shroff,	Banker.
Sipahi,	Hence 'sepoy', a soldier; usually an infantryman is intended; contrasts with 'sowar' a trooper in a cavalry regiment.
Sirdar,	A term with a very wide range of use; its basic meaning is leader.
Soobah,	Or Subah, a Province.
Sowar,	A trooper in a cavalry regiment, in contrast to sepoy which was a term usually applied only to an infantryman.
Subadar,	Or Subedar. 1. a viceroy; a rank superior to that of governor, in theory granted only by kings *eg* the Mughal emperors in Delhi. The governor of a province. 2. a rank of officer in the infantry regiments of the army, some times roughly translated as captain; superior to a Jemadar in rank.
Sudder,	Chief or principal, used in a variety of ways, hence sudderameen or superior court of law court, sudder bazar or chief market of the city *etc*. Sadr.
Sunnud,	A charter or grant made in an official form. Sanad.
Swadharma,	One's own religion.
Swaraj	One's own rule.
Syce,	A groom or servant whose particular function was to look after horses.
Tahsil.	The office or court of a tahsildar or collector of revenue.
Tahsildar,	A revenue collector, a government official junior to a collector or magistrate.
Thana,	*See* Thannah.
Thanadar,	The rural policeman - later equated with the rank of inspector or sub-inspector, in charge of the police station known as a thannah.
Thannah,	A rural police station.
Tilangas,	Soldiers, by which was meant regular soldiers of the Bengal army. The term originated in Madras.

Topi-wallahs,	Literally 'hat-wearers', and generally taken to mean Europeans at this time.
Tulwar,	A sword, particularly the extremely sharp and slightly curved weapon favoured in northern India at this time.
Umlah,	*See* Amla.
Urzee,	A Petition.
Vakil,	Or vakeel. An agent, in particular of a ruler; *cf* karinda. Sometimes, lawyer.
Vazier,	Or vizier, or wazir; a minister of a ruling house, usually the prime or chief minister.
Vilaities,	Foreigners, often Europeans were meant, but sometimes it simply meant people from a great distance away.
Wazir,	Or wazeer etc; a minister of a ruling family, usually the chief or prime minister.
Wahhabi,	A Muslim purist sect originating in Saudi Arabia: members of the Tarikh-i-Muhamadiyya who sought to purify Islam in India, and were suspected of hostility to British rule.
Wilayati,	Foreigner. Afghan(?) mercenaries.
Zamini,	Was the grant of a portion of the land of a village.
Zemindar,	Or zamindar; a term not easy to translate. A landlord who *held* the land but did not *own* it.
Zemindari,	A landed property or estate.
Zillah,	Or zilla. An official description for a sub-division of a province *etc.*; a district.

BIBLIOGRAPHY
PUBLISHED WORKS

ABERIGH-MACKAY, G.R. ed. *The Times of India Handbook of Hindustan*, Bombay, 1875.

ADAM, H.L., *The Indian Criminal*, London, 1909. A book on Indian Penal Colonies. Contains a fascinating chapter on Maulvi Liaquat Ali of Allahabad, and even a photograph of him in captivity. The caption calls him 'Mulvi Ala-ud-din', but that was evidently an honorific title.

ADAMS, W.H..DAVENPORT, *The Makers of British India*, London, c1900. Contains helpful thumbnail treatments of the Mutiny/rebellion, station by station, including some of the more obscure places.

ADYE, Lt.Col.John, *The Defence of Cawnpore*, London, 1858.

AHMAD, Qeyamuddin, *Maulavi Ali Karim*, a scholar soldier of Bihar during the movement of 1857-58, in Indian Historical Records Commission Proceedings, 33, 9-15, 1958. Describes the activities of an unusual rebel.

AITCHISON, Sir Chas., *Collection of Treaties Engagements and Sunnuds etc, 7 vols.* Calcutta, 1876. Essential reference material for pre-Mutiny background political information.

AITCHISON, Sir Chas, *Lord Lawrence*, Oxford, 1892.

AITKEN, E.H., ie 'EHA', *Behind the Bungalow*. Simla, 1929. Written humorously but really a social document on British attitudes.

ALAVI, Seema, *Sepoys and the Company: Tradition and Transition in Northern India, 1770-1830*, Delhi, 1995. An essential, scholarly, reference work for students of the mutiny of the Bengal Army.

ALLEN, Charles, ed., *A Glimpse of the Burning Plain*, Leaves from the Indian Journals of Charlotte Canning, London, 1986. Lady Canning was the wife of the Governor-General (and first Viceroy) and was a perceptive, and liberal, observer of the Indian scene.

ALLEN, Charles, ed., *A Soldier of the Company: life of an Indian Ensign 1833-43: Captain Albert Hervey*, London, 1988.

ALLEN, Charles, *Raj: A Scrapbook of British India*, London, 1979. An excellent, profusely illustrated, summary of the British period in India.

ALTER, James Payne, *In the Doab and Rohilkhand: North Indian Christianity 1815-1915*, Delhi 1986. Contains a useful commentary on the period of the rebellion.

ANDERSON, George, and SUBADAR, Manilal Bhagandas, *Development of an Indian Policy, last days of the Company: a source book of Indian History 1818-1858*, London, 1918. Particularly useful for analysis of the permanent policy results of the mutiny/rebellion.

ANDERSON, George, and SUBADAR, Manilal Bhagandas, *Expansion of British India 1818-1858*, London, 1918. A collection of excerpts from original sources dealing with the last years of E.I.C.'s rule in India. An interesting and informative work with an unusual format designed, it is said, primarily for students.

ANDERSON, Robert P., *Personal Journal of the Siege of Lucknow*, London, 1858.

ANDERSON, T.C., *Ubique: War service of all the officers of HM Bengal Army*, Calcutta, 1863. Limited to factual and accurate short personal statements.

ANDREWS, C.F., *Maulvi Zaka Ullah of Delhi*, London, 1928. Contains an interesting chapter on the 'Mutiny' at Delhi, and the terrible events that followed the British assault on the city on 14 September 1857.

ANNALS OF THE INDIAN REBELLION 1859-9, Calcutta, 1959. See CHICK, N.

ANNAND, A.McKenzie, *Cavalry Surgeon: the Recollections of J.H.Sylvester*, London, 1971.

ANSON, Major Octavius, *With HM 9th Lancers during the Indian Mutiny*, London, 1896. A collection of letters written to his wife by a serving officer in the years 1857-8. Interesting as representing the many such first-hand accounts available but also because the shrewdness of the writer allows him to sum up some of the essential truths of the struggle; eg he writes of the Sikhs ...'they were very active and, *as usual*, got an enormous deal of plunder.' Deplores the brutality of the British retribution.

ANTHONY, Frank, *Britain's Betrayal in India: the Story of the Anglo/Indian Community*, Bombay, 1969. The Eurasians and 'native Christians' were linked by the mutineers with the British and suffered accordingly. But the British, is the argument, did not show either recognition of the sacrifice, or gratitude for the loyalty.

ARNOLD, Edwin, *The Marquis of Dalhousie's Adminis-*

tration of British India, 2 vols, London, 1862. The rectitude of Dalhousie's policy of annexation is hardly questioned here, but for a factual account of the events this is useful.

ATKINS, J.B., *Life of Sir W.H.Russell (2 vols)*, London, 1911. Vol I includes the Crimea and the Indian Mutiny in much detail. Russell was the highly thought of correspondent of the London 'Times', who came to India and covered the latter part of the campaign of the Commander-in-Chief, Lord Clyde, to conquer Awadh and pacify Rohilkhand. etc.

ATKINSON, G.F., *Curry and Rice on Forty Plates*, London, 1856, subtitled *The Ingredients of Social Life at 'Our' station in India*, by Captn Geo.F. Atkinson, Bengal Engineers. The Forty Plates are, including a vignette Title, delicately coloured steel engravings, drawings by Atkinson, engraved by A.Laby, and printed by Day & Coy, Lithographers to the Queen. Hardly necessary to add that the social life depicted is (almost) exclusively British. Each official and his wife, in the station is the subject of a semi-caricature drawing, to which is added a page of amusing text. The book, particularly the first edition, is now a collector's item. Reprinted Delhi, 1993.

ATKINSON, G.F., *The Campaign in India 1857-58*, London, 1858. By the same author as above, but with an altogether more serious approach. The illustrations are of greater significance than the text which is minimal.

AUBER, P., *Rise and Progress of the British Power in India, 2 vols*, London, 1837.

BADEN-POWELL, B.H., *Land Systems of British India*, 3 vols, Oxford 1892. An important and frequently consulted authority.

BAGLEY, F.R., *A Small Boy in the Indian Mutiny*, in Blackwoods Magazine, March 1930. Not unfortunately particularly revealing, but does tell his memories of life in Agra Fort, during the mutiny. Soldiers told him they'd 'run likes hares' from the rebels at Sussiah[Sasia].

BAHADUR SHAH II PROCEEDINGS OF THE TRIAL OF, Calcutta, 1895. The bare facts/evidence etc. Shows neither sympathy nor understanding for the King of Delhi.

BAIRD-SMITH, Lt Col Richard, *Delhi Journal*, in Professional Papers of the Corps of Royal Engineers, 1897. Baird-Smith was the leading Engineer on the British side during the siege and assault of Delhi in 1857, and nominally at least, was responsible for the final Plan of Attack.

BAKER, David, *Colonial Beginnings and the Indian Response: the Revolt of 1857-58 in Madhya Pradesh*, Modern Asian Studies, vol 25, No 3, 1991.

BALDWIN, Rev J.R., *Indian Gup*, London, 1897. Useful background (gossipy) material as the title implies. No sympathy for Indian aspirations.

BALL, Chas, *History of the Indian Mutiny (2 vols)*, London, n.d. (1858?). A very readable, superbly illustrated, but highly jingoistic account. The author was not an historian but he could write, and he knew his public. An interesting feature is that he quotes from many letters written by protagonists in the great struggle. One of the standard histories, but somewhat dated as it makes little effort to be objective.

BAMFIELD, Veronica, *On the Strength: the Story of the British Army Wife,* London, 1974. Possibly the best 'recent' book on the role of those 'memsahibs' who were army wives.

BANERJEE, Brojendra Nath, *The Last Day's of Nana Sahib,* Indian Historical Records Commission Proceedings 1929. Quotes the British Resident in Nepal (Colonel Ramsay) who gives his reasons for believing that Nana Sahib had obtained refuge in Nepal.

BANERJEE, Sudhansu Mohan, *Some Facts about the Sepoy Mutiny Cawnpore Massacre from Contemporary Records,* Modern Review, Calcutta, June 1957. The title implies revelations, but unfortunately there are none.

BARA, Mahendra, *1857 in Assam*, Gauhati, 1957. Important because one of the few summaries of the rebellion in this area.

BARAT, Amiya, *The Bengal Native Infantry; its organisation and discipline 1796-1852,* Calcutta, 1962. Excellent background material, includes details of mutinies prior to 1857. There is a Foreword (October 1962) by Dr Surendra Nath Sen: 'The time has come when we should seriously consider the causes of the mutiny of 1857, but that is not possible if we do not know the development of the Bengal Army....in 1857 the British thought that the sepoys can be cowed down by bullets and they were wrong.' An important book of reference.

BARKER, Gen. Sir G.D., *Letters from Persia & India 1857-59: A Subaltern's Experiences in War.* Edinburgh, 1915. Written by a subaltern in the 78th Highlanders in Persia in 1857 and with Havelock's column during the relief of Lucknow (mentioned in Campbell's despatches). His letters were edited by his wife after his death in 1914. Mentioned in LADENDORF.

BARNES, George Carnac, and others, *Mutiny Records: Reports.* , 2 vols, Lahore, 1911.

BARNES, George Carnac, and others, *Mutiny Report from the Punjab and NWFP*, Lahore, 1970. A reprint from Lahore, 1911, *Selections from the Records, Vol 8, Mutiny Records.*

BARR, Pat, *Memsahibs*, London, 1976. An attempt to set the record straight with regard to the English women who lived in India in Victorian days. Usually portrayed as wives, sisters etc of serving officers and singularly idle and incapable, they could sometimes, as Pat Barr shows, exert themselves in unexpected fields and make a memorable mark. Includes some well-researched passages on the mutiny/rebellion as it affected some British 'memsahibs.'

BARTHORP, Michael, *The British Troops in the Indian Mutiny,* London, 1994. A pictorial statement of weapons, accoutrements and uniforms of those units of the British Army involved.

BARTRUM, Mrs Katherine, *A Widow's Reminiscences of the Siege of Lucknow,* London,1858. A particularly tragic personal history makes this a poignant and revealing document.

BAS, C.T. Le, *How we escaped from Delhi,* Fraser's Magazine, February 1858. A civilian's story.

BASU, Purnendo, *Oudh and the East India Company, 1785-1801,* Lucknow, 1943. Covers the period before the 1801 Treaty non-attendance to which was made the excuse for annexation by the British.

BAYLY, C.A., *Rulers, Townsmen and Bazaars, North Indian Society in the Age of British Expansion, 1770-1870,* Cambridge, 1983.

BAYLY, C.A., ed., *The Raj: India and the British:: 1600-1947,* London, 1990. Published for the exhibition at the National Portrait Gallery, London, from 19 October 1990 to 17 March 1991. An extremely well illustrated history of the period, with an emphasis on art and artefacts.The struggle for Indian freedom is well documented and illustrated.

BAYLY, C.A and KOLFF, D.H.A., eds, *Two Colonial Empires,* Lancaster, 1986. A description/comparison of the Dutch East Indies and British India etc.

BAYLEY, John, *The Assault of Delhi,* London, 1876

BEAMES, John, *Memoirs of a Bengal Civilian,* Columbia, 1984

BEAUMONT, Roger, *Sword of the Raj,* The British Army in India, 1747-1947, New York, 1977

BECHER, Augusta, ed. H.G.RAWLINSON. *Personal Reminiscences in India and Europe 1830-1888,* London, 1930. This is based on Mrs Becker's diary, and several passages deal with her escape from the mutineers at Ferozepur.

BELL, Thomas Evan, *Holkar's Appeal: papers relating to his conduct during the Mutiny,* London, 1881. Maharaja Holkar of Indore was suspected, despite his protestations, of being disloyal to the British in 1857. Sir Henry Marion Durand in particular doubted Holkar's trustworthiness, although the latter claimed that appearances were against him as his troops had rebelled and he had been unable to bring them back under control.

BELL, Major Evans, *The English in India: letters from Nagpore, 1857-58,* London, 1859.

BELL, Major Evans, *The Annexation of the Punjaub and the Maharaja Dulleep Singh,* London, 1882.

BELLASIS, Margaret, *Honourable Company,* London, 1952. A good background account, written in human rather than strictly scholarly terms, of the manner in which the H.E.I.C. built up 'British India' prior to 1858. No attention specifically directed to the events of 1857-58.

BELLEW, F.J., *Memoirs of a Griffin,* London, 1843. A 'griffin was a young British newcomer to service in the army of the HEIC: a novice.

BENNETT, Mrs Amelia, *Ten Months' Captivity etc.,* in The Nineteenth Century, June 1913. Mrs Bennett is somewhat better known as Amy Horne, an Eurasian girl who was abducted at the Satichaura Ghat by a trooper of the 2nd Cavalry. She wrote two accounts of her experiences: this is the second, written in old age, and is perhaps more frank and accurate than the first.

BENSON, A.C., ed. *Letters of Queen Victoria, 1837-61, Vol III,* London, 1907. The letters contain numerous references to the rebellion, which caused Victoria much genuine distress

BERNSTEIN, Henry T. *Steamboats on the Ganges,* Hyderabad, 1960

BEVERIDGE, Henry, *A Comprehensive History of India: Civil, Military & Social,* London, 1858. Perhaps substitute *concise* for *comprehensive* ?

BHARGAVA, K.D., and PRASAD, S.N., *Descriptive List of Mutiny Papers in the National Archives of India, Bhopal,* 5 vols, New Delhi, 1960-79. Prepared by the Head of the Archives. Volume 1 gives much information on the mutiny/rebellion in Central India, and includes the correspondence of the ruler of Bhopal who remained loyal to the British.

BHARGAVA, K.P. *Two unpublished proclamations of Nana Sahib,* Indian Historical Records Commission Proceedings, 1948. Provides documentary evidence of Nana Sahib's rudimentary military arrangements.

BHARGAVA, Moti Lal, *Architects of Indian Freedom Struggle,* New Delhi, 1981.

BHARGAVA, M.L., (ed. with RIZVI, S.A.A.), *Freedom Struggle in Uttar Pradesh (6vols),* Source Material, Lucknow, 1957-61. The six volumes cover I. 1857-59: Nature and Origin; II. Awadh: 1857-59; III. Bundelkhand and Adjoining Territories: 1857-59; IV. Eastern and Adjoining Districts: 1857-59; V. Western Districts and Rohilkhand:

1857-59; VI. Index to all volumes. While this is indisputably an essential reference tool for all students of the events of 1857-59 in India, it is not very easy to use. In their laudable desire to miss out nothing of significance the editors have had to sacrifice some order and method. While chronological sequence is observed in the main, there is much overlap and starting again. It is worth persevering however for the work is a veritable mine of information, and quite unique. It covers primary material from a wide variety of sources, and is not restricted by any means to the territorial limits of modern Uttar Pradesh: this is of course a bonus, not a cause for complaint, but it does make the title somewhat misleading. The editors have been meticulous in their reading of manuscript items, giving helpful interpretations/explanations wherever the text is suspect or illegible.

BHATNAGAR, G.D., *The Annexation of Oude,* Uttaara Bharati, vol 3, 1956.

BHATNAGAR, G.D., *Awadh under Wajid Ali Shah,* Varanasi, 1968. An exhaustive and scholarly re-appraisal of the character of King Wajid Ali Shah. 'On the basis of contemporary evidence some of the current notions regarding Wajid Ali Shah and his government are misleading, and merit re-examination.' The author achieves this aim.

BHATNAGAR, O.P., ed., *Private Correspondence of J.W.Sherer, Collector of Fatehpur,* (19May-28July 1857), Allahabad, 1968. Sherer was a man of liberal views, kindly and perspicacious. His strictly contemporaneous writings are therefore of great value.

BLUE PAMPHLET, The, by an Officer once in the Bengal Artillery, London, 1858

BLUNT, E.A.H., *Christian Tombs and Monuments in the United Provinces,* Lucknow, 1911. Evidence is obtainable from the scrutiny of epitaphs.

BOASE, Frederic, *Modern English Biography,* London, 1965. Containing many thousand concise memoirs of persons who have died between the years 1851-1900.

BOLT, Christine, *Victorian Attitudes to Race,* London, 1971.

BONADONNA, G., *Il Vento del Diavolo,* Milan, 1994. The story of the sepoy rebellion by an Italian scholar.

BONHAM, John, *Oudh in 1857, some memories of the Indian Mutiny,* London, 1928. This is a small book but he gives an excellent, succinct, account of the mutiny/rebellion in Oude[Awadh], from the British point of view, station by station. It is a soldier's account (he was a Lieutenant in the Bengal Artillery) and has the merit of soldierly directness and detail.

BOST, Isabella, *Incidents in the Life of,* Glasgow,1913.

BOURCHIER, Colonel George, *Eight Months' Campaign against the Bengal Sepoy Army,* London, 1858. Even though he was present during the struggle, his account relies largely upon the work of other writers notably Henry Norman's journal *re* the siege of Delhi, and John Walter Sherer's story of Cawnpore. The book is valuable nevertheless for its succinct account of the suppression of the mutiny in the Punjab.

BRADSHAW, Lieutenant J.H., *Letters on the Delhi Campaign, Ox.& Bucks Lt.Inf.Chronicle,* 1909. This regiment was included in the Punjab Movable Column which marched on Delhi and took part in the assault on 14 September 1857.

BRASYER, Jeremiah, *The Memoirs of Jeremiah Brasyer,* London, 1892. This is the man who raised 'Brasyer's Sikhs', paramilitaries and police in the first place, who played a decisive role in the suppression of the mutiny. His influence with the Sikhs was paramount in this success.

BROCK, Rev William, *Biographical Sketch of Sir Henry Havelock,* London,1858. Havelock enjoyed an enormous popularity in England after his 'relief' of the Lucknow Residency, not least because he was seen as some kind of Christian knight fighting the good fight against an irreligious foe. There is hardly a town in England that has not got its 'Havelock Road', though few today have any idea of the origin of the name. This book was dedicated to Lady Havelock who helped by providing many of the General's private letters, although the 'official' biography was to be written by her brother, Mr J.C.Marshman, the Editor of the 'Friend of India', in Calcutta.

BRODKIN, E.I., *The Struggle for Succession: Rebels and Loyalists in the Indian Mutiny of 1857,* Modern Asian Studies, Vol 6, part 3, Cambridge, July 1972. Represents new thinking, and therefore significant.

BROEHL, Wayne G., Jnr, *Crisis of the Raj,* Boston, 1986. The revolt of 1857 seen through the eyes of four British Lieutenants: G.Cracklow, A.Lang, Thomas Watson and Fred Sleigh Roberts. A compilation of contemporary letters. But breaks no new ground.

BROOME, Captain Arthur, *History of the Rise and Progress of the Bengal Army,* Calcutta & London, 1850. Any pre-Mutiny commentary of this nature is valuable to the student of the mutiny of the Bengal army.

BROWN, J., *Capture of Lucknow,* in Calcutta Review, June 1860. Has the merit of spontaneity.

BROWN, J., *Havelock's Indian Campaign,* in Calcutta Review, March 1859. As above.

BROWNE, John, *Cawnpore and the Nana of Bithur,* Cawnpore,1890. Contributes surprisingly little to what is already known.

BROWNE, John, *The Lucknow Guide*, Lucknow, 1874.

BRUCE, J.F., *The Mutiny at Cawnpore*, in Punjab University History Society Journal, April 1934.

BUCKLAND, Charles, *Dictionary of Indian Biography*, London, 1906. Obviously the date of imprint is significant.

BUCKLAND, Charles, *Bengal under the Lieutenant Governors*, 2 vols, reprint, orig. 1902, New Delhi, 1976. Contains useful background information on the administration of Bengal during the mutiny/rebellion.

BUCKLE, E., *Memoir of the services of the Bengal Artillery, from the formation of the Corps to the present time*, London, 1852. Detailed information on the artillery not to be found in any other source. Especially valuable as it is pre-Mutiny.

BUCKLER, F.W., *The Political Theory of the Indian Mutiny*, in Royal Historical Society Transactions, Series IV Vol V, 1922. This was a paper read before the Society on 12 January 1922, by Mr Buckler, in which he demonstrated that if there was a mutiny in 1857, it was not by the sepoys but by the East India Company, against their suzerain, the Mughal Emperor. Caused, at the time, furore in academic circles.

BURGESS, James, *The Chronology of Modern India: 1494 to 1894*, Edinburgh, 1913. Contains a chronological statement on the mutiny/rebellion and is therefore a useful reference work.

BURNE, Sir Owen, *Clyde and Strathnairn*, Oxford, 1891. Definitive biographies of Sir Colin Campbell and Sir Hugh Rose.

BURTON, R.G., *Indian Military Leaders,* in United Services Magazine, Vol 174, 1916. A useful and reasonably objective view of the faults and virtues of the Indian leaders of the mutiny/rebellion.

BURTON, R.G., *Women Warriors in India*, in United Services Magazine, Vol 175, 1916.

BUSTEED, H.E., *Echoes of old Calcutta*, London 1908.

BUTLER, Spencer Harcourt, *Oudh Policy considered historically*, Allahabad, 1896. Published for private circulation only. Contains an interesting short summary of the situation in Oude[Awadh] before, during and after the rebellion of 1857, with particular reference to the Talooqdars.

BUTLER, William, D.D., *The Land of the Veda*, New York, 1872. Butler was a Methodist missionary who escaped from Bareilly to Naini Tal during the rebellion. His book contains transcriptions of other fugitives' statements, and much interesting background material, indeed is one of the best souces of information on the events in, and affecting, Naini Tal.

BUTLER, Lt.Colonel Lewis, *Annals of the King's Royal Rifle Corps*, Vol iii, London, 1926.

CADELL, Sir Patrick, *History of the Bombay Army*, London, 1938. Includes a chapter on the Indian Mutiny with an account of the affair at Kolhapur, and the unrest in the Southern Maratha country.

CAMBRIDGE HISTORY OF INDIA, contains general standard summaries of events by historians considered pre-eminent in their field.

CAMBRIDGE SOUTH ASIAN ARCHIVE: RECORDS of THE BRITISH PERIOD in SOUTH ASIA HELD in the CENTRE of SOUTH ASIAN STUDIES, UNIVERSITY of CAMBRIDGE, COMPILED and EDITED by M..THATCHER and L.CARTER, London, 1973, Second Collection, 1980, Third Collection, 1983, Fourth Collection, 1987(?). An extremely valuable resource. Available also on microfiche.

CAMPBELL, Sir George, *Memories of my Indian Career*, (2 vols) London, 1893. Campbell was married to one of the Vibart (killed at Cawnpore) daughters. His views are sometimes surprisingly anti-establishment, *eg* 'Neill did things worse than the massacre (*ie* Satichaura).

CAMPBELL, Robert, *India: its government, misgovernment and future considered*, London, 1858.

CAMPBELL, W., *My Indian Journal*, Edinburgh, 1864. Contains useful background material.

CAPUCHIN MISSION UNIT, *India and its Missions*, New York, 1923.

CARDEW, Major F.G., *Hodson's Horse, 1857-1922*, Edinburgh, 1922. Unmistakably biassed in favour of the founder of the regiment - Captain W.S.R.Hodson.

CARDEW, Francis Gordon, *Sketch of the services of the Bengal Native Army to the year 1895,* Calcutta, 1903. The standard review of the Bengal army, including the period of the mutiny/rebellion.

CAREY, W.H., *The Good Old Days of Honourable John Company*, 2 vols. Calcutta, 1906. Being 'curious reminiscences illustrating Manners and Customs of the British in India during the rule of the E.I.C. from 1600 to 1858, with brief notices of places and people of those times. Compiled from newspapers and other publications.' The obvious partisanship of the work is proclaimed in the title, but it is a valuable source book none the less. The events of 1857 are termed 'The Mahomedan Rebellion of 1857', and attempts are made to pinpoint the complicity of the ex-King of Awadh, Wajid Ali Shah, *see* vol II p.322.

CAREY, W.H., *The Mahomedan Rebellion; its Premonitory Symptoms; the Outbreak and Suppression*, Roorkee, 1857. Despite its bias this is a valuable contemporary commentary.

CARNEGY, P., *Historical Sketch of Tahsil Fyzabad, Zillah Fyzabad,* Lucknow, 1870. An important review of the history (including 1857-8) of Faizabad in Awadh.

CASE, Mrs Adelaide, *Day by Day at Lucknow,* London, 1858. The diary of the widow of Colonel Case, commander of H.M.'s 32nd Regiment, killed at the Battle of Chinhat in June 1857.

CAUSE AND EFFECT, The Rebellion in India, by a Resident of the NWP, London, 1857.

CAVE-BROWNE, John, *The Punjab and Delhi in 1857,* 2 vols, Edinburgh, 1861. A Narrative account of the measures by which the Punjab was 'saved' for government. Very one-sided (British) but readable none the less. Originally published in instalments in Blackwoods Magazine. Has some original material as it is based on the author's own journal, and because it is contemporaneous may be regarded as a primary source.

CAVE-BROWNE, John, *Incidents of Indian Life,* Maidstone, 1895.

CAVENAGH, Sir Orfeur, *Reminiscences of an Indian Official,* London, 1884.

CHALMERS, John, *Letters written from India during the Mutiny and Waziri Campaign,* Edinburgh, 1904.

CHAMBERLAIN, General Sir Crauford, *Remarks on Captain Trotter's Biography of Major W.S.R.Hodson,* Edinburgh, 1901. Hodson was a controversial figure, overly loved and denigrated. This was Chamberlain's attempt to redress the balance

CHAND, Bool, *Urdu Journalism in the Punjab,* Punjab University History Society Journal, April 1933. A neglected source.

CHANDA, S.N., *1857, Some Untold Stories,* Delhi, 1976.

CHATTOPADHYA, Haraprasad, *The Sepoy Mutiny 1857, A Social Study and Analysis,* Calcutta, 1957. An eminent Indian historian of the period who argues that there would have been no popular revolt if no military mutiny first; moreover as it was regional in scope it cannot be called a national or freedom movement.

CHATTOPADHYA H.B., *Mutiny in Bihar,* in Bengal Past & Present, Vols 74-5, 1955-6.

CHATTOPADHYA H.B., *The Sepoy Army, its strength, composition and recruitment,* in Calcutta Review, May, July September 1956.

CHOUDHARY, *Growth of Nationalism in India: 1857-1918,* New Delhi, 1973.

CHAUDHRY, Nazir Ahmad, *The Great Rising of 1857 and the Repression of the Muslims,* Lahore, 1970.

CHAUDHURI, Sashi Bhusan, *Civil Disturbances during the British Rule in India (1765-1857),* Calcutta, 1955. An interesting account of affairs in India before the Great Rebellion. The author concludes 'The great revolution of 1857-8 was but the consummation of a political and social development which had been in process for a century.'

CHAUDHURI, Sashi Bhusan, *Civil Rebellion in the Indian Mutinies, 1857-58,* Calcutta, 1957. This historian emphasises the widespread nature of the disturbances, and the general passive support of the populace for the rebel forces.

CHAUDHURI, S.B., *Theories of the Indian Mutiny 1857-59,* Calcutta, 1965. The sub-title to this monograph is *A Study of the views of an Eminent Historian (R.C.Majumdar) on the Subject of the Indian Mutiny.* He is especially concerned to combat the interpretation of 1857 as the last desperate effort of the old order to maintain its privileges: the landlords who led the revolt in many areas were, he argues, unconscious tools of a nascent nationalism.

CHAUDHURI, S.B., *English Historical Writings on the Indian Mutiny, 1857-59,* Calcutta, 1979. Despite the title it also contains a valuable list of Works in Indian Languages on the subject of the mutiny/rebellion.

CHAUDHURI, Sibadas, *The Literature on the Rebellion in India in 1857: a Bibliography,* Calcutta, 1971. All bibliographies are useful.

CHICK, Noah, ed., *Annals of the Indian Rebellion,* Calcutta, 1859. Re-issued (abridged), London, 1974. Anecdotal evidence, but of considerable interest; much of the material is not only contemporaneous but is unique, *ie* can not be found elsewhere.

CHOLMELEY, J.E., *John Nicholson, the Lion of the Punjab,* London, 1909. Eulogistic and far from being the definitive biography.

CHOUDHURY, P.C.Roy, *1857 in Bihar,* Patna, 1959. Specifically in Chotanagpur and Santhal Parganas.

CHOUDHURY, Sujit, *The Mutiny Period in Cáchar,* Silchar, 1981. Compiled from the despatches of Captain Robert Stewart, Superintendent of Cachar, 1857-58.

CHUNDER, Bolanauth, *Travels of a Hindu,* 2 vols, London, 1869. An exceptionally well written work, containing many perceptive statements on the mutiny/rebellion.

CHURCHER, David G., *Episodes of the Indian Mutiny,* Blackwoods Magazine May 1900. A European survivor, one of the few, of the uprising at Fatehgarh, he gives a first hand account of the mutiny and siege of the Fort there.

CHURCHER, Emery J., *Some Reminiscences of Three quarters of a Century in India,* London, 1909.

CLARK, F., *East India Register and Army List for 1858,* Second edition, London, 1858. A standard reference work.

CLEMONS, Mrs Major, *The Manners and Customs of Society; including scenes in the Moffusil stations interspersed with characteristic tales and anecdotes; and reminiscences of the late Burmese War, to which are added Instructions for the guidance of cadets and other young gentlemen, during their First Year's Residence in India,* London, 1841. The subtitle is a mouthful but is helpful. It needed hardly be said that the 'Society' referred to is of the expatriate English, but it is a very useful background work on life, particularly for the officer cadre, in the Company's army.

CLODE, C.M., *The Military Forces of the Crown, their administration and government,* London, 1869.

COHEN, Stephen P., *The Indian Army: Its Contribution to the Development of a Nation,* New Delhi, 1990. An objective view, by an American, of a complicated question. Useful for background material and for an understanding of the role of the army in India both in peace and war.

COLESWORTHY-GRANT, *Anglo/Indian Domestic Life, a letter from an artist in India to his mother in England,* Calcutta, 1862. A most detailed reference work on Anglo/Indian domestic life at the time of the rebellion.

COLLIER, Richard, *The Sound of Fury,* London, 1963. This is an attempt to recreate in dramatic language the actual events and the atmosphere of a selected number of major incidents in the course of the mutiny/rebellion, especially connected with Delhi, Cawnpore and Lucknow. It is not fiction but sometimes reads as though it were.

COLLINSON & WEBBER, *General Sir John Harkness, 1804-83,* for the Royal Engineers Institution, London 1903. Contains about seventy pages on his service during the mutiny/rebellion.

COLVIN, Sir Auckland, *Life of John Russell Colvin,* Oxford, 1895. At pains to rehabilitate Colvin the much criticised Lt Governor of the NWP, in Agra in 1857.

COLVIN, Sir Auckland, *Agra in 1857: a Reply to Chapter 21 of Lord Roberts 41 Years' Reminiscences in India,* in Nineteenth Century Magazine, April 1897. Discusses affairs in Agra in response to Lord Roberts's criticisms. The latter had suggested in his book *Forty-one Years in India,* that the authorities in Agra during the mutiny/rebellion had been incompetent and inept, for which opinion he produced a good argument. A reasoned and partially successful attempt to rehabilitate the reputation of Sir William Colvin, the Lt Governor of the NWP.

CONGREGATIONAL PUBLISHING SOCIETY, *History of Missions of American Board of Commissioners for Foreign Missions in India,* Boston, 1874.

CONRAN, H.M., *Memoir of Colonel Wheler,* London, 1866. Wheler, a fanatical Christian, became the centre of attention when the outbreak of the mutiny was blamed on his preaching to the sepoys of the regiment he commanded, the 34th BNI, to which Mangal Pande belonged. The book discusses the controversy surrounding his preaching, and his defence of his action.

COOPER, Frederick Henry, *Crisis in the Punjab from 10th May until the Fall of Delhi,* Lahore, 1858. Cooper was a Deputy Commissioner of the Punjab, based at Amritsar and, although a civilian, was responsible for the 'extermination', of the disarmed 26th BNI at Ujnalla, in particularly inhumane and unpleasant circumstances, for which he was praised and supported by his superiors but roundly condemned by many in the House of Commons when the news reached London. He wrote this book, to vindicate his conduct which he considered 'prompt, spirited and thorough'.

COOPER, Leonard, *Havelock,* London, 1957. A new biography of Sir Henry Havelock who 'relieved' Lucknow in September 1857.

COOPLAND, Mrs Ruth, *A Lady's Escape from Gwalior,* London, 1859. Ruth Coopland was the widow of the Reverend Coopland Chaplain at Gwalior and killed by the mutineers. His grave can still be seen at Morar. Her account is cold and factual and full of (understandable) animosity towards the mutineers who killed her husband, but perhaps is more interesting for her details of life in Agra Fort, which those who were there described as besieged, although it was nothing of the kind.

CORK, Barry Johnson, *Rider on a Grey Horse, a Life of Hodson of Hodson's Horse,* London, 1858. Readable, but biassed strongly in favour of this controversial character; undoubtedly influenced subsequent biographers.

COSENS, Lt.Colonel F.R. and WALLACE C.L., *Fatehgarh and the Mutiny,* Lucknow, 1933. Reprint Karachi, 1978. One of the best factual accounts of the mutiny at Fatehgarh, told by a soldier in soldierly language. It is particularly helpful in explaining the part played by the 10th N.I. in the mutiny, about which there is still in some quarters considerable confusion.

COTTON, E.S., *Indian and Home Memories,* London, 1907.

COTTON, Sir Sydney J., *Copies of Sundry despatches etc on subjects connected with the Mutiny and Rebellion in India during the years 1857 and 1858,* Roorkee, 1859. Has the merit of topicality and spontaneity.

COTTON, T.S *et al.* ed. *Imperial Gazetteer of India,* 26 vols, Oxford 1907-9. Based on the work of Sir W.W.Hunter. Gazetteers can provide surprisingly useful information for the student.

CROMB, James, *The Highland Brigade,* London, 1886. Includes details of the service of the three Highland Regiments of the British army involved in the suppression of the mutiny.

CROOKE, W., *The Northwestern Provinces of India,* London, 1897.

CROOKE, William, *Songs of the Mutiny,* in the Indian Antiquary, Vol XL, April & June 1911. Folksongs resulting from the struggle reveal a great deal of information on rural involvement.

CROOKE, William, *Tribes and Castes of North Western India, 4 vols,* 1896, reprint Delhi, 1974.

CROSSE, Lt.Col.Chas, *The Punjab Moveable Column under Nicholson,* United Services Magazine, June 1896.

CRUMP, Lieutenant C.W., *A Pictorial Record of the Cawnpore Massacre,* London, 1858. A disappointing (and inaccurate) record of the events at Kanpur in 1857.

CULROSS, James, *The Missionary Martyr of Delhi,* London, 1860.

CUNNINGHAM, Sir Henry, *Earl Canning,* Oxford, 1891. Useful, but not the definitive biography.

CUST, R.N., *Pictures of Indian Life,* London, 1881.

CUST, R.N., *A District during the Rebellion,* in the Calcutta Review, September 1858.

CUTHELL, Edith, *My Garden in the City of Gardens,* London, 1905. *re* Lucknow in its heyday.

DALHOUSIE, Lord, *Private Letters,* (ed.J.G.Baird) Edinburgh, 1910.

DANGERFIELD, George, *Bengal Mutiny,* London, 1933. Possibly the last of the general histories to be written wholly from the British point of view, although there is the admission that 'both sides did things in 1857 which they would gladly have seen undone, and which their successors do not care to think about'.

DANVERS, R.W., *Letters from India and China, 1854-58,* London, 1898. An officer of the 70th BNI, and interpreter to the 5th Fusiliers in the Lucknow Relief Force of Outram and Havelock. Excellent account of the relief.

DASHWOOD, A.F., *Untimely arrival at the Siege of Lucknow,* in The Listener, London, 2 Dec 1936. A brief account of the siege of Lucknow by one who was born there during the siege.

DATTA, K.K., *Biography of Kunwar Singh and Amar Singh,* Patna, 1957. An important addition to the story of the Bihari Rajput Raja who made such an honourable contribution to the rebel cause.

DATTA, K.K., *Some unpublished Papers relating to the Mutiny,* in Indian Historical Quarterly, March 1936.

DATTA, K.K., *Nature of the Indian Revolt of 1857-9,* in Bengal Past and Present, Vol 73, 1954.

DATTA, K.K., *Some newly discovered Records relating to the Bihar Phase of the Indian Movement,* in Patna University Journal, Vol 8, 1954.

DATTA, K.K., *Popular Discontent in Bihar,* in Bengal Past and Present, Vol 74, 1955.

DATTA, K.K., *Reflections on the Mutiny,* Calcutta, 1967.

DATTA, K.K., *Some original Documents relating to the Indian Mutiny,* in Indian Historical Records Commission Pro-ceedings, Vol 30, 1954.

DATTA, K.K., *Contemporary Account of the Indian Movement of 1857,* in Journal of the Bihar Record Society, Vol 36, 1950.

DATTA, K.K., ed. *Unrest against British Rule in Bihar, 1831-59,* Patna, 1957.

DAWSON, Captain (RN) Lionel, *Squires and Sepoys,* London, 1960. An account of the career at this time of his ancestor, 2nd Lieutenant George Blake of H.M.'s 84th Regiment, serving particularly at Kanpur and Lucknow. The anecdotes in the book are clearly part of family history, valuable none the less as being typical of the class from which the army officer of the time was then recruited

DEFICIENCY OF EUROPEAN OFFICERS IN THE ARMY OF INDIA, by One of Themselves, London, 1849. Interesting pre-Mutiny commentary.

De KANTZOW, Colonel C.A., *Record of Services in India, 1853-86,* Brighton, 1898. Useful for its account of events at Mainpuri which are not well chronicled elsewhere. A modest man, he was remarkably courageous in very difficult circumstances.

DELHI: A Man Hunt, in Blackwoods Magazine November 1930. An account of an Englishman's escape from the uprising in Delhi.

DEPOSITIONS TAKEN at CAWNPORE UNDER the DIRECTION of Lt.COLONEL G.W.WILLIAMS, COMMISSIONER of MILITARY POLICE, NWP, 1858. Essential source material but infuriatingly inconclusive.

DEWAR, Douglas, *A Handbook to the English pre-Mutiny Records, in the Government Records Room of the United Provinces of Agra and Oude,* Allahabad, 1919.

DEWAR, Douglas, *In the days of the Company,* Calcutta, 1920. Pleasant anecdotes but little else; picture of pre-mutiny life - for the British. Reprint, Calcutta, 1987.

DEWAR, Douglas, *Bygone Days in India,* London, 1922.

As above.

DEWAR, Douglas, and GARRETT, H.L.O., *Reply to Mr F.W.Buckler's Political Theory of the Indian Mutiny*, in Royal Historical Society Transactions, Vol 7, 1924.

DIGBY, W., *A Friend in Need*, London, 1890.

DIVER, Maud, *An Echo of Cawnpore*. in Cornhill Magazine June 1937. Discusses some victims of the massacre.

DIVER, Maud, *Honoria Lawrence*, London, 1936. Honoria Lawrence was the wife (she pre-deceased him) of Sir Henry Lawrence of Punjab and Lucknow fame.

DOBROLYUBOV, Nikolai, trans. Harish C.Gupta. *The Indian National Uprising of 1857*, Calcutta, 1988. This is a contemporary (1858) Russian account of the events of 1857 by a Russian liberal writer and historian. Necessarily reviewing the situation *in absentia* he nevertheless produces strong and convincing arguments.

DODD, G., *History of the Indian Revolt etc* London, 1860. Did not become widely accepted or read.

DODWELL, Edward, and MILES, James Samuel, *Alphabetical List of Officers of the Indian Army, 1760-1834*, corrected to 1837, London, 1838. Standard reference work; gives surprisingly little detail.

DODWELL, H., ed., *Cambridge History of India*, Cambridge, 1929.

DODGSON, Lt.Colonel David, *General Views and Special Points of Interest of the City of Lucknow*, London, 1860.

DOMIN, Dolores, *India in 1857-9: a Study in the role of the Sikhs in the People's Uprising*, Berlin, 1977.

D'OYLY, Charles Walter, *Eight Month's Experience of the Revolt in 1857*, Blandford, 1891.

DUBERLEY, Mrs Frances, *Campaigning in Rajputana and Central India during the suppression of the Mutiny*, London, 1859. One of the (rare) Englishwomen who, in order to stay near to their husbands, actually refused to be parted from them. Fanny was the wife of Captain Henry Duberley, an officer of the 8th Hussars, and accompanied him to the Crimea, and is also said to have 'ridden with the 8th Hussars in India for 11½ months, and been present when the **Rani of Jhansi** was killed by a Hussar of her husband's regiment.'

DUFF, Alexander, *The Indian Rebellion*, London 1858. Dr Duff was a clergyman but his account is bereft of Christian forgiveness, indeed contains uncorroborated sensationalist propaganda.

DUNLOP, Robert H.W., *Service and adventure with the Khakee Ressalah*, London, 1858. Reprint, Allahabad, 1974. A record taken from his own notes and letters, and as he was an active participant in the events he describes, the account is valuable. Useful as showing the nature of the civil rebellion in the Meerut District and the means used by the British to suppress it or at least keep it in check.

DURAND, H.Mortimer, *Central India in 1857*, London, 1876.

DURAND, H.Mortimer, *The Life of Major General Sir Henry Marion Durand*, 2 vols, London, 1883. Includes one chapter on Sir Henry Marion Durand's experiences during the mutiny/rebellion while he was i/c the Central Indian Agency.

DURAND, Sir H.Mortimer, *Life of the Rt Hon Sir Alfred Comyn Lyall*, Edinburgh 1913. Includes material, based on Lyall's own reports and diaries etc, of the progress of the rebellion in the Bulandshahr District.

DUTT, Romesh, *India in the Victorian Age*, London, 1904.

DUTT, Romesh, *Economic History of India*, 2 vols, Delhi, 1960. Significant for the background to rural revolt 1857-58.

EASTWICK, Edward B., ed., *Autobiography of Lutfullah*, a Mohamedan gentleman and his Transactions with his fellow creatures, London, 1858.

EDWARDES, Lady Emma, *Memoirs of the Life of Sir H.B.Edwardes*, 2 vols, London, 1886. There are many excerpts from diaries and letters. Herbert Edwardes played a significant part in the suppression of the Mutiny in the Punjab, and keeping that area under control until the successful end to the Siege of Delhi relieved the immediate fear of an overwhelming *emeute*. That is of course very much from the British point of view.

EDWARDES, H.B. and MERIVALE, H., *Life of Sir Henry Lawrence*, 2 vols, London, 1872. This is the definitive biography and is essential reading for those who wish to know more of this liberal-minded man who had a remarkable insight into Indian affairs.

EDWARDES, Michael, *The Necessary Hell*, London, 1958.

EDWARDES, Michael, *The Orchid House*, London, 1960.

EDWARDES, Michael, *Battles of the Indian Mutiny*, London, 1963. While not adding much to our knowledge of the subject this book is attractively written and produced and brings a refreshingly modern light to bear on old and vexed problems. In defiance of its title it is not overmuch pre-occupied with the battles.

EDWARDES, Michael, *British India 1792-1942: A survey of the nature and effects of alien rule*, London, 1967.

EDWARDES, Michael, *Red Year: the Indian Rebellion of 1857*, London, 1973.

EDWARDES, Michael, *A Season in Hell: the Defence of the Lucknow Residency,* London, 1973.

EDWARDES, Michael, *Bound to Exile: the Victorians in India,* London, 1969.

EDWARDES, S.M., *Babur: Diarist and Despot,* London, n.d. (1935?). This excellent sketch of Zahir-ud-din Muhammad Babur Padshah is based almost entirely upon the English translation of the Babur-nama (Memoirs of Babur) by Mrs A.S.Beveridge between 1912 and 1921. Mrs Beveridge translated direct from the original Turki.

EDWARDS, William, *Facts and Reflections connected with Rebellion,* Liverpool, 1859.

EDWARDS, William, *Personal Adventures during the Indian Rebellion,* London, 1858. Reprint, Allahabad, 1974. A typical memoir of personal experience, harrowing and dangerous, but in its very danger highlighting the major problem facing so many Indians in 1857 - the conflict of loyalties, which side to join? Edwards was a fugitive who, with a small party and with the aid of Indians loyal to government made his way from Budaon to safety in Kanpur.

EDWARDS, William, *Reminiscences of a Bengal Civilian,* London, 1886. Particularly intended to give credit to the influential Indians who had assisted him when he was a fugitive, and whom he believed to have been insufficiently rewarded, notably Raja Byjenath Misr and his son Gunga Pershad. He argues that there was a well-founded plot behind the mutiny/rebellion to restore the Mughal Empire.

EGERTON, Captain Francis, RN, *Journal of a Winter's Tour in India,* 2 vols, London, 1852.

ELECTRIC TELEGRAPH, The, in Calcutta Review, March 1858. Contemporaneous and significant.

ELLIOTT, C.A., *The Chronicles of Oonao,* Allahabad, 1862.

ELLIOT, Henry M., *Memoirs on the History, Folk-lore and Distribution of the races of the Northwestern Provinces of India, being an amplified edition of the Original Supplemental Glossary of Indian Terms.* 2 vols, edited, revised and re-arranged by John Beames, London, 1869.

EMBREE, Ainslie T., ed. *1857 in India: Mutiny or War of Independence,* Lexington, 1963. An interesting attempt by an American scholar to present the widely differing views of British and Indian historians (and of contemporaries of the events of 1857-9) in an attempt to decide the true nature of the struggle. The views of twenty-three historians are summarised and short illustrative extracts given. No conclusion is stated, the question in the title remains unanswered, but a full range of views has been presented. The bibliography is useful but not exhaustive.

EMBREE, Ainslie T., *India in 1857.The Revolt against Foreign Rule,* Delhi, 1992.

ENGLISH, Barbara, *John Company's Last War,* London, 1971. ie against Persia in January 1857.

ERSKINE, W.C., *Chapter of the Bengal Mutiny,* Edinburgh, 1871. Erskine was Commissioner of the Jabalpur Division and his evidence/opinion on the Rani of Jhansi was crucial.

EWART, Lt.General J.A., *Story of a Soldier's Life,* 2 vols, London, 1881.

EYRE, Sir Vincent, *Letters and Dispatches,* nd. Eyre was the hero of both 'Cabool' and Arrah.

FARRINGTON, A.J., *Guide to the Records of the India Office Military Department,* London, 1982.

FARWELL, Byron, *Armies of the Raj; from the Mutiny to Independence, 1858-1947,* London, 1989. Good historical/sociological study.

FARWELL, Byron, *Eminent Victorian Soldiers: Seekers of Glory,* New York, 1985. Includes Wolseley.

FARWELL, Byron, *Queen Victoria's Little Wars,* New York, 1972.

FAYRER, Sir Joseph, *Recollections of my Life,* Edinburgh, 1900. Fayrer was the civil surgeon in the Residency at Lucknow in 1857, and was one of the survivors of the siege of the Residency. Fayrer's House (or rather its preserved ruins) are still visible in Lucknow. Sir Henry Lawrence died there. Fayrer, like Gubbins, was afterwards criticised for having lived well during the siege. He was a well-known authority at the time on tropical diseases.

FESTING, Gabrielle, *Strangers within the Gates,* Edinburgh 1914. A History of India from the 1740s to the end of Company rule in 1858. Of no particular depth or scholarship, but eminently readable.

FIELD, C., *Sir John Field, KCB,* London, 1908, includes Scinde, Mutiny and Abyssinian campaigns.

FITCHETT, Rev. William H., *The Tale of the Great Mutiny,* London, 1903. Popular history, was much read when it was published. Of significance mainly for its inclusion of a diary, verbatim, of a Mrs Soppitt, an officer's wife, which she kept throughout the siege of the Lucknow Residency.

FITZGERALD Lee J., and RADCLIFFE, F.W., *The Indian Mutiny up to the Relief of Lucknow (17 Nov.1857),* for Staff College candidates, Rawalpindi, n.d.(1905?). Very much a soldier's perspective eg 'The first cause of the Mutiny was the relaxation of discipline in the Sepoy Army; the second cause was a question of religion and caste; the third and last, politics (sic)'. But it does contain an appendix entitled 'Synchronological Table of Events, Jan-November 19th 1857' which is of interest.

FORBES, Archibald, *Colin Campbell, Lord Clyde*, London, 1895.

FORBES, Archibald, *Havelock*, London, 1891.

FORBES, Archibald, *Glimpses through the Cannon Smoke*, London, 1880.

FORBES, Mrs Hamilton, *Some recollections of the Siege of Lucknow*, Axminster, 1905.

FORBES-LINDSAY, C.H., *India : Past and Present*, 2 vols, Philadelphia, 1903.

FORBES-MITCHELL, William, *Reminiscences of the Great Mutiny*, London, 1893. Including the relief, siege and capture of Lucknow and the campaigns in Rohilkhand and Oude[Awadh]. Forbes-Mitchell had been a Sergeant in the old 93rd Regiment (Latterly the Argyll & Sutherland Highlanders) and was both literate and observant. The result is a very readable book, written from the perspective of a serving non-commissioned officer; this has the disadvantage of containing much that was hearsay and ill-informed, but the bonus of being full of interesting or amusing anecdotes.

FORGUES, Paul Emile Dourand, *La Revolte des Cipayes*, Paris, 1861. A former editor of the Delhi Gazette he published a number of narrative histories of the rebellion, both in French and in English.

FORJETT, Charles, *Our real danger in India*, London, 1877. Written by a Commissioner of Police in Bombay and intended to explain the 'causes of the Mutiny which threatened to deprive us of our possession of India in 1857'. The author spends considerable time and trouble demolishing the familiar and widely accepted views of Sir John Kaye as to the causes of the rebellion.

FORREST, Sir George, *History of the Indian Mutiny*, 3 vols, Edinburgh, 1904 . Comprehensive survey of the military operations. Tries to be objective. A standard narrative history of the rebellion.

FORREST, Sir George, *Life of Field Marshal Sir Neville Chamberlain*, 1909.

FORREST, Sir George William, *Sepoy Generals, Wellington to Roberts*, Edinburgh, 1901.

FORREST, Sir George William, ed., *Selections from the Letters, Dispatches and other State Papers of the Military Department of the Government of India 1857-58*, 4 vols, Calcutta, 1893-1912. Although a selection, and not every paper is included, this work represents a conscientious attempt to put together, for the use of scholars and the general public, primary source material of particular significance to a study of the military operations of 1857-58. Thus political issues are side-lined and may appear only incidental. Vol I concerns the beginning of the uprising *viz* the outbreaks at Barrackpore, Berhampore , Meerut and Delhi, and the siege and capture of the latter city. Vols II and III are concerned with the operations against and around Kanpur and Lucknow, including the Residency siege. Vol IV continues Sir Colin Campbell's operations in Oude[Awadh], Sir Hugh Rose's campaign in Central India, and General Whitlock's campaign also, together with papers on smaller and less significant military operations. Courts of Enquiry, Courts martial, and Trials are included. As a collection of primary source material it is of equal significance to the *Freedom Struggle in Uttar Pradesh - See* Bhargava, K.D. etc above.

FORSYTH, Sir Douglas, *Autobiography and Reminiscences*, London, 1887. Useful especially for a chapter on his own personal experiences at Ambala.

FORTESCUE, J.W., *History of the British Army*, Vol xiii, London, 1930.

FOSTER, Sir William, *Guide to the India Office Records*, 1600-1858, London, 1966. Of general interest, it supplements Sutton.S.C., *Guide to the India Office Library*, London 1967, and Hill S.C., *Catalogue of the Home Miscellaneous Series of the India Office Records*, London, 1927. For more specific guidance, *see* Seton , Rosemary, *The Indian 'Mutiny' 1857-58*, a Guide to source material in the India Office Library and Records, London, 1986.

FOSTER, Sir William, *John Company*, London, 1926. A readable and comprehensive account of the history of the H.E.I.C. by an authority on the subject.

FOWLER, Marian, *Below the Peacock Fan*, London, 1988.

FOY, Rev W.H., *Buchanan's Christian Researches in India: with the rise suspension and probable future of England's rule as a Christian Power in India*, A detailed historical study. Buchanan's original work dates from the beginning of the nineteenth century.

FRANK, Dr Katherine, *Lucie Duff Gordon*, London, 1994. Provides details of **Azimullah Khan's** sojourn in England, when he went on the Nana Sahib's behalf to plead for the reversal of Dalhousie's decision not to grant him the pension of Baji Rao II his adoptive father, and the title of Peshwa. Valuable and original research and material.

FRASER, Antonia, *Boadicea's Chariot* (for Rani of Jhansi), London, 1988. The book is devoted to Warrior Queens, and other strong feminine characters like Mrs Thatcher, Indira Gandhi and Golda Meir. There is a chapter on the Valiant Rani, ie Lakshmibai.

FRASER, E., *The Pearl's Brigade in the Indian Mutiny*, Mariners' Mirror, Vol xii 1926.

FRASER, Captain Hastings, *Our Faithful Ally, the Nizam*, London, 1865.

FREELING, G.H., *Narrative of events connected with the*

Mutiny at Hamirpur, 1858.

FROST, Thomas, *A Complete Narrative of the Mutiny in India*, London, 1857. Nothing of the sort.

FUHRER, Rev A., *List of Christian Tombs and Monuments of Archaeological or Historical Interest and their Inscriptions in the North-Western Provinces and Oudh*, Allahabad, 1896. Epitaphs and Inscriptions can be useful sources of general as well as specific information.

GANDA SINGH, *The Indian Mutiny of 1857 and the Sikhs*, Delhi, 1969.

GANGULY, Anil Baran, *Guerilla Fighter of the First Freedom Movement*, Patna, 1980. The guerilla fighter in question is Tatya Tope 1813-59.

GANGULY, D.C., *Select Documents of British Period of Indian History,* in the collection of the Victoria Memorial, Calcutta, 1958.

GARDINER, Sir Robert, *A Military Analysis of the Remote and Proximate causes of the Indian Rebellion*, London, 1858.

GARDNER, Brian, *The East India Company*, New York, 1990.

GARDNER, Frank M, *The Indian Mutiny*, London, 1857.

GERMON, Maria, *Journal of the Siege of Lucknow*, London, 1958. She was the wife of Captain Germon an officer in the 13th Native Infantry, Bengal Army. They were among the besieged in the Residency at Lucknow and survived. She kept a Journal throughout the siege and the original is preserved in the Oriental and India Office collections of the British Library, *See MSS.Eur.B.134*. Maria Germon was a very ordinary woman and her journal concerns mundane matters, to do with survival and conditions generally: it is therefore valuable for detail of conditions within the Residency during the siege. First edition of her diary appeared as *A Diary kept by Mrs R.C.Germon*, London, 1870.

GHALIB, 1797-1869, *Life and Letters*, Vol I, transl. Ralph Russell & Khurshidul Islam, London, 1969, Reprinted as Oxford Paperback, 1994. Extremely valuable primary source material on the 'Siege' of Delhi. The poet Ghalib was present throughout, within the walls of the city, and he recalls not only his own experiences but the events of which news was brought to him.

GHOSE, Sailen, *Archives in India, history and assets*, Calcutta, 1958. Description of both organization and contents.

GHOSH, Kali Charan, *The Roll of Honour - Anecdotes of Indian Martyrs*, Calcutta, 1965. Princes chieftains etc are listed in 'those who participated in the great upheaval' on the side of the rebels' on p 21. Otherwise the events of 1857-59 are restricted to six pages only in a book 829 pages in length.

GIBBON, F.P., *The Lawrences of the Punjab*, biographies of the three brothers John, Henry and George. First Afghan War, Sikh Wars, Mutiny etc.

GIBBS, M.E., *The Anglican Church in India*, 1600-1970, Delhi, 1972.

GIBNEY, Robert D, *My Escape from the Mutineers in Oudh*, 2vols, London, 1858. Hardly warrants one volume.

GILBERT, Henry, *The Story of the Indian Mutiny*, London, 1916. Of little significance

GILLIATT, Edward, *Heroes of Modern India*, London, 1911.

GILLIATT, Edward, *Daring Deeds of the Indian Mutiny*, London, 1918. By which, given the date, it is hardly necessary to add, *British* daring deeds.

GILLIATT, Edward, *Heroes of the Indian Mutiny*, London, 1922. As above.

GIMLETTE, Lt Colonel George, *Postscript to the Records of the Indian Mutiny*, London, 1927. An attempt to trace the career of the sepoy regiments, after they had mutinied, through the whole course of the rebellion. In the absence of rebel documentation this is a most praiseworthy and painstaking attempt to solve many mysteries.

GOLANT, William, *The Long Afternoon; British India 1601-1947*, London, 1975.

GOLDSMID, F.J., *James Outram,* 2vols, London, 1880. An early biography of Sir James Outram, the 'Bayard of India', collaborator of Havelock and subsequently Chief Commissioner of Oude[Awadh].

GORDON, Rev. Andrew, *Our India Mission 1855-85*, New York, 1888. A thirty year history of the India Mission of the United Presbyterian Church of North America; particularly valuable for the story of Sialkot.

GORDON, C.A., *Recollections of thirty-nine years in the Army*, London, 1898. Includes the Mutiny

GORDON-ALEXANDER, Lt Colonel W., *Recollections of a Highland Subaltern under Sir Colin Campbell*, London, 1898.

GOUGH, Sir Hugh, *Old Memories*, Edinburgh, 1897. His career and reputation are justifiably prestigious, but this book adds little to our knowledge.

GOWING, Thomas, *A Soldier's Experience, or A Voice from the Ranks*, Nottingham, 1905. Gowing was a sergeant-major in the Royal Fusiliers, and served through the Crimean War and the Indian Mutiny campaigns. His book includes a 'Sketch of the Life of General Sir Henry Havelock, the Christian Soldier, together with SOME THINGS NOT

GENERALLY KNOWN': regrettably there are not many of the latter, but the book is useful in giving a non-commissioned officer's point of view on the struggle, and providing considerable statistical detail of the units engaged, casualties, wounded and killed etc.

GRAHAM, G.F., *Life and Work of Sir Syed Ahmed Khan,* London, 1909.

GRANT, James, *Cassell's Illustrated History of India,* 2 vols, London, 1890. Particularly useful for well deleivered accounts and sketchmaps of various military engagements during the mutiny/rebellion.

GRANT, Sir J.Hope, and KNOLLYS, Sir H., *Incidents in the Sepoy War,* Edinburgh, 1875. The work includes much primary material in the form of General Hope Grant's journal and letters, and the orders he received from the Commander-in-Chief Sir Colin Campbell. The scope (final march to Lucknow and subsequent operations) is limited but useful.

GRANT, Sir J.Hope, *Life and Selections from his correspondence,* 2 vols, Edinburgh, 1894. Revealing and useful.

GRAY, Lieutenant W.J., *Journal of the Siege Train from Ferozepore to Delhi,* in Society Army Hist. Research Journal, Vol X 1931. This essential constituent of the British plan for assaulting Delhi was very meagerly guarded on the journey from the Punjab, and was very vulnerable to rebel attack.

GREAT BRITAIN, PUBLIC RECORDS OFFICE, *Guide to the Contents of the Public Record Office,* 2 vols, HMSO, London, 1963. Essential reference work of first resort for those determined to use the P.R.O.

GREATHED, Harvey H, *Letters written during the Siege of Delhi, edited by his Widow,* London, 1858. Particularly useful for first-hand material on the events of May 1857 in Meerut.

GRETTON, Lt Colonel G.le M. *Campaigns and History of the Royal Irish Regiment,* Edinburgh, 1911.

GREY, Colonel L.J., *Tales of our grandfather,* London, 1912.

GRIFFITHS, Chas, *A Narrative of the Siege of Delhi,* London, 1910. Includes an account of the outbreak of mutiny at Firozpur. Has the merit of containing only detail of events personally experienced. Useful for the picture it gives of the occupation of Delhi after the assault in September 1857.

GROOM, William Tate, *With Havelock from Allahabad to Lucknow,* London, 1894. He died of wounds in the Lucknow Residency. Was an officer in the Madras Fusiliers.

AN ACCOUNT OF THE MUTINIES IN OUDH, AND OF THE SIEGE OF THE LUCKNOW RESIDENCY; WITH SOME OBSERVATIONS ON THE CONDITION OF THE PROVINCE OF OUDH, AND ON THE CAUSES OF THE MUTINY OF THE BENGAL ARMY. BY MARTIN RICHARD GUBBINS, OF THE BENGAL CIVIL SERVICE, FINANCIAL COMMISSIONER FOR OUDH. LONDON: RICHARD BENTLEY, NEW BURLINGTON STREET, Publisher in Ordinary to Her Majesty. 1858.

GUBBINS, M.R., *An Account of the Mutinies in Oudh and of the Siege of the Lucknow Residency,* London, 1858. This is a valuable source book, particularly for Oude[Awadh] before and during the initial stages of the struggle, and for details of the Residency siege. Martin Gubbins and the Officer i/c the garrison after Sir Henry Lawrence's death, Colonel Inglis, detested each other, with the result that the story of the siege is not told with any respect for the official 'line'. Gubbins was the Financial Commissioner in Oude[Awadh]. The book has a useful appendix, giving the names of all officers, their families, civilians and merchants who were in the Residency, and their fate.

GUHA, Ranajit, *Prose of Counter-Insurgency,* Subaltern Studies II, Delhi, 1983.

GUHA, Ranajit, *Elementary Aspects of Peasant Insur-*

gency in Colonial India, Delhi, 1983. Reprinted, Delhi, 1992. A scholarly and painstaking work, although it is far from easy for the layman to understand what positive conclusions are to be drawn from Dr Guha's investigations. One highly practical merit lies in his drawing attention to the dangers inherent in accepting, without question, the 'colonial terminology' in official documents of the time: the truth regarding events and attitudes can be obscured by too facile acceptance of the British use of 'lawlessness', 'dacoit', 'badmash', 'fanatical' etc.

GUPTA, Pratul Chandra, *The Last Peshwa and the English Commissioners, 1818-1853*, Calcutta, 1944. For good details of the background to the inheritance etc of Nana Sahib.

GUPTA, Pratul Chandra, *Baji Rao II and the East India Company*, Oxford, 1939. Eminently readable account by a distinguished historian of the background to Nana Sahib's claims to the title and pension of his adoptive father.

GUPTA, Pratul Chandra, *Nana Sahib at Bithur*, in Bengal Past and Present, Vol 77, 1958.

GUPTA, P.C., *Nana Sahib and the Rising at Cawnpore*, Oxford, 1963.

GUPTA, Rajni Kant, *Military Traits of Tatya Tope*, Delhi, 1987. The author is obviously an admirer of Tatya Tope whom he describes as a 'military genius - a legendary dynamic personality of the First War of Indian Independence (1857)'. Expresses doubts as to whether it was really Tatya Tope whom the British hanged.

HALDANE, Julia, *Story of our escape from Delhi in 1857*, Agra, 1888.

HALLOWAY, J., *Essays on the Indian Mutiny*, London, 1864.

HALLS, John, *Two months in Arrah in 1857*, London, 1860. see below.

HALLS, John, *Arrah in 1857, its defence and relief*, Dover, 1893. A short straightforward account of the defence of Arrah House in Bihar in July/August 1857 against the Danapur mutineers and Raja Kunwar Singh. Largely based on Halls's previous account, but includes illustrations and an appendix featuring Major Eyre's Despatch of 3 August 1857, and Mr Chas Kelly's supplementary account of the affair.

HANDCOCK, Colonel A.G., *A Short Account of the Siege of Delhi*, Simla, 1892.

HANDCOCK, Colonel A.G., *Siege of Delhi in 1857*, Allahabad, 1907.

HANDKIN, Nigel, *Hanklyn-Janklyn or a Stranger's Rumble tumble guide to some words, customs and quiddities, Indian and Indo-British*, Delhi, 1992. A useful companion and successor to Hobson-Jobson.

HAQ, Syed Moinul, *The Great Revolution of 1857*, Karachi, 1968.

HARDEN, Anne, ed., *A Diary of the Indian Mutiny*, in Notes and Queries 200:352-6, 1955.

HARE, Augustus, *The Story of Two Noble Lives*, vol ii, London, 1893.

HARRIS, Mrs G., *A Lady's Diary of the Siege of Lucknow*, London, 1858. Wife of the Reverend Harris, the officiating Chaplain in the Residency after the death of the Rev. Polehampton. Useful for a picture of conditions within the Lucknow Residency during the siege, and covers the period 15 May to December 1857.

HARRIS, J., *The Indian Mutiny*, London, 1973. It adds very little. Part of the 'British at War' series.

HARRISON, A.T., ed., *The Graham Indian Mutiny Papers*, being the diary and correspondence of the Graham family. Dr Graham was killed by the mutineers at Sialkot. His nephew James at Landour preserved and co-ordinated the correspondence. His son James committed suicide at Lucknow during the siege of the Residency which had ' a depressing effect on the whole garrison.' Intro. by A.T.Harrison, Public Record Office Northern Ireland, Belfast, 1980.

HARVEY, George Frederick, *Narrative of Events attending the outbreak of disturbances and the restoration of Authority in the Agra Division*, Agra, 1859.

HAY, Sidney, *Historic Lucknow*, Lucknow, 1939.

HEARSEY, Captain J.B., *Narrative of the Outbreak at Seetapore*, Sitapur, 1858.

HEATHCOTE, T.A., *The Indian Army. The Garrison of British Imperial India, 1822-1922*, London, 1974.

HEBER, Rev R., *Narrative of a Journey through the Upper Provinces of India from Calcutta to Bombay*, 2 vols, London, 1828. Bishop Heber was a remarkably observant man and a meticulous note-taker: although his books came out thirty years before the rebellion, they are still of relevance and significance. Recommended as a book providing excellent background information.

HEDAYAT ALI, *A Few words relative to the late Mutiny of the Bengal Army*, Calcutta, 1858.

HERBERT, David, *Great Historical Mutinies*, London, 1875.

HERFORD, Ivan, *Stirring Times under Canvas*, London, 1862.

HERVEY, Captain Albert, *Ten Years in India, or the Life of a Young Officer*, London, 1850.

HEWITT, James, ed., *Eye-witnesses to the Indian Mutiny*, London, 1972. A limited narrative account of the mutiny/rebellion in Delhi, Kanpur and Lucknow, based upon eye-witness accounts.

HIBBERT, Christopher, *The Great Mutiny, India 1857*, London, 1978. A recent and very readable account of the Great Rebellion, told from the particular viewpoint of those Europeans who suffered during the struggle. Scholarly and accurate so far as it goes, it is a narrative history incorporating 'as much hitherto unpublished material as I have been able to find'. It does not attempt the analysis of the military or political problems, nor enter into a discussion of the Mutiny as an early stage of India's struggle for independence. An excellent introduction to the events of 1857 in India.

HIGGINBOTHAM, J.J., *Men whom India has known*, Biographies of Eminent Indian Characters, Madras, 1874.

HILL, J.R., et alii, of the Society for the Propagation of the Gospel in Foreign Parts, *The Story of the Cawnpore Mission*, Westminster, 1909.

HILL, S.C., *Catalogue of the Home Miscellaneous Series of the India Office Records,* London, 1927. Includes the Mutiny Papers, 724a-727, of Sir John Kaye, and numerous other significant reference materials relevant to the rebellion. An essential resource.

HILTON, Edward, *The Tourist's Guide to Lucknow*, Lucknow, 1916.

HILTON, Edward H., *The Mutiny Records: Oudh and Lucknow (1857)*, reprint by Sheik Mubarak Ali, Lahore, 1911.

HILTON, Richard, *The Indian Mutiny*, London, 1957.

HINDU, A. (anon), *The Mutiny and the People, or Statements of Native Fidelity*, Calcutta, 1859, reprinted Calcutta 1969. The bookjacket of the reprint asserts that the author was Sambhu Chandra Mukherjee, a journalist, and one of the eventual founders of the India League in 1875. Contains a list of persons 'rewarded for loyalty' together with the nature of their reward, but the list is unfortunately not fully comprehensive.

HISTORY of the BRITISH SETTLEMENTS in INDIA to the CLOSE of the SEPOY REBELLION. London, 1861.

HISTORY of the INDIAN REVOLT & of the EXPEDITIONS to PERSIA CHINA JAPAN 1856,7,8, London, 1859. Sometimes known as 'Chambers' History', profusely illustrated.

HODSON, V.C., *List of the Officers of the Bengal Army, 1758-1834*, 3 vols, London, 1947. The majority of the senior officers of the Company's army in 1857 are to be found here, with career and background personal details.

HODSON, William, *Twelve Years of a Soldier's Life in India*, London, 1859. Captain W.S.R.Hodson was killed in Lucknow in 1858, so this is a book based on his letters, edited by his brother George H.Hodson. The latter does his best to defend his brothers' actions, which have always aroused controversy and fierce debate, especially the killing by Hodson of the Delhi Princes in September 1857.

HOEY, W., (trans), *Memoirs of Delhi and Faizabad*, Allahabad, 1887.

HOLLOWAY, John, *Essays on the Indian Mutiny*, London, 1863.

HOLMES, F.M., *Four Heroes of India*, London, 1892.

HOLMES, T.Rice, *History of the Indian Mutiny*, Fifth Edition, London, 1904. A 'heavyweight' English historian of the rebellion whose work has the merit of being distanced some forty years in time from the events he describes, which makes for an objective and historian's re-appraisal of the rebellion. The standard one volume work. The extension to the title reads 'and of the disturbances which accompanied it among the local population' and this indicates the true scope of the book.

HOLMES, T.Rice, *Four Famous Soldiers*, London, 1889.

HOSEASON, John Cochrane, RN., *Remarks on the rapid transmission of troops to India and the practicability of* *promptly establishing effectual means of conveyance between the sea coast and the interior by the navigation of the great rivers by steam*, London, 1858. From a collection of letters to be found in *Ames in Indian Affairs*, 2, Pamphlets 1857-60, No.8.

HOSKINS, H.L., *British Routes to India*, London, 1966. Useful in particular for description of the methods by which news of the mutiny/rebellion was conveyed from India to London.

HOW the ELECTRIC TELEGRAPH SAVED INDIA, in MACMILLAN'S MAGAZINE, 76, 401-6, October 1897, ie the warning telegram from Delhi which gave the British time to hold the Punjab.

HUGHES, Maj.Gen. B.P., *The Bengal Horse Artillery, 1800-1861*, London, 1971.

HUGHES, Derrick, *The Mutiny Chaplains*, Salisbury, 1991.

HUNTER, Chas, *Personal Reminiscences of an Indian Mutiny Veteran*, Brighton, 1911.

HUNTER, Sir William, *The Marquess of Dalhousie*, Oxford, 1890.

HUNTER, Sir W.W., *Imperial Gazetteer of India*, 9 vols, London 1881. The original gazetteer upon which all subsequent editions have been based. Contains useful little summaries of the events and subsequent developments *re* the outbreak, station by station. For access to quick, basic, information this gazetteer is ideal.

HUNTER, Sir W.W., *Imperial Gazetteer of India*, 2nd Edn., 14 vols, London, 1885-7.

HUNTER, W.W., *The Indian Mussulmans: Are they bound in Conscience to Rebel against the Queen?* London, 1871. Reprint, Delhi, 1969.

HUSAIN, Mahdi, *Bahadur Shah II and the War of 1857 in Delhi*, Delhi, 1958.

HUTCHINSON, David, *Annals of the Indian Rebellion:1857-58*, London 1974. This is a reprint of N.A.Chick's original version, and while it has the advantage of being more readily available, unfortunately it does not contain all of the original.

HUTCHINSON, Major General George, *Narrative of the Mutinies in Oudh*, London, 1859. Was, eventually, the senior engineer officer in the Lucknow Residency during the siege. An interesting and informative account, from the point of view of a soldier, but he was a shrewd and intelligent observer also.

IBBETSON, D., *Punjab Castes*, Lahore, 1916.

INDIA BEFORE and AFTER the MUTINY, by an Indian student, Edinburgh, 1886.

INDIA: The REVOLT and the HOME GOVERNMENT, London , 1857.

INDIA: MINISTRY of INFORMATION and BROADCASTING. 1857: A PICTORIAL REPRESENTATION, Delhi, 1957. This was a publication designed to give a pictorial account of the 'Indian Struggle of 1857'. There are seventy pages of annotated photographs, portraits or sketches, without formal text. Is of interest, but the relevance of many of the illustrations is obscure. The editors regret that reliance has had to be placed on British artists whose drawings are 'not in keeping with Indian sentiments.'

INDIAN MUTINY, See Spectator magazine, 1857, Issue 199, pp 156,185,212,243 and 271.

INDIA'S MUTINY & ENGLAND'S MOURNING, London, 1857.

INDIAN NATIONAL TRUST FOR ART AND CULTURAL HERITAGE (INTACH), *Preliminary Unedited Listing of Cawnpore, Uttar Pradesh, Unprotected Monuments, Buildings & Structures Listed for Conservation*, Kanpur, 1992. Useful details of buildings etc still standing in Kanpur which date from the 'mutiny' or before.

INGE, Lt Colonel D.M., *A Subaltern's Diary*, London, 1894. He served in the 8th Hussars in the Central India Field Force under Sir Hugh Rose.

INGLIS, Hon.Julia, *The Siege of Lucknow, a Diary*, London, 1892. She was a woman of considerable strength of character, and marked sense of duty who took her position seriously as senior lady within the British garrison. She survived the siege and, like her friend **Mrs Adelaide Case** published the diary she had kept through the siege. The book includes the extensive notes of **Colonel F.M.Birch**, her husband's A.D.C.

INNES, Lt General James J.McLeod, *Lucknow and Oudh in the Mutiny*, London, 1895. Outstandingly useful for the sketches he made of the Residency buildings and the defensive position taken up by the British, from various angles. His experience is first-hand; as an engineer he is naturally most interested in that side of the operations.

INNES, Lt General James, *The Sepoy Revolt*, London, 1897.

INNES, Lt General James, *Indian Mutiny Cuttings from Newspapers published during Mutinies*, nd. Disappointingly meagre and selective.

INNES, Lt Col P.R., *History of the Bengal European Regiment*, London, 1885. Detailed history 1756-1858.

INTACH, *See* Indian National Trust for Art and Cultural Heritage.

IRELAND, William W., *A History of the Siege of Delhi by an Officer who served there,* Edinburgh, 1861.

JACOB, Sir George le Grand, *Western India before and during the Mutinies,* London, 1871.

JAMES, David, *Life of Lord Roberts,* London, 1954.

JEFFREY, Robin, *People, Princes and Paramount Power,* Society and Politics in the Indian Princely States, Oxford, 1978. May help to explain the contradiction whereby many Princes remained pro-British while their people sided with the rebels.

JHA, Kameshwar, *A Study of some Mutiny Letters of Sohagpur,* Indian Historical Records Commission Proceedings, 1958.

JIVANLALA, RAI BAHADUR, *Short Account of the life and family of Rai Jeewan Lal Bahadur,* Delhi, 1902.

JOCELYN, Julian R.J., *History of the Royal and Indian Artillery in the Mutiny of 1857,* London, 1915. A truly detailed *military* history of the revolt with particular emphasis on the role of the artillery.

JOHNSON, W.T., *Twelve Years of a Soldier's Life,* London, 1897.

JONES, Gavin S., *My escape from Fatehgarh,* Cawnpore, 1913. Written by one of the few European survivors of Fatehgarh. He had knowledge and respect for Indian people and this probably saved his life.

JONES, Gavin Sibbald, *The Story of my Escape from Fatehgarh.* in Cornhill magazine, Jan. 1865.

JONES, Captain (RN) Oliver J., *Recollections of a Winter's campaign in India in 1857-58,* London, 1859.

JONES-PARRY, S.H., *An Old Soldier's Memories,* London, 1897.

JOSHI, Puran Chandra, ed., *Rebellion 1857. A Symposium,* Delhi, 1957. A Marxist, he apparently perceives the rebellion as a revolt by the working classes, a view not wholly shared by Karl Marx (*q.v.*) himself. Included is his small collection of Folk Songs on 1857 which are most interesting; included he says because 'In India folk-art forms have been the traditional media for approaching the masses.'

JOYCE, Michael, *Ordeal at Lucknow,* Lucknow, 1938. A careful and readable statement on the siege of the Lucknow Residency, but very much from the British point of view.

JUDICIAL TRIAL DURING the INDIAN MUTINY, in the Calcutta Review, Vol 115, 1902.

JWALA SAHAI, *The Loyal Rajputana,* Allahabad, 1902.

KAVANAGH, T.Henry, *How I won the VC,* London, 1860. Not just the self-glorification it suggests: there is a great deal of original material on the campaigns in Oude[Awadh].

KAYE, G.R., and JOHNSTON, E.H., *Minor Collections and Miscellaneous Manuscripts,* London, 1937.

KAYE, Sir John, *History of the Sepoy War,* 3 vols, London, 1864-7. Volume I covers the period 1846 to May 1857; Vols II and III are not distinguishable chronologically but carry the story, together, to the capture of the King of Delhi in September 1857. A fourth (and subsequent?) volume was clearly projected but the death of the author prevented this, and the work was undertaken by Colonel G.B.Malleson (*q.v.*). Kaye's history has remained the most read, the most quoted and therefore the most authoritative detailed account of the rebellion, although it contains inaccuracies, and occasionally opinion set down as fact. *See also* Pincott, Frederic, for an analytical Index.

KAYE, Sir John, *Lives of Indian Officers,* 2 vols, London, 1867.

KEENE, H.G., *Fifty Seven,* London, 1883. Of particular use for Saharanpur and Muzaffarnagar, Meerut, Bulandshahr Mainpuri and Etawa, Ghazipur Jaunpur and Mirzapur, and Rohilkhand.

KEENE, H.G., *Handbook for visitors to Allahabad, Cawnpore and Lucknow,* Calcutta, 1896.

KEENE, H.G., *A Servant of John Company,* London, 1897.

KEENE, H.G., *Here and There: Memories Indian and Other,* London, 1906.

KHADGAWAT, Nathu Ram, *Rajasthan's Role in the Struggle of 1857,* Jaipur, 1957. This was a grandiose project, but the author had to admit 'The vast population inhabiting this desert land though in a way dissatisfied against the British, failed to make a common cause with the mutineers.'

KINCAID, Dennis, *British Social Life in India, 1608-1937,* London, 1973. First published in 1938, it provides a reasonably accurate picture of the kind of social life enjoyed by the British expatriate in India, although the accent is upon the lighter side of that life: the illustrations, by Frank Wilson, are talented, perceptive and amusing.

KNIGHTON, William, *The Private Life of an Eastern King,* London,1855. Supposedly that of Wajid Ali Shah of Oude; certainly the detail would indicate as such.

KNOLLYS, Sir Henry, *See* Sir J Hope Grant.

KOLFF, D.H.A., and BAYLY, C.A. eds., *Two Colonial Empires,* Lancaster, 1986. *ie* the Dutch and British Empires in South and South East Asia compared.

KURSHIDUL ISLAM, transl. (with **RUSSELL, Ralph**) *Life and Letters of Ghalib,* Vol I, London, 1952, new ed.

1969. Reprinted as Oxford Paperback, 1994. Extremely valuable primary source material on the 'Siege' of Delhi. The poet Ghalib was present throughout, within the walls of the city, and he recalls not only his own experiences but the events of which news was brought to him .

LADENDORF, Janice M., *The Revolt in India, 1857-58*, Zug, 1966. A most valuable annotated bibliography of English language materials, with nearly 1,000 entries (including duplicate and triplicate). Arranged, however, rather fussily under headings which are not always fully justified eg 'Scholarly Studies', 'Narrative Histories', 'Biography', 'Diaries, Letters, Journals, Memoirs etc'. Decisions as to placing appear arbitrary in some cases, and it is possible to overlook valuable items as a result. Spelling mistakes are frequent. Nonetheless this remains one of the most important reference tools for the student.

LAL, Krishan, *The Sack of Delhi*, 1857-58, as witnessed by Ghalib. In Bengal Past and Present, July-December 1955.

LALE, R.H.B., *Historical Records of the 93rd Sutherland Highlanders*, London, 1892.

LANDON, Percival, *1857*, London, 1907.

LANDON, Percival, *Nepal*, 2 vols, London, 1928. Contains interesting speculation about the exile of the Nana Sahib's family, and the Begum of Oude.

LANDON, Percival, *Under the Sun*, London, 1906. Impressions of Indian cities with a chapter dealing with the later life of Nana Sahib.

LANG , Arthur Moffatt, *Diary and Letters 1857-58*, in Society of Army Historical Research Journal, 1930-32.

LANG, Arthur Moffatt, *Lahore to Lucknow*, The Indian Mutiny Journal of Arthur Moffatt Lang, ed. David Blomfield, London, 1992. The personal account, via his journal, by an intelligent and observant engineer officer, of the military campaign as he saw it and experienced it. Includes detail not available elsewhere and is altogether very readable.

LANG, John, *Wanderings in India*, London, 1859. Written by a lawyer and novelist whose wanderings preceded the mutiny/rebellion. In 1856 as a result of a successful case in an Indian court he found himself in demand to give advice to others who had quarrels with the East India Company, notably the **Nana Sahib** and the **Rani of Jhansi** . This book includes therefore details of a number of interesting meetings he had with major protagonists.

LAWRENCE, Lt General Sir George, *Forty Years' Service in India*, London, 1874. By the brother of John and Henry Lawrence.

LAWRENCE, Sir Henry, *Essays Military and Political*, London, 1859. Henry Lawrence was a man of exceptional talent and insight. His (pre-Mutiny) writings are therefore significant.

LAWRENCE, John, *Lawrence of Lucknow*, London, 1990. The story of Sir Henry Lawrence.

LAY THOUGHTS on the INDIAN MUTINY, by a Barrister, London, 1858.

L.E., *The Present Crisis in India*, London, 1857. Typical of much of the leaping into print by many in Britain with a pretension to knowledge of the East. Uses the familiar I-told-you-so and I-knew-it-would-happen indictment of the East India Company, but loses half its strength because the author chooses to remain anonymous.

LEASOR, James, *The Red Fort*, London, 1956.

LEATHER, G.F.T., ed., *Arrah in 1857*, Dover, 1893.

LEBRA-CHAPMAN, Dr Joyce, *The Rani of Jhansi, a Study in Female Heroism in India*, Honolulu, 1986. Includes original legends and folk-songs.

LECKEY, Edward, *Fictions connected with the Indian Outbreak of 1857*, Bombay, 1859. A courageous attempt to refute the deliberate sensationalism of much of the earlier (British media) reporting of the mutiny/rebellion.

LEE, Joseph, *The Indian Mutiny: a narrative of the events at Cawnpore*, Cawnpore, 1893. Written by a hotel keeper in Kanpur, who may, or may not, (opinions vary) have been a sergeant in the British army at the relief of Lucknow by Sir Colin Campbell. He certainly made a tidy living taking tourists around the 'sights' of Kanpur in the late nineteenth century.

LEE-WARNER, W., *Life of the Marquess of Dalhousie*, London, 1904.

LEHMANN, Joseph, *All Sir Garnet, a Life of F.M.Lord Wolseley*, London, 1968. Wolseley was a Captain in the British Army at the relief of Lucknow and subsequent campaign.

LETTER 235, dated 25 November 1857, from the Court of Directors, Honourable East India Company, setting up a Commission to inquire into the future of the Army in India, with a series of questions addressed to Officers in India, ORIENTAL & INDIA OFFICE Collections of the British LIBRARY.

LEWIN, Malcolm, ed., *Causes of the Indian Revolt, by a Hindu of Bengal*, London, 1857. A twenty-four page tract, neglected by historians but none the less of some interest. It is an appeal to the British, via Lewin who had been a Judge in Madras, to compel the East India Company to govern more in the interests of the governed than they had done of late.

LEWIN, Thomas L., *A Fly on the Wheel*, or How I helped govern India, London 1885.

LIND, Af Hagely, Axel Reinhold Ferdinand, *Reisebilder und Skissen aus Indien und den letzen indischen Kriege, 1857-9,* Leipzig, 1861. Written by an officer of the Royal Swedish Navy who served with the Naval Brigade (the crew of H.M.S. 'Shannon', and the 'Pearl') under Peel in the relief of the Lucknow Residency: interesting as a 'foreigner's account'.

LINDNUM, John, *Humour of the Mutiny,* in Chambers Journal 3 November 1917. Attempts without overmuch success to provide anecdotes and moments of humour, on the British side during the struggle.

LLEWELLYN, Alexander, *The Siege of Delhi,* London, 1977. A run of the mill account with nothing of moment added to what was already known. But it is a skilful compendium of contemporary memoirs and letters, reads well, and for the non-specialist is a satisfactory account.

LLEWELLYN-JONES, Dr Rosie, *A Fatal Friendship,* Delhi, 1985. Subtitled *The Nawabs, the British and the City of Lucknow,* this book gives an excellent physical description of the city in Nawabi days. Many of the glories of Lucknow architecture were lost in or as a result of the warfare in 1857-8. Dr Lllewellyn-Jones shows us, in this unique work, what was lost as well as explaining and describing what has been retained.

LLEWELLYN-JONES, Dr Rosie, *A Very Ingenious Man,* Claude Martin in Early Colonial India, Delhi, 1992. This is a first-rate biography of the builder of La Constantia, the palace in Lucknow which is also known as La Martinière, and is the home of the Martinière College, whose buildings, staff and pupils played parts in the relief of the Lucknow Residency.

LOVETT, A.C. & MACMUNN, F.G., The Armies of India., Edinburgh 1911. A standard work, superbly illustrated.

LOW, Chas R., *History of the Indian Navy 1613-1863,* 2 vols, London, 1877. Includes two chapters on the navy in the period of the mutiny/rebellion.

LOW, C.R., *A Memoir of Sir Garnet Wolseley,* London, 1878.

LOW, C.R., *Major General Sir F.S.Roberts,* London 1883. An early biography of 'Old Bobs'. Frederick Roberts was a Lieutenant at this time and served at Delhi and at Lucknow, and won the VC at Khudaganj.

LOW, D.A., ILTIS, J.C., and WAINWRIGHT, M.D., *Government Archives in South Asia: a Guide to National and State Archives in Ceylon, India and Pakistan,* Cambridge, 1969. A basic guide and essential as a reference tool for the beginner.

LOW, Ursula, *Fifty Years with John Company,* London, 1936.

LOWE, Thomas, *Central India during the Rebellion of 1857-8,* London, 1860. A standard history. Factual and concise.An excellent basic reference tool.

LUARD, C.E., *Contemporary Newspaper Accounts of Events during the Mutiny in Central India 1857-58,* Allahabad, 1912. An exciting title but a disappointing book. Slim and very selective. Of little practical use to the scholar.

LUCAS, J.J., *Memoir of Reverend Robert Stewart Fullerton,* Allahabad, 1928. Subtitled: *American Presbyterian missionary in North India, 1850-65: compiled from his letters during fifteen years in India, and his narratives of the trials faiths and constancy of Indian Christians during the Mutiny of 1857.*

LUCAS, Samuel, *Dacoitee in Excelsis or the Spoliation of Oude by the East India Company,* new edition, Lucknow, 1971.

LUCKNOW, GUIDE to, with HISTORICAL NOTES on MUTINY of 1857, Lucknow, 1911.

LUCKNOW, REMINISCENCES of 1857 by a MEMBER of the ORIGINAL RESIDENCY GARRISON, Lucknow, 1891.

LUNIYA, B.N., *Freedom Movement in Malwa,* nd (1957?). An interesting collection of letters 'discovered in the records of Madhya Bharat Government, translated from Marathi', which, it is claimed, throw a good deal of light on the events of the mutiny in this region. Contains also some unique illustrations.

LUNT, James, ed., *From Sepoy to Subedar: being the life and adventures of Subedar Sita Ram, a native officer of the Bengal Army written and related by himself,* translated by Lt Col Norgate, Lahore 1873, reprinted London, 1970.

LUSHINGTON, Stephen, *Banda and Kirwee Booty,* Judgement of the Hon.S.Lushington, delivered in the High Court of Admiralty of England, London, 1866.

LUTFULLAH, Syed, *Azimullah Khan Yusufzai: the man behind the War of Independence, 1857,* Karachi, 1970. Shows enthusiasm for the premiss that Azimullah Khan was more significant than most other observers consider him to have been: that is the basis of this little book, but unfortunately little direct evidence but a great deal of wishful conjecture is marshalled in defence of the author's argument.

LUTFULLAH, Syed, *The Men behind the War of Independence 1857,* Karachi, 1957.

MACCREA, R., *The Tablets in the Memorial Church Cawnpore,* Calcutta, 1894.

MACGREGOR, Sir Charles, *Life and Opinions,* Edinburgh, 1888.

MACKAY, Rev. James, *From London to Lucknow,* 2

vols, London, 1860.

MACKENZIE, Colonel Alfred R.D., *Mutiny Memoirs*, Allahabad, 1891. Originally published in the 'Pioneer'. According to Rebecca West this was a book much admired by Rudyard Kipling - when it was published by his father's press. He spoke of it once to Lord Beaverbrook. The book contains the author's personal reminiscences of the 'Great Sepoy Revolt of 1857', in particular in Oude[Awadh]. Mackenzie was Hon'y ADC to the Viceroy, and a most intelligent and literate man. The quality of the prose as well as the content are remarkable.

MACKENZIE, Mrs Colin, *Life in the Mission, the Camp and the Zenana*, 3 vols, London,1853.

MACKENZIE, Mrs Colin, *English women in the Rebellion*, in the Calcutta Review, September 1859.

MACLAGAN, Michael, *Clemency Canning*, London, 1962 A more recent biography of Earl Canning, the Governor General and first Viceroy.

MACLAGAN, Michael, *The White Mutiny*, London, 1964. An essay and summary on this somewhat neglected subject ie the refusal of the European soldiers in the Company's employ to accept automatic transfer (without re-enlistment bonus) to the British Army, and the subsequent mutiny which began, coincidentally, at Meerut.

MACMILLAN, Margaret, *Women of the Raj*, London, 1988. An excellent overview.

MACMULLEN, J., *Camp and Barrackroom*, London, 1846.

MACMUNN, Lt General Sir George, *The Indian Mutiny in Perspective*, London, 1931. Produced by a popular and prolific writer upon the mutiny/rebellion, but he has the advantage of sound military knowledge and the instinct for research. Very readable, and will encourage further investigation of the themes he deals with.

MACMUNN, Lt General Sir George, *Behind the Scenes in Many Wars*, London, 1930.

MACMUNN, Lt General Sir George, *Some new light on the Indian Mutiny*, in Blackwoods Magazine, April 1928.

MACMUNN, Lt General Sir George, *Dawn at Delhi, May 11th 1857*, in Cornhill Magazine, July 1913. Narrative account of the uprising.

MACMUNN, Lt General Sir George, *Devi Din, mutineer*, in Blackwoods magazine, April 1928. Stories of the Mutiny as told by an old sepoy to a British officer.

MACMUNN, Lt General Sir George, *1857, Delhi after the Storming*, in Cornhill magazine, May 1910. Events in Delhi after its fall in September 1857, from several original manuscripts.

MACMUNN, Lt General Sir George, *Jan Kampani Kee Jai*, in the Cornhill magazine, May 1910. The story of the mutiny at Mian Mir near Lahore.

MACMUNN, Sir George, *Indian States and Princes*, London 1936. Includes the Princes during the Indian Mutiny.

MACPHERSON, A.G., *The Siege of Lucknow*, in the Calcutta Review, September 1858.

MACPHERSON, William, ed., *Memorials of Service in India from the correspondence of Major S.C.MacPherson*, London, 1865.

MADRAS STAFF OFFICER, *What is history and what is Fact?* or *Three Days at Cawnpore in November 1857 under the command of Major General C.A.Windham*, Madras, 1866. Written by or for Windham, certainly in his support. He had been heavily criticised for his defeat at the hands of the Gwalior Contingent.

MAISEY, Lt General F.C., *An Account by an eye-witness of the taking of the Delhi Palace*, in Royal United Services Institution Journal, 1930. Maisey, then a Captain, had been Deputy Assistant Judge Advocate General at the time of the siege of Delhi in 1857. He was one of the (many) critics of Captain Hodson of Hodson's Horse.

MAJENDIE, Lieutenant Vivian, *Up among the Pandies, or a year's service in India*, London, 1859.

MAJUMDAR, Ramesh Chandra, *The Sepoy Mutiny and the Revolt of 1857*, Calcutta, 1957. He argues against regarding the rebellion as a national war of independence, and puts the sepoys' mutiny down to fear of loss of caste and religion. He says there was no plot, and points out the lack of a great leader. His conclusion may be summarised in his own words: 'To regard the outbreak of 1857 simply as a mutiny of sepoys is probably as great an error as to look upon it as a national war of independence.' One of the most respected of the historians of this period, and essential reading.

MAJUMDAR, Ramesh Chandra, *Some 10 unpublished documents regarding the Mutiny of 1857*, in Bengal Past and Present, 1957.

MAJUMDAR, Ramesh Chandra, *Some unpublished Records regarding the Sepoy Mutiny*, in Indian Historical Records Commission Proceedings, 1958. Includes correspondence revealing intrigues between Bahadur Shah's entourage and the British during the siege of Delhi.

MAJUMDAR, R.C., *History of the Freedom Movement in India* Vol I, Calcutta 1962.

MALCOLM, Thomas, *Barracks and Battlefields in India, or the Experiences of a soldier of the 10th Foot (North Lincoln) in the Sikh Wars and the Sepoy Mutiny*, Caesar Caine, ed., Punjab, 1971.

MALET, J.P., *Lost links in the Indian Mutiny,* London, 1867.

MALGONKAR, Manohar, *The Devil's Wind: Nana Saheb's Story,* London, 1972. Reprint New Delhi,1988. Although technically a novel, and undoubtedly a work of fiction and imagination, this very readable book is based largely upon fact. Care must be taken however not to accept conjecture as evidence.

MALIK, Hafeez, and Dembo, Morris, *Sir Sayyid Ahmad Khan's History of the Bijnor Rebellion,* East Lansing, 1972. A valuable translation of a rare and important book.

MALLESON, Colonel George, *History of the Indian Mutiny,* 6 vols, London, 1878-80. Commencing from the close of the second volume of Sir John Kaye's *History of the Sepoy War,* Vol I is described as 'contemporaneous' with Kaye's third volume, and yet since the two are by no means identical in approach they complement rather than duplicate each other. Malleson and Kaye are two names almost impossible to disassociate from each other: massive, verbose, all-embracing histories: authoritative and dogmatic; entirely noble in sentiment, entirely British in attitude and viewpoint. *See also* Pincott, Frederic, for an analytical Index.

MALLESON, Colonel George, *The Indian Mutiny,* London, 1891. There is also a facsimile reprint of the 4th edition, (1892), of this book, London, 1993. An abbreviated version of the six volume work, of no particular merit.

MALLESON, Colonel George, *The Mutiny of the Bengal Army,* 2 parts, London, 1857-8.

MALLESON, Colonel George, *Recreations of an Indian Official,* London, 1872.

MANGIN, Arthur, *La Révolte au Bengale en 1857 et 1858, Souvenirs d'un officier Irlandais,* Tours, 1867. A most interesting statement in French, but written by an Irish officer who displays little love for the English who he regards as the conquerors and oppressors of Ireland (as well as India). He was present in Bengal in 1857-58. The death of Macdowell and Hodson perhaps 'had redeemed their evil deeds: God's mercy is infinite'. (Leurs morts á tous deux a peut-etre racheté leurs fautes: la clemence de Dieu est infinie!'

MARIAM (Marry), A Story of the Indian Mutiny of 1857, JF Fanthome, 1896.

MARRIOTT, Sir John, *The English in India, A Problem of Politics,* Oxford, 1932. An authoritative work; includes the Anglo/French duel, the HEIC, The Problem of the Frontier, Indian Mutiny etc.

MARRYAT, Florence, *Gup, Sketches of Anglo-Indian Life,* London, 1868.

MARSHMAN, J.C., *Memoir of Major General Sir Henry Havelock,* London, 1860. Marshman was the brother-in-law of Havelock. He was also the editor of that most influential English language newspaper in Calcutta, The 'Friend of India' (still in uninterrupted existence, but now known as the 'Statesman'). Some of the eulogy expended upon Havelock may be traced to the reports that appeared in the 'Friend' at the time, which were possibly based upon family loyalty rather than calm appraisal of the facts.

MARTIN, R.Montgomery, *The Indian Empire,* 3 vols, London, 1858-61. Volume II is concerned exclusively with the mutiny/rebellion, and is indispensable reading. Succinct and far more objective than its contemporary British rivals, it not only reads well, but has the advantage of an excellent index, and is well illustrated.

MARX, Karl, and ENGELS, Frederic, *The First Indian War of Independence,* Moscow, 1960. Contains reprints of the two political theorists' newspaper articles (in the American press); they were in general very sympathetic to the Indian struggle although the general tone of their articles does not completely vindicate the title given to this volume by the Soviet government in 1960: they are both particularly critical of the role played by the princes and other feudal leaders in the struggle.

MASANI, R.P., *The Aftermath of Revolt, India 1858-70,* Princeton, 1965.

MASIH-UD-DIN KHAN, *Oude: its Princes and Government Vindicated,* edited by S.Ahmad, Meerut, 1969.

MASON, Philip, *A matter of Honour,* an account of the Indian army, its officers and men, Jonathan Cape, London, 1974. The definitive historical account covering the years from 1746 to 1947 ie the HEIC to Independence.

MASON, Philip, *The Men who ruled India,* (2 vols, The Founders, The Guardians), London, 1953. An absorbing account of the men who served in the I.C.S. The mutiny/rebellion is covered with rare insight, both as to the causes and the results. Philip Mason also wrote under the name Philip Woodruff.

MATHUR, L.P., *Kala Pani, History of Andaman and Nicobar Islands, with a study of India's Freedom Struggle,* Delhi, 1985.

MAUDE, Colonel E., *Oriental Campaigns and European Furloughs: the autobiography of a veteran of the Indian Mutiny,* London 1900.

MAUDE, Colonel F.C., and SHERER, J.W., *Memories of the Mutiny,* 2 vols, London, 1894. Very well and sympathetically (to all concerned) written by two men who were very close to the action in Kanpur, Lucknow etc; the one was an artillery officer with Havelock and then Campbell, the other was a magistrate in Fatehpur and then Kanpur. Both write well.

MAUNSELL, F.R., *The Siege of Delhi,* London, 1912.

MAYNE, F.O., *Narrative of Events attending the Outbreak of Disturbances and the Restoration of Authority in the District of Banda,* 1858.

McRAE, Colonel H.StG.M., *Regimental History of the 45th Rattray's Sikhs. Vol !, 1856-1914,* Glasgow, 1933. Has detailed statements on the rebellion in Patna, Danapur, Arrah, Jugdishpur, Chota Nagpur and, above all, the campaign against Kunwar Singh.

MD, *Scenes from the late Indian Mutinies,* (in verse) London, 1858.

MEAD, Henry, *The Sepoy Revolt,* London, 1857. Much quoted by early commentators, but rarely read today.

MECHAM, C.H., and COUPER, G., *Sketches and Incidents of the Siege of Lucknow,* London, 1858. Copies of this are now valuable collector's items, but the drawings/engravings are very frequently reproduced in books on the rebellion/mutiny.

MEDLEY, Julius, *A Year's Campaigning in India,* London, 1858.

MEEK, Rev Robert, *The Martyr of Allahabad,* London, 1857.

MEHTA, Ashoka, *1857, The Great Rebellion,* Bombay, 1946.

METCALF, Thomas R, *The Aftermath of Revolt,* Princeton, 1965. A scholarly work. He sees the sepoy revolt as 'little more than the spark which touched off a smouldering mass of combustible material'. As a result of agrarian grievances arising from British over-assessment (of land-revenue), and the passage of landed property to the money-lender, the bulk of the people in the NW Provinces gave their support to the rebel cause.

METCALF, Thomas R, *Land Landlords & the British Raj,* Berkeley, 1979.

METCALFE, Charles, *Two native narratives of the Mutiny of Delhi,* London, 1898. The author (editor?) has given us a translation of two accounts, by Mainodin Hussain and Munshi Jiwan Lal, written from *inside* Delhi during the siege in May-September 1857, which go some way to redress the balance: practically everything else written and in print on the subject of the Siege of Delhi is written from the British point of view. There are some criticisms of the quality of his translation.

METCALFE, Private Henry, *Chronicle of the 32nd Foot,* London, 1953.

MILES, A.H. and POTTLE, A.J., *Fifty two stories of the Indian Mutiny,* London, 1895. Of little value to the serious student; written for children in a series of 'Fifty two stories of etc'.

MILLS, Arthur, *India in 1858,* London, 1858. A review of the Aftermath of the Mutiny.

MISHRA, J.P., *The Bundela Rebellion,* Delhi, 1982

MISRA, Anand Sarup, *Nana Sahib Peishwa and the fight for freedom,* Lucknow 1961. A large rambling volume containing much more than the story of Nana Sahib: it is a wordy defence of the view that the mutiny/rebellion was a planned National War of Independence. It contains a great deal of useful information; local lore from the Kanpur area, transcripts of proclamations, genealogies, and a great deal more; very useful.

MISRA, B.R., *Land Revenue Policy in the United Provinces under British Rule,* Benares, 1942.

MISTRY, H.D., *Rebels of Destiny,* Bombay, 1959.

MOLLO, Boris, *The Indian Army,* Poole, 1981. A statistical survey rather than a history.

MONCRIEFF, Ascot Robert Hope, *The Story of the Indian Mutiny,* Edinburgh, 1898.

MONSON, J.L., *My Garden in Lucknow,* 1905.

MONTGOMERY, F.J., *The Mutiny at Sialkot,* Sialkot? 1894.

MONTGOMERY, Robert, *Selections from the Public Correspondence of the Administration for the Affairs of the Punjab.* v.4 No.1. Punjab Mutiny Report, Lahore, 1859. Gives an account of the suppression of the rebellion in the Punjab. Includes official reports and letters. Very useful on the whole, but whenever the word 'selected' appears in a Government publication one wonders what has been left out.

MOORE, Kate, *Meerut during the Mutiny.* In Nineteenth Century magazine, November 1903. Personal recollections of an eighteen year old English girl.

MOORE, Rev T., *Guide to the Residency,* Lucknow, 1885.

MORRISON, J.L., *Lawrence of Lucknow,* London, 1934. Another account of the career of Sir Henry Lawrence.

MUBARAK ALI, SHEIK, *The Mutiny Records: Oudh & Lucknow 1856-57, Lahore, 1975.* A reprint of the 7th (1911) edition published in Lucknow under the title the 'Tourists' Guide to Lucknow'.

MUDFORD, Peter, *Birds of a different Plumage: a study of British Indian relations from Akbar to Curzon,* London, 1974. An attempt, says the author, to see whether Kipling's famous dictum 'East is East and West is West and never the twain shall meet', is really true. This is not a history of British India but rather a study of relationships.

MUIR, John R.B., Making of British India 1756-1858

Documents selected., Manchester, 1915.

MUIR, Sir William, *Records of the Intelligence Dept of the NWP of India during the Mutiny of 1857*, including correspondence with the Supreme Government, Delhi, Kanpur and other places, ed. by William Coldstream 2 vols, Edinburgh, 1902. Muir was in charge of intelligence at Agra during the mutiny/rebellion. Excellent for an appraisal of conditions during the outbreak, especially for detail of the civil unrest which accompanied the military mutiny.

MUKHERJEE, Hirendra Nath, *India Struggles for Freedom*, 1946.

MUKHERJEE, Haridas and Uma, *The Growth of Nationalism in India 1857-1905*,

MUKHERJEE, Rudrangshu, *Awadh in Revolt 1857-58; a study of Popular Resistance*, Delhi, 1984. The author sets out to explore the popular character of the uprising, and in doing so draws attention to hitherto unemphasised aspects: certain features of the immediately pre-annexation agrarian scene are shown to be causally linked to the specific characteristics of the revolt in Oude[Awadh] - ie the fact that the general populace , especially talukdars and peasants, fought together against a common foe. An excellent and scholarly work, likely to be long regarded as the definitive statement on the subject.

MUKHERJEE, Rudrangshu, *The Azamgarh Proclamation and Some Questions on the Revolt of 1857 in the NWP*, Essays in honour of Professor S.C.Sarkar, Delhi, 1976.

MUKHERJEE, Rudrangshu, *'Satan let loose upon Earth': the Kanpur Massacres in India in the Revolt of 1857*, in Past & Present, No.128, Oxford, August 1990, and No.142, Oxford, 1994. A refreshingly new, and overdue, look at the massacres in Cawnpore [Kanpur] in 1857, connected in particular with the name of Dhondu Pant, the Nana Sahib.The author stresses context rather than cause, and his argument is strong and convincing, although his conclusions are sometimes open to question by other academics.

MUKHERJEE, Rudrangshu, *Trade and Empire in Awadh 1765-1804*, in Past & Present, No.94. Oxford, Feb.1982.

MUKHERJEE, Sambhu Chandra, see HINDU, A.

MUKHOPADHYA, S.C., *The Mutinies and People*, Calcutta, 1905.

MUNRO, Surgeon William, *Reminiscences of Military Service with the 93rd Highlanders*, London, 1883.

MURRAY, et al., *Murray's Handbook for travellers in India, Burma and Ceylon*, London, 1859, Useful reference tool for background to the mutiny/rebellion.

MUSSEEHOOD-deen Khan Bahadoor, Maulvie Moh., *Oude: its Princes and its Government vindicated*, London, 1857. A book 'suppressed immediately after publication' (but by whom is unclear), and of considerable interest, especially for its summary of the condition of Oude[Awadh] in the months after annexation, -'a measure to which the natives themselves were so averse that a petition was signed against it by not less than 50,000 persons of all ranks'. The author describes himself as an 'hereditary native of Oude', and produced his book to support Wajid Ali Shah's appeal to London: he may well have been a member of the Begum's entourage.

MUTER, Mrs Dunbar, *My recollections of the Sepoy Revolt*, London, 1911. The wife of a serving officer, she describes events from the beginning of the mutiny in Meerut in May to the fall of Delhi in September 1857. Included is a description of life in the city and conditions obtaining there after it fell to the British.

MUTER, Mrs Dunbar, *Travels and Adventures of an Officer's wife in India China and N.Z*, 2 vols, London, 1864.

MUTINY of the BENGAL ARMY, an HISTORICAL NARRATIVE, by ONE WHO SERVED under Sir CHAS.. NAPIER, London, 1857.

NAHAL, J.S., *Sialkot: Reminiscences of 1857 Mutiny*, in Pakistan Review, September 1962.

NARRATIVE of EVENTS ATTENDING the OUTBREAK of DISTURBANCES and the RESTORATION of AUTHORITY in ALL the DISTRICTS of the NWP in 1857-8, 3 vols, Calcutta, 1881. Every district of the NWP produced an official narrative of events, prepared by the Commissioner or District Officer for each area. The resulting record is generally believed to be essentially accurate and to be the best source for detailed information. Bhargava, M.L., ed., with Rizvi, S.A.A., *Freedom Struggle in Uttar Pradesh*, 6 vols, Lucknow, 1957-61 makes extensive use of these narratives from which large extracts are taken.

NARRATIVE of an ESCAPE FROM GWALIOR , London, n.d.

NARRATIVE of the Indian Revolt from the Outbreak to the Capture of Lucknow, London, 1858. NB This is occasionally wrongly ascribed to Sir Colin Campbell, the Commander-in-Chief, because of the ambiguous wording of the title page. It was issued in 'penny numbers', and then made available, bound, in 1858. While much is sensationalist and produced with a view to capitalizing on the intense interest in England in 1857-8 on all things to do with the 'mutiny', it is still worthy to be consulted, if only because (a) it quotes from a large range of contemporary letters not available elsewhere, and (b) it has a large number of illustrations, and while the majority are of little value, there are a few quite unique sketches.

> INDIAN REVOLT
>
> FROM ITS OUTBREAK
>
> TO THE
>
> CAPTURE OF LUCKNOW BY SIR COLIN CAMPBELL.

NASH, J.T., *Volunteering in India*, London, 1893.

NEALE, Walter C., *Economic Change in Rural India's Land Tenure and Reform in Uttah Pradesh*, 1800-1955. New Haven, 1962.

NEVILL, H.R., *Cawnpore: a Gazetteer*, (Vol XIX of the District Gazetteers of the United Provinces of Agra and Oudh), Allahabad, 1929. Full of fascinating material relevant to the mutiny/rebellion including a listing of which landlords joined and which did not, and the effect of the struggle on the district.

NEWBOULT, A.W., *Padre Elliot of Fyzabad*, London, 1906.

NEWTON, John, *Hist. Sketches of the India Missions of the Presbyterian Church in the USA, 1834-1884*, Allahabad, 1886.

NICHOLL, Lt General T., *Saugor, a story of 1857*, Woolwich, 1894.

NIGAM, N.K., *Delhi in 1857*, Delhi, 1957. For a graphic description of the city *after* the assault see Chapter V, 'The City of the Dead', pp 148-171.

NIGAM, S., *Disciplining and Policing the Criminals by Birth: the making of a Colonial Stereotype - the Criminal Tribes and Castes of North India*, IESHR, vol xxvii, No 2, April-June 1990.

NOEL, Baptist Wriothesley, *England and India*, an essay on the Duty of Englishmen towards the Hindoos, London, 1859.

NORMAN, Sir Henry, *Narrative of the Campaign of the Delhi Army*, London, 1858. Norman was a young staff officer thrust into a highly responsible post by the sickness of his superiors; he observed much, knew more than most, and could write.

O'CALLAGHAN, Daniel, *Scattered Chapters: the fatal falter at Meerut*, Calcutta, 1861.

OLDENBURG, Veena Talwar, *The Making of Colonial Lucknow*, Delhi, 1984. Covering the years 1856-77, the author is particularly concerned, in examining the history of Lucknow, to show how the results of its transformation after the Mutiny of 1857 continue to pervade the city even today.

OMISSI, Dr David, *The Sepoy and the Raj*, The Indian Army 1860-1940, London, 1994. The sepoys conquered India for the British, then protected the Raj from its enemies without and within. His conclusion is that the sepoys fought for pay, but were much more than mere mercenaries. Their letters reveal a strong sense of honour, a loyalty to kin, caste and regiment, and, eventually, a deep devotion to the King-Emperor. A scholarly work by a lecturer in Imperial and Military history in the Department of History and Centre for Indian Studies at the University of Hull.

OUTRAM, Sir James, *The Campaign in India, 1857-8*, London, 1860. Contains General Outram's letters and despatches relating to the defence and relief of the Lucknow Garrison and the capture of the city. The defence of the Alam Bagh is detailed. Valuable as a source of military detail but otherwise throws little light on events.

OUDH, in Blackwood's Magazine, May 1858.

OUTRAM AT ALAMBAGH, in the Calcutta Review, March 1860.

OUVRY, Henry, *Cavalry Experiences and Leaves from my Journal*, Lymington, 1892.

OUVRY, Mrs Matilda, *A lady's diary before and during the Indian Mutiny*, Lymington, 1892.

OWEN, Arthur, *Recollections of a Veteran of the days of the great Indian Mutiny*, Simla, 1914. (Lucknow, 1916?)

OWEN, Rev. William, *Memorials of Christian Martyrs in the Indian Rebellion*, London, 1859.

PAGET, Mrs Leopold, *Camp and Cantonment*, London, 1865.

PAL, Dharam, *The Poorbeah Soldier; the Hero of India's War of Independence, 1857*, Delhi, 1957.

PAL, Dharam, *Tatya Tope*, Delhi, 1957.

PALMER, J.A.B., *The Mutiny outbreak at Meerut in 1857*, Cambridge, 1966. An exhaustive account based on the accessible sources containing eye-witness accounts or reminiscences or statements and reports or despatches, of the events of 10 May 1857 at Meerut and 11 May 1857 at Delhi. The authoritative and standard work?

PANIKKAR, Kavalam Madhava, *In 1857*, Ahmedabad, 1957.

PANIKKAR, K.M., *The Evolution of British Policy towards Indian States, 1774-1858,* Calcutta, 1929

PARKER, N.T., *Memories of the Indian mutiny in Meerut,* Meerut, 1914.

PARLIAMENTARY REPORTS, Mutiny in the East Indies.1857: Vols 29,30; 1857-8: 42,43,44; 1859: 11,23,25,27; 1860: Vol 50, 1. Mutinies in the East Indies 2.Papers relating to the East India Mutinies. 3.Further Papers relative to the Mutinies in the East Indies. An invaluable reference tool. Letters despatches reports, includes causes as well as events. Not easy however to find one's way through the great mass of material. For this it is possible to get help from Percy Ford, A Guide to Parliamentary Papers: what they are; how to find them: how to use them, Oxford, 1955.

PARRY, S.H.J., *An Old Soldier's Memories,* London, 1897.

PARSONS, Lieutenant Richard, *A Story of Jugdespore,* London, 1904. An absorbing account of an action in which, for once, the British were badly beaten.

PEARSE, Hugh, *The Hearseys,* Edinburgh, 1905.

PEARSON, J.D., *Guide to Manuscripts and Documents in the British Isles Relating to South and South-East Asia,* London, 1989. Volume 1 is of particular significance, giving a complete account of archive holdings in many libraries and collections. It is an essential reference tool for the serious investigator.

PEARSON, Hesketh, *The Hero of Delhi: John Nicholson and his Wars,* London, 1939.

PELLY, Captain Lewis, ed., *The Views and Opinions of Brig.Gen.John Jacob.* London, 1858.

PEMBLE, John, *The Raj, the Indian Mutiny, and the Princely State of Oudh, 1801-59,* Delhi, 1976.

PEILE, Mrs Fanny, *The Delhi Massacre: a Narrative by a Lady,* Calcutta, 1870.

PERKINS, Roger, *The Kashmir Gate,* Chippenham, 1983. The story of Lieutenant Home and the 'Delhi VCs'.

PHILLIPS, Alfred, *Anecdotes and Reminiscences of Service in Bengal,* Inverness, 1878.

PINCOTT, Frederic, *Analytical Index to Sir John Kaye's History of the Sepoy War, and Colonel G.B.Malleson's History of the Indian Mutiny,* London, 1880. This is an indispensable and an invaluable guide to the two major English works on the mutiny/rebellion. Indeed without Pincott much time is wasted in tracing events and references in these two major works.

PINKNEY, J.W., *Narrative of Events attending the Outbreak of Disturbances and the Restoration of Authority in the Division of Jhansi,* 1858.

PITT, F.W., *Incidents in India and Memories of the Mutiny,* London, 1896.

POLEHAMPTON, Rev Henry, *A Memoir, Letters and Diary,* London, 1858. Polehampton was the Chaplain in the Residency at Lucknow, but he did not survive the siege. His widow edited these letters etc.

POLLOCK, J.C., *Way to Glory: the Life of Havelock of Lucknow,* London, 1957.

PRESS-LIST of 'MUTINY PAPERS' 1857, being a COLLECTION of the CORRESPONDENCE of the MUTINEERS at DELHI, REPORTS of SPIES to ENGLISH OFFICIALS and other MISCELLANEOUS PAPERS, Calcutta, 1921. A most revealing collection, painting a picture of life within Delhi, during the siege.

PRITCHARD, Iltudus T., *The Mutinies in Rajpootana, being a personal narrative of the mutiny at Nusseerabad,* London, 1864. This includes the outbreak at Neemuch, the mutiny of the Jodhpore Legion at Erinpura, and the attack on Mount Abu. In an appendix is Dr Murray's narrative of his escape from Neemuch on the night of the mutiny, 3 June 1857.

PUNJAB GOVERNMENT, Mutiny Records, Correspondence and Reports, 4 Vols, 2 vols ed. by John Lawrence, two vols ed. by G.C.Barnes, Lahore, 1911.

QEYAMMUDDIN AHMED, *The Wahhabi Movement in India,* Calcutta 1966. Useful reading for background to the so-called Wahhabis of Patna, in Commissioner Tayler's time.

RADCLIFFE, F.W., *see* FITZGERALD, Lee J.

RAIKES, Charles, *Notes on the Northwest Provinces,* London, 1852. Useful background material.

RAIKES, Charles, *Notes on the Revolt in the NW Provinces,* London, 1858. Raikes was the Judge in Agra, and was in a good position to know exactly what was going on.

RAJ, Jagdish, *The Mutiny and British Land Policy in North India,* London, 1965.

RAMESAN, N., ed., *Freedom Struggle in Hyderabad,* (1857-85), 4 vols, Hyderabad, 1956. Vol II contains material dealing with unrest in the Contingent.

RAMSAY, Lt Colonel Balcarres, *Rough recollections of military service and society,* 2 vols, Edinburgh, 1882.

RAWLINS, J.S., *The autobiography of an old Soldier, 1843-79,* nd, privately printed.

READ, Rev. Hollis, *India and its People,* Columbus, 1858. Contains much 'atrocity' material.

RED PAMPHLET, London, 1858.

REES, L.E.R., *A personal narrative of the Siege of Lucknow*, London, 1858. The author was a merchant, and his non-official attitude to the events and decisions of the time is refreshing and illuminating, although the impression is given that he exaggerates his own and his friends' contributions to the outcome, *via* anecdote.

REEVES, P.D., *Sleeman in Oudh: an Abridgement of W.H.Sleeman's 'A Journey through the Kingdom of Oude in 1849-50'*, Cambridge, 1971.

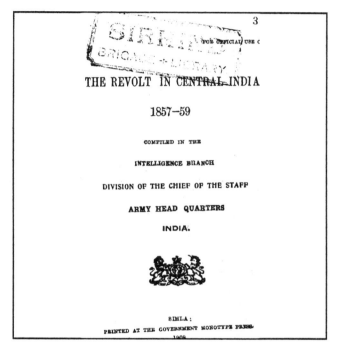

REVOLT in CENTRAL INDIA, 1857-59, Simla, 1908. Produced by the Intelligence Branch Army HQ India. A very detailed account of the military operations in Central India including Jhansi, Gwalior. Banda etc. With useful appendices including the names of the Europeans murdered at Jhansi, and the full statement of Tatya Tope, after his capture, 10 April 1859.

REYNOLDS, Reginald, *The White Sahibs in India*, London, 1937.

RICH, Captain Gregory, *The Mutiny in Sialkot*, Sialkot, 1924. Rich was the Cantonment Magistrate of Sialkot in 1924, and decided to produce this undistinguished little book to preserve knowledge of events that might otherwise have been forgotten.

RICKETTS, George H., *Extracts from the Diary of a Bengal Civilian 1857-59*, 1893.

RISLEY, H.H., *The Tribes and Castes of Bengal*, 2 vols, Calcutta, 1891.

RIVETT-CARNAC, Colonel H., *Many memories of life in India, at home and abroad*, Edinburgh, 1910.

RIZVI, S.A.A., ed., <u>with</u> BHARGAVA, M.L., *Freedom Struggle in Uttar Pradesh*, 6 vols, Lucknow, Source Material, 1957-61. The six volumes cover I 1857-59: Nature and Origin; II Awadh: 1857-59; III Bundelkhand and Adjoining Territories 1857-59; IV Eastern and Adjoining Districts 1857-59; V Western Districts and Rohilkhand 1857-59; VI Index to all volumes. While this is indisputably an essential reference tool for all students of the events of 1857-59 in India, it is not very easy to use. In their laudable desire to miss out nothing of significance the editors have had to sacrifice some order and method. While chronological sequence is observed in the main, there is much overlap and starting again. It is worth persevering however for the work is a veritable mine of information, and quite unique. It covers primary material from a wide variety of sources, and is not restricted by any means to the territorial limits of modern Uttar Pradesh: this is of course a bonus, not a cause for complaint, but it does make the title somewhat misleading. The editors have been meticulous in their reading of manuscript items, giving helpful interpretations/explanations wherever the text is suspect or illegible.

ROBERTS, Emma, *Sketches of Hindustan*, London, 1837.

ROBERTS, Field Marshal Lord, *Letters written during the Indian Mutiny*, London, 1924. Reprint New Delhi, 1979. Fred Roberts was a subaltern during the mutiny/rebellion, young, passionately adventurous and patriotic. His letters, thirty in number, addressed to his father, mother and sister, during his service at this time are most revealing. 'Old Bobs' as he later became, popular with both Indian and British troops, mellowed over the years, but in 1857 the rebellion filled him with the indignation which was typical of his age and class.

ROBERTS, Field Marshal Lord, *Forty-one Years in India*, 2 vols, London, 1897. This book ran to many editions and many reprints. It is, as might be expected, written from the point of view of a British Officer who rose to the highest rank, ie Commander-in-Chief. But it is noteworthy that Roberts was a *Sepoy General* that is he was from the old Indian Army of the East India Company, not one of the Queen's officers. The result, equally predictable, is that he shows love and respect for India and things Indian. This is a fascinating work, by an honest and straightforward soldier doing what he believed to be his duty, through a long and hard life.

ROBERTS, N.J.P., *Personal Adventures and Anecdotes of an old soldier*, London, 1906.

ROBERTSON, H.Dundas, *District Duties during the Revolt in the NW Provinces*, London, 1859. Robertson was joint magistrate at Saharanpur in May 1857, becoming officiating Collector in September, and administering the district with skill and courage, and very little outside help, from then onwards. He was then appointed (with J.Cracroft Wilson and T.D.Forsyth) on a commission for the investigation and trial of cases connected with 'the mutiny and rebellion'. This book reflects both of these activities. He could write and his views were both perceptive and dispassionate.

ROBERTSON, James P., *Personal Adventures*, London, 1906.

ROBINSON, Jane., *Angels of Albion,* Viking Penguin, London, 1996. A sensitive and refreshingly honest look at the 'memsahibs' during the 'mutiny', by this brilliant young writer.

ROHATGI, P., *Portraits in the India Office Library and Records*, London, 1982.

ROPER-LAWRENCE, Walter, *The India we served*, London, 1927.

ROSEBERRY III, J.Royal, *Imperial Rule in Punjab, 1818-1881*, Delhi, 1987. Contains an interesting note on Multan during the period 1857-58.

ROTTON, Rev John, The Chaplain's narrative of the Siege of Delhi., London, 1858.

ROUTLEDGE, James, *English Rule and Native Opinion in India*, London, 1878. Of interest in that it was compiled from interview notes painstakingly taken in 1870-74, and includes chapters specifically on the rebellion, particularly in Awadh.

ROWBOTHAM, William, ed., *The naval brigades in the Indian Mutiny*, London, 1947. Standard history of the three brigades, Shannon and Pearl in Bengal and the little-known Pelorus on the Irrawaddy River.

ROY, Tapti, *Sepoy Mutiny and the Uprising of 1857 in Bundelkhand*, Calcutta, 1991. An excellent factual statement by a gifted historian. Required reading for those particularly concerned with Bundelkhand and Jhansi.

ROY, Tapti, *The Politics of a Popular Uprising: Bundelkhand in 1857*, Delhi, 1994. An outstanding study of the outbreak in Bundelkhand, arranged under the headings of Sepoys; Rajas; Thakurs; The People, together with a helpful bibliography and glossary.

RUGGLES, John, *Recollections of a Lucknow Veteran*, London, 1906.

RUSSELL, Ralph, transl. with Kurshidul Islam, *Life and Letters of Ghalib*, Vol I, London, 1952, new ed. 1969. Reprinted as Oxford Paperback, 1994. Extremely valuable primary source material on the 'Siege' of Delhi. The poet Ghalib was present throughout, within the walls of the city, and he recalls not only his own experiences but the events of which news was brought to him.

RUSSELL, William Howard, *My Diary in India, 1858-9*, 2 vols London, 1860. Russell was the 'Times' correspondent who had just returned to England with enormous prestige from covering the Crimean War as one of the world's first war correspondents. His diary is concerned with the events of 1858 not 1857, ie the military campaign in suppression of the revolt rather than the mutiny itself. His liberal attitude leads him to find many reasons for the Indian rejection of British rule. A very shrewd observer: this work is invaluable to the student of the rebellion.

RUSSELL, William Howard, . *My Indian Mutiny Diary.* , abridged version. Michael Edwardes, . ed. ., London, 1957

SAMADDAR, J.N., *Two forgotten Mutiny Heroes*, in Indian Historical Records Commission Proceedings,Vol x, 1927.

SARAN, N., *Some Mutiny Telegrams in the Bihar State Archives*, in Indian Historical Records Commission Proceedings, 1960.

SARKAR, J.N., *Fall of the Mughal Empire ,4 vols*, 1938-50, reprinted Delhi, 1971.

SAVARKAR, Vinayak, *The Indian War of Independence of 1857*, Calcutta, 1930 (and Bombay 1947). Banned originally by the British government as seditious. Patriotism emphasised perhaps at the expense of scholarship.

SCHOLBERG, Henry, *The Indian Literature of the Great Rebellion*, New Delhi, 1993. A valuable source book, divided into three parts. Part 1: Published Literature in Assamese Bengali Gujarati Hindi Kannada Malyalam Marathi Oriya Persian Punjabi Sindhi Tamil Telugu and Urdu. Part 2: The Primary Sources, National Archives of India, New Delhi, Urdu Journals, Residency Vernacular Records, UP Regional Archives Bhopal State Archives, Delhi State Archives, the Western Archives, the Central Archives. Part 3: Folk Songs of the Great Rebellion, The Avadhi Songs, Songs of the Ranchi Mutiny, Bahadur Shah the Poet, the William Crooke Collection, the P.C.Joshi Collection, the Joyce Lebra Collection.

SEATON, Major General Sir Thomas, *From Cadet to Colonel*, 2 vols, London, 1866.

SEDGWICK, Francis R, *The Indian Mutiny of 1857, A Sketch of the Principal Military Events*, Foster Groom, London, 1909. An undistinguished little book in many ways, that fails to live up to its pretensious title, but it is valuable as a ready source of statistics relative to the number and designation of the troops involved, eg Appendix I gives the distribution of troops, by military station, British and Indian, at the outbreak of the mutiny/rebellion.

SELECTIONS from the RECORDS of the GOVT. of INDIA MILITARY DEPT., Calcutta, 1861.

SEN, Ashoka Kumar, *The Popular Uprising and the Intelligentsia: Bengal between 1855-73*, Calcutta, 1992.

SEN, Surendra Nath, *Eighteen Fifty-Seven*, Delhi, 1957. Dr Sen was commissioned by the Government of India to write the 'official' history of the mutiny/rebellion to coincide with its centennial. Although there must have been strong pressure to present the events in a strongly nationalist light, Dr Sen was judicious and free from rancour and produced a most scholarly and very readable narrative. He showed how conflicting much of the evidence is, and expresses the view that the struggle cannot be characterised as a great national uprising. A useful bibliography is included.

SEN, S.N., *A New Account of the Siege of Delhi*, in Bengal Past and Present, 1957.

SEN, S.N., *The Military System of the Marathas*, Calcutta, 1928 reprint 1958.

SEN GUPTA, S., ed., *War of Independence: centenary souvenir*, Calcutta, 1957.

SENGUPTA, K.K., *Recent Writings on the Revolt of 1857: a Survey*, New Delhi, 1975.

SETON, Rosemary, *The Indian 'Mutiny' 1857-58*, a guide to source material in the India Office Library and Records, London, The British Library, 1986. An invaluable guide to all relevant material in the Oriental and India Office Collectiions of the British Library, currently at 197, Blackfriars Road, London SE1 8NG, scheduled eventually to be moved to the new British Library building at St.Pancras, London. She has brought together and described the vast collection of archival source materials available in the India Office Library and Records as it was then called. Whether or not the student is able to visit the collection in person, this guide is an essential reference tool for all interested in the subject of the mutiny/rebellion of 1857 in India.

SHACKLETON, Robert, *A Soldier of Delhi*, in Harpers magazine October 1909. This is the story told by a private soldier, James Irvine, who survived the siege and the assault; it has the merit of showing events from the viewpoint of a humble private soldier.

SHADWELL, Lt General L., *Life of Colin Campbell, Lord Clyde,* 2 vols, Edinburgh, 1881.

SHARAR, Abdul Harim, *Last Phase of an Oriental Culture*, London, 1975.

SHARMA, Benudhar, *The Rebellion of 1857 vis-à-vis Assam*, Calcutta, 1958.

SHARPE, Jenny, *Allegories of Empire: the figure Woman in the Colonial Text*, Minnesota, 1993. An examination of the question of rape during and after the mutiny.

SHASTITKO, Pyotr Mikhailovitch, *Nana Sahib: an Account of the People's Revolt in India 1857-58*, Pune, 1980. An intriguing view of Nana Sahib and the mutiny/rebellion from a Russian (Marxist) standpoint. Translated by Savitri Shahani.

SHEPHERD, William Jonah, *A Personal Narrative of the Outbreak and Massacre at Cawnpore*, Lucknow, 1886. Reprint, New Delhi, 1980. In addition to his personal experiences the book contains a complete list of all the English (and Anglo/Indian?) residents of Cawnpore and their ultimate fate so far as he knew it. He writes accurately and fastidiously, and can be relied upon to have got his facts right before committing himself to paper.

SHEPHERD, William Jonah, *Guilty Men of 1857*, Reprint of 1879 edition, Delhi 1987.

SHERER, John W., *Daily Life during the Indian Mutiny*, London, 1898.

SHERER, John W., *At home and in India*, London, 1883.

SHERER, John W., *Havelock's march on Cawnpore*, A civilian's notes. Edinburgh c. 1910. Mutiny outbreak at Fatehpur where Sherer was the Collector and Magistrate; re-occupation of Kanpur with Havelock's force and subsequent events. He writes extremely well.

SHERRING, M.A., *The Indian Church during the Great Rebellion*, an authentic narrative of the Disasters that Befell it; its sufferings; and Faithfulness until death of many of its European and Native members, London, 1859.

SHOWERS, Lt General Chas.Lionel, *A Missing Chapter of the Indian Mutiny*, London, 1888. His post as Political Resident in the Meywar States of Rajputana during the period in question gives considerable interest and significance to the book, although the author's combative and far from lovable personality is strongly in evidence.

SIEVEKING, L.G., *A Turning Point in the Indian Mutiny*, London, 1910. Author purports to show that the siege of Arrah House in Bihar in August 1857 was the watershed of the mutiny/rebellion. Full of praise for English heroes, little detail of the rebel forces. Contains letters by John Nicholson

SINGH, Ganda, *The Indian Mutiny of 1857 and the Sikhs*, Delhi, 1969.

SINGH, Madan Paul, *Indian Army under the East India Company*, Delhi, 1976. Written by a serving officer of the Indian Army, well-researched and detailed examination of the Indian Army under the Company in the years 1746 to 1858. An excellent reference book.

SINGH, Sheo Bahadur, ed., *Letters of Sir Henry Lawrence*, New Delhi, 1978.

SINHA, Shyam Narain, *Rani Lakshmibai of Jhansi,* Allahabad, 1980.

SINHA, Shyam Narain, *The Revolt of 1857 in Bundelkhand,* Lucknow, 1982. One of the first attempts to study the insurrection specifically in this territory 'so rich in the activities and exploits of leaders and the people'. The author has succeeded in producing a comprehensive account of events in Bundelkhand 1857-9, including the story of the Rani of Jhansi.

SITA RAM, trans. Lt.Col J.T.Norgate, *From Sepoy to Subedar,* 1873. Latest reprint ed. James Lunt, London, 1970.

SLEEMAN, W.H., *Rambles and Recollections of an Indian Official,* London, 1843, reprinted Karachi, 1973.

SLEEMAN, W.H., *A Journey through the Kingdom of Oude in 1849-50,* 2 vols, London, 1858.

SMITH, R.Bosworth, *Life of Lord Lawrence,* 2 vols. London, 1883. This is the classic, partisan, biography, of John Lawrence, brother of Henry and George (who also served in India in distinguished capacities). It is in all senses a 'Life', telling the story of the man from childhood to old age, and including considerable detail of his activity during the mutiny/rebellion as Chief Commissioner of the Punjab, and later as Viceroy of India. It is partisan in the sense that there is little adverse criticism, either direct, or reported, of a man who, in Indian eyes, might well be considered controversial.

SMITH, V.A., *Popular Songs of the Hamirpur District in Bundelkhand, North-Western Provinces,* JASB, vol 44, 1875.

SMITH, V.A., *History of Bundelkhand,* IA, March 1908.

SMYTH, Sir John, *The Rebellious Rani,* London, 1966. This account of the career of Lakshmibai, Rani of Jhansi produces little new evidence to shed further light on the Rani's activities. Much remains conjecture, but at least the range of conjecture is here defined.

SONGHAL, D.P., *A History of the Indian People,* London, 1983.

SOURCE MATERIAL for a HISTORY of the FREEDOM MOVEMENT in INDIA, VOL I, Bombay, 1957.

SPEAR, Perceval, *The Twilight of the MUGHULS,* Cambridge, 1951.

SPROT, Lt General John, *Incidents and Anecdotes,* 2 vols, Edinburgh, 1906.

SRIVASTAVA, Khushhalial, *The Revolt of 1857 in Central India-Malwa,* Bombay, 1966. Contains an excellent and most helpful 'Mutiny Papers' section and Bibliography.

STANFORD, J.K. *Ladies in the sun: memsahibs in India 1790-1860,* London, 1962.

STARK, Herbert Alick, *Hostages to India, or the Life story of the Anglo/Indian race,* Calcutta, 1936.

STEEL, Flora Annie, *India Through the Ages,* London, 1908. Better known perhaps for her novels on the subject of India, including the Indian Rebellion, Mrs Steel also wrote this popular 'history'. It has a short section on the mutiny/rebellion, remarkable only for a number of interesting quotations which are found nowhere else.

STEUART, Mary, ed., *The Reminiscences of Lt.Col Thomas Ruddiman Steuart,* privately printed, 1900. Taken from his diaries and including service during the Indian Mutiny, with the Bombay Native Infantry.

STEWART, Charles Edward, *Through Persia in disguise with Reminiscences of the Indian Mutiny,* London, 1911. The author was in Peshawar on the outbreak of the struggle and gives detail of the military movements and mobilization in the Punjab which made possible the Siege of Delhi by the British. He also took part in the campaigns against Lucknow and in Oude[Awadh].

STEWART, Lt Colonel Charles, *The Victoria Cross,* London, 1916.

STOKES, Eric, *Agrarian Relations, Northern and Central India,* pp 36-85 in D.Kumar & M.Desai, eds, Cambridge Economic History of India, Vol 2, Cambridge, 1983.

STOKES, Eric, *The Peasant and the Raj: Studies in Agrarian Society and Peasant Rebellion in Colonial India,* Cambridge, 1978.

STOKES, Eric, *The Peasant Armed,* the Indian Revolt of 1857, Oxford, 1986. It was the loss of land rights, so the argument runs, that supplied the force behind the rural explosion of 1857. Unfortunately this work was not finished in the author's lifetime, although C.A.Bayly has produced this edited version which goes far to achieve what Prof.Stokes had intended. Bayly tells us 'Inevitably the Peasant Armed is an unbalanced book. The connection between the military mutiny so interestingly treated in the first two chapters, and the detailed social analysis which exists thereafter is implicit rather than explicit. This makes the book difficult for the non-specialist to follow. Clearly this is intended as a detailed treatment of the social origins of the revolt, but the lack of Eric Stokes's own conclusion is nevertheless a severe problem.'

STORY of the NORTH-WEST FRONTIER PROVINCE, Peshawar, 1929.

STRANG, Herbert, ed., *STORIES OF THE INDIAN MUTINY,* London, 1931. A book containing nine stories of the mutiny/rebellion: the stories have in common a high charge of inflammatory emotion; the mutineers are condemned as ignorant, cruel and despicable savages with no excuse for their conduct. Perversely because it is so biassed this book is useful - it is historical propaganda *par excellence,* reminiscent of Foxe's Book of Martyrs in the degree of its prejudice.

STUBBS, F.W., *Regiment of Bengal Artillery,* Vol I, London, 1877.

SUNDERLAL, P., *British Rule in India,* Bombay, 1972. Includes much on the mutiny/rebellion.

SURTEES, Virginia, *Charlotte Canning,* London, 1972.

SUTTON, S.C., *A Guide to the India Office Library,* with a note on the India Office Records, London, 1967. A brief but useful guide, for the beginner, to the India Office Library, issued by the Foreign and Commonwealth Office which was then responsible for it. S.C.Sutton was the Librarian and Keeper of the Records. The archives now need 13 kilometres of shelf space.

SWANSTON, W.O., *My Journal by a Volunteer,* Calcutta, 1858.

SWINEY G.C., *Historical Records of the 32nd (Duke of Cornwall's) Light Infantry,* London, 1893. The 32nd was the regiment that provided the British part of the garrison at both Kanpur and Lucknow.

SYED AHMAD KHAN, Sir, *An Essay on the Causes of the Indian Revolt,* trans. by Captain W.N.Lees, Calcutta, 1860. *See also* translation by Hafeez Malik and Morris Dembo, Sir Sayyid Ahmad Khan's History of the Bijnor Rebellion, East Lansing, 1972.

SYLVESTER, J.H., *Recollections of the campaign in Malwa and Central India,* Bombay, 1860.

TAHMANKAR, D.V., *The Ranee of Jhansi,* London, 1958. The author believes that with the death of the Ranee the resistance of the rebels became symbolic rather than real, and that her disappearance from the scene is rightly regarded as a significant landmark in Indian history, and the beginning of a new phase in Indo-British relations. He attempts, with some success, to show that the Ranee was innocent of the grave crime she was charged with, ie the massacre at Jhansi of English men women and children.

TAIMURI, M.H.R., *Some unpublished Documents on the death of the Rani of Jhansi and the Mutiny in Central India,* in Indian Historical Records Commission Proceedings, 1953. Nothing, unfortunately, very new.

TARACHAND, *History of the Freedom Movement in India, 4 vols,* Delhi, 1961-72.

TARIKH-E-BADSHAH BEGUM, Trans. By Muh Taqi Ahmad, Allahabad, 1938.

TAVENDER, I.T., *Casualty Roll for the Indian Mutiny 1857-59,* London, 1983.

TAYLER, William, *Thirty-eight Years in India,* 2 vols, London 1881/2.

TAYLER, William, *The Patna Crisis*: or *Three months at Patna, during the insurrection of 1857,* London, 1882. Yet another broadside from the one-time Commissioner of Patna against his old enemy, Mr Halliday the Lieutenant Governor, who removed him from office. It certainly seems to indicate that Tayler was very badly treated, which was the impression of many others, including Colonel Malleson.

TAYLOR, Alicia Cameron, *Life of General Sir Alex.Taylor,* London, 1913.

TAYLOR, Bayard, *A Visit to India China and Japan in the year 1858,* Edinburgh, 1859.

TAYLOR, Meadows, *Letters during the Indian Rebellion,* London, 1857. Better known for his novels, the author nevertheless provided succinct contemporary comment via his letters.

TAYLOR, P.J.O., *Chronicles of the Mutiny, and other Historical Sketches,* HarperCollins, Delhi, 1992. Reprints of articles originally published in the 'Statesman' of New Delhi. Written by a former officer in the Mahratta Light Infantry (1944-47) and a lifelong Indophile.

TAYLOR, P.J.O., *A Star Shall Fall ; India 1857,* HarperCollins, Delhi, 1993. This book attempts to look again at the facts of the great rebellion of 1857, but objectively. To give equal weight to the role, the performance, the credibility of all the actors upon that stage is not easy, since most contemporary source material is British, but it can be done, and has been done, in this book - or so the author claims. Although written in a popular history style it is not a work of fiction; all the tales are based on fact: if the truth cannot be established for want of evidence then this is stated.

TAYLOR, P.J.O., *A Sahib Remembers,* HarperCollins, Delhi, 1994. Further reprints of articles originally published in the New Delhi 'Statesman'.

TAYLOR, P.J.O., *A Feeling of Quiet Power; the Siege of Lucknow 1857,* HarperCollins, Delhi, 1994. A new look at the historic siege of the Lucknow Residency.

TEMPLE, Sir Richard, *Lord Lawrence,* London, 1889.

TEMPLE, Sir Richard, *Men and events of my time in India,* London, 1882.

TEMPLE, Richard, *The story of my life.,* London 1896. Includes description of Delhi after its capture.

TERNAN, A.H., *Narrative of Events attending the Outbreak of Disturbances and the Restoration of Authority in the District of Jalaun,* 1858.

TERRILL, Richard, ed., *John Chalmers. Letters from the Indian Mutiny 1857-59,* Norwich, 1992. Of interest in that

Chalmers served in the Punjab in a newly raised corps of Sikh Pioneers; useful for background detail of events in the Punjab, and for the assault on Delhi, also Lucknow.

THACKERAY, C.B., *A Dark and Fateful Sunday,* in the Journal of the Royal Artillery 1934.

THACKERAY C.B., *Four Men on a Ridge,* in the Army Quarterly, October 1933.

THACKERAY, Colonel Sir Edward, *Two Indian Campaigns 1857-58,* Chatham, 1896.

THACKERAY, Colonel Sir Edward Talbot, *Recollections of the Siege of Delhi,* in Cornhill Magazine Sept. 1913.

THACKERAY, Colonel Sir Edward Talbot, *A Subaltern in the Indian Mutiny,* in Royal Engineers Journal, 1930.

THACKERAY, Colonel Sir Edward, *Reminiscences of the Indian Mutiny,* London, 1916.

THACKERAY, Colonel Sir Edward, *Biographical Notices of Officers of the Royal (Bengal) Engineers,* London, 1900.

THICKNESSE, H.J.A., *The Indian Mutiny at Mhow and Indore,* in Journal of the Royal Artillery, 1934.

THOMPSON, Edward J., *The Other side of the Medal,* London, 1925. He deals with English atrocities against Indians during the rebellion, and claims that native memories of the Mutiny are a heavy obstacle to friendship and understanding. Perhaps goes too far in his denunciation to be considered truly objective, but his book is valuable, and, almost, unique in its approach.

THOMPSON, Edward, *Suttee,* London, 1928.

THOMPSON, Edward, and GARRATT, G.T., *Rise and fulfilment of British rule in India,* London, 1934.

THOMSON, Captain Mowbray, *The Story of Cawnpore,* London, 1859. The author was one of the four survivors of the Satichaura Ghat massacre in June 1857. His account of the events at Kanpur in that year, and in particular his own escape are written in a straightforward and soldierly manner without too much embellishment. He admits to having had a 'friend' assist him with the narrative: the ghost writer has managed to reflect Thomson's personality. It is possible that this friend was J.W.Sherer, *q.v.*

THORNHILL, Mark, *Personal Adventures of a Magistrate during the Indian Mutiny,* London, 1884. This member of the Thornhill family was the Magistrate at Muttra, 34 miles from Agra. Generally liberal attitude. A particularly useful attempt to explain rural disturbance in terms of the Indian peasants' attitude to land and land-holding, especially their resentment of the sale of land to pay debts to the state (native governments had collected revenues by siezing standing crops).

THORNTON, E., *A Gazetteer of the Territories under the Government of the East India Company, and the Native States on the Continent of India,* London, 1857.

THOUGHTS of a NATIVE of NORTHERN INDIA on the REBELLION, London, 1858.

THURBURN, E.A., *Reminiscences of the Indian Rebellion by a Staff Officer,* London, 1889.

TISDALL, E.E.P., *Mrs Duberley's Campaigns,* London, 1963. It was Mrs Duberley who refused to be parted from her husband, and went through both the Crimean War and the Indian 'Mutiny' at his side. He was an officer in the 8th Hussars, a trooper of which regiment is said to have slain the Rani of Jhansi.

TOD, James, *Annals and Antiquities of Rajasthan,* 2 vols, 1914, reprinted Delhi, 1978.

TOOMEY, T.E., *Heroes of the Victoria Cross,* London, 1895.

TRACEY, Louis, *Red Year,* London, 1908.

TREVELYAN, Sir George Otto, *Cawnpore,* London, 1886. Reprint, Delhi, 1992. A very readable version, or rather versions, of the events in Kanpur, based on the Depositions of 63 witnesses taken by Colonel Williams Police Commissioner of NWP, plus a narrative by Nanak Chand, plus Mowbray Thomson's account, plus the official narratives. Has been criticised for the resultant 'unscholarly account', but he is not himself to blame for the contradictory nature of the evidence he adduces.

TREVELYAN, Sir George, *The Competitition Wallah,* London, 1864. Reprint, Delhi, 1992. Ostensibly letters written by 'Mr Henry Broughton', the successful candidate at the Indian Civil Service Examination, to his friend 'Charles Simkins' who had failed the said competition, in furtherance of a pact that they had made that if only one passed he would write regularly to the other about his Indian experiences. Letter IV is 'A Story of the Great Mutiny', and concerns the siege of Arrah House: factually correct it is very well written, by a master of his craft.

TREVELYAN, Raleigh, *The Golden Oriole,* London, 1988. The author belongs to a family that had served with distinction in India for many generations, and included Sir Charles Trevelyan the Governor of Madras, and Lord Macauley. This book is very sensitively written and is far more than personal reminiscence.

TROTTER, Captain Lionel, *Life of John Nicholson,* London, 1898. The standard biography of Brigadier General John Nicholson, the 'Hero of Delhi', the man credited more than anyone else with the successful British assault on the city in September 1857. The book is however so dedicated to praising the idol that it is difficult to take an objective look at Nicholson the real man.

TROTTER, Captain Lionel, *Life of Hodson of Hodson's Horse*, London, 1910.

TROTTER, Captain Lionel, *The Bayard of India*, London, 1910.

TULLOCH, J., *Volunteering in India*, London, 1893.

TURNBULL, Lt Colonel John, *Letters written during the Siege of Delhi*, Torquay, 1886.

TURNBULL, Lt Colonel John, *Sketches of Delhi*, London, 1850.

TWEEDIE, Major General W., *A Memory and a Study of the Indian Mutiny*, in Blackwoods Magazine, August 1904. This account is useful as providing first hand evidence of the outbreak at Benares.

TYRRELL, Isaac, *From England to the Antipodes and India*, Madras, 1902.

TYTLER, Harriet C., *Through the Sepoy Mutiny and the Siege of Delhi*, in Chambers Journal 1931. This is Harriet Tytler's original journal. She was the only British woman present throughout the siege, and her detailed account, plus her own experiences are therefore unique.

TUKER, Lt General Sir Francis, ed., *The Chronicle of Private Henry Metcalfe, HM 32nd Regt of Foot*, London,1953. Metcalfe's memories are uninformed and inaccurate, but there is much here of interest nevertheless. The 32nd served especially at Lucknow in the garrison of the Residency.

VALBEZEN, E. de, *The English and India*, London, 1883. The interest of this particular work is that it was written by one who might be able to bring an entirely new viewpoint to the subject: the author was the French Consul-General in Calcutta. He describes the dissension among government officials in Calcutta, a matter not discussed in depth in any other publication.

VAUGHAN, Sir J.L., *My service in the Indian Army and after*, London, 1904.

VERNEY, Lieutenant Edmund, RN, *The Shannon's Brigade in India*, London, 1862. Although a naval officer he served in the force that attacked Lucknow in March 1858.

VERNEY, Major General G.L., *The Devil's Wind: the story of the Naval Brigade at Lucknow*, London, 1956. A book dealing with the contribution made by the naval brigade from HMS Shannon, at Lucknow in 1858, based upon the letters of his great-uncle Midshipman Edmund Verney who served with Captain Sir William Peel VC . The book has the merit of confining itself strictly to its chosen field - the military campaign.

VIBART, Colonel Edward, *The Sepoy Mutiny*, London, 1898. As seen by a subaltern from Delhi to Lucknow. With the 54th BNI at Delhi, he gives an extensive account of the Mutiny outbreak and the siege. Also Rohilkhand and Oude[Awadh].

VIBART, Colonel Henry, *Addiscombe its heroes and men of note*, London, 1894. Addiscombe was the College, in Hertfordshire, set up by the HEIC for the training of its army officers.

VIBART, Colonel Henry, *Richard Baird-Smith, Leader of the Delhi Heroes*, London, 1897.

VOLUNTEER, A, [pseud], *My Journal or What I saw and What I Did between 9th June and 25th November 1857*, Cal-cutta, 1858.

WAGENTREIBER, Florence, *The Story of our escape from Delhi*, Delhi, 1894. The matriarch of the Wagentreiber family was a remarkable woman and this account demonstrates both the courage and resolve of this Eurasian family.

WAINWRIGHT, M.D. and MATHEWS, Noel, *A Guide to Western Manuscripts and Documents in the British Isles relating to South and Southeast Asia*, London, 1965. A comprehensive listing of manuscript materials. But see Pearson, J.D., whose work to some extent supersedes this.

WAJID ALI SHAH, Nawab, *Reply to the charges against the King of Oude*, Calcutta, nd.

WALKER, T.N., *Through the Mutiny*, London, 1907. Run of the mill soldier's memoirs.

WALLACE, C.L., *Fatehgarh Camp, 1777-1857*, Lucknow, 1934.

WALLACE-DUNLOP, M.A. & R., *The timely retreat*, 2 vols, London, 1858. Two British sisters depart from N.India before the mutiny begins.

WARD, Andrew, *Our Bones are Scattered*, the Cawnpore Massacres in the Indian Mutiny of 1857, New York, 1996. The author is an acknowledged expert on the affairs of Kanpur in 1857-58 which he has meticulously researched over a long period of time.

WATERFIELD, Arthur, *Children of the Mutiny*, Worthing, 1935. A record of those still living (1935) who were in India during the Sepoy War 1857-59.

WATERFIELD, Robert, *The Memoirs of Private Waterfield, Soldier in HM's 32nd Regt of Foot, 1842-57*, London,1968. While any contemporary account is interesting, few add subatantially to what is already known.

WATSON, Bruce, *The Great Indian Mutiny, Colin Campbell and the Campaign at Lucknow*, New York, 1991. An eulogy of Campbell, 'the last British C-inC to lead in the field'.

WATSON, Edmund S., *Journal with HMS Shannon's Naval Brigade*, Kettering, 1858.

WHEELER, Harold F.B., *The Story of Lord Roberts*, London, 1922.

WHITE, S.Dewe, *Indian Reminiscences*, London, 1880.

WHITE, Samuel Dewe, *My First Battle in the Indian Mutiny*, in the Westminster Review, 1896. Describes the Battle of Shahganj(Sasia) outside Agra, and is one of the few detailed accounts of that battle as a result of which Brigadier Polwhele was suspended and the British forced to take refuge within Agra Fort.etc.

WHITE, Samuel Dewe, *Reminiscences of the great Sepoy Revolt*, in Westminister Review, October 1898. Gives an account of the experiences of an officer in Agra, not afraid to be critical of his superiors.

WILBERFORCE, Reginald, *An unrecorded chapter of the Indian Mutiny*, London, 1894.

WILDER, R.G., *Mission Schools in India of the American Board of Commissioners for Foreign Missions*, New York, 1861.

WILKINSON, JOHNSON & OSBORN, *The Memoirs of the Gemini Generals*, London 1896. Contains mutiny anecdotes and reminiscences.

WILKINSON-LATHAM, Christopher, *The India Mutiny*, London, 1977. With colour plates by G.A.Embleton.

WILLIAMS, Rev Edward, *Cruise of the Pearl round the World*, London, 1859. The British ship, HMS Pearl, provided a naval brigade, *ie* its sailors served on land during the rebellion.

WILLIAMS, George Walter, *Depositions taken at Cawnpore under the direction of Lt.Col G.W.W.*, Allahabad, 1859. Often quoted primary source material of great value : see TREVELYAN, Sir George Otto, *Cawnpore*.

WILLIAMS, George Walter, *Depositions taken at Meerut under the direction of LtCol G.W.W..*, Allahabad, 1858. Valuable source material, but it leaves the reader with the strange feeling that the list of witnesses is far from complete, and that other depositions should be there, to complete the picture.

WILLIAMSON, Dr George, *Notes on the wounded from the Mutiny in India*, London, 1859.

WILSON, Major Thomas F., *The Defence of Lucknow*, London, 1858.

WILSON, W.C. ed., *The Soldier's Cry from India*, London, 1858.

WINTRINGHAM, Thomas, *Mutiny*, London, 1936.

WISE, James, *The Diary of a medical officer during the Great Indian Mutiny of 1857*, Cork, 1894.

WOLSELEY, Field Marshal Viscount, *The story of a Soldier's Life*, London, 1903. Wolseley was a young officer in India during the rebellion: his attitude and outlook are those of a single-minded professional soldier, with room neither for sentiment nor understanding of the other side's point of view.

WOOD, Sir Evelyn, *The Revolt in Hindustan*, London, 1908. A good workmanlike account by a man who was a Lieutenant at the time but rose to the rank of Field Marshal. It is writen from the soldier's point of view and is concerned mainly with military operations, and avoids jingoistic distortion or overpraise for British achievements.

WOODRUFF, Philip, *The Men who ruled India, (2 vols, The Founders, The Guardians)*, London, 1953. A definitive account of the men who served in the I.C.S. The mutiny/rebellion is covered with rare insight, both as to the causes and the results. Philip Woodruff also wrote under the name Philip Mason.

WORSWICK, Clark, & EMBREE, Ainslie, *The Last Empire: Photography in British India, 1855-1911*, New York, 1976.

WYLIE, Macleod, *The English Captives in Oudh*, London, 1858.

WYLLY, H.C., *Neill's Blue caps*, Aldershot, 1925. The story of the Madras Fusiliers.

YADAV, K.C., *The Revolt of 1857 in Haryana*, Delhi, 1977.

YALLAND, Zoë, *Traders and Nabobs, the British in Cawnpore 1765-1857*, Wilton, Salisbury, 1987. The author takes the story of Cawnpore (Kanpur) up to the mutiny: she was a first-rate historian, a master of her craft. The book contains much original material never before published, and is likely to be the definitive history of early Kanpur.

YALLAND, Zoë, *Box Wallahs*, Norwich, 1994. The second volume of Mrs Yalland's excellent history of the city of Kanpur (Cawnpore) in which she was born and where she headed a distinguished school; this continues the story after the mutiny/rebellion.

YALLAND, Zoë, *Kacheri Cemetery, Kanpur*, London, 1985. Much more than just a list of graves: considerable historical detail is included.

YEOWARD, George, *An Episode of the Rebellion and Mutiny in Oudh*, Lucknow, 1871.

YOUNG, Colonel Keith, *Delhi, 1857*, London, 1902. The most comprehensive and detailed account of the siege assault and capture of Delhi, from the diary and correspondence of Colonel Keith Young, Judge Advocate General, Bengal, presented by General Sir Henry Wylie Norman. With nine useful (in terms of information) appendices, plus illustrations and maps not found elsewhere.

YOUNGHUSBAND, Colonel G.J., *The Story of the Guides*, London, 1909. The Guides came to set the standard of excellence for the Indian Army, and took precedence over other regiments.

YOUNGHUSBAND, Maj-Gen. Sir George, *A Soldier's Memories in Peace and War*, London, 1917.

YULE, H. ed., *Life and service of Major General W.H.Greathed*, London, 1879.

YULE, H. & BURNELL, A., *Hobson-Jobson: a glossary of colloquial Anglo-Indian words and phrases*, London,1908. Reissued London, 1968, reprinted 1985.

BIBLIOGRAPHY
PUBLISHED WORKS
in INDIAN LANGUAGES

'It should be stated that the most serious Indian historians of the Revolt (R.C.Majumdar, S.B.Chaudhuri, S.N.Sen etc) have written in English' - Henry Scholberg, 1993. see below.

BANDOPADHAYA, Durga Das, *Bidrohe Bengali,or Amar Jiban Charit,* Calcutta, 1925. BENGALI.

BHATTACHARYA, Mahashweta, *Jhansir Rani*, 1956. BENGALI.

GODSE, *Majha Pravas,* 1948. (*see* NAGAR, A.L. below.), MARATHI.

GUPTA, Rajani Kanta, *Sipahi Guddher Itihas,*or *History of Sepoy War,* 5 vols, Calcutta, 1886-1900, BENGALI.

HARDIKAR, Shriniwas Balaji, *Tatya Tope*, Delhi, 1969. HINDI.

JAFRI, Rais Ahmad, *Hazrat Mahal,* Lahore: Sheikh Ghulam Ali, 1969. URDU.

JHA, Kamal Narain, *Nana Sahaba,* [Lekhana] Kamalanarayana Jha 'Kamelsa' Prakkathana-lekhana, Patana, 1962.

JUNOON. , (Eng. 'Frenzy'), a film , Hindi, based upon Ruskin Bond's novel (*q.v.*).

KALYAN SINGH, *Lakshmi Bai ka Raso*, nd, HINDI.

KAMAL-ud-DIN HAIDAR HUSAINI, *Tarikh-i-Awadh.* or . *Qaisar-ut-Tawarikh*, 2 vols. Lucknow 1897. URDU.

MAULANA FAZLI-HAQ KHAIRABADI, *Saurat-ul-Hindiya*, Bijnor, 1947. ARABIC.

MIRZA GHULAM HUSAIN, *Siyyar-ul-Muntaquirin,* URDU, available also in English, trans. Briggs.

MIRZA MOHAMED HADI RUSVA, *Umrao Ada Jan,* URDU.

NAGAR, A.L., *Ankhon Dekha Ghadar,* (being translation of Majha Pravas *see* GODSE above), Lucknow, 1957. HINDI.

NAJMUL GHANI, *Tarikh-i-Awadh,* 5 vols, URDU.

NAJMUL GHANI, *Begamat-i-Awadh ke Khutut,* Delhi, URDU.

PARASNIS, D.B., *Maharani Lakshmi Bai Saheb,*

Hyanche Charita, 1894. MARATHI.

RIZVI, S.A.A., *Swatantra Dilli,*Varanasi, 1957. HINDI.

RUSVA, Mirza Muhammad Hadi, *Umrao Jan Jada*, Delhi, 1899. URDU.

SAIYID AHMAD KHAN, Sir, *Resalah Asbab-i-Baghawat-i-Hind*, 1858. URDU.

SAIYID KAMAL-ud-Din HAIDER, *See* Kamal-ud-Din Haidar above.

SCHOLBERG, Henry, *The Indian Literature of the Great Rebellion,* New Delhi, 1993. A valuable source book, divided into three parts. Part 1: Published Literature in Assamese Bengali Gujarati Hindi Kannada Malyalam Marathi Oriya Persian Punjabi Sindhi Tamil Telugu and Urdu. Part 2: The Primary Sources. National Archives of India, New Delhi, Urdu Journals, Residency Vernacular Records, UP Regional Archives Bhopal State Archives, Delhi State Archives, the Western Archives, the Central Archives. Part 3: Folk Songs of the Great Rebellion, The Avadhi Songs, Songs of the Ranchi Mutiny, Bahadur Shah the Poet, the William Crooke Collection, the P.C.Joshi Collection, the

Joyce Lebra Collection.

SIMHAQ, Mahendra P. ed. *Chattra Prakash*, by Goretala Lalkavi Purohita, Delhi, 1973. HINDI

TAGORE, Jyotirindranath, *Jhansir Rani*, 1900. BENGALI.

VERMA , Bindraban Lal, *Maharani Lakshmi Bai*, nd. HINDI.

BIBLIOGRAPHY
NOVELS & PLAYS

ALLARD HAFIZ ,(pseud), *Nirgis, a Tale of the Indian Mutiny*, London, 1869.

BONCIAULT, Dion, *Jessie Brown or The Relief of Lucknow*, A drama in three Acts. New York, 1858.

BOND, Ruskin, *A Flight of Pigeons*, Delhi, 1972.

CHAN-TOON, Mabel Mary Agnes, *Love Letters of an English Peeress to an Indian Prince*, London, 1912. Supposedly based on **Azimullah Khan**'s love letters.

CHESNEY, George T., *Dilemma: a Tale of the Mutiny*, London, 1908.

CHILDHOOD in INDIA, by the WIFE of an OFFICER, London, 1898.

COTES, Evarard, *The Story of Sonny Sahib*, London, 1894.

COX, Philip, *Rani of Jhansi: a historical play in 4 Acts.*, London, 1933.

FANTHOME, J.F., *Mariam: a story of the Indian Mutiny*, Benares, 1896.

FENN, C.R., *For the old Flag: a tale of the Indian Mutiny*, London, 1899.

FIELD, Mrs E.M., *BRYDA, a story of the Indian Mutiny*, London, 1888.

FORREST, Robert E., *Eight Days*, a novel, 3 vols, London, 1891.

FORREST, Robert E., *The Sword of Azreel*, a chronicle of the Great Mutiny, London, 1903.

FORREST, Robert E., *The Touchstone of Peril*, London, 1886.

FRASER, George MacDonald, *Flashman in the Great Game*, London, 1975.

GRANT, James, *First Love and Last Love*, London, 1908.

HENTY, George Alfred, *In Times of Peril, A Tale of India*, London, 1881.

HOLLIDAY, Mary, *Open Season for Fury*, Delhi, 1991 .

'JFF', *see* FANTOME, J.F.

JUNOON, a film based upon Ruskin Bond's novel *A Flight of Pigeons* , see above.

KAYE, M.M., *Shadow of the Moon*, New York, 1956.

KNIGHTON, William, *The Private Life of an Eastern King*, New York, 1855. Said to reflect the life and times of Wajid Ali Shah, last King of Awadh.

LANGTON, Jarvis, *A Foster Son: a tale of the Indian Mutiny*, London, 1896.

MALET, H.P., *Lost Links in the Indian Mutiny*, London, 1867.

MALGONKAR, Manohar, *The Devil's Wind: Nana Saheb's Story*, London, 1972. Reprint New Delhi,1988.

MASTERS, John, *Night runners of Bengal*, London, 1951.

MISTRY, Homi D., *Rebels of Destiny*, a Play, Bombay, 1859.

MUDDOCK, J.E., *The Great White Hand, or the Tiger of Cawnpore*, London, 1896.

MUJEEB, Mohammed, *Ordeal 1857*, a historical play, Bombay 1958.

NISBET, Hume, *The Queen's Desire*, London, 1893.

PEARCE, Chas E., *Red Revenge: a romance of Cawnpore*, London, 1912.

RAINES, G.Percy, *Terrible Times: a tale of the Sepoy Revolt*, London, 1898.

REID, C.L., *Masque of Mutiny*, London, 1947.

ROUSSILET, Louis, *A Tale of the Indian Mutiny*, New York, 1888.

RUSWA, Muhammad Hadi, *Umrao Jan Ada*, the Courtesan of Lucknow, (trans. Khushwant Singh). Calcutta, 1961.

SCHOLBERG, Henry, *Another Time, Another Country*, New York, 1965.

SELWYN, Francis, *Sergeant Verity and the Imperial Diamond*, London, 1975.

SINGH, Khushwant, *Delhi: a novel*, Delhi, 1990.

STEEL, Mrs Flora, *The Garden of Fidelity*, London, 1929.

STEEL, Mrs Flora, *On the Face of the Waters*, London, 1897.

TAYLOR, Meadows, *SEETA*, London, 1880.

VERNE, Jules, *The Steam House: Demon of Cawnpore: Tigers and Traitors*, London, 1881.

VIDROH, A seven-part serial broadcast by Doordarshan in 1993; the story of Babu Kunwar Singh.

WARD, Andrew, *The Blood Seed*, New York, 1985. A novel about the aftermath at Kanpur.

WHITE, Michael, *Lachmi Bai, Rani of Jhansi; the Jeanne d'Arc of India*, New York, 1901.

WOOD, Charles, *'H' or Monologues at front of Burning Cities*, First performed National Theatre, London, Feb.1969.

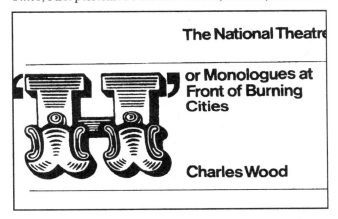

See also **INDIAN MUTINY FICTION** in Blackwoods Magazine, February 1897.

ARCHIVAL SOURCES

There is inevitable duplication here of material available both in India and in the United Kingdom, and students and researchers are advised to consult not only the Archives of 'first resort' so to speak, but also the very helpful Guides and Bibliographies of primary source material that have been produced in recent years and of which mention is made where appropriate in this section, and in the main Bibliography above.

NATIONAL ARCHIVES OF INDIA, NEW DELHI.

MILITARY DEPARTMENT PAPERS

MUTINY PAPERS

FOREIGN DEPARTMENT POLITICAL CONSULTATIONS

FOREIGN DEPARTMENT POLITICAL PROCEEDINGS

FOREIGN DEPARTMENT SECRET CONSULTATIONS

LUCKNOW UNIVERSITY LIBRARY.

MANUSCRIPTS (URDU), Muh. Azmat Alvi and Mirza Muh. Taqi

REGIONAL ARCHIVES, ALLAHABAD

CAWNPORE COLLECTORATE RECORDS

MEERUT AND BAREILLY COMMISSIONER'S OFFICE MUTINY RECORDS

UTTAR PRADESH STATE ARCHIVES, LUCKNOW

DAILY BULLETINS ISSUED BY E.A.READE, AGRA

TELEGRAMS SENT TO E.A.READE, AGRA

PROCEEDINGS OF THE FOREIGN, POLITICAL, GENERAL, JUDICIAL AND MILITARY DEPARTMENTS OF THE NORTH-WESTERN PROVINCES

INCLUDED ALSO ARE THE 'NARRATIVE OF EVENTS' FOR EACH DISTRICT (AVAILABLE ALSO ELSEWHERE)

BRITISH LIBRARY, LONDON.

Housed at the moment in the British Museum in Bloomsbury, London, but due to move to new premises at St Pancras in West London, where for the first time most of the Library's collection will be housed under one roof. The Library is probably the finest in the world and as it includes, by statute, a copy of every book published in England (this right is shared with the Bodleian at Oxford) it contains a truly comprehensive collection of published monographs, in addition to unpublished primary source material. The separate collection known in the past as the **India Office Library and Records**, and now known as the **'Oriental and India Office collections of the British Library'**, will eventually also be moved to St Pancras. No student of the mutiny/rebellion can afford to ignore the India Office material: it contains a complete collection of the Official Records of the Government of India at the time. Major series held include FOREIGN DEPARTMENT, PROCEEDINGS, POLITICAL; FOREIGN DEPARTMENT, PROCEEDINGS, SECRET; HOME DEPARTMENT, PUBLIC PROCEEDINGS; MILITARY DEPARTMENT PROCEEDINGS; DESPATCHES FROM INDIA AND BENGAL 1856-58; POLITICAL LETTERS FROM INDIA, 1856-58; HOME MISCELLANEOUS SERIES 724A-727 ie **Sir John Kaye's** Mutiny Papers etc. A complete and detailed list of the material available is included in the invaluable little book by **Rosemary Seton**, *The Indian 'Mutiny' 1857-58*, London, 1986.

BHARGAVA, M.L., (ed. with RIZVI, S.A.A.),

Freedom Struggle in Uttar Pradesh (6vols), . . Source Material, Lucknow, 1957-61. The six volumes cover I. 1857-59: Nature and Origin; II. Awadh: 1857-59; III. Bundelkhand and Adjoining Territories 1857-59; IV. Eastern and Adjoining Districts 1857-59; V. Western Districts and Rohilkhand 1857-59; VI. Index to all volumes. This is archival material in the sense that it publishes without comment (other than editorial headlines) original source material. But it is , necessarily, selective.

PEARSON, J.D.,

Guide to Manuscripts and Documents in the British Isles

Relating to South and South-East Asia,.. London, 1989. Volume 1 is of particular significance, giving a complete account of archive holdings in many libraries and collections. It would be difficult to overvalue this guide since it leads the student expertly to the material he seeks: if the latter was in the United Kingdom in 1989 it would be most likely to be listed here.

SCHOLBERG, Henry, *The Indian Literature of the Great Rebellion,* New Delhi, 1993. A valuable source book, divided into three parts. Part 1: Published Literature in Assamese Bengali Gujarati Hindi Kannada Malyalam Marathi Oriya Persian Punjabi Sindhi Tamil Telugu and Urdu. **Part 2: The Primary Sources. National Archives of India, New Delhi, Urdu Journals, Residency Vernacular Records, UP Regional Archives Bhopal State Archives, Delhi State Archives, the Western Archives, the Central Archives. Part 3: Folk Songs of the Great Rebellion, The Avadhi Songs, Songs of the Ranchi Mutiny, Bahadur Shah the Poet, the William Crooke Collection, the P.C.Joshi Collection, the Joyce Lebra Collection.**

THATCHER, M., and CARTER, L.,

Cambridge South Asian Archive: Records of the British Period in South Asia Held in the Centre of South Asian Studies, University of Cambridge,.. London, 1973, Second Collection, 1980, Third Collection, 1983, Fourth Collection , 1987(?). An extremely valuable resource listing much unique archival and manuscript material. Now available on microfiche.

There is of course also much material in private hands, particularly contemporary letters and unpublished memoirs, and much still remains no doubt to be discovered. Books of general interest on the mutiny/rebellion eg Christopher Hibbert's *The Great Rebellion,* London, 1978 or Richard Collier's *The Sound of Fury,* London, 1963 often contain lists of such material, particularly if it has been used in the book, but a mention of the existence of privately owned material is no guarantee of course that it is available to the student.

NEWSPAPERS, JOURNALS & PERIODICALS

(English language unless otherwise stated)

AGRA UKHBAR

ALLEN'S INDIAN MAIL

ARMY & NAVY GAZETTE

ARMY QUARTERLY

ATHÆNEUM & DAILY NEWS (MADRAS)

BENGAL HURKARU AND INDIA GAZETTE

BENGAL PAST AND PRESENT

BLACKWOOD'S MAGAZINE

BOMBAY CALENDAR AND ALMANAC

BOMBAY GUARDIAN AND BOMBAY GAZETTE

BOMBAY TIMES (1844)

CALCUTTA REVIEW

CALCUTTA WEEKLY CHRONICLE

CONTRIBUTIONS TO INDIAN SOCIOLOGY

DELHI GAZETTE

DELHI NEWS

DELHI SKETCH BOOK

DELHI URDU AKHBAR

EAST INDIA UNITED SERVICE JOURNAL

The ENGLISHMAN (CALCUTTA)

The FRIEND OF INDIA

HABIB-UL-AKHBAR-BADAUN (URDU)

The HINDU INTELLIGENCER

The HINDU PATRIOT

ILLUSTRATED LONDON NEWS

IMPERIAL AND ASIATIC QUARTERLY REVIEW

INDIAN ANTIQUARY

INDIAN ECON. & SOCIAL HISTORY REVIEW

INDIAN HISTORICAL QUARTERLY

INDIAN HISTORICAL RECORDS COMMISSION, PROCEEDINGS

INDIAN PLANTERS' GAZETTE

INDIAN SPORTING REVIEW

JOURNAL OF BIHAR & ORISSA RESEARCH SOC.

JOURNAL OF INDIAN HISTORY

LAHORE CHRONICLE

MADRAS CHURCH MISSIONARY RECORD

MOFFUSILITE (MEERUT)

MORNING CHRONICLE (LONDON)

NAVAL AND MILITARY GAZETTE

The OVERLAND BOMBAY TIMES

PAST & PRESENT (OXFORD)

PHOENIX

PUNCH

PUNJAB UNIVERSITY HISTORICAL SOCIETY JOURNAL

ROYAL ARTILLERY JOURNAL

ROYAL ENGINEERS' JOURNAL

ROYAL HISTORICAL SOCIETY TRANSACTIONS

ROYAL UNITED SERVICES INSTITUTE - JOURNAL

SADIQ-UL-AKHBAR- DELHI (URDU)

SIHR-I-SAMRI- LUCKNOW (URDU)

SIRAJ-UL AKHBAR- DELHI (PERSIAN)

SOCIETY OF ARMY HISTORICAL RESEARCH: JOURNAL

SOUTH ASIA REVIEW

TAIT'S EDINBURGH MAGAZINE

TELEGRAPH AND COURIER (BOMBAY)

TILISMI-I-LAKHNAU- LUCKNOW (URDU)

UNITED SERVICES INSTITUTE OF INDIA - JOURNAL

The WITNESS in the EAST

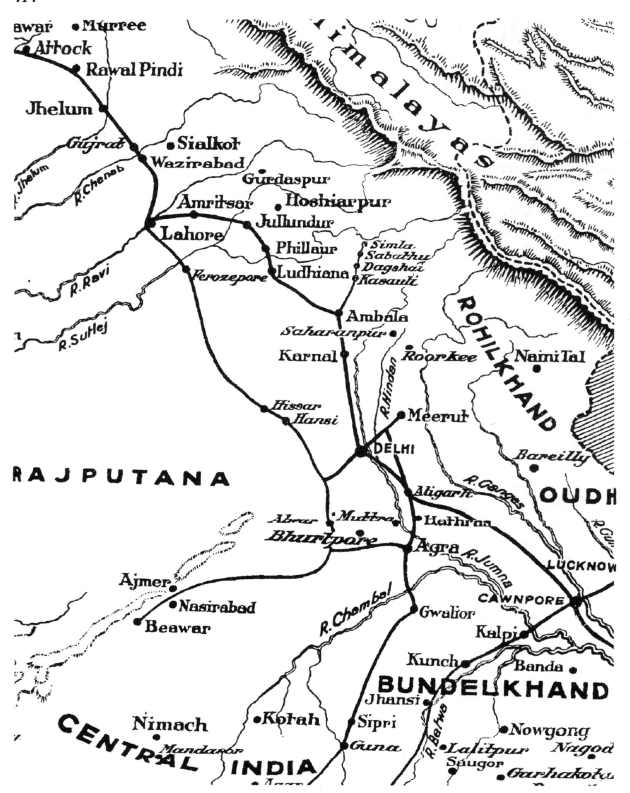

NORTHERN INDIA - WEST

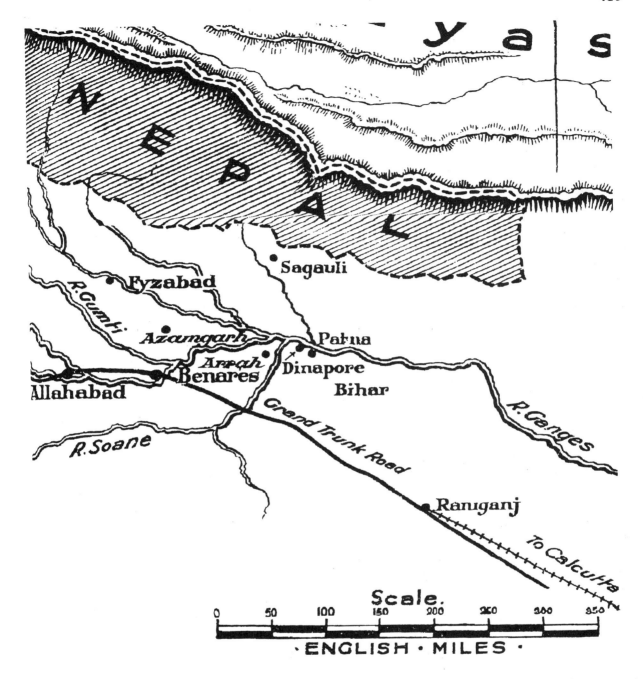

NORTHERN INDIA - EAST